Modern Birkhäuser Classics

Many of the original research and survey monographs in pure and applied mathematics published by Birkhäuser in recent decades have been groundbreaking and have come to be regarded as foundational to the subject. Through the MBC Series, a select number of these modern classics, entirely uncorrected, are being re-released in paperback (and as eBooks) to ensure that these treasures remain accessible to new generations of students, scholars, and researchers.

Optimal Control and Viscosity Solutions of Hamilton–Jacobi–Bellman Equations

Martino Bardi
Italo Capuzzo-Dolcetta

Reprint of the 1997 Edition

Birkhäuser
Boston • Basel • Berlin

Martino Bardi
Dipartimento di Matematica P. ed. A.
Università di Padova
35131 Padova
Italy

Italo Capuzzo-Dolcetta
Dipartimento di Matematica
Universià di Roma "La Sapienza"
00185 Roma
Italy

Originally published in the series *Systems & Control: Foundations & Applications*

ISBN-13: 978-0-8176-4754-4   e-ISBN-13: 978-0-8176-4755-1
DOI: 10.1007/978-0-8176-4755-1

Library of Congress Control Number: 2007940263

Mathematics Subject Classification (2000): 49L20, 49L25, 35F20, 90D25

©2008 Birkhäuser Boston
All rights reserved. This work may not be translated or copied in whole or in part without the written permission of the publisher (Birkhäuser Boston, c/o Springer Science+Business Media LLC, 233 Spring Street, New York, NY 10013, USA), except for brief excerpts in connection with reviews or scholarly analysis. Use in connection with any form of information storage and retrieval, electronic adaptation, computer software, or by similar or dissimilar methodology now known or hereafter developed is forbidden.
The use in this publication of trade names, trademarks, service marks and similar terms, even if they are not identified as such, is not to be taken as an expression of opinion as to whether or not they are subject to proprietary rights.

Cover design by Alex Gerasev.

Printed on acid-free paper.

9 8 7 6 5 4 3 2 1

www.birkhauser.com

Martino Bardi
Italo Capuzzo-Dolcetta

Optimal Control and Viscosity Solutions of Hamilton-Jacobi-Bellman Equations

Birkhäuser
Boston • Basel • Berlin

Martino Bardi
Dipartimento di Matematica P. ed A.
Università di Padova
35131 Padova
Italy
bardi@math.unipd.it

Italo Capuzzo-Dolcetta
Dipartimento di Matematica
Università di Roma "La Sapienza"
00185 Roma
Italy
capuzzo@mat.uniroma1.it

Library of Congress Cataloging-in-Publication Data

Bardi, M. (Martino)
   Optimal control and viscosity solutions of Hamilton-Jacobi-Bellman
equations / Martino Bardi, Italo Capuzzo Dolcetta.
     p. cm. -- (Sytems and control)
   Includes bibliographical references.
   ISBN 0-8176-3640-4 (hardcover : alk. paper)
   1. Viscosity solutions.   2. Control theory.   3. Differential
games.   I. Capuzzo Dolcetta, I. (Italo), 1948-   . II. Title.
III. Series: Systems & control.
QA316.B33   1997
515' .64--dc21   97-23275
                                                                   CIP

AMS Classifications: 49L20, 49L25, 35F20, 90D25

Printed on acid-free paper
© 1997 Birkhäuser Boston   *Birkhäuser*

Copyright is not claimed for works of U.S. Government employees.
All rights reserved. No part of this publication may be reproduced, stored in a retrieval system, or transmitted, in any form or by any means, electronic, mechanical, photocopying, recording, or otherwise, without prior permission of the copyright owner.

Permission to photocopy for internal or personal use of specific clients is granted by Birkhäuser Boston for libraries and other users registered with the Copyright Clearance Center (CCC), provided that the base fee of $6.00 per copy, plus $0.20 per page is paid directly to CCC, 222 Rosewood Drive, Danvers, MA 01923, U.S.A. Special requests should be addressed directly to Birkhäuser Boston, 675 Massachusetts Avenue, Cambridge, MA 02139, U.S.A.

ISBN 0-8176-3640-4
ISBN 3-7643-3640-4
Camera-ready text provided by the authors in LATEX.
Printed and bound by Edwards Brothers, Ann Arbor, MI.
Printed in the U.S.A.

9 8 7 6 5 4 3 2 1

Contents

Preface   xi

Basic notations   xv

Chapter I. Outline of the main ideas on a model problem   1
1. The infinite horizon discounted regulator 1
2. The Dynamic Programming Principle 2
3. The Hamilton-Jacobi-Bellman equation in the viscosity sense 3
4. Comparison, uniqueness and stability of viscosity solutions 6
5. Synthesis of optimal controls and verification theorems 10
6. Pontryagin Maximum Principle as a necessary and sufficient condition of optimality 13
7. Discrete time Dynamic Programming and convergence of approximations 17
8. The viscosity approximation and stochastic control 20
9. Bibliographical notes 21

Chapter II. Continuous viscosity solutions of Hamilton-Jacobi equations   25
1. Definitions and basic properties 25
2. Some calculus and further properties of viscosity solutions 34
3. Some comparison and uniqueness results 50
4. Lipschitz continuity and semiconcavity 61
   4.1. Lipschitz continuity 62
   4.2. Semiconcavity 65
5. Special results for convex Hamiltonians 77
   5.1. Semiconcave generalized solutions and bilateral supersolutions 77
   5.2. Differentiability of solutions 80
   5.3. A comparison theorem 82
   5.4. Solutions in the extended sense 84

    5.5. Differential inequalities in the viscosity sense 86
    5.6. Monotonicity of value functions along trajectories 90
  6. Bibliographical notes 95

Chapter III. Optimal control problems with continuous value functions: unrestricted state space   97

  1. The controlled dynamical system 97
  2. The infinite horizon problem 99
    2.1. Dynamic Programming and the Hamilton-Jacobi-Bellman equation 99
    2.2. Some simple applications: verification theorems, relaxation, stability 110
    2.3. Backward Dynamic Programming, sub- and superoptimality principles, bilateral solutions 119
    2.4. Generalized directional derivatives and equivalent notions of solution 125
    2.5. Necessary and sufficient conditions of optimality, minimum principles, and multivalued optimal feedbacks 133
  3. The finite horizon problem 147
    3.1. The HJB equation 147
    3.2. Local comparison and unbounded value functions 156
    3.3. Equivalent notions of solution 160
    3.4. Necessary and sufficient conditions of optimality and the Pontryagin Maximum Principle 170
  4. Problems whose HJB equation is a variational or quasivariational inequality 183
    4.1. The monotone control problem 184
    4.2. Optimal stopping 193
    4.3. Impulse control 200
    4.4. Optimal switching 210
  5. Appendix: Some results on ordinary differential equations 218
  6. Bibliographical notes 223

Chapter IV. Optimal control problems with continuous value functions: restricted state space   227

  1. Small-time controllability and minimal time functions 227
  2. HJB equations and boundary value problems for the minimal time function: basic theory 239
  3. Problems with exit times and non-zero terminal cost 247
    3.1. Compatible terminal cost and continuity of the value function 248
    3.2. The HJB equation and a superoptimality principle 254
  4. Free boundaries and local comparison results for undiscounted problems with exit times 261
  5. Problems with state constraints 271
  6. Bibliographical notes 282

Chapter V. Discontinuous viscosity solutions and applications   285
1. Semicontinuous sub- and supersolutions, weak limits, and stability .   286
2. Non-continuous solutions   295
    2.1. Definitions, basic properties, and examples   295
    2.2. Existence of solutions by Perron's method   302
3. Envelope solutions of Dirichlet problems   308
    3.1. Existence and uniqueness of e-solutions   309
    3.2. Time-optimal problems lacking controllability   313
4. Boundary conditions in the viscosity sense   316
    4.1. Motivations and basic properties   317
    4.2. Comparison results and applications to exit-time problems
         and stability   326
    4.3. Uniqueness and complete solution for time-optimal control . .   333
5. Bilateral supersolutions   342
    5.1. Problems with exit times and general targets   342
    5.2. Finite horizon problems with constraints on the endpoint
         of the trajectories   348
6. Bibliographical notes   357

Chapter VI. Approximation and perturbation problems   359
1. Semidiscrete approximation and ε-optimal feedbacks   359
    1.1. Approximation of the value function and construction
         of optimal controls   360
    1.2. A first result on the rate of convergence   369
    1.3. Improving the rate of convergence   372
2. Regular perturbations   376
3. Stochastic control with small noise and vanishing viscosity   383
4. Appendix: Dynamic Programming for Discrete Time Systems . . .   388
5. Bibliographical notes   395

Chapter VII. Asymptotic problems   397
1. Ergodic problems   397
    1.1. Vanishing discount in the state constrained problem   397
    1.2. Vanishing discount in the unconstrained case:
         optimal stopping   402
2. Vanishing switching costs   411
3. Penalization   414
    3.1. Penalization of stopping costs   414
    3.2. Penalization of state constraints   417
4. Singular perturbation problems   420
    4.1. The infinite horizon problem for systems with fast
         components   420
    4.2. Asymptotics for the monotone control problem   426
5. Bibliographical notes   429

Chapter VIII. Differential Games   **431**
1. Dynamic Programming for lower and upper values 433
2. Existence of a value, relaxation, verification theorems 442
3. Comparison with other information patterns and other
   notions of value 448
   3.1. Feedback strategies 448
   3.2. Approximation by discrete time games 457
4. Bibliographical notes 468

**Appendix A. Numerical Solution of Dynamic Programming
Equations** *by Maurizio Falcone*   **471**
1. The infinite horizon problem 472
   1.1. The Dynamic Programming equation 473
   1.2. Synthesis of feedback controls 478
   1.3. Numerical tests 481
2. Problems with state constraints 484
3. Minimum time problems and pursuit-evasion games 488
   3.1. Time-optimal control 488
   3.2. Pursuit-evasion games 494
   3.3. Numerical tests 496
4. Some hints for the construction of the algorithms 499
5. Bibliographical notes 502

Appendix B. Nonlinear $\mathcal{H}_\infty$ control *by Pierpaolo Soravia*   **505**
1. Definitions 506
2. Linear systems 510
3. $\mathcal{H}_\infty$ control and differential games 512
4. Dynamic Programming equation 515
5. On the partial information problem 521
6. Solving the problem 527
7. Exercises 530
8. Bibliographical notes 531

Bibliography   **533**

Index   **565**

To Alessandra and Patrizia

Preface

The purpose of the present book is to offer an up-to-date account of the theory of viscosity solutions of first order partial differential equations of Hamilton-Jacobi type and its applications to optimal deterministic control and differential games. The theory of viscosity solutions, initiated in the early 80's by the papers of M.G. Crandall and P.L. Lions [CL81, CL83], M.G. Crandall, L.C. Evans and P.L. Lions [CEL84] and P.L. Lions' influential monograph [L82], provides an extremely convenient PDE framework for dealing with the lack of smoothness of the value functions arising in dynamic optimization problems.

The leading theme of this book is a description of the implementation of the viscosity solutions approach to a number of significant model problems in optimal deterministic control and differential games. We have tried to emphasize the advantages offered by this approach in establishing the well-posedness of the corresponding Hamilton-Jacobi equations and to point out its role (when combined with various techniques from optimal control theory and nonsmooth analysis) in the important issue of feedback synthesis.

The main ideas are introduced in Chapter I where the infinite horizon discounted regulator problem is taken as a model. After the derivation of the Hamilton-Jacobi-Bellman equation from the Dynamic Programming optimality principle we cover, in a rather informal way, such topics as uniqueness, stability and necessary and sufficient conditions for optimality. A quick review of discrete time and stochastic approximations to the value function is given in the last two sections.

Chapter II is devoted to the basic theory of continuous viscosity solutions. A long section of this chapter deals with the case of Hamilton-Jacobi-Bellman equations, corresponding to Hamiltonians which are convex with respect to the gradient variable. This is the relevant case in connection with optimal control problems. In particular we discuss the connections between viscosity solutions and some different notions, such as Lipschitz continuous functions solving the equation almost everywhere, Barron and Jensen's bilateral supersolutions, and solutions in the extended sense of Clarke.

In the next two chapters the basic theory of Chapter II is specialized and developed with reference to various optimal control problems with continuous value function. We made the choice to present in Chapter III some problems with unrestricted state space, corresponding to the simpler case of Hamilton-Jacobi equations without boundary conditions. We note that Section 4 deals with some quasivariational inequalities and systems arising in connection with stopping times, impulse control, or switching costs. On the other hand, Chapter IV is dedicated to problems involving exit times from a given domain or constraints on the state variables for systems having suitable controllability properties, leading to boundary value problems.

Chapter V is dedicated to the case of discontinuous value functions, a typical motivation coming from minimum time problems lacking controllability. Various notions of discontinuous viscosity solutions, including those of Barles and Perthame, Ishii, Barron and Jensen, and Subbotin, are discussed and compared. Section 1 of this chapter is of independent interest since there we develop the so-called weak limits technique of Barles and Perthame, a very useful tool which we adopt to tackle the various limit problems in Chapters VI and VII.

In particular, in Chapter VI we consider an approximation scheme for value functions based on discrete time Dynamic Programming. This is an important issue in the applications because it also provides a method to construct almost optimal feedback controls. The convergence of the scheme and some estimates on the rate of convergence are proved by viscosity solutions methods. A section is dedicated to regular perturbation problems where similar techniques can be employed.

Chapter VII deals with the analysis of some asymptotic problems such as singular perturbations, penalization of state constraints, vanishing discount and vanishing switching costs. The limiting behavior of the associated Hamilton-Jacobi equations is analyzed in a simple way by the viscosity solutions approach and the weak limit technique.

The last chapter is intended as an introduction to the theory of two-person zero sum differential games. Different notions of value are discussed together with the derivation of the relevant Hamilton-Jacobi-Isaacs equations. It is worth noting that the viscosity solutions method appears to be highly privileged for the treatment of this kind of nonconvex, nonlinear PDE's.

Finally, two appendices deal with some additional topics which are important for the applications. The first one, by M. Falcone, describes in some detail a computational method based on the approximation theory developed in Chapter VI. The second, by P. Soravia, gives a fairly complete account of some recent results on the viscosity solutions approach to $\mathcal{H}_\infty$ control.

Our main goal in planning this work was to provide a self-contained but, at the same time, rather comprehensive presentation of the topic to scientists in the areas of optimal control, system theory, and partial differential equations. To this end, each chapter is enriched with a section of bibliographical and historical notes.

As is often the case, the present book originated from lecture notes of courses taught by the authors. We believe that the style of presentation (in particular, the set of exercises proposed at the end of each section) reflects this pedagogically oriented origin, and so selected parts could easily be used for a graduate course in optimal control. In this regard, a possible path for a one semester course, which we have tested in lectures at Paris-Dauphine, Rome, Padua, and Naples, might include the first three sections of Chapter II, a choice of a few sections from Chapters III and IV (for example, selected parts of sections 1–3 of Chapter III and the first two sections of Chapter IV) and the initial sections of Chapters V and VI.

The prerequisites for reading most of the book are just advanced calculus, some very basic functional analysis (uniform convergence and the Ascoli-Arzelà theorem) as well as, of course, the fundamental facts about nonlinear ordinary differential equations which are recalled in the appendix to Chapter III. These are also the only prerequisites for the course outlined above. More sophisticated mathematical tools such as relaxed controls, 2nd order PDE's, stochastic control and the geometric theory of controllability appear in the book but are not essential for understanding its core.

We are happy to thank L.C. Evans and P.L. Lions who introduced us to viscosity solutions in the very early stages of the theory and also the other colleagues and friends with whom we have collaborated, in particular H. Ishii, B. Perthame, M. Falcone, P. Soravia. We had the occasion to discuss the overall project of the book with W.H. Fleming, M.H. Soner and G. Barles. Additional thanks are due to M. Falcone and P. Soravia for contributing the Appendices. Several people read parts of the manuscript. We are particularly grateful to O. Alvarez, S. Bortoletto and A. Cutrì for reading the book from cover to cover and for stimulating remarks, and to F. Da Lio, P. Goatin, F. Gozzi, S. Mirica, M. Motta, F. Rampazzo, C. Sartori, F. Sullivan. We also thank A.I. Subbotin, N.N. Subbotina, and A.M. Tarasyev for translating some of their papers in Russian into English for us. Finally, thanks are due to G. Bertin for skilled and sensitive typing.

April 1997 *Martino Bardi, Italo Capuzzo-Dolcetta*

Basic notations

$\mathbb{R}^N$	the euclidean N-dimensional space
$x \cdot y$	the scalar product $\sum_{i=1}^{N} x_i y_i$ of vectors $x = (x_1, \ldots, x_N)$ and $y = (y_1, \ldots, y_N)$
span$\{\ldots\}$	the vector space generated by the vectors $\{\ldots\}$
$|x|$	the euclidean norm of $x \in \mathbb{R}^N$, $|x| = (x \cdot x)^{1/2}$
$B(x_0, r)$	the open ball $\{x \in \mathbb{R}^N : |x - x_0| < r\}$
$\bar{B}(x_0, r)$	the closed ball $\{x \in \mathbb{R}^N : |x - x_0| \leq r\}$
∂E	the boundary of the set E
int E	the interior of the set E
$\overline{E}$	the closure of the set E
E^c	the complement of the set E
co E	the convex hull of the set E
$\overline{\text{co}}E$	the closed convex hull of the set E
$\mathcal{P}(X)$	the set of all subsets of X
$\Omega' \subset\subset \Omega$	means $\overline{\Omega'} \subseteq \Omega$
$|E| = \text{meas } E$	the Lebesgue measure of the set E
$d(x, E) = \text{dist}(x, E)$	the distance from x to E (i.e., $d(x, E) = \inf_{y \in E} |x - y|$)
diam E	the diameter of the set E (i.e., diam $E = \sup\{|x - y| : x, y \in E\}$)
$d_H(E, S)$	the Hausdorff distance between the sets E and S (§III.2.2)
$n(x)$	the outward normal unit vector to a set E at $x \in \partial E$
$\gamma \wedge \sigma$	$\min\{\gamma, \sigma\}$ for $\gamma, \sigma \in \mathbb{R}$
$\gamma \vee \sigma$	$\max\{\gamma, \sigma\}$ for $\gamma, \sigma \in \mathbb{R}$
γ^+, γ^-	the positive and negative part of $\gamma \in \mathbb{R}$ (i.e., $\gamma^+ = \gamma \vee 0$, $\gamma^- = \gamma \wedge 0$)
$[r]$	the integer part of $r \in \mathbb{R}$

BASIC NOTATIONS

sgn r	the sign of $r \in \mathbb{R}$ (1 if $r > 0$, -1 if $r < 0$, 0 if $r = 0$)
$\arg\min_E u$	the set of minimum points of $u : E \to \mathbb{R}$
$\|u\|_\infty$	the supremum norm $\sup_{x \in E} |u(x)|$ of a function $u : E \to \mathbb{R}$
$u_n \searrow u$	the sequence of functions u_n is nonincreasing and tends to u
$u_n \nearrow u$	the same as the preceding, but u_n is nondecreasing
$u_n \rightrightarrows u$	uniform convergence of the functions u_n to u
supp u	the support of the function u, i.e., the closure of the set $\{x : u(x) \neq 0\}$
χ_X	the characteristic (or indicator) function of the set X (1 in X, 0 outside)
ω	a modulus, i.e., a function $\omega : [0, +\infty[\to [0, +\infty[$ continuous, nondecreasing, and such that $\omega(0) = 0$; or, more generally, $\omega : [0, +\infty[^2 \to [0, +\infty[$ such that for all $R > 0$ $\omega(\,\cdot\,, R)$ has the preceding properties
$o(t)$ as $t \to a$	a function u such that $\lim_{t \to a} u(t)/t = 0$
$Du(x)$	the gradient of the function u at x, i.e., $Du(x) = \left(\dfrac{\partial u}{\partial x_1}(x), \ldots, \dfrac{\partial u}{\partial x_1}(x)\right)$
$D^+u(x)$, $D^-u(x)$	the super- and subdifferential of u at x (§I.3)
$\partial_c u(x)$	the subdifferential of the convex function of u at x (§II.1)
$\partial u(x)$	the Clarke's gradient of u at x (§II.4.1)
$D^*u(x)$	$\{p \in \mathbb{R}^N : p = \lim_{n \to +\infty} Du(x_n),\ x_n \to x\}$ (§II.4.1)
$\partial u(x; q)$	the (one-sided) directional derivative of u at x in the direction q (§II.4.1)
$\partial^+ u(x; q)$, $\partial^- u(x; q)$	the (generalized) Dini directional derivatives (§II.4.1 and §III.2.4)
$u_0(x; q)$, $u^0(x; q)$	the regularized directional derivatives (§II.4.1)
Δu	the laplacian of the function u, i.e., $\Delta u = \sum\limits_{i=1}^{N} \dfrac{\partial^2 u}{\partial x_i^2}$
$\underline{u}(x) = \liminf\nolimits_{\varepsilon \to 0^+} {}_* u_\varepsilon(x)$	the lower weak limit of u_ε at x as $\varepsilon \to 0^+$ (§V.1)
$\overline{u}(x) = \limsup\nolimits_{\varepsilon \to 0^+} {}^* u_\varepsilon(x)$	the upper weak limit of u_ε at x as $\varepsilon \to 0^+$ (§V.1)
u_*, u^*	the lower and the upper semicontinuous envelopes of u (§V.2)
$\mathcal{A}$	the set of controls, i.e., (Lebesgue) measurable functions $\alpha : [0, +\infty[\to A$
$\mathcal{A}^{\mathrm{r}}$	the set of relaxed controls (§III.2.2)
$y_x(t, \alpha)$	the state at time t of a control system (§III.1)
$t_x(\alpha)$	the first entry time of the state in some given closed set $\mathcal{T}$ (§I.5.6, §III.2.3, and §IV.1)
$\widehat{t}_x(\alpha)$	the first entry time of the state in int $\mathcal{T}$ (§IV.1)

$f(x, A)$	$\{q \in \mathbb{R}^N : q = f(x, a) \text{ for some } a \in A\}$
$D_x f$	the Jacobian matrix of f with respect to the x variable
$B(E)$	the space of functions $u : E \to \mathbb{R}$ with $\|u\|_\infty < +\infty$
$C(E)$	the space of continuous functions $u : E \to \mathbb{R}$
$BC(E)$	the space $B(E) \cap C(E)$
$UC(E)$	the space of uniformly continuous functions $u : E \to \mathbb{R}$
$BUC(E)$	the space $B(E) \cap UC(E)$
$\text{Lip}(E)$	the space of Lipschitz continuous functions $u : E \to \mathbb{R}$, i.e., such that for some $L \geq 0$ $|u(x) - u(y)| \leq L |x - y|$ for all $x, y \in E$
$\text{Lip}_{\text{loc}}(E)$	the space of locally Lipschitz continuous functions $u : E \to \mathbb{R}$, i.e., functions whose restriction to any compact subset of E is Lipschitz continuous
$C^{0,\gamma}(E)$	the space of γ-Hölder continuous functions $u : E \to \mathbb{R}$, $\gamma \in\ ]0, 1[$, i.e., $\displaystyle\sup_{x,y \in E} \frac{|u(x) - u(y)|}{|x - y|^\gamma} < +\infty$
$C^k(\Omega)$	for $k \geq 1$ and Ω open subset of $\mathbb{R}^N$, the subspace of $C(\Omega)$ of functions with continuous partial derivatives in Ω up to order k
$C^k(E)$	for $E \subseteq \mathbb{R}^N$, the space of functions that are the restrictions to E of some $u \in C^k(\Omega)$ for some open set $\Omega \supseteq E$
$USC(E), LSC(E)$	the spaces of lower and upper semicontinuous functions $u : E \to \mathbb{R}$
$BUSC(E), BLSC(E)$	the spaces $USC(E) \cap B(E)$ and $LSC(E) \cap B(E)$
$BC_\partial(\overline{\Omega})$	the subspace of $B(\overline{\Omega})$ of the functions continuous at all points of $\partial\Omega$
$L^1([t_0, t_1])$	the Lebesgue space of integrable function $[t_0, t_1] \to \mathbb{R}$
$L^\infty([t_0, t_1])$	the Lebesgue space of essentially bounded function $[t_0, t_1] \to \mathbb{R}$
$L^1(0, T; E)$	the space of integrable function $[0, T] \to E$
$L^\infty(0, T; E)$	the space of essentially bounded function $[0, T] \to E$
STC$\mathcal{T}$	small-time controllability on $\mathcal{T}$ (§IV.1)
STLC	small-time local controllability (§IV.1)
STC$\overset{\circ}{\mathcal{T}}$	small-time controllability on int $\mathcal{T}$ (§IV.1)
Γ, Δ	nonanticipating strategies for the first and the second player (§VIII.1)
$\mathcal{F}, \mathcal{G}$	feedback strategies for the first and the second player (§VIII.3.1)
◀	end of a proof
◁	end of a definition, end of a remark, end of an example

CHAPTER I

Outline of the main ideas on a model problem

The purpose of this introductory chapter is to motivate the relevance of the notion of *viscosity solution* of partial differential equations of the form

$$F(x, u(x), Du(x)) = 0$$

in a *Dynamic Programming* approach to *deterministic optimal control* theory.

The important role played in this respect by the theory of viscosity solutions is exhibited with reference to the classical *infinite horizon discounted regulator* problem.

The presentation is rather informal in its aim of pointing out the main ideas. For detailed treatment of all topics discussed here the reader is referred to subsequent chapters.

1. The infinite horizon discounted regulator

Consider a *control system* governed by the *state equation*

(1.1)
$$y'(t) = f(y(t), \alpha(t)), \qquad t > 0,$$
$$y(0) = x .$$

Here, the *control* α is any measurable function of $t \in [0, +\infty[$ with values in A, the *control space* (typically, A is a closed bounded set in $\mathbb{R}^M$ or, more generally, A is a topological space).

Assume that the *dynamics* $f : \mathbb{R}^N \times A \to \mathbb{R}^N$ is such that, for any choice of the control α and of the initial position $x \in \mathbb{R}^N$, the state equation (1.1) has a unique solution defined for all $t \in [0, +\infty[$, denoted by $y_x(t, \alpha)$. The model also includes a *running cost*, associated with this controlled evolution, described by a given function $\ell : \mathbb{R}^N \times A \to \mathbb{R}$.

The *cost functional* to be minimized is

(1.2)
$$J(x, \alpha) = \int_0^{+\infty} \ell(y_x(t, \alpha), \alpha(t)) e^{-\lambda t} dt,$$

where $\lambda > 0$ represents a (fixed) *discount factor*. The first step in the standard dynamic programming approach to the optimal control problem described above is to introduce the *value function* v, defined by

$$(1.3) \qquad v(x) := \inf_{\alpha \in \mathcal{A}} J(x, \alpha),$$

where $\mathcal{A}$ denotes the set of all measurable functions $\alpha : [0, +\infty[\to A$. The fundamental idea of dynamic programming is that v satisfies a functional equation, often called the *Dynamic Programming Principle*, and when v is smooth enough, its infinitesimal version, the *Hamilton-Jacobi-Bellman* (HJB) *equation*. These equations contain all the relevant information needed to design, through the so-called *synthesis procedure*, an *optimal feedback map* for the problem.

In the subsequent Sections 2–5 we outline this program in the framework of the theory of viscosity solutions.

2. The Dynamic Programming Principle

Assume temporarily that an optimal control α_x^* exists for each x, so that

$$v(x) = J(x, \alpha_x^*) = \int_0^{+\infty} \ell(y_x(t, \alpha_x^*), \alpha_x^*(t)) e^{-\lambda t} \, dt \, .$$

Observe that, for any fixed $T > 0$,

$$J(x, \alpha_x^*) = \int_0^T \ell(y_x(t, \alpha_x^*), \alpha_x^*(t)) e^{-\lambda t} \, dt + \int_T^{+\infty} \ell(y_x(t, \alpha_x^*), \alpha_x^*(t)) e^{-\lambda t} \, dt$$

A simple argument based on the semigroup property

$$(2.1) \qquad y_x(t+s, \alpha_x^*) = y_{y_x(t,\alpha_x^*)}(s, \alpha_x^*(\cdot + t)) \qquad \forall t, s > 0,$$

leads to the following identity

$$(2.2) \qquad v(x) = \int_0^T \ell(y_x(t, \alpha_x^*), \alpha_x^*(t)) e^{-\lambda t} \, dt + v(y_x(T, \alpha_x^*)) e^{-\lambda T}$$

which holds for all $T > 0$ and $x \in \mathbb{R}^N$. In the general case (i.e., when the existence of optimal controls is not assumed), (2.2) is replaced by

$$(2.3) \qquad v(x) = \inf_{\alpha \in \mathcal{A}} \left\{ \int_0^T \ell(y_x(t, \alpha), \alpha(t)) e^{-\lambda t} \, dt + v(y_x(T, \alpha)) e^{-\lambda T} \right\} .$$

The functional equation (2.3) comprises the statement of the *Dynamic Programming Principle* for our problem.

An important fact to be stressed at this point is that (2.3) holds for all $x \in \mathbb{R}^N$ and $T > 0$ under very general conditions on the data. When ℓ and, consequently, v is bounded, (2.3) characterizes the value function v in the sense that if u is a bounded function satisfying (2.3) for all $x \in \mathbb{R}^N$ and $T > 0$, then $u \equiv v$.

Let us observe, finally, that the semigroup property (2.1) has a counterpart at the level of the dynamic optimization problem that leads to a more abstract viewpoint on the Dynamic Programming Principle. More precisely, consider the one parameter family of nonlinear operators $\{G_t\}$ defined by

(2.4)
$$(G_t\varphi)(x) = \inf_\alpha\left\{\int_0^t \ell(y_x(s,\alpha),\alpha(s))\,e^{-\lambda s}\,ds + \varphi(y_x(t,\alpha))\,e^{-\lambda t}\right\}$$
$$=: \inf_\alpha (G_t^\alpha\varphi)(x)\,.$$

It is not hard to check that $\{G_t\}$ has the following properties

(2.5)
$$G_t\varphi \in B(\mathbb{R}^N)$$
$$G_0\varphi = \varphi, \qquad G_{t+s}\varphi = G_t(G_s\varphi) \qquad \forall\, t,s > 0$$

for all $\varphi \in B(\mathbb{R}^N)$, the space of bounded functions on $\mathbb{R}^N$. Moreover, from the inequality

$$|(G_t^\alpha\varphi)(x) - J(x,\alpha)| \leq (\lambda^{-1}\sup|\ell| + \sup|\varphi|)\,e^{-\lambda t},$$

which holds for any fixed α, we get

(2.6)
$$\lim_{t\to+\infty} \|G_t\varphi - v\|_\infty = 0, \qquad \forall\,\varphi \in B(\mathbb{R}^N),$$

where $\|\cdot\|_\infty$ denotes the sup-norm. On the other hand, the semigroup property in (2.5) yields

$$\lim_{t\to+\infty} G_{t+T}\varphi = G_T \lim_{t\to+\infty} G_t\varphi\,.$$

Hence, by (2.6), v satisfies

(2.7)
$$v(x) = (G_T v)(x) \qquad \forall\, x \in \mathbb{R}^N,\ T > 0\,.$$

The preceding says that the value function v is invariant for the dynamical system defined on $B(\mathbb{R}^N)$ by the semigroup $\{G_t\}$. Of course, (2.7) is an equivalent formulation of the Dynamic Programming Principle (2.3).

3. The Hamilton-Jacobi-Bellman equation in the viscosity sense

In this section we derive an infinitesimal version of the Dynamic Programming Principle (2.3) or (2.7) in the form of a first order nonlinear partial differential equation satisfied by the value function v, the *Hamilton-Jacobi-Bellman equation*:

(HJB)
$$\lambda v(x) + \sup_{a\in A}\{-f(x,a)\cdot Dv(x) - \ell(x,a)\} = 0\,.$$

Here, $Dv(x)$ denotes the gradient of v at the point x.

The derivation of (HJB) requires, of course, some smoothness of v. In this event, (HJB) holds at all points where v is differentiable. This follows from (2.3) by standard arguments (divide both sides of (2.3) by $T > 0$ and let $T \to 0 \ldots$). This classical result can be found in many books (see the notes at the end of the chapter for references). A complete proof appears in Chapter III.

It is well known, however, that everywhere differentiability is a too restrictive assumption on v. A simple explicit example is the following.

EXAMPLE 3.1. Consider the infinite horizon problem with $N = 1$, $A = \{-1, 1\}$, $f(x, a) = a$. Suppose that $\ell(x, a) = \ell(x)$ is a smooth function such that

$$\ell(x) = \ell(-x), \qquad \ell \equiv 0 \quad \text{for } |x| > R,$$
$$\max \ell = \ell(0) > 0, \qquad x\ell'(x) < 0 \quad \text{for } |x| < R.$$

It is easy to realize that an optimal control is $\alpha_x^*(t) = \operatorname{sgn} x$ for $x \neq 0$, whereas, for $x = 0$, both $\alpha^*(t) \equiv 1$ and $\alpha^*(t) \equiv -1$ are optimal since ℓ is even. A simple computation then leads to the following expression for the value function,

$$v(x) = \begin{cases} \int_0^{+\infty} \ell(x-t) e^{-\lambda t} \, dt, & x < 0, \\ \int_0^{+\infty} \ell(-t) e^{-\lambda t} \, dt = \int_0^{+\infty} \ell(t) e^{-\lambda t} \, dt, & x = 0, \\ \int_0^{+\infty} \ell(x+t) e^{-\lambda t} \, dt, & x > 0. \end{cases}$$

Hence

$$v'_+(0) = \lim_{x \to 0^+} \frac{v(x) - v(0)}{x} = \int_0^{+\infty} \ell'(t) e^{-\lambda t} \, dt$$

$$v'_-(0) = \lim_{x \to 0^-} \frac{v(x) - v(0)}{x} = \int_0^{+\infty} \ell'(-t) e^{-\lambda t} \, dt \, .$$

Since $\ell'(-x) = -\ell'(x)$, v is not differentiable at $x = 0$. The (HJB) equation for this example is

$$\lambda v(x) + |v'(x)| - \ell(x) = 0;$$

of course the above has no classical meaning for $x = 0$ since v is not differentiable at $x = 0$. ◁

The preceding discussion shows that the derivation of (HJB) should be performed in some weak sense. Observe that under rather general conditions on the data, including the boundedness of ℓ, the value function is continuous and bounded in $\mathbb{R}^N$.

Let us describe then how to interpret (HJB) in the *viscosity sense* assuming that v is continuous in $\mathbb{R}^N$. For this purpose, we consider the sets

$$D^+v(x) = \left\{ p \in \mathbb{R}^N : \limsup_{y \to x} \frac{v(y) - v(x) - p \cdot (y - x)}{|y - x|} \leq 0 \right\}$$

$$D^-v(x) = \left\{ q \in \mathbb{R}^N : \liminf_{y \to x} \frac{v(y) - v(x) - q \cdot (y - x)}{|y - x|} \geq 0 \right\},$$

called, respectively, the *superdifferential* and the *subdifferential* of v at x. Note that if both $D^+v(x)$ and $D^-v(x)$ are nonempty at some x, then $D^+v(x) = D^-v(x) = \{Dv(x)\}$ and v is differentiable at x.

Following [CL83] and [CEL84], a continuous function u is a *viscosity solution* of the nonlinear partial differential equation

$$(3.1) \qquad F(x, u(x), Du(x)) = 0, \qquad x \in \mathbb{R}^N,$$

if the following conditions are satisfied:

(i) $\qquad F(x, u(x), p) \leq 0 \qquad\qquad \forall x \in \mathbb{R}^N, \ \forall p \in D^+u(x)$

(ii) $\qquad F(x, u(x), q) \geq 0 \qquad\qquad \forall x \in \mathbb{R}^N, \ \forall q \in D^-u(x)$.

Any u satisfying (i) will be called a *viscosity subsolution* of (3.1) whereas, if (ii) holds, then u is a *viscosity supersolution*.

An equivalent way to formulate conditions (i) and (ii) is in terms of *test functions*. Namely, (i) holds if and only if

(j) for any $\varphi \in C^1$, if x is a local maximum point for $u - \varphi$, then

$$F(x, u(x), D\varphi(x)) \leq 0;$$

similarly, (ii) holds if and only if

(jj) for any $\varphi \in C^1$, if x is a local minimum point for $u - \varphi$, then

$$F(x, u(x), D\varphi(x)) \geq 0.$$

We refer to Chapter II for the proof of the equivalence.

Let us now sketch the proof that if the value function v is continuous, then it satisfies (HJB) in the viscosity sense. We first check the subsolution condition (i) which, in the present case where

$$F(x, u, p) = \lambda u + \sup_{a \in A}\{-f(x, a) \cdot p - \ell(x, a)\},$$

reads as

$$(3.2) \quad \lambda v(x) + \sup_{a \in A}\{-f(x, a) \cdot p - \ell(x, a)\} \leq 0, \qquad \forall p \in D^+v(x), \ \forall x \in \mathbb{R}^N.$$

To see this, choose $\alpha(t) \equiv a$ with arbitrary fixed $a \in A$ in the Dynamic Programming Principle (2.3) to obtain

$$(3.3) \quad v(y_x(T, a)) - v(x) + \int_0^T \ell(y_x(t, a), a) e^{-\lambda t} \, dt + v(y_x(T, a))(e^{-\lambda T} - 1) \geq 0.$$

For any $p \in D^+v(x)$ we have, for small enough $T > 0$,

$$v(y_x(T, a)) - v(x) \leq p \cdot (y_x(T, a) - x) + o(|y_x(T, a) - x|),$$

since $y_x(T,a) \to x$ as $T \to 0^+$. Hence, from (3.3), after dividing by $T > 0$,

$$(3.4) \quad p \cdot \frac{y_x(T,a) - x}{T} + \frac{o(|y_x(T,a) - x|)}{T} + \frac{1}{T}\int_0^T \ell(y_x(t,a), a)\, e^{-\lambda t}\, dt$$
$$+ v(y_x(T,a))\frac{e^{-\lambda T} - 1}{T} \geq 0 \,.$$

Observe now that if f and ℓ are continuous, then

$$\frac{y_x(T,a) - x}{T} = \frac{1}{T}\int_0^T f(y_x(t,a), a)\, dt \longrightarrow f(x,a),$$

$$\frac{1}{T}\int_0^T \ell(y_x(t,a), a)\, e^{-\lambda t}\, dt \longrightarrow \ell(x,a),$$

as $T \to 0^+$. Therefore, if v is continuous, letting $T \to 0^+$ in (3.4) yields

$$p \cdot f(x,a) + \ell(x,a) - \lambda v(x) \geq 0 \,.$$

Since a was arbitrary in A, this proves (3.2).

To check the supersolution condition, note that for any $\varepsilon > 0$ and $T > 0$ there exists a control $\overline{\alpha}$ (depending on ε, T) such that

$$(3.5) \quad v(x) \geq \int_0^T \ell(y_x(t,\overline{\alpha}), \overline{\alpha}(t))\, e^{-\lambda t}\, dt + v(y_x(T,\overline{\alpha}))\, e^{-\lambda T} - T\varepsilon \,.$$

This follows easily from (2.3). A similar argument as above, with an additional technical difficulty due to the fact that $\overline{\alpha}$ is not necessarily constant, allows us to use (3.5) to conclude that

$$\lambda v(x) + \sup_{a \in A}\{-f(x,a) \cdot q - \ell(x,a)\} \geq 0 \quad \forall x \in \mathbb{R}^N,\ \forall q \in D^- v(x) \,.$$

Therefore, v is a viscosity solution of equation (HJB).

Note finally that the above proof shows that (HJB) holds in the classical sense at all points where v is differentiable.

For the reader who knows the classical derivation of the HJB equation at points of differentiability of v, we point out that proving that v is a viscosity subsolution is very easy by means of the second definition based on test functions. In fact, at a point x where $v - \varphi$ has a maximum we can replace the term $v(y_x(T,a)) - v(x)$ in (3.3) by $\varphi(y_x(T,a)) - \varphi(x)$, preserving the inequality. Then, the passage to the limit can be performed because φ is C^1.

4. Comparison, uniqueness and stability of viscosity solutions

The topics we discuss in this section form, in a sense, the core of the theory of viscosity solutions, specifically in the direction of applications to optimal control.

4. COMPARISON, UNIQUENESS AND STABILITY

On the other hand, from a PDE point of view, they relate to the notion of well-posedness. We have seen in Section 3 that the value function of the infinite horizon problem satisfies, when it is continuous, the HJB equation

(HJB) $$\lambda v(x) + \sup_{a \in A}\{-f(x,a) \cdot Dv(x) - \ell(x,a)\} = 0, \quad x \in \mathbb{R}^N,$$

in the viscosity sense, that is

(4.1) $\lambda v(x) + \sup_{a \in A}\{-f(x,a) \cdot p - \ell(x,a)\} \leq 0 \quad \forall x \in \mathbb{R}^N, \, p \in D^+ v(x)$

(4.2) $\lambda v(x) + \sup_{a \in A}\{-f(x,a) \cdot q - \ell(x,a)\} \geq 0 \quad \forall x \in \mathbb{R}^N, \, q \in D^- v(x).$

It is of major importance to point out that v is in fact characterized by conditions (4.1), (4.2) under rather general assumptions on the data f, ℓ, A (see §II.3 and §III.2). This is a key point since it provides the basis for rigorous justification of the synthesis procedure in the nonsmooth case (see Sections 5–7).

Let us present here a prototype *uniqueness result* for (HJB). Assume that $\lambda = 1$, $f(x,a) \equiv f(a)$, $\ell(x,a) \equiv \ell(x)$ is uniformly continuous and, in order to avoid technicalities, that v satisfies the "boundary condition"

(4.3) $$\lim_{|x| \to +\infty} v(x) = 0.$$

The claim is that if u is any bounded continuous function in $\mathbb{R}^N$ satisfying (HJB) in the viscosity sense and (4.3), then

(4.4) $$u \equiv v.$$

In the present simplified setting (HJB) becomes

(4.5) $$v(x) + H(Dv(x)) - \ell(x) = 0 \quad \text{in } \mathbb{R}^N,$$

with

(4.6) $$H(p) = \sup_{a \in A}\{-f(a) \cdot p\}.$$

To prove the claim, we exploit the information contained in the *subsolution* condition (i.e., (4.1)) for u and in the *supersolution* condition (i.e., (4.2)) for v, to obtain the *comparison principle*

(4.7) $$u(x) \leq v(x) \quad \forall x \in \mathbb{R}^N.$$

The technical tool to this end is to consider the auxiliary function

$$\Phi(x,y) := u(x) - v(y) - \frac{|x-y|^2}{2\varepsilon} \quad x, y \in \mathbb{R}^N$$

where $\varepsilon > 0$ is a small parameter. Assume by contradiction that (4.7) is false so that $(u-v)(\widetilde{x}) > 0$ for some $\widetilde{x}$. Then, by virtue of (4.3) there exists a global maximum point $(x_\varepsilon, y_\varepsilon)$ for Φ. Then x_ε is a maximum for

$$\varphi(x) := u(x) - v(y_\varepsilon) - \frac{|x - y_\varepsilon|^2}{2\varepsilon},$$

while y_ε is a minimum for

$$\psi(y) := v(y) - u(x_\varepsilon) + \frac{|x_\varepsilon - y|^2}{2\varepsilon}.$$

Consequently, as easily seen from the definition of D^+, D^-,

(4.8) $$0 \in D^+\varphi(x_\varepsilon), \qquad 0 \in D^-\psi(y_\varepsilon).$$

Now it is not hard to realize (note that the terms $|x-y_\varepsilon|^2/2\varepsilon$ and $|x_\varepsilon - y|^2/2\varepsilon$ are differentiable) that

$$D^+\varphi(x_\varepsilon) = D^+u(x_\varepsilon) - \frac{x_\varepsilon - y_\varepsilon}{\varepsilon}, \qquad D^-\psi(y_\varepsilon) = D^-v(y_\varepsilon) - \frac{x_\varepsilon - y_\varepsilon}{\varepsilon}$$

(see §II.2 for these calculus properties). Therefore, using (4.8), we see that

$$\frac{x_\varepsilon - y_\varepsilon}{\varepsilon} \in D^+u(x_\varepsilon) \cap D^-v(y_\varepsilon).$$

Since u and v are, respectively, viscosity sub- and supersolution of (4.5), we then obtain

$$u(x_\varepsilon) + H\left(\frac{x_\varepsilon - y_\varepsilon}{\varepsilon}\right) - \ell(x_\varepsilon) \leq 0, \qquad v(y_\varepsilon) + H\left(\frac{x_\varepsilon - y_\varepsilon}{\varepsilon}\right) - \ell(y_\varepsilon) \geq 0.$$

By addition the above yields $u(x_\varepsilon) - v(y_\varepsilon) \leq \ell(x_\varepsilon) - \ell(y_\varepsilon)$. Since, for any $x \in \mathbb{R}^N$, $\Phi(x_\varepsilon, y_\varepsilon) \geq \Phi(x, x) = u(x) - v(x)$, we deduce the inequality

(4.9) $$u(x) - v(x) \leq \ell(x_\varepsilon) - \ell(y_\varepsilon) \qquad \forall x \in \mathbb{R}^N.$$

From the inequality $\Phi(x_\varepsilon, y_\varepsilon) \geq \Phi(x_\varepsilon, x_\varepsilon)$ we get

$$\frac{|x_\varepsilon - y_\varepsilon|^2}{2\varepsilon} \leq v(x_\varepsilon) - v(y_\varepsilon).$$

Since v is bounded, the preceding yields the estimate

$$|x_\varepsilon - y_\varepsilon| \leq C\varepsilon^{1/2}.$$

Therefore, from (4.9) and the uniform continuity of ℓ,

(4.10) $$u(x) - v(x) \leq \omega(C\varepsilon^{1/2}) \qquad \forall x \in \mathbb{R}^N,$$

where ω is a modulus of continuity for ℓ. The above inequality holds for any $\varepsilon > 0$, and the this contradicts $(u-v)(\widetilde{x}) > 0$ for small ε. Hence (4.7) is proved. Since u

and v are, respectively, also a viscosity super and subsolution of (4.5), we can exchange the roles of u and v in the previous proof to obtain the inequality $u \geq v$. This completes the proof of (4.4).

Let us observe explicitly that the preceding proof shows that the following comparison principle holds:

(4.11) if u_1, u_2 are, respectively, viscosity sub- and supersolution of (4.5), then $u_1 \leq u_2$.

Essentially the same proof works when one adopts the test functions definition. The test functions one uses are

$$x \longmapsto v(y_\varepsilon) + \frac{|x - y_\varepsilon|^2}{2\varepsilon} := \varphi_1(x)$$

and

$$y \longmapsto u(x_\varepsilon) - \frac{|x_\varepsilon - y|^2}{2\varepsilon} := \varphi_2(y);$$

note that

$$D\varphi_1(x_\varepsilon) = \frac{x_\varepsilon - y_\varepsilon}{\varepsilon} = D\varphi_2(y_\varepsilon) .$$

The second issue that we touch upon here is that of stability of viscosity solutions. In order to illustrate this point, consider viscosity solutions v_n of

(HJB)$_n$ $\qquad \lambda v_n(x) + H_n(x, Dv_n(x)) = 0 \qquad$ in $\mathbb{R}^N$

with

$$H_n(x, p) := \sup_{a \in A} \{ -f_n(x, a) \cdot p - \ell_n(x, a) \} .$$

Assume we know that, as $n \to +\infty$,

(4.12) $\qquad\qquad v_n \longrightarrow u, \qquad H_n \longrightarrow H$

with

$$H(x, p) = \sup_{a \in A} \{ -f(x, a) \cdot p - \ell(x, a) \} .$$

The natural question is: does u satisfy (HJB) in the viscosity sense? The answer is yes, provided the convergence in (4.12) is uniform on compact sets. Of course, if uniqueness holds for the limit equation (HJB), this is a stability property of the value function v with respect to suitable perturbations of the data f, ℓ.

The proof is particularly simple with the test function definition. Let x_0 be a local maximum, which we can always assume to be strict, for $u - \varphi$, $\varphi \in C^1$. By uniform convergence, there exists a sequence $\{x_n\}$ such that

(4.13) $\qquad x_n$ is a local maximum for $v_n - \varphi$, $x_n \to x_0$, $v_n(x_n) \longrightarrow u(x_0)$.

By definition of viscosity solution we have

$$\lambda v_n(x_n) + H_n(x_n, D\varphi(x_n)) \leq 0;$$

letting $n \to +\infty$ in the above we obtain since $D\varphi$ is continuous

(4.14) $$\lambda u(x_0) + H(x_0, D\varphi(x_0)) \leq 0$$

that is, u is a viscosity subsolution of (HJB). The proof that u is also a supersolution is completely similar.

Let us point out that the so-called *generalized solutions* of (HJB), i.e., locally Lipschitz continuous functions satisfying the equation almost everywhere, do not share these uniqueness and stability properties (see §II.2). In some situations it is not easy to use the stability property as previously formulated since equicontinuity estimates ensuring the uniform convergence of v_n may be hard to get. In fact, a stronger stability property holds. Consider the following *weak limits in the viscosity sense*:

$$\underline{u}(x) := \lim_{\delta \to 0^+} \inf\{ v_n(y) : |x - y| \leq \delta,\ n \geq 1/\delta \}$$

$$\overline{u}(x) := \lim_{\delta \to 0^+} \sup\{ v_n(y) : |x - y| \leq \delta,\ n \geq 1/\delta \}.$$

It is not hard to prove that if x_0 is a strict local maximum for $\overline{u} - \varphi$, then there exists $\{x_n\}$ satisfying (4.13). Hence, the argument above shows that $\overline{u}$ satisfies the subsolution inequality (4.14). Note, however, that $\overline{u}$ is, in general, just upper semicontinuous.

Similarly, the lower semicontinuous function $\underline{u}$ satisfies the supersolution inequality. This leads naturally to extending the notion of subsolution to upper semicontinuous functions and that of supersolution to lower semicontinuous functions.

A remarkable fact is that the comparison principle (4.11) still holds for these extended definitions. Therefore, $\overline{u} \leq \underline{u}$ and since the reverse inequality is obvious, we obtain $\underline{u} = \overline{u}$. This implies the continuity of the weak limits $\underline{u}, \overline{u}$, their coincidence with the value function v and the local uniform convergence of v_n to v. This topic is discussed in Chapter V, and several applications are given in Chapters VI and VII.

5. Synthesis of optimal controls and verification theorems

Let us recall the classical *synthesis procedure* when the value function v is smooth. In the derivation of the Dynamic Programming Principle we observed that the function

$$h(t) := v(y^*(t)) e^{-\lambda t} + \int_0^t \ell(y^*(s), \alpha^*(s)) e^{-\lambda s}\, ds$$

is constant for all $t > 0$ if and only if α^*, y^* is an optimal control-trajectory pair for the initial position x; see (2.2). Therefore, if v is smooth, the optimality condition is $h' \equiv 0$, that is

$$e^{-\lambda t}[\lambda v(y^*(t)) - f(y^*(t), \alpha^*(t)) \cdot Dv(y^*(t)) - \ell(y^*(t), \alpha^*(t))] \equiv 0 .$$

Since in this case v is a classical solution of the HJB equation

$$\lambda v + H(x, Dv) = 0 \quad \text{in } \mathbb{R}^N,$$

where

$$H(x, p) := \sup_{a \in A} \{ -f(x, a) \cdot p - \ell(x, a) \},$$

we conclude that the control α^* is optimal for the initial state x if and only if

(5.1) $$\alpha^*(t) = S(y^*(t)) \quad \text{for a.e. } t > 0,$$

for any choice of $S(z)$ such that

(5.2) $$S(z) \in \arg\max_{a \in A} \{ -f(z, a) \cdot Dv(z) - \ell(z, a) \},$$

that is, if and only if

$$H(y^*(t), Dv(y^*(t))) = -f(y^*(t), \alpha^*(t)) \cdot Dv(y^*(t)) - \ell(y^*(t), \alpha^*(t)),$$

for a.e. $t > 0$. This characterization of optimal open loop controls provides a method for constructing an optimal pair control-trajectory for every initial condition. The first step is to find a map $S : \mathbb{R}^N \to A$ with the property (5.2); if v is known this is a static, finite dimensional, mathematical programming problem. Such a map S is called an *optimal feedback map*. The second step is solving

(5.3) $$\begin{cases} y' = f(y, S(y)) & t > 0, \\ y(0) = x, \end{cases}$$

and a solution $y^*(t)$ generates a control $\alpha^*(t) := S(y^*(t))$, which is optimal for the initial state x.

The applicability of this method requires the regularity of the value function for the characterization of optimal controls and some regularity of the feedback map S for the solvability of (5.3). A typical example when this favorable situation occurs is linear systems with quadratic costs. In Chapter III we present several necessary and sufficient conditions of optimality in the spirit of this method holding for nondifferentiable value function.

We consider, for instance, the case when the value function v is *semiconcave*; that is, for some $C > 0$

(5.4) $$v(x - h) - 2v(x) + v(x + h) \leq C |h|^2$$

for all x and h. A simple situation where this is true occurs for f and ℓ smooth and interest rate λ large enough. In this case a necessary and sufficient condition of optimality is that (5.1) holds for a.e. $t > 0$, where S satisfies

$$(5.5) \quad S(z) \in \arg\max_{a \in A}\{-f(z,a) \cdot p - \ell(z,a)\} \quad \text{for all } p \in D^+v(z) \cup D^-v(z)$$

instead of (5.2). We remark that a key property of semiconcave functions in this respect is that $D^+v(z) \neq \emptyset$ for all z. For merely Lipschitz value functions even $D^+v(z) \cup D^-v(z)$ may be empty at some point, so that condition (5.5) does not make sense. However, it can be replaced by a condition involving generalized directional derivatives in the sense of Dini (see §III.2.5).

If the value function is merely continuous, (5.1) and (5.5) are still necessary conditions of optimality and the problem of finding sufficient conditions in this form is open to our knowledge.

A slightly different approach to sufficient conditions of optimality for merely continuous value functions is a nonsmooth version of the classical Hamilton-Jacobi-Carathéodory *verification technique*, also named "the Basic Algorithm" by L.C. Young [You69]. This is based on the notion of *verification function*. Classically, such a function u is assumed to be bounded and C^1 and satisfy

$$(5.6) \quad \lambda u(x) + \sup_{a \in A}\{-f(x,a) \cdot Du(x) - \ell(x,a)\} \leq 0 \quad \forall x \in \mathbb{R}^N.$$

Let α^* be an admissible control with corresponding trajectory y^* starting at x. A simple verification result is that α^* is optimal if there exists a verification function u such that

$$(5.7) \quad \lambda u(y^*(t)) - f(y^*(t), \alpha^*(t)) \cdot Du(y^*(t), \alpha^*(t)) - \ell(y^*(t), \alpha^*(t)) = 0$$

for $t \in [0, +\infty[$. In fact, for arbitrary $\alpha \in \mathcal{A}$ and corresponding trajectory $y(t)$ starting at x, we have actually

$$\lambda u(y(t)) - f(y(t), \alpha(t)) \cdot Du(y(t)) \leq \ell(y(t), \alpha(t))$$

by the very definition of verification function. Observe now that the left-hand member of the above inequality is precisely

$$-e^{\lambda t} \frac{d}{dt}(e^{-\lambda t} u(y(t))).$$

Hence, integration of both sides of the inequality on $[0, +\infty[$ yields

$$u(x) \leq \int_0^{+\infty} \ell(y(t), \alpha(t)) e^{-\lambda t} dt = J(x, \alpha), \quad \forall \alpha \in \mathcal{A}.$$

For the particular control α^* the same argument gives

$$u(x) = J(x, \alpha^*),$$

so that α^* is optimal.

Conversely, it is immediate to check that if the *control law* $x \mapsto \alpha_x \in \mathcal{A}$ is optimal for all x and $u(x) := J(x, \alpha_x)$ is differentiable, then u is a verification function. Indeed, by its very definition, u coincides with the value function v, which in this case is a classical solution of (HJB) and, a fortiori, of (5.6).

Viscosity subsolutions provide a quite natural tool to extend the verification technique to a more general setting. Let us stipulate that $u \in C(\mathbb{R}^N)$ is a generalized verification function if inequality (5.6) is satisfied in the viscosity sense. The verification result in this framework is as follows: a control $\alpha^* \in \mathcal{A}$ is optimal for the initial state x if and only if there exists a generalized verification function u such that $u(x) \geq J(x, \alpha^*)$. The proof is very simple: the comparison principle yields $u(x) \leq v(x)$, so that

$$J(x, \alpha^*) \leq u(x) \leq v(x) \leq J(x, \alpha^*) \,.$$

To prove the reverse implication, just observe that the optimality of α^* implies $v(x) = J(x, \alpha^*)$ and recall that v itself is a viscosity subsolution of (HJB).

A computationally oriented approach to the synthesis problem is reviewed in §7.

6. Pontryagin Maximum Principle as a necessary and sufficient condition of optimality

The Pontryagin Maximum Principle is the most classical and useful necessary condition of optimality. It is well-known that if the value function is smooth, there is a formula linking its gradient with the costate vector. Next we show how to interpret this relation in terms of generalized gradients when the value is not differentiable. A relevant by-product is a formulation of the Principle as a necessary and sufficient condition.

To illustrate these ideas we choose the simplest problem, that is, the finite horizon Mayer problem whose cost functional is

(6.1) $$J(x, t, \alpha) := g(y_x(t, \alpha)), \qquad g \in C^1(\mathbb{R}^N),$$

where t is the fixed horizon. Later in this section we outline the extension of the result to the infinite horizon problem. The value function is now

$$v(x, t) := \inf_{\alpha \in \mathcal{A}} J(x, t, \alpha) \,.$$

We fix $x_0 \in \mathbb{R}^N$, $T > 0$ and $\alpha \in \mathcal{A}$, and investigate the optimality of the corresponding trajectory

$$y(t) := y_{x_0}(t, \alpha), \qquad 0 \leq t \leq T \,.$$

Assume that the dynamics f is differentiable with respect to the state variable and $D_x f$ is bounded. The *costate* (or *adjoint vector*) associated with x_0, T, α is the unique solution of the adjoint system

(6.2) $$\begin{cases} p'(t) = -p(t) D_x f(y(t), \alpha(t)) & 0 < t < T, \\ p(T) = Dg(y(T)) . \end{cases}$$

The classical Pontryagin Maximum Principle (shortly, PMP) states that the following condition is necessary for optimality:

(6.3) $$-p(t) \cdot f(y(t), \alpha(t)) = \max_{a \in A}\{ -p(t) \cdot f(y(t), a) \} .$$

Moreover, if v is smooth, the following relations between the costate and the value function are well known (see also §III.3.4)

(6.4) $$p(t) = D_x v(y(t), T - t),$$

(6.5) $$p(t) \cdot f(y(t), \alpha(t)) = \frac{\partial v}{\partial t}(y(t), T - t) .$$

The Hamilton-Jacobi-Bellman equation for this problem is

$$\frac{\partial v}{\partial t} + H(x, D_x v) = 0 \quad \text{in } \mathbb{R}^N \times]0, +\infty[,$$

where

$$H(x, q) := \sup_{a \in A}\{ -q \cdot f(x, a) \} .$$

With this notation (6.5) can be written as

(6.6) $$\frac{\partial v}{\partial t}(y(t), T - t) = -H(y(t), p(t)) .$$

In the case where v is not differentiable we consider the following nonsmooth counterpart of (6.4) and (6.6):

(6.7) $$(p(t), -H(y(t), p(t))) \in D^+ v(y(t), T - t) .$$

Here, D^+ denotes the superdifferential of v with respect to the $N+1$ dimensional vector (x, t). Now we can formulate the PMP as follows: *the control α is optimal if and only if (6.3) and (6.7) are verified for a.e. $t \in]0, T[$.*

To prove the necessity it is enough to show that, for a.e. t,

(6.8) $$(p(t), p(t) \cdot f(y(t), \alpha(t))) \in D^+ v(y(t), T - t) .$$

In fact, v being a subsolution of the HJB equation, this implies

$$p(t) \cdot f(y(t), \alpha(t)) + \sup_{a \in A}\{ -p(t) \cdot f(y(t), a) \} \leq 0,$$

which immediately gives (6.3) and (6.7). Next we sketch the proof of (6.8) (see Proof 3 of Theorem III.3.42 for more details). First note that, if α is optimal, the Dynamic Programming Principle yields

$$v(y(t), T-t) = J(y(t), T-t, \alpha^{(t)}),$$

for all $t \in [0, T]$, where

$$\alpha^{(t)}(s) := \alpha(t+s) .$$

Therefore,

(6.9) $\quad v(x, s) - v(y(t), T-t) \leq J(x + h(x, s), T-t, \alpha^{(t)}) - J(y(t), T-t, \alpha^{(t)})$

where

(6.10) $$h(x, s) = \int_{T-s}^{t} [f(y(\tau), \alpha(\tau)) + o(1)] \, d\tau$$

as $s \to T-t$, $x \to y(t)$. By the differentiability of g and of solutions of ordinary differential equations with respect to the initial data, with some computations one can see that J is differentiable with respect to the state variable with

$$D_x J(y(t), T-t, \alpha^{(t)}) = p(t) .$$

Hence, by (6.9) and (6.10),

$$v(x, s) - v(y(t), T-t) \leq p(t) \cdot (x + h(x, s) - y(t)) + o(|x - y(t)| + |s - T + t|) .$$

Now, the validity of (6.8) for a.e. t follows easily from (6.10) and Lebesgue derivation theorem.

For the sufficiency, we set

$$p_0(t) := -H(y(t), p(t))$$

and note that, by (6.3),

$$p(t) \cdot y'(t) - p_0(t) = 0$$

for all t where y' exists. Then, by (6.7),

(6.11) $\quad \varepsilon^{-1}[v(y(t+\varepsilon), T-t-\varepsilon) - v(y(t), T-t)] \leq p \cdot y'(t) - p_0 + o(1)$
$\hspace{6cm} = o(1) \qquad \text{for } \varepsilon \to 0 .$

Now we recall that

$$h(t) := v(y(t), T-t)$$

is nonincreasing if and only if the trajectory is optimal. Since v is locally Lipschitz because the data f and g are differentiable, h is locally Lipschitz as well, therefore differentiable almost everywhere. Thus for almost every t (6.11) gives $h'(t) \leq 0$, which completes the proof.

It is useful to point out that the previous formulation of the PMP remains valid even if g is not differentiable, provided $D^+g(y(t))$ is nonempty. Indeed, it is easy to check that everything works if the terminal condition in the definition of the costate (6.2) is replaced by

$$p(T) \in D^+g(y(T)) \,.$$

Next we show how to formulate the PMP as a necessary and sufficient condition of optimality for the model problem of this chapter, that is, the infinite horizon problem. To this end we introduce

$$(6.12) \qquad V(x,t) := \inf_{\alpha \in \mathcal{A}} \left\{ \int_t^T \ell(y(s), \alpha(s)) e^{-\lambda s} \, ds + e^{-\lambda T} v(y(T)) \right\},$$

where v is the value function of the infinite horizon problem defined in §1 and y is the trajectory of the system such that $y(t) = x$. Note that

$$V(x, 0) = v(x)$$

by the Dynamic Programming Principle (2.3), and it is not hard to show that

$$(6.13) \qquad V(x,t) = e^{-\lambda t} v(x) \qquad \text{for all } t \in [0,T] \,.$$

Observe also that, by its very definition, V is the value function of a finite horizon Bolza problem where v represents the terminal cost. By standard tricks based on the introduction of new state variables and by reversing time this problem can be reformulated as a Mayer problem such as the one previously discussed in this section. Then we can apply the PMP, provided ℓ is C^1 with respect to the state variables, and $D^+v(z)$ is nonempty for all $z \in \mathbb{R}^N$. This is the case, for example, if v is semiconcave; that is, it satisfies (5.4).

For problems with running cost, the costate equation is

$$(6.14) \qquad p'(t) = -p(t) \, D_x f(y(t), \alpha(t)) - e^{-\lambda t} D_x \ell(y(t), \alpha(t)) \,.$$

For the Bolza problem in (6.12) one defines the costate $p_T(t)$, $t \in [0,T]$, as a solution of (6.14) with the terminal condition

$$(6.15) \qquad p_T(T) \in e^{-\lambda T} D^+ v(y(T)) \,.$$

For this problem, the control α is optimal if and only if

$$(6.16) \qquad -p_T(t) \cdot f(y(t), \alpha(t)) - e^{-\lambda t} \ell(y(t), \alpha(t)) = \widetilde{H}(t, y(t), p_T(t))$$

where

$$\widetilde{H}(t, z, q) := \sup_{a \in A} \{ -q \cdot f(z, a) - e^{-\lambda t} \ell(z, a) \},$$

which replaces (6.3), and

$$(p_T(t), \tilde{H}(t, y(t), p_T(t))) \in D^+V(y(t), t),$$

which replaces (6.7). Observe that, by (6.13), this inclusion can be split as

(6.17) $$p_T(t) \in e^{-\lambda t} D^+ v(y(t))$$

and

(6.18) $$\tilde{H}(t, y(t), p_T(t)) = -\lambda e^{-\lambda t} v(y(t)) .$$

Suppose now that α is an optimal control for x in the infinite horizon problem. Then by the Dynamic Programming Principle in the form (2.2), it is also optimal for the finite horizon $T > 0$. The necessary condition of optimality can therefore be obtained by letting $T \to +\infty$ in (6.16), (6.17), (6.18) and it is the following: there exists a solution $p(t)$, $t \in [0, +\infty[$ of the costate equation (6.14) such that for a.e. $t > 0$

$$-p(t) \cdot f(y(t), \alpha(t)) - e^{-\lambda t} \ell(y(t), \alpha(t)) = \tilde{H}(t, y(t), p(t)),$$
$$p(t) \in e^{-\lambda t} D^+ v(y(t)),$$
$$\tilde{H}(t, y(t), p(t)) = -\lambda e^{-\lambda t} v(y(t)) .$$

This result holds as soon as one knows that $p_T(t)$ has a limit $p(t)$ as $T \to +\infty$ which satisfies the costate equation. This is true, for instance, if v is Lipschitz continuous and, for all t, $\lambda > \operatorname{Re} \mu(t)$ for all eigenvalues $\mu(t)$ of the symmetric part of the matrix $D_x f(y(t), \alpha(t))$. The assumption on v is used to get $p_T(T) \to 0$ from (6.15), and that on λ allows us to pass to the limit in the variation of constants formula for $p_T(t)$. The proof of the sufficiency of these optimality conditions is essentially the same as that for Mayer's problem.

7. Discrete time Dynamic Programming and convergence of approximations

In §5 we discussed the synthesis algorithm in the classical as well as in some "semiclassical" setting (i.e., for semiconcave or Lipschitz continuous value functions). As pointed out, sufficient conditions in the form described there fail for merely continuous value functions.

A natural way to overcome the difficulties related to the lack of regularity is to look at *approximate value functions* v_h satisfying pointwise a finite increment approximate version of the HJB equation. This can be done in several ways. The approach outlined below, which relies on *discrete time Dynamic Programming*, produces closed loop controls which are *suboptimal* for the original problem by virtue of a convergence result of v_h to v.

Let us also point out that the convergence of v_h to v, supplemented by quantitative error estimates, is the basis for *computational methods* of solution of the

HJB equation. We refer to Chapter VI and Appendixes A and B for developments of the themes touched here.

In order to illustrate the approach sketched above, assume that the evolution described by the control system (1.1) is observed only at a sequence of times $t_j = jh$ ($j = 0, 1, \ldots$), where h is a fixed positive real number. Assume as well that the dynamics f and the running cost ℓ remain unchanged between two subsequent observations. More precisely, suppose that, for $t \in [t_j, t_{j+1}[$,

$$f(y(t), \alpha(t)) = f(y_j, a_j), \qquad \ell(y(t), \alpha(t)) = \ell(y_j, a_j),$$

where $a_j = \alpha(t_j)$ and $y_j = y(t_j)$ is given by the recursion

$$(7.1) \qquad y_0 = x, \quad y_{j+1} = y_j + hf(y_j, a_j), \qquad j = 0, 1, \ldots$$

The infinite horizon cost functional associated with initial position x and control α is chosen in this discretized model as the series

$$(7.2) \qquad h \sum_{j=0}^{+\infty} \beta^j \ell(y_j, a_j), \qquad a_j \in A \text{ for } j = 0, 1, \ldots$$

with $\beta = 1 - \lambda h$. The corresponding value function is

$$(7.3) \qquad v_h(x) := \inf_{\alpha \in \mathcal{A}_h} h \sum_{j=0}^{+\infty} \beta^j \ell(y_j, a_j),$$

where $\mathcal{A}_h$ is the subset of $\mathcal{A}$ consisting of those controls that take on a constant value on each time interval, namely, $\alpha(t) = a_j$ for $t \in [t_j, t_{j+1}[$.

It is not hard to check that v_h satisfies a discrete time version of the Dynamic Programming Principle, namely

$$(7.4) \qquad v_h(x) = \inf_{\alpha \in \mathcal{A}_h} \left\{ h \sum_{j=0}^{k-1} \beta^j \ell(y_j, a_j) + \beta^k v_h(y_k) \right\}$$

for all $x \in \mathbb{R}^N$ and integer k. For the choice $k = 1$ the above reads as

$$\text{(HJB)}_h \qquad v_h(x) + \sup_{a \in A} \{ -\beta v_h(x + hf(x, a)) - h\ell(x, a) \} = 0.$$

This equation for v_h can be thought as a discrete version of (HJB).

Let us examine now the limiting behavior of v_h as $h \to 0^+$. Assume we know that v_h converges locally uniformly in $\mathbb{R}^N$ as $h \to 0^+$ to some continuous function u. As shown below, it turns out that u is a viscosity solution of (HJB). Hence u coincides with v, the value function of the infinite horizon problem, provided uniqueness holds for (HJB) (see §4). The fact that $u = \lim_{h \to 0^+} v_h$ is a viscosity subsolution of (HJB) is easily verified. For this purpose, let $\varphi \in C^1(\mathbb{R}^N)$ and x be a strict local maximum point for $u - \varphi$. By uniform convergence, there exist

$x_h \to x$ as $h \to 0^+$ such that x_h is a local maximum for $v_h - \varphi$. Therefore, for small h,

$$v_h(x_h) - v_h(x_h + hf(x_h, a)) \geq \varphi(x_h) - \varphi(x_h + hf(x_h, a)), \qquad \forall a \in A.$$

From (HJB)$_h$ we then obtain

$$0 \geq \varphi(x_h) - \varphi(x_h + hf(x_h, a)) + \lambda h v_h(x_h + hf(x_h, a)) - h\ell(x_h, a)$$

for all $a \in A$. Since $\varphi \in C^1$ this yields, after dividing by $h > 0$,

$$0 \geq -D\varphi(\xi_h) \cdot f(x_h, a) + \lambda v_h(x_h + hf(x_h, a)) - \ell(x_h, a),$$

for some $\xi_h \to 0$ as $h \to 0^+$ By uniform convergence we can pass to the limit as $h \to 0^+$ in the preceding and take the supremum with respect to a to get

$$0 \geq \sup_{a \in A}\{ -D\varphi(x) \cdot f(x, a) + \lambda u(x) - \ell(x, a) \}.$$

Hence, u is a viscosity subsolution of (HJB); the proof that u is a supersolution as well is similar.

Let us observe that the previous argument shows that the weak notion of viscosity solution allows us to pass to the limit as $h \to 0^+$ in the nonlinear equations (HJB)$_h$ under uniform convergence and just continuity assumptions on f and ℓ. Note also that the same result could have been obtained by the weak limits technique (see §4 and Chapter VI).

Let us come back now to the synthesis problem. It is quite natural in the present setting to consider the mapping

(7.5) $$S_h(z) \in \arg\max_{a \in A}\{ -\beta v_h(z + hf(z, a)) - h\ell(z, a) \}.$$

This is easily seen to be an optimal feedback map for the discrete time problem (7.3), that is

(7.6) $$v_h(x) = h \sum_{j=0}^{+\infty} \beta^j \ell(y_j^*, S_h(y_j^*)) \qquad \forall x \in \mathbb{R}^N,$$

with

(7.7) $$y_0^* = x, \qquad y_{j+1}^* = y_j^* + hf(y_j^*, S_h(y_j^*)).$$

Of course, no regularity of v_h or S_h is required for defining the feedback map S_h and the optimal discrete trajectory $\{y_j^*\}$. By the foregoing, the optimal cost $v(x)$ for the original continuous time problem can be approximately computed by (7.5), (7.6), (7.7). We refer to Chapter VI for more complete convergence results.

8. The viscosity approximation and stochastic control

A different insight to viscosity solutions is provided by the so-called *viscosity approximation* of equation (HJB), namely,

$$(\text{HJB})^\varepsilon \qquad -\varepsilon \Delta u^\varepsilon + \lambda u^\varepsilon + \sup_{a \in A}\{-f(x,a) \cdot Du^\varepsilon - \ell(x,a)\} = 0, \qquad x \in \mathbb{R}^N.$$

Here, ε is a positive real parameter and

$$\Delta = \sum_{i=1}^N \frac{\partial^2}{\partial x_i^2}$$

is the Laplace operator. Equation (HJB)$^\varepsilon$ can be interpreted in terms of *stochastic optimal control* theory. A rigorous explanation of this assertion is outside the scope of this book. Therefore we just outline it and refer to [FR75, FS93] for details. Consider the function

$$(8.1) \qquad v^\varepsilon(x) := \inf E_x \int_0^{+\infty} \ell(y_x^\varepsilon(t), \alpha(t)) e^{-\lambda t} \, dt \, .$$

In the above formula, y^ε is the solution of the *stochastic differential equation*

$$(8.2) \qquad \begin{aligned} dy &= f(y(t), \alpha(t)) \, dt + \sqrt{2\varepsilon} \, dw(t), \\ y(0) &= x, \end{aligned}$$

where w is a N-dimensional standard Brownian motion, E_x denotes expectation and the infimum is taken on the class of progressively measurable functions α with values in A.

Function v^ε appears then as the value function of a stochastic version of the infinite horizon problem where one tries to control in an optimal way the trajectories of (8.2), a random perturbation of system (1.1). Under suitable conditions on the data, v^ε happens to be a smooth, say C^2, solution of (HJB)$^\varepsilon$; this follows from the Dynamic Programming Principle and Ito's stochastic calculus. It is therefore natural to guess that when the randomness parameter ε is sent to zero, the value functions v^ε should converge to the value function v, the viscosity solution of (HJB). Let us next sketch a PDE proof of this guess.

Observe that passing to the limit in (HJB)$^\varepsilon$ as $\varepsilon \to 0^+$ in order to recover equation (HJB) is not an easy task. This is due to the nonlinearity of the equation and the fact that the required estimates on u^ε, which can be assumed to be smooth, explode as $\varepsilon \to 0^+$ since the regularizing effect of the additive term $-\varepsilon \Delta$ becomes weaker and weaker. On the other hand, one does not expect (HJB) to have smooth solutions.

Assume therefore that $u^\varepsilon \in C^2(\mathbb{R}^N)$ converge locally uniformly as $\varepsilon \to 0^+$ to some $u \in C(\mathbb{R}^N)$. Now let φ be a C^2 function and x be a strict maximum point for $u - \varphi$. By uniform convergence, $u^\varepsilon - \varphi$ attains a local maximum at some point x^ε and $x^\varepsilon \to x$ as $\varepsilon \to 0^+$. Hence, by elementary calculus,

$$D(u^\varepsilon - \varphi)(x^\varepsilon) = 0, \qquad -\Delta(u^\varepsilon - \varphi)(x^\varepsilon) \geq 0 \, .$$

By $(HJB)^\varepsilon$, then

$$-\varepsilon\Delta\varphi(x^\varepsilon) + \lambda u^\varepsilon(x^\varepsilon) - f(x^\varepsilon,a)\cdot D\varphi(x^\varepsilon) - \ell(x^\varepsilon,a) \leq 0$$

for all $a \in A$. Since $x^\varepsilon \to x$ as $\varepsilon \to 0^+$, we can easily pass to the limit in the preceding using the continuity of $\Delta\varphi, D\varphi, f, \ell$ and the uniform convergence of u^ε to u. The conclusion is that

$$\lambda u(x) + \sup_{a \in A}\{-f(x,a)\cdot D\varphi(x) - \ell(x,a)\} \leq 0,$$

therefore u is a viscosity subsolution of (HJB). In a similar way one can show that u is also a supersolution in the viscosity sense.

The limiting procedure described above to deal with the *singular perturbation problem* $(HJB)^\varepsilon \to (HJB)$ can be thought as a way to define weak solutions of (HJB). This is actually the motivation for the terminology viscosity solutions in the original paper of M.G. Crandall and P.L. Lions [CL81].

9. Bibliographical notes

The Dynamic Programming method in optimal control theory was developed in the '50s by Richard Bellman and his school [Bel57, Ho60, BelD62, Bel71], mainly for discrete time dynamics; for systems governed by differential equations it appears in a paper of Kalman [Kal63] and in Isaacs' monograph on differential games [Is65]. However, the central role played by value functions and Hamilton-Jacobi equations in the classical setting of the Calculus of Variations was recognized as early as in 1935 by C. Carathéodory [Ca35], see [PB94] for a recent survey.

In the '60s, Dynamic Programming became a standard topic in deterministic optimal control theory; see, for example, the books [Hes66, LM67, You69] and the more recent [FR75, Ber76]. However, the lack of smoothness of the value function, even for simple optimal control problems of continuous time processes, caused a considerable restriction to the range of applicability of the Hamilton-Jacobi theory. The main reason for this limitation was the very basic difficulty to give an appropriate sense to the Hamilton-Jacobi equation, a fully nonlinear first order partial differential equation, satisfied at points of differentiability by the value function, and to characterize the value as the unique solution of that equation.

Several non-classical notions of solution of Hamilton-Jacobi type equations have been therefore proposed in the literature. Kružkov built a rather complete theory for semiconvex functions solving the equation almost everywhere in the case of convex (and sufficiently smooth) Hamiltonians [Kr60, Kr67, Kr75] (see also [Dou61]), including the vanishing viscosity approximation [Kr64, Kr66a], and the convergence of finite difference schemes [Kr66b]. More recent references are [Bt77, Hr78, Cla83]; see also [Go76] and, for second order equations, [Me80a, Me80b, BL82]. All these papers dealt with equations convex in the derivatives of the unknown function, whereas Subbotin [Su80, Su84] gave some results for the nonconvex Hamiltonians arising in differential games. His ideas evolved at the end of the '80s into the theory of minimax solutions [Su91b, Su95] which has several deep connections with viscosity solutions (at least for first order equations); see §2.4 of Chapter III, as well as Chapter VIII for some more remarks in connection with the theory of differential games.

A new impulse to the rigorous mathematical justification of Dynamic Programming was originated by the introduction in the early '80s of the notion of viscosity solution by M.G. Crandall and P.L. Lions [CL81, CL83] for general Hamiltonians. Their original definition is related to Kružkov's theory of entropy solutions for scalar conservation laws [Kr70] (see [Cas92] for a direct proof of the connection between the two theories and [CDP96] for related results). In [CL83] some equivalent ways to define viscosity solutions were indicated; the one based on test functions is intimately connected with earlier work of Evans on weak passages to the limit in fully nonlinear PDEs [E78, E80]; see also [E90]. This property was already noted in [Kr75] in the case of convex Hamiltonians. Using this equivalent notion of solution Crandall, Evans and Lions gave rather simple proofs of the good comparison and stability properties of viscosity solutions [CEL84]. The theory was extended to second order degenerate elliptic HJB equations by P.L. Lions [L83a, L83b] with a proof of uniqueness that relied on stochastic control arguments. A major breakthrough was achieved by R. Jensen [J88] who proved a comparison principle for nonconvex degenerate elliptic equations by purely PDE methods; see also [J89]. A comprehensive account of the theory of viscosity solutions at this level of generality can be found in [CIL92, C95], where the reader can find more information and references on its history and developments. For some more recent advances of the theory of second order equations and several applications to various fields we refer to the lecture notes [CDL97], to [AGLM93, ESp95, CafCa95, CCKS96, CKSS96], and the references therein.

Starting from P.L. Lions' remark in the book [L82] that value functions do in fact satisfy Hamilton-Jacobi equations in the viscosity sense, a considerable amount of work has been subsequently devoted to applications of viscosity solutions to optimal control and differential games. Detailed references are given at the end of each chapter of the present book (see also the bibliographies of [L92, Ba94, FS93]). Let us just mention here that applications to switching and impulse control problems were considered, respectively, by Capuzzo Dolcetta and Evans [CDE84] and Barles [Ba85a], differential games were treated first by Evans and Souganidis [ES84] and Barron, Evans and Jensen [BEJ84], that boundary value problems arising in time-optimal control were studied by Bardi [B89] and Evans and James [EJ89], and those arising in state-constrained problems by Soner [S86] and Capuzzo Dolcetta and Lions [CDL90]. Relevant extensions of the theory, motivated by exit time, constrained and differential games problems, to include discontinuous solutions, were proposed by Barles and Perthame [BaP87], Ishii [I87a, I89], Barron and Jensen [BJ90], Bardi and Soravia [BS91a], Subbotin [Su93a]. The book of Elliott [El87] is devoted to some applications of viscosity solutions to deterministic optimal control, whereas Barles' book [Ba94] gives a wider and more recent account.

9. BIBLIOGRAPHICAL NOTES

The problem of the synthesis of optimal feedbacks received important contributions from Boltyanskii [Bol66], Brunovsky [Bru80] and Sussman (see [FR75, Sus89a, Sus90, Mi90a, PSus96] for some recent presentations) but it remains largely open. The approach we present here develops some results of Berkovitz [Be89] and Frankowska [Fr89b], whereas the use of viscosity subsolutions as verification functions gives verification theorems as a very easy consequence of the comparison principle.

For the connections of viscosity solutions with the Pontryagin Maximum Principle we refer to the papers [BJ86, Sua89, Z90, CF91], while the result for the infinite horizon problem presented in §6 is adapted from [Te95].

The relevance of comparison and stability properties of viscosity solutions in connection with the convergence of approximations and computational schemes for value functions and optimal feedbacks has been recognized since the early stages of the theory by Capuzzo Dolcetta [CD83] and Crandall and Lions [CL84], see also [CDI84, Sou85a, Fa87] and Appendix A of the present book.

Among the more recent applications of viscosity solutions in optimal control theory we mention those concerning risk-sensitive and $\mathcal{H}_\infty$ control (see [BJ89, FM92a, Jam92, Sor96a]) and Appendix B of this book.

The generalization of the theory of viscosity solutions to optimal control problems for diffusion processes, initiated by P.L. Lions [L83a, L83b, L83c], goes beyond the scope of the present work. We refer to the recent book by Fleming and Soner [FS93] for a comprehensive presentation of this important topic, to [S95] for its applications to mathematical finance and to [FSo89] for stochastic differential games.

Another issue that is not touched in this book is the viscosity solutions approach to optimal control problems for distributed parameter systems and the corresponding theory of infinite dimensional Hamilton-Jacobi equations; see, for example, [CL85, CL94a, CL94b, L88, L89, CDaP89, CDaP90, CGS93, CT96, I92a, T94, LoS93, KSw95, KSS97], the books [BDaP83, BDDM93, Bar93a, LiY95] and the references therein.

CHAPTER II

Continuous viscosity solutions of Hamilton-Jacobi equations

This chapter is devoted to the basic theory of continuous viscosity solutions of the Hamilton-Jacobi equation

(HJ) $$F(x, u(x), Du(x)) = 0 \quad x \in \Omega,$$

where Ω is an open domain of $\mathbb{R}^N$ and the Hamiltonian $F = F(x, r, p)$ is a continuous real valued function on $\Omega \times \mathbb{R} \times \mathbb{R}^N$.

Special attention will be dedicated in §4,5 to the case where $p \mapsto F(x, r, p)$ is convex and, more particularly, of the form

(0.1) $$F(x, r, p) = r + H(x, p) = r + \sup_{a \in A} \{ -f(x, a) \cdot p - \ell(x, a) \}.$$

Hamiltonian functions of this form arise naturally in connection with optimal control problems as indicated in Chapter I.

1. Definitions and basic properties

In this section we recall the two equivalent definitions of viscosity solutions of (HJ) introduced in Chapter I and discuss their relations with a comparison principle as well as some connections with classical notions of solutions of (HJ).

DEFINITION 1.1. A function $u \in C(\Omega)$ is a *viscosity subsolution* of (HJ) if, for any $\varphi \in C^1(\Omega)$,

(1.1) $$F(x_0, u(x_0), D\varphi(x_0)) \le 0$$

at any local maximum point $x_0 \in \Omega$ of $u - \varphi$. Similarly, $u \in C(\Omega)$ is a *viscosity supersolution* of (HJ) if, for any $\varphi \in C^1(\Omega)$,

(1.2) $$F(x_1, u(x_1), D\varphi(x_1)) \ge 0$$

at any local minimum point $x_1 \in \Omega$ of $u - \varphi$. Finally, u is a *viscosity solution* of (HJ) if it is simultaneously a viscosity sub- and supersolution. ◁

Let us mention explicitly that the definition applies to evolutionary Hamilton-Jacobi equation of the form

$$u_t(t,y) + F(t, y, u(t,y), D_y u(t,y)) = 0, \qquad (t,y) \in \,]0, T[\, \times D\,.$$

Indeed, the equation above is reduced to the form (HJ) by the positions

$$x = (t,y) \in \Omega = \,]0, T[\, \times D \subseteq \mathbb{R}^{N+1}, \qquad \widetilde{F}(x, r, q) = q_{N+1} + F(x, r, q_1, \ldots, q_N)$$

with

$$q = (q_1, \ldots, q_N, q_{N+1}) \in \mathbb{R}^{N+1}\,.$$

REMARK 1.2. In the definition of subsolution we can always assume that x_0 is a local strict maximum point for $u - \varphi$ (otherwise, replace $\varphi(x)$ by $\varphi(x) + |x - x_0|^2$). Moreover, since (1.1) depends only on the value of $D\varphi$ at x_0, it is not restrictive to assume that $u(x_0) = \varphi(x_0)$. Similar remarks apply of course to the definition of supersolution. Geometrically, this means that the validity of the subsolution condition (1.1) for u is tested on smooth functions "touching from above" the graph of u at x_0.

We note also that the space $C^1(\Omega)$ of test functions in Definition 1.1 can be replaced by $C^\infty(\Omega)$, see Exercise 2.1. ◁

The following proposition explains the local character of the notion of viscosity solution and its consistency with the classical pointwise definition.

PROPOSITION 1.3. (a) *If $u \in C(\Omega)$ is a viscosity solution of* (HJ) *in Ω, then u is a viscosity solution of* (HJ) *in Ω', for any open set $\Omega' \subset \Omega$;*
(b) *if $u \in C(\Omega)$ is a classical solution of* (HJ), *that is, u is differentiable at any $x \in \Omega$ and*

(1.3) $$F(x, u(x), Du(x)) = 0 \qquad \forall x \in \Omega,$$

then u is a viscosity solution of (HJ);
(c) *if $u \in C^1(\Omega)$ is a viscosity solution of* (HJ), *then u is a classical solution of* (HJ).

PROOF. (a) If x_0 is a local maximum (on Ω') for $u - \varphi$, $\varphi \in C^1(\Omega')$, then x_0 is a local maximum (on Ω) for $u - \widetilde{\varphi}$, for any $\widetilde{\varphi} \in C^1(\Omega)$ such that $\widetilde{\varphi} \equiv \varphi$ on $\overline{B}(x_0, r)$ for some $r > 0$. By (1.1)

$$0 \geq F(x_0, u(x_0), D\widetilde{\varphi}(x_0)) = F(x_0, u(x_0), D\varphi(x_0))$$

showing that u is a viscosity subsolution of (HJ) on Ω'. The same argument applies to prove that u is also a supersolution on Ω'.
(b) Take any $\varphi \in C^1(\Omega)$. By the differentiability of u, at any local maximum or minimum $x \in \Omega$ of $u - \varphi$ we have $Du(x) = D\varphi(x)$. Hence (1.3) yields

$$0 = F(x_0, u(x_0), D\varphi(x_0)) \leq 0$$

if x_0 is a local maximum for $u - \varphi$ and

$$0 = F(x_1, u(x_1), D\varphi(x_1)) \geq 0$$

if x_1 is a local minimum for $u - \varphi$.

(c) If $u \in C^1(\Omega)$, then $\varphi \equiv u$ is a feasible choice in the definition of viscosity solution. With this choice, any $x \in \Omega$ is simultaneously a local maximum and minimum for $u - \varphi$. Hence, by (1.1) and (1.2),

$$F(x, u(x), Du(x)) = 0 \quad \forall x \in \Omega. \quad \blacktriangleleft$$

Statement (a) says that the notion of viscosity solution is a local one. Consequently, one can take the test functions in (1.1) and (1.2) in $C^1(\mathbb{R}^N)$ or in any sufficiently small ball $B(x, r)$ centered at $x \in \Omega$.

The definition of viscosity solution is closely related to two properties that are typical in the theory of elliptic and parabolic equations, namely the *maximum principle* (MP) and the *comparison principle* (CP). For equation (HJ) these properties can be respectively formulated as follows.

DEFINITION 1.4. A function $u \in C(\Omega)$ satisfies the comparison principle with smooth strict supersolutions, briefly (CP), if for any $\varphi \in C^1(\Omega)$ and $\mathcal{O}$ open subset of Ω,

$$F(x, \varphi(x), D\varphi(x)) > 0 \quad \text{in } \mathcal{O}, \qquad u \leq \varphi \quad \text{on } \partial\mathcal{O}$$

implies $u \leq \varphi$ in $\mathcal{O}$.

We say that $u \in C(\Omega)$ satisfies the maximum principle (MP) if for any $\varphi \in C^1(\Omega)$ and $\mathcal{O}$ open subset of Ω the inequality

$$F(x, \varphi(x), D\varphi(x)) > 0 \quad \text{in } \mathcal{O},$$

implies that $u - \varphi$ cannot have a nonnegative maximum in $\mathcal{O}$. $\triangleleft$

It is quite clear that (MP) implies (CP). The connections with the notion of viscosity subsolution of (HJ) are expressed by the next result.

PROPOSITION 1.5. *If $u \in C(\Omega)$ satisfies* (CP), *then u is a viscosity subsolution of* (HJ). *Conversely, if u is a viscosity subsolution of* (HJ) *and $r \mapsto F(x, r, p)$ is nondecreasing for all x, p, then u satisfies* (MP) *and* (CP).

PROOF. Assume that $u \in C(\Omega)$ satisfies (CP). If, by contradiction, u is not a subsolution of (HJ) there exist $x_0 \in \Omega$, $\varphi \in C^1(\Omega)$, such that x_0 is a strict maximum point for $u - \varphi$, $(u - \varphi)(x_0) = 0$, and

$$F(x_0, u(x_0), D\varphi(x_0)) > 0.$$

For n large enough we have

$$a_n := \sup_{\partial B(x_0, 1/n)} (u - \varphi) < 0.$$

Observe also that
$$u - (\varphi + a_n) \leq 0 \quad \text{on } \partial B(x_0, 1/n),$$
$$u(x_0) - \varphi(x_0) - a_n > 0 .$$

By (CP) for any n there exists $x_n \in \mathcal{O}_n := B(x_0, 1/n)$ such that
$$F(x_n, \varphi(x_n) + a_n, D\varphi(x_n)) \leq 0 .$$

Since $a_n \to 0$ and $x_n \to x_0$ we obtain the contradiction
$$F(x_0, u(x_0), D\varphi(x_0)) \leq 0 .$$

Conversely, let u be a viscosity subsolution of (HJ) and take any $\varphi \in C^1(\Omega)$ such that
$$F(x, \varphi(x), D\varphi(x)) > 0 \quad \text{for all } x \in \mathcal{O} .$$

If $u - \varphi$ attains a local maximum at some $x_0 \in \mathcal{O}$ with $u(x_0) - \varphi(x_0) \geq 0$, then the monotonicity assumption on F implies the contradiction
$$0 < F(x_0, \varphi(x_0), D\varphi(x_0)) \leq F(x_0, u(x_0), D\varphi(x_0)) \leq 0 .$$

Therefore, u satisfies (MP) and, a fortiori, (CP). ◄

A similar result holds for viscosity supersolutions, provided all inequalities are reversed in (CP), (MP) and nonnegative maximum is replaced by nonpositive minimum.

A perhaps striking fact to be stressed here is that viscosity solutions are not preserved by change of sign in the equation. Indeed, since any local maximum of $u - \varphi$ is a local minimum of $-u - (-\varphi)$, u is a viscosity subsolution of (HJ) if and only if $v = -u$ is a viscosity supersolution of $-F(x, -v(x), -Dv(x)) = 0$ in Ω; similarly, u is a viscosity supersolution of (HJ) if and only if $v = -u$ is a viscosity subsolution of $-F(x, -v(x), -Dv(x)) = 0$. An explicit example is as follows.

EXAMPLE 1.6. The function $u(x) = |x|$ is a viscosity solution of the 1-dimensional equation
$$-|u'(x)| + 1 = 0 \quad x \in {]-1, 1[} .$$

To check this, notice first that if $x \neq 0$ is a local extremum for $u - \varphi$, then $u'(x) = \varphi'(x)$. Therefore, at those points both the supersolution and the subsolution conditions are trivially satisfied. Also, if 0 is a local minimum for $u - \varphi$, a simple calculation shows that $|\varphi'(0)| \leq 1$ and the supersolution condition holds. To conclude it is enough to observe that 0 cannot be a local maximum for $u - \varphi$ with $\varphi \in C^1(]-1, 1[)$ (this would imply $-1 \geq \varphi'(0) \geq 1$).

On the other hand, $u(x) = |x|$ is not a viscosity solution of
$$|u'(x)| - 1 = 0, \quad x \in {]-1, 1[} .$$

Actually, the supersolution condition is not fulfilled at $x_0 = 0$ which is a local minimum for $|x| - (-x^2)$. ◁

1. DEFINITIONS AND BASIC PROPERTIES

We describe now an alternative way of defining viscosity solutions of equation (HJ) and prove the equivalence of the new definition with the one given previously (see Exercise 1.11 for another equivalent definition). Let us associate with a function $u \in C(\Omega)$ and $x \in \Omega$ the sets

$$D^+u(x) := \Big\{ p \in \mathbb{R}^N : \limsup_{y \to x,\, y \in \Omega} \frac{u(y) - u(x) - p \cdot (y - x)}{|x - y|} \leq 0 \Big\}$$

$$D^-u(x) := \Big\{ p \in \mathbb{R}^N : \liminf_{y \to x,\, y \in \Omega} \frac{u(y) - u(x) - p \cdot (y - x)}{|x - y|} \geq 0 \Big\}.$$

These sets are called, respectively, the *super-* and the *subdifferential* (or *semidifferentials*) of u at x.

The next lemma provides a description of $D^+u(x)$, $D^-u(x)$ in terms of test functions.

LEMMA 1.7. *Let $u \in C(\Omega)$. Then,*

(a) *$p \in D^+u(x)$ if and only if there exists $\varphi \in C^1(\Omega)$ such that $D\varphi(x) = p$ and $u - \varphi$ has a local maximum at x;*

(b) *$p \in D^-u(x)$ if and only if there exists $\varphi \in C^1(\Omega)$ such that $D\varphi(x) = p$ and $u - \varphi$ has a local minimum at x.*

PROOF. Let $p \in D^+u(x)$. Then, for some $\delta > 0$,

$$u(y) \leq u(x) + p \cdot (y - x) + \sigma(|y - x|)|y - x| \quad \forall y \in B(x, \delta),$$

where σ is a continuous increasing function on $[0, +\infty[$ such that $\sigma(0) = 0$. Now define a C^1 function ϱ by

$$\varrho(r) = \int_0^r \sigma(t)\, dt\,.$$

The following properties of ϱ

$$\varrho(0) = \varrho'(0) = 0, \qquad \varrho(2r) \geq \sigma(r)r$$

imply, as it is easy to check, that the function φ defined by

$$\varphi(y) = u(x) + p \cdot (y - x) + \varrho(2|y - x|)$$

belongs to $C^1(\mathbb{R}^N)$ and $D\varphi(x) = p$. Moreover, for $y \in B(x, \delta)$,

$$(u - \varphi)(y) \leq \sigma(|y - x|)|y - x| - \varrho(2|y - x|) \leq 0 = (u - \varphi)(x)\,.$$

For the opposite implication it is enough to observe that

$$u(y) - u(x) - D\varphi(x) \cdot (y - x) \leq \varphi(y) - \varphi(x) - D\varphi(x) \cdot (y - x)$$

for $y \in B(x, \delta)$ and the proof of (a) is complete.

Since $D^-u(x) = -(D^+(-u)(x))$, the proof of (b) follows from the above argument when applied to $-u$. ◂

Some properties of the sub- and superdifferential are collected in Lemma 1.8.

LEMMA 1.8. *Let $u \in C(\Omega)$ and $x \in \Omega$. Then,*

(a) $D^+u(x)$ *and* $D^-u(x)$ *are closed convex (possibly empty) subsets of* $\mathbb{R}^N$;

(b) *if u is differentiable at x, then* $\{Du(x)\} = D^+u(x) = D^-u(x)$;

(c) *if for some x both $D^+u(x)$ and $D^-u(x)$ are nonempty, then*

$$D^+u(x) = D^-u(x) = \{Du(x)\};$$

(d) *the sets $A^+ = \{x \in \Omega : D^+u(x) \neq \emptyset\}$, $A^- = \{x \in \Omega : D^-u(x) \neq \emptyset\}$ are dense.*

PROOF. The convexity of $D^+u(x)$ and $D^-u(x)$ is a straightforward consequence of the definition of lim sup and lim inf.

To prove that $D^+u(x)$ is closed we take a sequence $p_n \to p$ such that $p_n \in D^+u(x)$ for all n, and assume by contradiction that

$$(1.4) \qquad \lim_n \frac{u(y_n) - u(x) - p \cdot (y_n - x)}{|y_n - x|} = \alpha > 0$$

for a sequence $y_n \to x$. For k large enough we have $|p_k - p| \leq \alpha/2$. Then, by adding and subtracting $p_k \cdot (y_n - x)/|y_n - x|$ to (1.4) we get

$$\limsup_n \frac{u(y_n) - u(x) - p_k \cdot (y_n - x)}{|y_n - x|} \geq \frac{\alpha}{2},$$

a contradiction to $p_k \in D^+u(x)$.

To proceed in the proof observe that for any $x, y \in \Omega$ and $p, q \in \mathbb{R}^N$ we have

$$(1.5) \qquad (p - q) \cdot \frac{y - x}{|y - x|} = \frac{u(y) - u(x) - q \cdot (y - x)}{|y - x|} - \frac{u(y) - u(x) - p \cdot (y - x)}{|y - x|}.$$

For any $n \in \mathbb{N}$, set $y_n := x + (1/n)(p - q)$ and take $y = y_n$ in (1.5) to obtain

$$|p - q| = \frac{u(y_n) - u(x) - q \cdot (y_n - x)}{|y_n - x|} - \frac{u(y_n) - u(x) - p \cdot (y_n - x)}{|y_n - x|};$$

by definition of lim sup and lim inf this yields

$$(1.6) \qquad |p - q| \leq \limsup_{y \to x,\, y \in \Omega} \frac{u(y) - u(x) - q \cdot (y - x)}{|y - x|} - \liminf_{y \to x,\, y \in \Omega} \frac{u(y) - u(x) - p \cdot (y - x)}{|y - x|}.$$

If u is differentiable at x, then $D^+u(x) \cap D^-u(x) \neq \emptyset$ since it contains $Du(x)$. In this case $D^+u(x)$ and $D^-u(x)$ reduce to singletons as a consequence of (1.6). Conversely, if for some x one has $D^+u(x) \neq \emptyset$, $D^-u(x) \neq \emptyset$, then by (1.6) $D^+u(x) = D^-u(x)$ is a singleton. This means that u is differentiable at x with $\{Du(x)\} = D^+u(x) = D^-u(x)$.

1. DEFINITIONS AND BASIC PROPERTIES

In order to prove (d), let $\bar{x} \in \Omega$ and consider the smooth function $\varphi_\varepsilon(x) = |x - \bar{x}|^2/\varepsilon$. For any $\varepsilon > 0$, $u - \varphi_\varepsilon$ attains its maximum over $\bar{B} = \bar{B}(\bar{x}, R)$ at some point x_ε. From the inequality

$$(u - \varphi_\varepsilon)(x_\varepsilon) \geq (u - \varphi_\varepsilon)(\bar{x}) = u(\bar{x})$$

we get, for all $\varepsilon > 0$,

$$|x_\varepsilon - \bar{x}|^2 \leq 2\varepsilon \sup_{x \in \bar{B}} |u(x)| \ .$$

Thus x_ε is not on the boundary of $\bar{B}$ for ε small enough, and by Lemma 1.7 (a), $D\varphi_\varepsilon(x_\varepsilon) = 2(x_\varepsilon - \bar{x})/\varepsilon$ belongs to $D^+u(x_\varepsilon)$. This proves that A^+ is dense, and similar arguments show that A^- is dense too. ◀

As a direct consequence of Lemma 1.7 the following new definition of viscosity solution turns out to be equivalent to the initial one: a function $u \in C(\Omega)$ is a viscosity subsolution of (HJ) in Ω if

(1.7) $\qquad F(x, u(x), p) \leq 0 \qquad \forall\, x \in \Omega, \ \forall\, p \in D^+u(x);$

a viscosity supersolution of (HJ) in Ω if

(1.8) $\qquad F(x, u(x), p) \geq 0 \qquad \forall\, x \in \Omega, \ \forall\, p \in D^-u(x) \ .$

Of course, u will be called a viscosity solution of (HJ) in Ω if (1.7) and (1.8) hold simultaneously.

The above definition, which is more in the spirit of nonsmooth analysis, is sometimes easier to handle than the previous one. We employ it in the proofs of some important properties of viscosity solutions.

As a first example we present a consistency result that improves Proposition 1.3.

PROPOSITION 1.9. (a) *If $u \in C(\Omega)$ is a viscosity solution of* (HJ), *then*

$$F(x, u(x), Du(x)) = 0$$

at any point $x \in \Omega$ where u is differentiable;
(b) *if u is locally Lipschitz continuous and it is a viscosity solution of* (HJ), *then*

$$F(x, u(x), Du(x)) = 0 \quad \text{almost everywhere in } \Omega \ .$$

PROOF. If x is a point of differentiability for u then by Lemma 1.8 (b) $\{Du(x)\} = D^+u(x) = D^-u(x)$. Hence, by definitions (1.7), (1.8)

$$0 \geq F(x, u(x), Du(x)) \geq 0,$$

which proves (a). Statement (b) follows immediately from (a) and the Rademacher's theorem on the almost everywhere differentiability of Lipschitz continuous functions (see [EG92]). ◀

REMARK 1.10. Part (b) of Proposition 1.9 says that any viscosity solution of (HJ) is also a *generalized solution*, i.e., a locally Lipschitz continuous function u such that
$$F(x, u(x), Du(x)) = 0 \quad \text{a.e. in } \Omega.$$
The converse is false in general: there are many generalized solutions which are not viscosity solutions. As an example, observe that $u(x) = |x|$ satisfies
$$|u'(x)| - 1 = 0 \quad \text{in }]-1,1[\setminus \{0\},$$
but it is not a viscosity supersolution of the same equation in $]-1,1[$ (see Example 1.6 or, alternatively, just observe that $p = 0$ belongs to $D^-u(0)$ and (1.8) is not satisfied at $x = 0$). In Remark 2.3 we define infinitely many generalized solutions of this equation which are not viscosity solutions. We shall come back to this point in §5. ◁

Exercises

1.1. Check that
$$u(x) = \begin{cases} x, & 0 < x \leq 1/2 \\ 1 - x, & 1/2 < x < 1 \end{cases}$$
is a viscosity solution of $|u'(x)| - 1 = 0$, $x \in (0,1)$. Is u a viscosity solution of equation $-|u'(x)| + 1 = 0$ in $]0,1[$?

1.2. Let
$$u(x) = \begin{cases} 0, & x \leq 0 \\ \frac{1}{2}bx^2 + ax, & x > 0. \end{cases}$$
Compute $D^+u(0)$.

1.3. If $u : \mathbb{R}^N \to \mathbb{R}$ is convex (i.e., $u(\lambda x + (1-\lambda)y) \leq \lambda u(x) + (1-\lambda)u(y)$, for any x, y in $\mathbb{R}^N$, $\lambda \in [0,1]$), then its subdifferential at x in the sense of convex analysis is the set
$$\partial_c u(x) := \{ p \in \mathbb{R}^N : u(y) \geq u(x) + p \cdot (y - x), \forall y \in \mathbb{R}^N \}.$$
Show that if u is convex then $\partial_c u(x) = D^-u(x)$ at any x.

1.4. Let $u \in C(\Omega)$. Prove that $D^-u(x_0) \neq \emptyset$ if $x_0 \in \Omega$ is a local minimum of u and that $D^+u(x_0) \neq \emptyset$ if x_0 is a local maximum.

1.5. Show that $D^+u(0) = D^-u(0) = \emptyset$ where u is given by
$$u(x) = |x|^{1/2} \sin 1/x^2, \quad x \neq 0, \quad u(0) = 0$$
while $D^+v(0) = \emptyset$, $D^-v(0) = \{0\}$ for
$$v(x) = |x \sin 1/x|, \quad x \neq 0, \quad v(0) = 0.$$

1. DEFINITIONS AND BASIC PROPERTIES

1.6. Let $u \in C([a,b])$. Prove the mean value property: there exists $\xi \in (a,b)$ such that
$$u(b) - u(a) = p(b-a)$$
for some $p \in D^-u(\xi) \cup D^+u(\xi)$.

1.7. Check that both $u_1(t,x) \equiv 0$ and $u_2(t,x) = (t-|x|)^+$ are viscosity supersolutions of
$$u_t - |u'(x)| = 0 \quad \text{in } [0,+\infty[\times \mathbb{R}$$
$$u(0,x) = 0, \quad x \in \mathbb{R}.$$
Is u_2 a subsolution? [Hint: look at $D^+u_2(1,0)$.]

1.8. Consider for $x \in \mathbb{R}$ the function $u(x) := Ce^{-E|x|}$ with $C, E > 0$. Check that
$$D^-u(x) = \begin{cases} -CEe^{-E|x|}x/|x| & \text{if } x \neq 0 \\ \emptyset & \text{if } x = 0. \end{cases}$$

1.9. Let $F(x,r,p) := r + H(p) - \ell(x)$ with H, ℓ such that, for some constants L, A, B,
$$|H(p)| \leq L|p|, \quad |\ell(x)| \leq Ae^{-B|x|}, \quad \forall x, p \in \mathbb{R}^N.$$
Show that if $E < \min\{B; 1/L\}$ and $C = A/(1-LE)$, then u given in Exercise 1.8 is a viscosity supersolution of
$$u(x) + H(Du(x)) - \ell(x) = 0, \quad x \in \mathbb{R}^N.$$

1.10. Assume that $u_n \in C(\Omega)$, $u_n \to u$ as $n \to +\infty$ locally uniformly in Ω. Show that for any $x_0 \in \Omega$ the following holds
$$D^+u(x_0) \subseteq \limsup_{\substack{n \to +\infty \\ x \to x_0}} D^+u_n(x)$$
(i.e., for any $p \in D^+u(x_0)$ there exist $x_n \in \Omega$, $p_n \in D^+u_n(x_n)$ such that $x_n \to x_0$, $p_n \to p$ as $n \to +\infty$).

1.11. Show that $u \in C(\Omega)$ is a viscosity subsolution of (HJ) if and only if the following holds: for all $\varphi \in C_0^1(\Omega)$, $\varphi \geq 0$ and $k \in \mathbb{R}$, if $\max_\Omega \varphi(u-k) > 0$ then there exists $x_0 \in \Omega$ satisfying $\varphi(u-k)(x_0) = \max_\Omega \varphi(u-k)$ and
$$F\left(x_0, u(x_0), -\frac{D\varphi(x_0)}{\varphi(x_0)}(u(x_0)-k)\right) \leq 0.$$
Also find the corresponding property of supersolutions.

1.12. Let u be a Lipschitz continuous function in Ω with Lipschitz constant L. Prove that $D^+u(x)$ and $D^-u(x)$ are contained in $\bar{B}(0,L)$ for all $x \in \Omega$, and that u is a (viscosity) subsolution of $|Du| - L = 0$ and supersolution of $-|Du| + L = 0$ in Ω.

2. Some calculus and further properties of viscosity solutions

In the first part of this section we collect some important stability properties of viscosity solutions and basic rules of calculus (change of unknown in (HJ), chain rule, ...). In the second part we establish some useful formulas for the semidifferentials of continuous functions of the form $u(x) = \inf_{b \in B} g(x,b)$. This is an important class of nonsmooth functions, sometimes called *marginal functions* which includes the *distance function* and is closely related to Hamilton-Jacobi equations.

The first result is on the stability with respect to the lattice operations in $C(\Omega)$:

$$(u \vee v)(x) = \max\{\, u(x), v(x)\,\}$$
$$(u \wedge v)(x) = \min\{\, u(x), v(x)\,\}\,.$$

PROPOSITION 2.1.

(a) *Let $u, v \in C(\Omega)$ be viscosity subsolutions of (HJ); then $u \vee v$ is a viscosity subsolution of (HJ).*

(b) *Let $u, v \in C(\Omega)$ be viscosity supersolutions of (HJ); then $u \wedge v$ is a viscosity supersolution of (HJ).*

(c) *Let $u \in C(\Omega)$ be a viscosity subsolution of (HJ) such that $u \geq v$ for any viscosity subsolution $v \in C(\Omega)$ of (HJ); then u is a viscosity supersolution and therefore a viscosity solution of (HJ).*

PROOF. Let x_0 be a local maximum for $u \vee v - \varphi$ with $\varphi \in C^1(\Omega)$ and assume without loss of generality that $(u \vee v)(x_0) = u(x_0)$. Then x_0 is a local maximum for $u - \varphi$; so

$$F(x_0, u(x_0), D\varphi(x_0)) \leq 0$$

and (a) is proved. An analogous argument can be used for (b).

In order to prove (c), let us suppose by contradiction that

$$h := F(x_0, u(x_0), D\varphi(x_0)) < 0$$

for some $\varphi \in C^1(\Omega)$ and $x_0 \in \Omega$ such that

$$u(x_0) - \varphi(x_0) \leq u(x) - \varphi(x) \qquad \forall\, x \in \overline{B}(x_0, \delta_0) \subseteq \Omega,$$

for some $\delta_0 > 0$. Consider next the function $w \in C^1(\Omega)$ defined by

$$w(x) = \varphi(x) - |x - x_0|^2 + u(x_0) - \varphi(x_0) + \frac{1}{2}\delta^2$$

for $0 < \delta < \delta_0$. It is immediate to check that

(2.1) $\qquad (u - w)(x_0) < (u - w)(x), \qquad \forall\, x \text{ such that } |x - x_0| = \delta\,.$

Let us prove now that, for sufficiently small δ,

(2.2) $\qquad F(x, w(x), Dw(x)) \leq 0 \quad \forall x \in B(x_0, \delta)$.

For this purpose, a local uniform continuity argument shows that, for $0 < \delta < \delta_0$,

(2.3) $\qquad \begin{cases} |\varphi(x) - \varphi(x_0)| \leq \omega_1(\delta), \\ |D\varphi(x) - 2(x - x_0) - D\varphi(x_0)| \leq \omega_2(\delta) + 2\delta, \end{cases}$

for any $x \in \bar{B}(x_0, \delta)$, where ω_i ($i = 1, 2$) are the moduli of continuity of φ and $D\varphi$. Hence

$$|w(x) - u(x_0)| \leq \omega_1(\delta) + \delta^2 \quad \forall x \in \bar{B}(x_0, \delta) .$$

Now,

(2.4) $F(x, w(x), Dw(x))$
$\qquad = h + F(x, w(x), D\varphi(x) - 2(x - x_0)) - F(x_0, u(x_0), D\varphi(x_0))$.

If ω is a modulus of continuity for F, then

$$F(x, w(x), Dw(x)) \leq h + \omega(\delta, \omega_1(\delta) + \delta^2, \omega_2(\delta) + 2\delta),$$

for all $x \in \bar{B}(x_0, \delta)$. Since $h < 0$, the preceding proves the validity of (2.2) for small enough $\delta > 0$. Fix any such δ and set

$$\widehat{v}(x) = \begin{cases} u \vee w & \text{on } B(x_0, \delta) \\ u & \text{on } \Omega \setminus B(x_0, \delta) . \end{cases}$$

It is easy to check that $\widehat{v} \in C(\Omega)$ (see (2.1)) and, by Propositions 1.3 (a) and 2.1 (a), $\widehat{v}$ is a subsolution of (HJ) in Ω. Since $\widehat{v}(x_0) > u(x_0)$, the statement is proved. ◀

The next result is a stability result in the uniform topology of $C(\Omega)$.

PROPOSITION 2.2. *Let $u_n \in C(\Omega)$ ($n \in \mathbb{N}$) be a viscosity solution of*

(HJ)$_n$ $\qquad F_n(x, u_n(x), Du_n(x)) = 0 \quad \text{in } \Omega$.

Assume that

$$\begin{aligned} u_n &\longrightarrow u & \text{locally uniformly in } \Omega \\ F_n &\longrightarrow F & \text{locally uniformly in } \Omega \times \mathbb{R} \times \mathbb{R}^N . \end{aligned}$$

Then u is a viscosity solution of (HJ) in Ω.

PROOF. Let $\varphi \in C^1(\Omega)$ and x_0 be a local maximum point of $u - \varphi$. As observed before, it is not restrictive to assume that

$$u(x_0) - \varphi(x_0) > u(x) - \varphi(x)$$

for $x \neq x_0$ in a neighborhood of x_0. By uniform convergence, $u_n - \varphi$ attains, for large n, a local maximum at a point x_n close to x_0 (see Lemma 2.4). Then,

$$F_n(x_n, u_n(x_n), D\varphi(x_n)) \leq 0 .$$

Since $x_n \to x_0$, passing to the limit as $n \to +\infty$ in the above yields

$$F(x_0, u(x_0), D\varphi(x_0)) \leq 0 .$$

A similar argument proves that u is also a viscosity supersolution. ◄

REMARK 2.3. Proposition 2.2 does not hold for generalized solutions of (HJ)$_n$. As an example, consider the *saw-tooth* like functions u_n defined by $u_1(x) = 1 - x$ and for $n \geq 2$ by

$$u_n(x) = \begin{cases} x - \dfrac{2j}{2^n} & x \in \,]2j/2^n, (2j+1)/2^n[\\ \dfrac{2j+2}{2^n} - x & x \in \,](2j+1)/2^n, (2j+2)/2^n[\end{cases} \quad j = 0, 1, \ldots, 2^{n-1} - 1$$

for $x \in \,]0, 1[$. It is evident that $|u_n'(x)| - 1 = 0$ almost everywhere in $]0, 1[$, but although u_1 is a classical (and therefore a viscosity) solution, u_n is not a viscosity solution for $n \geq 2$. The uniform limit of the sequence $\{u_n\}$ is identically zero and does not satisfy the equation at any point. ◁

In the proof of Proposition 2.2 we used the following elementary fact which is useful in many situations.

LEMMA 2.4. *Let $v \in C(\Omega)$ and suppose that $x_0 \in \Omega$ is a strict maximum point for v in $\bar{B}(x_0, \delta) \subseteq \Omega$. If $v_n \in C(\Omega)$ converges locally uniformly to v in Ω, then there exists a sequence $\{x_n\}$ such that*

(2.5) $\qquad\qquad x_n \longrightarrow x_0, \qquad v_n(x_n) \geq v_n(x) \quad \forall x \in \bar{B}(x_0, \delta) .$

PROOF. Let x_n be a maximum point for v_n on $\bar{B}(x_0, \delta)$ and let $\{x_{n_k}\}$, $k \in \mathbb{N}$, be any converging subsequence of $\{x_n\}$, $n \in \mathbb{N}$. By uniform convergence,

$$v_{n_k}(x_{n_k}) \longrightarrow v(\tilde{x}) \qquad \text{as } k \to +\infty,$$

where $\tilde{x} = \lim x_{n_k}$ as $k \to +\infty$. The choice of $\{x_n\}$ yields

$$v(\tilde{x}) \geq v(x) \qquad \forall x \in \bar{B}(x_0, \delta),$$

so that, in particular,

$$v(\tilde{x}) \geq v(x_0) .$$

This implies $\tilde{x} = x_0$ and the convergence of the whole sequence. ◄

2. FURTHER PROPERTIES OF VISCOSITY SOLUTIONS

The next result is on the change of unknown in (HJ).

PROPOSITION 2.5. *Let $u \in C(\Omega)$ be a viscosity solution of* (HJ) *and* $\Phi \in C^1(\mathbb{R})$ *be such that* $\Phi'(t) > 0$. *Then* $v = \Phi(u)$ *is a viscosity solution of*

$$(2.6) \qquad F(x, \Psi(v(x)), \Psi'(v(x))Dv(x)) = 0 \qquad x \in \Omega,$$

where $\Psi = \Phi^{-1}$.

PROOF. Since $G(x, r, p) := F(x, \Psi(r), \Psi'(r)p)$ is defined only for $r \in \Phi(\mathbb{R})$, here by viscosity solution of (2.6) we mean a function taking its values in $\Phi(\mathbb{R})$ and satisfying the properties of Definition 1.1.

Let $x \in \Omega$ and $p \in D^+v(x)$. Then

$$v(y) \leq v(x) + p \cdot (y - x) + o(|y - x|) \quad \text{as } y \to x.$$

Since Ψ is increasing,

$$\Psi(v(y)) \leq \Psi(v(x) + p \cdot (y - x) + o(|y - x|))$$
$$= \Psi(v(x)) + \Psi'(v(x))p \cdot (y - x) + o(|y - x|).$$

By definition of v, this amounts to saying that

$$\Psi'(v(x))p \in D^+u(x).$$

Therefore,

$$F(x, u(x), \Psi'(v(x))p) \leq 0,$$

showing that v is a viscosity subsolution of (2.6).

In a completely similar way one can prove that v is also a viscosity supersolution of (2.6). ◂

A slight generalization which is useful when dealing with evolution equations (see Exercise 2.3) is as follows.

PROPOSITION 2.6. *Let $u \in C(\Omega)$ be a viscosity solution of* (HJ) *and* $\Phi : \Omega \times \mathbb{R} \to \mathbb{R}$ *a C^1 function such that*

$$\Phi_r(x, r) > 0 \qquad \forall (x, r) \in \Omega \times \mathbb{R}.$$

Then the function $v \in C(\Omega)$ defined implicitly by

$$\Phi(x, v(x)) = u(x),$$

is a viscosity solution of

$$(2.7) \qquad \widetilde{F}(x, v(x), Dv(x)) = 0 \qquad \text{in } \Omega,$$

where

$$\widetilde{F}(x, r, p) = F(x, \Phi(x, r), D_x\Phi(x, r) + \Phi_r(x, r)p).$$

PROOF. Let us detail only the subsolution part. Let $x \in \Omega$ and $p \in D^+v(x)$. Then,
$$v(y) \leq v(x) + p \cdot (y - x) + o(|y - x|),$$
as $y \to x$. Since Φ is nondecreasing with respect to r, we have
$$\Phi(y, v(y)) \leq \Phi(y, v(x) + p \cdot (y - x) + o(|y - x|)),$$
and consequently
$$\Phi(y, v(y)) \leq \Phi(x, v(x)) + D_x\Phi(x, v(x)) \cdot (y - x)$$
$$+ \Phi_r(x, v(x))p \cdot (y - x) + o(|y - x|).$$
By definition of v this means that $D_x\Phi(x, v(x)) + \Phi_r(x, v(x))p \in D^+u(x)$. Therefore
$$F(x, u(x), D_x\Phi(x, v(x)) + \Phi_r(x, v(x))p) \leq 0,$$
that is, v is a viscosity subsolution of (2.7). ◀

The next proposition collects the semidifferential versions of some useful facts in elementary calculus.

PROPOSITION 2.7. *Let $u \in C(\Omega)$. Then*
(i) *for $v(x, r) = \varphi(r)u(x)$ ($x \in \Omega$, $r \in \mathbb{R}$) we have*
$$D^+v(x, r) = \{ (q, \varrho) \in \mathbb{R}^{N+1} : q \in \varphi(r)D^+u(x),\ \varrho = \varphi'(r)u(x) \},$$
provided $\varphi \in C^1(\mathbb{R})$, $\varphi(r) \geq 0$ for all $r \in \mathbb{R}$;
(ii) *for $u(x) = v(T(x))$, with $v \in C(\widehat{\Omega})$, we have*
$$p \in D^+v(y_0) \qquad \textit{if and only if} \qquad (DT(x_0))^t p \in D^+u(x)$$
(where ^{t} denotes transposition) provided $T : \Omega \to \widehat{\Omega}$ is a C^1-diffeomorphism and $y_0 = T(x_0)$;
(iii) *for $\eta(r) = u(y(r))$ we have*
$$D^+\eta(r) \supseteq D^+u(y(r)) \cdot \dot{y}(r),$$
provided $y \in C^1(\mathbb{R}, \Omega)$.

PROOF. Let us give some details only for parts (i) and (ii) and leave the proof of (iii) as an exercise. Since $\varphi \in C^1(\mathbb{R})$ we have
$$v(y, s) - v(x, r) = \varphi'(r)u(x)(s - r) + \varphi(s)(u(y) - u(x)) + o(|s - r|);$$
hence $(q, \varrho) \in D^+v(x, r)$ if and only if
$$(\varphi'(r)u(x) - \varrho)(s - r) + \varphi(s)(u(y) - u(x)) - q \cdot (y - x) \leq o(|y - x| + |s - r|)$$

2. FURTHER PROPERTIES OF VISCOSITY SOLUTIONS

for all (y,s) in a neighborhood of (x,r). This easily implies $\varrho = \varphi'(r)u(x)$.

Assume now $\varphi(r) \neq 0$ and that $q/\varphi(r) \notin D^+u(x)$. This implies

$$\varphi(r)(u(y) - u(x)) - q \cdot (y - x) > \varphi(r)o(|y - x|),$$

a contradiction to the above inequality with $s = r$. If $\varphi(r) = 0$ the choice $s = r$ gives $-q \cdot (y - x) \leq o(|y - x|)$, thus $q = 0$. Hence (i) is proved.

In order to prove (ii), let $y_0 \in \widehat{\Omega}$ and $p \in D^+v(y_0)$. Then

$$v(y) \leq v(y_0) + p \cdot (y - y_0) + o(|y - y_0|)$$

as $y \to y_0$. Set $x_0 = T^{-1}(y_0)$ and $x = T^{-1}(y)$. Then

$$u(x) \leq u(x_0) + p \cdot (T(x) - T(x_0)) + o(|x - x_0|)$$
$$= u(x_0) + p \cdot DT(x_0)(x - x_0) + o(|x - x_0|),$$

which proves (ii). ◀

REMARK 2.8. Similar results hold for D^-. The sign condition on φ is essential in (i) (see Exercise 2.4). The equality in the "chain rule" (iii) does not hold in general. The "change of variables" formula (ii) implies that u is a viscosity subsolution of (HJ) in Ω if and only if $v(\widehat{x}) = u(T^{-1}(\widehat{x}))$ is a viscosity subsolution of

$$F(T^{-1}(\widehat{x}), v(\widehat{x}), DT(T^{-1}(\widehat{x}))Dv(\widehat{x})) = 0, \qquad \widehat{x} \in \widehat{\Omega},$$

see Exercise 2.6. ◁

Let us now discuss the following quite natural question: when is a piecewise C^1 function satisfying (HJ) at all points of differentiability a viscosity solution of the same equation?

The answer is very simple in the one-dimensional case. Assume $u \in C(]-a,a[) \cap C^1(]-a,0[\cup]0,a[)$ is a classical and therefore a viscosity solution of

$$F(x, u(x), u'(x)) = 0 \quad \text{in }]-a, 0[\cup]0, a[\,.$$

It is immediate, by Taylor's expansion, to check that for small enough $|y|$,

$$u(y) = \begin{cases} u(0) + u'_+(0)y + o(|y|) & \text{if } y > 0, \\ u(0) + u'_-(0)y + o(|y|) & \text{if } y < 0, \end{cases}$$

where $u'_+(0), u'_-(0)$ denote, respectively, the right and the left derivative of u at 0. Hence, $p \in D^+u(0)$ if and only if

$$u'_+(0) \leq p \leq u'_-(0) \,.$$

Similarly, $p \in D^-u(0)$ is equivalent to $p \in [u'_-(0), u'_+(0)]$. Therefore u will be a viscosity solution in $]-a, a[$ provided

$$F(0, u(0), p) \leq 0 \quad \forall p \in I_+ := [u'_+(0), u'_-(0)] = D^+u(0)$$
$$F(0, u(0), p) \geq 0 \quad \forall p \in I_- := [u'_-(0), u'_+(0)] = D^-u(0) \,.$$

Of course, in the case where $u'_+(0) \neq u'_-(0)$, only one of the intervals I_+, I_- is nonempty.

The preceding simple remark provides a different explanation of the phenomenon described in Example 1.6.

In order to extend these considerations to higher dimensions, let us assume that

$$\Omega = \Omega^1 \cup \Omega^2 \cup \Gamma,$$

where Ω^i ($i = 1, 2$) are open subsets of Ω and Γ is a smooth surface. Let us denote by $n(x)$ and $T(x)$ ($x \in \Gamma$) the unit normal vector to Γ pointing inward Ω^1 and the tangent space to Γ at x, respectively. Denote finally by P_T and P_N the orthogonal projections of $\mathbb{R}^N$ onto $T(x)$ and, respectively, the space spanned by $n(x)$. It is an immediate consequence of the definition that if $u \in C(\Omega^i)$ is a viscosity solution of (HJ) in Ω^i ($i = 1, 2$), then u is a viscosity solution of (HJ) in $\Omega^1 \cup \Omega^2 = \Omega \setminus \Gamma$. As expected, such a u will be a viscosity solution in Ω if some interface condition on Γ is satisfied.

PROPOSITION 2.9. *Let $u \in C(\Omega)$ and assume that its restrictions u^i to $\Omega^i \cup \Gamma$ belong to $C^1(\Omega^i \cup \Gamma)$ ($i = 1, 2$). Then u is a viscosity solution of (HJ) in Ω if and only if the following conditions hold*

(a) u^i is a classical solution of (HJ) in Ω^i ($i = 1, 2$)

(b_1)
$$F(x, u(x), P_T Du^i(x) + \xi n(x)) \leq 0$$
$$\forall \xi \in [Du^1(x) \cdot n(x), Du^2(x) \cdot n(x)], \quad \forall x \in \Gamma,$$

(b_2)
$$F(x, u(x), P_T Du^i(x) + \xi n(x)) \geq 0$$
$$\forall \xi \in [Du^2(x) \cdot n(x), Du^1(x) \cdot n(x)], \quad \forall x \in \Gamma.$$

PROOF. By Proposition 1.3, u is a viscosity solution of (HJ) in $\Omega \setminus \Gamma$ if and only if (a) holds. So we are left to prove that (b_1), (b_2) are equivalent to the fulfillment of (1.7), (1.8) at any $x \in \Gamma$. Since $u_1 \equiv u_2$ on Γ we have $(u_1 - u_2)(x(t)) = 0$ for all $t \in [0, 1]$, where $x(\cdot)$ is a parametrization of a curve lying on Γ such that $x(0) = x$. Hence, by the smoothness of Γ,

$$0 = D(u_1 - u_2)(x(t)) \cdot \dot{x}(t), \quad \forall t \in [0, 1].$$

In particular for $t = 0$ this gives $D(u_1 - u_2)(x) \cdot \dot{x}(0) = 0$ so

$$D(u_1 - u_2)(x) \cdot \tau = 0 \quad \forall \tau \in T(x).$$

whence it follows that

$$P_T Du^1(x) = P_T Du^2(x), \quad x \in \Gamma.$$

Let $x \in \Gamma$ and observe that

$$u(y) = u(x) + (P_T Du^i(x) + P_N Du^i(x)) \cdot (y - x) + o(|y - x|)$$

for any $y \in B(x,\delta) \cap \Omega^i$, $(i=1,2)$. Hence, $p \in D^+u(x)$ if and only if

(2.8) $(P_T Du^i(x) + P_N Du^i(x)) \cdot (y-x) \leq (P_T p + P_N p) \cdot (y-x) + o(|y-x|)$,

for any $y \in B(x,\delta)$. If we choose now $y = x + t\tau$, with $\tau \in T(x)$, we get

$$P_T p = P_T Du^1(x) = P_T Du^2(x) \,.$$

On the other hand, the choice $y = x + tn(x)$ with $t > 0$ in (2.8) gives

$$P_N Du^1(x) \cdot n(x) \leq P_N p \cdot n(x) + o(t)/t,$$

since $y \in \Omega^1$. Similarly, $y = x - tn(x) \in \Omega^2$ for $t > 0$ and therefore

$$P_N p \cdot n(x) \leq P_N Du^2(x) \cdot n(x) + o(t)/t \,.$$

Now, $P_N q \cdot n(x) = q \cdot n(x)$, so that

$$D^+u(x) \subseteq \{\, p \in \mathbb{R}^N : p = P_T Du^i(x) + \xi n(x);$$
$$\xi \in [Du^1(x) \cdot n(x), Du^2(x) \cdot n(x)] \,\}$$

for any $x \in \Gamma$.

It is not hard to show that this inclusion is actually an equality by taking an element of the set on the right-hand side and proving by contradiction that it satisfies (2.8). In a completely analogous way one shows that

$$D^-u(x) = \{\, q \in \mathbb{R}^N : q = P_T Du^i(x) + \xi n(x);\ \xi \in [Du^2(x) \cdot n(x), Du^1(x) \cdot n(x)] \,\}$$

and the proof is complete. ◀

The following technical lemma is useful in the study of evolution equations.

LEMMA 2.10. *Let us assume that*

(2.9) $$p_1 \longmapsto F(\overline{x}, \overline{r}, p_1, \overline{p}_2, \ldots, \overline{p}_N)$$

is nondecreasing for any fixed $\overline{x}$, $\overline{r}$, $\overline{p}_2 \ldots, \overline{p}_N$. *Assume also that* $\Omega =]a,b[\times \Omega'$, *with* Ω' *an open subset in* $\mathbb{R}^{N-1}$. *If* $u \in C(\overline{\Omega})$ *is a viscosity subsolution (respectively, a supersolution) of* (HJ), *then*

(2.10) $$F(\overline{x}, u(\overline{x}), D\varphi(\overline{x})) \leq 0, \qquad (\text{respectively}, \geq 0)$$

at any local maximum (respectively, minimum) $\overline{x}$ *of* $u - \varphi$ *over* $]a,b] \times \Omega'$, *for any* $\varphi \in C^1(]a,b] \times \Omega')$.

PROOF. The only thing to check is that (2.10) holds true at strict local maxima $\overline{x}$ of the form $\overline{x} = (b, \overline{x}_2, \ldots, \overline{x}_N)$. Let x_n be a maximum point of

$$\psi_n(x) = u(x) - \left(\varphi(x) + \frac{1}{n(b-x_1)}\right)$$

in $[b-r,b[\,\times\bar{B}(\bar{x},r)\subseteq\Omega$. We claim that $x_n\to\bar{x}$. To prove the claim we assume that

$$x_n\longrightarrow\tilde{x},\qquad \psi_n(x_n)\longrightarrow\ell$$

after taking a subsequence if necessary. Since $\psi_n(x_n)\leq(u-\varphi)(x_n)$, then $\ell\leq(u-\varphi)(\tilde{x})$. On the other hand, we take $x^n=(b-1/\sqrt{n},\bar{x}_2,\ldots,\bar{x}_N)$ to get

$$\psi_n(x_n)\geq\psi_n(x^n)=(u-\varphi)(x^n)-\frac{\sqrt{n}}{n};$$

therefore

$$\ell\geq(u-\varphi)(\bar{x}).$$

Since $\bar{x}$ was a strict maximum point for $u-\varphi$ we conclude that $\tilde{x}=\bar{x}$. By the definition of subsolution,

$$F(x_n,u(x_n),p_n)\leq 0$$

where

$$p_n=\left(\partial_{x_1}\varphi(x_n)+\frac{1}{n(b-(x_n)_1)^2},\partial_{x_2}\varphi(x_n),\ldots,\partial_{x_N}\varphi(x_n)\right).$$

By the monotonicity assumption on F this implies

$$F(x_n,u(x_n),D\varphi(x_n))\leq 0$$

and the claim follows by letting $n\to+\infty$. ◀

In the final part of this section we consider functions that are expressible as pointwise infimum of a family of functions:

(2.11) $$u(x)=\inf_{b\in B}g(x,b),$$

where $g:\Omega\times B\to\mathbb{R}$, $\Omega\subseteq\mathbb{R}^N$ is open and B is a topological space. Functions of this kind are often called *marginal functions*. In the context of this book the main examples are, of course, value functions of optimal control problems,

$$V(x)=\inf_{\alpha\in\mathcal{A}}J(x,\alpha)$$

(see Chapter I and, for a detailed investigation, Chapter III, §3). A more basic example is the *distance function* from a set $S\subseteq\mathbb{R}^N$,

$$d(x,S):=\inf_{s\in S}|x-s|.$$

2. FURTHER PROPERTIES OF VISCOSITY SOLUTIONS

Let us also mention that in §4 we consider the *inf-convolution* of a function u, defined as

$$u_\varepsilon(x) := \inf_{y \in \Omega} [\, u(y) + |x-y|^2/2\varepsilon\,], \qquad \varepsilon > 0,$$

another relevant example of marginal function.

We denote by $M(x)$ the (possibly empty) set

$$M(x) = \arg\min_{b \in B} g(x,b) := \{\, b \in B : u(x) = g(x,b)\,\}$$

and assume that $g(x, B)$ is bounded for all $x \in \Omega$ as well as that $x \mapsto g(x,b)$ is continuous in x uniformly with respect to $b \in B$; that is

(2.12) $$|g(x,b) - g(y,b)| \leq \omega(|x-y|, R),$$

for all $|x|, |y| \leq R$, $b \in B$, for some modulus ω.

A first elementary result is on the relations between the semidifferentials of u and the semidifferentials of g with respect to the variable x, denoted by $D_x^\pm g$.

LEMMA 2.11. *Let us assume* (2.12). *Then* $u \in C(\Omega)$ *and*

$$\begin{aligned} D^+ u(x) &\supseteq D_x^+ g(x,b) \\ D_x^- g(x,b) &\supseteq D^- u(x), \end{aligned} \qquad \text{for all } b \in M(x).$$

PROOF. The continuity of u is a simple consequence of definition 2.11 and assumption (2.12). Now let $p \in D_x^+ g(x,b)$; then

$$g(x+h, b) \leq g(x,b) + p \cdot h + o(|h|) \qquad \text{as } |h| \to 0.$$

For $b \in M(x)$, we deduce

$$u(x+h) \leq u(x) + p \cdot h + o(|h|) \qquad \text{as } |h| \to 0,$$

and thus $p \in D^+ u(x)$. If $p \in D^- u(x)$, then

$$u(x+h) \geq u(x) + p \cdot h + o(|h|) \qquad \text{as } |h| \to 0.$$

Hence, for $b \in M(x)$ we deduce

$$g(x+h, b) \geq g(x,b) + p \cdot h + o(|h|) \qquad \text{as } |h| \to 0,$$

showing that $p \in D_x^- g(x,b)$, for all $b \in M(x)$. ◀

REMARK 2.12. The result shows in particular that $D^+ u(x) \neq \emptyset$ at those points x where g is differentiable with respect to x and $M(x) \neq \emptyset$. ◁

To proceed further we assume $g(\,\cdot\,,b)$ differentiable at x uniformly in b; that is, for some modulus ω_1,

(2.13) $$|g(x+h,b) - g(x,b) - D_x g(x,b)\cdot h| \leq |h|\omega_1(|h|)$$

for all $b \in B$ and h small. We assume also that

(2.14) $\qquad\qquad b \longmapsto D_x g(x,b)\ $ is continuous,

(2.15) $\qquad\qquad b \longmapsto g(x,b)\ $ is lower semicontinuous.

We will use the following notations:

$$Y(x) := \{\, D_x g(x,b) : b \in M(x)\,\}$$

and, for the (one-sided) *directional derivative* of u in the direction q,

$$\frac{\partial u}{\partial q}(x) := \lim_{t \to 0^+} \frac{u(x+tq) - u(x)}{t}$$

(in §4 and Chapter III this directional derivative will be also denoted with $\partial u(x; q)$).

PROPOSITION 2.13. *Let us assume* (2.12), (2.13), (2.14), (2.15) *and that B is compact. Then*

(2.16) $\qquad\qquad Y(x) \neq \emptyset,$

(2.17) $\qquad\qquad D^+ u(x) = \overline{\mathrm{co}}\, Y(x),$

(2.18) $$D^- u(x) = \begin{cases} \{y\} & \text{if } Y(x) = \{y\}, \\ \emptyset & \text{if } Y(x) \text{ is not a singleton}. \end{cases}$$

In particular, u is differentiable at x if and only if $Y(x)$ is a singleton. Moreover, u has the (one-sided) directional derivative in any direction q, given by

$$\frac{\partial u}{\partial q}(x) = \min_{y \in Y(x)} y \cdot q = \min_{p \in D^+ u(x)} p \cdot q.$$

PROOF. By the assumptions made, $M(x)$ and $Y(x)$ are nonempty. The inclusion $\overline{\mathrm{co}}\, Y(x) \subseteq D^+ u(x)$ follows from Lemma 2.11 and the fact that $D^+ u(x)$ is closed and convex (see Lemma 1.8). To prove the opposite inclusion, let us assume by contradiction the existence of some $p \in D^+ u(x)$ such that $p \notin \overline{\mathrm{co}}\, Y(x)$. Then there exists a unit vector v such that

(2.19) $\qquad\qquad p \cdot v \leq 0, \quad q \cdot v > 0 \qquad \text{for all } q \in \overline{\mathrm{co}}\, Y(x).$

Let $x_n = x + v/n$ and $b_n \in B$ be such that

(2.20) $$g(x_n, b_n) \leq u(x_n) + 1/n^2.$$

2. FURTHER PROPERTIES OF VISCOSITY SOLUTIONS

From the definition of $D^+u(x)$ and (2.13) we get

$$g(x,b_n) + D_x g(x,b_n) \cdot \frac{v}{n} + \frac{1}{n}\omega_1(1/n) \leq u(x) + p \cdot \frac{v}{n} + o(1/n),$$

as $n \to +\infty$, and then, by (2.19) and the inequality $g(x,b_n) \geq u(x)$,

(2.21) $$D_x g(x,b_n) \cdot v \leq \frac{o(1/n)}{1/n} + \omega_1(1/n) \quad \text{as } n \to +\infty.$$

We extract now from b_n a subsequence converging to some $\bar{b} \in B$ and use (2.14) to obtain

$$D_x g(x,\bar{b}) \cdot v \leq 0,$$

a contradiction to (2.19) if $\bar{b} \in M(x)$. This is indeed the case; actually (2.13) implies

$$g(x,b_n) \leq g(x_n,b_n) - D_x g(x,b_n) \cdot v/n + o(1/n).$$

Thus, using (2.15), (2.20), the continuity of u and assuming without loss of generality that $b_n \to \bar{b}$,

$$g(x,\bar{b}) \leq \liminf_n g(x,b_n) \leq u(x).$$

This completes the proof of (2.17).

In order to prove (2.18), let us assume first that $Y(x)$ is not a singleton and, by contradiction, that $D^-u(x) \neq \emptyset$. Since we know by (2.16), (2.17) that $D^+u(x) \neq \emptyset$, then u is differentiable at x with $D^+u(x) = \{Du(x)\}$ (see Lemma 1.8). Since $Y(x)$ and, a fortiori, $\overline{co}Y(x)$ is not a singleton, we obtain a contradiction.

Consider now the case $Y(x) = \{y\}$. By Lemma 2.11, $D^-u(x) \subseteq \{y\}$. To prove the opposite inclusion, assume by contradiction that

(2.22) $$u(x+h_n) - u(x) - y \cdot h_n \leq -\varepsilon |h_n|$$

for some $\varepsilon > 0$ and $h_n \to 0$. Choose $b_n \in B$ such that

(2.23) $$g(x+h_n, b_n) \leq u(x+h_n) + |h_n|^2,$$

and use (2.13) and (2.22) to get

$$g(x,b_n) + D_x g(x,b_n) \cdot h_n - u(x) - y \cdot h_n \leq -\varepsilon |h_n| + o(|h_n|),$$

and, consequently,

$$(D_x g(x,b_n) - y) \cdot \frac{h_n}{|h_n|} \leq -\varepsilon + o(1) \quad \text{as } n \to +\infty.$$

Now we extract subsequences such that, after renaming, $b_n \to \bar{b} \in B$ and $h_n/|h_n| \to q \neq 0$. Then

$$(D_x g(x, \bar{b}) - y) \cdot q \leq -\varepsilon,$$

which gives a contradiction if $\bar{b} \in M(x)$. This is indeed the case. Actually, from

$$g(x, b_n) \leq g(x + h_n, b_n) - D_x g(x, b_n) \cdot h_n + o(|h_n|),$$

(2.23), the continuity of u and the lower semicontinuity of g with respect to b, it follows that

$$g(x, \bar{b}) \leq \liminf_{n \to +\infty} g(x, b_n) \leq u(x) \, .$$

To prove the statements about the directional derivatives, we first observe that for all $p \in D^+ u(x)$

$$\limsup_{t \to 0} \frac{u(x + tq) - u(x)}{t} \leq p \cdot q \, .$$

Then we pick $t_n \to 0$ such that

$$\lim_n \frac{u(x + t_n q) - u(x)}{t_n} = \liminf_{t \to 0} \frac{u(x + tq) - u(x)}{t}$$

and choose $b_n \in B$ such that $g(x + t_n q, b_n) \leq u(x + t_n q) + t_n^2$. Next we extract a subsequence converging to $\bar{b} \in B$ and deduce, by the same arguments as before, that

$$\lim_n \frac{u(x + t_n q) - u(x)}{t_n} \geq D_x g(x, \bar{b}) \cdot q$$

and that $D_x g(x, \bar{b}) \in Y(x)$, since $\bar{b} \in M(x)$. Hence, for all $p \in D^+ u(x)$,

(2.24)
$$p \cdot q \geq \limsup_{t \to 0^+} \frac{u(x + tq) - u(x)}{t} \geq \liminf_{t \to 0^+} \frac{u(x + tq) - u(x)}{t}$$
$$\geq D_x g(x, \bar{b}) \cdot q \, .$$

With the choice $p = D_x g(x, \bar{b})$, the above proves the existence of $\dfrac{\partial u}{\partial q}(x)$. From (2.24) it also follows that

$$\inf_{p \in D^+ u(x)} p \cdot q \geq \frac{\partial u}{\partial q}(x) \geq \inf_{y \in Y(x)} y \cdot q \, .$$

By (2.14) and (2.17), $\overline{\text{co}} Y(x) = D^+ u(x)$ is a convex compact set and the proof is complete. ◀

Next we apply Proposition 2.13 to compute the semidifferentials of the distance function from an arbitrary set $S \subseteq \mathbb{R}^N$, $S \neq \emptyset$, that is

(2.25) $$d(x) = d(x, S) := \inf_{z \in S} |x - z| = \min_{z \in \overline{S}} |x - z|.$$

The last expression for $d(x)$ is more convenient and leads to consider the set of *projections* of x onto $\overline{S}$, that is

$$P(x) := \{ z \in \partial S : d(x) = |x - z| \} \neq \emptyset.$$

PROPOSITION 2.14. *For any $S \neq \emptyset$, $d \in C(\mathbb{R}^N)$ and for all $x \notin \overline{S}$ and unit vector q,*

$$D^+ d(x) = \overline{\mathrm{co}} \left\{ \frac{x - z}{|x - z|} : z \in P(x) \right\}$$

$$D^- d(x) = \begin{cases} \dfrac{x - p(x)}{|x - p(x)|} & \text{if } P(x) = \{p(x)\} \\ \emptyset & \text{if } P(x) \text{ is not a singleton} \end{cases}$$

$$\frac{\partial d}{\partial q}(x) = \min_{z \in P(x)} \frac{x - z}{|x - z|} \cdot q.$$

PROOF. The continuity of d is easy to prove. If we restrict ourselves to those x such that $\varepsilon \leq d(x) \leq \varepsilon^{-1}$, $|x| \leq C$ for some $\varepsilon, C > 0$, then the search for the minimum in (2.25) can be restricted to $z \in S$ such that $|z| \leq C + \varepsilon^{-1}$. For x and z satisfying these restrictions, $x \mapsto |x - z|$ has bounded second derivatives, thus $g(x, z) := |x - z|$ fulfills (2.13) as well as the other assumptions of Proposition 2.13, and the conclusions follow from the arbitrariness of ε. ◂

REMARK 2.15. Note that, by Proposition 2.14 and Lemma 1.7, d is differentiable at x if and only if the projection of x onto $\overline{S}$ is unique. A well-known result in convex analysis (Motzkin's theorem, see [Val64]) asserts that the set of projections $P(x)$ is a singleton for all x if and only if $\overline{S}$ is convex. Then $\overline{S}$ is convex if and only if d is differentiable in $\mathbb{R}^N \smallsetminus \overline{S}$. In this case the projection $p(x)$ depends continuously on x, so that $d \in C^1(\mathbb{R}^N \smallsetminus \overline{S})$. ◁

It is well known that for smooth ∂S, say C^2, the distance function is smooth near ∂S and satisfies the *eikonal equation*

(2.26) $$|Du| = 1 \quad \text{in } \mathbb{R}^N \smallsetminus \overline{S},$$

locally around ∂S. If this equation is interpreted in the viscosity sense then it holds globally for any S.

COROLLARY 2.16. *The distance function d from S is a viscosity solution of (2.26). It is also a viscosity subsolution, but not a supersolution, in all $\mathbb{R}^N$.*

PROOF. The first statement follows immediately from Proposition 2.14. Since $d(x) \geq 0$, we have $0 \in D^- d(x)$ for all $x \in \overline{S}$, so d is not a supersolution at any such point. Moreover, for $x \in \overline{S}$ either $D^+ d(x) = \emptyset$ or $D^+ d(x) = \{0\}$, so d is a subsolution. ◂

In §5 we prove that d is actually the unique viscosity solution of (2.26), continuous up to ∂S and null there, provided $\mathbb{R}^N \smallsetminus \overline{S}$ is bounded (see Remark 5.10). A uniqueness result for the case when this set is unbounded can be found in Chapter IV.

Another notion which is strongly related to the properties of the distance function is that of exterior normal $n(x)$ to the set S at the point $x \in \partial S$. In the case of smooth ∂S, the function n is the unique unique extension of Dd to ∂S, and for $x \notin \overline{S}$ sufficiently close to ∂S we have

$$(2.27) \qquad x = p(x) + d(x)\, n(p(x)), \qquad Dd(x) = n(p(x)),$$

where $p(x)$ indicates the (unique) projection of x onto $\overline{S}$. This motivates the following definition.

DEFINITION 2.17. Let $S \subseteq \mathbb{R}^N$ be a general nonempty set. A unit vector ν is a *(generalized) exterior normal* to $\overline{S}$ at $z \in \partial S$ (and we denote $\nu \in N(z)$) if there exists $x \notin \overline{S}$ such that

$$x = z + d(x)\nu \qquad \text{and} \qquad \{z\} = P(x) \,. \quad \triangleleft$$

Equivalently, z is the projection onto $\overline{S}$ of some point off $\overline{S}$ in the direction of ν, i.e., $\exists t > 0$ such that

$$z + t\nu \notin \overline{S}, \qquad P(z + t\nu) = \{z\} \,.$$

Equivalently, there is a ball outside $\overline{S}$, centered at $z + t\nu$ for some $t > 0$, touching $\overline{S}$ precisely at z, that is $\overline{B}(z + t\nu, t) \cap \overline{S} = \{z\}$. Clearly $N(z) = \{n(z)\}$ if ∂S is C^2, but in general $N(z)$ may be empty or may not be a singleton, see Exercise 2.11. Note that, by Proposition 2.14, the second formula in (2.27) becomes

$$(2.28) \qquad Dd(x) \in N(p(x)), \qquad P(x) = \{p(x)\},$$

at every point x where d is differentiable.

For other characterization of the exterior normal and the connection with the notion of *proximal normal* see Exercises 2.10 and 2.12.

Exercises

2.1. Show by a density argument that an equivalent definition of viscosity solution of (HJ) can be given by using $C^\infty(\Omega)$ instead of $C^1(\Omega)$ as the "test function space".

2.2. Show by exhibiting an example that is false in general that if u, v are viscosity solutions of (HJ) the same is true for $u \vee v$, $u \wedge v$.

2.3. Let u be a viscosity solution of

$$u_t + F(t, x, u(t,x), D_x u(t,x)) = 0 \,.$$

Use Proposition 2.6 to show that $v(t,x) := e^{-\lambda t} u(t,x)$, $\lambda \in \mathbb{R}$, is a viscosity solution of

$$v_t + \lambda v + e^{-\lambda t} F(t, x, e^{\lambda t} v(t,x), e^{\lambda t} D_x v(t,x)) = 0 \,.$$

2.4. Let $u \in C(\Omega)$. Prove the following statements:

(a) $\quad D^+(\alpha u)(x) = \alpha D^+ u(x) \qquad$ if $\alpha > 0$,

(b) $\quad D^+(\alpha u)(x) = \alpha D^- u(x) \qquad$ if $\alpha < 0$,

(c) $\quad D^+(u+\varphi)(x) = D^+ u(x) + D\varphi(x) \qquad$ if $\varphi \in C^1(\Omega)$,

(d) $\quad D^+ u_\alpha(x) = \alpha D^+ u(x) + (1-\alpha)D\varphi(x)$,

where $\varphi \in C^1(\Omega)$ and $u_\alpha := \alpha u + (1-\alpha)\varphi$, $\alpha \in [0,1]$.

2.5 (ON THE CHAIN RULE). For $u \in C(\Omega)$ and $y \in C^1(\mathbb{R}, \Omega)$ let $\eta(r) := u(y(r))$. Prove that

$$\{\rho \in \mathbb{R} : \rho = \dot{y}(r) \cdot p, \ p \in D^+ u(y(r))\} \subseteq D^+ \eta(r), \qquad \forall r \in \mathbb{R}.$$

The two-dimensional example

$$u(x) = u(x_1, x_2) = |x_1|^{1/2}, \qquad y(r) = (y_1(r), y_2(r)) = (0, r)$$

shows that a strict inclusion may occur in the above formula.

2.6. Let $T : \Omega \to T(\Omega) := \widehat{\Omega}$ be a C^1 diffeomorphism. Prove that $u \in C(\Omega)$ is a viscosity solution of

$$F(x, u(x), Du(x)) = 0, \qquad x \in \Omega$$

if and only if $v(\widehat{x}) := u(T^{-1}(\widehat{x}))$ is a viscosity solution of

$$F(T^{-1}(\widehat{x}), v(\widehat{x}), DT(T^{-1}(\widehat{x}))Dv(\widehat{x})) = 0, \qquad \widehat{x} \in \widehat{\Omega}.$$

Here DT denotes the jacobian matrix of T. Work out in detail the linear case $F(x, r, p) = \lambda r - f(x) \cdot p - \ell(x)$, $\lambda \geq 0$.

2.7. Suppose that equation $F_n(x, u_n(x), Du_n(x)) = 0$ has a classical solution $u_n \in C^1(\Omega)$ for $n = 1, 2, \ldots$. Show that, under the assumptions of Proposition 2.2, $u = \lim_{n \to +\infty} u_n$ is a viscosity solution of

$$-F(x, u(x), Du(x)) = 0.$$

2.8. Let $u \in C(\mathbb{R}^N)$ ($N \geq 1$), and e^i be the ith vector of the standard basis in $\mathbb{R}^N$. The set

$$D_i^+ u(x) := \left\{ \alpha \in \mathbb{R} : \limsup_{t \to 0} \frac{u(x + te^i) - u(x) - \alpha t}{|t|} \leq 0 \right\}$$

is called the ith partial superdifferential of u at x. Prove that

(a) $p = (p_1 \ldots p_N) \in D^+ u(x)$ if and only if $p_i \in D_i^+ u(x)$;

(b) $D^+ \eta(x_1) = D_1^+ u(x_1, \overline{x}_2, \ldots, \overline{x}_N)$, for any fixed $(\overline{x}_2, \ldots, \overline{x}_N) \in \mathbb{R}^{N-1}$, where $\eta(x_1) := u(x_1, \overline{x}_2, \ldots, \overline{x}_N)$;

(c) deduce from the above that u is a viscosity subsolution of $(\partial u)/(\partial x_1)(x) \leq 0$, $x \in \mathbb{R}^N$, if and only if
$$\eta'(x_1) \leq 0, \qquad x_1 \in \mathbb{R}$$
in the viscosity sense, for any $(x_2, \ldots, x_N) \in \mathbb{R}^{N-1}$.

2.9. Consider the distance function d from a set S. Under which conditions does $u := -d$ solve $|Du| = 1$ in the viscosity sense?

2.10. Prove that ν is a (generalized) exterior normal to $\bar{S}$ at z if and only if there exists $\varepsilon > 0$ such that
$$B(z + \varepsilon\nu, \varepsilon) \cap \bar{S} = \emptyset \,.$$

2.11. Compute the (generalized) exterior normal vectors to S in the following cases:

(a) $\qquad S = \{\,0\,\}$;
(b) $\qquad S = \{\,x \in \mathbb{R}^N : x = tx_0 + (1-t)x_1,\ t \in [0,1]\,\}$;
(c) $\qquad S = \{\,x \in \mathbb{R}^N : g(x) \leq 0\,\} \cap \{\,x \in \mathbb{R}^N : f(x) \leq 0\,\}$;
(d) $\qquad S = \{\,x \in \mathbb{R}^N : g(x) \leq 0\,\} \cup \{\,x \in \mathbb{R}^N : f(x) \leq 0\,\} \,.$

Here, f and g are C^1 functions. Finally, let
$$S = \{\,x = (x_1, x_2) \in \mathbb{R}^2 : |x_2| \leq |x_1|^{3/2}\,\} \,.$$
Show that $N(0) = \emptyset$ although the exterior normal at $x = 0$ exists (namely, $n(0) = (0, 1)$). Observe that ∂S is C^1 but not C^2.

2.12. Define a proximal normal to a closed set $S \subseteq \mathbb{R}^N$ at $x \in \partial S$ as any vector $p \in \mathbb{R}^N$ such that $d(x + rp, S) = r|p|$, for some $r > 0$. Prove the following characterization of generalized exterior normals to S:
$$N(x) = \{\nu = p/|p| : p \text{ proximal normal to } S \text{ at } x\} \,.$$

3. Some comparison and uniqueness results

In this section we address the problem of comparison and uniqueness of viscosity solutions. This is a major issue in the theory, also in view of its relevance in connection with sufficient conditions in optimal control problems. The results presented here are not the most general; in fact they are selected to show the main ideas involved in the proofs with as few technicalities as possible. In Chapters III, IV and V we prove many other comparison theorems especially designed for the Hamilton-Jacobi-Bellman equations arising in optimal control theory, and we refer also to the literature quoted in the bibliographical notes for more general results.

3. SOME COMPARISON AND UNIQUENESS RESULTS

As an introduction to the subject, suppose that $u_1, u_2 \in C(\overline{\Omega}) \cap C^1(\Omega)$, satisfy the inequalities

(3.1)
$$u_1(x) + H(x, Du_1(x)) \leq 0$$
$$u_2(x) + H(x, Du_2(x)) \geq 0$$

for $x \in \Omega$ and

(3.2) $$u_1 \leq u_2 \quad \text{on } \partial\Omega.$$

Suppose Ω bounded and let x_0 be a maximum point for $u_1 - u_2$ on $\overline{\Omega}$. If $x_0 \in \Omega$, then $Du_1(x_0) = Du_2(x_0)$ and (3.1) gives by subtraction

$$u_1(x) - u_2(x) \leq u_1(x_0) - u_2(x_0) \leq 0 \quad \forall x \in \overline{\Omega}.$$

If, on the other hand, $x_0 \in \partial\Omega$, then

$$u_1(x) - u_2(x) \leq u_1(x_0) - u_2(x_0) \leq 0 \quad \forall x \in \overline{\Omega},$$

as a consequence of (3.2). Hence, $u_1 \leq u_2$ in $\overline{\Omega}$. Reversing the role of u_1 and u_2 in (3.1), (3.2), we get, of course, the uniqueness of a classical solution of the Dirichlet problem

$$u(x) + H(x, Du(x)) = 0, \quad x \in \Omega,$$
$$u = \varphi, \quad \text{on } \partial\Omega.$$

The preceding elementary proof fails if u_1, u_2 are continuous functions satisfying the inequalities (3.1) in the viscosity sense since Du_i may not exist at x_0. However, the information contained in the notion of viscosity sub- and supersolution is strong enough to allow the extension of comparison and uniqueness results to continuous viscosity solutions of equation (HJ) with rather general F.

In the following we present some comparison theorems between viscosity sub- and supersolutions in the cases Ω bounded, $\Omega = \mathbb{R}^N$, and for the Cauchy problem. As a simple corollary, each comparison result produces a uniqueness theorem, as indicated in the remarks. For simplicity we restrict our attention to the case $F(x, r, p) = r + H(x, p)$; the results hold, however, for a general F provided $r \mapsto F(x, r, p)$ is strictly increasing (see Exercises 3.6 and 3.7), and for some special F independent of r, see Theorem 3.7, §5.3 and Chapter IV, §2.

THEOREM 3.1. *Let Ω be a bounded open subset of $\mathbb{R}^N$. Assume that $u_1, u_2 \in C(\overline{\Omega})$ are, respectively, viscosity sub- and supersolution of*

(3.3) $$u(x) + H(x, Du(x)) = 0, \quad x \in \Omega$$

and

(3.4) $$u_1 \leq u_2 \quad \text{on } \partial\Omega.$$

Assume also that H satisfies

(H$_1$) $$|H(x,p) - H(y,p)| \leq \omega_1(|x-y|(1+|p|)),$$

for $x, y \in \Omega$, $p \in \mathbb{R}^N$, where $\omega_1 : [0,+\infty[\to [0,+\infty[$ is continuous nondecreasing with $\omega_1(0) = 0$. Then $u_1 \leq u_2$ in $\overline{\Omega}$.

PROOF. Define, for $\varepsilon > 0$, a continuous function Φ_ε on $\overline{\Omega} \times \overline{\Omega}$ by setting

$$\Phi_\varepsilon(x, y) = u_1(x) - u_2(y) - \frac{|x-y|^2}{2\varepsilon}$$

and let $(x_\varepsilon, y_\varepsilon)$ be a maximum point for Φ_ε on $\overline{\Omega} \times \overline{\Omega}$. Then, for any $\varepsilon > 0$

(3.5) $$\max_{x \in \overline{\Omega}}(u_1 - u_2)(x) = \max_{x \in \overline{\Omega}} \Phi_\varepsilon(x, x) \leq \max_{(x,y) \in \overline{\Omega} \times \overline{\Omega}} \Phi_\varepsilon(x, y) = \Phi_\varepsilon(x_\varepsilon, y_\varepsilon).$$

We claim that

(3.6) $$\liminf \Phi_\varepsilon(x_\varepsilon, y_\varepsilon) \leq 0 \quad \text{as } \varepsilon \to 0.$$

This, together with (3.5), proves the theorem.

In order to prove (3.6), let us observe first that the inequality

$$\Phi_\varepsilon(x_\varepsilon, x_\varepsilon) \leq \Phi_\varepsilon(x_\varepsilon, y_\varepsilon)$$

amounts to

$$\frac{|x_\varepsilon - y_\varepsilon|^2}{2\varepsilon} \leq u_2(x_\varepsilon) - u_2(y_\varepsilon).$$

This implies

$$|x_\varepsilon - y_\varepsilon| \leq (C\varepsilon)^{1/2},$$

where C depends only on the maximum of $|u_2|$ in $\overline{\Omega}$. Therefore,

(3.7) $$|x_\varepsilon - y_\varepsilon| \longrightarrow 0 \quad \text{as } \varepsilon \to 0;$$

and, by continuity of u_2,

(3.8) $$\frac{|x_\varepsilon - y_\varepsilon|^2}{2\varepsilon} \longrightarrow 0 \quad \text{as } \varepsilon \to 0.$$

Now there are two possible cases:
(i) $(x_{\varepsilon_n}, y_{\varepsilon_n}) \in \partial(\Omega \times \Omega)$ for some sequence $\varepsilon_n \to 0^+$;
(ii) $(x_\varepsilon, y_\varepsilon) \in \Omega \times \Omega$ for all $\varepsilon \in]0, \overline{\varepsilon}[$.

In case (i) either $x_{\varepsilon_n} \in \partial\Omega$, and then, by (3.4),

$$u_1(x_{\varepsilon_n}) - u_2(y_{\varepsilon_n}) \leq u_2(x_{\varepsilon_n}) - u_2(y_{\varepsilon_n}),$$

or $y_{\varepsilon_n} \in \partial\Omega$ and then

$$u_1(x_{\varepsilon_n}) - u_2(y_{\varepsilon_n}) \leq u_1(x_{\varepsilon_n}) - u_1(y_{\varepsilon_n}).$$

Note that the right-hand sides of both these inequalities tend to 0 as $n \to \infty$ by (3.7) and the uniform continuity of u_1 and u_2. Therefore

$$\Phi_{\varepsilon_n}(x_{\varepsilon_n}, y_{\varepsilon_n}) \leq u_1(x_{\varepsilon_n}) - u_2(y_{\varepsilon_n}) \longrightarrow 0 \quad \text{as } n \to \infty,$$

and the claim (3.6) is proved in this case.

Assume now that $(x_\varepsilon, y_\varepsilon) \in \Omega \times \Omega$ and set

$$\varphi_2(x) = u_2(y_\varepsilon) + \frac{|x - y_\varepsilon|^2}{2\varepsilon}, \qquad \varphi_1(y) = u_1(x_\varepsilon) - \frac{|x_\varepsilon - y|^2}{2\varepsilon}.$$

It is immediate to check that $\varphi_i \in C^1(\Omega)$ ($i = 1, 2$) and x_ε is a local maximum point for $u_1 - \varphi_2$, whereas y_ε is a local minimum point for $u_2 - \varphi_1$. Moreover,

$$D\varphi_1(y_\varepsilon) = \frac{x_\varepsilon - y_\varepsilon}{\varepsilon} = D\varphi_2(x_\varepsilon).$$

By the definition of viscosity sub- and supersolution, then,

$$u_1(x_\varepsilon) + H\left(x_\varepsilon, \frac{x_\varepsilon - y_\varepsilon}{\varepsilon}\right) \leq 0, \qquad -u_2(y_\varepsilon) - H\left(y_\varepsilon, \frac{x_\varepsilon - y_\varepsilon}{\varepsilon}\right) \leq 0.$$

By (H_1), this implies

$$u_1(x_\varepsilon) - u_2(y_\varepsilon) \leq \omega_1\left(|x_\varepsilon - y_\varepsilon|\left(1 + \frac{|x_\varepsilon - y_\varepsilon|}{\varepsilon}\right)\right)$$

and, a fortiori

$$\Phi_\varepsilon(x_\varepsilon, y_\varepsilon) \leq \omega_1\left(|x_\varepsilon - y_\varepsilon|\left(1 + \frac{|x_\varepsilon - y_\varepsilon|}{\varepsilon}\right)\right).$$

Taking (3.7) and (3.8) into account, (3.6) follows and the proof is complete. ◀

REMARK 3.2. If u_1, u_2 are both viscosity solutions of (3.3) with $u_1 = u_2$ on $\partial\Omega$, from Theorem 3.1 it follows that $u_1 = u_2$ in $\overline{\Omega}$. ◁

REMARK 3.3. The statement of Theorem 3.1 is true also for the equation

$$\lambda u(x) + H(x, Du(x)) = 0 \quad x \in \Omega$$

with $\lambda > 0$. On the other hand, for the equation $H(x, Du(x)) = 0$ non-uniqueness phenomena may appear. An extreme case is $H(x, p) = 0$ for all x and p: in this case any $u \in C(\Omega)$ is a viscosity solution. ◁

REMARK 3.4. When the Hamiltonian has the form (0.1), the assumption (H$_1$) of the theorem is satisfied, for example, if f is continuous, A is compact, $x \mapsto f(x,a)$ is Lipschitz continuous uniformly with respect to $a \in A$ and $x \mapsto \ell(x,a)$ is continuous in $\overline{\Omega}$, with a modulus of continuity independent of $a \in A$ (see Exercise 3.2). ◁

We consider now the case $\Omega = \mathbb{R}^N$ and prove a comparison result in the space $BC(\mathbb{R}^N)$ of bounded continuous functions on $\mathbb{R}^N$. The scheme of the proof is similar to that of Theorem 3.1, with more technicalities due to the lack of compactness. We assume the following condition on H:

(H$_2$)
$$H(y, \lambda(x-y)+p) - H(x, \lambda(x-y)+q) \leq \omega_2(|x-y| + \lambda|x-y|^2, R) + \omega_3(|p-q|)$$
$$\text{for all } \lambda \geq 1,\ p,q \in \bar{B}(0,1),\ x,y \in \bar{B}(0,R),\ \forall R > 0,$$

where ω_2, ω_3 are moduli. It is easy to realize that (H$_1$) and

(H$_3$) $\qquad |H(x,p) - H(x,q)| \leq \omega(|p-q|) \qquad \forall x, p, q \in \mathbb{R}^N,$

imply (H$_2$) with $\omega_3 = \omega$ and $\omega_2(r, R) = \omega_1(2r)$ for all $r, R > 0$.

THEOREM 3.5. *Assume that $u_1, u_2 \in BC(\mathbb{R}^N)$ are, respectively, viscosity sub- and supersolutions of*

(3.9) $\qquad u(x) + H(x, Du(x)) = 0, \qquad x \in \mathbb{R}^N$

with H satisfying (H$_2$). *Then, $u_1 \leq u_2$ in $\mathbb{R}^N$.*

PROOF. Set $g(x) = \frac{1}{2}\log(1 + |x|^2)$ and
$$\Phi(x,y) = u_1(x) - u_2(y) - \frac{|x-y|^2}{2\varepsilon} - \beta(g(x) + g(y)),$$
where ε, β are positive parameters to be chosen later. Observe that for $\beta = 0$, Φ coincides with the auxiliary function employed in the proof of Theorem 3.1. Suppose by contradiction that $\delta := u_1(\tilde{x}) - u_2(\tilde{x}) > 0$ for some $\tilde{x}$ and choose β such that $\beta g(\tilde{x}) \leq \delta/4$ and $\omega_3(2\beta) < \delta/6$. Hence,

(3.10) $\qquad \delta/2 \leq \delta - 2\beta g(\tilde{x}) = \Phi(\tilde{x}, \tilde{x}) \leq \sup_{\mathbb{R}^N \times \mathbb{R}^N} \Phi(x,y).$

Since Φ is continuous and $\lim_{|x|+|y| \to +\infty} \Phi(x,y) = -\infty$, there exists $(x_\varepsilon, y_\varepsilon)$ such that
$$\Phi(x_\varepsilon, y_\varepsilon) = \sup_{\mathbb{R}^N \times \mathbb{R}^N} \Phi(x,y).$$

The claim is that $\Phi(x_\varepsilon, y_\varepsilon) < \delta/2$ for small enough $\varepsilon > 0$, leading to a contradiction to (3.10). In order to prove the claim, let us observe first that $\Phi(x_\varepsilon, y_\varepsilon) \geq \delta/2 > 0$ implies
$$\beta(g(x_\varepsilon) + g(y_\varepsilon)) \leq \sup_{\mathbb{R}^N} u_1 + \sup_{\mathbb{R}^N}(-u_2).$$

Therefore there exists $R > 0$ such that x_ε and y_ε belong to $\overline{B}(0,R)$ for all $\varepsilon > 0$. In fact, if this were not true, the above inequality would contradict the boundedness of u_1, u_2 since $\lim_{|x| \to +\infty} g(x) = +\infty$ and $\beta > 0$.

Next we observe that the inequality $\Phi(x_\varepsilon, x_\varepsilon) + \Phi(y_\varepsilon, y_\varepsilon) \leq 2\Phi(x_\varepsilon, y_\varepsilon)$ amounts to

$$(3.11) \qquad \frac{|x_\varepsilon - y_\varepsilon|^2}{\varepsilon} \leq u_1(x_\varepsilon) - u_1(y_\varepsilon) + u_2(x_\varepsilon) - u_2(y_\varepsilon)$$

and this gives

$$(3.12) \qquad |x_\varepsilon - y_\varepsilon| \leq (\varepsilon C)^{1/2},$$

where C is a constant depending on $\sup_{\mathbb{R}^N} |u_1|$ and $\sup_{\mathbb{R}^N} |u_2|$. By plugging (3.12) into (3.11) we obtain

$$(3.13) \qquad \frac{|x_\varepsilon - y_\varepsilon|^2}{\varepsilon} \longrightarrow 0 \qquad \text{as } \varepsilon \to 0^+$$

because of the uniform continuity of u_1, u_2 in $\overline{B}(0,R)$.

Let us consider now the C^1 functions

$$\varphi_1(y) = u_1(x_\varepsilon) - \frac{|x_\varepsilon - y|^2}{2\varepsilon} - \beta(g(x_\varepsilon) + g(y)),$$

$$\varphi_2(x) = u_2(y_\varepsilon) + \frac{|x - y_\varepsilon|^2}{2\varepsilon} + \beta(g(x) + g(y_\varepsilon));$$

observe that x_ε is a maximum for $u_1 - \varphi_2$, y_ε is a minimum for $u_2 - \varphi_1$ and that

$$D\varphi_1(y_\varepsilon) = \frac{x_\varepsilon - y_\varepsilon}{\varepsilon} - \beta Dg(y_\varepsilon),$$

$$D\varphi_2(x_\varepsilon) = \frac{x_\varepsilon - y_\varepsilon}{\varepsilon} + \beta Dg(x_\varepsilon).$$

By the definition of viscosity sub- and supersolution,

$$u_1(x_\varepsilon) + H\left(x_\varepsilon, \frac{x_\varepsilon - y_\varepsilon}{\varepsilon} + \beta Dg(x_\varepsilon)\right) \leq 0$$

$$u_2(y_\varepsilon) + H\left(y_\varepsilon, \frac{x_\varepsilon - y_\varepsilon}{\varepsilon} - \beta Dg(y_\varepsilon)\right) \geq 0$$

so that

$$u_1(x_\varepsilon) - u_2(y_\varepsilon) \leq H(y_\varepsilon, \lambda(x_\varepsilon - y_\varepsilon) + p) - H(x_\varepsilon, \lambda(x_\varepsilon - y_\varepsilon) + q)$$

with $\lambda = 1/\varepsilon$, $p = -\beta Dg(y_\varepsilon)$, $q = \beta Dg(x_\varepsilon)$. Since $|Dg| \leq 1$, choosing $\beta \leq 1$ allows us to apply (H$_2$). This gives

$$u_1(x_\varepsilon) - u_2(y_\varepsilon) \leq \omega_2\left(|x_\varepsilon - y_\varepsilon| + \frac{|x_\varepsilon - y_\varepsilon|^2}{\varepsilon}, R\right) + \omega_3(2\beta).$$

Thanks to (3.12) and (3.13), the preceding gives

$$u_1(x_\varepsilon) - u_2(y_\varepsilon) \leq \delta/3 + \omega_3(2\beta)$$

for sufficiently small $\varepsilon > 0$; a fortiori,
$$\Phi(x_\varepsilon, y_\varepsilon) \leq \delta/3 + \omega_3(2\beta).$$

Since β has been chosen such that $\omega_3(2\beta) < \delta/6$ the desired contradiction is obtained. ◂

It is easy to generalize Theorem 3.5 to cover the case of a general unbounded open set $\Omega \subset \mathbb{R}^N$. Indeed, one can prove that if (H_2) holds and $u_1, u_2 \in BC(\overline{\Omega})$ are, respectively, a viscosity sub- and supersolution of
$$u(x) + H(x, Du(x)) = 0, \quad x \in \Omega,$$
with $u_1 \leq u_2$ on $\partial\Omega$, then $u_1 \leq u_2$ in Ω. The proof is similar to that of Theorem 3.5; the case where the maximum point $(x_\varepsilon, y_\varepsilon)$ of the auxiliary function belongs to $\partial(\Omega \times \Omega)$ is dealt with as in the proof of Theorem 3.1 (see also Exercise 3.7).

REMARK 3.6. A useful variant of Theorem 3.5 is obtained when the boundedness assumption on u_1 and u_2 is replaced by their uniform continuity (which implies that they grow at most linearly at infinity). One can prove indeed that if $u_1, u_2 \in UC(\mathbb{R}^N)$ are, respectively, viscosity sub- and supersolution of (3.9) with H satisfying (H_1) and (H_3), then $u_1 \leq u_2$ in $\mathbb{R}^N$. ◃

The next comparison result concerns an evolutionary case. As usual it gives a uniqueness result for the Cauchy problem
$$\begin{cases} u_t(t,x) + H(t, Du(t,x)) = 0 & (t,x) \in \,]0, T[\, \times \mathbb{R}^N \\ u(0,x) = u_0(x) & x \in \mathbb{R}^N, \end{cases}$$
with initial condition $u_0 \in UC(\mathbb{R}^N)$. See Remark 3.8 for more general Hamiltonians.

THEOREM 3.7. *Assume* $H \in C([0,T] \times \mathbb{R}^N)$. *Let* $u_1, u_2 \in UC([0,T] \times \mathbb{R}^N)$ *be, respectively, viscosity sub- and supersolution of*
$$u_t(t,x) + H(t, D_x u(t,x)) = 0 \quad \text{in }]0, T[\, \times \mathbb{R}^N.$$
Then,
$$\sup_{[0,T] \times \mathbb{R}^N} (u_1 - u_2) \leq \sup_{\mathbb{R}^N}(u_1(0, \cdot) - u_2(0, \cdot)).$$

PROOF. We may assume that
$$A := \sup_{\mathbb{R}^N}(u_1(0, \cdot) - u_2(0, \cdot)) < +\infty.$$
Set, for $r \geq 0$ and $i = 1, 2$,
$$\omega_i(r) = \sup\{|u_i(t,x) - u_i(s,y)| : |t-s| + |x-y| \leq r\};$$

by uniform continuity we have

$$\sup_{r\geq 0} \frac{\omega_i(r)}{1+r} < +\infty. \tag{3.14}$$

A straightforward computation shows that

$$u_1(t,x) - u_2(s,y) = u_1(t,x) - u_1(0,x) + u_1(0,x) - u_2(0,x) + u_2(0,x) - u_2(s,y)$$
$$\leq \omega_1(t) + A + \omega_2(s + |x-y|),$$

for all $(t,x,s,y) \in Q^2$ where $Q := [0,T] \times \mathbb{R}^N$. Hence (3.14) yields

$$|u_1(t,x) - u_2(s,y)| \leq C(1 + |x-y|), \quad \forall (t,x,s,y) \in Q^2, \tag{3.15}$$

for some $C > 0$.

The proof proceeds by contradiction; we assume therefore that

$$\sup_Q (u_1 - u_2) > A + \sigma_0 \tag{3.16}$$

for some $\sigma_0 > 0$. Consider for $\alpha, \varepsilon > 0$ the auxiliary function

$$\Phi(t,x,s,y) = u_1(t,x) - u_2(s,y) - \frac{|x-y|^2}{\varepsilon} - \frac{|t-s|^2}{\alpha}.$$

Thanks to (3.15), $\sup_{Q^2} \Phi < +\infty$ so that, for each $0 < \delta < \sigma_0/2$, there exists $(t_0, x_0, s_0, y_0) \in Q^2$ with

$$\Phi(t_0, x_0, s_0, y_0) + \delta > \sup_{Q^2} \Phi.$$

Choose now $\zeta \in C_0^\infty(\mathbb{R}^N)$ such that

$$\zeta(x_0, y_0) = 1, \quad 0 \leq \zeta \leq 1, \quad |D\zeta(x,y)| \leq 1.$$

For $0 < \sigma < \sigma_0/2T$ we set

$$\Psi(t,x,s,y) = \Phi(t,x,s,y) + \delta\zeta(x,y) - \sigma t.$$

It is easy to realize that there exists $(\bar{t}, \bar{x}, \bar{s}, \bar{y}) \in Q^2$ such that

$$\Psi(\bar{t}, \bar{x}, \bar{s}, \bar{y}) = \sup_{Q^2} \Psi.$$

The inequality

$$2\Psi(\bar{t}, \bar{x}, \bar{s}, \bar{y}) \geq \Psi(\bar{t}, \bar{x}, \bar{t}, \bar{x}) + \Psi(\bar{s}, \bar{y}, \bar{s}, \bar{y})$$

and (3.15) yield after some computations

$$\frac{|\bar{x} - \bar{y}|^2}{\varepsilon} + \frac{|\bar{t} - \bar{s}|^2}{\alpha} \leq C(1 + |\bar{x} - \bar{y}|) + 4\delta + \sigma(\bar{s} - \bar{t})$$
$$\leq C(1 + |\bar{x} - \bar{y}|) + 2\sigma_0,$$

for $0 < \delta < \sigma_0/4$ and $0 < \sigma < \sigma_0/2T$.

The above implies that for each $\varepsilon, \alpha \in \,]0,1[$ there exists constants $C_\varepsilon, C_\alpha > 0$ such that

(3.17) $\qquad |\overline{x} - \overline{y}| \leq C_\varepsilon, \quad |\overline{t} - \overline{s}| \leq C_\alpha, \quad \lim_{\alpha \to 0} C_\alpha = 0 .$

If $(\overline{t}, \overline{x}), (\overline{s}, \overline{y}) \in \,]0, T] \times \mathbb{R}^N$, using the definition of viscosity solution and Lemma 2.10 we obtain

$$\sigma + \frac{1}{\alpha}(\overline{t} - \overline{s}) + H(\overline{t}, \frac{1}{\varepsilon}(\overline{x} - \overline{y}) - \delta D_x \zeta(\overline{x}, \overline{y})) \leq 0$$

and

$$\frac{1}{\alpha}(\overline{t} - \overline{s}) + H(\overline{s}, \frac{1}{\varepsilon}(\overline{x} - \overline{y}) + \delta D_y \zeta(\overline{x}, \overline{y})) \geq 0 .$$

Therefore, by addition,

$$\sigma \leq H(\overline{s}, \frac{1}{\varepsilon}(\overline{x} - \overline{y}) + \delta D_y \zeta(\overline{x}, \overline{y})) - H(\overline{t}, \frac{1}{\varepsilon}(\overline{x} - \overline{y}) + \delta D_x \zeta(\overline{x}, \overline{y})) .$$

Hence, by (3.17),

(3.18) $\qquad \sigma \leq \omega_H\left(|\overline{t} - \overline{s}| + \delta, \frac{C_\varepsilon}{\varepsilon} + \sigma_0\right)$

where the local modulus ω_H is given by

$$\omega_H(r, R) := \sup\{|H(t, p) - H(s, q)| : |t - s| + |p - q| \leq r, \ p, q \in B(0, R)\} .$$

Now we choose α and δ small enough to make the right-hand side of (3.18) strictly less than σ and get the desired contradiction.

In order to complete the proof we must examine the case where $\overline{t} = 0$ or $\overline{s} = 0$. We show that this cannot occur if α is small enough. Assume by contradiction there is a sequence $\alpha_n \to 0^+$ such that the corresponding Ψ attains the maximum at (t_n, x_n, y_n, s_n) with $t_n = 0$ or $s_n = 0$ for all n. If $t_n = 0$, by (3.17) and the definition of A,

$$u_1(x_n, 0) - u_2(y_n, s_n) \leq A + u_2(y_n, 0) - u_2(y_n, s_n) \leq A + \omega_2(C_{\alpha_n}) .$$

Similarly, if $s_n = 0$

$$u_1(x_n, t_n) - u_2(y_n, 0) \leq A + \omega_1(C_{\alpha_n}) .$$

Therefore

$$\sup_{(t,x) \in Q} \Psi(t, x, t, x) \leq \sup_{Q^2} \Psi \leq A + o(1) + \delta$$

as $n \to \infty$. On the other hand, by (3.16),

$$\sup_{(t,x) \in Q} \Psi(t, x, t, x) \geq \sup_Q (u_1 - u_2) - \sigma T > A + \sigma_0 - \sigma T .$$

Then $\sigma_0 < \delta + \sigma T + o(1)$ as $n \to \infty$, which contradicts the choice of δ and σ for n large enough. ◀

REMARK 3.8. The comparison Theorem 3.7 can be extended to the equation

$$u_t + H(t, x, Du) = 0$$

if the Hamiltonian $H \in UC([0, T] \times \mathbb{R}^N \times B(0, R))$ for every $R > 0$ and H satisfies (H$_1$) with a modulus ω_1 independent of $t \in [0, T]$. Further extension can be found in Exercises 3.9 and 3.10, also to Hamiltonians depending on u. ◁

REMARK 3.9. The comparison principles can be used to obtain bounds in the supremum norm for viscosity solutions of (HJ). To show this in a simple setting, let $u_n \in BC(\mathbb{R}^N)$, $n \in \mathbb{N}$, be a viscosity solution of

(3.19) $\qquad u_n(x) + H_n(x, Du_n(x)) = 0, \qquad x \in \mathbb{R}^N,$

where H_n satisfies (H$_1$), (H$_3$) for each $n \in \mathbb{N}$. Assume also that

$$\sup_{x \in \mathbb{R}^N} |H_n(x, 0)| \leq C < +\infty$$

for some constant C, uniformly in n. It is obvious that C and $-C$ are, respectively, a super- and a subsolution of (3.19) for any $n \in \mathbb{N}$. Hence, by Theorem 3.5

$$-C \leq u_n(x) \leq C,$$

for any $n \in \mathbb{N}$ and $x \in \mathbb{R}^N$. Several applications of this idea will be considered in subsequent chapters. ◁

Exercises

3.1. Observe that, if (3.4) does not hold, the proof of Theorem 3.1 shows actually that

$$\max_{\overline{\Omega}}(u_1 - u_2)^+ \leq \max_{\partial \Omega}(u_1 - u_2)^+.$$

Compare this with Proposition 1.5. Observe also that the result of Theorem 3.1 remains valid if one replaces assumption (H$_1$) with the requirement that either u_1 or u_2 is Lipschitz continuous.

3.2. Let $H(x, p) = \sup_{a \in A}\{-f(x, a) \cdot p - \ell(x, a)\}$, with A compact f and ℓ continuous. Assume also that, for all x, y,

$$|f(x, a) - f(y, a)| \leq L|x - y|, \qquad |\ell(x, a) - \ell(y, a)| \leq \omega(|x - y|)$$

where the constant L and the modulus ω are independent of $a \in A$. Show that H satisfies assumption (H$_1$) in Theorem 3.1.

3.3. Prove that the result of Theorem 3.1 remains true if assumption (H$_1$) is replaced by the coercivity condition

$$r + H(x, p) \longrightarrow +\infty \qquad \text{as } |p| \to +\infty,$$

uniformly with respect to $x \in \overline{\Omega}$, $|r| \leq R$, ($\forall R > 0$). [Hint: observe that the inequality

$$u_1(x_\varepsilon) + H\left(x_\varepsilon, \frac{x_\varepsilon - y_\varepsilon}{\varepsilon}\right) \leq 0$$

in the proof of Theorem 3.1 and coercivity imply that $|(x_\varepsilon - y_\varepsilon)/\varepsilon|$ is uniformly bounded as $\varepsilon \to 0^+$.]

3.4. Take $H(x,p) = |A(x)p|^2 - \ell(x)$, where $A(x) = (a_{ij}(x))$ is a $M \times N$ matrix, $M \leq N$. Assume that ℓ, a_{ij} are bounded and Lipschitz continuous. Observe that $|H(x,p) - H(y,p)| \simeq C|x-y| |p|^2$ and consequently (H_1) is not satisfied. Observe also that if $M = N$ and $A(x)$ is invertible at all $x \in \overline{\Omega}$, then H satisfies the coercivity condition in Exercise 3.3. Prove that if $u_1, u_2 \in C(\overline{\Omega})$ are viscosity solutions of

$$u(x) + |A(x)Du(x)|^2 - \ell(x) = 0 \quad \text{in } \Omega,$$

with $\Omega \subseteq \mathbb{R}^N$ open and bounded, then $u_1 \equiv u_2$ provided $u_1 = u_2$ on $\partial \Omega$.

3.5. Let Ω be a bounded open set with smooth boundary $\partial \Omega$. Assume that

$$t \longmapsto F(x, r, p + tn(x))$$

is nondecreasing for each fixed r, p and for all $x \in \Gamma \subseteq \partial \Omega$. Extend Lemma 2.10 to this setting. Generalize then Theorem 3.1 to show that if u and v are viscosity sub- and supersolution of (3.3) with $u \leq v$ on $\partial \Omega \setminus \Gamma$, then $u \leq v$ in $\overline{\Omega}$.

3.6. Prove the following generalization of Theorem 3.1. Assume that for any $R > 0$ there exists $\gamma_R > 0$ such that

$$F(x, r, p) - F(x, s, p) \geq \gamma_R(r - s)$$

for all $x \in \Omega$, $p \in \mathbb{R}^N$ and $-R \leq s \leq r \leq R$. Assume also that

$$|F(x, r, p) - F(y, r, p)| \leq \omega_R(|x - y|(1 + |p|)),$$

with $\omega_R(t) \to 0$ as $t \to 0^+$, for all $x, y \in \Omega$ and $|r| \leq R$. Show that if $u_1, u_2 \in C(\overline{\Omega})$ are, respectively, viscosity sub- and supersolution of

$$F(x, u(x), Du(x)) = 0, \quad x \in \Omega,$$

then,

$$\max_{\overline{\Omega}}(u_1 - u_2)^+ \leq \max_{\partial \Omega}(u_1 - u_2)^+.$$

3.7. Assume $u_1, u_2 \in BC(\overline{\Omega})$ are, respectively, a sub- and a supersolution of

$$\lambda(x)u(x) + H(x, Du(x)) = 0 \quad \text{in } \Omega,$$

such that $u_1 \leq u_2$ on $\partial \Omega$, where $\Omega \subseteq \mathbb{R}^N$ is open, H satisfies (H_2), and $\lambda \in C(\overline{\Omega})$, $\lambda(x) > 0$ for all $x \in \Omega$. Prove that $u_1 \leq u_2$ in Ω.

3.8. The one-dimensional equation
$$u(x) + \frac{1}{2}|u'(x)|^2 = 0, \qquad x \in \mathbb{R},$$
has the classical solutions $u_1 \equiv 0$ and $u_2(x) = -\frac{1}{2}(x - x_0)^2$ (for any fixed x_0). Prove that u_1 is the unique viscosity solution in $BC(\mathbb{R})$ and in $UC(\mathbb{R})$.

3.9. Under the assumptions of Remark 3.8, namely, $H \in UC([0,T] \times \mathbb{R}^N \times B(0,R))$ $\forall R > 0$ and, for a modulus ω_1,
$$|H(t,x,p) - H(t,y,p)| \leq \omega_1(|x-y|(1+|p|))$$
for all t, x, y, p, consider solutions $u_i \in UC([0,T] \times \mathbb{R}^N)$ of
$$u_t + H(t,x,Du) = f_i(t,x) \qquad \text{in }]0,T[\times \mathbb{R}^N,$$
where $f_i \in UC([0,T] \times \mathbb{R}^N)$, $i = 1, 2$. Prove that
$$\sup_{\mathbb{R}^N}(u_1(t,\cdot) - u_2(t,\cdot)) \leq \sup_{\mathbb{R}^N}(u_1(0,\cdot) - u_2(0,\cdot)) + \int_0^t \sup_{\mathbb{R}^N}(f_1(s,\cdot) - f_2(s,\cdot))\,ds$$
for all $t \in [0,T]$.

3.10. Prove comparison results for solutions in BUC and in UC of
$$u_t + H(t,u,Du) = 0 \qquad \text{in }]0,T[\times \mathbb{R}^N,$$
with H continuous such that $r \mapsto H(t,r,p)$ is nondecreasing. Then extend it to the case where $r \mapsto H(t,r,p) + \lambda r$ is nondecreasing for some $\lambda > 0$. [Hint: make the change of variables $v = e^{-\lambda t}u$.]

3.11. Under the assumptions of Theorem 3.7 let u_1, u_2 be, respectively, solutions of
$$u_t + \lambda u + H(t,Du) = e^{-t}f_i(t,x) \qquad \text{in }]0,+\infty[\times \mathbb{R}^N,$$
for $i = 1, 2$, where $\lambda > 0$ and $f_i \in UC([0,T] \times \mathbb{R}^N)$ $\forall T > 0$. Prove that, if $u_1(0,\cdot) - u_2(0,\cdot)$ and $f_1 - f_2$ are bounded, then $\sup_{\mathbb{R}^N}|u_1(t,\cdot) - u_2(t,\cdot)|$ decays exponentially to 0 as $t \to +\infty$. [Hint: make the change of variables $v = e^{-\lambda t}u$ and use Exercise 3.9.]

4. Lipschitz continuity and semiconcavity

In the first part of this section we present two results (Propositions 4.1 and 4.2) which show that, under suitable assumptions on H, viscosity solutions of

(HJ) $\qquad \lambda u(x) + H(x,Du(x)) = 0, \qquad x \in \mathbb{R}^N$

are actually Lipschitz continuous. The section includes a quick overview of some differentiability properties of Lipschitz functions.

Section 4.2 is devoted to the basic theory of semiconcave functions and the related topic of inf-convolution. The main results in this direction are Theorem 4.9 which asserts that, under suitable conditions on H, Lipschitz continuous viscosity solutions of (HJ) are semiconcave and Proposition 4.13, showing that a viscosity subsolution of (HJ) can be uniformly approximated from below by a semiconcave subsolution u_ε of an approximate equation.

4.1. Lipschitz continuity

We assume for the next result that H satisfies the coercivity condition

(H$_4$) $\qquad H(x,p) \longrightarrow +\infty \qquad \text{as } |p| \to +\infty,$

uniformly with respect to x. When H is given by

(4.1) $\qquad H(x,p) = \sup_{a \in A}\{-f(x,a) \cdot p - \ell(x,a)\},$

sufficient conditions for (H$_4$) to hold are the boundedness of ℓ together with the controllability assumption

(4.2) $\qquad \exists r > 0 : B(0,r) \subseteq \overline{\text{co}} f(x,A), \qquad \forall x \in \mathbb{R}^N,$

(see Exercise 4.1).

PROPOSITION 4.1. *Let us assume condition* (H$_4$). *Then any viscosity subsolution* $u \in BC(\mathbb{R}^N)$ *of* (HJ) *is Lipschitz continuous.*

PROOF. For fixed $x \in \mathbb{R}^N$ consider the function

$$\varphi(y) = u(y) - C|y - x|$$

where $C > 0$ is a constant to be chosen later. The boundedness of u implies the existence of $\overline{y} \in \mathbb{R}^N$ such that

$$\varphi(\overline{y}) = \max_{y \in \mathbb{R}^N} \varphi(y).$$

The claim is that $\overline{y} = x$ for C large. If not we would have

(4.3) $\qquad \lambda u(\overline{y}) + H\left(\overline{y}, C \dfrac{\overline{y} - x}{|\overline{y} - x|}\right) \leq 0,$

since u is a viscosity subsolution of (HJ) and $y \mapsto C|y - x|$ is differentiable at $y = \overline{y} \neq x$. For sufficiently large C, independent of x, (4.3) is in contradiction to (H$_4$). Therefore, for such C,

$$u(y) - C|y - x| \leq u(\overline{y}) - C|\overline{y} - x| = u(x), \qquad \forall y \in \mathbb{R}^N.$$

By interchanging the roles of x, y, the proof is complete. ◀

A different condition on H guaranteeing the Lipschitz continuity of viscosity solutions of (HJ) is

(H$_5$) $\quad \exists C > 0: H\left(x, C\dfrac{x-y}{|x-y|}\right) - H\left(y, C\dfrac{x-y}{|x-y|}\right) \geq -C|x-y|, \quad \forall x,y \in \mathbb{R}^N.$

For H given by (4.1), condition (H$_5$) holds with $C \geq M/(1-L)$, provided that f, ℓ satisfy

$$(f(x,a) - f(y,a)) \cdot (x-y) \leq L|x-y|^2, \qquad L < 1,$$
$$|\ell(x,a) - \ell(y,a)| \leq M|x-y|, \qquad \forall x, y, a$$

(see Exercise 4.2).

PROPOSITION 4.2. *Let us assume* (H_1), (H_3), (H_5), $\lambda \geq 1$ *and let* $u \in UC(\mathbb{R}^N)$ *be a viscosity solution of* (HJ). *Then*

$$|u(x) - u(y)| \leq C|x - y|.$$

PROOF. It is easy to check that $v(x, y) = u(x) - u(y)$ is a viscosity solution of

$$\lambda v(x,y) + \widehat{H}(x, y, D_x v(x,y), D_y v(x,y)) = 0 \quad \text{in } \mathbb{R}^{2N},$$

where $\widehat{H}(x, y, p, q) := H(x, p) - H(y, -q)$. On the other hand, $w(x, y) := C|x-y|$ satisfies

$$\lambda w(x,y) + \widehat{H}(x, y, D_x w(x,y), D_y w(x,y)) - g(x,y) = 0 \quad \text{in } \mathbb{R}^{2N},$$

where

$$g(x, y) = \lambda C |x - y| + H\left(x, C\frac{x-y}{|x-y|}\right) - H\left(y, C\frac{x-y}{|x-y|}\right).$$

Since $g \geq 0$ by (H_5), $v, w \in UC(\mathbb{R}^{2N})$ and $\widehat{H}$ satisfies (H_1), (H_3), as it is easy to check, the comparison result in Remark 3.6 gives $v \leq w$ in $\mathbb{R}^{2N}$, which proves the statement. ◀

Let us recall briefly now some relevant differential properties of locally Lipschitz continuous functions. By the classical Rademacher theorem (see [EG92]) such functions are differentiable almost everywhere with locally bounded gradient. Hence, if $u \in \text{Lip}_{\text{loc}}(\Omega)$, the set

$$D^*u(x) = \left\{ p \in \mathbb{R}^N : p = \lim_{n \to +\infty} Du(x_n),\ x_n \to x \right\}$$

is nonempty and closed for any $x \in \Omega$. Denote by $\text{co}\,D^*u(x)$ its convex hull. It is a well-known result in nonsmooth analysis (see [Cla83]) that

(4.4) $$\text{co}\,D^*u(x) = \partial u(x), \quad \forall x \in \Omega,$$

where $\partial u(x)$ is the *generalized gradient* or *Clarke's gradient* of u at x, defined by

$$\partial u(x) := \{ p \in \mathbb{R}^N : u^0(x; q) \geq p \cdot q,\ \forall q \in \mathbb{R}^N \}$$
$$= \{ p \in \mathbb{R}^N : u_0(x; q) \leq p \cdot q,\ \forall q \in \mathbb{R}^N \}.$$

Here $u^0(x; q)$ and $u_0(x; q)$ are the *generalized* (or *regularized*) *directional derivatives* defined by

$$u^0(x; q) := \limsup_{\substack{y \to x \\ t \to 0^+}} \frac{u(y + tq) - u(y)}{t}$$

$$u_0(x; q) := \liminf_{\substack{y \to x \\ t \to 0^+}} \frac{u(y + tq) - u(y)}{t}.$$

A related notion is that of *Dini directional derivatives*, namely,

$$\partial^+ u(x;q) := \limsup_{t \to 0^+} \frac{u(x+tq) - u(x)}{t}$$

$$\partial^- u(x;q) := \liminf_{t \to 0^+} \frac{u(x+tq) - u(x)}{t}.$$

From the definitions it follows easily that

(4.5) $\quad u_0(x;q) \leq \partial^- u(x;q) \leq \partial^+ u(x;q) \leq u^0(x;q), \qquad \forall x \in \Omega, \ q \in \mathbb{R}^N,$

and this implies that for $u \in \text{Lip}_{\text{loc}}(\Omega)$,

(4.6) $\quad D^- u(x) \cup D^+ u(x) \subseteq \partial u(x), \qquad \forall x \in \Omega$

(see Exercise 4.4). Observe also that $D^+ u(x), D^- u(x)$ are bounded sets (see Exercise 1.12).

The next result is about the existence of classical (one-sided) directional derivatives of locally Lipschitz continuous functions, namely

$$\frac{\partial u}{\partial q}(x) := \partial u(x;q) := \lim_{t \to 0^+} \frac{u(x+tq) - u(x)}{t}, \qquad |q| = 1.$$

PROPOSITION 4.3. *Let $u \in \text{Lip}_{\text{loc}}(\Omega)$. Then, for all q with $|q| = 1$, there exists*

(4.7) $$\frac{\partial u}{\partial q}(x) = \min_{p \in D^+ u(x)} p \cdot q = u_0(x;q)$$

at any $x \in \Omega$ where $D^+ u(x) = \partial u(x)$.

PROOF. Let $p \in D^+ u(x)$ and $|q| = 1$. Then

$$u(x + tq) - u(x) - tp \cdot q \leq o(|t|),$$

for sufficiently small t. Hence

$$p \cdot q \geq \frac{u(x+tq) - u(x)}{t} - \frac{o(t)}{t}, \qquad \text{for small } t > 0.$$

Consequently,

$$\inf_{p \in D^+ u(x)} p \cdot q \geq \partial^+ u(x;q).$$

Combining this with (4.5) we get

$$u_0(x;q) \leq \partial^- u(x;q) \leq \partial^+ u(x;q) \leq \inf_{p \in D^+ u(x)} p \cdot q.$$

It is not hard to show (see Exercise 4.4) that

$$u_0(x;q) = \min_{p \in \partial u(x)} p \cdot q;$$

hence (4.7) holds if x is such that $D^+ u(x) = \partial u(x)$. ◀

The preceding proposition allows us to prove a useful variant to Proposition 2.13 on semidifferential and directional derivatives of a marginal function u defined by

$$u(x) := \inf_{b \in B} g(x, b) .$$

PROPOSITION 4.4. *Let us assume B compact, g continuous on $\Omega \times B$, differentiable with respect to x with $D_x g$ continuous on $\Omega \times B$. Then $u \in \text{Lip}_{\text{loc}}(\Omega)$, $D^+ u(x) = \partial u(x)$ for all x and the same conclusions of Proposition 2.13 hold.*

PROOF. By the assumptions made, $x \mapsto g(x, b)$ is locally Lipschitz, uniformly in $b \in B$. Then it is easy to conclude that $u \in \text{Lip}_{\text{loc}}(\Omega)$. Next we show that $D^+ u(x) = \overline{\text{co}} Y(x)$, where

$$Y(x) = \{\, D_x g(x, b) : b \in M(x) \,\} \quad \text{and} \quad M(x) = \{\, b \in B : u(x) = g(x, b) \,\} .$$

By Lemma 2.11, it is enough to prove that $D^+ u(x) \subseteq \overline{\text{co}} Y(x)$; in view of (4.4) and (4.6) this amounts to show that $D^* u(x) \subseteq Y(x)$.

Let $p \in D^* u(x)$ and take $x_n \to x$ such that $Du(x_n) \to p$. Then pick $b_n \in M(x_n)$ and suppose, without loss of generality, that $b_n \to \overline{b} \in B$. Since $g(x_n, b_n) \leq g(x_n, b)$ for all $n \in \mathbb{N}$ and $b \in B$, we conclude by continuity that $\overline{b} \in M(x)$. By differentiability of u at x_n and Lemma 2.11, we have $Du(x_n) = D_x g(x_n, b_n)$. Hence, letting $n \to +\infty$, we get $p = D_x g(x, \overline{b})$; that is, $p \in Y(x)$. The existence of directional derivatives and the formula

$$(4.8) \qquad \frac{\partial u}{\partial q}(x) = \min_{y \in Y(x)} y \cdot q, \qquad \forall q \text{ with } |q| = 1,$$

follow directly from Proposition 4.3. It remains to prove that $D^- u(x) = \{y\}$, whenever $Y(x)$ reduces to the singleton $\{y\}$. Observe that in this case (4.8) implies $\partial u(x; q) = y \cdot q$ for all q, and this ensures $y \in D^- u(x)$ (this is not hard to check, see Lemma III.2.37 for a more general result). The opposite inclusion follows from Lemma 2.11. ◂

4.2. Semiconcavity

We say that $u : \Omega \to \mathbb{R}$ is *semiconcave* on the open convex set Ω if there exists a constant $C \geq 0$ such that

$$(4.9) \qquad \mu u(x) + (1 - \mu) u(y) \leq u(\mu x + (1 - \mu) y) + \frac{1}{2} C \mu (1 - \mu) |x - y|^2$$

for all $x, y \in \Omega$ and $\mu \in [0, 1]$.

This amounts to the concavity of $x \mapsto u(x) - \frac{1}{2} C |x|^2$, as it is easy to check. If u is continuous, an equivalent way to express condition (4.9) is to require that

$$(4.10) \qquad u(x + h) - 2u(x) + u(x - h) \leq C |h|^2,$$

for any $x \in \Omega$ and $h \in \mathbb{R}^N$, with sufficiently small $|h|$ (see Exercise 4.5). Concave functions are, of course, semiconcave. A less trivial class of semiconcave functions is that of C^1 functions with locally Lipschitz gradient.

A relevant class of nondifferentiable semiconcave functions is that of marginal functions $u(x) = \inf_{b \in B} g(x,b)$, provided $x \mapsto g(x,b)$ satisfies (4.9) uniformly in b (see Exercise 4.6). Consequently, value functions of optimal control problems are semiconcave under suitable conditions on the data (see later chapters for developments in this direction). An immediate application of this remark gives the following examples.

EXAMPLE 4.5. Let $S \subseteq \mathbb{R}^N$, $S \neq \emptyset$,

$$d(x) := \operatorname{dist}(x, S) = \inf_{s \in S} |x - s|.$$

Then d^2 is semiconcave in $\mathbb{R}^N$ because $x \mapsto |x - s|^2$ is C^∞ with constant second derivatives. On the other hand, d itself is semiconcave on any compact set having positive distance from S, because $x \mapsto |x - s|$ has bounded second derivatives in such a set, uniformly for s in a bounded set. ◁

The main properties of semiconcave functions which will be useful in the sequel are summarized in Propositions 4.6 and 4.7.

PROPOSITION 4.6. *Let u be semiconcave in Ω. Then u is locally Lipschitz continuous in Ω.*

PROOF. For fixed $x \in \Omega$ and any h such that $x + h \in \Omega$,

$$u(x + h) - u(x) = \psi(x + h) - \psi(x) + Cx \cdot h + \frac{C}{2}|h|^2,$$

where $\psi(x) = u(x) - \frac{1}{2}C|x|^2$ is concave and, consequently, locally Lipschitz continuous in Ω. Hence the statement. ◀

We know already from §4.1 that $D^+u(x) \subseteq \partial u(x) = \operatorname{co} D^*u(x)$ for any $u \in \operatorname{Lip}_{\operatorname{loc}}(\Omega)$. If in addition u is semiconcave, it turns out that $D^+u(x) = \partial u(x)$.

This and other differential properties of semiconcave functions are collected in the next statement.

PROPOSITION 4.7. *Let u be semiconcave in Ω. Then for all $x \in \Omega$*
(a) $D^+u(x) = \partial u(x) = \operatorname{co} D^*u(x)$;
(b) *either $D^-u(x) = \emptyset$ or u is differentiable at x;*
(c) *if $D^+u(x)$ is a singleton, then u is differentiable at x;*
(d) $\dfrac{\partial u}{\partial q}(x) = \min_{p \in D^+u(x)} p \cdot q$, *for all unit vectors q.*

PROOF. Let us start by showing that

(4.11) $$u(x + h) - u(x) - p \cdot h \leq \frac{C}{2}|h|^2,$$

for all $p \in \operatorname{co} D^*u(x)$ and sufficiently small $|h|$.

Assume first that u is differentiable at x. Since

$$u(x+h) = \lim_{\mu \to 0^+} u(x + (1-\mu)h)$$

and

$$-Du(x) \cdot h = Du(x) \cdot (-h) = \lim_{\mu \to 0^+} \frac{u(x - \mu h) - u(x)}{\mu}$$
$$= \lim_{\mu \to 0^+} (1 - \mu) \frac{u(x - \mu h) - u(x)}{\mu},$$

we have

(4.12) $\quad u(x+h) - u(x) - Du(x) \cdot h$
$$= \lim_{\mu \to 0^+} \frac{1}{\mu} [\mu u(x + (1-\mu)h) + (1-\mu)u(x - \mu h) - u(x)].$$

Now let us apply (4.9) with $x + (1-\mu)h$ and $x - \mu h$ playing the role of x and y, respectively, to obtain

$$\mu u(x + (1-\mu)h) + (1-\mu)u(x - \mu h) - u(x) \leq \frac{C}{2} \mu (1-\mu) |h|^2.$$

Taking (4.12) into account we get

(4.13) $$u(x+h) - u(x) - Du(x) \cdot h \leq \frac{C}{2} |h|^2,$$

which proves (4.11) in the present case.

Let us now take an arbitrary $x \in \Omega$ and $p \in D^*u(x)$. Next pick $x_n \to x$ such that $Du(x_n) \to p$. By (4.13),

$$u(x_n + h) - u(x_n) - Du(x_n) \cdot h \leq \frac{C}{2} |h|^2$$

and (4.11) follows after letting $n \to +\infty$. The validity of (4.11) for arbitrary $p \in \operatorname{co} D^*u(x)$ follows by a convexity argument.

In view of (4.4) and (4.6), the proof of (a) is reduced to show that $\operatorname{co} D^*u(x) \subseteq D^+u(x)$. This is an immediate consequence of (4.11). Since $D^*u(x) \neq \emptyset$, (a) implies $D^+u(x) \neq \emptyset$. Hence, if $D^-u(x) \neq \emptyset$, then u is differentiable at x (see Lemma 1.8) and (b) is proved.

Observe now that, in view of (a), the one-sided estimate (4.11) holds true for all $p \in D^+u(x)$. In order to prove (c), let $D^+u(x) = \{p_0\}$ and choose $x_n \to x$ such that

(4.14) $$\lim_{n \to +\infty} \frac{u(x_n) - u(x) - p_0 \cdot (x_n - x)}{|x_n - x|} = \liminf_{y \to x} \frac{u(y) - u(x) - p_0 \cdot (y - x)}{|y - x|}.$$

On the other hand, using (4.11) at $x = x_n$ with $h = x - x_n$,

(4.15) $\quad u(x_n) - u(x) - p_0 \cdot (x_n - x) \geq -\dfrac{C}{2}|x_n - x|^2 - |p_n - p_0| |x_n - x|,$

for all $p_n \in D^+u(x_n)$. Since $D^+u(x_n)$ is bounded uniformly in n (see Exercise 1.12), we can assume (passing to a subsequence if necessary) that $p_n \to \bar{p}$ as $n \to +\infty$. Estimate (4.11) gives

$$u(x_n + h) - u(x_n) - p_n \cdot h \leq \dfrac{C}{2}|h|^2$$

and therefore, by continuity,

$$u(x + h) - u(x) - \bar{p} \cdot h \leq \dfrac{C}{2}|h|^2 .$$

This shows that $\lim_{n \to +\infty} p_n = \bar{p} \in D^+u(x)$. Since we assumed $D^+u(x) = \{p_0\}$, we conclude that $\bar{p} = p_0$ and the whole sequence $\{p_n\}$ converges to p_0. From (4.14), (4.15) it then follows

$$\liminf_{y \to x} \dfrac{u(y) - u(x) - p_0 \cdot (y - x)}{|y - x|} \geq 0;$$

thus $p_0 \in D^-u(x)$. Therefore (c) is proved, taking Lemma 1.8 into account.

Finally, statement (d) is an immediate consequence of Proposition 4.3 in view of (a). ◀

As a simple consequence of the previous Proposition 4.7 we get a partial converse to statement (b) in Proposition 1.9 (see also §5.1 for further results in this direction).

PROPOSITION 4.8. *Let u be semiconcave and such that*

(4.16) $\quad\quad\quad\quad F(x, u(x), Du(x)) \geq 0 \quad$ *a.e. in* Ω,

where F is continuous. Then u is a viscosity supersolution of

(4.17) $\quad\quad\quad\quad F(x, u(x), Du(x)) = 0 \quad$ *in* Ω .

PROOF. By Proposition 4.7 (b), at any $x \in \Omega$ either

$$D^-u(x) = \emptyset \quad \text{or} \quad D^-u(x) = D^+u(x) = \{Du(x)\} .$$

In the first case the supersolution condition is automatically satisfied. Assume therefore that x is a point of differentiability of u. Then, there exists a sequence $x_n \to x$ such that u is differentiable at x_n and

(4.18) $\quad\quad\quad\quad F(x_n, u(x_n), Du(x_n)) \geq 0 \quad \forall n$

(one can take $x_n \equiv x$, provided (4.16) holds at x). Since u is locally Lipschitz continuous, by definition of $D^*u(x)$ we have

$$Du(x_n) \longrightarrow p \in D^*u(x) \qquad \text{as } n \to +\infty,$$

at least for a subsequence. In the present case,

$$D^*u(x) = D^+u(x) = D^-u(x) = \{Du(x)\},$$

so that, letting $n \to +\infty$ in (4.18), we get

$$F(x, u(x), p) \geq 0 \qquad \forall p \in D^-u(x)$$

and the statement is proved. ◀

The preceding proof shows that a semiconcave generalized solution of (4.17) satisfies $F(x, u(x), p) = 0$ for all $x \in \Omega$ and $p \in D^-u(x)$. A deeper result in this spirit is Theorem 5.5 in the following section.

The next result is on the semiconcavity of viscosity solutions of (HJ).

THEOREM 4.9. *Let $u \in BC(\mathbb{R}^N) \cap \mathrm{Lip}(\mathbb{R}^N)$ be a viscosity solution of*

(HJ) $$u(x) + H(x, Du(x)) = 0, \qquad x \in \mathbb{R}^N,$$

with Lipschitz constant L_u. Assume that H satisfies

(H3) $$|H(x, p) - H(x, q)| \leq \omega(|p - q|), \qquad \forall x, p, q \in \mathbb{R}^N,$$

and that, for some $C > 0$ and $L' > 2L_u$,

(H6) $$H(x + h, p + Ch) - 2H(x, p) + H(x - h, p - Ch) \geq -C|h|^2$$

holds for all $x, h \in \mathbb{R}^N$, $p \in B(0, L')$. Then u is semiconcave on $\mathbb{R}^N$.

PROOF. Consider the function $v : \mathbb{R}^{3N} \to \mathbb{R}$ defined by $v(x, y, z) = u(x) - 2u(y) + u(z)$. It is easy to check that

$$(p, q, r) \in D^+v(x, y, z)$$

if and only if

$$p \in D^+u(x), \quad -\frac{q}{2} \in D^-u(y), \quad r \in D^+u(z).$$

Since u is a viscosity solution of (HJ), then

$$u(x) + H(x, p) \leq 0 \qquad \forall p \in D^+u(x),$$
$$u(z) + H(z, r) \leq 0 \qquad \forall q \in D^-u(y).$$

Hence, v is a viscosity solution of

(4.19) $$v(x, y, z) + \widetilde{H}(x, y, z, D_xv, D_yv, D_zv) = 0, \qquad (x, y, z) \in \mathbb{R}^{3N},$$

where $\widetilde{H}(x,y,z,p,q,r) = H(x,p) - 2H(y,-q/2) + H(z,r)$.

Consider now the auxiliary function

$$\Phi(x,y,z) = v(x,y,z) - \frac{1}{\varepsilon}|x - 2y + z|^2 - \frac{C}{2}|x - y|^2 - \frac{C}{2}|y - z|^2 - \delta(1 + |y|^2)^{1/2}$$

where ε, δ are positive parameters and C is any fixed constant satisfying (H$_6$). From the boundedness of u it follows that Φ attains a maximum over $\mathbb{R}^{3N}$ at some point (x_0, y_0, z_0) depending on ε, δ.

Now a computation shows that

$$D_x(v - \Phi)(x_0, y_0, z_0) = p_0 + C(x_0 - y_0)$$
$$D_y(v - \Phi)(x_0, y_0, z_0) = -2p_0 + \delta\langle y_0\rangle^{-1}y_0 - C(x_0 - y_0) + C(y_0 - z_0)$$
$$D_z(v - \Phi)(x_0, y_0, z_0) = p_0 - C(y_0 - z_0),$$

where we set $\langle y\rangle = (1 + |y|^2)^{1/2}$ and $p_0 = \frac{2}{\varepsilon}(x_0 - 2y_0 + z_0)$.

Since v is a subsolution of (4.19) we obtain

(4.20) $v(x_0, y_0, z_0) \leq -H(x_0, p_0 + C(x_0 - y_0))$
$\quad + 2H(y_0, p_0 - \frac{1}{2}\delta\langle y_0\rangle^{-1}y_0 + \frac{1}{2}C(x_0 - 2y_0 + z_0)) - H(z_0, p_0 + C(z_0 - y_0))$.

The inequality $\Phi(y + h, y, y - h) \leq \Phi(x_0, y_0, z_0)$, which holds for any $y, h \in \mathbb{R}^N$, gives

(4.21) $\quad u(y + h) - 2u(y) + u(y - h) - C|h|^2 - \delta\langle y\rangle \leq \xi,$

where we set

$$\xi = v(x_0, y_0, z_0) - \frac{C}{2}|x_0 - y_0|^2 - \frac{C}{2}|y_0 - z_0|^2 - \delta\langle y_0\rangle.$$

By assumption (H$_3$),

$H(y_0, p_0 - \frac{1}{2}\delta\langle y_0\rangle^{-1}y_0 + \frac{1}{2}C(x_0 - 2y_0 + z_0))$
$$\leq H(y_0, p_0 + \frac{1}{2}C(x_0 - 2y_0 + z_0)) + \omega(\delta/2)$$

since $\langle y_0\rangle^{-1}|y_0| \leq 1$. Therefore, taking (4.20) into account, ξ is estimated from above as

(4.22) $\xi \leq -H(x_0, p_0 + C(x_0 - y_0))$
$\quad + 2H(y_0, p_0 + \frac{1}{2}C(x_0 - 2y_0 + z_0)) - H(z_0, p_0 + C(z_0 - y_0))$
$\quad + 2\omega(\delta/2) - \delta\langle y_0\rangle - \frac{C}{2}|x_0 - y_0|^2 - \frac{C}{2}|y_0 - z_0|^2$.

Let us proceed now to estimate various quantities using the Lipschitz continuity of u. The inequality $\Phi(0,0,0) \leq \Phi(x_0, y_0, z_0)$ yields

$$-\delta \leq L_u|x_0 - y_0| + L_u|y_0 - z_0| - \frac{C}{2}|x_0 - y_0|^2 - \frac{C}{2}|y_0 - z_0|^2 - \delta\langle y_0\rangle$$

and this implies that

(4.23) $$|x_0 - y_0| + |y_0 - z_0| + \delta\langle y_0\rangle \leq R$$

for some $R > 0$. On the other hand, $\Phi(x_0, \frac{1}{2}(x_0 + z_0), z_0) \leq \Phi(x_0, y_0, z_0)$ gives

$$\frac{1}{\varepsilon}|x_0 - 2y_0 + z_0|^2 \leq 2[u(\frac{1}{2}(x_0 + z_0)) - u(y_0)] + \delta\langle\frac{1}{2}(x_0 + z_0)\rangle - \delta\langle y_0\rangle,$$

so that

(4.24) $$\frac{1}{\varepsilon}|x_0 - 2y_0 + z_0|^2 \leq L_u|x_0 - 2y_0 + z_0| + \delta|\frac{1}{2}(x_0 + z_0 - 2y_0)|.$$

Here we applied the mean value theorem to $\psi(t) = (1 + t^2)^{1/2}$ and the fact that $|\psi'| \leq 1$. Consequently,

(4.25) $$|p_0| \leq 2L_u + \delta \leq L' \quad \text{for } 0 < \delta < L' - 2L_u.$$

We send now $\varepsilon \to 0^+$ for fixed δ as above; by (4.23), (4.24), (4.25), at least for a subsequence,

(4.26) $$\begin{aligned} x_0 - y_0 &\longrightarrow h, & y_0 &\longrightarrow \overline{y}, \\ y_0 - z_0 &\longrightarrow h', & p_0 &\longrightarrow \overline{p}. \end{aligned}$$

for some $h, h', \overline{y} \in \overline{B}(0, R)$, $\overline{p} \in \overline{B}(0, L')$. Also, from (4.24),

$$x_0 - 2y_0 + z_0 \longrightarrow 0$$

so that $h = h' = \overline{h}$. Hence, from (4.22),

$$\xi \leq -H(\overline{y} + \overline{h}, \overline{p} + C\overline{h}) + 2H(\overline{y}, \overline{p}) - H(\overline{y} - \overline{h}, \overline{p} - C\overline{h}) - C|\overline{h}|^2 + 2\omega(\delta/2) - \delta\langle\overline{y}\rangle.$$

By (H_6) then,

$$\limsup_{\varepsilon \to 0^+} \xi \leq 2\omega(\delta/2) - \delta\langle\overline{y}\rangle;$$

using this in (4.21), the claim follows after letting $\delta \to 0^+$. ◂

REMARK 4.10. If H is convex with respect to p, then (H_6) is satisfied under suitable assumptions on the x-dependence. Consider the simplest case $H(x, p) = H(p) - \ell(x)$, with H convex. Then, as it is easy to check, (H_6) holds if ℓ semiconcave (see Exercise 4.8).

A more general case is H of the form

$$H(x, p) = \sup_{a \in A}\{-f(x, a) \cdot p - \ell(x, a)\}.$$

Assume that, for some $L, M, F \geq 0$ the following hold

$$(f(x + h, a) - f(x, a)) \cdot h \leq L|h|^2$$
$$|f(x + h, a) - 2f(x, a) + f(x - h, a)| \leq F|h|^2$$
$$\ell(x + h, a) - 2\ell(x, a) + \ell(x - h, a) \leq M|h|^2,$$

for all $x, h \in \mathbb{R}^N$, $a \in A$. An easy computation shows that if $L < 1/2$, then (H_6) is satisfied with $C \geq (M + 2FL_u)/(1 - 2L)$ (see Exercise 4.9). ◁

A convenient way to build semiconcave approximations of a given function is provided by the method of *inf-convolution*, a standard tool in convex and non-smooth analysis. Let Ω be a subset of $\mathbb{R}^N$ and u a bounded function in Ω. Define, for any $\varepsilon > 0$,

$$(4.27) \qquad u_\varepsilon(x) := \inf\{ u(y) + \frac{1}{2\varepsilon}|x-y|^2 : y \in \Omega \}.$$

The function u_ε is called the ε inf-convolution of u. Similarly,

$$(4.28) \qquad u^\varepsilon(x) := \sup\{ u(y) - \frac{1}{2\varepsilon}|x-y|^2 : y \in \Omega \}$$

is the ε *sup-convolution* of u.

LEMMA 4.11. *Let u be continuous and bounded in Ω. Then*
(a) *u_ε and $-u^\varepsilon$ are semiconcave in Ω;*
(b) *$u_\varepsilon \nearrow u$, $u^\varepsilon \searrow u$, as $\varepsilon \to 0^+$, locally uniformly in Ω;*
(c) *the inf and the sup in (4.27) and (4.28) are attained if $\varepsilon < d^2(x, \partial\Omega)/(4\|u\|_\infty)$.*

PROOF. Let us observe that

$$u_\varepsilon(x) = \inf\left\{ u(y) + \frac{1}{2\varepsilon}|x|^2 - \frac{1}{\varepsilon}x \cdot y + \frac{1}{2\varepsilon}|y|^2 : y \in \Omega \right\}.$$

Hence

$$u_\varepsilon(x) - \frac{1}{2\varepsilon}|x|^2 = \inf\left\{ u(y) - \frac{1}{\varepsilon}x \cdot y + \frac{1}{2\varepsilon}|y|^2 : y \in \Omega \right\}.$$

Since

$$x \longmapsto u(y) - \frac{1}{\varepsilon}x \cdot y + \frac{1}{2\varepsilon}|y|^2$$

is affine for any fixed $y \in \Omega$, it turns out that $u_\varepsilon(x) - \frac{1}{2\varepsilon}|x|^2$ is concave, so that u_ε is semiconcave.

In order to prove (c) observe that if $\varepsilon < \dfrac{d^2(x, \partial\Omega)}{4\|u\|_\infty}$ then,

$$\bar{B} = \bar{B}(x, 2\sqrt{\varepsilon}\|u\|_\infty^{1/2}) \subset \Omega, \qquad \forall x \in \Omega.$$

Observe also that

$$u(y) + \frac{1}{2\varepsilon}|x-y|^2 \geq -\|u\|_\infty + 2\|u\|_\infty, \qquad \forall y \in \Omega \setminus \bar{B}.$$

Since $u_\varepsilon(x) \leq \|u\|_\infty$, the above implies

$$u_\varepsilon(x) = \min\{ u(y) + \frac{1}{2\varepsilon}|x-y|^2, y \in \bar{B} \}.$$

4. LIPSCHITZ CONTINUITY AND SEMICONCAVITY

In order to prove (b), let us observe that $0 < \varepsilon' < \varepsilon$ implies

$$u_\varepsilon(x) \leq u_{\varepsilon'}(x) \leq u(x) \qquad \forall\, x \in \Omega.$$

By (c), there exists $y_\varepsilon \in \Omega$ such that

$$u_\varepsilon(x) = u(y_\varepsilon) + \frac{1}{2\varepsilon}|x - y_\varepsilon|^2.$$

This implies $y_\varepsilon \to x$ as $\varepsilon \to 0$. Since u is continuous,

$$\lim_{\varepsilon \to 0^+} u_\varepsilon(x) \geq \lim_{\varepsilon \to 0^+} u(y_\varepsilon) = u(x);$$

therefore

$$u_\varepsilon(x) \nearrow u(x), \qquad \text{as } \varepsilon \to 0,$$

for any fixed $x \in \Omega$. The local uniform convergence now follows from Dini's Lemma.

The proofs for u^ε are completely similar. ◀

The next lemma collects some useful facts about u_ε and u^ε. These are simple consequences of results on marginal functions. By (c) of Lemma 4.11 we can set, for ε small,

$$M_\varepsilon(x) := \arg\min_{y \in \Omega} \{\, u(y) + |x-y|^2/2\varepsilon \,\},$$
$$M^\varepsilon(x) := \arg\max_{y \in \Omega} \{\, u(y) - |x-y|^2/2\varepsilon \,\}.$$

LEMMA 4.12. *Let $u \in C(\Omega)$ be bounded, $x \in \Omega$ and $\varepsilon < d^2(x, \partial\Omega)/(4\|u\|_\infty)$. Then, either $D^- u_\varepsilon(x) = \emptyset$ or $D^- u_\varepsilon(x) = \{(x - y_\varepsilon)/\varepsilon\}$, where $\{y_\varepsilon\} = M_\varepsilon(x)$ (respectively, either $D^+ u^\varepsilon(x) = \emptyset$ or $D^+ u^\varepsilon(x) = \{-(x - y_\varepsilon)/\varepsilon\}$ where $\{y_\varepsilon\} = M^\varepsilon(x)$). Moreover, for any $y_\varepsilon \in M_\varepsilon(x)$ (respectively, $M^\varepsilon(x)$),*

(i) $|x - y_\varepsilon| \leq 2\sqrt{\varepsilon}\|u\|_\infty^{1/2}$;

(ii) $|x - y_\varepsilon|^2/\varepsilon \to 0$ as $\varepsilon \to 0^+$, *uniformly on compact subsets of Ω*;

(iii) $(x - y_\varepsilon)/\varepsilon \in D^- u(y_\varepsilon)$ *(respectively, $-(x - y_\varepsilon)/\varepsilon \in D^+ u(y_\varepsilon)$).*

PROOF. Let us detail only the proof for u_ε, the argument for u^ε being completely similar. By Lemma 4.11 (c)

$$u_\varepsilon(x) = \min_{y \in \bar{B}}\{u(y) + \frac{1}{2\varepsilon}|x-y|^2\} = u(y_\varepsilon) + \frac{1}{2\varepsilon}|x - y_\varepsilon|^2,$$

where $\bar{B} = \bar{B}(x, 2\sqrt{\varepsilon}\|u\|_\infty^{1/2})$. The function

$$g(x, y) = u(y) + \frac{1}{2\varepsilon}|x-y|^2$$

satisfies on $\bar{B}$ the assumptions of Proposition 4.4. Since $D_x g(x, y_\varepsilon) = (x - y_\varepsilon)/\varepsilon$, the first part of the statement follows. Since $M_\varepsilon(x) \subseteq \bar{B}$, we get immediately (i).

If $y_\varepsilon \in M_\varepsilon(x)$, from the inequality

$$\frac{1}{2\varepsilon}|x - y_\varepsilon|^2 = u_\varepsilon(x) - u(y_\varepsilon) \le u(x) - u(y_\varepsilon),$$

and the continuity of u, we get (ii).

To prove (iii) it is enough to observe that

$$0 \in D_y^- g(x, y_\varepsilon) = D^- u(y_\varepsilon) - \frac{1}{\varepsilon}(x - y_\varepsilon),$$

because $g(x, \cdot)$ attains a minimum at y_ε. ◀

Lemmas 4.11 and 4.12 suggest that a continuous viscosity solution of (HJ) has a two-sided uniform approximation by locally Lipschitz continuous viscosity solutions of approximated equations. More precisely, we have

PROPOSITION 4.13. *Assume that H satisfies*

(H$_1$) $\qquad |H(x, p) - H(y, p)| \le \omega_1(|x - y|(1 + |p|)),$

for $x, y \in \Omega$, $p \in \mathbb{R}^N$, where ω_1 is a modulus. If $u \in C(\Omega)$ is a viscosity subsolution of (HJ) in Ω, then $u^\varepsilon \in \mathrm{Lip}_{loc}(\Omega)$ is a viscosity subsolution of

(HJ$_\varepsilon$) $\qquad \lambda u^\varepsilon(x) + H(x, Du^\varepsilon(x)) = \varrho^\varepsilon(x) \qquad \text{in } \Omega^\varepsilon,$

where

$$\Omega^\varepsilon = \{ x \in \Omega : d(x, \partial\Omega) > 2\sqrt{\varepsilon}\,\|u\|_\infty^{1/2}\}$$

and $\varrho^\varepsilon(x) \to 0$ as $\varepsilon \to 0$, uniformly on compact subsets of Ω.

PROOF. By Lemma 4.12 it is enough to check that

$$\lambda u^\varepsilon(x) + H(x, -(x - y_\varepsilon)/\varepsilon) \le \varrho^\varepsilon(x) \qquad \text{in } \Omega^\varepsilon,$$

where y_ε is such that $u^\varepsilon(x) = u(y_\varepsilon) - |x - y_\varepsilon|^2/2\varepsilon$.

Now by Lemma 4.12 (i), $y_\varepsilon \in \Omega$ for $x \in \Omega^\varepsilon$. Hence (HJ) gives

$$\lambda u(y_\varepsilon) + H(y_\varepsilon, -(x - y_\varepsilon)/\varepsilon) \le 0$$

by Lemma 4.12 (iii). Therefore we use $u^\varepsilon(x) \le u(y_\varepsilon)$ and (H$_1$) to get

$$\lambda u^\varepsilon(x) + H(x, -(x - y_\varepsilon)/\varepsilon) \le \omega_1(|x - y_\varepsilon|(1 + |x - y_\varepsilon|/\varepsilon)),$$

and the statement follows on the account of Lemma 4.12 (i), (ii). ◀

REMARK 4.14. A similar result holds of course for a supersolution u and its inf-convolution. ◁

Exercises

4.1. Take H as in (4.1) with f continuous on $\mathbb{R}^N \times A$. Assume also that (4.2) holds. Show that

(i) $$\sup_{a \in A}\{-f(x,a) \cdot p\} = \sup_{\xi \in \overline{co}f(x,A)}\{-\xi \cdot p\} \geq r|p|.$$

Prove then that H satisfies (H_4) provided ℓ is bounded.

4.2. Take H as in (4.1) with f, ℓ such that
$$(f(x,a) - f(y,a)) \cdot (x-y) \leq L|x-y|^2$$
$$|\ell(x,a) - \ell(y,a)| \leq M|x-y|, \qquad \forall x, y$$

with constants L, M independent of $a \in A$. Show that for any $C > 0$:

$$H\left(x, C\frac{x-y}{|x-y|}\right) - H\left(y, C\frac{x-y}{|x-y|}\right) \geq (-LC - M)|x-y|, \qquad \forall x, y.$$

4.3. Let $u : \Omega \to \mathbb{R}$ be locally Lipschitz continuous in the open set Ω. Prove the following statements:

(a) $D^+u(x)$ and $D^-u(x)$ are bounded for all $x \in \Omega$;

(b) $u_0(x; q) = -u^0(x, -q)$ for all $x \in \Omega$, $q \in \mathbb{R}^N$;

(c) for any $x \in \Omega$, $q \to u^0(x; q)$ is finite, positively homogeneous, subadditive, convex (and locally Lipschitz continuous);

(d) $(x, q) \to u^0(x; q)$ is upper semicontinuous.

4.4. For u as in Exercise 4.3 show that
$$D^+u(x) \subseteq \partial u(x), \qquad D^-u(x) \subseteq \partial u(x).$$

[Hint: use inequality (4.5) and choose $y = x + tq$ with $t > 0$ in the definition of $D^\pm u(x)$.] Prove also that

$$u_0(x; q) = \min_{p \in \partial u(x)} p \cdot q; \qquad u^0(x; q) = \max_{p \in \partial u(x)} p \cdot q.$$

4.5. Prove that the two definitions (4.9) and (4.10) of semiconcavity are equivalent for a continuous function.

4.6. Show that the marginal function $u(x) = \inf_{b \in B} g(x, b)$ is semiconcave provided that for each $b \in B$, $x \mapsto g(x, b)$ satisfies (4.9) with a constant independent of $b \in B$. [Hint: for any $\varepsilon > 0$ choose $b_\varepsilon \in B$ such that

$$u(\mu x + (1-\mu)y) \geq g(\mu x + (1-\mu)y, b_\varepsilon) - \varepsilon$$

and observe that

$$\mu u(x) + (1-\mu)u(y) \leq \mu g(x, b_\varepsilon) + (1-\mu)g(y, b_\varepsilon).]$$

4.7. Let u be locally Lipschitz continuous in Ω. Show that for all $x, y \in \Omega$ with $|x - y|$ sufficiently small then

$$|p - q| \leq L + \text{diam}\, D^+ u(y), \qquad \forall p \in D^+ u(y), q \in D^+ u(x)$$

where L is a (local) Lipschitz constant for u. [Hint: use inequality (1.6) in the proof of Lemma 1.8 to get

$$|p - q| + \liminf_{\substack{y \to x \\ x \in \Omega}} \frac{u(y) - u(x) - p \cdot (y - x)}{|y - x|} \leq 0 \qquad \forall q \in D^+ u(x) \,.]$$

4.8. Let $H : \mathbb{R}^N \to \mathbb{R}$ be convex. Show that

$$H(p + Ch) - 2H(p) + H(p - Ch) \geq 0$$

for all $p, h \in \mathbb{R}^N$ and $C \geq 0$. Deduce from this that $F(x, p) := H(p) - \ell(x)$, with ℓ semiconcave, satisfies (H_6).

4.9. Let H be as in Remark 4.10 and set

$$X(C) = H(x + h, p + Ch) - 2H(x, p) + H(x - h, p - Ch), \qquad C \geq 0\,.$$

Prove that for any $\varepsilon > 0$ there exists $a_\varepsilon \in A$ such that

$$X(C) \geq -X_1 |p| - C X_2 \cdot h - X_3 - \varepsilon$$

where

$$\begin{aligned} X_1 &= |f(x + h, a_\varepsilon) - 2f(x, a_\varepsilon) + f(x - h, a_\varepsilon)| \\ X_2 &= f(x + h, a_\varepsilon) - f(x - h, a_\varepsilon) \\ X_3 &= |\ell(x + h, a_\varepsilon) - 2\ell(x, a_\varepsilon) + \ell(x - h, a_\varepsilon)| \,. \end{aligned}$$

Deduce that if $L < 1/2$ then (H_6) holds for any $C \geq (2FL_u + M)/(1 - 2L)$.

4.10. Let $f(x, a) = Ex + Ba$, $\ell(x, a) = \frac{1}{2} Mx \cdot x + \frac{1}{2} Qa \cdot a$, where E, B, M, Q are matrices of appropriate dimensions. Prove that f, ℓ satisfy the conditions in Remark 4.10 with $L = \|E\|$, $F = 0$, $M = \|M\|$.

4.11. Let u be semiconcave in Ω, $\varphi \in C^1(\Omega)$ and $x_0 \in \Omega$ a local minimum for $u - \varphi$. Show that u is differentiable at x_0 with $Du(x_0) = D\varphi(x_0)$. [Hint: semiconcavity of u yields $u(x_0 + h) - u(x_0) \leq u(x_0) - u(x_0 - h) + C|h|^2$.]

4.12. Let u be Lipschitz continuous on $\mathbb{R}^N$. Show that $\sup |u - u_\varepsilon| \leq C\varepsilon$, where u_ε is the inf-convolution of u.

5. Special results for convex Hamiltonians

The convexity of the function $p \mapsto F(x, r, p)$ allows us to prove some additional results in the theory of viscosity solutions of Hamilton-Jacobi equations. On the other hand, the Hamiltonians associated with optimal control problems, always satisfy the convexity property (see the various examples in Chapter III and IV). For the sake of simplicity we shall consider only equations of the form

(HJ) $$\lambda u(x) + H(x, Du(x)) = 0$$

where $\lambda \geq 0$ and H satisfies

(H$_7$) $\qquad\qquad p \longmapsto H(x, p)$ is convex for any fixed x .

In the first subsection and in §5.4 we study the relations among viscosity solutions and other generalized notions of solution, namely semiconcave generalized solutions, bilateral supersolutions, and solutions in the extended sense of Clarke. The main result is Theorem 5.6 asserting that, when (H$_7$) holds, viscosity solutions of (HJ) are characterized by the single condition $\lambda u(x) + H(x, p) = 0$ for any $p \in D^- u(x)$; this is the starting point of the theory of lower semicontinuous viscosity solutions that we develop in §V.5.

Different issues where (H$_7$) plays a role are regularity, treated in §5.2, and comparison for the "degenerate" case $\lambda = 0$, see §5.3.

The last two subsections are devoted to monotonicity properties along trajectories of viscosity subsolutions of the Hamilton-Jacobi-Bellman equation

(HJB) $$\lambda u(x) + \sup_{a \in A}\{-f(x, a) \cdot Du(x) - \ell(x, a)\} = 0 \quad \text{in } \Omega$$

with $\lambda \geq 0$ (see Theorem 5.21). This is related to the method of characteristics and to some different characterizations of the value function.

5.1. Semiconcave generalized solutions and bilateral supersolutions

Let us go back first to the question about the relationship between generalized and viscosity solutions of (HJ).

PROPOSITION 5.1. *Let $u \in \text{Lip}_{\text{loc}}(\Omega)$. If*

$$\lambda u(x) + H(x, Du(x)) \leq 0 \qquad a.e. \text{ in } \Omega$$

and (H$_7$) *holds, then u is a viscosity subsolution of* (HJ) *in Ω.*

PROOF. Let $\varrho \in C^\infty(\mathbb{R}^N)$, $\varrho \geq 0$, with $\text{supp}\, \varrho \subset B(0,1)$ and $\int_{\mathbb{R}^N} \varrho(y)\, dy = 1$ be a standard regularizing kernel and set

$$\varrho_\delta(x) = \frac{1}{\delta^N} \varrho\left(\frac{x}{\delta}\right), \qquad \delta > 0 .$$

It is well-known that for each $\Omega' \subset\subset \Omega$,

$$u_\delta(x) := (u * \varrho_\delta)(x) := \int_{\mathbb{R}^N} u(y)\varrho_\delta(x-y)\,dy, \qquad x \in \Omega',$$

is well defined (for small δ) and C^1, with $Du_\delta(x) = (Du * \varrho_\delta)(x)$. Moreover, as $\delta \to 0^+$,

(5.1) $\qquad\qquad\qquad u_\delta \longrightarrow u \quad \text{locally uniformly in } \Omega.$

Since

$$\lambda u(y) + H(y, Du(y)) \leq 0 \quad \text{a.e. in } \Omega,$$

we have

(5.2) $\qquad \lambda u_\delta(x) + \int_{\mathbb{R}^N} H(y, Du(y))\varrho_\delta(x-y)\,dy \leq 0 \qquad \text{a.e. in } \Omega.$

Now, if H depends only on p, the above gives

(5.3) $\qquad\qquad \lambda u_\delta(x) + H(Du_\delta(x)) \leq 0 \qquad \forall x \in \Omega',$

as a consequence of (H$_7$) and the Jensen's inequality. Since $u_\delta \in C^1(\Omega')$, by Proposition 1.3 (b), it satisfies (5.3) in the viscosity sense. Thanks to (5.1) we can apply Proposition 2.2 to obtain that u is a viscosity subsolution of (HJ) in Ω', for any $\Omega' \subset\subset \Omega$ and the claim is proved.

In the general case, from (5.2) and Jensen's inequality it follows that

$$\lambda u_\delta(x) + H(x, Du_\delta(x)) \leq h_\delta(x) \qquad \forall x \in \Omega',$$

where

$$h_\delta(x) = \int_{\mathbb{R}^N} [H(x, Du(y)) - H(y, Du(y))]\varrho_\delta(x-y)\,dy.$$

By continuity of H, h_δ converge uniformly to zero as $\delta \to 0^+$ and the proof is completed as above. ◀

By putting together Propositions 4.8 and 5.1, and taking into account Theorem 4.9, we obtain

COROLLARY 5.2. *Assume* (H$_7$).
(i) *If u is a semiconcave generalized solution of* (HJ) *in Ω, then it is also a viscosity solution.*
(ii) *If, in addition, $u \in BC(\mathbb{R}^N) \cap \mathrm{Lip}(\mathbb{R}^N)$ and H satisfies the assumptions* (H$_3$) *and* (H$_6$) *of Theorem 4.9, then u is a viscosity solution of* (HJ) *in $\mathbb{R}^N$ if and only if it is a semiconcave generalized solution.*

REMARK 5.3. Corollary 5.2 has various extensions, e.g., to solutions of evolutive equations which are semiconcave in the space variables. The semiconcavity assumption of part (i) can be weakened to semi-superharmonicity, i.e., $\Delta u \leq C$ in the sense of distributions, and even to

$$\sum_{i=1}^{N}(u(x+te_i) - 2u(x) + u(x-te_i)) \leq Ct^{2-\alpha}, \qquad \alpha \in [0,1[,$$

for all $t \geq 0$ and $x \in \mathbb{R}^N$, where e_i are the vectors of the canonical basis of $\mathbb{R}^N$, but the proof that u is a viscosity supersolution requires a different argument. ◁

PROPOSITION 5.4. *Let $u \in \text{Lip}_{\text{loc}}(\Omega)$ be a viscosity subsolution of (HJ) and assume that (H_7) holds. Then u is a viscosity supersolution of*

(HJ)_ $\qquad -\lambda u(x) - H(x, Du(x)) = 0, \qquad x \in \Omega$.

PROOF. By Proposition 1.9 (b),

(5.4) $\qquad \lambda u(y) + H(y, Du(y)) \leq 0 \qquad$ a.e. in Ω.

With the same notations as in the proof of Proposition 5.1,

$$\lambda u_\delta(x) + H(x, Du_\delta(x)) \leq h_\delta(x) \qquad \forall x \in \Omega'.$$

Since $u_\delta \in C^1(\Omega')$, the foregoing implies

$$-\lambda u_\delta(x) - H(x, Du_\delta(x)) \geq -h_\delta(x)$$

at any $x \in \Omega'$ and a fortiori, by Proposition 1.3 (b), in the viscosity sense. Since u_δ converges locally uniformly to u as $\delta \to 0^+$ we obtain

$$-\lambda u(x) - H(x, Du(x)) \geq 0 \qquad \text{in } \Omega$$

in the viscosity sense (see Proposition 2.2). ◀

REMARK 5.5. The convexity of H is necessary for Proposition 5.4 to hold (see the Example 1.6). ◁

Theorem 5.6 shows that if H is convex and satisfies some technical assumption, then any viscosity solution of (HJ) satisfies

(5.5) $\qquad \lambda u(x) + H(x,p) = 0 \qquad \forall x \in \Omega, \ \forall p \in D^- u(x)$.

and, conversely, the fulfillment of the preceding implies automatically the subsolution condition. Any $u \in C(\Omega)$ satisfying (5.5) will be called a *bilateral supersolution* (see §III.2.3 for further study of this notion and §V.5 for its extension to lower semicontinuous functions). This rather surprising result can indeed be proved immediately for H smooth enough, because viscosity solutions coincide with semiconcave generalized solutions, by Corollary 5.2, and then the result follows at once from the property of semiconcave functions that either $D^- u(x) = \emptyset$ or u is differentiable at x (see Proposition 4.7 (b)). This observation can be used in the general case via regularization by inf and sup convolution and stability properties, yielding:

THEOREM 5.6. *Assume that* (H_1), (H_7) *hold. Then* $u \in C(\Omega)$ *is a viscosity solution of* (HJ) *if and only if*

(5.6) $$\lambda u(x) + H(x,p) = 0, \quad \forall x \in \Omega, \; \forall p \in D^- u(x).$$

PROOF. Let us assume that u is a viscosity subsolution of (HJ). By Propositions 4.13 and 5.4 its sup-convolution u^ε satisfies, in the viscosity sense,

(HJ_ε^-) $$-\lambda u^\varepsilon(x) - H(x, Du^\varepsilon(x)) \geq -\varrho^\varepsilon(x) \quad \text{in } \Omega^\varepsilon,$$

where $\varrho^\varepsilon(x) \to 0$ as $\varepsilon \to 0^+$ and Ω^ε is as in Proposition 4.13. Hence, by Proposition 2.2 and Lemma 4.11 (b),

$$-\lambda u(x) - H(x,p) \geq 0 \quad \forall x \in \Omega, \; \forall p \in D^- u(x).$$

If u is also a viscosity supersolution of (HJ), the reverse inequality is also true and (5.6) is proved.

Let us assume now that (5.6) holds. We can take, in particular, $x = y_\varepsilon$ where $u_\varepsilon(x) = u(y_\varepsilon) + |x - y_\varepsilon|^2/2\varepsilon$ and $p = (x - y_\varepsilon)/\varepsilon \in D^- u(y_\varepsilon)$ (see Lemma 4.12). Hence, (5.6) yields

$$\lambda u(y_\varepsilon) + H(y_\varepsilon, (x - y_\varepsilon)/\varepsilon) = 0$$

and, by (H_1),

(5.7) $$\lambda u_\varepsilon(x) + H(x, (x - y_\varepsilon)/\varepsilon) \leq \varrho_\varepsilon(x) \quad \text{in } \Omega^\varepsilon,$$

where $\varrho_\varepsilon(x) := (\lambda/2\varepsilon)|x - y_\varepsilon|^2 + \omega_1(|x - y_\varepsilon|(1 + |x - y_\varepsilon|/\varepsilon))$. Since u_ε is semiconcave, it is differentiable almost everywhere by Proposition 4.6, and $Du_\varepsilon(x) = (x - y_\varepsilon)/\varepsilon$ by Lemma 4.12. Therefore (5.7) reads, for almost every $x \in \Omega^\varepsilon$,

$$\lambda u_\varepsilon(x) + H(x, Du_\varepsilon(x)) \leq \varrho_\varepsilon(x),$$

and by Proposition 5.1 the same inequality holds in the viscosity sense. By Lemmata 4.11 and 4.12, u_ε and ϱ_ε converge, respectively, to u and 0 locally uniformly as $\varepsilon \to 0^+$. Therefore, using Proposition 2.2 we can conclude that u is a viscosity subsolution of (HJ) in Ω. Since (5.6) implies trivially that u is a viscosity supersolution, the proof is complete. ◂

5.2. Differentiability of solutions

A different issue where the convexity of H plays a role is regularity. The next proposition is a (local) regularity result.

PROPOSITION 5.7. *Let us assume that* $u \in C(\Omega)$ *is a viscosity solution of*

$$\lambda u(x) + H(x, Du(x)) = 0 \quad \text{in } \Omega,$$

with $\lambda \geq 0$. *Assume also that* $p \mapsto H(x,p)$ *is strictly convex for any fixed* $x \in \Omega$ *and that* $-u$ *is semiconcave. Then* $u \in C^1(\Omega)$.

PROOF. Let us first show that u is differentiable at any $x \in \Omega$. In view of Proposition 4.7 (c) we just have to prove that $D^+(-u)(x)$ is a singleton for any $x \in \Omega$; by Proposition 4.7 (a) this will be true, provided $D^*(-u)(x)$ is a singleton.

Suppose by contradiction that $p^1 \neq p^2 \in D^*(-u)(x)$. Then there exist sequences $\{x_n\}$, $\{y_m\}$ in Ω where u is differentiable and such that

$$x = \lim_{n \to +\infty} x_n = \lim_{m \to +\infty} y_m, \quad p^1 = \lim_{n \to +\infty} D(-u)(x_n), \quad p^2 = \lim_{m \to +\infty} D(-u)(y_m).$$

By Proposition 1.9 (a),

$$\lambda u(x_n) + H(x_n, Du(x_n)) = \lambda u(y_m) + H(y_m, Du(y_m)) = 0;$$

by continuity this yields

(5.8) $$\lambda u(x) + H(x, -p^1) = \lambda u(x) + H(x, -p^2) = 0.$$

Now let

$$\bar{p} = \frac{1}{2} p^1 + \frac{1}{2} p^2;$$

by strict convexity, (5.8) implies

(5.9) $$\lambda u(x) + H(x, -\bar{p}) < \lambda u(x) + \frac{1}{2} H(x, -p^1) + \frac{1}{2} H(x, -p^2) = 0.$$

On the other hand, by Proposition 4.7 (a), $\bar{p} \in \operatorname{co} D^*(-u)(x) = D^+(-u)(x) = -D^-u(x)$. Since u is a viscosity solution of (HJ), then

$$\lambda u(x) + H(x, -\bar{p}) \geq 0,$$

contradicting (5.9). Hence u is differentiable at all points. The continuity of Du is a consequence of the upper semicontinuity of the multifunction D^+u for semiconcave u, that is the following property:

$$x_n \to x, \quad p_n \in D^+u(x_n), \quad p_n \to p \implies p \in D^+u(x)$$

(the proof of this fact is implicit in that of Proposition 4.7 (c)). ◀

REMARK 5.8. The strict convexity assumption is essential. To realize this, observe that $u(x) = |x|$ is a viscosity solution of $a(x)(|Du(x)|^2 - 1) = 0$, where a is a continuous function such that $a(x) > a(0) = 0$, for any x. In this example $-u$ is semiconcave but $p \mapsto H(x,p) = a(x)(|p|^2 - 1)$ is not strictly convex for $x = 0$. Another remark is that strict convexity of H is a serious restriction for Hamiltonians of optimal control problem. In fact it is not satisfied in the very simple case

$$H(x, p) = \sup_{a \in A} \{-f(a)p - \ell(x)\} \qquad x, p \in \mathbb{R},$$

with A compact, $A \subset \mathbb{R}$, because H takes the explicit piecewise affine form

$$H(x, p) = \begin{cases} -p \min_{a \in A} f(a) - \ell(x), & p \geq 0, \\ -p \max_{a \in A} f(a) - \ell(x), & p \leq 0. \end{cases} \triangleleft$$

5.3. A comparison theorem

In this subsection we present a comparison and uniqueness result where the convexity of H with respect to the p variable plays a key role. As already pointed out in Remark 3.3, equations of the form

(5.10) $$H(x, Du(x)) = 0$$

may have more than one viscosity (and even classical) solution satisfying prescribed boundary data. The classical model problem for (5.10) is the eikonal equation of geometric optics

(5.11) $$|Du(x)| = \ell(x).$$

We recall that in §2 we proved that for $\ell \equiv 1$ a viscosity solution of this equation in the open set Ω is the distance function from $\partial \Omega$.

THEOREM 5.9. *Let Ω be a bounded open subset of $\mathbb{R}^N$. Assume $u_1, u_2 \in C(\overline{\Omega})$ are, respectively viscosity sub- and supersolution of (5.10) in Ω with $u_1 \leq u_2$ on $\partial \Omega$. Assume that H satisfies*

(H$_1$) $$|H(x,p) - H(y,p)| \leq \omega_1(|x-y|(1+|p|))$$

for $x, y \in \Omega$ and $p \in \mathbb{R}^N$. Assume also

(H$_7$) *$p \mapsto H(x,p)$ is convex on $\mathbb{R}^N$ for each $x \in \Omega$;*

(H$_8$) *there exists $\varphi \in C(\overline{\Omega}) \cap C^1(\Omega)$ such that $\varphi \leq u_1$ in $\overline{\Omega}$ and*

$$\sup_{x \in \Omega'} H(x, D\varphi(x)) < 0, \qquad \forall \Omega' \subset\subset \Omega.$$

Then $u_1 \leq u_2$ in Ω.

PROOF. Set, for $t \in \;]0,1[$;

$$u^t(x) = tu_1(x) + (1-t)\varphi(x), \qquad x \in \overline{\Omega}.$$

The following properties of u^t are easy to check (see also Exercise 2.4):

$$u^t \in C(\overline{\Omega}) \qquad u^t \leq u_1 \quad \text{in } \overline{\Omega};$$
$$D^+ u^t(x) = \{q \in \mathbb{R}^N : q = tp + (1-t)D\varphi(x),\ p \in D^+ u_1(x)\}$$

and

(5.12) $$u^t \longrightarrow u_1 \qquad \text{as } t \to 1.$$

By (H$_7$), for any $q \in D^+ u^t(x)$, we have

$$H(x,q) = H(x, tp + (1-t)D\varphi(x)) \leq tH(x,p) + (1-t)H(x, D\varphi(x)),$$

5. SPECIAL RESULTS FOR CONVEX HAMILTONIANS

for some $p \in D^+ u_1(x)$. Since u_1 is a subsolution of (5.10),

(5.13) $$H(x,q) \leq (1-t)H(x, D\varphi(x)), \qquad \forall q \in D^+ u^t(x);$$

this means that for any $t \in [0, 1]$, u^t is a viscosity subsolution of

(5.14) $$H(x, Du^t(x)) - (1-t)g(x) = 0, \qquad x \in \Omega,$$

where $g(x) := H(x, D\varphi(x)) \in C(\Omega)$.

We claim that this implies

(5.15) $$u^t \leq u_2 \quad \text{on } \Omega, \quad \forall t \in \,]0, 1[\,.$$

From (5.12) the thesis then follows. In order to prove (5.15), let us suppose by contradiction that, for some $\bar{t} \in \,]0, 1[$,

(5.16) $$\sup_{x \in \Omega}(u^{\bar{t}} - u_2)(x) = \delta > 0\,.$$

Since $u_1 \leq u_2$ on $\partial\Omega$, the open set

$$\Omega' = \{x \in \Omega : (u^{\bar{t}} - u_2)(x) > \delta/2\}$$

is such that $\overline{\Omega'} \subset \Omega$. Consider now the auxiliary function

$$\Phi_\varepsilon(x, y) = u^{\bar{t}}(x) - u_2(y) - \frac{|x - y|^2}{2\varepsilon}$$

and let $(x_\varepsilon, y_\varepsilon)$ be a maximum point for Φ_ε on $\overline{\Omega'} \times \overline{\Omega'}$. As in the proof of Theorem 3.1 one obtains

(5.17) $$|x_\varepsilon - y_\varepsilon| \longrightarrow 0, \qquad \frac{|x_\varepsilon - y_\varepsilon|^2}{2\varepsilon} \longrightarrow 0$$

as $\varepsilon \to 0^+$. Let us observe also that

$$\Phi_\varepsilon(x_\varepsilon, y_\varepsilon) \geq \sup_{x \in \Omega'} \Phi_\varepsilon(x, x) = \sup_{x \in \Omega'}(u^{\bar{t}}(x) - u_2(x)) = \delta\,.$$

This implies that $(x_\varepsilon, y_\varepsilon) \in \Omega' \times \Omega'$ for sufficiently small ε. Indeed, if this were not the case, then either x_ε or y_ε would belong to $\partial\Omega'$, yielding

$$\frac{\delta}{2} = u^{\bar{t}}(x_\varepsilon) - u_2(x_\varepsilon) \geq \Phi_\varepsilon(x_\varepsilon, y_\varepsilon) + u_2(y_\varepsilon) - u_2(x_\varepsilon) \geq \delta - \omega_2(|x_\varepsilon - y_\varepsilon|),$$

where ω_2 is a modulus of continuity for u_2. Taking (5.17) into account we obtain a contradiction.

Since x_ε is a local maximum for $x \mapsto u^{\bar{t}}(x) - (u_2(y_\varepsilon) + |x - y_\varepsilon|^2/2\varepsilon)$ and $u^{\bar{t}}$ is a viscosity subsolution of (5.14), we have

(5.18) $$H\left(x_\varepsilon, \frac{x_\varepsilon - y_\varepsilon}{\varepsilon}\right) - (1 - \bar{t})g(x_\varepsilon) \leq 0;$$

on the other hand, y_ε is a local minimum for $y \mapsto u_2(y) - \big(u^{\bar{t}}(x_\varepsilon) - |x_\varepsilon - y|^2/2\varepsilon\big)$, yielding

(5.19) $$H\left(y_\varepsilon, \frac{x_\varepsilon - y_\varepsilon}{\varepsilon}\right) \geq 0,$$

since u_2 is a viscosity supersolution of (5.10).

By (H$_8$), there exists $\alpha > 0$ such that, for all ε small,
$$(1 - \bar{t})g(x_\varepsilon) = (1 - \bar{t})H(x_\varepsilon, D\varphi(x_\varepsilon)) \leq -\alpha \,.$$

Hence, taking (H$_1$) into account, from (5.18), (5.19), and the preceding, it follows that
$$\alpha \leq H\left(y_\varepsilon, \frac{x_\varepsilon - y_\varepsilon}{\varepsilon}\right) - H\left(x_\varepsilon, \frac{x_\varepsilon - y_\varepsilon}{\varepsilon}\right) \leq \omega_1\left(|x_\varepsilon - y_\varepsilon|\left(1 + \frac{|x_\varepsilon - y_\varepsilon|}{\varepsilon}\right)\right),$$
a contradiction for ε small enough, in view of (5.17). ◄

REMARK 5.10. Theorem 5.9 applies to the eikonal equation (5.11) if ℓ is uniformly continuous in Ω and strictly positive. In fact (H$_8$) is satisfied by taking
$$\varphi(x) \equiv \min_{\overline{\Omega}} u_1 \,.$$

Therefore the Dirichlet problem for this equation has at most one viscosity solution in $C(\overline{\Omega})$ for any bounded open set Ω. In particular $u(x) = d(x, \partial\Omega)$ is the unique solution of
$$|Du| = 1 \quad \text{in } \Omega, \qquad u = 0 \quad \text{on } \partial\Omega \,.$$

The next remark shows that this uniqueness result is sharp because there can be many solutions if ℓ vanish at a single point in Ω. ◁

REMARK 5.11. Assumption (H$_8$) is not satisfied by $H(x,p) = |p| - 2|x|$ with $\Omega =]-1, 1[$. Indeed, the supremum of $H(x, \varphi'(x))$ in any neighborhood of $x = 0$ is at least $|\varphi'(0)|$ for any $\varphi \in C^1(\Omega)$. Uniqueness is not true for the associated Hamilton-Jacobi equation. Actually $u_1(x) = x^2 - 1$, $u_2(x) = 1 - x^2$ are both classical solutions of $|u'| = 2|x|$ in $]-1, 1[$ and $u_1 = u_2 = 0$ at $x = \pm 1$. A more careful investigation reveals the existence of a continuum of viscosity solutions connecting u_1 with u_2 (see Exercise 5.6). ◁

REMARK 5.12. An alternative approach to comparison and uniqueness for equation (5.10) under different assumptions is via the Kružkov transform: see Chapter IV for this approach (see also Exercise 5.4). ◁

5.4. Solutions in the extended sense

We briefly discuss here the relation between viscosity solutions and another concept of solution which is known in the Lipschitz continuous case. Let us consider the model Hamiltonian
$$H(x, p) = \sup_{a \in A}\{-f(x, a) \cdot p - \ell(x, a)\}$$

5. SPECIAL RESULTS FOR CONVEX HAMILTONIANS

with the usual continuity assumption. The following notion of solution of the Hamilton-Jacobi-Bellman equation is given in terms of the Clarke's gradient ∂u of u (see §4.1 for the definition).

A function $u \in \text{Lip}(\Omega)$ is said to be a *solution in the extended sense* of the equation

(HJ) $$\lambda u(x) + H(x, Du(x)) = 0, \quad x \in \Omega,$$

if

(5.20) $$\lambda u(x) + \max_{p \in \partial u(x)} H(x, p) = 0 \quad \forall\, x \in \Omega.$$

PROPOSITION 5.13. *Let $u \in \text{Lip}(\Omega)$. If u solves (HJ) in the extended sense, then u is a viscosity subsolution of (HJ) and a viscosity supersolution of*

$$-\lambda u(x) - H(x, Du(x)) = 0, \quad x \in \Omega.$$

Conversely, any viscosity solution of (HJ) *is a solution in the extended sense.*

PROOF. Since, by (4.6), $D^+u(x) \cup D^-u(x) \subseteq \partial u(x)$, from (5.20) we obtain directly

$$\lambda u(x) + H(x,p) \leq 0, \quad \forall\, x \in \Omega,\ \forall\, p \in D^+u(x)$$
$$\lambda u(x) + H(x,q) \leq 0, \quad \forall\, x \in \Omega,\ \forall\, q \in D^-u(x).$$

From the above inequalities the first part of the statement follows.

Now let $u \in \text{Lip}(\Omega)$ be a viscosity solution of (HJ). We know already that

$$\lambda u(x) + H(x, Du(x)) = 0 \quad \text{almost everywhere in } \Omega$$

(see Proposition 1.9). By definition of D^*, it follows that

$$\lambda u(x) + H(x,p) = 0 \quad \forall\, x \in \Omega,\ \forall\, p \in D^*u(x).$$

Hence

(5.21) $$\lambda u(x) + \max_{p \in D^*u(x)} H(x,p) = 0, \quad \forall\, x \in \Omega.$$

Now the convexity of H with respect to p yields

$$\max_{p \in \text{co}\, D^*u(x)} H(x,p) = \max_{p \in D^*u(x)} H(x,p)$$

and the thesis follows from (5.21) and the identity $\text{co}\, D^*u(x) = \partial u(x)$ (see (4.4)). ◀

REMARK 5.14. Lipschitz continuous viscosity solutions form in general a proper subset of solutions in the extended sense. To realize this, consider the one-dimensional equation $|u'| - 1 = 0$ in $]-1, 1[$ and the functions $u_1(x) = 1 - |x|$, $u_2(x) = |x| - 1$. Since $\partial u_1(0) = \partial u_2(0) = [-1, 1]$, it is easy to check that both u_1

and u_2 solve the equation in the extended sense. However, u_1 is a viscosity solution while u_2 is not. Observe also that the boundary value problem $|u'| - 1 = 0$ in $]-1, 1[$, $u(-1) = u(1) = 0$, has a unique viscosity solution by Remark 5.10, namely u_1.

A final remark is that solutions in the extended sense are not stable in the uniform topology. A simple example is provided by the saw-tooth function in Remark 2.3. ◁

5.5. Differential inequalities in the viscosity sense

The classical method of characteristics for first order partial differential equations starts from the observation that smooth subsolutions of the linear equation

(5.22) $$-f(x) \cdot Du(x) = 0$$

with $f \in C^1(\mathbb{R}^N)$ are nondecreasing along the trajectories of the characteristic system

(5.23) $$\dot{y}(t) = f(y(t))$$
$$y(0) = x \,.$$

Proposition 5.18 shows that the same result holds true for viscosity subsolutions of (5.22). We then consider the nonlinear model case

(HJB) $$\lambda u(x) + \sup_{a \in A} \{ -f(x, a) \cdot Du(x) - \ell(x, a) \} = 0$$

and prove that viscosity subsolutions of (HJB) have the same behavior on the trajectories of the controlled system

(5.24) $$\dot{y}(t) = f(y(t), \alpha(t))$$
$$y(0) = x$$

where $\alpha \in \mathcal{A} = \{\alpha : [0, +\infty[\to A, \ \alpha \text{ measurable}\}$.

The first step is the following one-dimensional monotonicity result.

LEMMA 5.15. *Let $u \in C(]0, T[)$, $T > 0$. Then the following statements are equivalent:*

(i)   *u is nondecreasing on $]0, T[$;*

(ii) *$u' \geq 0$ in $]0, T[$ in the viscosity sense;*

(iii) *$-u' \leq 0$ in $]0, T[$ in the viscosity sense.*

PROOF. Let $\varphi \in C^1(]0, T[)$ and t_0 be a local minimum for $u - \varphi$. If u is nondecreasing, then

$$\varphi(t) - \varphi(t_0) \leq u(t) - u(t_0) \leq 0$$

for $t < t_0$ close enough to t_0. This implies $\varphi'(t_0) \geq 0$, showing that (ii) holds. Let us assume now that (ii) holds and suppose that

(5.25) $\qquad u(t_1) > u(t_2) \qquad$ for some $0 < t_1 < t_2 < T$.

By continuity, there exists $\bar{t} \in [t_1, t_2]$ such that $u(t_2) < u(\bar{t}) < u(t_1)$. Choose now $\varphi \in C^1(]t_1, t_2[)$ (see Exercise 5.7) such that

$$\varphi'(t) < 0, \quad \forall t \in]t_1, t_2[, \qquad \varphi(\bar{t}) > u(\bar{t}), \qquad \varphi(t_i) = u(t_i), \quad i = 1, 2.$$

Since $(u - \varphi)(t_i) = 0$, $(u - \varphi)(\bar{t}) < 0$, $u - \varphi$ attains its minimum over $[t_1, t_2]$ at some interior point t_0. Hence $\varphi'(t_0) < 0$, a contradiction to (ii).

The equivalence between (i) and (iii) is proved by similar arguments. ◂

REMARK 5.16. It is easy to deduce from Lemma 5.15 that if $\ell \in C(]0, T[)$, then $t \mapsto u(t) + \int_0^t \ell(s)\, ds$ is nondecreasing if and only if $u' + \ell \geq 0$ or $-u' - \ell \leq 0$ in the viscosity sense.

Also, it can be proved similarly that u is nonincreasing if and only if $u' \leq 0$ in the viscosity sense. Hence, $u' = 0$ in the viscosity sense is equivalent to u being a constant. ◁

In order to generalize this result to a higher dimension, let Ω be an open subset of $\mathbb{R}^N$ and $u \in C(\Omega)$. For fixed $z = (x_2, \ldots, x_N) \in \mathbb{R}^{N-1}$ we set

$$\Omega_z := \{\, x_1 \in \mathbb{R} : x = (x_1, z) \in \Omega\,\}$$

and

$$u_z(x_1) := u(x_1, z), \qquad u_z : \Omega_z \longrightarrow \mathbb{R}.$$

LEMMA 5.17. *Let $u, \ell \in C(\Omega)$. Then the following statements are equivalent:*

(i) *for each $z \in \mathbb{R}^{N-1}$, u_z is a viscosity supersolution of*

$$u'_z(x_1) \geq \ell_z(x_1) \qquad in\ \Omega_z;$$

(ii) *u is a viscosity supersolution of*

$$\frac{\partial u}{\partial x_1}(x_1, x_2, \ldots, x_N) \geq \ell(x_1, x_2, \ldots, x_N) \qquad in\ \Omega.$$

PROOF. We first prove that (ii) implies (i). Let $z^0 \in \mathbb{R}^{N-1}$ be such that $\Omega_{z^0} \neq \emptyset$ and assume that x_1^0 is a strict local minimum for $u_{z_0} - \eta$ with $\eta \in C^1$. It is not restrictive to assume that $\eta \leq -1$ in $B(x_1^0, \delta)$ for some $\delta > 0$. Consider now

$$\varphi_\varepsilon(x_1, \ldots, x_N) := \eta(x_1)\Big(1 + \frac{|z - z^0|^2}{\varepsilon}\Big), \qquad \varepsilon > 0.$$

If $x^\varepsilon = (x_1^\varepsilon, z^\varepsilon)$ is a minimum point for $u - \varphi_\varepsilon$ in $\bar{B}(x^0, \delta)$ ($x^0 = (x_1^0, z^0)$), then

(5.26) $$u(x^\varepsilon) - \varphi_\varepsilon(x^\varepsilon) = u(x^\varepsilon) - \eta(x_1^\varepsilon) - \eta(x_1^\varepsilon)\frac{|z^\varepsilon - z^0|^2}{\varepsilon}$$
$$\leq u(x^0) - \varphi_\varepsilon(x^0) = u_{z^0}(x_1^0) - \eta(x_1^0).$$

Since $\eta \leq -1$ in $B(x_1^0, \delta)$, it follows that

$$\frac{|z^\varepsilon - z^0|^2}{\varepsilon} \leq u_{z^0}(x_1^0) - \eta(x_1^0) + \eta(x_1^\varepsilon) - u(x^\varepsilon).$$

Therefore

$$\frac{|z^\varepsilon - z^0|^2}{\varepsilon} \leq C, \quad z^\varepsilon \longrightarrow z^0, \quad \text{as } \varepsilon \to 0^+.$$

Then, at least for a subsequence,

$$x_1^\varepsilon \longrightarrow \bar{x}_1, \quad \frac{|z^\varepsilon - z^0|^2}{\varepsilon} \longrightarrow \alpha \geq 0, \quad \text{as } \varepsilon \to 0^+.$$

Letting $\varepsilon \to 0^+$ in (5.26) we obtain

$$u_{z^0}(x_1^0) - \eta(x_1^0) \geq u(\bar{x}_1, z^0) - \eta(\bar{x}_1) - \eta(\bar{x}_1)\alpha$$
$$\geq u_{z^0}(\bar{x}_1) - \eta(\bar{x}_1).$$

Since x_1^0 was a local strict minimum for $u_{z^0} - \eta$, the above implies $\bar{x}_1 = x_1^0$ and $\alpha = 0$. Now, assuming the validity of (ii),

$$\frac{\partial \varphi_\varepsilon}{\partial x_1}(x^\varepsilon) = \eta'(x_1^\varepsilon)\left(1 + \frac{|z^\varepsilon - z^0|^2}{\varepsilon}\right) \geq \ell(x^\varepsilon).$$

If we let $\varepsilon \to 0^+$ in the above inequality we conclude

$$\eta'(x_1^0) \geq \ell(x_1^0, z^0) = \ell_{z^0}(x_1^0);$$

this shows that (i) holds true. The proof of the reverse implication is straightforward. It is enough to observe that if $\bar{x} = (\bar{x}_1, \bar{z})$ is a local minimum for $u - \varphi$, $\varphi \in C^1(\Omega)$, then $\bar{x}_1$ is a local minimum for $u_{\bar{z}}(x_1) - \varphi(x_1, \bar{z})$. ◀

In the following we use the notation

$$t_x := \inf\{t \geq 0 : y_x(t) \notin \Omega\}, \quad x \in \Omega,$$

where $y_x(\cdot)$ is the solution of (5.23).

PROPOSITION 5.18. *Let us assume* $f \in C^1(\Omega, \mathbb{R}^N)$, $\ell \in C(\Omega)$, $\lambda \in \mathbb{R}$ *and* $u \in C(\Omega)$. *Then the following statements are equivalent:*

(i) $e^{-\lambda s}u(y_x(s)) - e^{-\lambda t}u(y_x(t)) \leq \int_s^t \ell(y_x(\tau))e^{-\lambda \tau}\,d\tau,$
for all $0 \leq s \leq t < t_x$, $x \in \Omega$;

(ii) $-\lambda u(x) + f(x) \cdot Du(x) + \ell(x) \geq 0$, $x \in \Omega$, in the viscosity sense;

(iii) $\lambda u(x) - f(x) \cdot Du(x) - \ell(x) \leq 0$, $x \in \Omega$, in the viscosity sense.

PROOF. Let us observe first that it is not restrictive to assume $\lambda = 0$. Indeed, u satisfies (ii) if and only if $\widehat{u}(x, x_{N+1}) := x_{N+1} u(x)$ is a viscosity supersolution of

$$\widehat{f} \cdot D\widehat{u} + \widehat{\ell} \geq 0 \qquad \text{in } \Omega \times \mathbb{R}_+,$$

where $\widehat{f}(x, x_{N+1}) := (f(x), -\lambda x_{N+1})$ and $\widehat{\ell}(x, x_{N+1}) := x_{N+1}\ell(x)$ (see Proposition 2.7). On the other hand, it is easy to check that (i) is equivalent to

$$\widehat{u}(\widehat{y}(s)) - \widehat{u}(\widehat{y}(t)) \leq \int_s^t \widehat{\ell}(\widehat{y}(\tau))\, d\tau, \qquad s \leq t,$$

where $\widehat{y}$ is the solution of

$$\dot{\widehat{y}}(t) = \widehat{f}(\widehat{y}(t)) = (f(y(t)), -\lambda y_{N+1}(t))$$
$$\widehat{y}(0) = (x, 1) .$$

Assume then that (i) holds with $\lambda = 0$. Then it is not hard to prove, for $x \in \Omega$ and $s < 0$, $|s|$ small enough, that

(5.27) $$u(y_x(s)) - u(x) \leq \int_s^0 \ell(y_x(\tau))\, d\tau .$$

Now, if $x \in \Omega$ is a local minimum for $u - \varphi$, $\varphi \in C^1(\Omega)$, then

(5.28) $$\varphi(y_x(s)) - \varphi(x) \leq u(y_x(s)) - u(x)$$

for $s < 0$ small enough. Combining (5.27) with (5.28) we obtain

$$\varphi(y_x(s)) - \varphi(x) \leq \int_s^0 \ell(y_x(\tau))\, d\tau .$$

Dividing this by $-s > 0$ and letting $s \to 0$ we conclude that $-f(x) \cdot D\varphi(x) \leq \ell(x)$ and (ii) is proved.

To prove the reverse implication, let us assume first that x_0 is a local minimum for $u - \varphi$ with $\varphi \in C^1(\Omega)$. Then (ii) gives

(5.29) $$f(x_0) \cdot Du(x_0) + \ell(x_0) \geq 0 .$$

Now if $f(x_0) = 0$, then $y_{x_0}(\tau) \equiv x_0$. Inequality (i) reduces in this case to

$$0 \leq \ell(x_0) \int_s^t d\tau = (t-s)\,\ell(x_0)$$

for $s \leq t$ which is clearly implied by (5.29).

Consider now the case $f(x_0) \neq 0$. Since $f \in C^1(\mathbb{R}^N)$, by classical results on ordinary differential equations, there exists a local diffeomorphism Φ such that in the new coordinates $\xi = \Phi(x)$ system (5.23) becomes

(5.30)
$$\dot{\xi}(t) = e_1 = (1, 0, \ldots, 0)$$
$$\xi(0) = x_0.$$

The change of coordinates in (ii) implies that $u(\Phi^{-1}(\xi))$ satisfies

$$f(\Phi^{-1}(\xi)) \cdot D\Phi(\Phi^{-1}(\xi))Du(\Phi^{-1}(\xi)) + \ell(\Phi^{-1}(\xi)) \geq 0$$

in the viscosity sense (see Exercise 2.6). Thanks to (5.30) this amounts to

$$\frac{\partial u}{\partial \xi_1}(\Phi^{-1}(\xi)) + \ell(\Phi^{-1}(\xi)) \geq 0.$$

Using now Lemmas 5.17 and 5.15 (see also Remark 5.16) we conclude that

$$u(\Phi^{-1}(s, 0, \ldots, 0)) - u(\Phi^{-1}(t, 0, \ldots, 0)) \leq \int_s^t \ell(\Phi^{-1}(\tau, 0, \ldots, 0)) \, d\tau$$

for $0 \leq s \leq t$, with s, t sufficiently small.

By definition of Φ this is the same as

$$u(y_{x_0}(s)) - u(y_{x_0}(t)) \leq \int_s^t \ell(y_{x_0}(\tau)) \, d\tau,$$

for $0 \leq s \leq t$, s, t small. A simple continuation argument shows the validity of (i) for any $0 \leq s \leq t < t_x$.

Up to now we have proved that (i) holds for any $x \in A^- = \{x \in \Omega : D^-u(x) \neq \emptyset\}$ (recall Lemma 1.7). Since A^- is dense in Ω, by Lemma 1.8 (d), we conclude the validity of (i) for all $x \in \Omega$ using the continuous dependence of the solution of (5.23) with respect to the initial datum.

The equivalence between (i) and (iii) can be proved similarly. ◀

REMARK 5.19. A symmetric result holds for viscosity supersolutions of (iii). Therefore u is a viscosity solution of $\lambda u - f \cdot Du = \ell$ if and only if

$$s \longmapsto e^{-\lambda s} u(y_x(s)) + \int_0^s \ell(y_x(\tau)) e^{-\lambda \tau} \, d\tau$$

is constant for any fixed $x \in \Omega$. ◁

5.6. Monotonicity of value functions along trajectories

Let us now consider the controlled system (5.24). We denote by $y_x(\cdot, \alpha)$ the solution of (5.24) corresponding to a given control $\alpha \in \mathcal{A}$ and set, for $x \in \Omega$,

$$t_x(\alpha) := \inf\{t \geq 0 : y_x(t, \alpha) \notin \Omega\}.$$

5. SPECIAL RESULTS FOR CONVEX HAMILTONIANS

If we use constant controls $\alpha(t) \equiv a$ in (5.24), then the equivalence between

(5.31)
$$e^{-\lambda s}u(y_x(s,a)) \leq e^{-\lambda t}u(y_x(t,a)) + \int_s^t \ell(y_x(\tau,a),a)\,e^{-\lambda \tau}\,d\tau$$
$$\forall x \in \Omega,\ 0 \leq s \leq t \leq t_x(a),\ \forall a \in A,$$

(5.32) $\quad \lambda u + H(x, Du) \leq 0 \quad$ in Ω (viscosity sense),

(5.33) $\quad -\lambda u - H(x, Du) \geq 0 \quad$ in Ω (viscosity sense),

with
$$H(x,p) = \sup_{a \in A}\{-f(x,a) \cdot p - \ell(x,a)\},$$

is a straightforward consequence of Proposition 5.18. A repeated application of Proposition 5.18 allows us to show that (5.32) and (5.33) are equivalent to

(5.34) $\quad e^{-\lambda s}u(y_x(s,\alpha)) \leq e^{-\lambda t}u(y_x(t,\alpha)) + \int_s^t \ell(y_x(\tau,\alpha),\alpha(\tau))\,e^{-\lambda \tau}\,d\tau$

$\forall x \in \Omega, 0 \leq s \leq t \leq t_x(\alpha), \forall \alpha \in \mathcal{P}$, where $\mathcal{P} \subset \mathcal{A}$ is the class of piecewise constant controls. To establish the equivalence between (5.32), (5.33), (5.34) for general controls $\alpha \in \mathcal{A}$ we need the following approximation lemma.

LEMMA 5.20. *Let $\alpha \in \mathcal{A}$, $x \in \Omega$, and $y(t) = y_x(t, \alpha)$ be the corresponding solution of (5.24). Assume*

(5.35) $\quad f$ *is continuous and bounded on* $\Omega \times A$,

(5.36) $\quad \exists L > 0 : |f(x,a) - f(y,a)| \leq L|x - y|, \quad \forall x, y \in \Omega, \forall a \in A$.

Then for every $T > 0$ there exists a sequence $\{\alpha_n\} \subset \mathcal{A}$ such that

(5.37)
$$\begin{cases} \alpha_n \text{ is piecewise constant on } [0, T], \\ |\alpha_n(t) - \alpha(t)| < 1/n, \quad \forall t \in E_n \subseteq [0, T],\ E_n \text{ compact},\ |[0, T] \setminus E_n| < 1/n \end{cases}$$

(5.38) $\quad\quad\quad\quad y_n \longrightarrow y \quad$ *uniformly in* $[0, T]$,

where $y_n(t) := y_x(t, \alpha_n)$ and $|E|$ denotes the Lebesgue measure of E.

PROOF. Assertion (5.37) is a consequence of Lusin's theorem (see Exercise 5.8). To prove (5.38) observe that

$$|y_n(t) - y(t)| \leq L \int_0^t |y_n(s) - y(s)|\,ds + \int_0^t |B_n(s)|\,ds,$$

where L is as in (5.36) and

$$B_n(s) = f(y(s), \alpha_n(s)) - f(y(s), \alpha(s)).$$

Hence, by Gronwall's inequality,

(5.39) $$|y_n(t) - y(t)| \leq \int_0^T |B_n(s)|\,ds + L\,e^{LT} \int_0^T \int_0^s |B_n(\tau)|\,d\tau\,ds,$$

for $t \in [0,T]$. From (5.37), $\alpha_n(t) \to \alpha(t)$ a.e. in $[0,T]$ (at least for a subsequence). Hence, by continuity

$$B_n(\tau) \longrightarrow 0 \quad \text{a.e. in } [0,T].$$

Moreover, by assumption (5.35),

$$|B_n(\tau)| \leq C \quad \forall \tau \in [0,T],\ n \in \mathbb{N}.$$

Assertion (5.38) now follows from (5.39) and the dominated convergence theorem. ◀

THEOREM 5.21. *Let us assume that f satisfies* (5.35), (5.36), *and*

(5.40) $$f(\,\cdot\,,a) \in C^1(\Omega) \quad \text{for every } a \in A.$$

Assume also

(5.41) $$\ell \in C(\Omega \times A), \quad \ell \text{ bounded}, \lambda \geq 0.$$

Then for $u \in C(\Omega)$ the following statements are equivalent:

(j) $\begin{cases} e^{-\lambda s}u(y_x(s,\alpha)) - e^{-\lambda t}u(y_x(t,\alpha)) \leq \int_s^t \ell(y_x(\tau,\alpha),\alpha(\tau))\,e^{-\lambda \tau}\,d\tau, \\ \forall \alpha \in \mathcal{A},\ x \in \Omega,\ 0 \leq s \leq t < t_x(\alpha); \end{cases}$

(jj) $\quad \lambda u(x) + \sup\limits_{a \in A}\{-f(x,a)\cdot Du(x) - \ell(x,a)\} \leq 0 \quad \text{in } \Omega;$

(jjj) $\quad -\lambda u(x) - \sup\limits_{a \in A}\{-f(x,a)\cdot Du(x) - \ell(x,a)\} \geq 0 \quad \text{in } \Omega;$

where (jj) *and* (jjj) *are understood in the viscosity sense.*

PROOF. The discussion before Lemma 5.20 shows that in order to prove the statement it is enough to show that (5.34) implies (j) for any $u \in C(\Omega)$ (the reverse implication being trivial). For this purpose let α be arbitrary in $\mathcal{A}$ and take α_n and y_n as in Lemma 5.20. By (5.34),

$$e^{-\lambda s}u(y_n(s)) - e^{-\lambda t}u(y_n(t)) \leq \int_s^t \ell(y_n(\tau),\alpha_n(\tau))\,e^{-\lambda \tau}\,d\tau$$

for $0 \leq s \leq t \leq t_x(\alpha_n)$. Since $\alpha_n \to \alpha$ almost everywhere and $y_n \to y(\,\cdot\,,\alpha)$ uniformly on compact intervals by Lemma 5.20, a passage to the limit in the previous inequality as $n \to +\infty$ gives the desired result (see also Exercise 5.10). ◀

REMARK 5.22. All the results of this section hold under more general assumptions (see Chapter III, §2.3). Let us mention here that the differentiability assumption (5.40) on f can be relaxed to $(f(x,a) - f(y,a))\cdot(x-y) \leq L|x-y|^2$. ◁

REMARK 5.23. Theorem 5.21 shows in particular that if u is a viscosity subsolution of (HJB) then

$$g(s) := e^{-\lambda s} u(y_x(s, \alpha)) + \int_0^s \ell(y_x(\tau, \alpha)) e^{-\lambda \tau} \, d\tau$$

is nondecreasing for any choice of $\alpha \in \mathcal{A}$. Of course one cannot expect g to be nonincreasing for all $\alpha \in \mathcal{A}$ when u is a viscosity (or even classical) supersolution of (HJB) (see Exercise 5.9). Indeed we will see in Chapter III, §2, that g nonincreasing is a sufficient condition of optimality for α. ◁

REMARK 5.24. The maximum bounded function u_M satisfying (j) has been proposed in the literature as a good notion of weak solution of (HJB) (see [BL82, Me80a]). Theorem 5.21 and Proposition 2.1 show that u_M is actually a viscosity solution of (HJB), provided $u_M \in C(\Omega)$. ◁

Exercises

5.1. Let $H(x, p) = \sup_{a \in A} \{ -f(x, a) \cdot p - \ell(x, a) \}$ where A is a topological space, $f : \mathbb{R}^N \times A \to \mathbb{R}^N$, $\ell : \mathbb{R}^N \times A \to \mathbb{R}$ are given functions.

(i) Show that $p \mapsto H(x, p)$ is convex for each x.

(ii) Assume that $x \mapsto f(x, a)$, $x \mapsto \ell(x, a)$ are continuous for each fixed $a \in A$ and show that $H(x, p)$ is lower semicontinuous.

(iii) Assume that for some moduli ω_1, ω_2 and $M > 0$ independent of $a \in A$

$$|f(x, a) - f(y, a)| \leq \omega_1(|x - y|), \qquad |f(x, a)| \leq M,$$
$$|\ell(x, a) - \ell(y, a)| \leq \omega_2(|x - y|), \qquad \forall x, y \in \mathbb{R}^N$$

and show that H is jointly continuous with respect to (x, p).

5.2. Suppose that (H$_7$) holds. Use Propositions 1.9 and 5.1 to show that the set

$$S := \{ u \in \text{Lip}(\Omega) : u(x) + H(x, Du(x)) \leq 0 \text{ in } \Omega, \text{ in the viscosity sense} \}$$

is convex.

5.3. Take $H(x, p)$ as in Exercise 5.1 with $f(x, a) \equiv a \in A$, a closed convex set in $\mathbb{R}^N$. Assume also that $a \mapsto \ell(x, a)$ is C^1 strictly convex and (in case A is unbounded) that $\ell(x, a)/|a| \to +\infty$ as $|a| \to +\infty$, for each fixed x. Show that

$$H(x, p) = -a^* \cdot p - \ell(x, a^*)$$

where $a^* = a^*(x, p) \in A$ is the unique solution of the variational inequality

$$(D_a \ell(x, a^*) + p) \cdot (a - a^*) \geq 0 \qquad \forall a \in A.$$

Compute explicitly a^* in the special case $A = \mathbb{R}^N$, $\ell(x, a) = \ell_0(x) + \frac{1}{2}|a|^2$.

5.4 (KRUŽKOV'S CHANGE OF VARIABLES). Show that u is a viscosity solution of $|Du(x)| = \ell(x)$ if and only if $v(x) = -e^{-u(x)}$ is a viscosity solution of

$$|Dv(x)| + \ell(x)v(x) = 0.$$

5.5. Let $u(x,b)$, $b \in B$ compact, be a family of C^1 solutions of

$$H(x, u(x,b), D_x u(x,b)) = 0$$

in Ω. Suppose that $b \mapsto u(x,b)$ is lower semicontinuous and that $p \mapsto H(x,r,p)$ is convex for all fixed x, r. Show that $u(x) := \min_{b \in B} u(x,b)$ is a viscosity solution of the same equation. [Hint: use Proposition 2.13.]

5.6. Show that for each $t \in [0,1]$ the function

$$v_t(x) = \begin{cases} 1 - x^2, & -1 \leq x < -t \\ x^2 - 1 + 2(1 - t^2), & |x| < t \\ 1 - x^2, & t < x \leq 1 \end{cases}$$

is a viscosity solution of $|u'(x)| - 2|x| = 0$, $x \in (-1,1)$ such that $v_t(-1) = v_t(1) = 0$.

5.7. Let $u \in C([t_1, t_2])$, $\bar{t} \in \,]t_1, t_2[$ with $u(t_2) < u(\bar{t}) < u(t_1)$. Fix $\delta > 0$ such that $\bar{t} + \delta < t_2$, and set

$$\varphi(t) = \begin{cases} mt + q, & t_1 \leq t \leq \bar{t} + \delta \\ at^2 + bt + c, & \bar{t} + \delta < t \leq t_2. \end{cases}$$

Find constants m, q, a, b, c and δ such that $\varphi \in C^1(t_1, t_2)$, $\varphi' < 0$, $\varphi(\bar{t}) > u(\bar{t})$, $\varphi(t_i) = u(t_i)$ ($i = 1, 2$).

5.8. Let $\alpha : [0, T] \to A$ be measurable. Recall that (Lusin's Theorem) for any $\varepsilon > 0$ there exist a compact set $E \subset [0, T]$ and $\delta(\varepsilon) > 0$ such that

$$|\alpha(t) - \alpha(s)| < \varepsilon, \quad \forall s, t \in E \text{ with } |s - t| < \delta(\varepsilon)$$
$$|[0, T] \setminus E| < \varepsilon.$$

Use this and the partition of $[0, T]$ into intervals $[k\delta, (k+1)\delta[$ ($k = 0, \ldots, [T/\delta]$) to prove assertion (5.37) of Lemma 5.20.

5.9. Observe that $u(x) = x$ is a classical solution of $|u'(x)| - 1 = 0$, $x \in \mathbb{R}$, or (equivalently) of

$$\sup_{a \in [-1,1]} \{-au'(x) - 1\} = 0.$$

For arbitrarily fixed $T > 0$ set

$$\bar{\alpha}(s) = \begin{cases} 1, & 0 \leq s \leq T \\ -1, & s > T \end{cases}$$

and show that $g(s) < g(t)$ for $s < T < t$, where g is as in Remark 5.23.

5.10. Give alternative proofs of Propositions 5.1 and 5.4 by means of (4.4) and (4.6) without using regularization procedures.

5.11. Use Theorem 5.21 to show that the value function of the infinite horizon problem in Chapter I is the maximum viscosity subsolution of the corresponding (HJB) equation.

5.12. With the assumptions and notations of Lemma 5.20 prove that

$$\liminf_n t_x(\alpha_n) \geq t_x(\alpha) \ .$$

6. Bibliographical notes

The presentation of the theory of viscosity solutions in Sections 1 to 3 is basically inspired by the papers [CEL84, CL83, I84] and the lecture notes [I88b]. We adopted the common habit, initiated by Crandall, Evans and Lions [CEL84], of working with either the "test functions" or the "semidifferentials" definitions of viscosity solutions rather than with the original one in [CL81, L82, CL83].

The connections with the Maximum Principle in elliptic theory are inspired by [Caf89, CafCa95]. The subdifferential D^-u defined here coincides with the classical subdifferential of convex analysis [R70] provided that function u is convex (see Exercise 1.3). The comparison result for unbounded solutions Theorem 3.7 as well as Proposition 4.2 on Lipschitz continuity and Theorem 4.9 on semiconcavity are due to Ishii [I84]. Among the many papers touching the important topic of comparison and uniqueness of continuous viscosity solutions let us mention here also [CDE84, CN85, Sou86, Ba85b, L85b, S86, I89, BaP87, BaP90, CDL90, DI90, BaL91, I91, CIL92, IK96] for bounded solutions, and [I85a, I86, CL86, CL87, CIL87, B89, CDL90, Ba90a, Ba90b, ME95, BDL96, Alv97a] for unbounded solutions, the book [Ba94], and the references therein. Our treatment of the basic theory of marginal functions in §2 seems to be new, as well as some results; the formula for directional derivatives in Proposition 2.13 is known in Operations Research as Danskin's theorem (under the assumptions of Proposition 4.4), see [DeM71, Cla75] and the references therein. The notion of (generalized) exterior normal is due to Bony [Bon69]; proximal normals are a standard tool in nonsmooth analysis [R81, Cla89].

A proof of the classical Rademacher theorem on differentiability of Lipschitz functions can be found for example in [EG92]. The generalized differential calculus for Lipschitz continuous functions and the related notion of solution of Hamilton-Jacobi equations were developed mainly by F.H. Clarke [Cla75, Cla83]. Some relations between Clarke's generalized gradient and the semidifferentials of the theory of viscosity solutions have been investigated in [CS87, CS89, Fr89a]. See also the recent survey [CLSW95].

The role of semiconcavity in Hamilton-Jacobi theory and optimal control was pointed out by Kružkov [Kr60, Kr67, Kr75] (see also [Bt77, Hr78]) and in connections with viscosity solutions in [L82, CDI84, CS87, CS89, CF91, CSi95a]. Proposition 4.7 follows [CS89]; the semiconcavity of viscosity solutions (Theorem 4.9) is taken from [I84] and Proposition 4.13 is from [Ba93]. For the related topic of inf-convolution, we refer to [Mo66] and [LL86].

Other results on the regularity of solutions and the regularizing effect in the Cauchy problem can be found in [L85d, Ba90b, Ba90c, ArT96].

The equivalence of semiconcave generalized solutions and viscosity solutions for convex and smooth Hamiltonians (Corollary 5.2) is adapted from [CL83]; it gives a further explanation of the uniqueness theory for semiconcave generalized solutions in [Kr60, Kr67, Kr75, Dou61, L82]. The proof of Remark 5.3 is in [LS95]. The characterization of viscosity solutions as bilateral supersolutions in the case of convex Hamiltonian (Theorem 5.6) is due to Barron and Jensen [BJ90], but our proof follows [Ba93]. Proposition 5.7 is a variant of a result in [CS89]. Theorem 5.9 is essentially a result of [Kr75], presented here following [I87b]. Most of the results in §5.5 are from [CL83], our presentation and §5.6 take advantage of some unpublished work by P. Soravia. The connection between the classical method of characteristics and viscosity solutions is studied in more detail in [Sua91, Ko93a, Ko93b].

Some important topics in the general theory of viscosity solutions of first order Hamilton-Jacobi equations are not covered in this chapter. First of all, we did not give any existence result. The reason is that in the next two chapters we always prove directly that the value function of a given control problem is a viscosity solution of the corresponding HJB equation. The best approach to the existence theory is the so-called Perron's method, settled by Ishii [I87a], which needs the notion of discontinuous solution and is presented in Chapter V, §2; for earlier results see also [L83d, I83, Ba84, EI84, I86] and the references in [CIL92, Ba94].

Representation formulas are a classical issue since the work of Hopf [H65] and Kružkov [Kr67]. The connection between Hopf's formulas and viscosity solutions was studied in [L82, BE84, LR86], see also [BOs91, Su95, BJL96, BFag96] for some recent extensions of these formulas, and [Car93] for their use in linking viscosity solutions with geometric solutions in the sense of symplectic mechanics. Of course any characterization of the value function in optimal control or differential games as the unique solution of a HJ equation gives a representation result for solutions of a class of equations whose Hamiltonian can be written in the Bellman or Isaacs form. This point of view was pursued, e.g., in [L82, ES84, I85b, I88a].

Among the other topics that we do not touch in this book we also mention the analysis of singularities of solutions in [CS87, CS89, ACS93, Ko93a, Ko93b, IzK95, Bad94, Wi96], geometric properties of solutions in [B85, BadB90, BadB92, Bad94, ALL97], propagation of fronts with curvature-dependent speed [OSe88, ESp91, ESp95, CGG91, Sor94a], the problem of shape-from-shading in vision theory [RT92, Tou92, LRT93], equations with singular coefficients [Si95, IR95], the connections with nonconvex problems in the Calculus of Variations [Ba90a, Cut93, DaM96].

CHAPTER III

Optimal control problems with continuous value functions: unrestricted state space

In this Chapter we consider several optimal control problems whose value function is defined and continuous on the whole space $\mathbb{R}^N$. This setting is suitable for those problems where no a priori constraint is imposed on the state of the control system. For all the problems considered we establish the Dynamic Programming Principle and derive from it the appropriate Hamilton-Jacobi-Bellman equation for the value function. This allows us to apply the theory of Chapter II, and some extensions of it, to prove that the value function can in fact be characterized as the unique viscosity solution of the corresponding Hamilton-Jacobi-Bellman equation.

This has some direct consequences such as *verification theorems*. More involved applications such as those to necessary and sufficient conditions of optimality require some additional tools from nonsmooth analysis like *generalized directional derivatives*. These themes are developed in Sections 2.3 to 2.5 with reference to the infinite horizon problem. The important issue of connections with the Pontryagin Maximum Principle (in a necessary and sufficient form) is covered in detail in Section 3.4 for the finite horizon problem.

1. The controlled dynamical system

We list here some basic assumptions on the nonlinear control system that we consider. We will assume:

(A$_0$) $\quad\begin{cases} A \text{ is a topological space,} \\ f: \mathbb{R}^N \times A \to \mathbb{R}^N \text{ is continuous;} \end{cases}$

(A$_1$) $\quad f$ is bounded on $B(0, R) \times A$ for all $R > 0$;

(the local boundedness of f, uniformly with respect to the control variable a)

(A$_2$) $\quad\begin{array}{l}\text{there is a modulus } \omega_f \text{ such that} \\ |f(y, a) - f(x, a)| \leq \omega_f(|x - y|, R), \\ \text{for all } x, y \in B(0, R) \text{ and } R > 0,\end{array}$

(the local uniform continuity of f, uniformly in a) where a *modulus* is a function $\omega : \mathbb{R}_+ \times \mathbb{R}_+ \to \mathbb{R}_+$ such that, for all $R > 0$, $\omega(\,\cdot\,, R)$ is continuous, nondecreasing, and $\omega(0, R) = 0$.

The reader can assume for simplicity that A is a subset of $\mathbb{R}^M$. In most applications A is compact and in this case (A_1) and (A_2) are consequences of (A_0).

We will also assume

(A$_3$) $\qquad (f(x, a) - f(y, a)) \cdot (x - y) \leq L |x - y|^2 \qquad$ for all $x, y \in \mathbb{R}^N$, $a \in A$;

that is, there is $L \in \mathbb{R}$ such that $f(\,\cdot\,, a) - LI$, where I is the identity, is a monotone "nonincreasing" map for all a. A more familiar condition which implies (A_3) (and (A_2)) is the global Lipschitz continuity of f in the state variable, uniformly in the control variable, that is,

(1.1) $\qquad |f(x, a) - f(y, a)| \leq L |x - y| \qquad$ for all $x, y \in \mathbb{R}^N$, $a \in A$.

We look at trajectories of the nonlinear system

(S) $\qquad \begin{cases} y'(t) = f(y(t), \alpha(t)), & t > 0, \\ y(0) = x, \end{cases}$

for control functions $\alpha(\,\cdot\,)$ ("open loop" controls) in the set[1]

$$\mathcal{A} := \{\text{measurable functions } [0, +\infty[\to A\}.$$

This means that $y(\,\cdot\,)$ solves the integral equation

(1.2) $\qquad y(t) = x + \int_0^t f(y(s), \alpha(s))\, ds, \qquad t > 0,$

so that, in particular, $y(\,\cdot\,)$ is absolutely continuous on compact intervals of $[0, +\infty[$ and it solves (S) almost everywhere.

Under the assumptions (A_0), (A_1), (A_3) the existence and uniqueness of a solution of (S) (i.e., of (1.2)), defined for all $t \in [0, +\infty[$, follows from the standard theory of ordinary differential equations and we recall the proof in the appendix of this chapter, Section 5. For given $\alpha \in \mathcal{A}$ and $x \in \mathbb{R}^N$, the solution will be denoted by $y_x(t, \alpha)$, or briefly by $y_x(t)$ if no confusion may arise. The basic estimates on y_x are the following (for the proofs see §5):

(1.3) $\qquad |y_x(t, \alpha) - x| \leq M_x t \qquad$ for all $\alpha \in \mathcal{A}$ and $t \in [0, 1/M_x]$,

where $M_x := \sup\{|f(z, a)| : |z - x| \leq 1,\ a \in A\}$;

(1.4) $\qquad |y_x(t, \alpha)| \leq (|x| + \sqrt{2Kt})e^{Kt} \qquad$ for all $\alpha \in \mathcal{A}$ and $t > 0$,

where $K := L + \sup\{|f(0, a)| : a \in A\}$;

---

[1] We recall that $\alpha : [0, +\infty[\to A$ is (Lebesgue) measurable if for any open set $B \subseteq A$ the set $\{t : \alpha(t) \in B\}$ is Lebesgue measurable.

(1.5) $\quad |y_x(t,\alpha) - y_z(t,\alpha)| \leq e^{Lt}|x-z| \quad$ for all $\alpha \in \mathcal{A}$ and $t > 0$.

We will also be interested in some subsets of $\mathcal{A}$, such as *piecewise constant controls*

$$\mathcal{P} := \{\, \alpha \in \mathcal{A} : \text{there is an increasing sequence } t_n \text{ such that }$$
$$\lim_n t_n = +\infty \text{ and } \alpha \text{ is constant on }]t_n, t_{n+1}[\text{ for all } n\,\},$$

and *monotone controls*, for $A \subseteq \mathbb{R}$,

$$\mathcal{A}_m := \{\, \alpha \in \mathcal{A} : \alpha \text{ is nondecreasing}\,\}.$$

2. The infinite horizon problem

2.1. Dynamic Programming and the Hamilton-Jacobi-Bellman equation

In this section the cost to minimize is

$$J(x,\alpha) := \int_0^\infty \ell(y_x(t), \alpha(t)) e^{-\lambda t}\, dt,$$

where the running cost $\ell : \mathbb{R}^N \times A \to \mathbb{R}$ and the constant interest rate λ satisfy

(A$_4$) $\quad \begin{cases} \ell \text{ is continuous;} \\ \text{there are a modulus } \omega_\ell \text{ and a constant } M \text{ such that} \\ |\ell(x,a) - \ell(y,a)| \leq \omega_\ell(|x-y|) \text{ and} \\ |\ell(x,a)| \leq M, \text{ for all } x,y \in \mathbb{R}^N \text{ and } a \in A; \\ \lambda > 0. \end{cases}$

The *value function* for this problem is

$$v(x) := \inf_{\alpha \in \mathcal{A}} J(x,\alpha).$$

It is also called Bellman's function, the optimal result function, the guaranteed outcome. The name "value" comes from an economic interpretation of the equivalent problem of maximizing the payoff $-J$: in this problem $-v(x)$ is the best we can get from the initial state x, thus it is the economic value of x.

We first look at the regularity of v.

PROPOSITION 2.1. *Assume* (A$_0$), (A$_1$), (A$_3$), *and* (A$_4$). *Then* $v \in BUC(\mathbb{R}^N)$. *If moreover* $\omega_\ell(r) = L_\ell r$ (*i.e.*, ℓ *is Lipschitz in y, uniformly in a), then v is Hölder continuous with the following exponent* γ:

$$\gamma = \begin{cases} 1 & \text{if } \lambda > L, \\ \text{any } \gamma < 1 & \text{if } \lambda = L, \\ \lambda/L & \text{if } \lambda < L. \end{cases}$$

PROOF. The estimate $|v(x)| \leq M/\lambda$ for all $x \in \mathbb{R}^N$ is an immediate consequence of (A$_4$). Next we fix $x, z \in \mathbb{R}^N$, $\varepsilon > 0$ and $\overline{\alpha} \in \mathcal{A}$ such that

$$v(z) \geq \int_0^\infty \ell(y_z(t), \overline{\alpha}(t)) e^{-\lambda t} \, dt - \varepsilon \, .$$

Then

(2.1) $v(x) - v(z)$
$$\leq \int_0^T e^{-\lambda t} |\ell(y_x(t), \overline{\alpha}(t)) - \ell(y_z(t), \overline{\alpha}(t))| \, dt + \int_T^\infty 2M \, e^{-\lambda t} \, dt + \varepsilon \, .$$

For T large enough we can make the second integral in the right-hand side less then ε and get

(2.2)
$$v(x) - v(z) \leq \int_0^T \omega_\ell(|y_x(t) - y_z(t)|) \, e^{-\lambda t} \, dt + 2\varepsilon$$
$$\leq \int_0^T \omega_\ell(e^{LT} |x - z|) \, e^{-\lambda t} \, dt + 2\varepsilon,$$

where we have used the estimate (1.5). Now the right-hand side of (2.2) can be made less than 3ε for $|x - z|$ small enough, which proves the uniform continuity of v.

Next we assume $\omega_\ell(r) = L_\ell r$ and get from (2.1) and (1.5)

(2.3)
$$v(x) - v(z) \leq L_\ell |x - z| \int_0^T e^{Lt} e^{-\lambda t} \, dt + \frac{2M}{\lambda} e^{-\lambda T} + \varepsilon$$
$$= L_\ell |x - z| \frac{e^{(L-\lambda)T} - 1}{L - \lambda} + \frac{2M}{\lambda} e^{-\lambda T} + \varepsilon,$$

where the last inequality holds only for $L \neq \lambda$.

If $\lambda > L$, it is easy to get from (2.3), letting $T \to +\infty$ and $\varepsilon \to 0$,

$$|v(x) - v(z)| \leq \frac{L_\ell}{\lambda - L} |x - z|,$$

which proves the statement in this case. If $\lambda < L$, we restrict ourselves (without loss of generality) to $|x - z| < 2M/L_\ell$ and choose

$$T = \frac{1}{L} \log \frac{2M}{L_\ell |x - z|} > 0,$$

which minimizes the last term of (2.3) as a function of T. We get

$$v(x) - v(z) \leq C_1 |x - z| \, |x - z|^{(\lambda - L)/L} + C_2 |x - z|^{\lambda/L} + \varepsilon,$$

where C_i depend on λ, L and L_ℓ, and then $|v(x) - v(z)| \leq C_3 |x - z|^{\lambda/L}$, which proves the statement in this case as well.

Finally if $\lambda = L$, we use the first inequality in (2.3) with the choice $T = -\lambda^{-1} \log |x - z|$ (and we restrict ourselves to $|x - z| < 1$) and get

$$v(x) - v(z) \leq \frac{-L_\ell}{\lambda} |x - z| \log |x - z| + \frac{2M}{\lambda} |x - z| + \varepsilon \, .$$

Then the modulus of continuity of v in this case is $\omega_v(r) = C_4 r \, |\log r|$, and the proof is complete. ◂

REMARK 2.2. In the study of ergodic problems in Chapter VII we need the following estimate. Under the assumptions of Proposition 2.1 with $\omega_\ell(r) = L_\ell r$, if $L < 0$ then for all $\lambda > 0$

$$|v(x) - v(z)| \leq \frac{L_\ell}{-L} |x - z| \qquad \text{for all } x, z \in \mathbb{R}^N \, . \quad \triangleleft$$

Lipschitz and Hölder continuity of v can be obtained by assuming some controllability of the system instead of a large interest rate. In the next result we assume, for instance, that in a given set $\mathcal{K}$ the following holds:

(2.4) $\quad$ there exist $C, \gamma > 0$ and, for each $x, z \in \mathcal{K}$, a control $\alpha_1 \in \mathcal{A}$ such that $y_x(t_1, \alpha_1) = z$ for some $t_1 \leq C |x - z|^\gamma$.

Roughly speaking (2.4) means that the system can reach any point of $\mathcal{K}$ in a time proportional to a power of the distance from the initial point. In the terminology of Chapter IV, (2.4) is equivalent to saying that, for any $z \in \mathcal{K}$, the minimum time function with target set $\{z\}$ is finite and γ-Hölder continuous in $\mathcal{K}$. In Section IV.1 we give some results on the Hölder continuity of minimal time functions for linear and nonlinear systems. Note that a trivial condition for (2.4) to hold with $\gamma = 1$ in $\mathcal{K}$ convex is that $f(x, A) \supseteq \bar{B}(0, 1/C)$. This condition for Lipschitz continuity can be weakened to $\overline{co} f(x, A) \supseteq \bar{B}(0, r)$ for some $r > 0$, as we show in Corollary 2.10.

PROPOSITION 2.3. *Assume* (A_0), (A_1), (A_3), (A_4), *and* (2.4) *in a set* $\mathcal{K} \subseteq \mathbb{R}^N$. *Then, for all* $x, z \in \mathcal{K}$,

$$|v(x) - v(z)| \leq MC|x - z|^\gamma \, .$$

PROOF. Since ℓ is bounded it is not restrictive to assume $v \geq 0$ because the value function of the problem with running cost $\ell - \inf \ell \geq 0$ is $v - \inf \ell$. For given $x, z \in \mathcal{K}$, we choose $\overline{\alpha}$ as in the previous proof and define

$$\alpha_2(t) = \begin{cases} \alpha_1(t) & t \leq t_1, \\ \overline{\alpha}(t - t_1) & t > t_1 \, . \end{cases}$$

Then $y_x(s + t_1, \alpha_2) = y_z(s, \overline{\alpha})$ for $s \geq 0$, and

$$v(x) \leq J(x, \alpha_2)$$
$$= \int_0^{t_1} e^{-\lambda t} \ell(y_x(t, \alpha_1), \alpha_1(t)) \, dt + e^{-\lambda t_1} \int_0^\infty e^{-\lambda s} \ell(y_x(s + t_1, \alpha_2), \overline{\alpha}(s)) \, ds$$
$$\leq M t_1 + e^{-\lambda t_1} (v(z) + \varepsilon) \leq v(z) + MC |x - z|^\gamma + \varepsilon,$$

where in the last inequality we used (2.4) and $v \geq 0$. The proof of the claim is easily completed by exchanging the roles of x and z. ◂

EXAMPLE 2.4. The following example shows that the regularity proved in Proposition 2.1 is optimal if we do not assume controllability conditions, and that $v(x)$ is not differentiable in general, even if $N = 1$, A is a singleton, and ℓ and f are C^∞.

Let $f \in C^\infty(\mathbb{R})$ be such that $f(x) = x$ for $|x| < 1$, $|f(x)| \leq 2$ and $0 \leq f'(x) \leq 1$ for all x. The system is $y' = f(y) = f(y, a)$. Let λ be the discount rate and take

$$\ell(x, a) = \ell(x) := \lambda |f(x)|^\lambda (1 - f'(x)).$$

Note that $\ell \in C^\infty(\mathbb{R})$ for all $\lambda > 0$ because $\ell \equiv 0$ in $[-1, 1]$. Now we compute, for $x > 0$ (so that $f(y_x(t)) > 0$ for all t)

$$v(x) = \int_0^\infty \ell(y_x(t)) e^{-\lambda t}\, dt = \int_0^\infty -\frac{d}{dt}\left[f^\lambda(y_x(t)) e^{-\lambda t}\right] dt = f^\lambda(x).$$

A similar calculation for $x < 0$ shows that $v(x) = |f(x)|^\lambda$. Then $v(x) = |x|^\lambda$ in $[-1, 1]$. Note that $L = 1$ in this example. ◁

The next result is fundamental. As is common nowadays, we use Bellman's terminology and call it the Dynamic Programming Principle, briefly DPP, even if our formulation is more akin to what Isaacs called the *tenet of transition*. See Remark 2.6 for Bellman's formulation. The DPP expresses as an equation the intuitive belief that to achieve the minimum cost function $v(x)$ it is necessary and sufficient to behave as follows:

(a) let the system evolve for a time t choosing an arbitrary control function α on the interval $[0, t]$;

(b) pay the corresponding cost, that is, $\int_0^t \ell(y_x(s), \alpha(s)) e^{-\lambda s}\, ds$;

(c) pay what remains to pay with the best possible controls, that is, $v(y_x(t)) e^{-\lambda t}$;

(d) finally minimize over all possible controls α.

PROPOSITION 2.5 (DYNAMIC PROGRAMMING PRINCIPLE). *Assume* (A_0), (A_1), (A_3), *and* (A_4). *Then for all* $x \in \mathbb{R}^N$ *and* $t > 0$

$$(2.5) \qquad v(x) = \inf_{\alpha \in \mathcal{A}} \left\{ \int_0^t \ell(y_x(s, \alpha), \alpha(s)) e^{-\lambda s}\, ds + v(y_x(t, \alpha)) e^{-\lambda t} \right\}.$$

PROOF. Let us name $w(x)$ the right-hand side of (2.5). We first show that $v(x) \geq w(x)$. For all $\alpha \in \mathcal{A}$ we have

$$(2.6) \qquad J(x, \alpha) = \int_0^t \ell(y_x(s), \alpha(s)) e^{-\lambda s}\, ds + \int_t^\infty \ell(y_x(s), \alpha(s)) e^{-\lambda s}\, ds$$

$$= I_1 + \int_0^\infty \ell(y_x(s+t), \alpha(s+t)) e^{-\lambda s} e^{-\lambda t}\, ds$$

$$= I_1 + e^{-\lambda t} J(y_x(t), \widetilde{\alpha}) \geq I_1 + e^{-\lambda t} v(y_x(t)) \geq w(x),$$

where I_1 is the first integral in the right-hand side of (2.6) and $\widetilde{\alpha}(s) := \alpha(s + t)$. Taking the infimum over $\mathcal{A}$ we get $v(x) \geq w(x)$.

2. THE INFINITE HORIZON PROBLEM

To prove the opposite inequality, we fix $\alpha \in \mathcal{A}$, set $z := y_x(t, \alpha)$, and fix $\varepsilon > 0$ and $\alpha^1 \in \mathcal{A}$ such that
$$v(z) \geq J(z, \alpha^1) - \varepsilon \, .$$

Define the control
$$\overline{\alpha}(s) = \begin{cases} \alpha(s) & s \leq t, \\ \alpha^1(s-t) & s > t, \end{cases}$$

and let $\overline{y}$ and y^1 be the trajectories corresponding to $\overline{\alpha}$ and α^1, respectively. Then

$$(2.7) \quad v(x) \leq J(x, \overline{\alpha}) = \int_0^t \ell(\overline{y}_x(s), \overline{\alpha}(s)) e^{-\lambda s} \, ds + \int_t^\infty \ell(\overline{y}_x(s), \overline{\alpha}(s)) e^{-\lambda s} \, ds$$
$$= I_1 + e^{-\lambda t} \int_0^\infty \ell(y_z^1(\tau), \alpha^1(\tau)) e^{-\lambda \tau} \, d\tau$$
$$= I_1 + e^{-\lambda t} J(z, \alpha^1) \leq I_1 + e^{-\lambda t} v(y_x(t, \alpha)) + \varepsilon,$$

where I_1 is the first integral in the right-hand side of (2.7). Since ε and α are arbitrary we get $v(x) \leq w(x)$. ◀

REMARK 2.6. If there is an optimal control α^* for x, that is, $v(x) = J(x, \alpha^*)$, then the infimum in (2.5) is attained at α^*:

$$(2.8) \quad v(x) = \int_0^t \ell(y_x(s, \alpha^*), \alpha^*(s)) e^{-\lambda s} \, ds + v(y_x(t, \alpha^*)) e^{-\lambda t}$$

for all $t > 0$. In fact the calculation after (2.6) gives

$$v(x) = \int_0^t \ell(y_x(s), \alpha^*(s)) e^{-\lambda s} \, ds + e^{-\lambda t} J(y_x(t, \alpha^*), \alpha^*(\cdot + t))$$

which shows the inequality "$\geq$" in (2.8). Therefore we also have

$$v(y_x(t, \alpha^*)) = J(y_x(t, \alpha^*), \alpha^*(\cdot + t)),$$

that is, an optimal control for x is optimal for each point of the corresponding trajectory, provided it is appropriately shifted in time. This is the way Bellman formulated the "Principle of Optimality" [Bel71]: "An optimal policy has the property that whatever the initial state and the initial decision are, the remaining decisions must constitute an optimal policy with regard to the state resulting from the first decision" (Bellman referred to discrete time processes. In our setting the "initial decision" is the choice of the control in the interval $[0, t]$).

Note also that (2.8) implies that the function

$$t \longmapsto \int_0^t \ell(y_x(s, \alpha), \alpha(s)) e^{-\lambda s} \, ds + v(y_x(t, \alpha)) e^{-\lambda t}$$

is constant if and only if α is optimal. It is also easy to show, using the DPP, that this function is *nondecreasing* for any $\alpha \in \mathcal{A}$: we leave the proof as an exercise for the reader. ◁

REMARK 2.7. Proposition 2.1 holds for $w(x) = \inf_{a \in \mathcal{B}} J(x, a)$ for any $\mathcal{B} \subseteq \mathcal{A}$. Proposition 2.5 holds for any $\mathcal{B}$ with the following properties:

(i) if $\alpha \in \mathcal{B}$, $\tau > 0$, then $t \mapsto \alpha(t + \tau)$ is in $\mathcal{B}$;

(ii) if $\alpha_1, \alpha_2 \in \mathcal{B}$, $s > 0$,

$$\alpha(t) := \begin{cases} \alpha_1(t) & t \leq s, \\ \alpha_2(t-s) & t > s, \end{cases}$$

then $\alpha \in \mathcal{B}$.

Hence, for instance, $v^\#(x) = \inf_{a \in \mathcal{P}} J(x, a)$ satisfies (2.5) with $\mathcal{A}$ replaced by $\mathcal{P}$. ◁

The Hamiltonian in the Hamilton-Jacobi-Bellman equation for the infinite horizon problem is $H : \mathbb{R}^N \times \mathbb{R}^N \to \mathbb{R}$ defined by

(2.9) $$H(x, p) := \sup_{a \in A} \{-f(x, a) \cdot p - \ell(x, a)\}.$$

Note that $H(x, p) < +\infty$ for all x and p by (A$_1$) and (A$_4$).

The main consequence of the Dynamic Programming Principle is the following.

PROPOSITION 2.8. *Assume* (A$_0$)–(A$_4$). *Then the value function v is a viscosity solution of*

(2.10) $$\lambda v + H(x, Dv) = 0 \quad \text{in } \mathbb{R}^N.$$

PROOF. Let $\phi \in C^1(\mathbb{R}^N)$ and x be a local maximum point of $v - \phi$, that is, for some $r > 0$ $v(x) - v(z) \geq \phi(x) - \phi(z)$ for all $z \in B(x, r)$. Fix an arbitrary $a \in A$ and let $y_x(t)$ be the solution corresponding to the constant control $\alpha(t) = a$ for all t. For t small enough $y_x(t) \in B(x, r)$ (by (1.3)) and then

$$\phi(x) - \phi(y_x(t)) \leq v(x) - v(y_x(t)) \quad \text{for all } 0 \leq t \leq t_0.$$

By using the inequality "$\leq$" in the DPP (2.5) we get

$$\phi(x) - \phi(y_x(t)) \leq \int_0^t \ell(y_x(s), a) e^{-\lambda s}\, ds + v(y_x(t))(e^{-\lambda t} - 1);$$

therefore, dividing by $t > 0$ and letting $t \to 0$, we obtain, by the differentiability of ϕ and the continuity of v, y_x, f, and ℓ,

$$-D\phi(x) \cdot y_x'(0) = -D\phi(x) \cdot f(x, a) \leq \ell(x, a) - \lambda v(x).$$

Since $a \in A$ is arbitrary we have proved that

$$\lambda v(x) + \sup_{a \in A} \{-f(x, a) \cdot D\phi(x) - \ell(x, a)\} \leq 0,$$

that is, v is a viscosity subsolution of (2.10).

Next assume x is a local minimum point of $v - \phi$, that is, for some $r > 0$

(2.11) $$v(x) - v(z) \leq \phi(x) - \phi(z) \quad \text{for all } z \in B(x, r).$$

2. THE INFINITE HORIZON PROBLEM

For each $\varepsilon > 0$ and $t > 0$, by the inequality "$\geq$" in the DPP (2.5), there exists $\overline{\alpha} \in \mathcal{A}$ (depending on ε and t) such that

$$(2.12) \qquad v(x) \geq \int_0^t \ell(\overline{y}_x(s), \overline{\alpha}(s)) e^{-\lambda s}\, ds + v(\overline{y}_x(t)) e^{-\lambda t} - t\varepsilon,$$

where $\overline{y}_x(s) = y_x(s, \overline{\alpha})$, the trajectory of (S) corresponding to $\overline{\alpha}$. Now observe that (1.3) and (A$_4$) imply

$$(2.13) \qquad |\ell(\overline{y}_x(s), \overline{\alpha}(s)) - \ell(x, \overline{\alpha}(s))| \leq \omega_\ell(M_x s) \qquad \text{for } 0 \leq s \leq t_0,$$

and (1.3) and (A$_2$) imply

$$(2.14) \qquad |f(\overline{y}_x(s), \overline{\alpha}(s)) - f(x, \overline{\alpha}(s))| \leq \omega_f(M_x s, |x| + M_x t_0)$$

for $0 \leq s \leq t_0$, where t_0 does not depend on $\overline{\alpha}, \varepsilon$ and t. By (2.13) the integral on the right-hand side of (2.12) can be written as

$$\int_0^t \ell(x, \overline{\alpha}(s)) e^{-\lambda s}\, ds + o(t) \qquad \text{as } t \to 0,$$

where $o(t)$ indicates a function $g(t)$ such that $\lim_{t \to 0+} |g(t)|/t = 0$, and in this case $|g(t)| \leq t\omega_\ell(M_x t)$. Then (2.11) with $z = \overline{y}_x(t)$ and (2.12) give

$$(2.15) \quad \phi(x) - \phi(\overline{y}_x(t)) - \int_0^t \ell(x, \overline{\alpha}(s)) e^{-\lambda s}\, ds + v(y_x(t))(1 - e^{-\lambda t}) \geq -t\varepsilon + o(t).$$

Moreover,

$$(2.16) \qquad \begin{aligned} \phi(x) - \phi(\overline{y}_x(t)) &= -\int_0^t \frac{d}{ds} \phi(\overline{y}_x(s))\, ds \\ &= -\int_0^t D\phi(\overline{y}_x(s)) \cdot f(\overline{y}_x(s), \overline{\alpha}(s))\, ds \\ &= -\int_0^t D\phi(x) \cdot f(x, \overline{\alpha}(s))\, ds + o(t), \end{aligned}$$

where we used (1.3), (2.14) and $\phi \in C^1$ in the last equality to estimate the difference between $D\phi \cdot f$ computed at $\overline{y}_x(s)$ and at x, respectively. Plugging (2.16) into (2.15) and adding $\pm \int_0^t \ell(x, \overline{\alpha}(s))\, ds$ we get

$$(2.17) \quad \int_0^t \{-D\phi(x) \cdot f(x, \overline{\alpha}(s)) - \ell(x, \overline{\alpha}(s))\}\, ds$$
$$+ \int_0^t \ell(x, \overline{\alpha}(s))(1 - e^{-\lambda s})\, ds + v(y_x(t))(1 - e^{-\lambda t}) \geq -t\varepsilon + o(t).$$

The term in brackets in the first integral is estimated from above by

$$\sup_{a \in A} \{ -D\phi(x) \cdot f(x, a) - \ell(x, a) \}$$

and the second integral is $o(t)$ because ℓ is bounded by (A$_4$), so we can divide (2.17) by $t > 0$ and pass to the limit to get

$$\sup_{a \in A}\{ -D\phi(x) \cdot f(x,a) - \ell(x,a) \} + \lambda v(x) \geq -\varepsilon,$$

where we have also used the continuity of v at x and of $\bar{y}_x$ at 0. Since ε is arbitrary, the proof that v is a supersolution is complete. ◀

REMARK 2.9. The proof of Proposition 2.8 shows that any function satisfying the DPP (2.5) with $\mathcal{A}$ replaced by any $\mathcal{B} \subseteq \mathcal{A}$, $\mathcal{B}$ containing the constant functions, is a viscosity solution of (2.10). Then, for instance, $v^\sharp$ satisfies (2.10) by Remark 2.7. ◁

COROLLARY 2.10. *Let us assume* (A$_0$)–(A$_4$) *and the existence of* $r > 0$ *such that* $\bar{B}(0,r) \subseteq \overline{\text{co}} f(x, A)$. *Then the value function v is Lipschitz continuous in* $\mathbb{R}^N$.

PROOF. Since v is a bounded solution of (2.10) we can apply Proposition II.4.1. Indeed, under the present assumptions we have

$$H(x,p) \geq \sup_{a \in A}\{ -f(x,a) \cdot p \} - M = \sup_{q \in \overline{\text{co}} f(x,a)}\{ -q \cdot p \} - M$$
$$\geq \sup_{|q| \leq r}\{ -q \cdot p \} - M = r|p| - M . \quad \blacktriangleleft$$

We know from the section on differential inequalities in the viscosity sense in Chapter II that the value function v satisfies $v(x) \geq u(x)$, at any x, for all subsolutions u of (2.10). Therefore, by Proposition 2.8, v is the maximal subsolution of (2.10), so it is characterized by the Hamilton-Jacobi-Bellman equation. Since v is also a supersolution of (2.10), we are now going to characterize v as the unique solution of (2.10). We do this via a comparison principle which also implies that v is the minimal supersolution and gives a completely different proof of the "maximal subsolution" characterization previously recalled. We premise a technical lemma on the regularity of the Hamiltonian H defined by (2.9).

LEMMA 2.11. *Assume* (A$_0$)–(A$_4$). *Then H is continuous and it satisfies*

(2.18) $H(y, \mu(x-y) - \tau y) - H(x, \mu(x-y) + \gamma x)$
$$\leq \mu L |x-y|^2 + \tau K(1+|y|^2) + \gamma K(1+|x|^2) + \omega_\ell(|x-y|)$$

for all $\mu, \tau, \gamma > 0$, $x, y \in \mathbb{R}^N$, *where* $K := L + \sup\{ |f(0,a)| : a \in A \}$.

Moreover, if (1.1) *holds, then H satisfies*

(2.19) $\qquad |H(x,p) - H(y,p)| \leq |p| L |x-y| + \omega_\ell(|x-y|)$

for all $x, y, p \in \mathbb{R}^N$. *If, in addition,*

(2.20) $\qquad \exists M' : \quad |f(x,a)| \leq M' \quad$ *for all* $x \in \mathbb{R}^N$, $a \in A$,

then

(2.21) $\qquad |H(x,p) - H(x,q)| \leq M'|p-q|$.

2. THE INFINITE HORIZON PROBLEM

PROOF. Observe first that (A$_3$) implies

(2.22) $$f(z,a) \cdot z \leq L|z|^2 + \sup_{a \in A}|f(0,a)||z| \leq K(1+|z|^2)$$

for all $z \in \mathbb{R}^N$, $a \in A$. Now fix $\varepsilon > 0$ and $\bar{a}$ such that

$$H(y, \mu(x-y) - \tau y) \leq -f(y, \bar{a}) \cdot (\mu(x-y) - \tau y) - \ell(y, \bar{a}) + \varepsilon.$$

Then the left-hand side of (2.18) is majorized by

$$\mu(f(x,\bar{a}) - f(y,\bar{a})) \cdot (x-y) + \tau f(y,\bar{a}) \cdot y + \gamma f(x,\bar{a}) \cdot x - \ell(y,\bar{a}) + \ell(x,\bar{a}) + \varepsilon,$$

which can be estimated by the right-hand side of (2.18) plus ε by (A$_3$), (A$_4$), and (2.22). Since ε is arbitrary the proof of (2.18) is complete. The proofs of the other statements are similar and are left to the reader. ◀

Lemma 2.11 implies that if f satisfies (A$_0$), (2.20), and (1.1), then the Hamiltonian H satisfies the assumptions of Theorem II.3.5, which states that a subsolution of (2.10) is nowhere larger than a supersolution, provided both belong to $BUC(\mathbb{R}^N)$. Since, by Proposition 2.8, the value function v is both a viscosity subsolution and a supersolution of (2.10), we conclude that v is below any supersolution in $BUC(\mathbb{R}^N)$ and above any subsolution in $BUC(\mathbb{R}^N)$. Hence v is the unique viscosity solution of (2.10) in $BUC(\mathbb{R}^N)$ and, more precisely, it is at the same time the maximal subsolution and the minimal supersolution, a property that we summarize by saying that v is the *complete solution* of (2.10) in $BUC(\mathbb{R}^N)$. All this holds under the more general assumptions (A$_0$)–(A$_4$) as well, as the next theorem states.

THEOREM 2.12. *Assume $H : \mathbb{R}^{2N} \to \mathbb{R}$ is continuous and satisfies (2.18). If u_1, $u_2 \in BUC(\mathbb{R}^N)$ are, respectively, a viscosity sub- and supersolution of*

(2.23) $$u + H(x, Du) = 0 \quad \text{in } \mathbb{R}^N,$$

then $u_1 \leq u_2$ in $\mathbb{R}^N$. In particular, if H is given by (2.9) and (A$_0$)–(A$_4$) hold, then the value function v is the complete solution of (2.10) in $BUC(\mathbb{R}^N)$ and thus it is also the unique viscosity solution in the same class.

PROOF. The second part of the theorem follows from Proposition 2.8 and Lemma 2.11: it is enough to divide (2.10) by $\lambda > 0$ and observe that H/λ satisfies the structural assumption (2.18) (with the constants and the modulus ω_ℓ divided by λ, which does not affect the following proof).

The proof of the comparison assertion is a variant of the proofs of Theorems II.3.1 and II.3.5; we consider $\Phi : \mathbb{R}^{2N} \to \mathbb{R}$,

$$\Phi(x,y) := u_1(x) - u_2(y) - \frac{|x-y|^2}{2\varepsilon} - \beta(\langle x \rangle^m + \langle y \rangle^m),$$

where $\langle x \rangle := (1 + |x|^2)^{1/2}$, and ε, β, m are positive parameters to be chosen conveniently. Note that for $\beta = 0$ we recover the auxiliary function Φ of Theorem II.3.1.

Let us assume by contradiction that there is $\delta > 0$ and $\tilde{x}$ such that $u_1(\tilde{x}) - u_2(\tilde{x}) = \delta$. We choose $\beta > 0$ such that $2\beta\langle\tilde{x}\rangle \leq \delta/2$, so that, for all $0 < m \leq 1$,

(2.24) $$\frac{\delta}{2} < \delta - 2\beta\langle\tilde{x}\rangle^m = \Phi(\tilde{x},\tilde{x}) \leq \sup\Phi(x,y) .$$

Since Φ is continuous and $\lim_{|x|+|y|\to\infty}\Phi(x,y) = -\infty$, there exist $\overline{x},\overline{y}$ such that

(2.25) $$\Phi(\overline{x},\overline{y}) = \sup\Phi(x,y) .$$

From the inequality $\Phi(\overline{x},\overline{x}) + \Phi(\overline{y},\overline{y}) \leq 2\Phi(\overline{x},\overline{y})$ we easily get

(2.26) $$\frac{|\overline{x}-\overline{y}|^2}{\varepsilon} \leq u_1(\overline{x}) - u_1(\overline{y}) + u_2(\overline{x}) - u_2(\overline{y}) .$$

Then the boundedness of u_1 and u_2 implies

(2.27) $$|\overline{x}-\overline{y}| \leq c\sqrt{\varepsilon}$$

for a suitable constant c. By plugging (2.27) into (2.26) and using the uniform continuity of u_1 and u_2 we get

(2.28) $$\frac{|\overline{x}-\overline{y}|^2}{\varepsilon} \leq \omega(\sqrt{\varepsilon}),$$

for some modulus ω.

Next define the C^1 "test functions"

$$\varphi(x) := u_2(\overline{y}) + \frac{|x-\overline{y}|^2}{2\varepsilon} + \beta(\langle x\rangle^m + \langle\overline{y}\rangle^m),$$

$$\psi(y) := u_1(\overline{x}) - \frac{|\overline{x}-y|^2}{2\varepsilon} - \beta(\langle\overline{x}\rangle^m + \langle y\rangle^m),$$

and observe that, by definition of $\overline{x},\overline{y}$, $u_1 - \varphi$ attains its maximum at $\overline{x}$ and $u_2 - \psi$ attains its minimum at $\overline{y}$. It is easy to compute

$$D\varphi(\overline{x}) = \frac{\overline{x}-\overline{y}}{\varepsilon} + \gamma\overline{x}, \quad \gamma = \beta m\langle\overline{x}\rangle^{m-2},$$

$$D\psi(\overline{y}) = \frac{\overline{x}-\overline{y}}{\varepsilon} - \tau\overline{y}, \quad \tau = \beta m\langle\overline{y}\rangle^{m-2} .$$

By definition of viscosity sub- and supersolution,

(2.29) $$u_1(\overline{x}) + H(\overline{x}, D\varphi(\overline{x})) \leq 0 \leq u_2(\overline{y}) + H(\overline{y}, D\psi(\overline{y})),$$

and by using (2.18) with $\mu = 1/\varepsilon$, τ,γ as above, and the definition of $\langle x\rangle$, we get

$$u_1(\overline{x}) - u_2(\overline{y}) \leq \frac{L}{\varepsilon}|\overline{x}-\overline{y}|^2 + \tau K\langle\overline{y}\rangle^2 + \gamma K\langle\overline{x}\rangle^2 + \omega_\ell(|\overline{x}-\overline{y}|)$$

$$\leq L\omega(\sqrt{\varepsilon}) + \beta mK(\langle\overline{x}\rangle^m + \langle\overline{y}\rangle^m) + \omega_\ell(c\sqrt{\varepsilon}),$$

where in the last inequality we have used (2.27) and (2.28). Therefore, by choosing $0 < m \leq 1/K$, we obtain

$$\Phi(\overline{x},\overline{y}) \leq u_1(\overline{x}) - u_2(\overline{y}) - \beta(\langle\overline{x}\rangle^m + \langle\overline{y}\rangle^m) \leq L\omega(\sqrt{\varepsilon}) + \omega_\ell(c\sqrt{\varepsilon}),$$

and the right-hand side of the last inequality can be made smaller than $\delta/2$ for ε small enough, a contradiction to (2.24) and (2.25). ◂

REMARK 2.13. The uniform continuity assumption on u_1 and u_2 in Theorem 2.12 can be easily replaced by mere continuity, so that v is the complete solution of (2.10) in $BC(\mathbb{R}^N)$ (bounded and continuous functions $\mathbb{R}^N \to \mathbb{R}$) as well. To prove this observe that, once β is chosen, we have the estimate

$$\langle \overline{x} \rangle^m + \langle \overline{y} \rangle^m < \frac{\sup u_1 - \inf u_2}{\beta} =: c_0,$$

for all $\varepsilon > 0$, $m \in \,]0,1]$, because the violation of this inequality leads to $\Phi(\overline{x}, \overline{y}) \leq 0$, a contradiction to (2.24) and (2.25). Then, for all $\varepsilon > 0$, $\overline{x}, \overline{y}$ are in the compact set $\overline{B}(0, c_0^{1/m})$, where $m = \min\{1, 1/K\}$, and so the uniform continuity of u_1, u_2 in such a set is enough to deduce (2.28) from (2.26) and (2.27), which is the only step where we used the uniform continuity of u_1, u_2. ◁

REMARK 2.14. Theorem 2.12 and Remark 2.13 are easily extended to a comparison result for a Dirichlet boundary value problem that is useful in Section 3 and Chapter IV. If H is continuous and satisfies (2.18), $\Omega \subseteq \mathbb{R}^N$ is an open set, $u_1, u_2 \in BC(\overline{\Omega})$ are, respectively, a viscosity sub- and supersolution of $u + H(x, Du)$ in Ω and $u_1 \leq u_2$ on $\partial \Omega$, then $u_1 \leq u_2$ in $\overline{\Omega}$. The only difference in the proof is that the maximum point of Φ, $(\overline{x}, \overline{y})$, may not belong to $\Omega \times \Omega$. If $\overline{x} \in \partial \Omega$, then the boundary inequality holds

$$\Phi(\overline{x}, \overline{y}) \leq u_1(\overline{x}) - u_2(\overline{x}) + u_2(\overline{x}) - u_2(\overline{y}) \leq \overline{\omega}(c\sqrt{\varepsilon}),$$

where $\overline{\omega}$ is the modulus of continuity of u_2 in $\overline{B}(0, c_0^{1/m})$ (see the proof of Remark 2.13), and we have used (2.27). By choosing ε small enough we get a contradiction to (2.24) and (2.25). If $\overline{x} \in \Omega$ and $\overline{y} \in \partial \Omega$ we add and subtract $u_1(\overline{y})$ and use the continuity of u_1 to conclude in the same way (see also the proof of Theorem II.3.1). ◁

REMARK 2.15. An immediate consequence of Theorem 2.12 and Remarks 2.7 and 2.9 is the equality $v = v^\sharp$, that is, measurable control functions and piecewise constant control functions generate the same value function. In §VII.4.1 we prove in a rather different way that also Lipschitz continuous controls generate the same value function. ◁

We end this section with some comments on the optimality of the assumptions of the comparison Theorem 2.12 for Hamiltonians of the form (2.9). First note that we cannot drop the assumption that u_1 and u_2 are both bounded because the linear 1-dimensional equation $u - x u_x = 0$ in $\mathbb{R}$ has the classical, uniformly continuous solutions $u_1 \equiv 0$ and $u_2 = x$. It has also the viscosity solutions $u_3(x) = |x|$ and $u_4(x) = -|x|$ which are, respectively, bounded below and bounded above and remain solutions in arbitrary dimension for

$$u - x \cdot Du = 0 \quad \text{in } \mathbb{R}^N.$$

We remark that there are comparison and uniqueness results in classes of unbounded solutions, typically under suitable restrictions on the rate of growth at infinity (see, e.g., Remark II.3.6).

The assumption of uniform continuity of ℓ in (A$_4$) can be relaxed to mere continuity, see Exercise 2.7.

Now let us focus on (A$_3$) by looking at dimension $N = 1$, A a singleton, so that the Hamilton-Jacobi-Bellman equation is linear, and $f(x)$ a power of x. The "decreasing case," $f(x) := -\operatorname{sgn} x\,|x|^b$, $x \in \mathbb{R}$, satisfies (A$_3$) for any $b > 0$ (with $L = 0$), so Theorem 2.12 applies to

$$u + \operatorname{sgn} x\,|x|^b u_x = 0 \qquad \text{in } \mathbb{R}$$

and $u \equiv 0$ is the unique bounded solution. The "increasing case", $f(x) := \operatorname{sgn} x\,|x|^b$, $b > 0$, satisfies (A$_3$) if and only if $b = 1$, and then $u \equiv 0$ is the only bounded solution of

$$u - \operatorname{sgn} x\,|x|^b u_x = 0 \qquad \text{in } \mathbb{R}$$

if $b = 1$. Here is a second classical solution in $BUC(\mathbb{R})$ when $b > 1$:

$$u(x) := \exp\left(\frac{|x|^{1-b}}{1-b}\right) \qquad \text{if } x \neq 0,\ u(0) = 0,$$

as it is easy to check since $u_x(x) = \operatorname{sgn} x |x|^{-b} u(x)$ if $x \neq 0$ and $u_x(0) = 0$. Note that for vector fields C^1 in the x variable the inequality in (A$_3$) always holds if x and y are restricted to a compact set $\mathcal{K}$, for a suitable constant $L_{\mathcal{K}}$. Thus, for smooth f, (A$_3$) is essentially a growth condition at infinity, saying, roughly speaking, that there is no restriction if the fields are monotone nonincreasing, whereas the growth has to be at most linear in the opposite case. The preceding example says that this assumption is essentially optimal for the comparison Theorem 2.12, at least for fields having a polynomial growth at infinity.

2.2. Some simple applications: verification theorems, relaxation, stability

In this subsection we draw some simple consequences on the infinite horizon optimal problem from the results on the associated Hamilton-Jacobi-Bellman equation of the previous section.

We begin with some necessary and sufficient conditions for optimality of a *control law* or *presynthesis* which is defined as follows.

DEFINITION 2.16. A *control law* or *presynthesis* on a set $\Omega \subseteq \mathbb{R}^N$ is a map $\mathrm{A} : \Omega \to \mathcal{A}$, that is, it associates with each point $x \in \Omega$ a control function $\mathrm{A}(x) =: \alpha_x$. It is *optimal* (on Ω) if the cost associated with it, that is, $J_\mathrm{A}(x) := J(x, \alpha_x)$, satisfies

$$J_\mathrm{A}(x) = \min_{\alpha \in \mathcal{A}} J(x, \alpha) = v(x) \qquad \text{for all } x \in \Omega\,. \quad \triangleleft$$

The most important examples of control laws are generated by feedback maps $\Phi : \Omega \to A$, provided the feedback is admissible in the following sense.

2. THE INFINITE HORIZON PROBLEM

DEFINITION 2.17. A feedback map on a set $\Omega \subseteq \mathbb{R}^N$, $\Phi : \Omega \to A$, is *admissible* if for all $x \in \Omega$ there exists a unique solution $y_x(\,\cdot\,, \Phi)$ on $[0, +\infty[$ of

$$\begin{cases} y' = f(y, \Phi(y)) \\ y(0) = x \end{cases}$$

such that $t \mapsto \Phi(y_x(t, \Phi))$ is measurable and $y_x(t, \Phi) \in \Omega$ for all $t \geq 0$. ◁

It is natural to associate the following control law with an admissible feedback map

$$\alpha_x(\,\cdot\,) := \Phi(y_x(\,\cdot\,, \Phi)) \in \mathcal{A},$$

and call it a *feedback law*; α_x in this case is called a *closed-loop* control. A feedback law has the property of being *memoryless*, that is, for any $z = y_x(t, \alpha_x)$, $\alpha_z(\,\cdot\,) = \alpha_x(\,\cdot\, + t)$. A memoryless presynthesis defined on all $\mathbb{R}^N$ is called a *synthesis*.

Before stating the conditions of optimality of a control law we recall the Hamilton-Jacobi-Bellman equation

(2.30) $$\lambda u + \sup_{a \in A}\{-f(x,a) \cdot Du - \ell(x,a)\} = 0.$$

In the next result by subsolution we mean a bounded and uniformly continuous viscosity subsolution, and we use the same convention for supersolutions and solutions. By Remark 2.13 "uniformly continuous" can be replaced by "continuous."

COROLLARY 2.18. *Assume* (A_0)–(A_4) *and let* $\mathbb{A}$ *be a control law on a set* Ω. *Then*

(i) (Necessary condition for optimality) *If* $\mathbb{A}$ *is optimal and* Ω *is open, then the associated cost* $J_\mathbb{A}$ *is a viscosity solution of* (2.30) *in* Ω.

(ii) (Sufficient condition for optimality) *If there exists a subsolution* u *of* (2.30) *on* $\mathbb{R}^N$, *such that* $J_\mathbb{A} \leq u$ *on* Ω, *then* $\mathbb{A}$ *is optimal.*

(iii) (Sufficient conditions for non-optimality) *If there exists a supersolution* w *of* (2.30) *on* $\mathbb{R}^N$, *such that* $J_\mathbb{A} > w$ *at some point of* Ω, *then* $\mathbb{A}$ *is not optimal.*

(iv) *In particular, for* $\Omega = \mathbb{R}^N$, *the optimality of* $\mathbb{A}$ *is equivalent to each of the following facts:*

$J_\mathbb{A}$ *is a subsolution of* (2.30) *in* $\mathbb{R}^N$;

$J_\mathbb{A}$ *is a solution of* (2.30) *in* $\mathbb{R}^N$;

$J_\mathbb{A} \leq w$ *for any supersolution* w *of* (2.30) *in* $\mathbb{R}^N$.

PROOF. All the statements follow from the equivalence of the optimality of $\mathbb{A}$ and the equality $J_\mathbb{A} = v$, where v is the value function. Thus (i) follows from Proposition 2.8 and the fact that being a viscosity solution is a local property; (ii) follows from Theorem 2.12 because

$$v \leq J_\mathbb{A} \leq u \leq v \quad \text{in } \Omega;$$

(iii) follows from the same theorem because

$$v \leq w < J_\mathbb{A};$$

and (iv) is also a trivial consequence of Theorem 2.12. ◂

Corollary 2.18 and the classical verification technique recalled in Chapter I suggest the name *(generalized) verification function* for any viscosity subsolution $u \in BUC(\mathbb{R}^N)$: to verify that a control law $\mathbb{A}$ is optimal in $\mathbb{R}^N$ it is enough to find such a function satisfying $u \geq J_\mathbb{A}$. By analogy we call a *falsification function* a viscosity supersolution $w \in BUC(\mathbb{R}^N)$, because the existence of such a function satisfying $w < J_\mathbb{A}$ implies the non-optimality of $\mathbb{A}$.

It is easy to give a similar verification theorem valid for open loop controls $\alpha \in \mathcal{A}$: we say that $\alpha_0 \in \mathcal{A}$ is optimal for $x_0 \in \mathbb{R}^N$ if $J(x_0, \alpha_0) = \min_{\alpha \in \mathcal{A}} J(x_0, \alpha) = v(x_0)$.

COROLLARY 2.19. *Assume* (A_0)–(A_4).

(i) *A control α_0 is optimal for x_0 if and only if there exists a verification function u such that $J(x_0, \alpha_0) \leq u(x_0)$;*

(ii) *α_0 is not optimal for x_0 if and only if there exists a falsification function w such that $J(x_0, \alpha_0) > w(x_0)$.*

PROOF. The very easy proof is left to the reader. ◀

★ ★ ★

Next we want to show that the value functions of two different "relaxed" problems coincide with v. These results are consequences of the following simple observation. Consider a control system $\widetilde{f} : \mathbb{R}^N \times \widetilde{A} \to \mathbb{R}^N$ and running cost $\widetilde{\ell}$ satisfying the structural conditions (A_0)–(A_4) (maybe with different constants), and the corresponding Hamiltonian

$$\widetilde{H}(x,p) := \sup_{a \in \widetilde{A}} \{ -\widetilde{f}(x,a) \cdot p - \widetilde{\ell}(x,a) \} .$$

If $\widetilde{H}(x,p) = H(x,p)$ for all $x, p \in \mathbb{R}^N$, then the corresponding value function $\widetilde{v}$ satisfies $\widetilde{v}(x) = v(x)$ for all $x \in \mathbb{R}^N$. In fact v and $\widetilde{v}$ satisfy the same Hamilton-Jacobi-Bellman equation (2.10) and then they must coincide by Theorem 2.12.

The first relaxation result concerns the *convexified problem*. Let us begin with a simple case. Assume the system is *affine*, that is,

$$f(x,a) = f_1(x) + f_2(x)a \qquad \text{for all } x \in \mathbb{R}^N, a \in A,$$

where $A \subseteq \mathbb{R}^M$ and $f_2(x)$ is an $N \times M$ matrix, and the running cost is concave in the control variable, that is, $-\ell$ can be extended to all of $\mathbb{R}^N \times \mathbb{R}^M$ as a convex function of a for each fixed x.

We consider the same system and cost, but for the larger control set $\widetilde{A} = \operatorname{co} A$ (i.e., the convex hull of A), and we call $\widetilde{v}$ the corresponding value function. We claim that $v = \widetilde{v}$, and to prove the claim we have only to check that $H \geq \widetilde{H}$ because the inequality $H \leq \widetilde{H}$ is trivial. Fix $x, p \in \mathbb{R}^N$, $\varepsilon > 0$, and $b \in \widetilde{A}$ such that

$$\widetilde{H}(x,p) - \varepsilon \leq -f_1(x) \cdot p - f_2(x)b \cdot p - \ell(x,b) .$$

We write $b = \sum_{i=1}^{n} \lambda_i a_i$ with $a_i \in A$, $\lambda_i \geq 0$, $\sum_{i=1}^{n} \lambda_i = 1$, and obtain, by the concavity of ℓ in the control variable,

$$\widetilde{H}(x,p) - \varepsilon \leq -f_1(x) \cdot p - \sum_{i=1}^{n} \lambda_i f_2(x) a_i \cdot p - \sum_{i=1}^{n} \lambda_i \ell(x, a_i)$$

$$\leq \sum_{i=1}^{n} \lambda_i \sup_{a \in A} \{ -f_1(x) \cdot p - f_2(x) a \cdot p - \ell(x, a) \} = H(x, p),$$

which completes the proof of the claim after letting $\varepsilon \to 0^+$.

The meaning of the result is clear: the value function does not change if we allow the controls to take values just on the vertices of a polyhedron rather than on the full polyhedron. This is a weak (nonlinear) version of the classical bang-bang principle.

For general system and cost, the convexified problem is the following. Let

$$\Lambda := \left\{ \lambda \in \mathbb{R}^{N+2} : \lambda_i \geq 0 \, \forall i, \sum_{i=1}^{N+2} \lambda_i = 1 \right\}, \quad A^c = \Lambda \times A^{N+2},$$

and define f^c and ℓ^c on $\mathbb{R}^N \times A^c$ by

$$f^c(x, \lambda, a_1, \ldots, a_{N+2}) = \sum_{i=1}^{N+2} \lambda_i f(x, a_i), \quad \ell^c(x, \lambda, a_1, \ldots, a_{N+2}) = \sum_{i=1}^{N+2} \lambda_i \ell(x, a_i).$$

It is easy to check that f^c, ℓ^c satisfy (A$_0$)–(A$_4$). Their crucial property is that

(2.31) $$\operatorname{co}(f(x, A) \times \ell(x, A)) = f^c(x, A^c) \times \ell^c(x, A^c),$$

for all $x \in \mathbb{R}^N$. In fact, by a classical theorem of Carathéodory (see [Ce83]), if $X \subseteq \mathbb{R}^n$, then every element of the convex hull $\operatorname{co} X$ can be written as a convex combination of at most $n+1$ elements of X. Using (2.31) it is easy to show, as previously, that the Hamiltonian corresponding to f^c, ℓ^c, A^c coincides with H, so the value functions of the original and of the convexified problem are the same. We recall that convex problems are interesting because they have some special properties; for example, if A is compact, the set of their trajectories is closed with respect to the uniform convergence over a bounded time-interval.

★ ★ ★

The second relaxation result concerns the enlargement of the set of measurable control functions $\mathcal{A}$ to the class of *relaxed* or *chattering* controls, that is,

$$\mathcal{A}^r := \{ \text{measurable functions } [0, +\infty[\to A^r \},$$

where

$$A^r := \{ \text{Radon probability measures on } A \}$$

and we are assuming that A is a compact subset of $\mathbb{R}^M$.[2] To give sense to the system (S) and the cost J for controls in $\mathcal{A}^r$ we extend f and ℓ to $\mathbb{R}^N \times \mathcal{A}^r$ by setting

$$f^r(x,m) := \int_A f(x,a)\,dm, \qquad \forall\, m \in \mathcal{A}^r,$$

$$\ell^r(x,m) := \int_A \ell(x,a)\,dm, \qquad \forall\, m \in \mathcal{A}^r,$$

and observing that $f^r(x,\delta_a) = f(x,a)$, $\ell^r(x,\delta_a) = \ell(x,a)$, where δ_a is the Dirac measure concentrated at a. If we identify $a \in A$ with $\delta_a \in \mathcal{A}^r$ we can view f^r and ℓ^r as extensions of f and ℓ, respectively.

We also endow $\mathcal{A}^r$ with a suitable topology by regarding it as a subset of $C(A)^*$, the dual space of the space $C(A)$ of all continuous functions $A \to \mathbb{R}$, with the weak star topology. It can be proved that $\mathcal{A}^r$ with this topology is metrizable and compact: see [Wa72].

We recall that a sequence m_n converges weak star to m if

$$\lim_{n\to\infty} \int_A \varphi(a)\,dm_n = \int_A \varphi(a)\,dm \qquad \text{for all } \varphi \in C(A)\,.$$

LEMMA 2.20. *If f, ℓ, A satisfy (A_0)–(A_4) and $A \subseteq \mathbb{R}^M$ is compact, then $f^r, \ell^r, \mathcal{A}^r$ satisfy (A_0)–(A_4).*

PROOF. It is enough to check the continuity of f^r and ℓ^r on sequences $(x_n, m_n) \to (x, m)$ in $\mathbb{R}^N \times \mathcal{A}^r$. By definition of weak star convergence

$$\int_A f(x,a)\,dm_n \longrightarrow \int_A f(x,a)\,dm = f^r(x,m),$$

whereas, for R such that $|x_n| \leq R$ for all n,

$$\left| f^r(x_n, m_n) - \int_A f(x,a)\,dm_n \right| = \left| \int_A [f(x_n,a) - f(x,a)]\,dm_n \right|$$
$$\leq \int_A \omega_f(|x_n - x|, R)\,dm_n$$
$$= \omega_f(|x_n - x|, R) \longrightarrow 0,$$

because $\int_A dm_n = 1$, thus $f^r(x_n, m_n) \to f^r(x, m)$. The same proof works for ℓ^r.

The other proofs are straightforward; for instance, (A_3) holds because

$$(f^r(x,m) - f^r(y,m)) \cdot (x-y) = \int_A (f(x,a) - f(y,a)) \cdot (x-y)\,dm$$
$$\leq \int_A L|x-y|^2\,dm = L|x-y|^2\,. \quad \blacktriangleleft$$

---

[2] A probability measure on A is a Radon measure if it is defined on all Borel sets of A (i.e., on the smallest σ-algebra of subsets of A that contains the open sets).

2. THE INFINITE HORIZON PROBLEM

We keep the notation $y_x(t)$ or $y_x(t,\mu)$ for the trajectory of the relaxed system, that is, (S) with f replaced by f^r and α by $\mu \in \mathcal{A}^r$, and note that, by definition, it is a solution of

$$y(t) = x + \int_0^t \int_A f(y(s), a)\, d\mu(s)\, ds;$$

similarly the cost and the value function of the relaxed problem are

$$J^r(x,\mu) := \int_0^\infty \ell^r(y_x(t), \mu(t))e^{-\lambda t}\, dt = \int_0^\infty \int_A \ell(y_x(t), a)\, d\mu(t) e^{-\lambda t}\, dt,$$
$$v^r(x) := \inf_{\mu \in \mathcal{A}^r} J^r(x, \mu).$$

Note that the relaxed trajectory and cost associated with $\mu(t) := \delta_{\alpha(t)}$ for all t, where $\alpha \in \mathcal{A}$, coincide with $y_x(\cdot, \alpha)$ and $J(x, \alpha)$, respectively.

COROLLARY 2.21. *If f, ℓ, A satisfy (A_0)-(A_4) and $A \subseteq \mathbb{R}^M$ is compact, then $v = v^r$; that is, the value functions of the original and the relaxed problem coincide.*

PROOF. Let

$$H^r(x, p) := \sup_{m \in A^r} \left\{ -f^r(x, m) \cdot p - \ell^r(x, m) \right\}.$$

By Lemma 2.20 and Propositions 2.1 and 2.8, $v^r \in BUC(\mathbb{R}^N)$ and it is a viscosity solution of $\lambda v + H^r(x, Dv) = 0$ in $\mathbb{R}^N$.

We prove that $H = H^r$ so that the conclusion follows immediately from Theorem 2.12.

For all $x, p \in \mathbb{R}^N$ we have

$$H^r(x,p) \geq \sup_{a \in A}\{-f^r(x, \delta_a) \cdot p - \ell^r(x, \delta_a)\} = H(x, p)$$

by the definitions of f^r, ℓ^r and H, see (2.9). On the other hand, for all $x, p \in \mathbb{R}^N$, $m \in A^r$,

$$-f^r(x,m) \cdot p - \ell^r(x,m) = \int_A (-f(x,a) \cdot p - \ell(x,a))\, dm$$
$$\leq \int_A \sup_{b \in A}\{-f(x,b) \cdot p - \ell(x,b)\}\, dm = H(x,p),$$

because $\int_A dm = 1$. Now we take $\sup_{m \in A^r}$ of both sides and get $H^r(x,p) \leq H(x,p)$. ◀

We point out that relaxed controls are interesting because the relaxed problem usually has an optimal control whereas the original one does not in general. In Chapter VI we construct an optimal relaxed control for the problem discussed here as the limit of piecewise constant controls computable via Dynamic Programming methods.

Finally we remark that considering relaxed controls is another way of convexifying the problem. In fact, it is not hard to prove that

(2.32) $$\overline{co}(f(x,A) \times \ell(x,A)) = f^{\mathrm{r}}(x, A^{\mathrm{r}}) \times \ell^{\mathrm{r}}(x, A^{\mathrm{r}}),$$

see Exercise 2.10.

$$\star \star \star$$

Next we prove a simple stability property of the value function v which is an immediate consequence of the stability of viscosity solutions with respect to the uniform convergence of the Hamiltonians.

We consider a family of systems

$$\begin{cases} y' = f_h(y, \alpha), & \alpha(t) \in A_h,\ t > 0, \\ y(0) = x \end{cases}$$

and costs

$$J_h(x, \alpha) = \int_0^\infty e^{-\lambda_h t} \ell_h(y_x(t), \alpha(t))\, dt,$$

where the parameter h varies in $[0, 1]$. We want to show that if the data $f_h, A_h, \lambda_h, \ell_h$ are continuous in a suitable sense for $h \to 0^+$, then the value function $v_h(x) := \inf_{\alpha \in A_h} J_h(x, \alpha)$ is also continuous as $h \to 0^+$.

We assume that for all $h \in [0, 1]$ f_h satisfy (A$_0$)–(A$_3$) with the same compact $A \subseteq \mathbb{R}^M$ and the same constant L in (A$_3$), and that ℓ_h, λ_h satisfy (A$_4$) with the same constant M and modulus ω_ℓ. We also assume that $f_h \to f_0$, $\ell_h \to \ell_0$, as $h \to 0^+$, uniformly on $\mathcal{K} \times A$ for each compact $\mathcal{K} \subseteq \mathbb{R}^N$, and that $\lim_{h \to 0} \lambda_h = \lambda_0$.

To define the convergence of the control sets A_h to A_0 we need the notion of *Hausdorff distance* of two compact sets $S, U \subseteq \mathbb{R}^M$:

$$d_{\mathrm{H}}(S, U) := \max\{\max_{s \in S} \mathrm{dist}(s, U),\ \max_{u \in U} \mathrm{dist}(u, S)\},$$

where $\mathrm{dist}(s, U) = \min\{|s - u| : u \in U\}$. Note that

$$d_{\mathrm{H}}(S, U) \leq r \quad \text{if and only if} \quad S \subseteq \bar{B}(U, r) \quad \text{and} \quad U \subseteq \bar{B}(S, r),$$

where $\bar{B}(X, r) = \{x \in \mathbb{R}^M : \mathrm{dist}(x, X) \leq r\}$, so in particular

$$d_{\mathrm{H}}(S, U) = 0 \quad \text{if and only if} \quad S = U.$$

We assume that:

$$A_h \subseteq A \text{ are compact for all } h \in [0, 1],$$
$$\lim_{h \to 0} d_{\mathrm{H}}(A_h, A_0) = 0.$$

COROLLARY 2.22. *Under the preceding assumptions v_h converge to v_0 as $h \to 0^+$ uniformly on each compact set of $\mathbb{R}^N$.*

PROOF. Fix $\bar{h} > 0$ such that $\lambda_h \geq \lambda_0/2$ for all $h \in [0, \bar{h}]$. By Proposition 2.1 $v_h \in BUC(\mathbb{R}^N)$ for all h, and it is easy to see from the proof of Proposition 2.1 that for $h \in [0, \bar{h}]$ there is a uniform bound and a common modulus of continuity for all v_h, depending only on $M, \lambda_0, L, \omega_\ell$.

By the Ascoli-Arzelà theorem, for every compact $\mathcal{K} \subseteq \mathbb{R}^N$ there is $h_n \to 0$ such that v_{h_n} converges uniformly on $\mathcal{K}$. We construct a sequence converging on all $\mathbb{R}^N$ by the following standard "diagonal procedure": let $h_n^{(1)}$ be such that $v_{h_n^{(1)}}$ converge uniformly to $u^{(1)}$ on $\bar{B}(0,1)$, and define by induction an $h_n^{(m)}$ subsequence of $h_n^{(m-1)}$ such that $v_{h_n^{(m)}}$ converge uniformly to $u^{(m)}$ on $\bar{B}(0,m)$. By construction $u^{(m)} = u^{(m-1)}$ on $\bar{B}(0, m-1)$, thus $u(x) := u^{(m)}(x)$ if $|x| \leq m$ is the uniform limit of $v_{h_n^{(n)}}$ on every compactum of $\mathbb{R}^N$. We claim that u is a viscosity solution of

$$\lambda_0 u + H_0(x, Du) = 0 \quad \text{in } \mathbb{R}^N,$$

where $H_h(x,p) = \max_{a \in A_h}\{-f_h(x,a) \cdot p - \ell_h(x,a)\}$, $h \in [0,1]$. This follows from Proposition II.2.2, once we have checked that H_h converge uniformly to H_0 on compacts of $\mathbb{R}^{2N}$. For this purpose we pick $a_h \in A_h$ such that

$$H_h(x,p) = -f_h(x, a_h) \cdot p - \ell_h(x, a_h)$$

and $a_0 \in A_0$ such that $|a_h - a_0| = d_H(A_h, A_0)$. Then

$$H_h(x,p) - H_0(x,p) \leq -f_h(x, a_h) \cdot p - \ell_h(x, a_h) + f_0(x, a_0) \cdot p + \ell_0(x, a_0)$$
$$\leq |p|(|f_0(x, a_h) - f_0(x, a_0)| + |f_h(x, a_h) - f_0(x, a_h)|)$$
$$+ |\ell_0(x, a_h) - \ell_0(x, a_0)| + |\ell_h(x, a_h) - \ell_0(x, a_h)|.$$

The right-hand side of this inequality tends to 0 as h goes to 0, uniformly for $|x|, |p| \leq R$, because

$$|f_0(x, a_h) - f_0(x, a_0)| \leq \omega_0(|a_h - a_0|) = \omega_0(d_H(A_h, A_0)),$$

where ω_0 is the modulus of continuity of f_0 on $\bar{B}(0, R) \times A$, whereas $|f_h(x, a_h) - f_0(x, a_h)|$ goes to 0 by the uniform convergence of f_h to f_0, and the same arguments hold for the terms involving ℓ. A similar argument on $H_0 - H_h$ proves the desired convergence $H_h \to H_0$.

Now the uniqueness Theorem 2.12 implies that $u = v_0$, so we have the compactness of v_h and the uniqueness of the limit of subsequences, and this implies that $v_h \to v_0$ by the following standard argument. Assume by contradiction there exists $\mathcal{K}$ compact such that

$$\limsup_{h \to 0} \|v_h - v_0\| = \delta > 0,$$

where $\|\cdot\|$ is the sup-norm in $\mathcal{K}$. Then there exists $h_n \to 0$ such that

(2.33) $$\lim_n \|v_{h_n} - v_0\| = \delta.$$

But the previous arguments imply the existence of a subsequence of v_{h_n} converging to v_0 uniformly on each compactum of $\mathbb{R}^N$, in particular $\mathcal{K}$, which contradicts (2.33). ◀

Corollary 2.22 is essentially a perturbation result: if we are aware of the possibility of errors in the knowledge of our data f, A, λ, ℓ, but the errors are small and do not affect the structure of the problem or the regularity of the data, then the value function is affected only by a small error. In Chapter V we also allow more general perturbations, and in Chapter VI we give estimates of the errors.

Corollary 2.22 can also be used to give approximation results. In the proof of the next corollary we smooth the data and "fatten" the control set A and get an approximation by more regular value functions.

COROLLARY 2.23. *Assume f, A, ℓ, λ satisfy (A_0)–(A_4), and that for all $x \in \mathbb{R}^N$ there exists $a_x \in A$ such that $f(x, a_x) = 0$. Then there exist sequences f_n, ℓ_n, λ_n, and a control set $\widetilde{A}$, such that the corresponding value functions v_n satisfy:*

(i) v_n *is locally Lipschitz continuous;*

(ii) v_n *is semiconcave (provided that f is bounded and, for L as in (A_3), $\lambda \geq 2L$);*

(iii) $v_n \to v$ *as $n \to +\infty$ locally uniformly in $\mathbb{R}^N$.*

PROOF. To get the Lipschitz approximation we set $\lambda_n = \lambda$, $\ell_n = \ell$ for all n, $\widetilde{A} := A \times \{b \in \mathbb{R}^N : |b| \leq 1\}$, and

$$f_n(x, a, b) = f(x, a) + \frac{1}{n} b.$$

By the assumption on f we have $f_n(x, \widetilde{A}) \supseteq \overline{B}(0, 1/n)$, therefore v_n is locally lipschitzean by Corollary 2.10. The semiconcave approximation is obtained by setting $\lambda_n = \lambda + 1/n$ and by smoothing f and ℓ with respect to x by means of the usual mollifiers, that is, nonnegative $\varrho_n \in C^\infty(\mathbb{R}^N)$ such that $\int_{\mathbb{R}^N} \varrho_n(x)\,dx = 1$ and $\mathrm{supp}\,\varrho_n = \overline{B}(0, 1/n)$. Then

$$f_n(x, a, b) := \int_{\mathbb{R}^N} \varrho_n(x - y) f(y, a)\,dy + \frac{1}{n} b,$$

$$\ell_n(x, a, b) := \int_{\mathbb{R}^N} \varrho_n(x - y) \ell(y, a)\,dy,$$

are $C^\infty(\mathbb{R}^N)$ for a and b fixed and it is easy to check that they satisfy (A_0)–(A_4) with the same constants and moduli. It is well known (and not hard to check) that f_n, ℓ_n converge uniformly to f, ℓ, respectively, on $\mathcal{K} \times \widetilde{A}$ for each compactum $\mathcal{K}$. The Hamiltonians

$$H_n(x, p) := \sup_{c \in A_n} \left\{ -f_n(x, c) \cdot p - \ell_n(x, c) \right\}$$

satisfy the assumptions of Theorem II.4.9 (with constants C depending on n), as it is easy to check (see also Remark II.4.10), thus v_n are semiconcave. The convergence of v_n to v follows from the uniform convergence on compacta of H_n to H, as in the previous corollary. ◂

2.3. Backward Dynamic Programming, sub- and superoptimality principles, bilateral solutions

In this subsection we discuss some further results about the Dynamic Programming optimality principle. First we derive from it a similar inequality concerning the backward trajectories of the system and prove that it implies that the value function v is also a viscosity supersolution of minus the Hamilton-Jacobi-Bellman equation (recall that in general it is not allowed to change the sign of a PDE in the viscosity sense, see Example II.1.6). Then we show that certain local versions of the DPP, named (local) sub- and superoptimality principles, are equivalent to the definition of viscosity solutions. This is the basic step to proving, in the next section, the equivalence of viscosity solutions with other concepts of weak solution of the Hamilton-Jacobi-Bellman equation.

Let us first define $\mathcal{A}^- = \{\alpha :]-\infty, 0] \to A, \text{ measurable}\}$ and recall that under the assumptions $(A_0), (A_1), (A_3)$ there exists for all negative times the *backward trajectory* of the system with initial point x, that is, the unique solution of

$$(S^-) \qquad \begin{cases} y'(s) = f(y(s), \alpha(s)) & s < 0, \\ y(0) = x, \end{cases}$$

or, more precisely, the unique solution of

$$y(s) = x - \int_s^0 f(y(\tau), \alpha(\tau))\,d\tau, \qquad s < 0.$$

The proof is the same as that for forward trajectories recalled in the appendix of this chapter, §5. We keep the notation $y_x(s, \alpha) = y_x(s)$ for the solution of (S^-).

DEFINITION 2.24. We say that x is an *optimal point*, and write $x \in \mathcal{O}$, if there exist $z \in \mathbb{R}^N$, $\alpha^* \in \mathcal{A}$ such that $v(z) = J(z, \alpha^*)$, and $t > 0$ such that $x = y_z(t, \alpha^*)$. ◁

PROPOSITION 2.25 (BACKWARD DYNAMIC PROGRAMMING PRINCIPLE). *Assume* $(A_0), (A_1), (A_3),$ *and* (A_4). *Then, for all* $x \in \mathbb{R}^N$, $t > 0$ *and* $\alpha \in \mathcal{A}^-$,

$$(2.34) \qquad v(x) \geq v(y_x(-t, \alpha))\,e^{\lambda t} - \int_0^t \ell(y_x(-s, \alpha), \alpha(-s))\,e^{\lambda s}\,ds;$$

moreover, for $x \in \mathcal{O}$ *and for all* $t > 0$ *small enough,*

$$(2.35) \qquad v(x) = \max_{\alpha \in \mathcal{A}^-} \left(v(y_x(-t, \alpha))\,e^{\lambda t} - \int_0^t \ell(y_x(-s, \alpha), \alpha(-s))\,e^{\lambda s}\,ds \right).$$

PROOF. Fix any x, t and $\alpha \in \mathcal{A}^-$, let $\alpha_1(s) := \alpha(s - t)$ for $0 \leq s \leq t$ and extend it arbitrarily for $s > t$ so that $\alpha_1 \in \mathcal{A}$. If $z = y_x(-t, \alpha)$, then $x = y_z(t, \alpha_1)$. Then the DPP Proposition 2.5 gives

$$(2.36) \qquad v(z) \leq \int_0^t \ell(y_z(s), \alpha_1(s))\,e^{-\lambda s}\,ds + v(y_z(t, \alpha_1))\,e^{-\lambda t}.$$

We multiply both sides by $e^{\lambda t}$, make the change of variable $s = t - \tau$ in the integral, and use $y_z(s, \alpha_1) = y_x(s - t, \alpha)$ to get (2.34).

To prove the second part we write $x = y_z(t, \alpha^*)$ with α^* optimal for z, define $\alpha_2(s) = \alpha^*(s + t)$ for $-t \leq s \leq 0$ and extend it arbitrarily for $s < -t$ so that $\alpha_2 \in \mathcal{A}^-$. By construction $z = y_x(-t, \alpha_2)$. Now we write the DPP in the form (2.8) (see Remark 2.6) and get the conclusion by the same calculations as in the first part of the proof. ◂

REMARK 2.26. It is easy to see that (2.34) is actually equivalent to the suboptimality principle (2.36), and both are equivalent to the fact that

$$t \longmapsto \int_0^t \ell(y_x(s, \alpha), \alpha(s)) e^{-\lambda s} \, ds + v(y_x(t, \alpha)) e^{-\lambda t}$$

is nondecreasing for any $\alpha \in \mathcal{A}$. ◁

We recall a definition from §5.1 of Chapter II.

DEFINITION 2.27. Let $\Omega \subseteq \mathbb{R}^N$ be open, $u \in C(\Omega)$, $x \in \Omega$. We say that u is a *bilateral* (viscosity) *subsolution* (respectively, *supersolution*) at x of

(2.37) $$F(x, u, Du) = 0$$

if it is a subsolution (respectively, supersolution) of (2.37) as well of

$$-F(x, u, Du) = 0;$$

that is, $F(x, u(x), p) = 0$ for all $p \in D^+ u(x)$ (respectively, $F(x, u(x), p) = 0$ for all $p \in D^- u(x)$).

If u is simultaneously a bilateral sub- and supersolution at x, we say that u is a *bilateral solution* at x. ◁

Of course, equivalent definitions can be given in terms of test functions. Note also that a viscosity solution is a bilateral solution at all points where it is differentiable, by Lemma II.1.8.

As a consequence of the Backward DPP we obtain

COROLLARY 2.28. *Assume* (A_0)–(A_4). *Then the value function v is a bilateral supersolution of*

$$\lambda v + H(x, Dv) = 0 \quad \text{in } \mathbb{R}^N,$$

where H is given by (2.9), and it is a bilateral subsolution of the same equation at any $x \in \mathcal{O}$.

PROOF. The argument is essentially the same as in the proof of Proposition 2.8. If $\varphi \in C^1(\mathbb{R}^N)$ and $v - \varphi$ has a local minimum point at x, then, by (2.34), for any $\alpha \in \mathcal{A}^-$ and $t > 0$

$$\varphi(x) - \varphi(y_x(-t)) \geq v(x) - v(y_x(-t))$$

$$\geq (e^{\lambda t} - 1) v(y_x(-t)) - \int_0^t \ell(y_x(-s), \alpha(-s)) e^{\lambda s} \, ds \,.$$

We choose $\alpha(s) \equiv a$, divide by t, and pass to the limit as $t \to 0^+$ to get
$$D\varphi(x) \cdot y'_x(0) = D\varphi(x) \cdot f(x,a) \geq \lambda v(x) - \ell(x,a) .$$
Then, by the arbitrariness of $a \in A$,
$$-\lambda v(x) + \inf_{a \in A} \{ f(x,a) \cdot D\varphi(x) + \ell(x,a) \} \geq 0$$
and the left-hand side is precisely $-\lambda v(x) - H(x, D\varphi(x))$.

The second statement is proved by the argument of the second part of the proof of Proposition 2.8, and it is actually easier because there exists an optimal control: we leave it as an exercise for the reader. ◀

Here is an application to the regularity of the value function.

COROLLARY 2.29. *Assume* (A_0)–(A_4) *and let x be a point where $H(x, \cdot)$ is strictly convex. Then $D^-v(x)$ has at most one element, and if $x \in \mathcal{O}$, $D^+v(x)$ also has at most one element.*

PROOF. The proof is easily obtained by contradiction using the convexity of $D^+v(x)$ and $D^-v(x)$ and Corollary 2.28. ◀

Next we give a simple example showing that the set of optimal points $\mathcal{O}$, though large, is not all of $\mathbb{R}^N$.

EXAMPLE 2.30. Let $N = 1$, $f(x,a) = a$, $A = [-1,1]$, $\ell(x,a) = \max\{1 - |x|, 0\}$, $\lambda = 1$. It is easy to guess that an optimal control for $x \geq 0$ is $\alpha(t) \equiv 1$ and, for $x \leq 0$, $\alpha \equiv -1$. By computing the cost corresponding to this choice we get
$$v(x) = \begin{cases} e^{|x|-1} - |x| & \text{for } |x| \leq 1, \\ 0 & \text{for } |x| \geq 1 . \end{cases}$$
The Hamilton-Jacobi-Bellman equation is

(2.38) $\qquad\qquad v + |v'| - \max\{1 - |x|, 0\} = 0 \qquad \text{in } \mathbb{R}$

and v is a classical, and therefore bilateral, solution for all $x \neq 0$. At $x = 0$ we have $D^-v(0) = \emptyset$ and $D^+v(0) = [e^{-1} - 1, 1 - e^{-1}]$ and it is immediate to check that v is a subsolution of (2.38) but not of "minus (2.38)." Indeed, 0 is the only point that does not belong to the relative interior of an optimal trajectory, thus $\mathcal{O} = \mathbb{R} \setminus \{0\}$. ◁

The Dynamic Programming Principle has been the crucial step in the proof that the value function v solves the Hamilton-Jacobi-Bellman equation; conversely, a solution of this equation in all of $\mathbb{R}^N$ has to be the value function, by the uniqueness Theorem 2.12, and therefore satisfy the DPP. Next we show that this equivalence is a local fact and that it can be split into two parts: subsolution is equivalent to a *local suboptimality principle* and supersolution to a *local superoptimality principle*.

DEFINITION 2.31. Let $\Omega \subseteq \mathbb{R}^N$ be open and bounded and $u \in C(\overline{\Omega})$. We say that u satisfies a *(local) suboptimality* (respectively, *superoptimality*) *principle*, if for every $x \in \Omega$ there exists $\tau > 0$ such that

$$(2.39) \qquad u(x) \leq \inf_{\alpha \in \mathcal{A}} \left(\int_0^t \ell(y_x(s,\alpha), \alpha(s)) e^{-\lambda s} \, ds + u(y_x(t,\alpha)) e^{-\lambda t} \right)$$

for all $0 < t \leq \tau$, (respectively,

$$u(x) \geq \inf_{\alpha \in \mathcal{A}} \left(\int_0^t \ell(y_x(s,\alpha), \alpha(s)) e^{-\lambda s} \, ds + u(y_x(t,\alpha)) e^{-\lambda t} \right)$$

for all $0 < t \leq \tau$). ◁

We recall that the Hamilton-Jacobi-Bellman equation is

$$(2.40) \qquad \lambda u + \sup_{a \in A} \{ -f(x,a) \cdot Du - \ell(x,a) \} = 0 \, .$$

THEOREM 2.32. *Assume* (A_0)–(A_4), $\Omega \subseteq \mathbb{R}^N$ *open and bounded, and* $u \in C(\overline{\Omega})$. *Then*

(i) *u is a viscosity subsolution of (2.40) in Ω if and only if it satisfies a suboptimality principle in Ω;*

(ii) *u is a viscosity supersolution of (2.40) in Ω if and only if it satisfies a superoptimality principle in Ω.*

PROOF. By the proof of Proposition 2.8, if u satisfies a suboptimality (respectively, a superoptimality) principle in Ω, then u is a viscosity subsolution (respectively, supersolution) of (2.40) in Ω.

Let us assume now that u is a viscosity supersolution of (2.40) and warn the reader that we are going to use a simple result from Section 3. For $\varepsilon > 0$ set $\Omega_\varepsilon := \{ x \in \Omega : \text{dist}(x, \partial\Omega) > \varepsilon \}$, and take $h_\varepsilon \in C^1(\mathbb{R}^N)$ such that $h_\varepsilon \equiv 1$ on Ω_ε and $h_\varepsilon \equiv 0$ on Ω^c, $0 \leq h_\varepsilon \leq 1$ in $\mathbb{R}^N$. Extend u off $\overline{\Omega}$ so that $u \in BC(\mathbb{R}^N)$ (this is possible by the classical Tietze-Urysohn extension theorem). By multiplying (2.40) by h_ε we get that u is a supersolution of

$$\lambda w + \sup_{a \in A} \{ -f_\varepsilon(x,a) \cdot Dw - \ell_\varepsilon(x,a) \} = 0 \qquad \text{in } \mathbb{R}^N,$$

where $f_\varepsilon := h_\varepsilon f$ and $\ell_\varepsilon := h_\varepsilon \ell + (1 - h_\varepsilon) \lambda u$. Then it is also a (time-independent) supersolution of the initial value problem

$$(2.41) \qquad \begin{cases} w_t + \lambda w + \sup_{a \in A}\{ -f_\varepsilon(x,a) \cdot Dw - \ell_\varepsilon(x,a) \} = 0 & \text{in } \mathbb{R}^N \times]0, +\infty[, \\ w(x,0) = u(x) & \text{in } \mathbb{R}^N \, . \end{cases}$$

It is easy to check that f_ε and ℓ_ε satisfy the structure assumptions (A_0)–(A_3) and (2.91), with different constants and moduli (depending on ε). Then by Corollary 3.6 and Remark 3.8 in the next section, the unique solution of (2.41) is

$$(2.42) \qquad w_\varepsilon(x,t) = \inf_{\alpha \in \mathcal{A}} \left\{ \int_0^t \ell_\varepsilon(y_x^{(\varepsilon)}(s), \alpha(s)) e^{-\lambda s} \, ds + u(y_x^{(\varepsilon)}(t)) e^{-\lambda t} \right\},$$

where $y_x^{(\varepsilon)}(\,\cdot\,)$ is the trajectory of the system $y' = f_\varepsilon(y, \alpha)$, $y(0) = x$, and

(2.43) $\qquad u(x) \geq w_\varepsilon(x, t) \qquad$ for all $x \in \mathbb{R}^N$, $t \geq 0$ and $\varepsilon > 0$.

Now we fix $x \in \Omega$ and choose $\varepsilon = \frac{1}{2}\operatorname{dist}(x, \partial\Omega)$. Next we pick τ such that $y_x(t) \in \overline{B}(x, \varepsilon)$ for all $t \leq \tau$, so that, for all such t, $h_\varepsilon(y_\varepsilon(t)) = 1$, $y_x^{(\varepsilon)}(t) = y_x(t)$ and $\ell_\varepsilon(y_x^{(\varepsilon)}(t), \alpha(t)) = \ell(y_x(t), \alpha(t))$. Then, combining (2.42) and (2.43), we obtain

$$u(x) \geq \inf_{\alpha \in \mathcal{A}} \left\{ \int_0^t \ell(y_x(s), \alpha(s)) e^{-\lambda s} \, ds + u(y_x(t)) e^{-\lambda t} \right\}$$

for all $0 < t \leq \tau$. Therefore u satisfies a superoptimality principle.

The proof of the "only if" part in (i) can be done in a completely analogous way. However, a more direct proof of a more general result can be obtained by extending to the present assumptions Theorem II.5.21 on differential inequalities. We state it separately as Theorem 2.33. ◀

The next result is the extension to the current hypotheses of the main theorem on differential inequalities in the viscosity sense, Theorem II.5.21.

We recall the definition of *exit time* from Ω:

$$t_x(\alpha) := \begin{cases} +\infty, & \text{if } \{t : y_x(t, \alpha) \in \partial\Omega\} = \emptyset, \\ \min\{t : y_x(t, \alpha) \in \partial\Omega\}, & \text{otherwise}. \end{cases}$$

THEOREM 2.33. *Assume* (A_0)–(A_4), Ω *open,* $u \in C(\overline{\Omega})$. *Then the following statements are equivalent:*

(i) u *is a subsolution of* (2.40) *in* Ω;

(ii) *for all* $\alpha \in \mathcal{A}$, $x \in \Omega$, *and* $0 \leq s \leq t < t_x(\alpha)$

(2.44) $\qquad e^{-\lambda s} u(y_x(s, \alpha)) - e^{-\lambda t} u(y_x(t, \alpha)) \leq \int_s^t \ell(y_x(\tau, \alpha), \alpha(\tau)) e^{-\lambda \tau} \, d\tau;$

(iii) u *is a supersolution of*

(2.45) $\qquad -\lambda u - \sup_{a \in A}\{-f(x, a) \cdot Du - \ell(x, a)\} = 0 \qquad$ in Ω.

PROOF. We only explain how to relax the assumptions of Theorem II.5.21. In Lemma II.5.20 the proof of the uniform convergence of a trajectory y_n corresponding to a piecewise constant control α_n to the trajectory y corresponding to α, where $\alpha_n \to \alpha$ a.e. in $[0, T]$, makes use of the uniform Lipschitz continuity of f with respect to x, i.e., (1.1). This assumption can be replaced by (A_3) by writing

$$\frac{1}{2}\frac{d}{dt}|y_n(t) - y(t)|^2 = (y_n(t) - y(t)) \cdot [f(y_n(t), \alpha_n(t)) - f(y(t), \alpha(t))]$$
$$\leq L|y_n(t) - y(t)|^2 + B_n(t), \qquad \text{a.e. } t \in [0, T],$$

where $B_n := (y_n - y) \cdot [f(y, \alpha_n) - f(y, \alpha)]$, integrating on $[0, s]$ and applying Gronwall's inequality as before.

Moreover, the continuous differentiability of $x \mapsto f(x, a)$ can be removed in Proposition II.5.18 by the following approximation argument. Let ϱ_n be a sequence of mollifiers, i.e., ϱ_n are smooth and nonnegative, $\int_{\mathbb{R}^N} \varrho_n \, dx = 1$, $\operatorname{supp} \varrho_n = \bar{B}(0, 1/n)$, and define $f_n(x, a) := \int_{\mathbb{R}^N} f(y, a) \varrho_n(x - y) \, dy$. Then f_n satisfy (A3) with the same constant L for all n, and $f_n(\cdot, a)$ converge uniformly to $f(\cdot, a)$ on compact subsets of $\mathbb{R}^N$. If y_n is the solution of $y' = f_n(y, a)$, $y(0) = x$, for $a \in A$ fixed, $y(t) = y_x(t, a)$, then $|y_n(t)|, |y(t)| \leq C$ for all $t \in [0, T]$ and all n. Moreover,

$$\frac{1}{2}\frac{d}{dt}|y_n(t) - y(t)|^2 = (y_n - y) \cdot [f_n(y_n, a) - f(y, a)]$$
$$\leq L|y_n - y|^2 + 2C|f_n(y(t)) - f(y(t))| \quad \text{on } [0, T],$$

and by Gronwall's inequality we get the uniform convergence of $y_n(\cdot)$ to $y(\cdot)$ on $[0, T]$. Since (2.44) is valid for $y_n(\cdot)$ and $\alpha \equiv a$ we can pass to the limit as $n \to \infty$ and get it for $y(\cdot)$. ◀

REMARK 2.34. It follows from Theorem 2.33 that any subsolution $u \in C(\overline{\Omega})$ of (2.40) satisfies for all $x \in \overline{\Omega}$ and $t \geq 0$

$$u(x) \leq \inf_{a \in \mathcal{A}} \left\{ \int_0^{t \wedge t_x} \ell(y_x(s), \alpha(s)) e^{-\lambda s} \, ds + e^{-\lambda(t \wedge t_x)} u(y_x(t \wedge t_x)) \right\},$$

which is a *global principle of suboptimality*. The local suboptimality principle is immediately obtained by observing that $T(x) := \inf_{\alpha \in \mathcal{A}} t_x(\alpha) > 0$ if $x \in \Omega$. A *global principle of superoptimality* is proved in §3.2 of Chapter IV by means of the inequality

$$(2.46) \qquad u(x) \geq \inf_\alpha \left\{ \int_0^t \ell(y_x, \alpha) e^{-\lambda s} \, ds + u(y_x(t)) e^{-\lambda t} \right\}$$

for all $x \in \Omega$ and $0 \leq t < T_\varepsilon(x)$ for some $\varepsilon > 0$, where

$$T_\varepsilon(x) := \inf\{ t : \operatorname{dist}(y_x(t, \alpha), \partial\Omega) \leq \varepsilon, \ \alpha \in \mathcal{A} \}.$$

This inequality follows immediately from the proof of Theorem 2.32.

Note that the assumption $\lambda > 0$ can be replaced by $\lambda \geq 0$ in Theorem 2.32 and Theorem 2.33: the only change in the proof of Theorem 2.32 is the use of Theorem 3.7 instead of Corollary 3.6. Also the boundedness of Ω can be dropped and replaced by $u \in BC(\overline{\Omega})$. In the proof we redefine Ω_ε as

$$\{ x \in \Omega : \operatorname{dist}(x, \partial\Omega) > \varepsilon \text{ and } |x| < 1/\varepsilon \},$$

the other minor changes are obvious. ◁

We end this subsection with the extension to the current weaker assumptions of one of the main results of §5.1 in Chapter II, which extends also the first statement of Corollary 2.28.

COROLLARY 2.35. *Assume* (A_0)-(A_4), $\Omega \subseteq \mathbb{R}^N$ *open. Then* $u \in C(\Omega)$ *is a viscosity solution of* (2.40) *in* Ω *if and only if it is a bilateral supersolution.*

PROOF. It is a straightforward consequence of Theorem 2.33: it is enough to apply it to open subsets of Ω where u is continuous up to the boundary. ◂

2.4. Generalized directional derivatives and equivalent notions of solution

In this subsection we recall some definitions of generalized directional derivatives (Dini derivatives, contingent derivatives) and use them to give an alternative definition of weak sub- and supersolutions of the Hamilton-Jacobi-Bellman equation

$$\lambda u + \sup_{a \in A}\{-f(x,a) \cdot Du - \ell(x,a)\} = 0,$$

the so-called *Dini solutions*, or *minimax solutions*, and prove the equivalence with viscosity solutions. This characterization of the value function is used in §2.5 to give necessary and sufficient conditions for optimality and to construct optimal multivalued feedback maps.

DEFINITION 2.36. *Let* $u : \Omega \to \mathbb{R}$, $\Omega \subseteq \mathbb{R}^N$ *an open set. The* lower (generalized) Dini derivative, *or* upper contingent derivative, *of* u *at the point* $x \in \Omega$ *in the direction* $q \in \mathbb{R}^N$ *is*

$$\partial^- u(x;q) = \liminf_{\substack{t \to 0^+ \\ p \to q}} \frac{u(x+tp) - u(x)}{t}$$

$$:= \sup_{\delta > 0} \inf \left\{ \frac{u(x+tq+tz) - u(x)}{t} : 0 < t \leq \delta, |z| \leq \delta \right\} \geq -\infty;$$

similarly the upper Dini derivative, *or* lower[3] contingent derivative, *of* u *at* x *in the direction* q *is*

$$\partial^+ u(x;q) := \limsup_{\substack{t \to 0^+ \\ p \to q}} \frac{u(x+tp) - u(x)}{t} \leq +\infty.\quad \triangleleft$$

Note that

(2.47)
$$\partial^- u(x;q) \leq \liminf_{t \to 0^+} \frac{u(x+tq) - u(x)}{t},$$
$$\partial^+ u(x;q) \geq \limsup_{t \to 0^+} \frac{u(x+tq) - u(x)}{t},$$

where the right-hand sides of these inequalities are the classical lower and upper Dini derivatives. It is straightforward to check that the inequalities (2.47) are both

---

[3]The switching of the adjectives upper and lower for contingent derivatives may seem a bit weird. However, this terminology was introduced in the field of set-valued analysis and it is natural in that context; see the notes at the end of the chapter for some references on the subject.

equalities if u is locally Lipschitz. This is no longer true if we drop the Lipschitz assumption, and u may even have the classical (one-sided) directional derivative

$$\partial u(x;q) := \frac{\partial u}{\partial q}(x) := \lim_{t \to 0^+} \frac{u(x+tq) - u(x)}{t},$$

despite $\partial^- u(x;q) < \partial^+ u(x;q)$ (for instance, $u(x_1, x_2) = |x_1|^{1/2}$ has $\partial u(0;(0,1)) = \partial^- u(0;(0,1)) = 0$ and $\partial^+ u(0;(0,1)) = +\infty$). The link between the sub- and superdifferentials and generalized directional derivatives is given by the following lemma.

LEMMA 2.37. *Let Ω be an open set, $u : \Omega \to \mathbb{R}$. Then*

$$D^- u(x) = \{ p : p \cdot q \leq \partial^- u(x;q) \text{ for all } q \in \mathbb{R}^N \},$$
$$D^+ u(x) = \{ p : p \cdot q \geq \partial^+ u(x;q) \text{ for all } q \in \mathbb{R}^N \}.$$

PROOF. We prove only the first equality, the other being analogous. We begin with the inclusion "$\subseteq$". Fix $p \in D^- u(x)$, $q \in \mathbb{R}^N$, $q \neq$, and $t_n \to 0^+$, $q_n \to q$ such that

$$\partial^- u(x;q) = \lim_n \frac{u(x + t_n q_n) - u(x)}{t_n}.$$

Then

$$\partial^- u(x;q) - p \cdot q = \lim_n |q| \frac{u(x + t_n q_n) - u(x) - p \cdot t_n q_n}{t_n |q_n|}$$
$$\geq |q| \liminf_{z \to 0} \frac{u(x+z) - u(x) - p \cdot z}{|z|} \geq 0.$$

To prove the inclusion "$\supseteq$", we fix p in the set on the right-hand side and choose $z_n \to 0$ such that

$$(2.48) \qquad \liminf_{z \to 0} \frac{u(x+z) - u(x) - p \cdot z}{|z|} = \lim_n \frac{u(x+z_n) - u(x) - p \cdot z_n}{|z_n|}$$

and $\lim_n z_n/|z_n| = q$. Since $z_n = |z_n| q_n$ with $q_n \to q$, the right-hand side of (2.48) is larger than or equal to

$$\liminf_{\substack{t \to 0^+ \\ \xi \to q}} \frac{u(x+t\xi) - u(x)}{t} - p \cdot q \geq 0$$

by the definition of ∂^- and the choice of p. ◂

Let us introduce the notation

$$(FL)(x) := \overline{co}\{ (f(x,a), \ell(x,a)) : a \in A \},$$

2. THE INFINITE HORIZON PROBLEM

and note that $(FL)(x)$ is compact by (A_2) and (A_4). With a little abuse of notation we call (f, ℓ) the generic element of $(FL)(x)$. The Hamiltonian of the Hamilton-Jacobi-Bellman equation

$$(2.49) \qquad \lambda u + H(x, Du) = 0$$

can be rewritten as

$$(2.50) \qquad H(x,p) := \sup_{a \in A}\{-f(x,a)\cdot p - \ell(x,a)\} = \max_{(f,\ell)\in(FL)(x)}\{-f\cdot p - \ell\},$$

as it is straightforward to check. Motivated by this, and by the classical formula $Du(x) \cdot q = \partial u(x; q)$ for differentiable functions we introduce the following definition.

DEFINITION 2.38. Let Ω be an open set and $u \in C(\Omega)$. Then u is a (*generalized*) *Dini subsolution* of (2.49) in Ω, with H given by (2.50), if for all $x \in \Omega$

$$(2.51) \qquad \lambda u(x) + \sup_{(f,\ell)\in(FL)(x)}\{-\partial^+ u(x; f) - \ell\} \leq 0;$$

similarly it is a *Dini supersolution* if

$$(2.52) \qquad \lambda u(x) + \sup_{(f,\ell)\in(FL)(x)}\{-\partial^- u(x; f) - \ell\} \geq 0,$$

and it is a (*generalized*) *Dini solution* (or minimax solution) if it is a sub- and supersolution. ◁

Let us make some comments on this definition. The first comment is that the choice of the particular representation formula for the Hamiltonian, i.e., the last one in (2.50), is crucial in Definition 2.38, whereas this choice is irrelevant for viscosity solutions. The reason is that, for instance, $f \mapsto \partial^- u(x; f)$ is not concave in general, so that $\sup_{a \in A}\{-\partial^- u(x; f(x,a)) - \ell(x,a)\}$ can be strictly smaller than the convexified counterpart in (2.52). This occurs in the next example.

EXAMPLE 2.39. Take $\lambda = 1$, $\ell \equiv 0$, $f(x,a) = a$, $A = \{a : |a| = 1\}$, so that the Hamiltonian is

$$H(x,p) = \max_{|a|=1}\{-a\cdot p\} = |p| = \max_{|a|\leq 1}\{-a\cdot p\}.$$

It is easy to check that $u(x) = |x|$ is a Dini supersolution of $u + |Du| = 0$ in $\mathbb{R}^N$, since $\partial^- u(0, f) = |f|$, but

$$u(0) + \max_{|a|=1}\{-\partial^- u(0; f(x,a))\} = -1 < 0 . \quad \triangleleft$$

The second comment is that if u is locally lipschitzean, if it has directional derivatives in the directions f for all $(f, \ell) \in (FL)(x)$, and if it is a Dini solution of (2.49), then

$$\lambda u(x) + \sup_{(f,\ell)\in(FL)(x)}\left\{-\frac{\partial u}{\partial f}(x) - \ell\right\} = 0 .$$

In this case we say that u is an *intrinsically classical solution* of the HJB equation (2.49), since no generalized derivatives are necessary in this formulation, even if u may not be differentiable and then it may not be a truly classical solution of (2.49). All this occurs, for instance, if u is semiconcave, see Chapter II, §4.2. We give two explicit applications of this remark at the end of the subsection.

The third comment is that some results in this section look nicer in the special case

(2.53) $\qquad$ (FL)$(x) = f(x, A) \times \ell(x, A) \qquad$ for all $x \in \mathbb{R}^N$.

In this case the definition of Dini solution of (2.49) can be written as

$$\lambda u(x) + \sup_{a \in A} \left\{ -\partial^+ u(x; f(x, a)) - \ell(x, a) \right\} \leq 0$$

$$\lambda u(x) + \sup_{a \in A} \left\{ -\partial^- u(x; f(x, a)) - \ell(x, a) \right\} \geq 0.$$

This is a very natural generalization of the equation

$$\lambda u(x) + \sup_{a \in A} \left\{ -\frac{\partial u(x)}{\partial f(x, a)} - \ell(x, a) \right\} = 0$$

which is equivalent to the HJB equation (2.49) when the solution u is smooth. The property (2.53) is obviously satisfied if $f(x, A)$ and $\ell(x, A)$ are closed convex sets for all x. Two more examples are the convexified problem f^c, ℓ^c, A^c, and the relaxed problem f^r, ℓ^r, A^r introduced in §2.2, provided A is compact, see (2.31) and (2.32).

The main result of this subsection is the following.

THEOREM 2.40. *Assume* (A_0)–(A_4), Ω *is a bounded open set, and* $u \in C(\overline{\Omega})$. *For the equation* (2.49) *in* Ω, *with Hamiltonian given by* (2.50), *the following statements are equivalent:*

(i) $\quad u$ *is a viscosity subsolution (respectively, supersolution);*

(ii) $\quad u$ *satisfies a suboptimality (respectively, superoptimality) principle;*

(iii) $\quad u$ *is a Dini subsolution (respectively, supersolution).*

PROOF. The equivalence of (i) and (ii) has already been proved in Theorem 2.32. We prove first that (iii) implies (i). By Lemma 2.37 we have

$$p \cdot f \leq \partial^- u(x; f) \leq \partial^+ u(x; f) \leq q \cdot f$$

for all $p \in D^- u(x)$, $q \in D^+ u(x)$, $f \in \mathbb{R}^N$. Thus, if u is a Dini supersolution of (2.49),

$$\lambda u(x) + \sup_{(f, \ell) \in (\mathrm{FL})(x)} \{-f \cdot p - \ell\} \geq \lambda u(x) + \sup_{(f, \ell) \in (\mathrm{FL})(x)} \{-\partial^- u(x; f) - \ell\} \geq 0,$$

for all $p \in D^- u(x)$. From (2.50) it follows that u is a viscosity supersolution of (2.49). The proof for subsolutions is just the same.

In order to complete the proof we need to show that (ii) implies (iii). For this purpose we make use of the following result in convex analysis.

2. THE INFINITE HORIZON PROBLEM

LEMMA 2.41. *For all $x \in \mathbb{R}^N$ we have*

(2.54) $\quad (FL)(x) = \Big\{ g = (f, \ell) \in \mathbb{R}^{N+1} :$

$g = \lim\limits_{n \to +\infty} \dfrac{1}{t_n} \displaystyle\int_0^{t_n} (f(x, \alpha_n(s)), \ell(x, \alpha_n(s))) \, ds \text{ for some } t_n \to 0^+ \text{ and } \alpha_n \in \mathcal{A} \Big\};$

PROOF. By the mean value theorem, for any $t_n \to 0^+$ and $\alpha_n \in \mathcal{A}$,

$$\frac{1}{t_n} \int_0^{t_n} (f(x, \alpha_n(s)), \ell(x, \alpha_n(s))) \, ds \in \overline{\operatorname{co}}(f(x, A) \times \ell(x, A)) = (FL)(x) \, .$$

Hence, the set on the right-hand side of (2.54) is contained in $(FL)(x)$.

For the opposite inclusion, set $G(x, a) := (f(x, a), \ell(x, a))$; if $g \in \operatorname{co} G(x, A)$, that is,

$$g = \sum_{i=1}^{j} \lambda_i G(x, a_i), \qquad a_i \in A, \quad \lambda_i \geq 0, \quad \sum_{i=1}^{j} \lambda_i = 1,$$

then

(2.55) $$g = \frac{1}{t} \int_0^t G(x, \alpha_t(s)) \, ds \qquad \text{for any } t > 0,$$

where

$$\alpha_t(s) = a_k \qquad \text{for } s \in \left[t \sum_{i=1}^{k-1} \lambda_i, t \sum_{i=1}^{k} \lambda_i \right[, \, k = 1, 2, \ldots, j \, .$$

Take now $g \in \overline{\operatorname{co}} G(x, A)$ and choose $g_n \in \operatorname{co} G(x, A)$ such that $g_n \to g$ and $t_n \to 0^+$. Since g_n can be written in the form (2.55) with $t = t_n$, the conclusion follows easily. ◀

END OF THE PROOF OF THEOREM 2.40. Fix $x \in \Omega$, and $(f, \ell) \in (FL)(x)$. Use Lemma 2.41 to pick $t_n \to 0^+$ and $\alpha_n \in \mathcal{A}$ such that

$$\frac{1}{t_n} \int_0^{t_n} f(x, \alpha_n(s)) \, ds \longrightarrow f,$$
$$\frac{1}{t_n} \int_0^{t_n} \ell(x, \alpha_n(s)) \, ds \longrightarrow \ell, \qquad \text{as } n \to \infty \, .$$

If u satisfies the suboptimality principle (2.39), then, for $0 < t \leq \tau$ and all $\alpha \in \mathcal{A}$,

(2.56) $\quad \dfrac{u(x) - u(y_x(t))}{t} + u(y_x(t)) \dfrac{1 - e^{-\lambda t}}{t} - \dfrac{1}{t} \displaystyle\int_0^t \ell(y_x(s), \alpha(s)) \, e^{-\lambda s} \, ds \leq 0 \, .$

By (A_0)–(A_4) (see (2.13) and (2.14)) we have

$$y_x(t_n, \alpha_n) = x + \int_0^{t_n} f(x, \alpha_n(s)) \, ds + o(t_n) = x + t_n f + o(t_n)$$

$$\frac{1}{t_n} \int_0^{t_n} \ell(y_x(s, \alpha_n), \alpha_n(s)) \, e^{-\lambda s} \, ds = \ell + o(1) \qquad \text{as } n \to \infty \, .$$

Then, using α_n and t_n in (2.56), we get
$$\frac{u(x) - u(x + t_n f + o(t_n))}{t_n} + \lambda u(x) - \ell \leq o(1) \qquad \text{as } n \to \infty$$
and then
$$\lambda u(x) - \partial^+ u(x; f) - \ell = \lambda u(x) + \liminf_{(t,g) \to (0^+ \cdot f)} \frac{u(x) - u(x + tg)}{t} - \ell \leq 0,$$
which shows that (ii) implies (iii) for subsolutions.

To prove that (ii) implies (iii) for supersolutions, fix $x \in \Omega$. By the superoptimality principle, for every $n \in \mathbb{N}$ we can find $\alpha_n \in \mathcal{A}$ and the corresponding trajectory $y^{(n)}(t) = y_x(t, \alpha_n)$ such that

$$(2.57) \quad \frac{u(x) - u(y^{(n)}(t))}{t} + u(y^{(n)}(t))\frac{1 - e^{-\lambda t}}{t} - \frac{1}{t}\int_0^t \ell(x, \alpha_n(s))\,ds \geq \frac{1}{n} + o(1),$$

as $t \to 0^+$, where we have already used (2.13) in the integral. By the compactness of (FL)(x) we can find $t_n \to 0^+$, a subsequence of α_n, still denoted α_n, and $(\bar{f}, \bar{\ell}) \in$ (FL)(x) such that

$$\frac{1}{t_n}\int_0^{t_n} \ell(x, \alpha_n(s))\,ds \longrightarrow \bar{\ell},$$
$$\frac{1}{t_n}\int_0^{t_n} f(x, \alpha_n(s))\,ds \longrightarrow \bar{f}, \qquad \text{as } n \to \infty.$$

Moreover, by (2.14),
$$y^{(n)}(t_n) = x + \int_0^{t_n} f(x, \alpha_n(s))\,ds + o(t_n) = x + t_n \bar{f} + o(t_n) \qquad \text{as } n \to \infty,$$
and thus, by (2.57),
$$0 \leq \lambda u(x) + \liminf_n \frac{u(x) - u(x + t_n \bar{f} + o(t_n))}{t_n} - \bar{\ell}$$
$$\leq \lambda u(x) + \limsup_{(t,g) \to (0^+ \cdot \bar{f})} \frac{u(x) - u(x + tg)}{t} - \bar{\ell} = \lambda u(x) - \partial^- u(x; \bar{f}) - \bar{\ell}$$
$$\leq \lambda u(x) + \sup_{(f,\ell) \in (\text{FL})(x)} \{-\partial^- u(x; f) - \ell\}. \quad \blacktriangleleft$$

REMARK 2.42. If u satisfies a suboptimality principle, then it satisfies a local backward suboptimality principle, that is, inequality (2.34) for all $0 < t \leq \tau$. By the arguments of the previous proof one can get
$$\lambda u(x) + \sup_{(f,\ell) \in (\text{FL})(x)} \{\partial^- u(x; -f) - \ell\} \leq 0.$$

Note that this inequality does not follow directly from either (2.51) or (2.52), and that it implies immediately, via Lemma 2.37, that u is a viscosity supersolution of $-\lambda u - H(x, Du) = 0$ (see Corollary 2.28). See also Exercise 2.11. ◁

REMARK 2.43. Some stronger results can be proved if u is locally lipschitzean because the generalized Dini derivatives take the simpler form of classical Dini derivatives (i.e., the right-hand sides of (2.47)). In the assumptions of Theorem 2.40, if u is locally Lipschitz and satisfies a suboptimality principle, then

$$(2.58) \qquad \lambda u(x) + \limsup_{t \to 0^+} \sup_{(f,\ell) \in (\text{FL})(x)} \left\{ \frac{u(x) - u(x+tf)}{t} - \ell \right\} \leq 0,$$

uniformly for x in a compact set of Ω, which easily implies

$$(2.59) \qquad \lambda u(x) + \sup_{(f,\ell) \in (\text{FL})(x)} \{ -\partial^- u(x;f) - \ell \} \leq 0 .$$

Note that (2.59) is stronger than (2.51), and it implies that u locally lipschitzean is a viscosity solution in Ω of (2.49) if and only if it is a *lower Dini solution*, that is,

$$\lambda u(x) + \sup_{(f,\ell) \in (\text{FL})(x)} \{ -\partial^- u(x;f) - \ell \} = 0 .$$

Here is a proof of (2.58). Choose $(f,\ell) \in \text{co}(f(x,A) \times \ell(x,A))$ and, for any $t > 0$, $\alpha_{(t)} \in \mathcal{A}$ such that

$$f = \frac{1}{t} \int_0^t f(x, \alpha_{(t)}(s)) \, ds, \qquad \ell = \frac{1}{t} \int_0^t \ell(x, \alpha_{(t)}(s)) \, ds .$$

Then $y_x(t, \alpha_{(t)}) = x + tf + o(t)$ as $t \to 0^+$ and $u(y_x(t, \alpha_{(t)})) = u(x + tf) + o(t)$ as $t \to 0^+$ by the local Lipschitz continuity of u. Then from (2.56) we get, as in the previous proof,

$$(2.60) \qquad \frac{u(x) - u(x+tf)}{t} + \lambda u(x) - \ell \leq o(1) \qquad \text{as } t \to 0^+ .$$

Since $(f,\ell) \in \text{co}(f(x,A) \times \ell(x,A))$ is arbitrary, the left-hand side of (2.60) is continuous in (f,ℓ), and the right-hand side tends to 0 independently of (f,ℓ) and uniformly for x in a compact set of Ω, we get

$$\lambda u(x) + \sup_{(f,\ell) \in (\text{FL})(x)} \left\{ \frac{u(x) - u(x+tf)}{t} - \ell \right\} \leq o(1) \qquad \text{as } t \to 0^+,$$

and (2.58) follows immediately.

See Exercise 2.12 for a result on locally Lipschitzean supersolutions.   ◁

REMARK 2.44. A direct proof of the equivalence of viscosity and Dini supersolutions (respectively, subsolutions) can be given by means of the following property linking Dini derivatives and subdifferentials: if u is continuous in a neighborhood of x, $\mathcal{C}$ is a closed convex set, and

$$\partial^- u(x;q) > 0 \qquad \text{for all } q \in \mathcal{C},$$

then there exist sequences $x_n \to x$ and $p_n \in D^- u(x_n)$ such that

$$p_n \cdot q \geq 0 \qquad \text{for all } q \in \mathcal{C} .$$

See Exercise 2.13.   ◁

We end this subsection with two special cases where, under stronger assumptions, the value function is an intrinsically classical solution of the Hamilton-Jacobi-Bellman equation. The first case is semiconcavity of v, see §4.2 in Chapter II for the definition. We just recall that sufficient conditions for semiconcavity are $\lambda > 2L$ (L is the constant in (A$_3$)) and suitable smoothness (e.g., C^2) of f and ℓ in the state variable uniformly in the control variable: see Remark II.4.10.

PROPOSITION 2.45. *Assume* (A$_0$)–(A$_4$) *and that the value function v is semiconcave. Then v is locally lipschitzean, it has all (classical one-sided) directional derivatives $\partial v(x;q)$ everywhere, and it solves*

$$(2.61) \qquad \lambda v(x) + \sup_{(f,\ell) \in (\mathrm{FL})(x)} \{-\partial v(x;f) - \ell\} = 0 \qquad \text{for all } x \in \mathbb{R}^N.$$

In particular, the formulation (2.61) *is equivalent to the viscosity and the Dini formulations of the Hamilton-Jacobi-Bellman equation* (2.49).

PROOF. The Lipschitz continuity of v is proved in Proposition II.4.6 and the existence of the directional derivatives in Proposition II.4.7 (d). All the other statements are immediate consequences of the theory of this subsection. ◀

REMARK 2.46. We proved in Corollary II.5.2 that for a semiconcave function being a viscosity solution is also equivalent to solving the Hamilton-Jacobi-Bellman equation a.e. We recall also that if $\lambda \geq 2L$, v can be approximated uniformly on compacta of $\mathbb{R}^N$ by semiconcave value functions of similar infinite horizon problems: see Corollary 2.23. ◁

The second case has a different proof but its assumptions are essentially of the same type: large λ and smooth f and ℓ.

PROPOSITION 2.47. *In addition to* (A$_0$)–(A$_4$) *suppose that f and ℓ are differentiable with respect to the state variable x and*

$$|f(x+q,a) - f(x,a) - D_x f(x,a)q| \leq |q|\,\omega(|q|)$$
$$|\ell(x+q,a) - \ell(x,a) - D_x \ell(x,a) \cdot q| \leq |q|\,\omega(|q|),$$

for all $x, q \in \mathbb{R}^N$, $a \in A$, where ω is a modulus and

$$\sup\{\,|D_x f(x,a)| : x \in \mathbb{R}^N,\ a \in A\,\} < \lambda\,.$$

Then all the conclusions of Proposition 2.45 *hold.*

SKETCH OF THE PROOF. The Lipschitz continuity of v follows from Proposition 2.1. All the statements follow easily from the theory of this section once we prove the existence of the directional derivatives. This follows from the next lemma, which is a variant of the results on derivatives of marginal functions of Chapter II, §§2 and 4.1. ◀

LEMMA 2.48. *Let B be a set, $w^b : \mathbb{R}^N \to \mathbb{R}$, $b \in B$, be equibounded, equicontinuous functions with directional derivatives such that*

$$\left| \frac{w^b(x+hq) - w^b(x)}{h} - \partial w^b(x;q) \right| \leq \omega(h)$$

for all $x, q \in \mathbb{R}^N$, $b \in B$, $h > 0$, where ω is a modulus. If $w(x) := \inf_{b \in B} w^b(x)$ is continuous, then it has all directional derivatives everywhere and

$$\partial w(x;q) = \inf \left\{ \liminf_n \partial w^{b_n}(x;q) : \lim_n w^{b_n}(x) = w(x) \right\}.$$

The proof of this lemma is left to the reader, see Exercise 2.15. To complete the proof one has to show that $w^b(x) := J(x, \alpha)$, with $b = \alpha$ and $B = \mathcal{A}$, satisfy the assumptions of Lemma 2.48. This is a tedious but straightforward computation that we leave to the reader.

2.5. Necessary and sufficient conditions of optimality, minimum principles, and multivalued optimal feedbacks

In this Section we apply the theory of §2.4 to give several necessary and sufficient conditions for optimality of open loop controls in terms of Dini derivatives or sub- and superdifferentials of the value function. Some of them take the form of a Minimum Principle, and we explain the connection with the classical Pontryagin Maximum Principle. We also study the optimality of feedback maps and construct optimal multivalued feedbacks which are a counterpart for problems with nonsmooth value functions of the classical calculations of optimal feedbacks in linear-quadratic problems.

Before stating the main result we introduce some notations and give some motivations. Let us fix $x_0 \in \mathbb{R}^N$ and for a given $\alpha \in \mathcal{A}$ let us write for brevity

$$y(t) := y_{x_0}(t, \alpha).$$

We know from the DPP and, in particular, from Remark 2.6, that the control α is optimal for x_0, i.e., $v(x_0) = J(x_0, \alpha)$, if and only if the function

(2.62) $$g(t) := v(y(t)) e^{-\lambda t} + \int_0^t \ell(y(s), \alpha(s)) e^{-\lambda s} \, ds$$

is constant, and that in this case the control $s \mapsto \alpha(s+t)$ is optimal for $y(t)$, for all $t \geq 0$, so that α generates, by shifting, an optimal control for each point of the trajectory. Moreover, since the function g is nondecreasing for any control, the true condition of optimality is that g be nonincreasing. Thus, if everything is smooth (v differentiable and α continuous), the condition of optimality is $g'(t) \leq 0$ for all t, that is,

$$e^{-\lambda t} \left[Dv(y(t)) \cdot y'(t) - \lambda v(y(t)) + \ell(y(t), \alpha(t)) \right] \leq 0 \qquad \text{for all } t.$$

There are several ways to rewrite this necessary and sufficient condition of optimality in an equivalent way which makes sense in the general nonsmooth case. Here are some possibilities: for a.e. $t > 0$

$$p \cdot y'(t) + \ell(y(t), \alpha(t)) \leq \lambda v(y(t)), \tag{2.63}$$

either for all or for some $p \in D^{\pm}v(y(t))$, where we denote

$$D^{\pm}v(x) := D^+v(x) \cup D^-v(x);$$

$$\partial^- v(y(t); y'(t)) + \ell(y(t), \alpha(t)) \leq \lambda v(y(t)) \tag{2.64}$$
$$-\partial^+ v(y(t); -y'(t)) + \ell(y(t), \alpha(t)) \leq \lambda v(y(t)). \tag{2.65}$$

Since v is a viscosity subsolution and a bilateral supersolution of the HJB equation

$$\lambda v + \sup_{a \in A}\{-Dv \cdot f(x, a) - \ell(x, a)\} = 0,$$

(2.63) is equivalent to

$$p \cdot y'(t) + \ell(y(t), \alpha(t)) = \lambda v(y(t)) \tag{2.66}$$

and it implies

$$p \cdot f(y(t), \alpha(t)) + \ell(y(t), \alpha(t)) = \min_{a \in A}\{p \cdot f(y(t), a) + \ell(y(t), a)\}. \tag{2.67}$$

Note that the right-hand side of (2.67) is $-H(y(t), p)$, where H is the Hamiltonian given by (2.50). Similarly, if v is locally lipschitzean then it is a lower Dini solution of the HJB equation (see Remark 2.43), that is,

$$\lambda v(x) + \sup_{(f, \ell) \in (FL)(x)} \{-\partial^- v(x; f) - \ell\} = 0,$$

so in this case (2.64) is equivalent to

$$\partial^- v(y; y') + \ell(y, \alpha) = \lambda v(y) \tag{2.68}$$

and to

$$\partial^- v(y; f(y, \alpha)) + \ell(y, \alpha) = \min_{(f, l) \in (FL)(y)} \{\partial^- v(y; f) + \ell\}. \tag{2.69}$$

We first prove the necessity of these conditions.

THEOREM 2.49 (NECESSARY CONDITIONS OF OPTIMALITY). *Assume* (A_0)–(A_4). *For* $\alpha \in \mathcal{A}$ *let* $y(\cdot)$ *be the corresponding trajectory with initial point* $x_0 \in \mathbb{R}^N$. *If* α *is optimal for* x_0, *then conditions* (2.64), (2.65), *and* (2.66), (2.67) *hold for all* $p \in D^{\pm}v(y(t))$, *for a.e.* $t > 0$.

2. THE INFINITE HORIZON PROBLEM

The necessity of condition (2.67) for all $p \in D^{\pm}v(y(t))$ is called the *Minimum Principle* (briefly MP) or the maximum principle if one changes sign to both sides of (2.67). It is a weak form of the classical Pontryagin MP because we are not assuming any differentiability on the data. Note that condition (2.67) can be rewritten as

(2.70) $$\alpha(t) \in \arg\min_{A} \{\, p \cdot f(y(t), \cdot) + \ell(y(t), \cdot) \,\}.$$

Also condition (2.69) can be viewed as a minimum principle expressed in terms of directional derivatives. When v is locally lipschitzean it implies the following further necessary condition of optimality

(2.71) $$\alpha(t) \in A_*(y(t)) \qquad \text{for a.e. } t,$$

where

(2.72) $$A_*(x) := \arg\min_{A} \{\, \partial^- v(x; f(x, \cdot)) + \ell(x, \cdot) \,\}.$$

In the proof of the theorem we need the following chain rule for Dini derivatives.

LEMMA 2.50. *Let $y : [0, T] \to \Omega$ be differentiable at s and $u \in C(\Omega)$. Then*

$$\partial^-(u \circ y)(s; 1) \geq \partial^- u(y(s); y'(s)),$$
$$\partial^+(u \circ y)(s; 1) \leq \partial^+ u(y(s); y'(s)).$$

If, moreover, u is locally lipschitzean, then both these inequalities are equalities.

PROOF. Since $y(s+h) = y(s) + h\, y'(s) + o(h)$ as $h \to 0$,

$$\partial^-(u \circ y)(s; 1) = \liminf_{h \to 0^+} \frac{u(y(s+h)) - u(y(s))}{h} \geq \partial^- u(y(s); y'(s)).$$

If u has a Lipschitz constant C in a neighborhood of $y(s)$, then

$$\frac{u(y(s+h)) - u(y(s))}{h} \leq \frac{u(y(s) + h\, y'(s)) - u(y(s)) + C\, o(h)}{h},$$

so we get the desired equality by taking the $\liminf_{h \to 0^+}$ of both sides, since in this case the generalized Dini derivative coincides with the usual Dini derivative. The proof for ∂^+ is analogous. ◂

PROOF OF THEOREM 2.49. If α is optimal, the function g defined by (2.62) is constant. The integral in the definition of g is differentiable for a.e. t, and therefore so is $v(y(\cdot))\, e^{-\lambda \cdot}$. We fix any such t where $y(\cdot)$ is differentiable as well, and choose $h_n \to 0^+$ such that

$$\lim_{n} \frac{v(y(t + h_n)) - v(y(t))}{h_n} = \partial^-(v \circ y)(t; 1).$$

Then

$$(v(y(t))e^{-\lambda t})' = \lim_n \frac{[v(y(t+h_n)) - v(y(t))]e^{-\lambda t} + v(y(t+h_n))\left[e^{-\lambda(t+h_n)} - e^{-\lambda t}\right]}{h_n}$$
$$= -\lambda e^{-\lambda t} v(y(t)) + e^{-\lambda t} \partial^-(v \circ y)(t;1)$$
$$\geq e^{-\lambda t}\left[-\lambda v(y(t)) + \partial^- v(y(t); y'(t))\right],$$

where the inequality comes from Lemma 2.50. Thus

$$0 = g'(t) \geq e^{-\lambda t}\left[-\lambda v(y(t)) + \partial^- v(y(t); y'(t)) + \ell(y(t), \alpha(t))\right]$$

for a.e. t, and so (2.64) holds.

The proof of the necessity of (2.65) is similar and we omit it.

Now (2.64) and Lemma 2.37 give (2.63) for all $p \in D^- v(y(t))$. Since v is a bilateral supersolution of the HJB equation, for all x and a

$$\lambda v(x) \leq p \cdot f(x,a) + \ell(x,a) \qquad \text{for all } p \in D^- v(x),$$

thus the inequality in (2.63) is an equality and we get (2.66) and (2.67) for all $p \in D^- v(y(t))$. Similarly, (2.65) and Lemma 2.37 give (2.63) for all $p \in D^+ v(y(t))$, and then (2.66) and (2.67) hold for such p because v is a subsolution of the HJB equation. ◀

REMARK 2.51. It follows immediately from Lemma 2.50 that for locally lipschitzean u and y the classical directional derivative $\partial u(y(s); y'(s))$ exists for all s such that both y and $u \circ y$ are differentiable, therefore a.e. in $[0,T]$. In this case the Dini derivatives in (2.64) and (2.65) are in fact classical (one-sided) derivatives. ◁

Next we study the sufficiency of the necessary conditions of optimality listed previously.

THEOREM 2.52 (SUFFICIENT CONDITIONS FOR OPTIMALITY). *Assume* (A_0)–(A_4) *and* $y(\cdot) = y_{x_0}(\cdot, \alpha)$. *Suppose that there exists a subsolution* $u \in BC(\mathbb{R}^N)$ *of the HJB equation which is locally lipschitzean in a neighborhood of* $\{y(t): t \geq 0\}$. *Then* α *is optimal for* x_0 *if for a.e.* $t > 0$ *at least one of the following conditions holds:*

(2.73) $\quad \partial^- u(y(t); y'(t)) + \ell(y(t), \alpha(t)) \leq \lambda u(y(t));$

(2.74) $\quad -\partial^+ u(y(t); -y'(t)) + \ell(y(t), \alpha(t)) \leq \lambda u(y(t));$

(2.75) $\quad \exists p \in D^\pm u(y(t)): p \cdot y'(t) + \ell(y(t), \alpha(t)) \leq \lambda u(y(t));$

(2.76) $\quad u$ *is also a supersolution of the HJB equation in a neighborhood of* $y(t)$ *and there is* $p \in D^- u(y(t))$ *or* $p \in D^* u(y(t)) \cap D^+ u(y(t))$ *such that* (2.67) *holds*.

By Remark 2.43, (2.73) is equivalent to

(2.77) $\quad \lambda u(y) = \partial^- u(y; y') + \ell(y, \alpha) = \min_{(f,\ell) \in (\mathrm{FL})(y)} \{\partial^- u(y; f) + \ell\}.$

PROOF. Let us consider the function
$$h(t) := u(y(t))\,e^{-\lambda t} + \int_0^t \ell(y(s),\alpha(s))\,e^{-\lambda s}\,ds, \qquad t \geq 0\,.$$
Our goal is to prove that h is nonincreasing, because in such a case
$$u(x_0) = h(0) \geq \lim_{t\to+\infty} h(t) = J(x_0,\alpha)\,.$$
By the comparison principle (Theorem 2.12 and Remark 2.13) $u(x) \leq v(x)$ and then $J(x_0,\alpha) = v(x_0)$, showing that α is optimal for x_0.

Since $\ell(y,\alpha)$ and y' are locally bounded and u is locally lipschitzean, h is locally lipschitzean and
$$h'(t) = e^{-\lambda t}\left[(u\circ y)'(t) - \lambda u(y(t)) + \ell(y(t),\alpha(t))\right],$$
for a.e. t. The right-hand side of this equality is ≤ 0 if, for a.e. t, either (2.73) or (2.74) holds, because at every point of differentiability of both $u \circ y$ and y
$$(u\circ y)'(t) = \partial^-(u\circ y)(t;1) = \partial^- u(y(t);y'(t))$$
and
$$(u\circ y)'(t) = -\partial^+(u\circ y)(t;-1) = -\partial^+ u(y(t);-y'(t)),$$
by Lemma 2.50. Thus h is nonincreasing in this case. On the other hand, if (2.75) holds with $p \in D^-u(y(t))$, then, by Lemma 2.37,
$$\lambda u - \ell \geq -p\cdot(-y') \geq -\partial^- u(y;-y') \geq -\partial^+ u(y;-y')$$
and thus (2.74) holds. Similarly (2.75) with $p \in D^+u(y(t))$ implies (2.73).

Now let us assume (2.76) with $p \in D^-v(y(t))$. Then (2.75) holds because v is a bilateral supersolution of the HJB equation. Finally let $p \in D^*u(y(t))\cap D^+u(y(t))$. By definition of D^*, see §II.4.1, $p = \lim_n Du(x_n)$ for a sequence $x_n \to y(t)$ of points where u is differentiable. Since u is a viscosity solution of the HJB equation near $y(t)$, by Proposition II.1.9 this equation is satisfied in the classical sense at points of differentiability, thus
$$\lambda u(x_n) + H(x_n, Du(x_n)) = 0\,.$$
By letting $n \to \infty$ we get
$$\lambda u(y(t)) + H(y(t),p) = 0,$$
and then (2.67) implies (2.75). ◂

REMARK 2.53. From the proof of the theorem one sees that the two Dini derivatives in (2.73) and (2.74) coincide. Moreover they are classical (one-sided) derivatives by Remark 2.51, thus each of the conditions (2.73) and (2.74) is equivalent to
$$\partial u(y;y') + \ell(y,\alpha) \leq \lambda u(y)$$
and also to
$$-\partial u(y;-y') + \ell(y,\alpha) \leq \lambda u(y)\,. \quad \triangleleft$$

The previous results can be combined to get several characterizations of optimal controls which are listed in the next theorem. Recall that the (local) semiconcavity of v implies $D^+v(x) = \partial v(x) \neq \emptyset$ for all x by Proposition II.4.7, where ∂v denotes Clarke's generalized gradient.

THEOREM 2.54 (NECESSARY AND SUFFICIENT CONDITIONS OF OPTIMALITY).
Assume (A_0)–(A_4) and $y(\,\cdot\,) = y_{x_0}(\,\cdot\,, \alpha)$. Suppose the value function v is locally lipschitzean in a neighborhood of $\{y(t) : t \geq 0\}$. Then

(i) α is optimal for x_0 if and only if for a.e. $t > 0$ one of the following conditions holds:

(2.78) $$\begin{aligned} \partial v(y(t); y'(t)) + \ell(y(t), \alpha(t)) &= \lambda v(y(t)), \\ \partial^- v(y; f(y, \alpha)) + \ell(y, \alpha) &= \min_{(f,\ell) \in (\mathrm{FL})(y)} \{\partial^- v(y; f) + \ell\}; \end{aligned}$$

(ii) if in addition $D^\pm v(y(t)) \neq \emptyset$ for a.e. t, then the optimality of α for x_0 is equivalent to each of the following conditions
- for a.e. t there exists $p \in D^\pm v(y(t))$ such that

(2.79) $$p \cdot y' + \ell(y, \alpha) = \lambda v(y),$$

- for a.e. t and for all $p \in D^\pm v(y(t))$ (2.79) holds;

(iii) if, moreover, $D^+v(y(t)) = \partial v(y(t))$ for a.e. t, then α is optimal for x_0 if and only if for a.e. t

(2.80) $$p \cdot f(y, \alpha) + \ell(y, \alpha) = \min_{a \in A}\{p \cdot f(y, a) + \ell(y, a)\}$$

for all $p \in D^\pm v(y(t))$.

PROOF. (i) The statement about the first condition follows from Theorems 2.49 and 2.52 and Remark 2.53. To prove the statement on the second condition it is enough also to use Remark 2.43.

(ii) is an obvious consequence of Theorem 2.49 and 2.52, and so is the "only if" statement in (iii). To complete the proof of (iii) we recall that $D^*v(z) \neq \emptyset$ for all z because v is locally lipschitzean. Since $\partial v(z) = \mathrm{co}\, D^*v(z)$ for all z, we have $D^*v(y(t)) \subseteq D^+v(y(t))$ for a.e. t. Therefore for a.e. t there exists $p \in D^*v(y(t)) \cap D^+v(y(t))$, and it satisfies (2.80) by assumption. Then the conclusion follows from Theorem 2.52. ◀

REMARK 2.55 (THE PONTRYAGIN MAXIMUM PRINCIPLE). A second interesting case where $D^\pm v(y(t)) \neq \emptyset$ along an optimal trajectory occurs when ℓ and f are C^1 with bounded derivatives and λ is large enough. In Section I.6 we saw that there exists a solution $p(t)$ of the *adjoint system* (I.6.14) such that, for a.e. t,

(2.81) $$e^{\lambda t}p(t) \in D^+v(y(t)), \qquad e^{\lambda t}p(t) \cdot y'(t) + \ell(y(t), \alpha(t)) = \lambda v(y(t))$$

2. THE INFINITE HORIZON PROBLEM

if y is the trajectory associated with an optimal control α. By Theorem 2.49 the maximum in the definition of the Hamiltonian $H(y(t), e^{\lambda t} p(t))$ is attained at $\alpha(t)$; that is,

$$(2.82) \quad e^{\lambda t} p(t) \cdot f(y(t), \alpha(t)) + \ell(y(t), \alpha(t))$$
$$= \min_{a \in A} \{ e^{\lambda t} p(t) \cdot f(y(t), a) + \ell(y(t), a) \},$$

for a.e. t. This is the classical necessary condition of the PMP.

If v is locally lipschitzean around a trajectory $y(\,\cdot\,) = y_{x_0}(\,\cdot\,, \alpha)$ we can use Theorem 2.54 to extend the PMP to a necessary and sufficient condition of optimality: α is optimal for x_0 if and only if the costate $p(t)$ satisfies (2.81) and (2.82) for a.e. $t > 0$.

Note that even if the PMP is a special case of the conditions of optimality derived from Dynamic Programming, it has the advantage that the necessary condition (2.67) can be checked without knowing the value function: it is enough to know the trajectory y and to solve the adjoint equation. ◁

REMARK 2.56. If in addition to the assumptions of Theorem 2.49 we have

$$(FL)(x) = f(x, A) \times \ell(x, A),$$

then the necessary and sufficient condition of optimality (2.78) becomes

$$\partial^- v(y; f(y, \alpha)) + \ell(y, \alpha) = \min_A \{ \partial^- v(y; f(y, \,\cdot\,)) + \ell(y, \,\cdot\,) \}$$

for a.e. t, that is,

$$(2.83) \qquad \alpha(t) \in A_*(y(t)) \qquad \text{for a.e. } t > 0,$$

where A_* is defined by (2.72). If the running cost does not depend explicitly on the control (i.e., $\ell(x, a) = \ell(x)$ for all x and a) then

$$A_*(x) = \arg\min_A \partial^- v(x; f(x, \,\cdot\,)) .$$

In this case the optimality condition (2.83) has a rather intuitive geometric interpretation: at almost every time t the controller must choose among all the available directions $f(y(t), a)$ at the current state $y(t)$ any one providing the steepest descent of the value function at the point $y(t)$, in the generalized sense of lower Dini derivatives. ◁

REMARK 2.57. We saw in §2.2 that

$$(FL)(x) = f^r(x, A^r) \times \ell^r(x, A^r) .$$

Then from the previous remark one gets, under the assumptions of Theorem 2.49, that a relaxed control $\mu \in \mathcal{A}^r$ is optimal for x_0, i.e., $J^r(x_0, \mu) = v(x_0)$, if and only if the corresponding relaxed trajectory $y(\,\cdot\,) = y_{x_0}(\,\cdot\,, \mu)$ satisfies

$$\partial^- v(y; f^r(y, \mu)) + \ell^r(y, \mu) = \min_{A^r} \{ \partial^- v(y; f^r(y, \,\cdot\,)) + \ell^r(y, \,\cdot\,) \}$$

for a.e. t. ◁

Let us make a brief comparison between the conditions of optimality in this subsection (Theorems 2.49, 2.52 and Corollary 2.54) and those in subsection 2.2, the so-called verification theorems. In both cases one needs first to know a global object, a subsolution of the HJB equation or the value function itself, but then the criteria are different. In a verification theorem (e.g., Corollary 2.19) one has to know the whole open loop control function, compute the payoff, and then compare it with the verification function. On the contrary, the conditions (2.64) or (2.66) relate the control at a given time t only with the state at the same time, so they are fit to design an optimal policy prescribing the instantaneous choice of the control according to the actual state.

This leads to the problem of the optimality of feedback maps and closed loop controls, which we discuss next. The classical synthesis procedure recalled in Section I.5 states that if the value function v is differentiable, then an admissible feedback $\Phi : \mathbb{R}^N \to A$ (see Definition 2.17) is optimal if

$$\Phi(z) \in \arg\min_{A} \{ Dv(z) \cdot f(z, \cdot) + \ell(z, \cdot) \}$$

for all $z \in \mathbb{R}^N$, and by the HJB equation this condition can also be written as

$$\Phi(z) \in \left\{ a \in A : \frac{\partial v(z)}{\partial f(z,a)} + \ell(z,a) \leq \lambda v(z) \right\}.$$

To obtain the generalization of this condition to the case of a merely continuous value function we introduce the sets

$$S_{\mathrm{D}}(z) := \{ a \in A : \partial^- v(z; f(z,a)) + \ell(z,a) \leq \lambda v(z) \},$$

$$S_{\mathrm{v}}(z) := \begin{cases} \bigcap_{p \in D^\pm v(z)} \arg\min_A \{ p \cdot f(z, \cdot) + \ell(z, \cdot) \} & \text{if } D^\pm v(z) \neq \emptyset, \\ A & \text{if } D^\pm v(z) = \emptyset, \end{cases}$$

where the subscripts D and v stand for Dini and viscosity, respectively. Under the assumptions (A_0)-(A_4), a necessary condition for the optimality of an admissible feedback Φ is the existence of a set $X \subseteq \mathbb{R}^N$ such that

(2.84) $$\Phi(z) \in S_{\mathrm{D}}(z) \quad \text{for all } z \in X,$$

and X is "large on the trajectories of Φ" in the following sense

(2.85) $$\forall x \in \mathbb{R}^N, \quad y_x(t, \Phi) \in X \quad \text{for a.e. } t > 0.$$

To prove this statement it is enough to define $X := \{ z : \Phi(z) \in S_{\mathrm{D}}(z) \}$ and use Theorem 2.49. By the same argument we see that another necessary condition of optimality for Φ is

(2.86) $$\Phi(z) \in S_{\mathrm{v}}(z) \quad \text{for all } z \in X,$$

for some set X satisfying (2.85).

Assume now v is locally lipschitzean. Then (2.84) with (2.85) is also a sufficient condition for the optimality of Φ by Theorem 2.52. If we assume in addition $D^+v(z) = \partial v(z)$ for all $z \in \mathbb{R}^N$, then (2.86) with (2.85) becomes also a sufficient condition of optimality for Φ, by Theorem 2.54 (iii).

It is natural to try to use the sufficient conditions (2.84) and (2.86) to construct optimal admissible feedback maps. An easy example is the case when $S_D(x)$ (or S_v) is nonempty and single valued for all x, and for each initial point there is a unique solution of $y' = f(y, S_D(y))$ such that $\alpha^*(\,\cdot\,) = S_D(y(\,\cdot\,))$ is measurable, for instance when S_D is a Lipschitz function. Then α^* is an optimal open loop control for the initial point $y(0)$, the unique one if we identify controls coinciding a.e., and S_D is an optimal feedback map. This situation occurs, for instance, in linear-quadratic problems, but it is very peculiar. Typically S_D and S_v are multivalued, or empty, or discontinuous. Therefore an optimal admissible feedback may not exist, as in the following simple example.

EXAMPLE 2.58. Consider the one-dimensional problem of Example 2.30. The value function v is globally lipschitzean and differentiable for $x \neq 0$. Since $f(x, a) = a \in [-1, 1]$ we compute $\partial^- v(0; a) = |a| (e^{-1} - 1)$. Now it is easy to check that

$$S_D(x) = S_v(x) = \begin{cases} 1, & x > 0, \\ -1, & x < 0, \end{cases} \qquad S_D(0) = \{-1, 1\}, \qquad S_v(0) = \emptyset.$$

If Φ were an optimal feedback map, it should be $\Phi(x) = 1$ for a.e. $x > 0$ and $\Phi(x) = -1$ for a.e. $x < 0$. Therefore, no matter how we define $\Phi(0)$, there are two distinct solutions of $y' = f(y, \Phi(y))$, $y(0) = 0$, namely, $y_1(t) = t$ and $y_2(t) = -t$. Thus Φ is not admissible. ◁

The previous discussion leads naturally to the introduction of the notion of a *multivalued feedback map*.

DEFINITION 2.59. A map $S : \mathbb{R}^N \to \mathcal{P}(A)$ is a multivalued admissible feedback if for all $x \in \mathbb{R}^N$ there is a pair (y, α) such that $\alpha \in \mathcal{A}$ and

$$\begin{cases} y(t) = x + \int_0^t f(y(s), \alpha(s))\, ds, & t \geq 0 \\ \alpha(s) \in S(y(s)), & \text{a.e. } s > 0. \end{cases}$$

Such a pair (y, α) is called a solution of

(2.87)
$$\begin{cases} y' \in f(y, S(y)), & t > 0, \\ y(0) = x. \end{cases} \quad \triangleleft$$

Note that any solution (y, α) of (2.87) is such that $y(\,\cdot\,) = y_x(\,\cdot\,, \alpha)$ in our standard notation. Our definition of solution of (2.87) is slightly different from the usual one in the theory of differential inclusions, because we explicitly mention the "selection" α as part of the solution, and we require it to be measurable. We do this because we want the cost functional to be defined on every solution of (2.87). However, this does not restrict the class of admissible feedbacks, by a classical theorem of Filippov.

DEFINITION 2.60. A multivalued admissible feedback S is *weakly optimal* if for all $x \in \mathbb{R}^N$ there exists an optimal solution (y, α) of (2.87), that is, a solution such that $J(x, \alpha) = \min_{\mathcal{A}} J(x, \cdot)$; S is *fully optimal* if for all $x \in \mathbb{R}^N$ every solution of (2.87) is optimal. ◁

The first obvious remark on this definition is that a necessary condition for the existence of a weakly optimal multivalued feedback is the existence of an optimal open loop control for every initial point $x \in \mathbb{R}^N$, and in this case $S(x) = A$ for all x is a trivial weakly optimal feedback, certainly not fully optimal in general. The next result says that this condition is actually sufficient for the existence of nontrivial weakly optimal feedbacks; it gives formulas for some of them in terms of generalized derivatives of the value function, and shows that they can be fully optimal.

THEOREM 2.61. *Assume* (A_0)–(A_4) *and suppose that for all* $x \in \mathbb{R}^N$ *there is a corresponding optimal control in* $\mathcal{A}$. *Then*

(i) *the multivalued feedbacks* S_D *and* S_v *are weakly optimal;*

(ii) *if in addition the value function v is locally lipschitzean, then S_D is fully optimal;*

(iii) *if moreover $D^+v(z) = \partial v(z)$ for all $z \in \mathbb{R}^N$, then S_v is fully optimal as well.*

PROOF. Given $x \in \mathbb{R}^N$ let $\alpha^* \in \mathcal{A}$ be a corresponding optimal control and set $y^*(\cdot) = y(\cdot, \alpha^*)$. We claim that the pair (y^*, α^*) is a solution of the differential inclusion (2.87) with $S = S_D$. In fact, by Theorem 2.49, for a.e. $t > 0$

$$\partial^- v(y^*(t); f(y^*(t), \alpha^*(t))) + \ell(y^*(t), \alpha^*(t)) \leq \lambda v(y^*(t)),$$

thus $\alpha^*(t) \in S_D(y^*(t))$ for a.e. t. Then the feedback S_D is weakly optimal, and a similar argument works for S_v.

To prove the full optimality of S_D (respectively, S_v) it is enough to observe that any solution of the differential inclusion (2.87) with $S = S_D$ (respectively, $S = S_v$) is optimal because it satisfies the sufficient condition (2.73) of Theorem 2.52 with $u = v$ (respectively, condition (2.80) of Theorem 2.54 (iii)), for a.e. t. ◀

REMARK 2.62. It would be nice to have a representation for S_D in the form "arg max of the Hamiltonian," as we have for S_v. It is easy to deduce from Remark 2.43 that $a \in S_D(z)$ if and only if

$$\partial^- v(z; f(z, a)) + \ell(z, a) = \min_{(f, \ell) \in (\mathrm{FL})(z)} \{\partial^- v(z; f) + \ell\},$$

for any $z \in \mathbb{R}^N$. In particular, if $(\mathrm{FL})(z) = f(z, A) \times \ell(z, A)$ for all z, then

$$S_D(z) = A_*(z) = \arg\min_A \{\partial^- v(z; f(z, \cdot)) + \ell(z, \cdot)\}, \qquad \text{for all } z. \quad ◁$$

The crucial assumption of Theorem 2.61 is the existence of an optimal open loop control for every initial point. There is a large literature on this issue. The most classical and simple sufficient condition for this property is a convexity assumption on the problem. Next we give two explicit applications of this observation, beginning with the relaxed problem.

COROLLARY 2.63. *Assume* (A$_0$), *A compact,* (1.1), (A$_4$) *with* $\omega_\ell(r) = Mr$. *Then the following multivalued feedback for the relaxed problem,*

$$A^r_*(z) := \arg\min_{A^r}\{\partial^- v(z; f^r(z, \cdot)) + \ell^r(z, \cdot)\},$$

is weakly optimal, and it is fully optimal if, in addition, v is locally lipschitzean.

PROOF. In Section VI.1 we will construct for any initial point an optimal relaxed control $\mu^* \in \mathcal{A}^r$, under the current assumptions. Moreover, by (2.32), we have

$$((\text{FL}))(z) = f^r(z, A^r) \times \ell^r(z, A^r),$$

for all z. Therefore the conclusions follow from Theorem 2.55 and Remark 2.62. ◀

COROLLARY 2.64. *Assume the hypotheses of Corollary 2.63 and suppose* $f(x, A) \times \ell(x, A)$ *is a convex set for all* $x \in \mathbb{R}^N$. *Then*

(i) *the multivalued feedbacks* $S_D = A_*$ *and* S_v *are weakly optimal;*

(ii) *if, in addition, v is locally lipschitzean, then* $S_D = A_*$ *is fully optimal;*

(iii) *if, moreover, v is locally semiconcave, then* S_v *is fully optimal.*

PROOF. In Section VI.1 we prove the existence of an optimal control for any initial point, under the current assumptions. Therefore all the conclusions follow from Theorem 2.61, Remark 2.62, and Proposition II.4.7. ◀

We end this subsection with a few remarks on the problem of finding an optimal synthesis as defined in Definition 2.17. In analogy to §2.2 we can associate with a multivalued admissible feedback S a *multivalued control law* or *multivalued presynthesis* $\mathbb{A}^S : \mathbb{R}^N \to \mathcal{P}(\mathcal{A})$ as follows

$$\mathbb{A}^S(x) = \{\alpha \in \mathcal{A} : (y, \alpha) \text{ is a solution of } (2.87)\}.$$

Under all the assumptions of Theorem 2.61, the presynthesis generated by S_D and S_v satisfy

$$\mathbb{A}^{S_D}(x) = \mathbb{A}^{S_v}(x) = \mathbb{A}^*(x) \qquad \text{for all } x \in \mathbb{R}^N,$$

where $\mathbb{A}^*$ is the maximal optimal presynthesis

$$\mathbb{A}^*(x) := \{\alpha \in \mathcal{A} : \alpha \text{ is optimal for } x\},$$

as it is easy to see from the proof of the theorem. By Remark 2.6, $\mathbb{A}^*$ is *memoryless*, in the sense that

$$\mathbb{A}^*(z) = \{\alpha(\cdot + t) : t \geq 0, \alpha \in \mathbb{A}(x), z = y_x(t, \alpha)\},$$

and we call it a *multivalued synthesis*. Therefore we have a method for generating the *maximal optimal multivalued synthesis* $\mathbb{A}^*$. To our knowledge there are no general results on the existence of single-valued optimal syntheses.

Exercises

2.1. Prove directly that there is at most one bounded function satisfying the DPP, that is equation (2.5).

2.2. Denote by $\mathcal{F}$ the set of admissible feedback maps on $\mathbb{R}^N$, and set, for $\Phi \in \mathcal{F}$,

$$J_{\mathcal{F}}(x, \Phi) := \int_0^\infty e^{-\lambda t} \ell(y_x(t, \Phi), \Phi(y_x(t, \Phi))) \, dt$$

$$v_{\mathcal{F}}(x) = \inf_{\Phi \in \mathcal{F}} J_{\mathcal{F}}(x, \Phi).$$

(i) Prove that $v_{\mathcal{F}} = v$. [Hint: one inequality is trivial, the other is easily obtained by adding time as a state variable.]

(ii) Prove directly (without using (i)) that $v_{\mathcal{F}}$ is continuous and satisfies the Dynamic Programming Principle.

2.3. Let $H : \mathbb{R}^{2N} \to \mathbb{R}$ be continuous and satisfy (2.18), and assume $u_1, u_2 \in BUC(\mathbb{R}^N)$ solve, respectively,

$$u_1 + H(x, Du_1) + g(x) \leq 0 \quad \text{in } \mathbb{R}^N,$$
$$u_2 + H(x, Du_2) + h(x) \geq 0 \quad \text{in } \mathbb{R}^N,$$

with $g, h \in BUC(\mathbb{R}^N)$. Show that

$$\sup_{\mathbb{R}^N}(u_1 - u_2) \leq \sup_{\mathbb{R}^N}(h - g).$$

2.4 (VARIABLE INTEREST RATE). Let $\lambda : \mathbb{R}^N \times A \to \mathbb{R}$ satisfy $0 < \lambda_0 \leq \lambda(x, a) \leq M'$ and $|\lambda(x, a) - \lambda(y, a)| \leq \omega_\lambda(|x - y|)$, where ω_λ is a modulus, for all $x, y \in \mathbb{R}^N$ and $a \in A$. Consider the payoff

$$J(x, \alpha) := \int_0^\infty \exp\left(-\int_0^t \lambda(y_x(s), \alpha(s)) \, ds\right) \ell(y_x(t), \alpha(t)) \, dt$$

under the hypotheses (A_0)–(A_4).

(i) Prove that the value function $v = \inf_\alpha J$ satisfies the following DPP:

$$v(x) = \inf_{\alpha \in \mathcal{A}} \left\{ \int_0^t \ell(y_x(s), \alpha(s)) \exp\left(-\int_0^s \lambda(y_x(\tau), \alpha(\tau)) \, d\tau\right) ds \right.$$
$$\left. + v(y_x(t)) \exp\left(-\int_0^t \lambda(y_x(\tau), \alpha(\tau)) \, d\tau\right) \right\}.$$

(ii) Prove that v is a viscosity solution of

$$\sup_{a \in A} \{\lambda(x, a) v - f(x, a) \cdot Dv - \ell(x, a)\} = 0 \quad \text{in } \mathbb{R}^N.$$

(iii) Rewrite the Bellman equation in (ii) in the form

(2.88) $$v + H(x, v, Dv) = 0 \quad \text{in } \mathbb{R}^N$$

2. THE INFINITE HORIZON PROBLEM

with H continuous and such that

(2.89) $\quad r \mapsto H(x,r,p)$ is nondecreasing in $\mathbb{R}$, for all $x, p \in \mathbb{R}^N$,

$$H(y,r,\mu(x-y) - \tau y) - H(x,r,\mu(x-y) + \gamma x)$$
(2.90) $\quad \leq \mu L'|x-y|^2 + \tau K(1+|y|^2) + \gamma K'(1+|x|^2) + \varrho(|x-y|,R)$

for all $\mu, \tau, \gamma > 0$, $x, y \in \mathbb{R}^N$, $r \in \mathbb{R}$ such that $|r| \leq R$

where ϱ is a modulus, $L' = L/\lambda_0$, $K' = K/\lambda_0$.

2.5. Prove the analogue of Theorem 2.12 for equation (2.88) and H satisfying the structure conditions (2.89) and (2.90). (Hint: follow closely the proof of Theorem 2.12, but treat separately the case $u_1(\overline{x}) \leq u_2(\overline{x})$ and the case $u_1(\overline{x}) > u_2(\overline{x})$.)

2.6. Discuss the possibility of dropping some assumptions on λ in the previous exercises.

2.7. Prove that $v \in BC(\mathbb{R}^N)$ also when the uniform continuity of ℓ in (A4) is weakened to

(2.91) $\qquad |\ell(x,a) - \ell(y,a)| \leq \omega_\ell(|x-y|, R)$

for all $a \in A$ and $x, y \in \mathbb{R}^N$ such that $|x|, |y| \leq R$, and all $R > 0$, and ω_ℓ is a modulus. Prove that Theorem 2.12 keeps holding if ω_ℓ is allowed to depend on R as above and with BUC replaced by BC. [Hint: see Remark 2.13.]

2.8. Prove that the value function v in Example 2.1 is semiconcave if and only if $\lambda \geq 2$. Note that the sufficient condition for semiconcavity in Remark II.4.10 (i.e., $\lambda > 2L$) is almost optimal.

2.9. Prove that $\pm |x|^b$ and the constant 0 are viscosity solutions of $u - (1/b)x \cdot Du = 0$ in $\mathbb{R}^N$ for any $b > 0$. In dimension $N = 1$ we also have the solutions $\pm \operatorname{sgn} x |x|^b$.

2.10. Under the assumptions of Lemma 2.20 prove (2.32). [Hint: for the inclusion "$\supseteq$" prove the following mean value theorem: for a probability measure m on A and an m-measurable $G : A \to \mathbb{R}^M$, $\int_A G(a)\,dm \in \overline{co} G(A)$.]

2.11. Prove the assertion in Remark 2.42. Using the backward DPP prove that the value function v satisfies, at any $x \in \mathcal{O}$,

$$\lambda v + \sup_{(f,\ell) \in (\mathrm{FL})(x)} \{\partial^+ u(x; -f) - \ell\} \geq 0,$$

and deduce from this that v is a subsolution of $-\lambda v - H(x, Dv) = 0$ at points of $\mathcal{O}$.

2.12. In the assumptions of Theorem 2.40 prove that a locally lipschitzean supersolution satisfies

$$\lambda u(x) + \liminf_{t \to 0^+} \sup_{(f,\ell) \in (\mathrm{FL})(x)} \left\{ \frac{u(x) - u(x+tf)}{t} - \ell \right\} \geq 0\,.$$

2.13. Extend the property of Dini derivatives given in Remark 2.44 to the following: if, for some $b \in \mathbb{R}$, $\partial^- u(x; q) > b$ for all $q \in \mathcal{C}$, then there exist sequences $x_n \to x$, $p_n \in D^- u(x_n)$, such that $p_n \cdot q \geq b$ for all $q \in \mathcal{C}$. Note that this implies

$$\partial^- u(x; q) \leq \limsup_{z \to x} \sup\{ p \cdot q : p \in D^- u(z) \}$$

for all $q \in \mathbb{R}^N$, which complements Lemma 2.37.

2.14. Give a direct proof of the statement (i) $\Rightarrow$ (iii) in Theorem 2.40 by means of Exercise 2.13. [Hint: consider the function $U(z, s) := u(z) + s$, $s \in \mathbb{R}$, and its derivatives in $(x, 0)$.]

2.15. Prove Lemma 2.48. [Hint: to prove "$\leq$" choose b_n such that $w^{b_n}(x) \to w(x)$ and show that

$$\frac{w(x + hq) - w(x)}{h} \leq \liminf_n \{ \partial w^{b_n}(x; q) + \omega(h) \} .$$

To prove "$\geq$" for $h_n \to 0^+$ choose b_n such that $w^{b_n}(x + h_n q) \leq w(x + h_n q) + h_n^2$, show that $w^{b_n}(x) \to w(x)$ and $\partial w^{b_n}(x; q) - \omega(h_n) \leq \dfrac{w(x + h_n q) - w(x)}{h_n} + h_n.$]

2.16. Fill in the details of the proof of Proposition 2.47.

2.17. Extend the chain rule Lemma 2.50 to the case y has just right or left-hand derivatives.

2.18. Prove that a necessary condition of optimality holding for all $t \geq 0$ is that

$$(2.92) \qquad (f, \ell) := \lim_n \frac{1}{t_n} \int_0^{t_n} \bigl(f(y(t), \alpha(t + s)), \ell(y(t), \alpha(t + s)) \bigr) \, ds$$

satisfies $\partial^- v(y(t); f) + \ell \leq \lambda v(y(t))$, for all $t_n \to 0^+$ such that the limit in (2.92) exists. [Hint: follow the last part of the proof of Theorem 2.40.] Note that, by the compactness of $(FL)(x)$ for all x, the limit in (2.92) exists for some $t_n \to 0^+$.

2.19. For what values of $\lambda > 0$ is the value function of Example 2.4 an intrinsically classical solution of the HJB equation in $]-1, 1[$?

2.20. Derive a necessary condition for optimality from the Backward DPP Proposition 2.25, and discuss its sufficiency.

2.21. Consider the multivalued feedbacks

$$S_\sharp(z) := \{ a \in A : p \cdot f(z, a) + \ell(z, a) \leq \lambda v(z) \quad \forall p \in D^\pm v(z) \},$$
$$S_{D+}(z) := \{ a \in S_D(z) : -\partial^+ v(x; -f(z, a)) + \ell(z, a) \leq \lambda v(z) \},$$

and prove that

(i) they are weakly optimal under the assumptions of Theorem 2.61 (i);

(ii) $S_{D+}(z) \subseteq S_D(z) \subseteq S_\sharp(z) \subseteq S_v(z)$ for all $z \in \mathbb{R}^N$;

(iii) $S_\sharp$ is a fully optimal feedback if $D^\pm v(x) \neq \emptyset$ for all $x \in \mathbb{R}^N$ and v is locally lipschitzean.

3. The finite horizon problem

3.1. The HJB equation

In this section the cost to minimize is

$$J(x,t,\alpha) := \int_0^t e^{-\lambda s}\ell(y_x(s,\alpha),\alpha(s))\,ds + e^{-\lambda t}g(y_x(t,\alpha)),$$

where the running cost ℓ and the constant interest rate λ satisfy

(A$_4'$) ℓ satisfies the hypotheses in (A$_4$); $\lambda \geq 0$;

and the *terminal cost* g satisfies, in this subsection,

(A$_5$) $g \in BUC(\mathbb{R}^N)$; that is, there exists a modulus ω_g and a constant G such that $|g(x) - g(y)| \leq \omega_g(|x-y|)$ and $|g(x)| \leq G$ for all $x,y \in \mathbb{R}^N$.

In §3.2 we consider the case of unbounded terminal cost. Following the terminology of the calculus of variations, the problem of minimizing $J(x,t,\cdot)$ is called a *Mayer problem* if $\ell \equiv 0$ and $\lambda = 0$, a *Lagrange problem* if $g \equiv 0$, and a *Bolza problem* in the general case. If $\lambda = 0$ and ℓ is Lipschitz continuous with respect to the state variables uniformly in the control variables, then the Bolza problem can be converted into a Mayer problem by adding a scalar state variable y_{N+1} to the vector y, with dynamics

$$y'_{N+1}(s) = \ell(y(s),\alpha(s)), \quad s > 0,$$
$$y_{N+1}(0) = 0,$$

and minimizing the payoff $y_{N+1}(t) + g(y_x(t))$. (This new terminal cost is unbounded but we can replace it with $\psi(y_{N+1} + g(y))$ where ψ is any bounded and strictly increasing function, e.g., $\psi(r) = \arctan r$, without changing the optimal control problem.) Therefore in the sequel the reader may assume for simplicity that ℓ is null, at least when $\lambda = 0$.

There are more general formulations of finite horizon problems. It is often important to incorporate constraints on the state variables, and this can be done by the methods of Chapter IV. It is also important to consider the nonautonomous case, that is, the case of dynamics and running cost depending explicitly on time $s > 0$:

$$f = f(s,y,a), \qquad \ell = \ell(s,y,a).$$

If the dependence of f on the variable s is Lipschitz continuous uniformly in y and a, then this case can be reduced to the standard autonomous one by treating time as an extra state variable, that is, by adding

$$y'_{N+1}(s) = 1, \quad s > 0,$$
$$y_{N+1} = 0,$$

to the system (S). When f depends on time in a less regular way we have to treat the nonautonomous problem separately, but this extension is not hard: see Remark 3.10 and Exercise 3.7.

The value function of the finite horizon problem is
$$v(x,t) := \inf_{\alpha \in \mathcal{A}} J(x,t,\alpha) .$$

We begin with a regularity result for v. As for the infinite horizon problem, the value function fails to be differentiable even for linear systems and smooth costs: the reader can easily construct some examples or see Exercises 3.3 and 3.4.

PROPOSITION 3.1. *Assume* (A_0), (A_1), (A_3), (A_4'), *and* (A_5). *Then*

(i) v *is bounded and continuous in* $\mathbb{R}^N \times [0,T]$ (*i.e.*, $v \in BC(\mathbb{R}^N \times [0,T])$), *for all* $T > 0$;

(ii) $\lambda > 0$ *implies* $v \in BC(\mathbb{R}^N \times [0,+\infty[)$;

(iii) g *Lipschitz continuous and* $\omega_\ell(r) = L_\ell r$ (*i.e.*, ℓ *is lipschitzean in* y *uniformly in* a) *imply that* v *is lipschitzean on* $\mathcal{K} \times [0,T]$ *for all compact* $\mathcal{K} \subseteq \mathbb{R}^N$ *and* $T > 0$.

PROOF. Let us fix $x, z \in \mathbb{R}^N$, $t, \tau \in [0,T]$, $\varepsilon > 0$ and $\overline{\alpha} \in \mathcal{A}$ such that
$$v(z,\tau) \geq \int_0^\tau e^{-\lambda s} \ell(y_z(s), \overline{\alpha}(s)) \, ds + e^{-\lambda \tau} g(y_z(\tau, \overline{\alpha})) - \varepsilon .$$

Then

$$\begin{aligned}
(3.1) \quad v(x,t) &- v(z,\tau) \\
&\leq \int_0^t e^{-\lambda s} \ell(y_x, \overline{\alpha}) \, ds + e^{-\lambda t} g(y_x(t, \overline{\alpha})) \\
&\quad - \int_0^\tau e^{-\lambda s} \ell(y_z, \overline{\alpha}) \, ds - e^{-\lambda \tau} g(y_z(\tau, \overline{\alpha})) + \varepsilon \\
&\leq M |t - \tau| + \int_0^{t \wedge \tau} e^{-\lambda s} \omega_\ell(|y_x(s) - y_z(s)|) \, ds \\
&\quad + |t - \tau| G + e^{-\lambda \tau} \omega_g(|y_x(t) - y_z(\tau)|) + \varepsilon \\
&\leq (M + G)|t - \tau| + T \omega_\ell(e^{LT} |x - z|) + \omega_g(C |t - \tau| + e^{LT} |x - z|) + \varepsilon,
\end{aligned}$$

where C is a Lipschitz constant for $y_x(\,\cdot\,, \overline{\alpha})$ on $[0,T]$ and we have used (A_4'), (A_5) and (1.5). This proves the continuity of v. The boundedness of v follows from
$$|J(x,t,\alpha)| \leq TM + G, \quad \text{for all } x \in \mathbb{R}^N, t \in [0,T], \alpha \in \mathcal{A} .$$

The estimate (3.1) proves (iii) as well, because C can be chosen independently of x in a compact set by (A_1).

To prove (ii) it is enough to observe that
$$|J(x,t,\alpha)| \leq M \int_0^\infty e^{-\lambda s} \, ds + G \quad \text{for all } x, t, \alpha . \quad \blacktriangleleft$$

3. THE FINITE HORIZON PROBLEM

PROPOSITION 3.2 (DYNAMIC PROGRAMMING PRINCIPLE). *Assume* (A_0), (A_1), (A_3), (A'_4), *and* (A_5). *Then for all* $x \in \mathbb{R}^N$ *and* $0 < \tau \leq t$

$$(3.2) \quad v(x,t) = \inf_{\alpha \in \mathcal{A}} \left\{ \int_0^\tau e^{-\lambda s} \ell(y_x(s), \alpha(s))\, ds + e^{-\lambda \tau} v(y_x(\tau), t - \tau) \right\}.$$

PROOF. Since $v(z,0) = g(z)$ by definition, (3.2) reduces to the definition of v when $\tau = t$. For $\tau < t$ and $\alpha \in \mathcal{A}$ fixed, we define $\widetilde{\alpha}(s) := \alpha(s+\tau)$ and observe that

$$(3.3)$$
$$J(x,t,\alpha) = \int_0^\tau e^{-\lambda s}\ell(y_x(s),\alpha(s))\,ds + \int_\tau^t e^{-\lambda s}\ell(y_x(s),\alpha(s))\,ds + e^{-\lambda t}g(y_x(t))$$
$$= I_1 + \int_0^{t-\tau} e^{-\lambda r}e^{-\lambda \tau}\ell(y_x(r+\tau),\alpha(r+\tau))\,dr + e^{-\lambda \tau}e^{-\lambda(t-\tau)}g(y_x(t))$$
$$= I_1 + e^{-\lambda \tau}J(y_x(\tau,\alpha),t-\tau,\widetilde{\alpha}) \geq I_1 + e^{-\lambda \tau}v(y_x(\tau,\alpha),t-\tau),$$

where I_1 denotes the first integral in (3.3). By taking the inf over $\mathcal{A}$ we get the inequality "$\geq$" in (3.2).

To prove the opposite inequality we fix $\alpha \in \mathcal{A}$, set $z := y_x(\tau,\alpha)$, fix $\varepsilon > 0$ and $\alpha^1 \in \mathcal{A}$ such that $v(z,t-\tau) + \varepsilon \geq J(z,t-\tau,\alpha^1)$. We define the control

$$\overline{\alpha}(s) = \begin{cases} \alpha(s) & s \leq \tau, \\ \alpha^1(s-\tau) & s > \tau, \end{cases}$$

and let $\overline{y}$ and y^1 be the trajectories corresponding to $\overline{\alpha}$ and α^1, respectively. Then

$$(3.4) \quad v(x,t) \leq J(x,t,\overline{\alpha})$$
$$= \int_0^\tau e^{-\lambda s}\ell(\overline{y}_x(s),\overline{\alpha}(s))\,ds + \int_\tau^t e^{-\lambda s}\ell(\overline{y}_x(s),\overline{\alpha}(s))\,ds + e^{-\lambda t}g(\overline{y}_x(t))$$
$$= I_1 + e^{-\lambda \tau}\int_0^{t-\tau} e^{-\lambda r}\ell(y_z^1(r),\alpha^1(r))\,dr + e^{-\lambda \tau}e^{-\lambda(t-\tau)}g(y_z^1(t-\tau))$$
$$= I_1 + e^{-\lambda \tau}J(z,t-\tau,\alpha^1) \leq I_1 + e^{-\lambda \tau}v(z,t-\tau) + \varepsilon,$$

where I_1 is the first integral in (3.4). Since α and ε are arbitrary we have obtained the desired inequality. ◀

REMARK 3.3. It is not hard to see from the arguments of the proof of Proposition 3.2 that if α^* is optimal for (x,t) (i.e., $v(x,t) = J(x,t,\alpha^*)$), then the infimum in the DPP (3.2) is attained at α^* for all τ, and that $\alpha^*(\cdot + \tau)$ is optimal for $(y_x(\tau,\alpha^*), t - \tau)$ for all $0 \leq \tau \leq t$.

Moreover the function

$$\tau \mapsto \int_0^\tau e^{-\lambda s}\ell(y_x(s),\alpha(s))\,ds + e^{-\lambda \tau}v(y_x(\tau),t-\tau)$$

is nondecreasing for all $\alpha \in \mathcal{A}$, and it is constant if and only if α is optimal for (x,t). See Remark 2.6 for the corresponding facts in the infinite horizon case and their interpretation in terms of Bellman's Principle of Optimality. ◁

REMARK 3.4. Proposition 3.1 and the DPP still hold if we replace the set of admissible control functions $\mathcal{A}$ in the definition of the value function with any $\mathcal{B} \subseteq \mathcal{A}$ with the properties of being closed with respect to the operations of translation in time and piecing together two different controls: see Remark 2.7. Thus for instance $v^\sharp(x,t) := \inf_{\alpha \in \mathcal{P}} J(x,t,\alpha)$, the value function associated with piecewise constant controls, satisfies (3.2) with $\mathcal{A}$ replaced by $\mathcal{P}$. ◁

As in Section 2 the DPP implies that v satisfies the Hamilton-Jacobi-Bellman equation. We recall the definition of the Hamiltonian H

$$(3.5) \qquad H(x,p) := \sup_{a \in A} \{ -f(x,a) \cdot p - \ell(x,a) \} .$$

PROPOSITION 3.5. *Assume* (A_0)–(A_3), (A_4'), (A_5). *Then the value function* $v(x,t)$ *is a viscosity solution of*

$$(3.6) \qquad v_t + \lambda v + H(x, Dv) = 0 \quad \text{in } \mathbb{R}^N \times \,]0, +\infty[,$$

where v_t denotes the partial derivative with respect to the last variable and $Dv = D_x v$ the gradient with respect to the first N variables.

PROOF. Let $\phi \in C^1(\mathbb{R}^N \times \,]0, +\infty[)$ and (x,t) be a local maximum point of $v - \phi$; that is, for some $r > 0$

$$(3.7) \qquad v(x,t) - v(z,s) \geq \phi(x,t) - \phi(z,s), \qquad \text{if } |x-z| < r,\ |t-s| < r .$$

Fix an arbitrary $a \in A$ and let $y_x(\,\cdot\,)$ be the solution corresponding to the constant control $\alpha(\tau) = a$ for all τ. For τ small enough $|y_x(\tau) - x| < r$ by (1.3), and then

$$(3.8) \qquad \phi(x,t) - \phi(y_x(\tau), t - \tau) \leq v(x,t) - v(y_x(\tau), t - \tau) \qquad \text{for } 0 \leq \tau \leq t_0 .$$

By using the inequality "$\leq$" in the DPP (3.2) we get

$$(3.9) \quad \phi(x,t) - \phi(y_x(\tau), t - \tau)$$
$$\leq \int_0^\tau e^{-\lambda s} \ell(y_x(s), \alpha(s))\, ds + v(y_x(\tau), t - \tau)(e^{-\lambda \tau} - 1) .$$

Next we divide by $\tau > 0$ and pass to the limit as $\tau \to 0$, to obtain

$$\phi_t(x,t) - D\phi(x,t) \cdot y_x'(0) = \phi_t - D\phi \cdot f(x,a) \leq \ell(x,a) - \lambda v(x,t),$$

where we have used the differentiability of ϕ and the continuity of v, y_x, f, and ℓ. Since $a \in A$ is arbitrary we get

$$\phi_t(x,t) + \lambda v(x,t) + \sup_{a \in A} \{ -f(x,a) \cdot D\phi(x,t) - \ell(x,a) \} \leq 0,$$

which means that v is a viscosity subsolution.

To check that v is a supersolution we assume now that (x,t) is a minimum point of $v - \phi$, so that (3.7) holds with the inequality reversed. For each $\varepsilon > 0$ and $\tau > 0$,

3. THE FINITE HORIZON PROBLEM

by the inequality "$\geq$" in the DPP (3.2), there exists $\alpha \in \mathcal{A}$, depending on ε and τ, such that

$$v(x,t) \geq \int_0^\tau e^{-\lambda s}\ell(y_x(s),\alpha(s))\,ds + e^{-\lambda \tau}v(y_x(\tau),t-\tau) - \tau\varepsilon\,.$$

Then (3.8) holds with the reversed inequality, and so does (3.9) with an extra term $-\tau\varepsilon$ in the right-hand side. As before we divide by $\tau > 0$ and let $\tau \to 0$: we have reduced to the classical proof because we have replaced v by the differentiable test function ϕ. However the details are a bit more lengthy than in the first part of the proof because α depends on τ and it is not a constant function: we leave them as an exercise for the reader since they are analogous to those of the proof of Proposition 2.8. The conclusion is

$$\phi_t(x,t) + \lambda v(x,t) + \sup_{a\in A}\{-f(x,a)\cdot D\phi(x,t) - \ell(x,a)\} \geq -\varepsilon,$$

which completes the proof by the arbitrariness of ε. ◀

Next we characterize the value function v as the unique viscosity solution of the HJB equation (3.6) with the natural initial condition

(3.10) $\qquad\qquad v(x,0) = g(x) \qquad \text{for all } x \in \mathbb{R}^N\,.$

Indeed, as in Section 2, we want to compare v with any subsolution of the initial value problem

(3.11) $\qquad \begin{cases} u_t + \lambda u + H(x, Du) = 0 & \text{in } \mathbb{R}^N \times \,]0, T[, \\ u(x,0) = g(x) & \text{in } \mathbb{R}^N, \end{cases}$

for $T \leq +\infty$, namely, a subsolution u_1 of the HJB equation such that $u_1(x,0) \leq g(x)$ in $\mathbb{R}^N$, and any supersolution of (3.11), that is, a supersolution u_2 of the HJB equation such that $u_2(x,0) \geq g(x)$ in $\mathbb{R}^N$. By analogy with Section 2 a subsolution and a supersolution of (3.11) may be called, respectively, a verification function and a falsification function. We call v a *complete solution* of (3.11) in a set of functions if it is the maximal subsolution and the minimal supersolution in such a set. Of course a complete solution in a set is the unique solution in the same set.

A first result for the case $\lambda > 0$ is an easy consequence of the theory in Section 2.

COROLLARY 3.6. *Assume* (A$_0$)–(A$_3$), (A$_4'$), (A$_5$) *and* $\lambda > 0$. *Then the value function v is the complete solution of* (3.11) *with* $T = +\infty$ *in* $BC(\mathbb{R}^N \times [0,\infty[)$.

PROOF. The value function v has the desired regularity by Proposition 3.1 (ii). The proof follows from Remark 2.14. It is enough to observe that $\widetilde{H} : \mathbb{R}^{2N+2} \to \mathbb{R}$, $\widetilde{H}(x,t,p,p_0) := \lambda^{-1}(p_0 + H(x,p))$ satisfies the structural assumption (2.18), rewrite the PDE in (3.11) as

$$u(z) + \widetilde{H}(z, D_z u) = 0 \qquad \text{in } \Omega := \mathbb{R}^N \times\,]0,\infty[,$$

where $z := (x,t)$, $D_z u := (D_x u, u_t)$, and note that the boundary condition $u_1 \leq u_2$ on $\partial\Omega$ (see Remark 2.14) follows from $u_1(x,0) \leq g(x) \leq u_2(x,0)$ in $\mathbb{R}^N$. ◀

Next we drop the assumption $\lambda > 0$. We recall that a comparison result for (3.11) with $\lambda = 0$ was already proved in Chapter II under different assumptions on H and on the sub- and the supersolution (see Theorem II.3.7).

The main result of this section is the following.

THEOREM 3.7. *Assume $H : \mathbb{R}^{2N} \to \mathbb{R}$ is continuous and satisfies (2.18), $T \in]0,+\infty[$, and $\lambda \geq 0$. If $u_1, u_2 \in BC(\mathbb{R}^N \times [0,T])$ are, respectively, a viscosity sub- and supersolution of*

(3.12) $$u_t + \lambda u + H(x, Du) = 0 \quad \text{in } \mathbb{R}^N \times]0, T[,$$

then

(3.13) $$\sup_{\mathbb{R}^N \times [0,T]} (u_1 - u_2) \leq \sup_{\mathbb{R}^N \times \{0\}} (u_1 - u_2)^+ \, .$$

In particular, if H is given by (2.9) and (A_0)–(A_3), (A_4'), and (A_5) hold, then the value function v is the complete solution of (3.11) in $BC(\mathbb{R}^N \times [0,T])$.

PROOF. The second part of the theorem follows easily from the first because for any subsolution u of (3.11) $u - v \leq 0$ on $\mathbb{R}^N \times \{0\}$, by definition, and then $u \leq v$ on $\mathbb{R}^N \times [0, T]$, and similarly v is below any supersolution.

The proof of the comparison assertion is based on the arguments in the proofs of Theorem 2.12, and Remarks 2.13 and 2.14, and makes use of Lemma II.2.10. Define

$$\Phi(x, y, t, s) := u_1(x,t) - u_2(y,s) - \frac{|x-y|^2 + |t-s|^2}{2\varepsilon} - \beta(\langle x \rangle^m + \langle y \rangle^m) - \eta(t+s),$$

where $\varepsilon, \beta, m, \eta$ are positive parameters, $\langle x \rangle = (1+|x|^2)^{1/2}$. Set

$$A := \sup_{\mathbb{R}^N \times \{0\}} (u_1 - u_2)^+$$

and assume by contradiction there exists $\delta > 0$, $\widetilde{x}, \widetilde{t}$ such that $(u_1 - u_2)(\widetilde{x}, \widetilde{t}) = A + \delta$. Choose $\beta > 0$ and $\eta > 0$ such that $2\beta\langle \widetilde{x} \rangle^m + 2\eta\widetilde{t} \leq \delta/2$ for all $m \leq 1$, so that

(3.14) $$A + \frac{\delta}{2} \leq A + \delta - 2\beta\langle \widetilde{x} \rangle^m - 2\eta\widetilde{t} = \Phi(\widetilde{x}, \widetilde{x}, \widetilde{t}, \widetilde{t}) \leq \sup_{\mathbb{R}^{2N} \times [0,T]^2} \Phi \, .$$

Since Φ is continuous and tends to $-\infty$ as $|x| + |y| \to +\infty$, there exist $\overline{x}, \overline{y}, \overline{t}, \overline{s}$ such that

(3.15) $$\sup_{\mathbb{R}^{2N} \times [0,T]^2} \Phi = \Phi(\overline{x}, \overline{y}, \overline{t}, \overline{s}) \, .$$

Since $\Phi(\overline{x}, \overline{y}, \overline{t}, \overline{s}) \geq A$, we have

(3.16) $\beta(\langle \overline{x} \rangle^m + \langle \overline{y} \rangle^m) \leq \sup u_1 - \inf u_2 - A =: c_1$, for all $\varepsilon > 0$, $m \in]0,1]$.

Now we fix $m := (\eta/Kc_1) \wedge 1$, where K is the constant appearing in assumption (2.18). Observe that $\overline{x}, \overline{y} \in \overline{B}(0, (c_1/\beta)^{1/m}) =: B_1$ for all $\varepsilon > 0$, by (3.16), and let ω be a modulus such that

(3.17) $$u_i(x,t) - u_i(y,s) = \omega(|x-y| + |t-s|)$$

for $i = 1, 2$, $x, y \in B_1$, $t, s \in [0, T]$. From the inequality $\Phi(\overline{x}, \overline{x}, \overline{t}, \overline{t}) + \Phi(\overline{y}, \overline{y}, \overline{s}, \overline{s}) \leq 2\Phi(\overline{x}, \overline{y}, \overline{t}, \overline{s})$ we easily get

(3.18) $$\frac{|\overline{x} - \overline{y}|^2 + |\overline{t} - \overline{s}|^2}{\varepsilon} \leq u_1(\overline{x}, \overline{t}) - u_1(\overline{y}, \overline{s}) + u_2(\overline{x}, \overline{t}) - u_2(\overline{y}, \overline{s}),$$

and the boundedness of u_1, u_2 gives, for a suitable constant C,

(3.19) $$|\overline{x} - \overline{y}| + |\overline{t} - \overline{s}| \leq C\sqrt{\varepsilon}.$$

By combining (3.17),(3.18) and (3.19) we get

(3.20) $$\frac{|\overline{x} - \overline{y}|^2 + |\overline{t} - \overline{s}|^2}{\varepsilon} \leq 2\omega(C\sqrt{\varepsilon}).$$

Now we show that neither $\overline{t}$ nor $\overline{s}$ can be 0. In fact, in $\overline{t} = 0$

$$\Phi(\overline{x}, \overline{y}, 0, \overline{s}) \leq u_1(\overline{x}, 0) - u_2(\overline{x}, 0) + u_2(\overline{x}, 0) - u_2(\overline{y}, \overline{s}) \leq A + \omega(C\sqrt{\varepsilon})$$

and we get a contradiction to (3.14) and (3.15) by choosing ε such that $\omega(C\sqrt{\varepsilon}) < \delta/2$. The proof that $\overline{s} > 0$ is analogous.

Next define the test functions

$$\varphi(x,t) := u_2(\overline{y}, \overline{s}) + \frac{|x - \overline{y}|^2 + |t - \overline{s}|^2}{2\varepsilon} + \beta(\langle x \rangle^m + \langle \overline{y} \rangle^m) + \eta(t + \overline{s}),$$

$$\psi(y,s) := u_1(\overline{x}, \overline{t}) - \frac{|\overline{x} - y|^2 + |\overline{t} - s|^2}{2\varepsilon} - \beta(\langle \overline{x} \rangle^m + \langle y \rangle^m) - \eta(\overline{t} + s),$$

and observe that $u_1 - \varphi$ attains its maximum at $(\overline{x}, \overline{t})$ and $u_2 - \psi$ attains its minimum at $(\overline{y}, \overline{s})$. Moreover

$$\varphi_t(\overline{x}, \overline{t}) = \frac{\overline{t} - \overline{s}}{\varepsilon} + \eta, \qquad D_x\varphi(\overline{x}, \overline{t}) = \frac{\overline{x} - \overline{y}}{\varepsilon} + \gamma\overline{x},$$

$$\psi_s(\overline{y}, \overline{s}) = \frac{\overline{t} - \overline{s}}{\varepsilon} - \eta, \qquad D_y\psi(\overline{y}, \overline{s}) = \frac{\overline{x} - \overline{y}}{\varepsilon} - \tau\overline{y},$$

where $\gamma = m\beta\langle \overline{x} \rangle^{m-2}$ and $\tau = m\beta\langle \overline{y} \rangle^{m-2}$. Since $\overline{t}, \overline{s} \in \,]0, T]$, by the definition of viscosity sub- and supersolution and Lemma II.2.10, we have

(3.21) $$\varphi_t(\overline{x}, \overline{t}) + \lambda u_1(\overline{x}, \overline{t}) + H(\overline{x}, D_x\varphi(\overline{x}, \overline{t})) \leq 0$$
$$\psi_s(\overline{y}, \overline{s}) + \lambda u_2(\overline{y}, \overline{s}) + H(\overline{y}, D_y\psi(\overline{y}, \overline{s})) \geq 0.$$

We use

(3.22) $$u_1(\overline{x}, \overline{t}) - u_2(\overline{y}, \overline{s}) \geq \Phi(\overline{x}, \overline{y}, \overline{t}, \overline{s}) \geq A + \delta/2 \geq 0$$

and the structural assumption (2.18) on H to obtain from (3.21)

$$
\begin{aligned}
(3.23) \quad 2\eta &\leq \frac{L}{\varepsilon}|\bar{x}-\bar{y}|^2 + \tau K\langle\bar{y}\rangle^2 + \gamma K\langle\bar{x}\rangle^2 + \omega_\ell(|\bar{x}-\bar{y}|) \\
&\leq 2L\omega(C\sqrt{\varepsilon}) + mK\beta(\langle\bar{y}\rangle^m + \langle\bar{x}\rangle^m) + \omega_\ell(C\sqrt{\varepsilon}) \\
&\leq o(1) + mKc_1 \leq o(1) + \eta \qquad \text{as } \varepsilon \to 0^+,
\end{aligned}
$$

where in the second inequality we have used (3.19) and (3.20), in the third inequality we have used (3.16) (and $o(1)$ indicates a function of ε going to 0 as $\varepsilon \to 0^+$), and in the last inequality we have used $m \leq \eta/Kc_1$. Now we reach the desired contradiction by taking ε small enough. ◀

REMARK 3.8. If $\lambda = 0$ in (3.12), we can drop the symbol of the positive part "+" in (3.13) and conclude that

$$\sup_{\mathbb{R}^N \times [0,T]} (u_1 - u_2) \leq \sup_{\mathbb{R}^N \times \{0\}} (u_1 - u_2).$$

In fact in the proof of the theorem we use that $A \geq 0$ only in the estimate (3.22), which is useless if $\lambda = 0$.

The assumption (2.18) can be weakened by replacing the term $\omega_\ell(|x-y|)$ by $\omega(|x-y|, R)$, where $\omega(\cdot, \cdot)$ is a modulus, and the inequality is assumed to hold for all $|x|, |y| \leq R$ and all $R > 0$. In fact, in the proof $\bar{x}, \bar{y} \in B(0, \overline{R})$, $\overline{R} = (c_1/\beta)^{1/m}$ for all $\varepsilon > 0$, and so in (3.23)

$$\omega(|\bar{x}-\bar{y}|, \overline{R}) \leq \omega(C\sqrt{\varepsilon}, \overline{R}) \longrightarrow 0 \qquad \text{as } \varepsilon \to 0^+.$$

Therefore, if H is given by (3.5), the assumption of uniform continuity of the running cost in (A$_4'$) can be weakened to

$$|\ell(x,a) - \ell(y,a)| \leq \omega(|x-y|, R)$$

for all $a \in A$, $|x|, |y| \leq R$ and all $R > 0$. ◁

REMARK 3.9. As in §2.2 we can apply the theory of this section to give verification theorems, relaxation results, and to prove stability for perturbations and approximations of the problem. We encourage the reader to figure out how to state these results. ◁

REMARK 3.10 (BACKWARD HJB EQUATION AND NONAUTONOMOUS SYSTEMS). Since we have assumed our system f and running cost ℓ to be autonomous (independent of time) the problem of minimizing $J(x, T-t, \alpha)$ is equivalent to that of minimizing

$$I(x,t,\alpha) := \int_t^T e^{\lambda(t-s)} \ell(y(s;x,t,\alpha), \alpha(s))\,ds + e^{\lambda(t-T)} g(y(T;x,t,\alpha)),$$

where $y(s; x, t, \alpha)$ indicates the position at time $s \geq t$ of the solution of

$$\begin{cases} y' = f(y, \alpha), \\ y(t) = x. \end{cases}$$

The value function of this problem
$$w(x,t) = \inf_{\alpha \in \mathcal{A}} I(x,t,\alpha)$$
is related to the previous one by the formula

(3.24) $$v(x,t) = w(x, T - t).$$

It is not hard to check the Dynamic Programming Principle for w:

(3.25)
$$w(x,t) = \inf_{\alpha \in \mathcal{A}} \left\{ \int_t^\tau e^{\lambda(t-s)} \ell(y(s; x, t, \alpha), \alpha(s))\, ds + e^{\lambda(t-\tau)} w(y(\tau; x, t, \alpha), \tau) \right\}$$

for all $\tau \in [t, T]$, and to deduce from this, or from (3.24), that w is a viscosity solution of
$$\begin{cases} -w_t + \lambda w + H(x, Dw) = 0 & \text{in } \mathbb{R}^N \times\,]-\infty, T[, \\ w(x, T) = g(x), \end{cases}$$
under the same assumptions as the corresponding statements for v.

If one is really interested only in minimizing the cost functional over the interval $[0, T]$, the choice between the intermediate functionals I and J is a matter of taste. We made the first choice for simplicity of notation and because initial value problems for forward evolution equations are much more usual in the theory of PDEs than terminal value problems for backward equations. However, the second choice better fits the direct treatment of nonautonomous problems (see the comments at the beginning of this section), and that is probably the reason why it is more usual in the control theoretic literature. In fact, if the system f depends explicitly on the time variable, a notation such as $y(s; x, t, \alpha)$ is mandatory because the trajectory depends on the initial time, and the Dynamic Programming Principle is (3.25) (with a possible explicit dependence of ℓ on the time variable s). The proof is essentially the same, once we replace the *semigroup property* for autonomous systems
$$y_x(t+s, \alpha) = y_{y(x,t)}(s, \alpha(\cdot + t)) \qquad t, s > 0$$
with the *unique evolution property*
$$y(t+s; x, \tau, \alpha) = y(t+s; y(t; x, \tau, \alpha), t, \alpha) \qquad s \geq 0,\, t \geq \tau.$$

The Hamiltonian in the HJB equation depends also on the t variable in this case (see Exercise 3.7), but it is not hard to extend the theory of the present section to cover this situation. In this setting it is also possible to consider time-dependent constraints on the controls. ◁

REMARK 3.11. By Remark 3.4 and the proof of Proposition 3.5, $v^\sharp \in BC(\mathbb{R}^N \times [0, T])$ and solves (3.11). Thus, under the assumptions of Theorem 3.7, $v = v^\sharp$; that is, measurable controls and piecewise constant controls generate the same value function. ◁

3.2. Local comparison and unbounded value functions

In this subsection we prove a local comparison result typical of evolutive HJB equations, saying that two viscosity solutions u_1, u_2 of

(3.26) $$u_t + H(x, Du) = 0$$

in the cone

$$\mathcal{C} := \{(x,t) : 0 < t < T \text{ and } |x| < C(T-t)\},$$

such that $u_1 \leq u_2$ at the initial time $t = 0$ in the ball $|x| < CT$, satisfy the same inequality in the whole cone $\mathcal{C}$, provided C is a Lipschitz constant for the Hamiltonian $H(x,p)$ in the variable p. This is usually called a *cone of dependence* result, and it is a typical property of hyperbolic PDEs, whereas global comparison results are typical properties of elliptic and parabolic PDEs. Of course the result has an easy control-theoretic interpretation.

Then we apply this result to prove global comparison and uniqueness of merely continuous solutions of (3.26). This allows us to extend the theory of the previous subsection to the case of unbounded running and terminal costs, paying a price in the strengthening of the structural conditions on the system. In particular, we assume the existence of a constant $K > 0$ such that

(3.27) $$|f(x,a)| \leq K(|x| + 1) \quad \text{for all } x \in \mathbb{R}^N, a \in A,$$

which is true, for instance, for linear systems with a bounded control set A.

THEOREM 3.12. *Let $H : \bar{B}(0, CT) \times \mathbb{R}^N \to \mathbb{R}$ be continuous and satisfy*

(3.28) $$|H(x,p) - H(x,q)| \leq C|p - q|$$
(3.29) $$|H(x,p) - H(y,p)| \leq \omega(|x-y|) + \omega(|x-y||p|)$$

for all $x, y \in B(0, CT)$, $p, q \in \mathbb{R}^N$, where ω is a modulus and $C, T > 0$.

If $u_1, u_2 \in C(\overline{\mathcal{C}})$ are, respectively, a viscosity sub- and supersolution of (3.26) in $\mathcal{C}$ and $u_1(x, 0) \leq u_2(x, 0)$ for all $x \in B(0, CT)$, then $u_1 \leq u_2$ in $\mathcal{C}$.

PROOF. We modify the auxiliary function in the proof of Theorem 3.7 in order to localize the problem in the cone $\mathcal{C}$. Assume by contradiction there exists $0 < \delta < T$ and $(\tilde{x}, \tilde{t})$ such that

(3.30) $$(u_1 - u_2)(\tilde{x}, \tilde{t}) = \delta \quad \text{and} \quad |\tilde{x}| \leq C(T - \tilde{t}) - 2\delta.$$

Take $M > \sup\{|u_1(x,t) - u_2(y,s)| : (x,t,y,s) \in \mathcal{C}^2\} \geq \delta$ and $h \in C^1(\mathbb{R})$ such that $h' \leq 0$, $h(r) = 0$ for $r \leq -\delta$, $h(r) = -3M$ for $r \geq 0$. Now define $\langle x \rangle_\beta := (|x|^2 + \beta^2)^{1/2}$ and

$$\Phi(x,y,t,s) := u_1(x,t) - u_2(y,s) - \frac{|x-y|^2 + |t-s|^2}{2\varepsilon}$$
$$- \eta(t+s) + h(\langle x \rangle_\beta - C(T-t)) + h(\langle y \rangle_\beta - C(T-s)),$$

where ε, η, β are positive parameters.

Let $(\bar{x}, \bar{y}, \bar{t}, \bar{s}) \in \overline{\mathcal{C}}^2$ be such that

$$\max_{\overline{\mathcal{C}}^2} \Phi = \Phi(\bar{x}, \bar{y}, \bar{t}, \bar{s}) .$$

We claim that either $\bar{t} = 0$ or $\bar{s} = 0$ or $(\bar{x}, \bar{y}, \bar{t}, \bar{s}) \in \mathcal{C}^2$, for β and η small enough. In fact, if $|\bar{x}| = C(T - \bar{t})$ or $|\bar{y}| = C(T - \bar{s})$ we have $\Phi(\bar{x}, \bar{y}, \bar{t}, \bar{s}) \leq M - 3M = -2M$ by the definition of h, because $\langle x \rangle_\beta > |x|$ for all x and β. On the other hand, by (3.30) and the definition of h, for any $\beta < \delta$ and $\eta < \delta/4\tilde{t}$,

$$(3.31) \qquad \max_{\overline{\mathcal{C}}^2} \Phi \geq \Phi(\tilde{x}, \tilde{x}, \tilde{t}, \tilde{t}) = \delta - 2\eta\tilde{t} + 2h(\langle \tilde{x} \rangle_\beta - C(T - \tilde{t})) \geq \frac{\delta}{2},$$

which proves the claim.

From the inequality $\Phi(\bar{x}, \bar{x}, \bar{t}, \bar{t}) + \Phi(\bar{y}, \bar{y}, \bar{s}, \bar{s}) \leq 2\Phi(\bar{x}, \bar{y}, \bar{t}, \bar{s})$ we get

$$(3.32) \qquad \frac{|\bar{x} - \bar{y}|^2 + |\bar{t} - \bar{s}|^2}{\varepsilon} \leq u_1(\bar{x}, \bar{t}) - u_1(\bar{y}, \bar{s}) + u_2(\bar{x}, \bar{t}) - u_2(\bar{y}, \bar{s}),$$

and thus

$$(3.33) \qquad |\bar{x} - \bar{y}|^2 + |\bar{t} - \bar{s}|^2 \leq 2M\varepsilon .$$

By the continuity of u_1 and u_2, (3.32) and (3.33), we obtain

$$(3.34) \qquad \frac{|\bar{x} - \bar{y}|^2 + |\bar{t} - \bar{s}|^2}{\varepsilon} \leq \omega_1(\varepsilon),$$

for a suitable modulus ω_1.

Now we show that neither $\bar{t}$ nor $\bar{s}$ can be zero if ε is chosen appropriately. In fact if $\bar{t} = 0$, since $u_1(\bar{x}, 0) \leq u_2(\bar{x}, 0)$,

$$\Phi(\bar{x}, \bar{y}, 0, \bar{s}) \leq u_1(\bar{x}, 0) - u_2(\bar{x}, 0) + u_2(\bar{x}, 0) - u_2(\bar{y}, \bar{s})$$
$$\leq \omega_2(\sqrt{2M\varepsilon}),$$

where ω_2 is the modulus of continuity of u_2 in $\overline{\mathcal{C}}$, which gives a contradiction to (3.31) for ε small enough. The proof that $\bar{s} > 0$ is analogous.

Next we define the test functions

$$\varphi(x, t) := \frac{|x - \bar{y}|^2 + |t - \bar{s}|^2}{2\varepsilon} + \eta(t + \bar{s}) - h(\langle x \rangle_\beta - C(T - t))$$

$$\psi(y, s) := -\frac{|\bar{x} - y|^2 + |\bar{t} - s|^2}{2\varepsilon} - \eta(\bar{t} + s) + h(\langle y \rangle_\beta - C(T - s)),$$

so that $u_1 - \varphi$ has a maximum at $(\bar{x}, \bar{t})$ and $u_2 - \psi$ has a minimum at $(\bar{y}, \bar{s})$. We compute the partial derivatives of φ and ψ, set $X = \langle \bar{x} \rangle_\beta - C(T - \bar{t})$, $Y = \langle \bar{y} \rangle_\beta - C(T - \bar{s})$, and use the definition of viscosity sub- and supersolution to get

$$2\eta \leq C(h'(Y) + h'(X))$$
$$+ H\left(\bar{y}, \frac{\bar{x} - \bar{y}}{\varepsilon} + h'(Y)\frac{\bar{y}}{\langle \bar{y} \rangle_\beta}\right) - H\left(\bar{x}, \frac{\bar{x} - \bar{y}}{\varepsilon} - h'(X)\frac{\bar{x}}{\langle \bar{x} \rangle_\beta}\right).$$

Now we add and subtract $H\big(\overline{x},(\overline{x}-\overline{y})/\varepsilon + h'(Y)\overline{y}/\langle\overline{y}\rangle_\beta\big)$ to the right-hand side and use (3.28) and (3.29) to obtain

$$(3.35) \quad 2\eta \leq C(h'(Y) + h'(X)) + \omega(|\overline{x}-\overline{y}|)$$
$$+ \omega\Big(\frac{|\overline{x}-\overline{y}|^2}{\varepsilon} + |\overline{x}-\overline{y}|\,|h'(Y)|\Big) + C\Big|h'(Y)\frac{\overline{y}}{\langle\overline{y}\rangle_\beta} + h'(X)\frac{\overline{x}}{\langle\overline{x}\rangle_\beta}\Big|.$$

Now we use (3.33), (3.34), and $h' = -|h'|$ to get that the right-hand side of (3.35) tends to zero as $\varepsilon \to 0^+$, a contradiction which completes the proof. ◀

REMARK 3.13. By combining the proofs of Theorems 3.7 and 3.12 the reader can easily extend Theorem 3.12 to the case where the assumption $u_1 \leq u_2$ at $t=0$ does not hold, so that one proves the inequality

$$(3.36) \qquad \sup_{\mathcal{C}}(u_1 - u_2) \leq \sup_{B(0,CT)\times\{0\}}(u_1 - u_2).$$

As in Theorem 3.7 one can include a term λu in the equation, provided $\lambda > 0$, and obtain inequality (3.36) with $u_1 - u_2$ replaced by $(u_1 - u_2)^+$ in the right-hand side. By using (3.36) and an easy approximation argument we can relax the conditions on u_1 and u_2 by assuming their continuity on $\mathcal{C} \cup B(0,CT) \times \{0\}$ only.

Theorem 3.12 can also be extended to the case where H depends explicitly on time t, by just adding the assumption that $H : \mathcal{C} \times \mathbb{R}^N \to \mathbb{R}$ be continuous, see Exercise 3.5. ◁

REMARK 3.14. It is easy to check that Theorem 3.12 holds in the cone $\{(x,t) : 0 < t < T$ and $|x - x_0| < C(T-t)\}$ for any x_0, if (3.28) and (3.29) hold for all $x, y \in B(x_0, CT)$ and $u_1 \leq u_2$ in $B(x_0, CT) \times \{0\}$. ◁

THEOREM 3.15. *Let $H : \mathbb{R}^{2N} \to \mathbb{R}$ be continuous and satisfy*

$$(3.37) \qquad |H(x,p) - H(x,q)| \leq K(|x|+1)|p-q|$$

for all $x, p, q \in \mathbb{R}^N$ and

$$(3.38) \qquad |H(x,p) - H(y,p)| \leq \omega(|x-y|, R) + \omega(|x-y|\,|p|, R)$$

for all $p \in \mathbb{R}^N$, $x,y \in B(0,R)$, $R > 0$, where ω is a modulus. If $u_1, u_2 \in C(\mathbb{R}^N \times [0,T])$ are, respectively, a viscosity sub- and supersolution of (3.26) in $\mathbb{R}^N \times\,]0,T[$ and $u_1(x,0) \leq u_2(x,0)$ for all $x \in \mathbb{R}^N$, then $u_1 \leq u_2$ in $\mathbb{R}^N \times [0,T]$.

PROOF. We may assume without loss of generality that $T < 1/K$, because the proof of the general case is obtained by iterating the following argument on time intervals of fixed length smaller than $1/K$. We fix $x_0 \in \mathbb{R}^N$ and define $r := KT(|x_0|+1)/(1-KT) > 0$, so that

$$(3.39) \qquad r = KT(|x_0|+1+r) =: CT.$$

We define the cone
$$\mathcal{C}_{x_0} := \{ (x,t) : 0 < t < T \text{ and } |x - x_0| < K(|x_0| + 1 + r)(T - t) \}.$$

If $x \in B(x_0, CT)$, then $|x| < |x_0| + r$ by (3.39). Thus by (3.37) we can take $C = K(|x_0| + 1 + r)$ as a Lipschitz constant for H with respect to the variable p in $B(x_0, CT) \times \mathbb{R}^N$ and apply Theorem 3.12 and Remark 3.14 to get $u_1 \leq u_2$ in $\mathcal{C}_{x_0}$. Since $\mathbb{R}^N \times\,]0, T[\, = \bigcup_{x_0 \in \mathbb{R}^N} \mathcal{C}_{x_0}$ the proof is complete. ◀

REMARK 3.16. The extensions of Theorem 3.12 mentioned in Remark 3.13 apply to Theorem 3.15 as well. In particular, if u_1, u_2 are continuous in $\mathbb{R}^N \times [0, T[$ only, we get that $u_1 \leq u_2$ in the same set. See Exercise 3.6 for the extension of Theorem 3.15 to time-dependent Hamiltonians. ◁

Next we apply the previous theory to characterize the value function
$$v(x,t) := \inf_{\alpha \in \mathcal{A}} \left\{ \int_0^t \ell(y_x(s), \alpha(s))\, ds + g(y_x(t)) \right\}$$
as the unique solution of the associated HJB equation under the conditions that g is continuous, ℓ satisfies

(3.40) $\quad \begin{cases} \ell \text{ is continuous and bounded on } B(0, R) \times A; \\ |\ell(x,a) - \ell(y,a)| \leq \omega_\ell(|x-y|, R),\ \forall R > 0,\ x, y \in B(0, R),\ a \in A, \end{cases}$

where ω_ℓ is a modulus, and the system f is locally lipschitzean in the state uniformly in the control and satisfies the growth condition (3.27).

THEOREM 3.17. *Assume* (A_0), (A_1), (A_2) *with* $\omega_f(r, R) = L_R r$, (3.27), (3.40), $g \in C(\mathbb{R}^N)$, *and* $0 < T < +\infty$. *Then the value function v is the complete solution in* $C(\mathbb{R}^N \times [0, T])$ *of the HJB equation* (3.26) *in* $\mathbb{R}^N \times\,]0, T[$ *with H given by* (3.5), *plus the initial condition* $u(x, 0) = g(x)$ *in* $\mathbb{R}^N$.

PROOF. The existence, uniqueness, and basic estimates on the trajectories under the current assumptions are given in Remark 5.7 of the appendix to this chapter. The proof of the continuity of v in Proposition 3.1 is still valid because the trajectories with initial points in a bounded set are uniformly bounded, and g and ℓ are bounded on bounded sets of the state space. Also, v is a viscosity solution of (3.26) because this is a local property and the proofs of Propositions 3.2 and 3.5 remain valid. Moreover, the argument of the proof of Lemma 2.11 and (3.27) show that the Hamiltonian H satisfies (3.37), and (3.38) with $\omega(r, R) = L_R r + \omega_\ell(r, R)$. Therefore, by Theorem 3.15, v is below (respectively, above) any continuous supersolution (respectively, subsolution) u such that $u \geq g$ (respectively, $u \leq g$) at $t = 0$. Thus v is the complete solution of the initial value problem (3.26) with g as initial data. ◀

REMARK 3.18. Theorem 3.17 can be extended without difficulties to cost functionals involving terms $e^{-\lambda t}$ with interest rate $\lambda > 0$, as in §3.1, by Remarks 3.13 and 3.16, and to nonautonomous problems, see Remark 3.10 and Exercise 3.7. ◁

3.3. Equivalent notions of solution

In this subsection we study for the finite horizon problem the properties discussed in §§2.3 and 2.4 for the infinite horizon problem: Backward Dynamic Programming, sub- and superoptimality principles, and the equivalence of viscosity solutions with bilateral supersolutions and with (generalized) Dini solutions, semiconcavity of the value function and existence of its directional derivatives. As a rule we leave the proofs as exercises for the reader as long as they are straightforward adaptations of those of §§2.3 and 2.4. For simplicity we restrict ourselves to a Mayer problem, that is, to the case of null running cost and interest rate, and we assume the conditions ensuring the well-posedness of the HJB equation according to the theory of §§3.1 and 3.2:

(3.41) $\begin{cases} 0 < T < +\infty, \quad \ell(x,a) \equiv 0, \quad \lambda = 0; \\ \text{either (i): } (A_0)\text{-}(A_3) \text{ and } g \in BUC(\mathbb{R}^n); \\ \text{or (ii): } (A_0), (A_1), (A_2) \text{ with } \omega_f(r,R) = L_R r, (3.27) \text{ and } g \in C(\mathbb{R}^n). \end{cases}$

Since some results are of a local nature these assumptions can be weakened: the reader can search for more general conditions for each result.

We recall that in this case the value function is

$$v(x,t) := \inf_{\alpha \in \mathcal{A}} g(y_x(t,\alpha))$$

and the HJB equation is

(3.42) $$u_t + \sup_{a \in A}\{-f(x,a) \cdot Du\} = 0,$$

where $Du = D_x u$ denotes the gradient with respect to space variables x. We refer to §2.3 for the definitions of backward trajectories, bilateral solution, and exit time from Ω, $t_x(\alpha)$. We begin with the Backward Dynamic Programming Principle.

DEFINITION 3.19. We say that $x \in \mathbb{R}^N$ is an *optimal point* for the horizon $t > 0$, and write $(x,t) \in \mathcal{O}$, if x belongs to an optimal trajectory for some larger horizon, that is, there exist $z \in \mathbb{R}^N$, $s > t$, and $\alpha^* \in \mathcal{A}$ such that $v(z,s) = J(z,s,\alpha^*)$ and $x = y_z(s-t,\alpha^*)$. ◁

PROPOSITION 3.20 (BACKWARD DPP). *Assume* (3.41). *Then, for all* $x \in \mathbb{R}^N$, $t, \tau > 0$, *and* $\alpha \in \mathcal{A}^-$

$$v(x,t) \geq v(y_x(-\tau,\alpha), t+\tau);$$

for $(x,t) \in \mathcal{O}$ *and for all* $\tau > 0$ *small enough*

$$v(x,t) = \max_{\alpha \in \mathcal{A}^-} v(y_x(-\tau,\alpha), t+\tau).$$

COROLLARY 3.21. *Assume* (3.41). *Then the value function v is a bilateral supersolution of* (3.42) *in* $\mathbb{R}^N \times \,]0,\infty[$, *and it is a bilateral subsolution of the same equation at any* $(x,t) \in \mathcal{O}$. *If in addition $H(x,\cdot)$ is strictly convex for some x, then $D^-v(x,t)$ has at most one element for all t, and if $(x,t) \in \mathcal{O}$, then $D^+v(x,t)$ has at most one element also.*

Next we study sub- and superoptimality principles.

THEOREM 3.22. *Assume* (3.41), $\Omega \subseteq \mathbb{R}^N$ *open and* $u \in C(\overline{\Omega} \times [0,T])$. *Then the following statements are equivalent:*

(i) u *is a subsolution of* (3.42) *in* $\Omega \times {]}0,T{[}$;

(ii) *for all* $\alpha \in \mathcal{A}$ *and* $0 \leq s \leq t < t_x(\alpha) \wedge T$ *the function* $s \mapsto u(y_x(s,\alpha),t-s)$ *is nondecreasing;*

(iii) u *is a supersolution of*

$$-u_t - \sup_{a \in A}\{-f(x,a) \cdot Du\} = 0 \qquad \text{in } \Omega \times {]}0,T{[};$$

(iv) u *satisfies a local suboptimality principle, that is, for all* $x \in \Omega$, $0 < t \leq T$, *there exists* $\tau > 0$ *such that*

$$u(x,t) \leq \inf_{\alpha \in \mathcal{A}} u(y_x(s,\alpha),t-s) \qquad \text{for all } 0 < s \leq \tau;$$

(v) u *satisfies a global suboptimality principle, that is,*

$$u(x,t) \leq \inf_{\alpha \in \mathcal{A}} u(y_x(s \wedge t_x(\alpha),\alpha),t - s \wedge t_x(\alpha)) \qquad \text{for all } 0 < s \leq t\ .$$

PROOF. The proof of Proposition 3.5 gives (iv) $\Rightarrow$ (i); (iv) and (v) follow immediately from (ii). The equivalence of (i), (ii), and (iii) is a special case of Theorem 2.33 because we can consider time as an additional state variable by adding $\tau' = -1$, $\tau(0) = t$ to the system (S), and then rewrite (3.42) in the form (2.40) with $\ell \equiv 0$ and $\lambda = 0$ (this is allowed by Remark 2.34). ◂

REMARK 3.23. From Theorem 3.22 we get that $u \in C(\Omega \times {]}0,T{[})$ is a viscosity solution of (3.42) in $\Omega \times {]}0,T{[}$ if and only if it is a bilateral supersolution (it is enough to apply Theorem 3.22 to open subsets of $\Omega \times {]}0,T{[}$ where u is continuous up to the boundary). Therefore the value function v is also the unique bilateral supersolution of (3.42) in $\mathbb{R}^N \times {]}0,\infty{[}$ taking up the initial data g continuously. ◁

THEOREM 3.24. *Assume* (3.41), $\Omega \subseteq \mathbb{R}^N$ *open, and* $u \in C(\overline{\Omega} \times [0,T])$. *Then* u *is a supersolution of the HJB equation* (3.42) *if and only if it satisfies a local superoptimality principle, that is, for all* $x \in \Omega$ *and* $0 < t \leq T$ *there exists* $\tau > 0$ *such that*

$$u(x,t) \geq \inf_{\alpha \in \mathcal{A}} u(y_x(s,\alpha),t-s) \qquad \text{for all } 0 < s \leq \tau\ .$$

PROOF 1. As in the previous proof we add time as a state variable and we apply Theorem 2.32 with $\ell \equiv 0$ and $\lambda = 0$, which is allowed by Remark 2.34. ◂

PROOF 2. It is worthwhile presenting a simple direct proof if either $\Omega = \mathbb{R}^N$ or f is locally Lipschitzean in x uniformly in a. Assume first $\Omega = \mathbb{R}^N$. Fix $0 < \sigma < t \leq T$ and set

$$h(z) := u(z, t-\sigma), \qquad w(x,s) := \inf_{\alpha \in \mathcal{A}} h(y_x(s,\alpha))\ .$$

By Proposition 3.5 w solves the HJB equation (3.42) in $\mathbb{R}^N \times]0,\infty[$, and the initial condition $w(x,0) = h(x)$. On the other hand, the function $(x,s) \mapsto u(x,s+t-\sigma)$ satisfies the same condition at $s = 0$ and it is a supersolution of (3.42) in $\mathbb{R}^N \times]0, T + \sigma - t[$. By a comparison theorem (either Theorem 3.7 or Theorem 3.17) $w(x,s) \leq u(x,s+t-\sigma)$, and setting $s = \sigma$ gives the conclusion with $\tau = t$. Note that in this case we have proved a global superoptimality principle.

Next let Ω be any open set. Fix x_0 and let r be such that $B(x_0,r) \subseteq \Omega$, C be an upper bound on f in $\bar{B}(x_0,r) \times A$. Set $\tau := (r/C) \wedge t$ and

$$C = \{(x,s) : 0 < s < \tau,\ |x - x_0| < C(\tau - s)\}.$$

For $(x,s) \in \bar{C}$ we have $|y_x(s,\alpha) - x| \leq Cs$ and then $y_x(s,\alpha) \in \bar{\Omega}$. Thus w is well defined in $\bar{C}$ and it satisfies (3.42) in C. In order to invoke the local comparison Theorem 3.12 we need the condition $|f(x,a) - f(y,a)| \leq C_1|x-y|$ for all $x,y \in B(x_0,r)$ and $a \in A$. As before we get $w(x,s) \leq u(x,s+t-\sigma)$ for all $(x,s) \in \bar{C}$ and $s \leq T + \sigma - t$, which yields the conclusion by choosing $s = \sigma$. ◀

Now we consider Dini derivatives and Dini solutions. First we simplify a bit the notation for (generalized) Dini directional derivatives (see Definition 2.36):

$$\partial^- u(x,t;q,s) := \partial^- u((x,t);(q,s)) := \liminf_{\substack{h \to 0^+ \\ (p,\tau) \to (q,s)}} \frac{u(x+hp, t+h\tau) - u(x,t)}{h},$$

$$\partial^+ u(x,t;q,s) := \partial^+ u((x,t);(q,s)) := \limsup_{\substack{h \to 0^+ \\ (p,\tau) \to (q,s)}} \frac{u(x+hp, t+h\tau) - u(x,t)}{h},$$

and for the classical (one-sided) directional derivative

$$\partial u(x,t;q,s) := \frac{\partial u}{\partial (q,s)}(x,t) := \lim_{h \to 0^+} \frac{u(x+hq, t+hs) - u(x,t)}{h}.$$

We define the compact set

$$F(x) := \overline{co} f(x, A),$$

and note that the HJB equation (3.42) can be rewritten

(3.43) $$\max_{f \in F(x)} \{u_t - f \cdot D_x u\} = 0.$$

Definition 2.38 of (generalized) *Dini subsolution* becomes, for this equation,

$$\sup_{f \in F(x)} \{-\partial^+ u(x,t; f, -1)\} \leq 0,$$

and that of *Dini supersolution*,

$$\sup_{f \in F(x)} \{-\partial^- u(x,t; f, -1)\} \geq 0,$$

and u is a *Dini solution* if it is simultaneously a sub- and a supersolution. The analogy with (3.43) comes from the formula

$$\partial u(x,t;f,-1) = f \cdot Du(x,t) - u_t(x,t)$$

at points where u is differentiable. If a lipschitzean Dini solution u has all directional derivatives at a point (x,t), then it satisfies

(3.44) $$\sup_{f \in F(x)} \left\{ -\frac{\partial u}{\partial (f,-1)}(x,t) \right\} = 0$$

and we say that it is an *intrinsically classical* solution of (3.42) (even if it may not be a classical solution!). See §2.4 for some more comments on these definitions.

The connection of the two notions of weak solution is given by the next result.

THEOREM 3.25. *Assume* (3.41), $\Omega \subseteq \mathbb{R}^N$ *open, and* $u \in C(\overline{\Omega} \times [0,T])$. *Then*

(i) u *is a viscosity subsolution (respectively, supersolution) of* (3.42) *in* $\Omega \times \,]0,T[$ *if and only if it is a Dini subsolution (respectively, supersolution);*

(ii) *if, in addition, u is locally lipschitzean, then it is a viscosity solution of* (3.42) *in* $\Omega \times \,]0,T[$ *if and only if*

$$\sup_{f \in F(x)} \{-\partial^- u(x,t;f,-1)\} = 0 \qquad in\ \Omega \times \,]0,T[\,.$$

PROOF. As in the previous results of this section we can view (3.42) as an equation of the form studied in §2. Thus (i) is a special case of Theorem 2.40 and (ii) is a special case of Remark 2.43. ◄

★ ★ ★

In the second part of this subsection we give two regularityth results, one on the semiconcavity of the value function v, the other saying that v is a good marginal function, in the sense that it satisfies the assumptions of Proposition II.4.4, under some regularity assumptions on the data. In both cases, v turns out to have all classical directional derivatives, thus it is an intrinsically classical solution of the HJB equation, and to have a nonempty superdifferential at every point, a useful fact for the synthesis problem in next section.

We say that a function is *locally semiconcave* in a set X if it is semiconcave in any bounded open set whose closure is contained in X.

THEOREM 3.26. *Assume* (3.41) *and, for some* $L, M > 0$,

(3.45) $$|f(x,a) - f(y,a)| \le L\,|x-y|$$

(3.46) $$|f(x-h,a) - 2f(x,a) + f(x+h,a)| \le M\,|h|^2,$$

for all $x,y,h \in \mathbb{R}^N$, $a \in A$, *and g locally semiconcave. Then v is locally semiconcave in* $\mathbb{R}^N \times [0,T]$.

An example where (3.46) is satisfied with $M = 0$ is any system affine in the state variables for all choices of controls (see Exercise 3.9 for a stronger result in this case).

We split the proof in some lemmata.

LEMMA 3.27. *For a fixed $t > 0$ let $\tau \mapsto z_x(\tau; \alpha, t)$ be the solution of*

(3.47) $$\begin{cases} z'(\tau) = t\, f(z(\tau), \alpha(\tau)), & \tau > 0, \\ z(0) = x\,. \end{cases}$$

Then $y_x(t, \alpha) = z_x(1; \alpha_t, t)$ for $\alpha_t(\tau) := \alpha(t\tau)$ and

(3.48) $$v(x, t) = \inf_{\alpha \in \mathcal{A}} g(z_x(1; \alpha, t))\,.$$

PROOF. The first statement is straightforward to check and the map $G : \mathcal{A} \to \mathcal{A}$, $G(\alpha) := \alpha_t$ is a one-to-one correspondence. ◀

We say that $\varphi : \Omega \to \mathbb{R}^N$ is *semiconcave-convex* in the open set $\Omega \subseteq \mathbb{R}^k$ if for some constant C

$$|\varphi(x - h) - 2\varphi(x) + \varphi(x + h)| \leq C\, |h|^2$$

for all $x \in \Omega$ and h sufficiently small. The name comes from the fact that for $N = 1$, φ is semiconcave-convex if and only if both φ and $-\varphi$ are semiconcave. It is easy to see that φ is semiconcave-convex if it has a Lipschitz gradient, and the converse is also true (but less obvious, see Exercise 3.8). Note that assumption (3.46) means that f is semiconcave-convex in $\mathbb{R}^N$, uniformly for $a \in A$.

LEMMA 3.28. *In the notations of Lemma 3.27, the map $\varphi(x, t) := z_x(1; \alpha, t)$ is semiconcave-convex, Lipschitz, and bounded in any bounded set of $\mathbb{R}^N \times\,]0, T[$, uniformly for $\alpha \in \mathcal{A}$.*

PROOF. Fix $\alpha \in \mathcal{A}$, $\eta \in \mathbb{R}$, $x, h \in \mathbb{R}^N$ such that $|h| \leq 1$, and set $z(s) := z_x(s; \alpha, t)$,

$$z_+(s) := z_{x+h}(s; \alpha, t + \eta), \qquad z_-(s) := z_{x-h}(s; \alpha, t - \eta),$$
$$w := z_- - 2z + z_+, \qquad \xi := z_+ - z_-\,.$$

In the sequel we do not write α any more in the formulas. First note that $\xi(s)$ satisfies

(3.49) $$\begin{cases} \xi' = (t + \eta)f(z_+) - (t - \eta)f(z_-), & s > 0, \\ \xi(0) = 2h\,. \end{cases}$$

Since the trajectories of (3.47) are bounded in $[0, 1]$ uniformly for x in a bounded set and $t \in [0, T]$, there is $C_1 > 0$ depending only on $|x|$ such that $|f(z_+(s))|$, $|f(z_-(s))| \leq C_1$ for all $s \in [0, 1]$. Then

$$|(t + \eta)f(z_+) - (t - \eta)f(z_-)| \leq tL|\xi| + 2|\eta|\, C_1,$$

and applying Gronwall's inequality to (3.49) we get

(3.50) $$|\xi(s)| \leq C_2(|\eta| + |h|) \quad \text{for } 0 \leq s \leq 1 .$$

By a similar argument we also obtain

(3.51) $$|z_-(s) - z(s)| \leq C_3(|\eta| + |h|) \quad \text{for } 0 \leq s \leq 1$$

(note that this gives the Lipschitz continuity of φ).

Next observe that $w(s)$ satisfies

$$\begin{cases} w' = (t-\eta)f(z_-) - 2tf(z) + (t+\eta)f(z_+), & s > 0, \\ w(0) = 0, \end{cases}$$

and estimate the norm of the right-hand side of the differential equation by

(3.52) $$t|f(z_-) - 2f(z) + f(z_+)| + |\eta| L |\xi| .$$

Now we write $z_- = z + (z_- - z)$, $z_+ = z - (z_- - z) + w$ and use (3.45) and (3.46) to estimate the first term of (3.52) by $TM|z_- - z|^2 + TL|w|$. Then we use (3.50) and (3.51) to estimate (3.52) by $TL|w| + C_4(|\eta|^2 + |h|^2)$. By Gronwall's inequality this gives

$$|w(1)| \leq C_4(|\eta|^2 + |h|^2) e^{TL},$$

which completes the proof. ◀

LEMMA 3.29. *Assume $\varphi : \Omega \to \mathbb{R}^N$ is semiconcave-convex, Lipschitz, and bounded. If $g : \mathbb{R}^N \to \mathbb{R}$ is locally semiconcave, then $g \circ \varphi$ is locally semiconcave.*

PROOF. Set $\zeta := \varphi(x-h) - 2\varphi(x) + \varphi(x+h)$. By assumption $|\zeta| \leq C_\varphi |h|^2$ and by writing $\varphi(x-h) = \varphi(x) - (\varphi(x+h) - \varphi(x)) + \zeta$ we get

(3.53) $$\begin{aligned} g(\varphi(x-h)) - 2g(\varphi(x)) + g(\varphi(x+h)) &\leq C_g |\varphi(x+h) - \varphi(x)|^2 + L_g |\zeta| \\ &\leq C_g L_\varphi^2 |h|^2 + L_g C_\varphi |h|^2, \end{aligned}$$

where C_g and L_g are, respectively, the constant of semiconcavity and the Lipschitz constant of g on $\varphi(\Omega)$. ◀

PROOF OF THEOREM 3.26. By Lemma 3.27 v is the marginal function of the family $g(z_x(1; \alpha, t))$ with respect to the parameter $\alpha \in \mathcal{A}$, and all these functions are locally semiconcave uniformly with respect to α by Lemmas 3.28 and 3.29 (see the explicit estimate (3.53)). Then, v is locally semiconcave by Exercise II.4.6. ◀

Next we use relaxed controls (see §2.2) to represent the value function v as a good marginal function when A is compact. For a relaxed control function $\mu \in \mathcal{A}^r$ we denote by $z_x(\,\cdot\,;\mu,t)$ the solution of

(3.54) $$\begin{cases} z'(\tau) = tf^r(z(\tau), \mu(\tau)), & \tau > 0 \\ z(0) = x . \end{cases}$$

It is straightforward to adapt the arguments of §2.2 to the finite horizon problem and show that the value function of the relaxed problem coincides with v. Thus Lemma 3.27 gives $v(x,t) = \inf_{\mu \in \mathcal{A}^r} g(z_x(1;\mu,t)) = \inf_{\mu \in \mathcal{A}_1^r} G(x,t,\mu)$, where

(3.55)
$$G(x,t,\mu) := g(z_x(1;\mu,t)),$$
$$\mathcal{A}_1^r := \{ \text{ measurable functions } [0,1] \to A^r \} .$$

THEOREM 3.30. *Assume* (3.41)(ii) *and A compact. Then there is a topology on $\mathcal{A}_1^r$ such that $\mathcal{A}_1^r$ is a compact metric space and $G : \mathbb{R}^N \times [0,\infty[\times \mathcal{A}_1^r \to \mathbb{R}$ defined by* (3.55) *is continuous. In particular, for all (x,t) there exists an optimal relaxed control function.*

PROOF. By the theory of Warga [Wa72], $\mathcal{A}^r$ can be identified with a compact subset of the dual space of the space

$$\mathcal{B} = \{ \phi : [0,1] \times A \to \mathbb{R} : \phi(\cdot,a) \text{ is measurable for all } a, \, \phi(t,\cdot) \text{ is continuous}$$
$$\text{and } \sup_{a \in A} |\phi(t,a)| \leq \psi(t) \text{ for all } t, \text{ for some } \psi \in L^1([0,1]) \},$$

with the weak star topology. More explicitly, a sequence μ_n in $\mathcal{A}^r$ converges to μ if and only if for all $\phi \in \mathcal{B}$

(3.56)
$$\lim_n \int_0^1 \int_A \phi(\tau,a) \, d\mu_n(\tau) \, d\tau = \int_0^1 \int_A \phi(\tau,a) \, d\mu(\tau) \, d\tau .$$

Moreover, since $\mathcal{B}$ is separable, this topology is metrizable and so it is enough to check the continuity of G on sequences. We take $x_n \to x$, $t_n \to t$ and $\mu_n \to \mu$ and denote z_n, z the corresponding solutions of (3.54) and $w_n := z_n - z$. In the ODE for w_n we add and subtract $tf^r(z_n,\mu_n)$ and $tf(z,\mu_n)$ and we integrate on $[0,s]$ to get

(3.57)
$$w_n(s) = x_n - x + \psi_n(s) + \xi_n(s) + \int_0^s \zeta_n(\tau) \, d\tau,$$

where

$$\psi_n(s) := (t_n - t) \int_0^s f^r(z_n,\mu_n) \, d\tau$$
$$\xi_n(s) := t \int_0^s (f^r(z,\mu_n) - f^r(z,\mu)) \, d\tau = t \int_0^s \int_A f(z,a) \, d(\mu_n - \mu) \, d\tau$$
$$\zeta_n := t(f^r(z_n,\mu_n) - f^r(z,\mu_n)) .$$

It is easy to check, as in Lemma 2.20, that f^r satisfies the structural conditions (3.41)(ii) on f. Thus all trajectories of (3.54) are uniformly bounded on $[0,1]$ for initial conditions in a bounded set. Then

$$|\psi_n(s)| \leq C\,|t_n - t| \qquad \text{for all } 0 \leq s \leq 1,$$
$$|\zeta_n(\tau)| \leq tL_R\,|w_n(\tau)| \qquad \text{for all } 0 \leq \tau \leq 1 .$$

Moreover, by choosing $\phi(\tau, a)$ null for $\tau > s$ and equal to a component of $f(z(\tau), a)$ for $\tau \leq s$ in the definition of weak convergence (3.56), we obtain

$$\lim_n \xi_n(s) = 0 \quad \text{for all } s.$$

Therefore we apply Gronwall inequality to (3.57) and get that $|w_n(s)| \to 0$ for all $s \in [0,1]$. Since g is continuous the proof is complete. ◄

REMARK 3.31. If, in addition to the assumptions of Theorem 3.30, $f(x, A)$ is convex for all x, then $f(x, A) = f^r(x, A^r)$ by (2.32). Then any relaxed trajectory is a standard trajectory because a controlled system $y' = \varphi(y, b)$, $b \in B$, is equivalent to the differential inclusion $y' \in \varphi(y, B)$ by a theorem of Filippov. Therefore Theorem 3.30 implies the existence of an optimal standard control function for all (x,t) under the present assumptions. This is a well-known result that can be proved also by convexity methods (Carathéodory's theorem) instead of using relaxed controls. ◁

THEOREM 3.32. *Assume the hypotheses of Theorem 3.30, $g \in C^1(\mathbb{R}^N)$, and f differentiable with respect to x with $D_x f$ continuous. Then*

(i) *G defined by (3.55) is differentiable with respect to (x,t) with $D_{(x,t)}G$ continuous;*

(ii) *v is locally lipschitzean;*

(iii) *for all (x,t) the superdifferential of v coincides with Clarke's gradient, that is, $D^+v(x,t) = \partial v(x,t)$;*

(iv) *v is differentiable at all points where there is a unique optimal relaxed control.*

PROOF. Since A is compact, $D_x f$ is bounded on $B(x,r) \times A$; thus a standard theorem on differentiability and continuity of integrals depending on parameters shows that f^r is differentiable with respect to x with $D_x f^r(x,m) = \int_A D_x f(x,a)\, dm$, and it is easy to see that this derivative is continuous with respect to (x,m). By the differentiability of solutions of an ODE with respect to the initial data (Theorem 5.8), the i-th partial derivative of $x \mapsto z_x(1; \mu, t)$ at x_0 is the solution, at time $\tau = 1$, of

(3.58) $$\begin{cases} q'(\tau) = tD_x f^r(z(\tau), \mu(\tau))q(\tau), & \tau > 0, \\ q(0) = e_i, \end{cases}$$

where $z(\tau) = z_{x_0}(\tau; \mu, t)$, e_i has the i-th component 1 and the other components null, and $D_x f^r$ is the Jacobian matrix of f^r with respect to x (i.e., it has $\partial f_i^r / \partial x_j$ on the i-th row and j-th column). By an argument similar to the proof of Theorem 3.30, one can show that the solution of (3.58) at $\tau = 1$ depends continuously on (x_0, t, μ). The derivative of $t \mapsto z_x(1; \mu, t)$ at t_0 can be computed either by a theorem on the differentiability of solutions of an ODE with respect to parameters, or by adding t as an extra variable to the system (3.54) with trivial dynamics

$t'(\tau) = 0$ and initial condition $t(0) = t_0$, and then using again Theorem 5.8 on derivatives with respect to initial data. The derivative turns out to be the solution, at time $\tau = 1$, of

(3.59) $$\begin{cases} \varphi'(\tau) = t_0 D_x f^{\mathrm{r}}(z(\tau), \mu(\tau))\varphi(\tau) + f^{\mathrm{r}}(z(\tau), \mu(\tau)), & \tau > 0, \\ \varphi(0) = 0, \end{cases}$$

where $z(\tau) = z_x(\tau; \mu, t_0)$, and this can be shown to depend continuously on (x, μ, t_0) as above. Since $g \in C^1(\mathbb{R}^N)$, this proves the first statement of the theorem. Then v satisfies the assumptions of Proposition II.4.4 which gives $D^+v(x,t) = \partial v(x,t)$ and formulas for $D^+v(x,t)$ and $D^-v(x,t)$ in terms of minimizers of $G(x, t, \cdot)$ in $\mathcal{A}_1^{\mathrm{r}}$. In particular, the uniqueness of the minimizer implies the differentiability of v at (x,t), and such a uniqueness is equivalent to the uniqueness of the optimal relaxed control function by the proof of Lemma 3.27. ◂

REMARK 3.33. From the arguments of the proof of Theorem 3.32 one can compute D^+v and D^-v at a point (x,t) where all optimal controls are known. In fact, by Proposition II.4.4,

$$D^+v(x,t) = \overline{\mathrm{co}}\{ D_{(x,t)}G(x,t,\mu) : \mu \in \arg\min G(x, t, \cdot) \},$$

$\mu \in \mathcal{A}_1^{\mathrm{r}}$ minimizes $G(x, t, \cdot)$ if and only if $s \mapsto \mu(s/t)$ is an optimal control for (x,t), and we computed $D_{(x,t)}G$ in terms of solutions of certain linear differential equations. It is actually more useful to know explicitly some elements of D^+v along an optimal trajectory: this leads to necessary conditions of optimality in the form of the Pontryagin Maximum Principle, as we see in the next section. It is not hard to prove (see Lemma 3.43) that for any (relaxed) trajectory $y(\cdot) = y_x(\cdot, \mu)$ we have

$$D_x G(y(t), T - t, \mu_{T-t}) = p(t)$$

(with $\mu_{T-t}(\tau) := \mu((T-t)\tau)$) and the *costate* or *adjoint vector* $p(\cdot)$ satisfies

(3.60) $$\begin{cases} p'(s) = -p(s) D_x f^{\mathrm{r}}(y(s), \mu(s)), \\ p(T) = Dg(y(T)). \end{cases}$$

Therefore, if μ is an optimal (relaxed) control, we have

$$(p(t), p_0(t)) \in D^+v(y(t), T-t), \qquad p_0(t) := \frac{\partial G}{\partial t}(y(t), T-t, \mu_{T-t}) .$$

A formula for p_0 can be deduced from (3.59) and other formulas are given in the next section in connection with the PMP. ◁

The uniqueness of the optimal control is not necessary for the differentiability of v: it is enough to take $f \equiv 0$ to have a counterexample. A more subtle question is whether the uniqueness of the optimal trajectory for (x,t) is sufficient for the differentiability of v at (x,t). The answer is positive for some problems with smooth Hamiltonian, as happens in the calculus of variations, but in general it is negative, even if the system and the terminal cost are smooth and $f(x, A)$ is convex, as the next example shows.

3. THE FINITE HORIZON PROBLEM

EXAMPLE 3.34. Consider the system in $\mathbb{R}^2$
$$\begin{cases} y_1' = a, \\ y_2' = h(y_1) + a^2 y_2, \end{cases}$$
with $A = [-1, 1]$, h smooth such that $h(0) = 0$, $h(s) > 0$ for $s \neq 0$, and cost functional $g(y_2(t)) = y_2(t)$. For the initial condition $x = (y_1(0), y_2(0)) = (0, 0)$ the only optimal trajectory is $(y_1(\tau), y_2(\tau)) \equiv 0$ which is obtained by just one standard control, $\alpha = 0$ a.e., but infinitely many relaxed controls, all those such that
$$\int_A a\, d\mu(\tau) = 0 \qquad \text{for a.e. } \tau.$$
Thus $v(0, 0, t) = 0$ for all t. To prove that v is not differentiable at such points we show that $v(0, \varepsilon, t) = \varepsilon$ and $v(0, -\varepsilon, t) = -\varepsilon e^t$ for all $t, \varepsilon > 0$. If the starting point is $(0, \varepsilon)$, then $\alpha = 0$ a.e. is again the unique optimal control, and now it is also the unique optimal relaxed control (thus v is differentiable at $(0, \varepsilon, t)$ for all $t, \varepsilon > 0$). If the starting point is $(0, -\varepsilon)$, then
$$\mu(\tau) = \frac{1}{2}\delta_{-1} + \frac{1}{2}\delta_1 \qquad \text{for a.e. } \tau$$
(δ_a is the Dirac measure concentrated at a) is the unique optimal relaxed control whose corresponding optimal trajectory is $\tau \mapsto (0, -\varepsilon e^\tau)$. ◁

Next we apply Theorems 3.26 and 3.32 to prove the differentiability of the value function along an optimal trajectory under a strict convexity assumption on the Hamiltonian. By an *optimal trajectory* for (x, T) we mean $y^*(s) := y_x(s, \alpha^*)$, $s \in [0, T]$, where α^* is optimal for (x, T), that is $v(x, T) = g(y_x(T, \alpha^*))$.

COROLLARY 3.35. *Under the assumptions of either Theorem 3.26 or Theorem 3.32 let y^* be an optimal trajectory for (x, T). Then v is differentiable at $(y^*(T - t), t)$ and the classical gradient Dv is continuous at this point, for all $t \in]0, T[$ such that $H(y^*(T - t), \cdot)$ is strictly convex.*

PROOF. For all $t \in]0, T[$
$$Y := (y^*(T - t), t) \in \mathcal{O}.$$
Then $D^+v(Y)$ has at most one element by Corollary 3.21. However $D^+v(Y) \neq \emptyset$ because $D^+v(Y) = \partial v(Y)$ (if we are under the assumptions of Theorem 3.26 this follows from Proposition II.4.7). Therefore $D^+v(Y)$ is a singleton, and this implies the differentiability of v at Y by either Proposition II.4.7 or II.4.4. Finally,
$$Dv(Y) = \lim_n Dv(X_n)$$
for any sequence $X_n \to Y$ of points of differentiability of v because
$$D^*v(Y) = D^+v(Y) = \{Dv(Y)\}$$
and Dv is locally bounded, so we get the continuity of Dv at Y. ◂

The last result of this subsection is a straightforward consequence of Theorems 3.26 and 3.32.

COROLLARY 3.36. *Under the assumptions of either Theorem 3.26 or Theorem 3.32 the value function is an intrinsically classical solution of the HJB equation (3.42), that is, v satisfies (3.44) in $\mathbb{R}^N \times\,]0, +\infty[$. In particular, the formulation (3.44) of the HJB equation is equivalent to the viscosity and the Dini formulations.*

PROOF. Semiconcave functions have $D^+v = \partial v$ everywhere by Proposition II.4.7, and this implies the existence of all directional derivatives by Proposition II.4.3. ◂

3.4. Necessary and sufficient conditions of optimality and the Pontryagin Maximum Principle

In this subsection we apply the theory of §3.3 to prove various necessary and/or sufficient conditions for optimality of open loop controls in terms of Dini derivatives or sub- and superdifferentials of the value function. Under some additional differentiability assumptions on the data we then give an easy proof of the Pontryagin Maximum Principle where the classical necessary condition of optimality is slightly strengthened to become sufficient as well. The new formulation enlightens the connection between the adjoint variables and the gradient of the value function.

Throughout the subsection we keep the main assumption of §3.3, that is (3.41). We fix $x_0 \in \mathbb{R}^N$, $T > 0$, and $\alpha \in \mathcal{A}$ and study the optimality of the corresponding trajectory

$$y(t) := y_{x_0}(t, \alpha), \qquad 0 \le t \le T.$$

A very easy consequence of the DPP, Proposition 3.2, and of the assumptions $\ell \equiv 0$, $\lambda = 0$, is that the function

$$h(t) := v(y(t), T - t)$$

is nondecreasing for any α and is constant if and only if α is optimal for (x_0, T), see Remark 3.3. Therefore $h' \equiv 0$ is a necessary condition of optimality and $h'(t) \le 0$ for a.e. t is a sufficient condition of optimality if h is absolutely continuous (e.g., lipschitzean). If v is smooth we have the necessary and sufficient condition of optimality

(3.61) $\quad D_x v(y(t), T - t) \cdot f(y(t), \alpha(t)) - v_t(y(t), T - t) \le 0 \qquad$ for a.e. t,

and since the HJB equation gives

$$v_t(x, s) - D_x v(x, s) \cdot f(x, a) \le 0 \qquad \text{for all } x, s,$$

an equivalent condition is the *Minimum Principle* (briefly MP)

(3.62) $\quad D_x v(y(t), T - t) \cdot f(y(t), \alpha(t)) = \min_{a \in A} D_x v(y(t), T - t) \cdot f(y(t), a),$

for a.e. t. In the following we study conditions that generalize (3.61) and (3.62) to the case of nonsmooth value functions by replacing the classical derivatives either with (generalized) Dini derivatives or by elements of the sub- and superdifferentials. We denote

$$D^{\pm}u(x,t) := D^+u(x,t) \cup D^-u(x,t),$$

and the elements of this set are written (p, p_0) with $p \in \mathbb{R}^N$, $p_0 \in \mathbb{R}$.

THEOREM 3.37 (NECESSARY CONDITIONS OF OPTIMALITY). *Assume* (3.41). *If the control function α is optimal for (x_0, T), then the corresponding trajectory $y(\,\cdot\,)$ satisfies for a.e. t, $0 \leq t \leq T$,*

(3.63) $\qquad\qquad \partial^- v(y(t), T - t; y'(t), -1) \leq 0,$

(3.64) $\qquad\qquad -\partial^+ v(y(t), T - t; -y'(t), 1) \leq 0,$

(3.65) $\qquad\qquad p_0 - p \cdot y'(t) = 0 \quad \forall\, (p, p_0) \in D^{\pm}v(y(t), T - t),$

(3.66) $\qquad\qquad -p \cdot y'(t) = \max_{a \in A}\{-p \cdot f(y(t), a)\} \quad \forall\, (p, p_0) \in D^{\pm}v(y(t), T - t),$

(3.67) $\qquad\qquad \alpha(t) \in \arg\min_A p \cdot f(y(t), \,\cdot\,) \quad \forall\, (p, p_0) \in D^{\pm}v(y(t), T - t).$

PROOF. Since $h(t) = v(G(t))$, with $G(t) := (y(t), T - t)$, is constant, by Lemma 2.50 we have, for each t where $y(\,\cdot\,)$ is differentiable,

$$0 = h'(t) = \partial(v \circ G)(t; 1) \geq \partial^- v(y(t), T - t; y'(t), -1)$$

which is (3.63). Similarly, we get (3.64) by writing $h'(t) = -\partial(v \circ G)(t; -1)$.

From Lemma 2.37 we have

$$\partial^- v(y(t), T - t; y'(t), -1) \geq p \cdot y'(t) - p_0$$

for all $(p, p_0) \in D^- v(y(t), T - t)$, so that (3.63) implies

(3.68) $\qquad\qquad p_0 - p \cdot y'(t) \geq 0$

for such vectors. Similarly, (3.64) implies (3.68) for all $(p, p_0) \in D^+ v(y(t), T - t)$.

Since v is a viscosity subsolution of the HJB equation (3.42), at each point where $y(\,\cdot\,)$ is differentiable we have

(3.69) $\qquad\qquad p_0 - p \cdot y'(t) \leq p_0 + \sup_{a \in A}\{-p \cdot f(y(t), a)\} \leq 0$

for all $(p, p_0) \in D^+ v(y(t), T - t)$, so (3.65), (3.66) and (3.67) hold for such vectors. Moreover, v is a bilateral supersolution of the HJB equation by Corollary 3.21, thus (3.69) holds for all $(p, p_0) \in D^- v(y(t), T - t)$ as well and so (3.65), (3.66), (3.67) are verified. ◂

Next we state some sufficient conditions of optimality in terms of verification functions which are viscosity subsolutions u of the HJB equation (3.42) such that $u(x, 0) \leq g(x)$ and with the following regularity: either $u \in BC(\mathbb{R}^N \times [0, T])$ if we assume (3.41) (i), or $u \in C(\mathbb{R}^N \times [0, T])$ if we assume (3.41) (ii). By the comparison theorems in §3.1 and §3.2 the value function v is the maximal verification function.

THEOREM 3.38 (SUFFICIENT CONDITION OF OPTIMALITY). *Assume (3.41) and set* $y(\,\cdot\,) = y_{x_0}(\,\cdot\,,\alpha)$. *Suppose there exists a verification function u which is locally lipschitzean in a neighborhood of $\{\,y(t): 0 \leq t \leq T\,\}$ and such that $u(y(T),0) = g(y(T))$. Then α is optimal for (x_0,T) if for a.e. $t \in [0,T]$ at least one of the following conditions holds:*

(3.70) $\quad\quad\quad\quad \partial^- u(y(t), T-t; y'(t), -1) \leq 0,$

(3.71) $\quad\quad\quad\quad -\partial^+ u(y(t), T-t; -y'(t), 1) \leq 0,$

(3.72) $\quad\quad\quad\quad \exists (p, p_0) \in D^{\pm} u(y(t), T-t) : \ p_0 - p \cdot y'(t) \geq 0\,.$

PROOF. We consider $\varphi(t) := u(G(t)) := u(y(t), T-t)$. Our goal is to prove that φ is nonincreasing, because in such a case

$$g(y(T)) = u(y(T), 0) = \varphi(T) \leq \varphi(0) = u(x_0, T) \leq v(x_0, T),$$

where the last inequality follows from the comparison principle (Theorem 3.7 or Theorem 3.17), and then $g(y_{x_0}(T, \alpha)) = v(x_0, T)$.

Since u and y are locally lipschitzean, we can use Lemma 2.50 to get

$$\varphi'(t) = \partial^-(u \circ G)(t; 1) = \partial^- u(y(t), T-t; y'(t), -1), \quad \text{and}$$
$$\varphi'(t) = -\partial^+(u \circ G)(t; -1) = -\partial^+ u(y(t), T-t; -y'(t), 1),$$

for all t where φ and y are both differentiable. Then $\varphi'(t) \leq 0$ at such points if either (3.70) or (3.71) holds.

On the other hand, if (3.72) holds with $(p, p_0) \in D^- u(y(t), T-t)$, then we use Lemma 2.37 to get

$$-\partial^+ u(y, T-t; -y', 1) \leq -\partial^- u(y, T-t; -y', 1) \leq -p_0 + p \cdot y' \leq 0$$

and thus (3.71) holds. Similarly, (3.72) with $(p, p_0) \in D^+ u(y(t), T-t)$ implies (3.70), and this completes the proof. ◀

THEOREM 3.39 (NECESSARY AND SUFFICIENT CONDITIONS OF OPTIMALITY). *Assume (3.41) and set $y(\,\cdot\,) = y_{x_0}(\,\cdot\,,\alpha)$. Suppose the value function v is locally lipschitzean in a neighborhood of $\{\,y(t): 0 \leq t \leq T\,\}$. Then*

(i) *α is optimal for (x_0, T) if and only if for a.e. $t \in [0, T]$ one of the following conditions hold: (3.63), (3.64), or*

(3.73) $\quad\quad \partial^- v(y(t), T-t; y'(t), -1) = \min_{f \in F(y(t))} \partial^- v(y(t), T-t; f, -1),$

where $F(x) := \overline{\mathrm{co}} f(x, A)$;

(ii) *if, moreover, $D^{\pm} v(y(t), T-t) \neq \emptyset$ for a.e. t, then the optimality of α is equivalent to (3.65) for a.e. t and it is also equivalent to (3.72) for a.e. t;*

(iii) *if, in addition, $D^+ v(y(t), T-t) = \partial v(y(t), T-t)$ for a.e. t, then the optimality of α is equivalent to (3.66) and (3.67) for a.e. t.*

Note that the additional assumption in (iii) that the superdifferential D^+v coincides with Clarke's gradient ∂v is stronger that the assumption of (ii), $D^{\pm}v \neq \emptyset$, and it is satisfied if either v is locally semiconcave (and sufficient conditions for this are given in Theorem 3.26), or the data satisfy the differentiability assumptions of Theorem 3.32. Note also that the conditions (3.66), (3.67), and (3.73) have the form of a Minimum (or Maximum) Principle. They say that for a.e. t an optimal trajectory takes a direction of steepest descent of the value function among the available directions $F(y(t))$, in some generalized sense.

PROOF. Most of the statements follow immediately from Theorems 3.37 and 3.38, for example, those concerning (3.63), (3.64), and (ii). It is easy to deduce from Theorem 3.25 (ii) that (3.73) is equivalent to (3.63).

The statement about necessity in (iii) is in Theorem 3.37. If there exist $(p, p_0) \in D^-v(y(t), T-t)$ such that

$$(3.74) \qquad -p \cdot y'(t) = \max_{a \in A} \{-p \cdot f(y(t), a)\},$$

then (3.72) holds with $u = v$ because v is a viscosity supersolution of the HJB equation (3.42), and so α is optimal by Theorem 3.38. To complete the proof of the sufficiency of (3.66) for optimality we show the sufficiency of (3.74) for some $(p, p_0) \in D^*v(y(t), T-t)$, which is weaker than (3.66) because $D^+v = \partial v = \overline{co}D^*v$ (see §II.4.1). We recall that

$$p = \lim_n Dv(x_n, t_n), \qquad p_0 = \lim_n v_t(x_n, t_n)$$

for a sequence $(x_n, t_n) \to (y(t), T-t)$ of points where v is differentiable, so that the HJB equation (3.42) holds in the classical sense at the points (x_n, t_n). By taking the limit as $n \to \infty$ in the equation we get

$$p_0 + \sup_{a \in A}\{-p \cdot f(y(t), a)\} = 0,$$

so (3.74) implies (3.72) and we conclude by Theorem 3.38. ◀

REMARK 3.40. As in §2.5 for the infinite horizon problem, one can use the necessary and sufficient conditions of optimality for open loop controls to construct optimal multivalued feedbacks for the finite horizon problem as well. It is not a hard exercise for the reader to adapt Definition 2.60 of weak and full optimality and then prove results analogous to Theorem 2.61 and Corollaries 2.63 and 2.64. ◁

★ ★ ★

Next we prove the Pontryagin Maximum Principle (PMP in the sequel) in a necessary and sufficient form, under the (usual) additional assumptions that $g \in C^1(\mathbb{R}^N)$ and f is differentiable with respect to x with $D_x f$ continuous.

Let us first recall its classical derivation from the HJB equation under the additional assumption that the value function v is C^2. The HJB equation (3.42) gives

$$(3.75) \qquad v_t(x, s) - D_x v(x, s) \cdot f(x, a) \leq 0 \qquad \text{for all } x, s,$$

and we saw at the beginning of the section that the left-hand side of (3.75) is ≥ 0 at $x = y(t)$, $s = T - t$, $a = \alpha(t)$ if $y(\,\cdot\,) = y_{x_0}(\,\cdot\,, \alpha)$ is optimal for (x_0, T), see (3.61). Thus $y(t)$ maximizes $x \mapsto v_t(x, T-t) - D_x v(x, T-t) \cdot f(x, \alpha(t))$ and we can differentiate with respect to x to get

(3.76) $$D_x v_t - f D_{xx}^2 v - D_x v D_x f = 0$$

for $x = y(t)$, $s = T - t$, $a = \alpha(t)$, where $D_{xx}^2 v$ is the Hessian matrix of v and $(D_x f)_{ij} = \partial f_i / \partial x_j$ (so (3.76) is an equality between row vectors). We define the row vector

(3.77) $$p(t) := D_x v(y(t), T-t)$$

and differentiate with respect to t to get, by (3.76),

$$p' = f D_{xx}^2 v - D_x v_t = -p D_x f \,.$$

Thus the *costate* or *adjoint vector* $p(t)$ associated with (x_0, T) and $\alpha \in \mathcal{A}$ can be redefined, without mentioning v, as the unique solution of the linear system adjoint to the linearization of the controlled system around the trajectory $y(\,\cdot\,)$, that is,

(3.78) $$\begin{cases} p'(t) = -p(t) D_x f(y(t), \alpha(t)), & 0 < t < T, \\ p(T) = Dg(y(T)), \end{cases}$$

and necessary conditions of optimality are

(3.79) $$-p(t) \cdot f(y(t), \alpha(t)) = \max_{a \in A} \{ -p(t) \cdot f(y(t), a) \} = H(y(t), p(t)),$$

(3.80) $$v_t(y(t), T-t) = p(t) \cdot f(y(t), \alpha(t)) = -H(y(t), p(t)) \,.$$

The necessity of (3.79) for the optimality of α, without the extra regularity assumption on v, is the statement of the classical PMP, and it is particularly useful because one can try to solve (3.78) and (3.79) without knowing a priori the value function v. However, these two conditions alone are not sufficient for optimality, as the next example shows.

EXAMPLE 3.41. For $N = 2$ consider the system

$$(y_1', y_2') = (a, y_1^2), \quad (y_1, y_2)(0) = (x_1, x_2) = x, \quad A = [-1, 1],$$

and the problem of minimizing $-y_2(t, \alpha)$, that is, $g(y) = -y_2$. It is easy to guess that an optimal control is $\alpha_+ \equiv 1$ if $x_1 \geq 0$ and $\alpha_- \equiv -1$ if $x_1 < 0$, whose corresponding cost is $-x_2 - x_1^2 t - \frac{1}{3} t^3 - |x_1| t^2$. It is easy to see that this is indeed the value function $v(x, t)$ because it satisfies the HJB equation

$$v_t + |v_{x_1}| - x_1^2 v_{x_2} = 0 \quad \text{in } \mathbb{R}^2 \times {]}0, \infty[$$

and the initial condition $v(x, 0) = g(x)$. The control $\alpha_0 \equiv 0$ is not optimal for $(0, 0, t)$ because the trajectory remains in $(0, 0)$ and the cost is $0 > -t^3/3 = v(0, t)$. The costate is $(p_1, p_2)(t) \equiv (0, -1)$ because the matrix $D_x f$ is null along the null trajectory, and the maximality condition (3.79) is satisfied because both sides are null. ◁

If we do not assume the extra regularity of the value function v, which is false in general, a natural nonsmooth counterpart of the formulas (3.77) and (3.80) connecting the costate with the derivatives of v is

(3.81) $\qquad (p(t), -H(y(t), p(t))) \in D^+ v(y(t), T-t)$.

Note that in the right-hand side of this formula it is equivalent to write D^+ or $D^\pm$ because $D^- v \subseteq D^+ v$ everywhere by Theorem 3.32. The condition that the costate satisfy for a.e. t both (3.79) and (3.81) is sufficient for the optimality of α by Theorem 3.38, because (3.72) is satisfied and v is locally lipschitzean by Theorem 3.32. The next result states that this condition is also necessary.

THEOREM 3.42 (PONTRYAGIN MAXIMUM PRINCIPLE). *Assume (3.41) (ii), A compact, $g \in C^1(\mathbb{R}^N)$, f differentiable with respect to x and $D_x f$ continuous. Set $y(\,\cdot\,) := y_{x_0}(\,\cdot\,, \alpha)$ and let the costate $p(\,\cdot\,)$ be the corresponding solution of (3.78). Then α is optimal for (x_0, T) if and only if for a.e. $t \in \,]0, T[$*

$$-p(t) \cdot f(y(t), \alpha(t)) = \max_{a \in A} \{ -p(t) \cdot f(y(t), a) \} = H(y(t), p(t)), \quad \text{and}$$

$$(p(t), -H(y(t), p(t))) \in D^+ v(y(t), T-t).$$

We present three proofs of this important theorem; they make a decreasing use of the previous theory while their length is increasing. The first is an easy byproduct of the representation of v as a minimum over relaxed controls derived in §3.3 and of the necessary conditions of optimality of the present subsection. In particular, it gives the PMP for relaxed controls, see Remark 3.48. The second does not use relaxed controls but just the theory of necessary conditions. The third is self-contained, but still not too long, and it has an interesting extension, see Theorem 3.44. For all of them we need some notations and a lemma. For a given $\alpha \in \mathcal{A}$ we denote by S the corresponding *solution operator* for the system (S), that is,

$$S(\tau; r, x) := \zeta(\tau),$$

where ζ solves

$$\zeta' = f(\zeta, \alpha), \qquad \zeta(r) = x.$$

We also denote by $\alpha^{(t)} \in \mathcal{A}$ the control $\alpha \in \mathcal{A}$ shifted by a time t:

$$\alpha^{(t)}(s) := \alpha(t+s).$$

LEMMA 3.43. *Under the assumptions and notations of Theorem 3.42, the cost functional*

$$J(x, \tau, \beta) := g(y_x(\tau, \beta))$$

corresponding to any $\beta \in \mathcal{A}$ is differentiable with respect to x, and for all t

$$D_x J(y(t), T-t, \alpha^{(t)}) = p(t).$$

PROOF. The first statement follows from the chain rule and the theorem on differentiability of solutions of an ODE with respect to the initial data, Theorem 5.8. Set $A(s) := D_x f(y(s), \alpha(s))$ and denote by $M(s,t)$ the fundamental matrix associated with the linear system

(3.82) $$\xi' = A(s)\xi,$$

that is, the unique solution of

$$\begin{cases} \dfrac{\partial M}{\partial s}(s,t) = A(s)M(s,t), & s \in \mathbb{R}, \\ M(t,t) = I, \end{cases}$$

where I is the identity matrix. By Theorem 5.8 the Jacobian matrix of the map $x \mapsto S(t+\tau;t,x)$ at $y(t)$ is $M(t+\tau,t)$. Since

$$J(x,\tau,\alpha^{(t)}) = g(S(t+\tau;t,x)),$$

we obtain

$$D_x J(y(t), T-t, \alpha^{(t)}) = Dg(y(T))M(T,t).$$

On the other hand, by standard results on linear ODE, the solution of the system adjoint to (3.82)

$$\eta' = -\eta A(t)$$

with terminal datum $\eta(T) = \bar{\eta}$ is given by $\bar{\eta}M(T,t)$, as it is easy to check by using the following well-known property of the fundamental matrix:

$$\dfrac{\partial M}{\partial t}(s,t) = -M(s,t)A(t).$$

Therefore, the definition of costate (3.78) gives

$$p(t) = Dg(y(T))M(T,t). \quad \blacktriangleleft$$

PROOF 1 OF THEOREM 3.42. The sufficiency follows from Theorem 3.38, as we remarked earlier. To prove the necessity we assume α is optimal and observe that it is enough to show for a.e. t the existence of $p_0(t) \in \mathbb{R}$ such that

(3.83) $$(p(t), p_0(t)) \in D^+ v(y(t), T-t),$$

because then the necessary conditions of optimality (3.65) and (3.66) in Theorem 3.37 imply for a.e. t

$$p_0(t) = p(t) \cdot f(y(t), \alpha(t)) = -H(y(t), p(t))$$

which gives (3.79) and (3.81). In the present proof we use the results of §3.3 on the representation of v as a marginal function with respect to rescaled relaxed controls

$\mu \in \mathcal{A}_1^r$ of the smooth (with respect to (x,t)) functions $G(x,t,\mu)$, see (3.55) and, in particular, Theorem 3.32 and Remark 3.33. Since α is optimal for $(y(t), T-t)$, at such point the minimum of G with respect to μ is attained at the rescaled relaxed control associated with $\alpha(\cdot + t)$, that is,

$$\eta_{T-t}(\tau) := \delta_{\alpha((T-t)\tau+t)}$$

where δ_a indicates the Dirac measure concentrated at a. Then the gradient of G with respect to (x,t) at $(y(t), T-t, \eta_{T-t})$ belongs to D^+v at the same point, see Remark 3.33. Thus it is enough to show that

$$D_x G(y(t), T-t, \eta_{T-t}) = p(t),$$

and this follows immediately from Lemma 3.43 because

$$G(x,t,\mu_t) = J^r(x,t,\mu) \qquad \text{for all } \mu \in \mathcal{A}^r,$$

where $\mu_t(\tau) = \mu(t\tau)$, and for a relaxed control associated with a standard control $\beta \in \mathcal{A}$ the relaxed trajectories coincide with the standard ones, so $J^r = J$. ◀

The next proof of the necessary condition of optimality does not use the results of §3.3 involving relaxed controls, but it is still based on Theorem 3.37 on necessary conditions in the general case.

PROOF 2 OF THEOREM 3.42. We recall from the beginning of Proof 1 that we have to show the existence of $p_0(t)$ satisfying (3.83). Since

$$v(x,s) \le J(x,s,\alpha^{(t)}) =: \varphi(x,s)$$

for all x, s, and

$$v(y(t), T-t) = J(y(t), T-t, \alpha^{(t)}) = \varphi(y(t), T-t),$$

it is enough to prove that there exists p_0 such that

$$\varphi(x,s) - \varphi(y(t), T-t)$$
$$\le p(t) \cdot (x - y(t)) + p_0(t)(s - T + t) + o(|x - y(t)| + |s - T + t|)$$

as $x \to y(t)$, $s \to T-t$. By Lemma 3.43,

$$\varphi(x, T-t) - \varphi(y(t), T-t) = p(t) \cdot (x - y(t)) + o(|x - y(t)|)$$

as $x \to y(t)$. On the other hand, since $\varphi(x,s) = g(S(t+s;t,x))$ we can use the differentiability of g to get

$$\varphi(x,s) - \varphi(x, T-t) = Dg(S(T;t,x)) \cdot \int_T^{t+s} f(S(\tau;t,x), \alpha(\tau)) \, d\tau + o(|s - T + t|)$$
$$\le p_0(t)(s - T + t) + o(|s - T + t|) \qquad \text{as } s \to T-t,$$

where

$$p_0(t) := \limsup_{(x,s) \to (y(t), T-t)} Dg(S(T;t,x)) \cdot \frac{1}{t+s-T} \int_T^{t+s} f(S(\tau;t,x), \alpha(\tau)) \, d\tau$$

which is finite because f is locally bounded. This completes the proof. ◀

The last proof of the necessary condition in Theorem 3.42 is the most direct: it does not use the general Theorem 3.37 on necessary conditions, but only Lemma 3.43.

PROOF 3 OF THEOREM 3.42. For all x, s
$$v(x,s) \leq J(x, s, \alpha^{(T-s)}) = g(S(T; T-s, x)), \quad \text{and}$$
$$S(T; T-s, x) = S(T; t, x + h(x, s)),$$

where

(3.84) $$h(x,s) := \int_{T-s}^{t} f(S(\tau; T-s, x), \alpha(\tau))\, d\tau,$$

and so we get
$$v(x,s) \leq J(x + h(x, s), T-t, \alpha^{(t)}).$$

Since
$$v(y(t), T-t) = J(y(t), T-t, \alpha^{(t)}),$$

Lemma 3.43 gives

(3.85)
$$v(x,s) - v(y(t), T-t) \leq p(t) \cdot (x + h(x,s) - y(t)) + o(|s - T + t| + |x - y(t)|)$$

as $s \to T-t$, $x \to y(t)$. Now observe that $y(\tau) = S(\tau; t, y(t))$ gives
$$S(\tau; T-s, x) = S(\tau; t, x) + o(1) = y(\tau) + o(1)$$

as $s \mapsto T-t$, $x \to y(t)$, and so
$$h(x,s) = \int_{T-s}^{t} f(y(\tau), \alpha(\tau))\, d\tau + o(1).$$

Moreover, by Lebesgue derivation theorem, for a.e. t,
$$\frac{1}{s - T + t} \int_{T-s}^{t} f(y(\tau), \alpha(\tau))\, d\tau = f(y(t), \alpha(t)) + o(1)$$

as $s \to T-t$. Therefore, for such t, we get from (3.85)

$v(x,s) - v(y(t), T-t)$
$\leq p(t) \cdot (x - y(t)) + p(t) \cdot f(y(t), \alpha(t))(s - T + t) + o(|s - T + t| + |x - y(t)|)$

as $s \to T-t$, $x \to y(t)$, which proves that
$$(p(t), p(t) \cdot f(y(t), \alpha(t))) \in D^+ v(y(t), T-t).$$

On the other hand, the definition of viscosity subsolution of the HJB equation (3.42) gives
$$p(t) \cdot f(y(t), \alpha(t)) + \sup_{a \in A}\{-p(t) \cdot f(y(t), a)\} \leq 0$$

so we get (3.79) and (3.81). ◀

Now we give some extensions of the PMP, Theorem 3.42, and several remarks and comments. In the next result we allow the terminal cost g to be nondifferentiable and we replace the value function v with verification functions. It applies, in particular, if g is semiconcave.

THEOREM 3.44 (EXTENDED PMP). *Assume* (3.41) (ii) *and* $x \mapsto f(x,a)$ *continuously differentiable for all* $a \in A$. *Given* $y(\,\cdot\,) = y_{x_0}(\,\cdot\,,\alpha)$, *assume* $D^+g(y(T)) \neq \emptyset$ *and define* $p(\,\cdot\,)$ *as the unique solution of*

(3.86) $$\begin{cases} p'(t) = -p(t)\, D_x f(y(t),\alpha(t)), & 0 < t < T, \\ p(T) = \bar{p}, \end{cases}$$

with $\bar{p} \in D^+g(y(T))$. *If* α *is optimal for* (x_0,T), *then for almost every* $t \in \,]0,T[$ *the maximum condition* (3.79) *holds and*

(3.87) $$(p(t), -H(y(t),p(t))) \in D^+u(y(t), T-t)$$

for any verification function u *such that*

$$u(y(t), T-t) = g(y(T)) \qquad \text{for all } t\,.$$

Vice versa, α *is optimal for* (x_0,T) *if there exists a verification function* u *that is locally lipschitzean in a neighborhood of* $\{\,y(t) : 0 \leq t \leq T\,\}$ *and such that* (3.79) *and* (3.87) *hold for a.e.* $t \in \,]0,T[$.

PROOF. The sufficient condition of optimality is an immediate corollary of Theorem 3.38. To prove the necessary condition of optimality, first we replace Lemma 3.43 with the claim

$$p(t) \in D_x^+ J(y(t), T-t, \alpha^{(t)}),$$

where $D_x^+ J(y,\tau,\beta)$ denotes the superdifferential of $x \mapsto J(x,\tau,\beta)$ at y. In fact, under the current assumptions, the arguments in the proof of Lemma 3.43 give

(3.88) $$p(t) = \bar{p} M(T,t),$$
(3.89) $$S(T;t,x) - y(T) = M(T,t)(x - y(T)) + o(x - y(T))$$

as $x \to y(T)$, and $J(x, T-t, \alpha^{(t)}) = g(S(T;t,x))$. By plugging (3.89) into the following property of $\bar{p} \in D^+g(y(T))$:

$$g(S) - g(y(T)) \leq \bar{p} \cdot (S - y(T)) + o(S - y(T)) \qquad \text{as } S \to y(T),$$

and using (3.88) we get the claim.

Next we note that

$$u(x,s) \leq v(x,s) \qquad \text{for all } x,s,$$

by the comparison principle (Theorem 3.7 or 3.15), and

$$u(y(t), T-t) = v(y(t), T-t)$$

by assumption. Thus we can repeat Proof 3 of Theorem 3.42 with v replaced by u and get the conclusion. ◀

The next result improves a bit the necessary condition of optimality connecting the costate and the semidifferentials of the value function.

COROLLARY 3.45. *Under the assumptions of Theorem 3.42, if α is optimal for (x_0, T), then for all $t \in [0, T]$*

$$D^- v(y(t), T-t) \subseteq \{ (p(t), -H(y(t), p(t))) \} \subseteq D^+ v(y(t), T-t).$$

Moreover, if either one of these inclusions is an equality, then v is differentiable at $(y(t), T-t)$, and in this case both the inclusions are equalities.

PROOF. The first inclusion follows from the second by the properties of the sub- and superdifferentials of good marginal functions (by Theorem 3.32 and Proposition II.4.4 either $D^- v$ is empty or v is differentiable at $(y(t), T-t)$, and then $D^- v = D^+ v$ at such point).

By Theorem 3.42 the second inclusion holds for a.e. $t \in \,]0, T[$. Then it is very easy to prove it for all t because $D^+ v(z) = \partial v(z)$ for all $z = (x, t)$, by Theorem 3.32, the multivalued map $z \mapsto \partial v(z)$ is known to be upper semicontinuous (i.e., $z_n \to z$, $p_n \to p$, $p_n \in \partial v(z_n)$ imply $p \in \partial v(z)$) and $p(\,\cdot\,), y(\,\cdot\,)$, and H are continuous.

If the second inclusion is an equality, then v is differentiable at $(y(t), T-t)$ by Theorem 3.32 and Proposition II.4.4. All the other statements follow from the basic properties of sub- and superdifferentials. ◂

COROLLARY 3.46. *Assume the hypotheses of Theorem 3.44. If α is optimal for (x_0, T), then $t \mapsto H(y(t), p(t))$ is constant.*

PROOF. By the assumption (3.41) (ii) the Hamiltonian H is locally lipschitzean, so $t \mapsto H(y(t), p(t))$ is absolutely continuous. We are going to show that its derivative is null at almost every t where it is differentiable, therefore almost everywhere. We fix such a t. By Theorem 3.44 we can choose s arbitrarily close to t such that

$$H(y(s), p(s)) = -p(s) \cdot f(y(s), \alpha(s)).$$

By the definition of H

$$H(y(s), p(s)) - H(y(t), p(t))$$
$$\leq [p(t) - p(s)] \cdot f(y(t), \alpha(t)) + p(s) \cdot [f(y(t), \alpha(t)) - f(y(s), \alpha(s))]$$
$$= (t-s)[p'(t) \cdot f(y(t), \alpha(t)) + p(t) D_x f(y(t), \alpha(t)) y'(t)] + o(|t-s|)$$
$$= o(|t-s|) \quad \text{as } s \to t,$$

where in the last equality we have used the definition of costate (3.86). If we exchange the roles of t and s we obtain a similar inequality which completes the proof. ◂

REMARK 3.47. A more concise but equivalent formulation of the PMP is as follows: α is optimal for (x_0, T) if and only if for a.e. $t \in \,]0, T[$

$$(p(t), p(t) \cdot f(y(t), \alpha(t))) \in D^+ v(y(t), T-t). \quad \triangleleft$$

REMARK 3.48. Proof 1 of Theorem 3.42 also gives a PMP for relaxed controls (we recall that the value function associated with relaxed controls coincides with v, so $\mu \in \mathcal{A}^r$ is optimal for (x,t) if $J^r(x,t,\mu) = v(x,t)$). The formulation is exactly the same after replacing f with f^r in (3.79) and in the definition of costate, which now is (3.60). ◁

REMARK 3.49. If the trajectory $y(\,\cdot\,)$ is optimal and H is differentiable at $(y(t),p(t))$, then the differential equations satisfied by the costate $p(\,\cdot\,)$ and by $y(\,\cdot\,)$ take the form of the *Hamiltonian system*

$$\begin{cases} p'(t) = D_x H(y(t),p(t)), \\ y'(t) = -D_p H(y(t),p(t)) \,. \end{cases}$$

We recall that for H smooth this is the characteristic system of ODEs of the HJB equation (3.42) and the functions $p(\,\cdot\,), y(\,\cdot\,)$ are the characteristics of this PDE. ◁

Exercises

3.1. Prove that under the assumptions (A$_1$)–(A$_3$), (A$'_4$), and (A$_5$)

$$\lim_{t \to +\infty} v(x,t) = \inf_{\alpha \in \mathcal{A}} \int_0^{+\infty} \ell(y_x(s), \alpha(s)) e^{-\lambda s} \, ds \,.$$

3.2. In addition to the assumptions of Proposition 3.1, suppose f is bounded. Prove that

(i) $v \in BUC(\mathbb{R}^N \times [0,T])$ for all $T > 0$,

(ii) if $\lambda > 0$, then $v \in BUC(\mathbb{R}^N \times [0,\infty[)$.

3.3. Consider $N = 1$, $f(x,a) = a$, $A = [-1,1]$, $g(x) = e^{-x^2}$, $\ell(x,a) \equiv 0$. Compute $v(x,t)$ and show that $\partial v / \partial x$ has a jump along the half-line $x = 0$, $t > 0$.

3.4. Consider $N = 1$, $f(x,a) = x^a$, $A = [0,1]$, $g(x) = -x$, $\ell(x,a) \equiv 0$. Compute $v(x,t)$ and $D^+v(x,t)$, $D^-v(x,t)$.

3.5. Prove the following extension of Theorem 3.12. Let $u_1, u_2 \in C(\overline{\mathcal{C}})$ be, respectively, a subsolution of

$$u_t + H(x,t,Du) - h_1(x,t) = 0 \quad \text{in } \mathcal{C}$$

and a supersolution of

$$u_t + H(x,t,Du) - h_2(x,t) = 0 \quad \text{in } \mathcal{C},$$

where $h_1, h_2 \in C(\mathcal{C})$, $H \in C(\mathcal{C} \times \mathbb{R}^N)$ satisfies

$$|H(x,t,p) - H(x,t,q)| \le C|p-q|,$$
$$|H(x,t,p) - H(y,t,p)| \le \omega(|x-y|) + \omega(|x-y||p|),$$

for all $(x,t), (y,t) \in \mathcal{C}$, $p, q \in \mathbb{R}^N$. Then

$$\sup_{B(0,C(T-t))} (u_1(\,\cdot\,,t) - u_2(\,\cdot\,,t))$$
$$\leq \sup_{B(0,CT)} (u_1(\,\cdot\,,0) - u_2(\,\cdot\,,0)) + \int_0^t \sup_{B(0,C(t-s))} (h_1(\,\cdot\,,s) - h_2(\,\cdot\,,s))\,ds\,.$$

3.6 (COMPARISON THEOREM FOR HJB EQUATIONS WITH TIME-DEPENDENT H). Prove the following extension of Theorem 3.15. Let $u_1, u_2 \in C(\mathbb{R}^N \times [0,T])$ be, respectively, a sub- and a supersolution of

$$u_t + H(x,t,Du) = 0 \quad \text{in } \mathbb{R}^N \times\,]0,T[,$$

where $H \in C(\mathbb{R}^N \times [0,T] \times \mathbb{R}^N)$ satisfies

$$|H(x,t,p) - H(x,t,q)| \leq K(|x|+1)|p-q|$$

for all x, t, p, and

$$|H(x,t,p) - H(y,t,p)| \leq \omega(|x-y|,R) + \omega(|x-y||p|,R),$$

for all $p \in \mathbb{R}^N$, $t \in [0,T]$, $x,y \in B(0,R)$, $R > 0$. Then

$$\sup_{\mathbb{R}^N \times [0,T]} (u_1 - u_2) = \sup_{\mathbb{R}^N}(u_1(\,\cdot\,,0) - u_2(\,\cdot\,))\,.$$

3.7 (NONAUTONOMOUS PROBLEMS). Assume A is a compact set, $g \in C(\mathbb{R}^N)$, $0 < T < +\infty$, $f: \mathbb{R}^N \times [0,T] \times A \to \mathbb{R}^N$ and $\ell: \mathbb{R}^N \times [0,T] \times A \to \mathbb{R}$ are continuous, $|f(x,t,a)| \leq K(1+|x|)$ for all x, t, a, and for any $R > 0$ there is L_R such that

$$|f(x,t,a) - f(y,t,a)| \leq L_R|x-y|$$

for all $t \in [0,T]$, $a \in A$ and $x,y \in B(0,R)$. Define

$$H(x,t,p) := \max_{a \in A}\{-f(x,t,a)\cdot p - \ell(x,t,a)\}, \quad \text{and}$$

$$v(x,t) := \inf_{\alpha \in \mathcal{A}}\left\{\int_t^T \ell(y(s),s,\alpha(s))\,ds + g(y(T))\right\},$$

where $y(\,\cdot\,)$ is the unique solution of

$$\begin{cases} y'(s) = f(y(s),s,\alpha(s)), & t < s < T, \\ y(t) = x. \end{cases}$$

(i) Prove the DPP

$$v(x,t) = \inf_{\alpha \in \mathcal{A}}\left\{\int_t^\tau \ell(y(s),s,\alpha(s))\,ds + v(y(\tau),\tau)\right\},$$

for all $x \in \mathbb{R}^N$, $0 \leq t < \tau \leq T$ (cfr. Remark 3.10).

(ii) Prove that v is the complete solution in $C(\mathbb{R}^N \times [0,T])$ of the terminal value problem
$$\begin{cases} -v_t + H(x,t,Dv) = 0 & \text{in } \mathbb{R}^N \times \,]0,T[, \\ v(\,\cdot\,,T) = g(\,\cdot\,) & \text{in } \mathbb{R}^N. \end{cases}$$

[Hint: use Exercise 3.6.]

3.8. Assume $u : \Omega \to \mathbb{R}$ is semiconcave-convex. Prove that

(i) $u \in C^1(\Omega)$;

(ii) u has bounded pure second derivatives in the sense of distributions;

(iii) all the second derivatives of u are in $L^\infty(\Omega)$;

(iv) Du is Lipschitz continuous.

3.9. Show that if $x \mapsto f(x,a)$ is affine for all $a \in A$, then $x \mapsto z_x(1;\alpha,t)$ is affine (see Lemma 3.27 for the notations). Therefore, if in addition g is concave (respectively, convex), $x \mapsto v(x,t)$ is concave (respectively, convex).

3.10. Fill in the details in the proof of Theorem 3.32, that is, prove that the solution of (3.58) at a fixed time depends continuously on (x,t,μ).

3.11. Prove that the conclusion of Theorem 3.39 (iii) holds if the assumption $D^+v = \partial v$ is replaced by $D^+v(z) \neq \emptyset$ for a.e. t, where $z := (y(t), T-t)$, and $\partial^- v(z;q) = \min_{\zeta \in D^+v(z)} \zeta \cdot q$ for all $q \in F(y(t)) \times \{-1\}$.

3.12. Under the assumptions and notations of Theorem 3.44 prove that, if α is optimal for (x_0, T), then

(i) $p(t) \in D_x^+ u(y(t), T-t)$ for all $t \in [0,T]$;

(ii) if in addition u is semiconcave, then $(p(t), H(y(t), p(t))) \in D^+u(y(t), T-t)$ (cfr. Corollary 3.45).

4. Problems whose HJB equation is a variational or quasivariational inequality

In the previous sections we considered optimal control problems where the admissible controls were all the measurable functions of time with values in a given set A. Here we give some examples where either some nontrivial restriction on the class of admissible controls is imposed (the *monotone control problem* in §4.1) or extra control variables appear in the model in the form of a sequence of *stopping times* and *switchings* or *impulses* (see §4.2 to §4.4) with corresponding costs.

All these problems, which can be seen as variants of the infinite horizon problem of §2, share a common feature, namely, that the HJB equation satisfied by their value function takes the form of a *variational* or *quasivariational inequality*. We show that the viscosity solution approach is flexible enough to deal with these more complicated situations.

Throughout the section we assume that the dynamics and the running cost satisfy the standard assumptions of §1 and §2, namely:

(A$_0'$) A is compact, $f : \mathbb{R}^N \times A \to \mathbb{R}^N$ is continuous,
which implies (A$_0$) and (A$_1$);

(A$_3$) $(f(x,a) - f(y,a)) \cdot (x - y) \leq L |x - y|^2$, $\forall\, x, y \in \mathbb{R}^N$, $\forall\, a \in A$;

(A$_4$) $\ell : \mathbb{R}^N \times A \to \mathbb{R}^N$ is continuous and
$|\ell(x,a) - \ell(y,a)| \leq \omega_\ell(|x - y|)$, $|\ell(x,a)| \leq M$ for all $x, y \in \mathbb{R}^N$, $a \in A$; $\lambda > 0$,

where ω_ℓ is a modulus. The control system is

(4.1) $$\begin{cases} y'(t) = f(y(t), \alpha(t)), & t \geq 0, \\ y(0) = x \in \mathbb{R}^N, \end{cases}$$

$\alpha \in \mathcal{A}$, the set of measurable functions with values in A.

4.1. The monotone control problem

In this problem the control set A is the closed interval $[0,1]$ and the dynamical system (4.1) is controlled by choosing α in the class

(4.2) $$\mathcal{A}_m := \{\, \alpha : [0, +\infty[\to [0,1] : \alpha \text{ nondecreasing} \,\}.$$

With the evolution defined by (4.1) and (4.2), we associate the infinite horizon payoff functional

(4.3) $$J(x, \alpha) := \int_0^{+\infty} \ell(y_x(t,\alpha), \alpha(t))\, e^{-\lambda t}\, dt$$

and the value function

(4.4) $$v_m(x) := \inf_{\alpha \in \mathcal{A}_m} J(x, \alpha).$$

Let us observe that the class of admissible controls $\mathcal{A}_m$ does not satisfy property (ii) in Remark 2.7. This causes difficulty in implementing the dynamic programming procedure for v_m.

A way to overcome such a difficulty is to introduce, for each fixed $a \in [0,1]$, the subset

(4.5) $$\mathcal{A}_m^a := \{\, \alpha \in \mathcal{A}_m : \alpha(0) \geq a \,\},$$

and look at

(4.6) $$v(x, a) := \inf_{\alpha \in \mathcal{A}_m^a} J(x, \alpha),$$

called the *monotone value function*. It is clear by definition that

(4.7) $$v(x, 0) = \inf_{\alpha \in \mathcal{A}_m^0} J(x, \alpha) = v_m(x).$$

4. HJB INEQUALITIES

In the following we study the basic properties of v as a function of $x \in \mathbb{R}^N$ and $a \in [0, 1]$, leading to its characterization as the unique viscosity solution of the *variational inequality*

(VI) $\quad \max\left\{ \lambda u(x, a) - f(x, a) \cdot D_x u(x, a) - \ell(x, a); \; -\dfrac{\partial u}{\partial a}(x, a) \right\} = 0$

in $\mathbb{R}^N \times [0, 1[$, satisfying the boundary condition

$$u(x, 1) = J(x, 1), \qquad \forall x \in \mathbb{R}^N .$$

For this purpose we need to strengthen the standard conditions on f and ℓ. We assume in addition

(4.8) $\quad \begin{cases} |f(x, a) - f(x', a')| \leq L(|x - x'| + |a - a'|) \\ |\ell(x, a) - \ell(x', a')| \leq \omega_\ell(|x - x'| + |a - a'|) \end{cases} \quad \forall x, x' \in \mathbb{R}^N, \; a, a' \in [0, 1].$

PROPOSITION 4.1. *Assume (4.8). Then the value function v defined by (4.6) has the following properties:*

(i) *for each fixed $x \in \mathbb{R}^N$, the function $a \mapsto v(x, a)$ is nondecreasing;*

(ii) *if $v(x, a) = J(x, \widetilde{\alpha})$ for some $\widetilde{\alpha} \in \mathcal{A}_m^a$, then $v(x, a) = v(x, a')$ for all $a' \in [a, \widetilde{\alpha}(0^+)]$;*

(iii) $v(x, 1) = \displaystyle\int_0^{+\infty} \ell(y_x^1(t), 1) e^{-\lambda t} \, dt = J(x, 1)$, *for all $x \in \mathbb{R}^N$, where y_x^1 is the trajectory of (4.1) corresponding to the constant control $\alpha(t) \equiv 1$;*

(iv) $v \in BUC(\mathbb{R}^N \times [0, 1])$.

PROOF. Statement (i) follows immediately from the inclusion $\mathcal{A}_m^{a'} \subseteq \mathcal{A}_m^a$, $\forall a' \geq a$.

In order to prove (ii), observe first that by monotonicity

$$\widetilde{\alpha}(0^+) := \lim_{t \to 0^+} \widetilde{\alpha}(t) \geq \widetilde{\alpha}(0) \geq a .$$

If $\widetilde{\alpha}(0^+) = a$, there is nothing to prove. Hence consider the case $\widetilde{\alpha}(0^+) > a$. Define α^* by setting $\alpha^*(0) = \widetilde{\alpha}(0^+) =: \widetilde{a}^+$ and $\alpha^*(t) = \widetilde{\alpha}(t)$ for $t > 0$. Of course, $\alpha^* \in \mathcal{A}_m^{\widetilde{a}^+}$ and $\alpha^* \equiv \widetilde{\alpha}$ almost everywhere. As a consequence, the corresponding trajectories and payoff coincide, $J(x, \alpha^*) = J(x, \widetilde{\alpha})$, so that, by assumption and the definition of v,

(4.9) $\qquad v(x, a) = J(x, \widetilde{\alpha}) = J(x, \alpha^*) \geq v(x, \widetilde{a}^+) .$

On the other hand, by property (i)

$$v(x, a) \leq v(x, a') \leq v(x, \widetilde{a}^+) \qquad \forall a' \in [a, \widetilde{a}^+] .$$

This combined with (4.9), proves (ii).

Equality (iii) is immediate since $\mathcal{A}_m^1$ contains only the constant control $\alpha \equiv 1$.

In order to prove (iv), let us first fix $x \in \mathbb{R}^N$ and take $a, a' \in [0,1]$ with $a' > a$ (the case $a' < a$ is treated in a similar way). If $\widetilde{\alpha} \in \mathcal{A}_m^a$ is such that

$$v(x,a) = J(x, \widetilde{\alpha}), \tag{4.10}$$

we set

$$t_0 := \inf\{t \geq 0 : \widetilde{\alpha}(t) \geq 1 + a - a'\}$$

and $t_0 = +\infty$ if $\widetilde{\alpha}(t) < 1 + a - a'$ for all $t \geq 0$. Consider the control $\overline{\alpha}$ defined by

$$\overline{\alpha}(t) = \begin{cases} \widetilde{\alpha}(t) + a' - a & \text{for } 0 \leq t \leq t_0 \\ 1 & \text{for } t > t_0. \end{cases}$$

By definition of t_0 (and the monotonicity of $\widetilde{\alpha}$) we have

$$|\overline{\alpha}(t) - \widetilde{\alpha}(t)| \leq |a' - a| \qquad \forall t \geq 0. \tag{4.11}$$

Since $\overline{\alpha} \in \mathcal{A}_m^{a'}$ we have, because of (4.10) and property (i),

$$0 \leq v(x,a') - v(x,a) \leq J(x, \overline{\alpha}) - J(x, \widetilde{\alpha}). \tag{4.12}$$

Let us now show that

$$|J(x, \overline{\alpha}) - J(x, \widetilde{\alpha})| \leq \omega(|a - a'|), \tag{4.13}$$

for some modulus ω. For this purpose observe that assumption (4.8), Gronwall's Lemma 5.1, and (4.11) yield

$$|\overline{y}_x(t) - \widetilde{y}_x(t)| \leq Lt\, e^{Lt} |a - a'| \qquad \forall t \geq 0, \tag{4.14}$$

where $\overline{y}_x(\cdot), \widetilde{y}_x(\cdot)$ are the trajectories of (4.1) corresponding to the controls $\overline{\alpha}, \widetilde{\alpha}$. At this point observe that

$$|J(x, \overline{\alpha}) - J(x, \widetilde{\alpha})| \leq I_T + I_\infty \qquad \forall T \geq 0,$$

where

$$I_T = \int_0^T \omega_\ell(|\overline{y}_x(t) - \widetilde{y}_x(t)| + |\overline{\alpha}(t) - \widetilde{\alpha}(t)|)\, e^{-\lambda t}\, dt,$$

$$I_\infty = \int_T^{+\infty} 2M\, e^{-\lambda t}\, dt.$$

Using the estimates (4.14) and (4.11) we obtain

$$I_T \leq \int_0^T \omega_\ell(Lt\, e^{Lt}|a - a'| + |a - a'|)\, e^{-\lambda t}\, dt.$$

Since I_∞ and I_T can be made arbitrarily small by choosing, respectively, T large and $|a - a'|$ small, we have proved (4.13). The uniform continuity of v with respect to a follows from (4.12), (4.13).

The proof of the uniform continuity of v as a function of x is the same as that of Proposition 2.1.

Observing that the modulus of continuity of $a \mapsto v(x,a)$ does not depend on x (as well as that of $x \mapsto v(x,a)$ is independent of a) the proof of (iv) is complete. Indeed, the boundedness of v is a straightforward consequence of the inequalities

$$-M/\lambda \leq v(x,0) \leq v(x,a) \leq v(x,1) \leq M/\lambda \, . \blacktriangleleft$$

The next proposition is the statement of the Dynamic Programming Principle of Optimality.

PROPOSITION 4.2. *For any $x \in \mathbb{R}^N$, $a \in [0,1]$, and $t > 0$ we have*

(i) $$v(x,a) \leq \int_0^t \ell(y_x(s,a),a)\, e^{-\lambda s}\, ds + e^{-\lambda t} v(y_x(t,a),a) \, .$$

If $\widetilde{\alpha} \in \mathcal{A}_m^a$ is optimal, then

(ii) $$v(x,a) = \int_0^t \ell(y_x(s,\widetilde{\alpha}),\widetilde{\alpha}(s))\, e^{-\lambda s}\, ds + e^{-\lambda t} v(y_x(t,\widetilde{\alpha}),\widetilde{\alpha}(t)) \, .$$

PROOF. Fix $t > 0$ and $\alpha \in [0,1]$ and set $z := y_x(t,a)$. For any $\varepsilon > 0$ there exists $\alpha^1 \in \mathcal{A}_m^a$ such that

(4.15) $$v(z,a) \geq J(z,\alpha^1) - \varepsilon \, .$$

Define the control

$$\overline{\alpha}(s) = \begin{cases} a & \text{if } 0 \leq s \leq t, \\ \alpha^1(s-t) & \text{if } s > t \, . \end{cases}$$

It is easy to check that $\overline{\alpha} \in \mathcal{A}_m^a$. Then a simple computation (see the proof of Proposition 2.5) shows that

$$v(x,a) \leq J(x,\overline{\alpha})$$
$$= \int_0^t \ell(y_x(s,a),a)\, e^{-\lambda s}\, ds + e^{-\lambda t} \int_0^{+\infty} \ell(y_z(s,\alpha^1),\alpha^1(s))\, e^{-\lambda s}\, ds \, .$$

Hence

$$v(x,a) \leq \int_0^t \ell(y_x(s,a),a)\, e^{-\lambda s}\, ds + e^{-\lambda t} J(z,\alpha^1) \, .$$

Taking (4.15) into account and the arbitrariness of ε, (i) follows.

Now let $\widetilde{\alpha} \in \mathcal{A}_m^a$ be optimal, that is

$$v(x,a) = J(x,\widetilde{\alpha}) \, .$$

Observe that

$$J(x,\tilde{\alpha}) = \int_0^t \ell(y_x(s,\tilde{\alpha}),\tilde{\alpha}(s))\,e^{-\lambda s}\,ds + e^{-\lambda t} J(y_x(t,\tilde{\alpha}),\overline{\alpha}),$$

where $\overline{\alpha}(s) := \tilde{\alpha}(s+t)$. Since, as it is easy to check, $\overline{\alpha} \in \mathcal{A}_m^{\tilde{a}(t)}$ is optimal for the initial position $y_x(t,\tilde{\alpha})$, (ii) is proved. ◀

Let us deduce now from the Dynamic Programming Principle the HJB equation, that takes the form of the *variational inequality* (VI). Of course, we do not expect v to satisfy it in the classical pointwise sense but only in the viscosity sense. In the present case, by this we mean that $v \in C(\mathbb{R}^N \times [0,1])$ and that the following two conditions hold for any $\varphi \in C^1(\mathbb{R}^N \times [0,1])$:

(4.16) $$\max\{\lambda v - f \cdot D_x\varphi - \ell; -\partial\varphi/\partial a\} \leq 0,$$

at any local maximum point $(x_0, a_0) \in \mathbb{R}^N \times [0,1[$ of $v - \varphi$, and

(4.17) $$\max\{\lambda v - f \cdot D_x\varphi - \ell; -\partial\varphi/\partial a\} \geq 0$$

at any local minimum point $(x_1, a_1) \in \mathbb{R}^N \times [0,1[$ of $v - \varphi$.

REMARK 4.3. Observe that if $v \in C^1(\mathbb{R}^N \times [0,1])$ and (4.16) and (4.17) hold, then v satisfies

$$\max\{\lambda v(x,a) - f(x,a) \cdot D_x v(x,a) - \ell(x,a); -\frac{\partial v}{\partial a}(x,a)\} = 0$$

at all $(x,a) \in \mathbb{R}^N \times [0,1]$ or, equivalently,

$$\lambda v(x,a) - f(x,a) \cdot D_x v(x,a) - \ell(x,a) \leq 0, \qquad -\frac{\partial v}{\partial a}(x,a) \leq 0,$$
$$-\frac{\partial v}{\partial a}(x,a)[\lambda v(x,a) - f(x,a) \cdot D_x v(x,a) - \ell(x,a)] = 0. \quad \triangleleft$$

THEOREM 4.4. *Assume* (4.8). *Then v defined by* (4.6) *is a viscosity solution of* (VI). *Moreover, v satisfies the boundary condition*

(4.18) $$v(x,1) = J(x,1).$$

PROOF. By Proposition 4.1 (iii) and (iv), $v \in BUC(\mathbb{R}^N \times [0,1])$ and (4.18) holds. We must then check the validity of (4.16) and (4.17).

In order to prove that v satisfies (4.16) we show that the inequalities

(4.19) $$-\frac{\partial\varphi}{\partial a} \leq 0$$

and

(4.20) $$\lambda v - f \cdot D_x\varphi - \ell \leq 0$$

hold at any local maximum point $(x_0, a_0) \in \mathbb{R}^N \times [0,1[$ of $v - \varphi$, for all $\varphi \in C^1(\mathbb{R}^N \times [0,1])$. Assume by contradiction that $-(\partial \varphi / \partial a)(x_0, a_0) > 0$ for some φ. Then

$$\varphi(x_0, a_0 + h) < \varphi(x_0, a_0)$$
$$v(x_0, a_0 + h) - v(x_0, a_0) \leq \varphi(x_0, a_0 + h) - \varphi(x_0, a_0)$$

for sufficiently small $h > 0$. This is a contradiction to the fact that v is nondecreasing with respect to a (see Proposition 4.1 (i)). Hence (4.19) is proved.

Let us now prove (4.20). Since (x_0, a_0) is a local maximum for $v - \varphi$, we have for sufficiently small $t > 0$,

$$\varphi(x_0, a_0) - \varphi(y^0(t), a_0) \leq v(x_0, a_0) - v(y^0(t), a_0),$$

where $y^0(\cdot)$ is the trajectory of (4.1) starting at x_0 associated with the constant control $\alpha(t) \equiv a_0$. Hence, inequality (i) in Proposition 4.2 yields

$$\varphi(x_0, a_0) - \varphi(y^0(t), a_0) \leq \int_0^t \ell(y^0(s), a_0) e^{-\lambda s} \, ds + (e^{-\lambda t} - 1) v(y^0(t), a_0).$$

Divide by $t > 0$ and let $t \to 0^+$ to get

$$-D_x \varphi(x_0, a_0) \cdot f(x_0, a_0) \leq \ell(x_0, a_0) - \lambda v(x_0, a_0)$$

and (4.20) is established.

To check the supersolution condition (4.17), observe first that if $-\dfrac{\partial \varphi}{\partial a}(x_1, a_1) \geq 0$, then (4.17) is automatically fulfilled. Consider then the case

(4.21) $$-\frac{\partial \varphi}{\partial a}(x_1, a_1) < 0.$$

We have to prove in this case that

(4.22) $$\lambda v - f \cdot D_x \varphi - \ell \geq 0 \quad \text{at } (x_1, a_1).$$

In order to do so, observe preliminarily that if $\widetilde{\alpha} \in \mathcal{A}_m^{a_1}$ is optimal for x_1, then (4.21) implies

(4.23) $$\lim_{t \to 0^+} \widetilde{\alpha}(t) =: \widetilde{\alpha}(0^+) = a_1.$$

Indeed, were $\widetilde{\alpha}(0^+) > a_1$, it would follow from (4.21) that

(4.24) $$\varphi(x_1, a_1 + h) > \varphi(x_1, a_1)$$

for some $h > 0$ such that $a_1 + h \leq \widetilde{\alpha}(0^+)$. Since (x_1, a_1) is a local minimum for $v - \varphi$,

$$v(x_1, a_1 + h) - v(x_1, a_1) \geq \varphi(x_1, a_1 + h) - \varphi(x_1, a_1).$$

This, together with (4.24), leads to a contradiction to Proposition 4.1 (ii) and (4.23) is established.

Let us consider the constant control $\bar{\alpha}(t) \equiv a_1$ and denote by $\tilde{y}(\,\cdot\,)$ and $\bar{y}(\,\cdot\,)$ the trajectories of (4.1) with initial point x_1 associated, respectively, with $\tilde{\alpha}$ and $\bar{\alpha}$. By Propositions 4.2 (ii) and 4.1 and the fact that (x_1, a_1) is a local minimum for $v - \varphi$, for t small we have

$$(4.25) \quad \begin{aligned} \varphi(x_1, a_1) - \varphi(\tilde{y}(t), a_1) &\geq v(x_1, a_1) - v(\tilde{y}(t), a_1) \\ &\geq (e^{-\lambda t} - 1) v(\tilde{y}(t), a_1) + I_t^{\tilde{\alpha}} \end{aligned}$$

with

$$I_t^{\tilde{\alpha}} = \int_0^t \ell(\tilde{y}(s), \tilde{\alpha}(s)) e^{-\lambda s}\, ds \ .$$

Now, for some constant C independent of t,

$$-\varphi(\tilde{y}(t), a_1) \leq -\varphi(\bar{y}(t), a_1) + C |\tilde{y}(t) - \bar{y}(t)| \ .$$

Therefore (4.25) implies

$$(4.26) \quad \frac{\varphi(x_1, a_1) - \varphi(\bar{y}(t), a_1) + C |\tilde{y}(t) - \bar{y}(t)|}{t} \geq \frac{e^{-\lambda t} - 1}{t} v(\tilde{y}(t), \tilde{\alpha}(t)) + \frac{1}{t} I_t^{\tilde{\alpha}} \ .$$

By the uniform continuity of ℓ,

$$(4.27) \quad I_t^{\tilde{\alpha}} \geq \int_0^t \ell(\bar{y}(s), a_1) e^{-\lambda s}\, ds - \int_0^t \omega_\ell(|\tilde{y}(s) - \bar{y}(s)| + |\tilde{\alpha}(s) - a_1|) e^{-\lambda s}\, ds \ .$$

The Gronwall's inequality implies

$$|\tilde{y}(t) - \bar{y}(t)| \leq L e^{Lt} \int_0^t |\tilde{\alpha}(s) - a_1|\, ds \ .$$

Thanks to (4.23), this gives

$$(4.28) \quad \frac{|\tilde{y}(t) - \bar{y}(t)|}{t} \longrightarrow 0 \quad \text{as } t \to 0^+ \ .$$

Observe now that, as a consequence of (4.23)

$$(4.29) \quad \bar{y}'(0) = f(x_1, a_1)$$
$$(4.30) \quad v(\tilde{y}(t), \tilde{\alpha}(t)) \longrightarrow v(x_1, a_1) \quad \text{as } t \to 0 \ .$$

Let us now pass to the limit in (4.26) as $t \to 0^+$. Thanks to (4.27), (4.28), (4.29), (4.30) we obtain

$$-f(x_1, a_1) \cdot D_x \varphi(x_1, a_1) \geq -\lambda v(x_1, a_1) + \ell(x_1, a_1)$$

and (4.17) is proved. ◀

4. HJB INEQUALITIES

REMARK 4.5. In the proof that v is a viscosity supersolution of (VI) we implicitly assumed the existence of $\widetilde{\alpha} \in \mathcal{A}_m^{a_1}$ such that $v(x_1, a_1) = J(x_1, \widetilde{\alpha})$. The existence of optimal monotone controls for each $(x, a) \in \mathbb{R}^N \times [0, 1]$ can be proved by means of the Helly selection theorem and the Lebesgue dominated convergence theorem (see [LM67], Corollary 2, p. 281). ◁

The main result of this section is the next one, which states the comparison principle between viscosity sub- and supersolutions of (VI).

THEOREM 4.6. *Assume* (4.8). *If* $u, w \in BUC(\mathbb{R}^N \times [0,1])$ *are, respectively, a viscosity sub- and supersolution of* (VI) *such that*

(4.31) $$u(x, 1) \leq w(x, 1) \quad \forall x \in \mathbb{R}^N,$$

then

(4.32) $$u(x, a) \leq w(x, a) \quad \forall (x, a) \in \mathbb{R}^N \times [0, 1].$$

PROOF. Consider the auxiliary function $\Phi : \mathbb{R}^{2N} \times [0,1]^2 \to \mathbb{R}$ defined by

$$\Phi(x, y, a, b)$$
$$:= u(x, a) - w(y, b) - \frac{|x-y|^2 + |a-b|^2}{2\varepsilon} - \beta(\langle x \rangle^m + \langle y \rangle^m) - \eta((1-a) + (1-b))$$

where $\varepsilon, \beta, m, \eta$ are positive parameters to be chosen conveniently and $\langle x \rangle^m := (1 + |x|^2)^{m/2}$. Assume by contradiction that (4.32) is false; hence for some $(\widetilde{x}, \widetilde{a}) \in \mathbb{R}^N \times [0,1]$ and $\delta > 0$,

$$(u - w)(\widetilde{x}, \widetilde{a}) = \delta > 0.$$

The first part of the proof follows along the lines of Theorems 2.12 and 3.12. We therefore omit the details. A suitable choice of β and η, namely,

$$2\beta \langle \widetilde{x} \rangle + 2\eta(1 - \widetilde{a}) \leq \delta/2$$

implies that for all $m \in \,]0, 1]$

(4.33) $$\sup_{\mathbb{R}^{2N} \times [0,1]^2} \Phi \geq \Phi(\widetilde{x}, \widetilde{x}, \widetilde{a}, \widetilde{a}) > \delta/2.$$

Also, there exists some point $(\overline{x}, \overline{y}, \overline{a}, \overline{b}) \in \mathbb{R}^{2N} \times [0,1]^2$ (depending on the various parameters) such that

(4.34) $$\Phi(\overline{x}, \overline{y}, \overline{a}, \overline{b}) = \sup_{\mathbb{R}^{2N} \times [0,1]^2} \Phi.$$

Moreover, the following estimates hold:

(4.35) $$|\overline{x} - \overline{y}| + |\overline{a} - \overline{b}| \leq C\sqrt{\varepsilon} \quad \text{for some } C,$$
$$\frac{|\overline{x} - \overline{y}|^2 + |\overline{a} - \overline{b}|^2}{2\varepsilon} \leq \omega(\sqrt{\varepsilon})$$

(see again the proof of Theorem 2.12 for details). Observe at this point that

(4.36) $$\bar{a} < 1, \quad \bar{b} < 1 \quad \text{for sufficiently small } \varepsilon.$$

Indeed, if $\bar{a} = 1$ (a similar argument applies in case $\bar{b} = 1$), then

(4.37) $$\sup_{\mathbb{R}^{2N} \times [0,1]^2} \Phi = \Phi(\bar{x}, \bar{y}, 1, \bar{b}) \leq u(\bar{x}, 1) - w(\bar{x}, 1) + w(\bar{x}, 1) - w(\bar{y}, \bar{b}).$$

Now, using the boundary condition (4.31), estimate (4.35), and the uniform continuity of w, from (4.37) it follows that

$$\sup_{\mathbb{R}^{2N} \times [0,1]} \Phi \leq \omega(C\sqrt{\varepsilon}),$$

a contradiction to (4.33) for small $\varepsilon > 0$.

Thanks to (4.36), we can use the information provided by the fact that u, w are, respectively, a viscosity sub- and supersolution of (VI). Choose then the test functions

$$\varphi(x, a) := w(\bar{y}, \bar{b}) + \frac{|x - \bar{y}|^2 + |a - \bar{b}|^2}{2\varepsilon} + \beta(\langle x \rangle^m + \langle \bar{y} \rangle^m) + \eta((1 - a) + (1 - \bar{b}))$$

$$\psi(y, b) := u(\bar{x}, \bar{a}) - \frac{|\bar{x} - y|^2 + |\bar{a} - b|^2}{2\varepsilon} - \beta(\langle \bar{x} \rangle^m + \langle y \rangle^m) - \eta((1 - \bar{a}) + (1 - b)).$$

From (4.16), (4.17) we get through an elementary computation

(4.38) $$\max\{\lambda u - f \cdot (p_\varepsilon + \gamma \bar{x}) - \ell; \alpha_\varepsilon + \eta\} \leq 0 \quad \text{at } (\bar{x}, \bar{a}),$$

(4.39) $$\max\{\lambda w - f \cdot (p_\varepsilon - \tau \bar{y}) - \ell; \alpha_\varepsilon - \eta\} \geq 0 \quad \text{at } (\bar{y}, \bar{b}),$$

where

$$p_\varepsilon := \frac{\bar{x} - \bar{y}}{\varepsilon}, \quad \alpha_\varepsilon = -\frac{\bar{a} - \bar{b}}{\varepsilon},$$

$$\gamma = \beta m \langle \bar{x} \rangle^{m-2}, \quad \tau = \beta m \langle \bar{y} \rangle^{m-2}.$$

Now the parameter $\eta > 0$ plays a role. The subsolution condition (4.38) obviously yields $\alpha_\varepsilon + \eta \leq 0$, so that

$$\alpha_\varepsilon - \eta < 0.$$

Therefore, by the supersolution condition (4.39),

$$\lambda w - f \cdot (p_\varepsilon - \tau \bar{y}) - \ell \geq 0 \quad \text{at } (\bar{y}, \bar{b}).$$

Combine the preceding with

$$\lambda u - f \cdot (p_\varepsilon + \gamma \bar{x}) - \ell \leq 0 \quad \text{at } (\bar{x}, \bar{a})$$

(another immediate consequence of (4.38)) to get

$$\lambda(u(\overline{x},\overline{a}) - w(\overline{y},\overline{b}))$$
$$\leq (f(\overline{x},\overline{a}) - f(\overline{y},\overline{b})) \cdot p_\varepsilon + \gamma f(\overline{x},\overline{a}) \cdot \overline{x} + \tau f(\overline{y},\overline{b}) \cdot \overline{y} + \ell(\overline{x},\overline{a}) - \ell(\overline{y},\overline{b}) \, .$$

At this point we use assumption (4.8) and the estimates (4.35) to deduce from the preceding that

(4.40) $\quad \lambda(u(\overline{x},\overline{a}) - w(\overline{y},\overline{b})) \leq 2L\omega(\sqrt{\varepsilon}) + \omega_\ell(2C\sqrt{\varepsilon}) + K\beta m(\langle\overline{x}\rangle^m + \langle\overline{y}\rangle^m),$

where $K := L + \sup_{a\in[0,1]} |f(0,a)|$. Let us now choose $m \in \,]0, \lambda/K]$ to conclude, thanks to (4.40) and the definition of Φ, that

(4.41) $\quad\quad\quad \Phi(\overline{x},\overline{y},\overline{a},\overline{b}) \leq 2L\lambda^{-1}\omega(\sqrt{\varepsilon}) + \lambda^{-1}\omega_\ell(2C\sqrt{\varepsilon}) \, .$

Since the right-hand side of the previous inequality tends to 0 as $\varepsilon \to 0$, a contradiction arises to (4.33) and (4.34), and the proof is complete. ◂

The following statement gives the characterization of the value function of the monotone control problem. Its proof is a straightforward consequence of Theorems 4.6 and 4.4 and Proposition 4.1 (iii), (iv).

COROLLARY 4.7. *Assume* (4.8). *Then the value function v defined by* (4.6) *is the unique viscosity solution of* (VI) *such that*

$$v(x,1) = J(x,1) \quad \forall\, x \in \mathbb{R}^N \, .$$

4.2. Optimal stopping

In the optimal stopping problem the controller can act on the evolution defined by (4.1) by choosing the control α and also by stopping it at some *stopping time* $\vartheta \geq 0$. The cost associated with the controlled evolution is given by

(4.42) $\quad\quad J(x,\alpha,\vartheta) := \int_0^\vartheta \ell(y_x(t,\alpha),\alpha(t)) e^{-\lambda t}\, dt + \psi(y_x(\vartheta,\alpha)) e^{-\lambda\vartheta} \, .$

Here $\lambda > 0$ is the discount factor and ψ represents a *stopping cost*. The value function is

(4.43) $\quad\quad\quad\quad v(x) := \inf_{\substack{\alpha\in\mathcal{A} \\ \vartheta\geq 0}} J(x,\alpha,\vartheta),$

where $\mathcal{A} = \{\,\alpha : [0,+\infty[\, \to A : \alpha \text{ measurable}\,\}$.

In the following we develop the Dynamic Programming approach to the optimal stopping problem, leading to the characterization of v as the unique viscosity solution of the *variational inequality of obstacle type*

$$\max\{\lambda v(x) + H(x,Dv(x)); v(x) - \psi(x)\,\} = 0 \quad x \in \mathbb{R}^N \, .$$

Here H is the same Hamiltonian as in Section 2, namely,
$$H(x,p) = \sup_{a \in A}\{-f(x,a) \cdot p - \ell(x,a)\} \,.$$

For the sake of simplicity and in order to point out the typical features of the problem, we perform all proofs in the case $f(x,a) \equiv f(x)$, $\ell(x,a) \equiv \ell(x)$. The general case can be treated easily using the arguments of §2.

The first result is on the continuity of v.

PROPOSITION 4.8. *Assume* (A_0'), (A_3), (A_4), *and*

(4.44) $\qquad\qquad\qquad\qquad \psi \in BUC(\mathbb{R}^N) \,.$

Then $v \in BUC(\mathbb{R}^N)$.

PROOF. The proof is very similar to (and actually simpler than) that of Proposition 2.1. We leave it as an exercise. ◂

The next proposition exploits some consequences of the definition of v. These can be seen as the formulation of the Dynamic Programming Principle for the present problem.

PROPOSITION 4.9. *Assume* (A_0'), (A_3), (A_4) *and* (4.44). *Then for all* $x \in \mathbb{R}^N$ *the following inequalities hold:*

(i) $\quad v(x) \leq \psi(x),$

(ii) $\quad v(x) \leq \int_0^t \ell(y_x(s,\alpha),\alpha(s)) e^{-\lambda s}\, ds + v(y_x(t,\alpha)) e^{-\lambda t}, \quad \forall \alpha \in \mathcal{A},\ t \geq 0\,.$

Furthermore, for any x where the strict inequality occurs in (i) there exists $t_0 = t_0(x)$ such that

(iii) $\quad v(x) = \inf_{\alpha \in \mathcal{A}} \left\{ \int_0^t \ell(y_x(s,\alpha),\alpha(s)) e^{-\lambda s}\, ds + v(y_x(t,\alpha)) e^{-\lambda t} \right\} \quad \forall t \in [0, t_0]\,.$

PROOF. Assume $f(x,a) \equiv f(x)$, $\ell(x,a) \equiv \ell(x)$. Inequality (i) is obvious since, in particular, $v(x) \leq J(x,0) = \psi(x)$. To prove (ii), fix any $t \geq 0$ and think of $y_x(t)$ as the initial position. For each $\varepsilon > 0$ there exists, by definition of v, some $\vartheta_\varepsilon \geq 0$ such that

(4.45) $\qquad\qquad\qquad J(y_x(t), \vartheta_\varepsilon) \leq v(y_x(t)) + \varepsilon \,.$

Now set $\widehat{\vartheta} = \vartheta_\varepsilon + t$ and observe that

$$J(x, \widehat{\vartheta}) = \int_0^t \ell(y_x(s)) e^{-\lambda s}\, ds + \int_t^{\vartheta_\varepsilon + t} \ell(y_x(s)) e^{-\lambda s}\, ds + \psi(y_x(\widehat{\vartheta})) e^{-\lambda \widehat{\vartheta}} \,.$$

4. HJB INEQUALITIES

Since $y_x(\tau + t) = y_{y_x(t)}(\tau)$ for any $\tau \geq 0$, the second integral above can be written, by a change of variable, as

$$e^{-\lambda t} \int_0^{\vartheta_\varepsilon} \ell(y_{y_x(t)}(\tau)) e^{-\lambda \tau} d\tau;$$

also,

$$\psi(y_x(\widehat{\vartheta})) e^{-\lambda \widehat{\vartheta}} = e^{-\lambda t} \psi(y_{y_x(t)}(\vartheta_\varepsilon)) e^{-\lambda \vartheta_\varepsilon}.$$

Therefore

(4.46) $\quad J(x, \widehat{\vartheta}) = \int_0^t \ell(y_x(s)) e^{-\lambda s} ds + e^{-\lambda t} J(y_{y_x(t)}, \vartheta_\varepsilon) \quad \forall t \geq 0.$

Hence it follows, by (4.45),

$$v(x) \leq J(x, \widehat{\vartheta}) \leq \int_0^t \ell(y_x(s)) e^{-\lambda s} ds + e^{-\lambda t}(v(y_x(t)) + \varepsilon))$$

and (ii) is proved.

Assume now that x is such that

(4.47) $\quad v(x) < \psi(x)$

and let $\vartheta_n \geq 0$ be such that

(4.48) $\quad \lim_{n \to +\infty} J(x, \vartheta_n) = v(x).$

We claim that there exists $t_0 > 0$ such that

(4.49) $\quad \vartheta_n \geq t_0 > 0,$

for sufficiently large n. To see this, set $\delta_n = J(x, \vartheta_n) - v(x)$. Then, if $\omega(\cdot)$ is a modulus of continuity for ψ and M is as in (A$_4$),

$$v(x) + \delta_n \geq -M \int_0^{\vartheta_n} e^{-\lambda s} ds + e^{-\lambda \vartheta_n}(\psi(x) - \omega(|y_x(\vartheta_n) - x|)).$$

If for some subsequence $\vartheta_n \to 0$, the preceding would imply

$$v(x) \geq \psi(x),$$

a contradiction to (4.47). Hence (4.49) holds. Observe now that for $t \in [0, t_0]$

$$J(x, \vartheta_n) = \int_0^t \ell(y_x(s)) e^{-\lambda s} ds + e^{-\lambda t} J(y_{y_x(t)}, \vartheta_n - t).$$

Since $\vartheta_n - t \geq 0$, it follows by definition of v that

$$J(x, \vartheta_n) \geq \int_0^t \ell(y_x(s)) e^{-\lambda s} ds + e^{-\lambda t} v(y_x(t)) \quad \forall t \in [0, t_0].$$

Let now $n \to +\infty$ to obtain, by (4.48),

$$v(x) \geq \int_0^t \ell(y_x(s)) e^{-\lambda s} \, ds + e^{-\lambda t} v(y_x(t)), \qquad \forall t \in [0, t_0].$$

The preceding and (ii) show the validity of statement (iii). ◂

As for the problems previously analyzed in this chapter, it is possible to derive from the Dynamic Programming Principle the optimality condition for v in PDE form. In the present case the HJB equation turns out to be the following

(VIO) $\qquad \max\{\lambda u(x) + H(x, Du(x)); u(x) - \psi(x)\} = 0, \qquad x \in \mathbb{R}^N.$

This PDE problem is usually called a *variational inequality of obstacle type* or a *complementarity system*. This terminology is more transparent if one observes that (VIO) is equivalent to

(4.50) $\qquad \begin{cases} u(x) \leq \psi(x), \\ \lambda u(x) + H(x, Du(x)) \leq 0, \\ [u(x) - \psi(x)][\lambda u(x) + H(x, Du(x))] = 0, \quad x \in \mathbb{R}^N, \end{cases}$

(see the proof of Theorem 4.12). The fact that v satisfies (VIO) (or equivalently (4.50)) can be easily deduced from Proposition 4.9 assuming $v \in C^1(\mathbb{R}^N)$. However, since in general $v \notin C^1(\mathbb{R}^N)$, (VIO) will be satisfied only in the viscosity sense. By this we mean that $v \in C(\mathbb{R}^N)$ and, for all $\varphi \in C^1(\mathbb{R}^N)$,

(4.51) $\qquad \max\{ \lambda v(x_0) + H(x_0, D\varphi(x_0)); v(x_0) - \psi(x_0) \} \leq 0$

at any local maximum x_0 of $v - \varphi$;

(4.52) $\qquad \max\{ \lambda v(x_1) + H(x_1, D\varphi(x_1)); v(x_1) - \psi(x_1) \} \geq 0$

at any local minimum x_1 of $v - \varphi$.

THEOREM 4.10. *Assume* (A_0'), (A_3), (A_4), *and* (4.44). *Then v defined by* (4.43) *is a viscosity solution of* (VIO).

PROOF. We already proved that $v \in BUC(\mathbb{R}^N)$. Assume again $f(x,a) \equiv f(x)$, $\ell(x,a) \equiv \ell(x)$, so that $H(x,p) = -f(x) \cdot p - \ell(x)$. Let x_0 be a local maximum of $v - \varphi$, $\varphi \in C^1(\mathbb{R}^N)$. Using (ii) of Proposition 4.9 and the continuity of v and φ,

$$\varphi(x_0) - \varphi(y_{x_0}(t)) \leq v(x_0) - v(y_{x_0}(t))$$
$$\leq \int_0^t \ell(y_{x_0}(s)) e^{-\lambda s} \, ds + (e^{-\lambda t} - 1) v(y_{x_0}(t))$$

for sufficiently small $t > 0$. Divide now by t and let $t \to 0^+$ to obtain, thanks to the differentiability of φ, $-D\varphi(x_0) \cdot f(x_0) \leq \ell(x_0) - \lambda v(x_0)$. Since $v(x) \leq \psi(x)$ at all x by Proposition 4.9 (i), the subsolution condition (4.51) follows.

4. HJB INEQUALITIES

Now let x_1 be a local minimum for $v - \varphi$. If $v(x_1) = \psi(x_1)$, then obviously

$$\max\{ v(x_1) - \psi(x_1); \lambda v(x_1) - f(x_1) \cdot D\varphi(x_1) - \ell(x_1) \} \geq v(x_1) - \psi(x_1) = 0$$

and v is a supersolution of (VIO). Assume then $v(x_1) < \psi(x_1)$ (the only other possibility in view of Proposition 4.9 (i)). Thanks to identity (iii) in Proposition 4.9 we get

$$\frac{\varphi(x_1) - \varphi(y_{x_1}(t))}{t} \geq \frac{v(x_1) - v(y_{x_1}(t))}{t}$$
$$= \frac{1}{t}\int_0^t \ell(y_{x_1}(s)) e^{-\lambda s} ds + \frac{e^{-\lambda t} - 1}{t} v(y_{x_1}(t))$$

for sufficiently small $t > 0$ (recall that x_1 is a local minimum for $v - \varphi$). Now let $t \to 0^+$ in the preceding to obtain

$$-f(x_1) \cdot D\varphi(x_1) \geq \ell(x_1) - \lambda v(x_1) .$$

Hence the supersolution condition is satisfied as well. ◀

Let us show now that the value function v is indeed the unique viscosity solution of (VIO). The result is obtained as a particular case of a uniqueness theorem already established in §2.1. The optimal stopping problem can indeed be interpreted as an infinite horizon problem (and consequently the whole theory of Section 2 applies to it). To see this, introduce the control set $\tilde{A} = A \cup \{\omega\}$ and the dynamics $\tilde{f} : \mathbb{R}^N \times \tilde{A} \to \mathbb{R}^N$ by

$$\tilde{f}(x, \omega) \equiv 0, \quad \tilde{f}(x, a) \equiv f(x, a), \quad \text{for } a \in A .$$

Let us also consider the cost $\tilde{\ell} : \mathbb{R}^N \times \tilde{A} \to \mathbb{R}^N$ defined as

$$\tilde{\ell}(x, \omega) = \lambda \psi(x), \quad \tilde{\ell}(x, a) = \ell(x, a), \quad \text{for } a \in A .$$

The Hamiltonian of the infinite horizon problem associated with $\tilde{f}, \tilde{\ell}$ is

$$\tilde{H}(x, r, p) = \sup_{\tilde{a} \in \tilde{A}}\{ \lambda r - \tilde{f}(x, \tilde{a}) \cdot p - \tilde{\ell}(x, \tilde{a}) \}$$

as we know from §2.1. It is now simple to check that (VIO) is equivalent to the HJB equation

(4.53) $$\tilde{H}(x, u(x), Du(x)) = 0 \quad \text{in } \mathbb{R}^N$$

in the viscosity sense. Therefore, we have the following

THEOREM 4.11. *Assume* (A_0'), (A_3), (A_4), *and* (4.44). *Then v is the unique viscosity solution of* (VIO) *in* $BUC(\mathbb{R}^N)$.

PROOF. The statement is a corollary of Theorem 2.12. We leave the details to the reader. ◀

Let us conclude this subsection with a different proof of uniqueness. Assume again $f(x,a) \equiv f(x)$, $\ell(x,a) \equiv \ell(x)$. The idea is to show that if u is a viscosity solution of (VIO), then u is represented by

$$u(x) = J(x, \vartheta^*(x)),$$

for some $\vartheta^*(x)$ such that $J(x, \vartheta^*(x)) = \inf_{\vartheta \geq 0} J(x, \vartheta)$.

THEOREM 4.12. *Assume* (A_0'), (A_3), (A_4), (4.44), $f(x,a) \equiv f(x)$, $\ell(x,a) \equiv \ell(x)$, *and let* $u \in BUC(\mathbb{R}^N)$ *be a viscosity solution of* (VIO). *Then*

(4.54) $$u(x) = J(x, \vartheta^*(x)) = v(x) \qquad \forall\, x \in \mathbb{R}^N$$

where

(4.55) $$\vartheta^*(x) = \inf\{\, t \geq 0 : u(y_x(t)) = \psi(y_x(t))\,\}.$$

PROOF. Let $u \in BUC(\mathbb{R}^N)$ be a viscosity solution of (VIO) and consider the open set $C = \{\, x \in \mathbb{R}^N : u(x) < \psi(x)\,\}$. The first step is to prove that

(4.56) $$u \leq \psi \qquad \text{in } \mathbb{R}^N$$

and that

(4.57) $$\lambda u - f \cdot Du - \ell \leq 0 \qquad \text{in } \mathbb{R}^N,$$
(4.58) $$\lambda u - f \cdot Du - \ell = 0 \qquad \text{in } C$$

in the viscosity sense. Assume by contradiction that $u(x_0) > \psi(x_0)$ at some $x_0 \in \mathbb{R}^N$. Hence, by continuity,

(4.59) $$u(x) > \psi(x) \qquad \forall\, x \in B(x_0, \delta),\ \delta > 0.$$

It is not difficult to show the existence of $\varphi \in C^1(\mathbb{R}^N)$ such that $u - \varphi$ has a local maximum at some point $\overline{x} \in B(x_0, \delta)$, so that by (4.51)

$$\max\{\,\lambda u(\overline{x}) - f(\overline{x}) \cdot D\varphi(\overline{x}) - \ell(\overline{x});\, u(\overline{x}) - \psi(\overline{x})\,\} \leq 0.$$

This contradicts (4.59) and (4.56) is established.

In view of (4.56) the inequality (4.57) is a straightforward consequence of the fact that u is a viscosity solution of (VIO). To prove (4.58) it is enough to show that

(4.60) $$\lambda u - f \cdot Du - \ell \geq 0 \qquad \text{in } C$$

in the viscosity sense. At any local minimum $x_1 \in C$ of $u - \varphi$, $\varphi \in C^1(\mathbb{R}^N)$, (VIO) gives

$$\max\{\,\lambda u(x_1) - f(x_1) \cdot D\varphi(x_1) - \ell(x_1);\, u(x_1) - \psi(x_1)\,\} \geq 0$$

and (4.60) follows since $x_1 \in C$.

Now we apply Proposition II.5.18 (see also Remark II.5.22) with $\Omega = \mathbb{R}^N$. Because of (4.56) and (4.57) we obtain

$$u(x) \leq u(y_x(\vartheta)) e^{-\lambda \vartheta} + \int_0^\vartheta \ell(y_x(t)) e^{-\lambda t} \, dt$$
$$\leq \psi(y_x(\vartheta)) e^{-\lambda \vartheta} + \int_0^\vartheta \ell(y_x(t)) e^{-\lambda t} \, dt$$

for all $\vartheta \geq 0$. Hence,

$$u(x) \leq \inf_{\vartheta \geq 0} J(x, \vartheta) =: v(x).$$

To prove the reverse inequality let us deal first with $x \notin C$. In this case $u(x) = \psi(x)$ and $\vartheta^*(x) = 0$ by (4.55). Hence

$$u(x) = \psi(x) = J(x, \vartheta^*(x)) \geq \inf_{\vartheta \geq 0} J(x, \vartheta) =: v(x).$$

Assume now $x \in C$, so that (4.58) holds, and apply Proposition II.5.18 with $\Omega = C$ to get, after some calculations,

$$u(x) = u(y_x(\vartheta)) e^{-\lambda \vartheta} + \int_0^\vartheta \ell(y_x(t)) e^{-\lambda t} \, dt \qquad \forall \, 0 \leq \vartheta < \vartheta^*(x)$$

(note that this equality can also be obtained by Remark 2.34). Let $\vartheta \to \vartheta^*(x)$ in the preceding to obtain

$$u(x) = J(x, \vartheta^*(x)) \geq \inf_{\vartheta \geq 0} J(x, \vartheta) := v(x),$$

since $u(y_x(\vartheta^*(x))) = \psi(y_x(\vartheta^*(x)))$. The statement is proved. ◀

REMARK 4.13. It may happen that, for some x, $y_x(t) \in C$ for all $t > 0$. This means that $u(y_x(t)) < \psi(y_x(t))$ for all $t > 0$. From the previous arguments it follows that

$$u(x) < \psi(y_x(\vartheta)) e^{-\lambda \vartheta} + \int_0^\vartheta \ell(y_x(t)) e^{-\lambda t} \, dt$$

for all $\vartheta \geq 0$. Hence for those x no finite optimal stopping time exists. ◁

REMARK 4.14. Let us point out that (4.54) and (4.55) say that $x \mapsto \vartheta^*(x)$ is an optimal feedback map. The optimal time $\vartheta^* = \vartheta^*(x)$ is the first time when the trajectory $y_x(\,\cdot\,)$ hits the set $S = \{\, x \in \mathbb{R}^N : u(x) = \psi(x) \,\}$, usually called the *free boundary* of (VIO). ◁

4.3. Impulse control

In this problem the controller can act on the evolution described by (4.1) by choosing *stopping times* $\vartheta^i \geq 0$ at which the current state is abruptly modified by an additive *impulse* ξ^i. The state equation becomes then

(4.61)
$$\begin{cases} y'(t) = f(y(t), \alpha(t)), & t \in]\vartheta^i, \vartheta^{i+1}[\\ y(0) = x, \\ y(\vartheta^i_+) = y(\vartheta^i_-) + \xi^i, & i = 0, 1, 2, \ldots \end{cases}$$

The notation $y(\vartheta^i_-)$ stands for

$$y(\vartheta^{i-1}) + \int_{\vartheta^{i-1}}^{\vartheta^i} f(y(t), \alpha(t)) \, dt$$

and $y(\vartheta^i_+) := \lim_{t \to \vartheta^i_+} y(t)$. The triplet $\beta = (\alpha, \{\vartheta^i\}, \{\xi^i\})$ is the control; we denote by $\mathcal{B}$ the set of all β such that $\alpha \in \mathcal{A}$ and

(4.62)
$$\begin{aligned} &\vartheta^i \in [0, +\infty], \quad \vartheta^{i+1} \geq \vartheta^i, \quad \vartheta^i \to +\infty \text{ as } i \to +\infty, \\ &\xi^i \in \mathbb{R}^N_+ \end{aligned}$$

($\mathbb{R}^N_+$ is the set $\{\xi = (\xi_1, \ldots, \xi_N) \in \mathbb{R}^N : \xi_k \geq 0, \ k = 1, \ldots, N\}$). We associate the controlled evolution defined by (4.61) and (4.62) with the cost functional

(4.63)
$$J(x, \beta) = \int_0^{+\infty} \ell(y_x(s, \beta), \alpha(s)) e^{-\lambda s} \, ds + \sum_{i=0}^{\infty} c(\xi^i) e^{-\lambda \vartheta^i},$$

where $y_x(\,\cdot\,)$ is the solution of (4.61) (see Exercise 4.8) and $e^{-\lambda(+\infty)} := 0$.

The *impulse cost* c is assumed to satisfy

(4.64)
$$\begin{cases} c(\xi) = k + c_0(\xi) & k > 0 \text{ and } c_0 \in C(\mathbb{R}^N_+) \text{ with } c_0(0) = 0, \\ c_0(\xi + \tilde{\xi}) \leq c_0(\xi) + c_0(\tilde{\xi}) & \forall \xi, \tilde{\xi} \in \mathbb{R}^N_+, \\ c_0(\xi) \leq c_0(\tilde{\xi}) & \text{if } \tilde{\xi} - \xi \in \mathbb{R}^N_+, \\ c_0(\xi) \longrightarrow +\infty & \text{as } |\xi| \to +\infty. \end{cases}$$

The value function of the present problem is

(4.65)
$$v(x) := \inf_{\beta \in \mathcal{B}} J(x, \beta).$$

We have

PROPOSITION 4.15. *Assume* (A'_0), (A_3), (A_4), *and* (4.64). *Then* $v \in BUC(\mathbb{R}^N)$.

PROOF. Choose $\beta = (\alpha, \{\vartheta^i\}, \{\xi^i\})$ with arbitrary $\alpha \in \mathcal{A}$ and $\vartheta^0 = +\infty$. Then (A_4) gives

$$-\frac{M}{\lambda} \leq v(x) \leq \frac{M}{\lambda}.$$

The proof of uniform continuity is essentially the same as that of Proposition 2.1.

◀

4. HJB INEQUALITIES

Following the usual pattern we now exploit the Dynamic Programming Principle for the present problem.

PROPOSITION 4.16. *Assume* (A_0'), (A_3), (A_4), *and* (4.64). *Then* v *satisfies*

(4.66) $v(x) =$
$$= \inf_{\substack{\alpha \in \mathcal{A} \\ \vartheta^0 \geq 0}} \left\{ \int_0^{\vartheta^0} \ell(y_x(s,\beta),\alpha(s)) e^{-\lambda s} ds + e^{-\lambda \vartheta^0} \inf_{\xi \in \mathbb{R}_+^N} \left(v(y_x(\vartheta_-^0,\beta) + \xi) + c(\xi) \right) \right\}.$$

PROOF. For the sake of simplicity let us assume that f and ℓ are independent of a. By definition of v, for $\vartheta^0 < +\infty$,

(4.67) $$v(x) \leq \int_0^{\vartheta^0} \ell(y_x(s)) e^{-\lambda s} ds + \int_{\vartheta^0}^{+\infty} \ell(y_x(s)) e^{-\lambda s} ds + \sum_{i=0}^{+\infty} c(\xi^i) e^{-\lambda \vartheta^i}.$$

Set
$$Y_1 := \int_{\vartheta^0}^{+\infty} \ell(y_x(s)) e^{-\lambda s} ds, \qquad Y_2 := \sum_{i=0}^{\infty} c(\xi^i) e^{-\lambda \vartheta^i}.$$

The change of variable $\tau = s - \vartheta^0$ and the state equation (4.61) give
$$Y_1 = \int_0^{+\infty} \ell(y_x(\tau + \vartheta^0)) e^{-\lambda(\tau + \vartheta^0)} d\tau = e^{-\lambda \vartheta^0} \int_0^{+\infty} \ell(y_z(\tau)) e^{-\lambda \tau} d\tau,$$

where

(4.68) $$z = y_x(\vartheta_-^0) + \xi^0.$$

On the other hand it is obvious that
$$Y_2 = e^{-\lambda \vartheta^0} \left(\sum_{i=1}^{\infty} c(\xi^i) e^{-\lambda(\vartheta^i - \vartheta^0)} + c(\xi^0) \right).$$

Hence

(4.69) $$Y_1 + Y_2 = e^{-\lambda \vartheta^0} \left(\sum_{i=0}^{\infty} c(\widetilde{\xi}^i) e^{-\lambda \widetilde{\vartheta}^i} + c(\xi^0) + \int_0^{+\infty} \ell(y_z(\tau)) e^{-\lambda \tau} d\tau \right)$$
$$= e^{-\lambda \vartheta^0} (J(z, \widetilde{\beta}) + c(\xi^0)),$$

where $\widetilde{\beta} := (\{\widetilde{\vartheta}^i\}, \{\widetilde{\xi}^i\})$ with
$$\widetilde{\vartheta}^i = \vartheta^{i+1} - \vartheta^0, \quad \widetilde{\xi}^i = \xi^{i+1}, \qquad i = 0, 1, 2, \ldots$$

As a consequence of (4.67) and (4.69) we get
$$v(x) \leq \int_0^{\vartheta^0} \ell(y_x(s)) e^{-\lambda s} ds + e^{-\lambda \vartheta^0} J(z, \widetilde{\beta}) + e^{-\lambda \vartheta^0} c(\xi^0).$$

Since $\widetilde{\beta}$ is arbitrary in $\mathcal{B}$ it follows that

$$v(x) \leq \int_0^{\vartheta^0} \ell(y_x(s)) e^{-\lambda s} \, ds + e^{-\lambda \vartheta^0} v(z) + e^{-\lambda \vartheta^0} c(\xi^0)$$

for all $\vartheta^0 \geq 0$ and $\xi^0 \in \mathbb{R}_+^N$. The preceding and (4.68) yield

$$v(x) \leq \inf_{\vartheta^0 \geq 0} \left\{ \int_0^{\vartheta^0} \ell(y_x(s)) e^{-\lambda s} \, ds + e^{-\lambda \vartheta^0} \inf_{\xi^0 \in \mathbb{R}_+^N} \left(v(y_x(\vartheta_-^0) + \xi^0) + c(\xi^0) \right) \right\},$$

because the inequality is obvious for $\vartheta^0 = +\infty$.

To prove the validity of the reverse inequality, let $\varepsilon > 0$ and choose $\beta_\varepsilon \in \mathcal{B}$ such that

$$J(x, \beta_\varepsilon) \leq v(x) + \varepsilon.$$

The same arguments as in the first part of the proof show that

$$J(x, \beta_\varepsilon) \geq \int_0^{\vartheta^0} \ell(y_x(s)) e^{-\lambda s} \, ds + e^{-\lambda \vartheta^0} \left(v(y_x(\vartheta_-^0) + \xi^0) + c(\xi^0) \right).$$

Hence, denoting by $w(x)$ the right-hand side of (4.66), $v(x) + \varepsilon \geq w(x)$; since ε is arbitrary the proof is complete. ◀

REMARK 4.17. The Dynamic Programming Principle (4.66) can also be expressed as

(i) $\quad v(x) \leq \inf_{\xi \in \mathbb{R}_+^N} \left(v(x + \xi) + c(\xi) \right),$

(ii) $\quad v(x) \leq \int_0^t \ell(y_x(s, \beta), \alpha(s)) e^{-\lambda s} \, ds + e^{-\lambda t} v(y_x(t, \beta)),$

$\forall t \geq 0$, $x \in \mathbb{R}^N$, $\beta \in \mathcal{B}$. Furthermore, for any x where the strict inequality occurs in (i) there exists $t_0(x) > 0$ such that

(iii) $\quad v(x) = \int_0^t \ell(y_x(s, \beta), \alpha(s)) e^{-\lambda s} \, ds + e^{-\lambda t} v(y_x(t, \beta))$

$\forall t \in [0, t_0(x)]$. ◁

It is convenient to introduce the (nonlocal) operator $\mathcal{M}$ defined by

(4.70) $$(\mathcal{M}u)(x) := \inf_{\xi \in \mathbb{R}_+^N} \left(u(x + \xi) + c(\xi) \right);$$

with this notation the Dynamic Programming Principle (4.66) reads as

(4.71) $$v(x) = \inf_{\substack{\alpha \in \mathcal{A} \\ \vartheta \geq 0}} \left\{ \int_0^\vartheta \ell(y_x(s, \alpha), \beta(s)) e^{-\lambda s} \, ds + e^{-\lambda \vartheta} (\mathcal{M}v)(y_x(\vartheta_-, \beta)) \right\}.$$

If we go back to §4.2, the previous formula shows that v can be interpreted as the value function of an optimal stopping time problem with stopping cost $\psi = \mathcal{M}v$. The implicit character of the stopping cost (it depends through $\mathcal{M}$ on the value function itself) is due of course to the presence of impulse controls.

Let us now derive the infinitesimal version of (4.71); in this case the HJB equation for v takes the form of a *quasivariational inequality*, namely,

(QVI) $\quad \max\{\lambda v(x) + H(x, Dv(x)); v(x) - (\mathcal{M}v)(x)\} = 0 \quad x \in \mathbb{R}^N$.

From (4.71) and Theorem 4.10 it would follow directly that v is a viscosity solution of (QVI) (by this we mean that (4.51) and (4.52) hold with $\psi = \mathcal{M}v$) if we knew that $\mathcal{M}v \in BUC(\mathbb{R}^N)$.

This property of the operator $\mathcal{M}$, as well as some others, which are useful later, is the content of the next result.

LEMMA 4.18. *Assume* (A_0'), (A_3), (A_4), (4.64). *Then, for all* $u, w \in BUC(\mathbb{R}^N)$,

(i) $\quad (k - \|w\|_\infty) \leq \mathcal{M}w \leq (k + \|w\|_\infty)$,

(ii) $\quad \mathcal{M}w \in BUC(\mathbb{R}^N)$,

(iii) $\quad \mathcal{M}w \geq \mathcal{M}u \quad \text{if } w \geq u$,

(iv) $\quad \mathcal{M}(\mu w + (1-\mu)u) \geq \mu \mathcal{M}w + (1-\mu)\mathcal{M}u \quad \text{for all } \mu \in [0, 1]$.

PROOF. Assumption (4.64) implies $c(\xi) \geq k + c_0(0) = k$ for all $\xi \in \mathbb{R}_+^N$. Hence

$$w(x + \xi) + c(\xi) \geq -\|w\|_\infty + k \ .$$

On the other hand, the choice $\xi = 0$ gives

$$(\mathcal{M}w)(x) \leq w(x) + c(0) = w(x) + k \leq \|w\|_\infty + k,$$

and the proof of (i) is complete.

To prove (ii), fix arbitrary x, x' in $\mathbb{R}^N$. For any $\varepsilon > 0$ there exists $\xi_\varepsilon \in \mathbb{R}_+^N$ such that

$$w(x' + \xi_\varepsilon) + c(\xi_\varepsilon) \leq (\mathcal{M}w)(x') + \varepsilon \ .$$

Then

$$(\mathcal{M}w)(x) - (\mathcal{M}w)(x') \leq w(x + \xi_\varepsilon) + c(\xi_\varepsilon) - w(x' + \xi_\varepsilon) - c(\xi_\varepsilon) + \varepsilon;$$

if ω is a modulus for w, we obtain

$$(\mathcal{M}w)(x) - (\mathcal{M}w)(x') \leq \omega(|x - x'|) + \varepsilon \ .$$

Exchanging the roles of x and x' the reverse inequality is established, showing that $\mathcal{M}w \in BUC(\mathbb{R}^N)$.

The monotonicity property (iii) is straightforward.

Let us now prove that (iv) holds (i.e., $\mathcal{M}$ is concave). For any $\varepsilon > 0$ there exists $\xi_\varepsilon \in \mathbb{R}_+^N$ such that

(4.72) $$z(x + \xi_\varepsilon) + c(\xi_\varepsilon) \leq (\mathcal{M}z)(x) + \varepsilon,$$

where $z = \mu w + (1 - \mu)u$. Now, by definition of $\mathcal{M}$,

$$\mu(\mathcal{M}w)(x) \leq \mu w(x + \xi_\varepsilon) + \mu c(\xi_\varepsilon),$$
$$(1 - \mu)(\mathcal{M}u)(x) \leq (1 - \mu)u(x + \xi_\varepsilon) + (1 - \mu)c(\xi_\varepsilon)$$

for all $\mu \in [0, 1]$. Hence, adding the two inequalities and using (4.72),

$$\mu(\mathcal{M}w)(x) + (1 - \mu)(\mathcal{M}u)(x) \leq (\mathcal{M}z)(x) + \varepsilon$$

and (iv) is proved. ◀

THEOREM 4.19. *Assume* (A_0'), (A_3), (A_4), *and* (4.64). *Then v is a viscosity solution of* (QVI).

PROOF. By Proposition 4.15 and Lemma 4.18 both v and $\mathcal{M}v$ belong to $BUC(\mathbb{R}^N)$. The statement follows from Theorem 4.10 thanks to the Dynamic Programming Principle (4.71). ◀

The characterization of v in terms of (QVI) is completed by a uniqueness result. Its proof makes use of the nonlinear operator T defined as

(4.73) $$(Tw)(x) := \inf_{\substack{\alpha \in \mathcal{A} \\ \vartheta \geq 0}} \left\{ \int_0^\vartheta \ell(y_x(s, \alpha), \alpha(s)) e^{-\lambda s} \, ds + (\mathcal{M}w)(y_x(\vartheta, \alpha)) e^{-\lambda \vartheta} \right\},$$

where y_x is the solution of

$$\begin{cases} y'(t) = f(y(t), \alpha(t)), & t \geq 0, \\ y(0) = x, \end{cases}$$

for $\alpha \in \mathcal{A}$.

LEMMA 4.20. *Assume* (A_0'), (A_3), (A_4), (4.64). *Then, for all* $u, w \in BUC(\mathbb{R}^N)$,
(j) $\min\{k - \|w\|_\infty; -M/\lambda\} \leq Tw \leq M/\lambda$;
(jj) $Tw \in BUC(\mathbb{R}^N)$;
(jjj) $Tw \geq Tu$ *if* $w \geq u$;
(jv) $T(\mu w + (1 - \mu)u) \geq \mu Tw + (1 - \mu)Tu$, $\forall \mu \in [0, 1]$;
(v) *if* $w_i \in BUC(\mathbb{R}^N)$, $w_i \geq 0$ $(i = 1, 2)$ *satisfy*

(4.74) $$w_1 - w_2 \leq \gamma w_1 \quad \textit{for some } \gamma \in [0, 1],$$

then, if $\ell \geq 0$, we have

$$Tw_1 - Tw_2 \leq \gamma(1-\mu)Tw_1 \qquad \forall \mu \in \,]0, k/\|\overline{u}\|_\infty \wedge 1[$$

where

$$\overline{u}(x) := \inf_{\alpha \in \mathcal{A}} \int_0^{+\infty} \ell(y_x(s,\alpha), \alpha(s)) e^{-\lambda s}\,ds\ .$$

PROOF. By definition of T, Lemma 4.18 (i) gives

$$(Tw)(x) \leq M\int_0^\vartheta e^{-\lambda s}\,ds + e^{-\lambda\vartheta}(k + \|w\|_\infty), \qquad \forall\,\vartheta \geq 0\ .$$

Let $\vartheta \to +\infty$ to obtain the right-hand side estimate in (j). On the other hand, by Lemma 4.18 (i) again,

$$\int_0^\vartheta \ell(y_x(s,\alpha), \alpha(s))e^{-\lambda s}\,ds + e^{-\lambda\vartheta}(\mathcal{M}w)(y_x(\vartheta,\alpha))$$
$$\geq -M\int_0^\vartheta e^{-\lambda s}\,ds + e^{-\lambda\vartheta}(k - \|w\|_\infty)$$

for all $\alpha \in \mathcal{A}$, $\vartheta \geq 0$. A simple computation proves the validity of the first inequality (j).

To prove (jj), fix arbitrary $\varepsilon > 0$ and choose $\alpha_\varepsilon \in \mathcal{A}$, $\vartheta_\varepsilon \geq 0$ such that

$$\int_0^{\vartheta_\varepsilon} \ell(y_{x'}(s,\alpha_\varepsilon), \alpha_\varepsilon(s))e^{-\lambda s}\,ds + (\mathcal{M}w)(y_{x'}(\vartheta_\varepsilon,\alpha_\varepsilon))e^{-\lambda\vartheta_\varepsilon} \leq (Tw)(x') + \varepsilon\ .$$

Then

$$(Tw)(x) - (Tw)(x') \leq A + e^{-\lambda\vartheta_\varepsilon}\bigl((\mathcal{M}w)(y_x(\vartheta_\varepsilon,\alpha_\varepsilon)) - (\mathcal{M}w)(y_{x'}(\vartheta_\varepsilon,\alpha_\varepsilon))\bigr) + \varepsilon,$$

where

$$A = \int_0^{\vartheta_\varepsilon} \bigl(\ell(y_x(s,\alpha_\varepsilon)) - \ell(y_{x'}(s,\alpha_\varepsilon))\bigr)e^{-\lambda s}\,ds\ .$$

The statement follows from Lemma 4.18 (ii) and standard estimates on $|y_x(\,\cdot\,,\alpha) - y_{x'}(\,\cdot\,,\alpha)|$ (see Exercise 4.8).

The monotonicity and concavity properties (jjj) and (jv) are easy consequences of the analogous ones for $\mathcal{M}$. Let us now prove (v). If (4.74) holds, then (jjj) gives $Tw_2 \geq T((1-\gamma)w_1)$. Now use (jv) with $u = 0$, $w = w_1$, $\mu = 1-\gamma$, to obtain $T((1-\gamma)w_1) \geq (1-\gamma)Tw_1 + \gamma T0$. Hence

$$(4.75) \qquad Tw_2 \geq (1-\gamma)Tw_1 + \gamma T0\ .$$

Let us now show that

$$(4.76) \qquad T0 \geq \mu\overline{u} \qquad \forall \mu \in \,]0, k/\|\overline{u}\|_\infty \wedge 1[\ .$$

Observe that $\mathcal{M}0 \equiv k$ (it follows from Lemma 4.18 (i)); hence

$$(T0)(x) = \inf_{\substack{\alpha \in \mathcal{A} \\ \vartheta \geq 0}} \left\{ \int_0^\vartheta \ell(y_x(s,\alpha), \alpha(s)) e^{-\lambda s}\, ds + k e^{-\lambda \vartheta} \right\}.$$

Since, by definition, $k \geq \mu \overline{u}(y_x(\vartheta, \alpha))$ for all $\mu \in\,]0, k/\|\overline{u}\|_\infty \wedge 1[$ and all $\vartheta \geq 0$, we get

(4.77) $\quad (T0)(x) \geq \inf_{\alpha \in \mathcal{A}} \left\{ \int_0^\vartheta \mu\ell(y_x(s,\alpha), \alpha(s)) e^{-\lambda s}\, ds + \mu e^{-\lambda \vartheta} \overline{u}(y_x(\vartheta, \alpha)) \right\}.$

Here we used the fact that $\ell \geq \mu \ell$, since $\ell \geq 0$ and $0 < \mu < 1$. By definition of $\overline{u}$, the right-hand side in (4.77) is $\mu \overline{u}(x)$ (see Proposition 2.5) and (4.76) is proved.

Observe at this point that for any $\alpha \in \mathcal{A}$, $\vartheta \geq 0$ and $w \in BUC(\mathbb{R}^N)$,

$$(Tw)(x) \leq \int_0^\vartheta \ell(y_x(s,\alpha), \alpha(s)) e^{-\lambda s}\, ds + (\mathcal{M}w)(y_x(\vartheta, \alpha)) e^{-\lambda \vartheta}.$$

Since $w \in BUC(\mathbb{R}^N)$, we can let $\vartheta \to +\infty$ to obtain, by Lemma 4.18 (i), that

$$(Tw)(x) \leq \int_0^{+\infty} \ell(y_x(s,\alpha), \alpha(s)) e^{-\lambda s}\, ds \qquad \forall \alpha \in \mathcal{A}$$

and, consequently,

(4.78) $\qquad\qquad\qquad Tw \leq \overline{u} \qquad \forall w \in BUC(\mathbb{R}^N).$

Use this with $w = w_1$ to obtain from (4.75) and (4.76) that $Tw_2 \geq (1-\gamma)Tw_1 + \gamma\mu Tw_1$ to complete the proof of (v). ◀

We are now in position to prove the announced uniqueness result for (QVI).

THEOREM 4.21. *Assume* (A_0'), (A_3), (A_4), *and* (4.64). *If* $u \in BUC(\mathbb{R}^N)$ *is a viscosity solution of* (QVI) *such that*

$$-\frac{M}{\lambda} \leq u(x) \leq \frac{M}{\lambda} \qquad \text{for all } x \in \mathbb{R}^N,$$

then $u = v$, *where* v *is given by* (4.65).

PROOF. It is not restrictive to assume that u, ℓ, and v are nonnegative. In fact, if we define $\widetilde{\ell} = \ell + M$, $\widetilde{v} = v + M/\lambda$, $\widetilde{u} = u + M/\lambda$, where M is the constant in (A_4), it is easy to see that $\widetilde{\ell}, \widetilde{v}, \widetilde{u}$ are nonnegative, $\widetilde{v}$ is the value function associated with the running cost $\widetilde{\ell}$, and $\widetilde{v}, \widetilde{u}$ are solutions of (QVI) with H replaced by $H - M$.

Let us observe first that Tu and $\mathcal{M}u$ belong to $BUC(\mathbb{R}^N)$ (see Lemmata 4.18 and 4.20). Also, by its very definition and the theory of §4.2, Tu satisfies

$$\max\{\lambda Tu + H(x, DTu); Tu - \mathcal{M}u\} = 0 \qquad \text{in } \mathbb{R}^N$$

in the viscosity sense. On the other hand, we also have
$$\max\{\lambda u + H(x, Du); u - \mathcal{M}u\} = 0 \quad \text{in } \mathbb{R}^N$$
in the viscosity sense. Therefore, by the uniqueness result for (VIO) (see Theorem 4.11),

(4.79) $$u = Tu.$$

The same argument also shows that

(4.80) $$v = Tv.$$

Since $v \geq 0$, we have
$$u - v \leq u.$$

Hence, by Lemma 4.20 (v) with $\gamma = 1$,
$$Tu - Tv \leq (1 - \mu)Tu$$
for all $\mu \in {]}0, k/\|\overline{u}\|_\infty \wedge 1[$. But $u = Tu$, $v = Tv$, so that
$$u - v \leq (1 - \mu)u$$
for all $\mu \in {]}0, k/\|\overline{u}\|_\infty \wedge 1[$. Now fix μ in the same interval and apply Lemma 4.20 (v) with $w_1 = u$, $w_2 = v$, $\gamma = 1 - \mu$. Hence
$$Tu - Tv \leq \gamma^2 Tu$$
or, which is the same thanks to (4.79) and (4.80),
$$u - v \leq \gamma^2 u.$$

Iterating this procedure gives the inequality
$$u - v \leq \gamma^n u \qquad n = 1, 2, \dots$$

Since u is bounded and $\gamma \in {]}0, 1[$, we obtain $u \leq v$. Exchanging the roles of u and v, the opposite inequality is proved similarly. ◄

The proof of Theorem 4.21 shows that the value function v of the impulse control problem is a fixed point in $BUC(\mathbb{R}^N)$ of the nonlinear operator T. This suggests that the iterative scheme

(4.81) $$v_{n+1} = Tv_n, \quad n = 1, 2, \dots$$

should converge to v, at least for a suitably chosen starting point v_1. This is indeed the case if we choose

(4.82) $$v_1(x) = \overline{u}(x) := \inf_{\alpha \in \mathcal{A}} \left\{ \int_0^{+\infty} \ell(y_x(s, \alpha), \alpha(s)) \, e^{-\lambda s} \, ds \right\},$$

and the interest rate λ is large enough. Observe that $\overline{u}$ is the value function of the infinite horizon problem in Section 2. In view of the theory developed there, $\overline{u}$ is the unique viscosity solution of

$$\lambda \overline{u}(x) + H(x, D\overline{u}(x)) = 0 \qquad x \in \mathbb{R}^N.$$

PROPOSITION 4.22. *Assume* (A'_0), (A_3), (4.64), *and that* (A_4) *holds with* $\omega_\ell(r) = c_\ell r$ *and* $\lambda > L$. *Then the sequence* $\{v_n\}$ *defined by* (4.81) *and* (4.82) *converges locally uniformly in* $\mathbb{R}^N$ *to* v *given by* (4.65).

PROOF. We showed in the proof of Lemma 4.20 (v) (see (4.78)) that $Tw \leq \overline{u}$ for any $w \in BUC(\mathbb{R}^N)$. Hence

$$v_2 := Tv_1 = T\overline{u} \leq \overline{u}.$$

Since T is nondecreasing, the previous inequality easily implies that $\{v_n\}$ is nonincreasing. This also gives the bound

(4.83) $$v_n \leq \overline{u} \leq M/\lambda.$$

On the other hand, since $-M/\lambda \leq \overline{u} \leq M/\lambda$ we have

(4.84) $$v_2 = Tv_1 \geq T(-M/\lambda) \geq -M/\lambda,$$

where we used Lemma 4.20 (j) and the assumption $k > 0$. From (4.84) it easily follows that

$$v_n \geq -M/\lambda.$$

This and (4.83) show that $\|v_n\|_\infty$ is uniformly bounded.

The sequence $\{v_n\}$ is also equicontinuous, namely,

(4.85) $$|v_n(x) - v_n(z)| \leq \frac{c_\ell}{\lambda - L}|x - z| \quad \forall x, z \in \mathbb{R}^N.$$

Estimate (4.85) is immediately checked for $n = 1$. Let us proceed then by induction. By definition of T, for any $\varepsilon > 0$ there exists $\alpha_\varepsilon \in \mathcal{A}$ and $\vartheta_\varepsilon \geq 0$ such that

(4.86) $$v_{n+1}(x) - v_{n+1}(z)$$
$$\leq I(\vartheta_\varepsilon) + e^{-\lambda\vartheta_\varepsilon}\big((\mathcal{M}v_n)(y_x(\vartheta_\varepsilon, \alpha_\varepsilon)) - (\mathcal{M}v_n)(y_z(\vartheta_\varepsilon, \alpha_\varepsilon))\big) + \varepsilon,$$

where

$$I(\vartheta_\varepsilon) := \int_0^{\vartheta_\varepsilon} \big(\ell(y_x(s, \alpha_\varepsilon), \alpha_\varepsilon(s)) - \ell(y_z(s, \alpha_\varepsilon), \alpha_\varepsilon(s))\big)e^{-\lambda s}\, ds.$$

A standard computation reveals that

$$I(\vartheta_\varepsilon) \leq \frac{c_\ell}{\lambda - L}|x - z|\big(1 - e^{-(\lambda - L)\vartheta_\varepsilon}\big).$$

Assuming that (4.85) holds for n, the term depending on $\mathcal{M}$ in (4.86) is estimated by

$$\frac{c_\ell}{\lambda - L}|x - z|e^{-(\lambda - L)\vartheta_\varepsilon}.$$

Hence (4.85) is true for $n+1$.

By the Ascoli-Arzelà theorem and the monotonicity of $\{v_n\}$ we deduce the existence of some $\widetilde{u} \in BUC(\mathbb{R}^N)$ such that

$$v_n \longrightarrow \widetilde{u}$$

locally uniformly in $\mathbb{R}^N$. At this point, the inequality

$$\|Tw - Tu\|_\infty \leq \|w - u\|_\infty \quad \forall\, w, u \in BUC(\mathbb{R}^N)$$

(its easy proof is left as Exercise 4.9) yields

$$Tv_n \longrightarrow T\widetilde{u}$$

locally uniformly in $\mathbb{R}^N$. Hence, by (4.81),

(4.87) $$\widetilde{u} = T\widetilde{u}\,.$$

We observed already that $T\widetilde{u}$ is, by the very definition of T, a viscosity solution of (QVI). Therefore, by (4.87) and Theorem 4.21, $\widetilde{u} = v$. ◀

REMARK 4.23. The above result holds more generally for $\ell \in BUC(\mathbb{R}^N)$; in this case (4.85) is replaced by the equicontinuity of $\{v_n\}$. A different proof of the convergence of $\{v_n\}$ can be performed using stability and comparison properties of viscosity solutions (see Exercise 4.10). ◁

REMARK 4.24. Using Lemma 4.20 (v) it is possible to show that

$$\|v_n - v_m\|_\infty \leq \gamma^{m-n} \|\overline{u}\|_\infty \quad \text{for } m > n,$$

for any $\gamma < 1$ and $\gamma > (1 - k/\|\overline{u}\|_\infty) \vee 0$, whence the geometric rate of convergence

$$\|v_n - v\|_\infty \leq \gamma^n \|\overline{u}\|_\infty\,. \quad ◁$$

Let us conclude this section by sketching how the previous results allows us to synthesize an optimal impulse control $\{\widehat{\vartheta}^i, \widehat{\xi}^i\}$ in the simplified case where f and ℓ are independent of a (i.e., the controls α are not present in the model).

Let v be the unique viscosity solution of (QVI). Since v is bounded and $c(\xi) \to +\infty$ as $|\xi| \to +\infty$ by (4.64), the infimum in the definition of $\mathcal{M}v$ (4.70) is attained. By standard selection results (see [FR75]), there exists a bounded measurable mapping $\widehat{\xi} : \mathbb{R}^N \to \mathbb{R}^N_+$ such that

$$(\mathcal{M}v)(x) = v(x + \widehat{\xi}(x)) + c(\widehat{\xi}(x)) \quad \forall\, x \in \mathbb{R}^N\,.$$

Define by recursion $\{\widehat{\vartheta}^i, \widehat{\xi}^i\}$ as follows:

(4.88)
$$\widehat{\vartheta}^0 = 0,$$
$$\frac{d}{dt}\widehat{y}^0(t) = f(\widehat{y}^0(t)), \quad t \geq \widehat{\vartheta}^0, \quad \widehat{y}_x^0(0) = x,$$
$$\widehat{\vartheta}^{i+1} = \begin{cases} \inf\{t \geq \widehat{\vartheta}^i : v(\widehat{y}^i(t)) = (\mathcal{M}v)(\widehat{y}^i(t))\}, \\ +\infty \text{ if the set above is empty,} \end{cases}$$
$$\widehat{\xi}^i = \widehat{\xi}(\widehat{y}^{i-1}(\widehat{\vartheta}^i)),$$
$$\frac{d}{dt}\widehat{y}^i(t) = f(\widehat{y}^i(t)) \quad t \geq \widehat{\vartheta}^i,$$
$$\widehat{y}^i(\widehat{\vartheta}^i) = \widehat{y}^{i-1}(\widehat{\vartheta}^i + \widehat{\xi}^i),$$

for $i = 1, 2, \ldots$. Using a technique similar to that employed in the proof of Theorem 4.12 one can show the optimality of $\{\widehat{\vartheta}^i, \widehat{\xi}^i\}$. The details are left to the reader (see, in particular, Exercise 4.12).

4.4. Optimal switching

In this section we consider the problem of minimizing an infinite horizon discounted running cost as in §2.1, over the class $\mathcal{P}$ of piecewise constant controls, namely,

(4.89) $$\mathcal{P} := \{\alpha \in \mathcal{A} : \alpha(t) \text{ is constant for } t \in\,]\vartheta^i, \vartheta^{i+1}]\,\},$$

where $\{\vartheta^i\}$ is a sequence in $[0, +\infty]$ such that

(4.90)
$$0 = \vartheta^0 \leq \vartheta^1 < \cdots < \vartheta^i < \vartheta^{i+1} < \cdots,$$
$$\lim_{i \to +\infty} \vartheta^i = +\infty.$$

Unlike in §2.1, however, we assume here that for every *switching* from one control setting to a different one, the controller incurs a positive switching cost. We assume that the control set is finite, that is,

(4.91) $$A = \{1, \ldots, m\} \quad \text{for some } m \in \mathbb{N}.$$

Observe that an admissible control $\alpha \in \mathcal{P}$ can be identified with the couple of sequences $\{\vartheta^i\}, \{a^i\}$, where $a^0 = \alpha(0)$, $a^i = \alpha(t)$ for $t \in\,]\vartheta^i, \vartheta^{i+1}]$ and $a^i \neq a^{i-1}$ if $\vartheta^i < +\infty$.

The controlled dynamical system (4.1) then becomes

(4.92) $$\begin{cases} y'(t) = f(y(t), a^i), \quad t \in\,]\vartheta^i, \vartheta^{i+1}[, \quad (i = 0, 1, \ldots), \\ y(0) = x \in \mathbb{R}^N, \end{cases}$$

and the cost functional is

(4.93) $$J(x, \alpha) := \sum_{i=1}^\infty \left(\int_{\vartheta^{i-1}}^{\vartheta^i} \ell(y_x(s, \alpha), a^{i-1}) e^{-\lambda s}\, ds + k(a^{i-1}, a^i) e^{-\lambda \vartheta^i} \right).$$

Here we assumed for simplicity that the switching cost k depends only on the control settings involved but not on the state where the switching occurs. Let us observe explicitly that if $k \equiv 0$, then

$$\inf_{\alpha \in \mathcal{P}} J(x,\alpha) = \inf_{\alpha \in \mathcal{A}} J(x,\alpha)$$

(see Remark 2.15); the case of zero switching cost is therefore covered by the theory of Section 2.

Let us assume then that

(4.94) $$\begin{aligned} k(j,\ell) &> 0, & &\text{if } \ell \neq j, \\ k(j,j) &= 0, & & \\ k(j,\ell) &< k(j,i) + k(i,\ell), & &\text{if } \ell \neq j \neq i. \end{aligned}$$

We consider the subsets $\mathcal{P}^j$ of admissible controls defined by

$$\mathcal{P}^j := \{\alpha \in \mathcal{P} : \alpha(0) = j\};$$

correspondingly we define the value functions

(4.95) $$v^j(x) := \inf_{\alpha \in \mathcal{P}^j} J(x,\alpha), \qquad j = 1,\ldots,m.$$

The proof of the next result on the regularity of v^j is completely similar to that of Proposition 2.1, so we omit it.

PROPOSITION 4.25. *Assume* (A'_0), (A_3), (A_4), *and* (4.94). *Then* $v^j \in BUC(\mathbb{R}^N)$ *for all* $j = 1,\ldots,m$.

Following the usual pattern we derive next the Dynamic Programming optimality conditions.

PROPOSITION 4.26. *Assume* (A'_0), (A_3), (A_4), *and* (4.94). *Then, for each* $j \in \{1,\ldots,m\}$ *the value function* v^j *satisfies for all* $x \in \mathbb{R}^N$ *and* $t > 0$

(i) $$v^j(x) \leq \min_{\ell \neq j} \left(v^\ell(x) + k(j,\ell) \right),$$

(ii) $$v^j(x) \leq \int_0^t \ell(y_x(s),j) e^{-\lambda s}\, ds + v^j(y_x(t))\, e^{-\lambda t}.$$

Furthermore, for any x where the strict inequality occurs in (i) *there exists $t_0 = t_0(x)$ such that*

(iii) $$v^j(x) = \int_0^t \ell(y_x(s),j) e^{-\lambda s}\, ds + v^j(y_x(t))\, e^{-\lambda t} \qquad \forall t \in [0,t_0].$$

PROOF. Fix $\ell \in \{1,\ldots,m\}$, $\ell \neq j$, and choose any $\alpha = (\{\vartheta^i\},\{a^i\})$ in $\mathcal{P}^\ell$. Then define $\widetilde{\alpha} = (\{\widetilde{\vartheta}^i\},\{\widetilde{a}^i\})$ by setting

$$\widetilde{\vartheta}^0 = 0, \qquad \widetilde{a}^0 = j,$$

and, if $\vartheta^0 < \vartheta^1$,
$$\widetilde{\vartheta}^i = \vartheta^{i-1}, \qquad \widetilde{a}^i = a^{i-1}, \qquad i = 1, 2, \ldots,$$
and in the case $\vartheta^0 = \vartheta^1$
$$\widetilde{\vartheta}^i = \vartheta^i, \qquad \widetilde{a}^i = a^i, \qquad i = 1, 2, \ldots$$
By definition, $\widetilde{\alpha} \in \mathcal{P}^j$; also we have, if $\vartheta^0 < \vartheta^1$,
$$J(x, \widetilde{\alpha}) = k(j, \ell) + J(x, \alpha),$$
as a simple computation shows, while in the case $\vartheta^0 = \vartheta^1$,
$$J(x, \widetilde{\alpha}) = k(j, a^1) + J(x, \alpha) - k(\ell, a^1) < k(j, \ell) + J(x, \alpha)$$
by (4.94). Therefore, taking the infimum over $\alpha \in \mathcal{P}^\ell$,
$$v^j(x) \leq J(x, \widetilde{\alpha}) \leq k(j, \ell) + v^\ell(x) \qquad \ell \neq j$$
and (i) follows.

The statements (ii) and (iii) are proved in the same way as the corresponding ones in Proposition 4.9. ◀

In view of the previous proposition it is convenient to define operators M^j acting on vectors $u = (u^1, \ldots, u^m) \in BUC(\mathbb{R}^N)^m$ by setting

(4.96) $$(M^j u)(x) = \min_{\ell \neq j}(u^\ell(x) + k(j, \ell)) \qquad j = 1, \ldots, m.$$

The HJB equation for the vector valued value function $v = (v^1, \ldots, v^m)$ defined by (4.95) turns out to be the following *system of quasivariational inequalities*:

(SQVI)
$$\max\{\lambda u^j(x) - f^j(x) \cdot Du^j(x) - \ell^j(x); u^j(x) - (M^j u)(x)\} = 0 \qquad x \in \mathbb{R}^N,$$

$j = 1, \ldots, m$. (We used here the notations $f^j(x)$ and $\ell^j(x)$ to denote $f(x, j)$ and $\ell(x, j)$.) The coupling of the unknowns in (SQVI) appears only in the pointwise *implicit obstacle* condition

(4.97) $$u^j(x) \leq (M^j u)(x).$$

On the basis of theory developed in §§4.1–3 it is natural to consider the following notion of viscosity solution of (SQVI): a vector function $u = (u^1, \ldots, u^m) \in BUC(\mathbb{R}^N)^m$ is a viscosity solution of (SQVI) if, for each $j \in \{1, \ldots, m\}$ and each $\varphi \in C^1(\mathbb{R}^N)$,

(4.98) $$\max\{\lambda u^j(x) - f^j(x) \cdot D\varphi(x) - \ell^j(x); u^j(x) - (M^j u)(x)\} \leq 0$$
at all local maxima x of $u^j - \varphi$,

(4.99) $$\max\{\lambda u^j(x) - f^j(x) \cdot D\varphi(x) - \ell^j(x); u^j(x) - (M^j u)(x)\} \geq 0$$
at all local minima x of $u^j - \varphi$.

THEOREM 4.27. *Assume* (A'_0), (A_3), (A_4), *and* (4.94). *Then the value function* $v = (v^1, \ldots, v^m)$ *defined by* (4.95) *is a viscosity solution of* (SQVI).

PROOF. The proof is essentially the same as that of Theorem 4.10 and we leave it to the reader. ◂

The main result in this section is on the uniqueness of viscosity solutions of (SQVI).

THEOREM 4.28. *Assume* (A'_0), (A_3), (A_4), *and* (4.94). *If* $u = (u^1, \ldots, u^m)$, $w = (w^1, \ldots, w^m) \in BUC(\mathbb{R}^N)^m$ *are viscosity solutions of* (SQVI), *then* $u \equiv w$.

PROOF. The scheme of the proof is similar to that of the comparison and uniqueness properties for HJB equations (see Section II.3 and §2.1 in this chapter). We therefore give details only for the points concerning the typical features of (SQVI) and just outline the standard part.

As in the proof of Theorem 2.12 we consider the auxiliary functions

$$\Phi^j(x,y) := u^j(x) - w^j(y) - \frac{|x-y|^2}{2\varepsilon} - \beta(\langle x \rangle^\gamma + \langle y \rangle^\gamma)$$

for $j = 1, \ldots, m$, where $\langle x \rangle = (1 + |x|^2)^{1/2}$ and $\varepsilon, \beta, \gamma$ are positive parameters to be chosen conveniently (see the proof of Theorem 2.12). Let (x_0^j, y_0^j) be a global maximum point for Φ^j over $\mathbb{R}^N \times \mathbb{R}^N$. Then there exists $\ell \in \{1, \ldots, m\}$, say $\ell = 1$, such that

(4.100) $$\Phi^1(x_0^1, y_0^1) = \max_{1 \le j \le m} \max_{\mathbb{R}^N \times \mathbb{R}^N} \Phi^j(x,y) .$$

Moreover, the estimates

$$|x_0^1 - y_0^1| \longrightarrow 0, \quad \frac{|x_0^1 - y_0^1|^2}{\varepsilon} \longrightarrow 0 \quad \text{as } \varepsilon \to 0$$

hold. Since u, w are viscosity solutions of (SQVI), by standard arguments

(4.101) $\max\{\lambda u^1(x_0^1) - f^1(x_0^1) \cdot \xi - \ell^1(x_0^1); u^1(x_0^1) - (M^1 u)(x_0^1)\} \le 0$

(4.102) $\max\{\lambda w^1(y_0^1) - f^1(y_0^1) \cdot \eta - \ell^1(y_0^1); w^1(y_0^1) - (M^1 w)(y_0^1)\} \ge 0$

with

$$\xi = \frac{x_0^1 - y_0^1}{\varepsilon} + \beta\gamma x_0^1 \langle x_0^1 \rangle^{\gamma-2}, \quad \eta = \frac{x_0^1 - y_0^1}{\varepsilon} - \beta\gamma y_0^1 \langle y_0^1 \rangle^{\gamma-2} .$$

Consider now the two possibilities:

case 1: $\quad w^1(y_0^1) < (M^1 w)(y_0^1)$
case 2: $\quad w^1(y_0^1) = (M^1 w)(y_0^1)$

(no other case is possible because w^1 is a subsolution of (SQVI)). If case 1 occurs, then (4.101) and (4.102) imply, respectively,

$$\lambda u^1(x_0^1) - f^1(x_0^1) \cdot \xi - \ell(x_0^1) \le 0$$

and
$$\lambda w^1(y_0^1) - f^1(y_0^1) \cdot \eta - \ell(y_0^1) \geq 0.$$

Hence
$$\lambda(u^1(x_0^1) - w^1(y_0^1)) \leq f^1(x_0^1) \cdot \xi - f^1(y_0^1) \cdot \eta + \ell(x_0^1) - \ell(y_0^1)$$

and we conclude that $u^1 = w^1$ as in the proof of Theorem 2.12. By (4.100), this implies $u^j = w^j$ for each $j \in \{1, \ldots, m\}$ (see Exercise 4.14).

If case 2 holds, then, by definition of M^1, there exists $j \neq 1$, say $j = 2$, such that

(4.103) $$w^1(y_0^1) = w^2(y_0^1) + k(a^1, a^2).$$

By (4.100), $\Phi^1(x_0^1, y_0^1) \geq \Phi^2(x_0^1, y_0^1)$, so that

(4.104) $$u^1(x_0^1) - u^2(x_0^1) \geq w^1(y_0^1) - w^2(y_0^1).$$

It is easy to deduce from (SQVI) that

(4.105) $$u^1(x_0^1) \leq u^2(x_0^1) + k(a^1, a^2).$$

Thus (4.103), (4.104), and (4.105) yield
$$u^1(x_0^1) = u^2(x_0^1) + k(a^1, a^2).$$

Because of (4.103) the above gives
$$u^1(x_0^1) - w^1(y_0^1) = u^2(x_0^1) - w^2(y_0^2),$$

so that
$$\Phi^2(x_0^1, y_0^1) = \Phi^1(x_0^1, y_0^1),$$

by definition of Φ^j.

Now repeat the previous considerations with the index 2 replacing 1. Should case 1 occur we are done. Otherwise there exists $\ell_2 \in \{1, \ldots, m\}$, $\ell_2 \neq 2$, such that
$$w^2(y_0^1) = w^{\ell_2}(y_0^1) + k(a^2, a^{\ell_2}).$$

Observe that necessarily $\ell_2 \neq 1$. Indeed, were $\ell_2 = 1$, the preceding and (4.103) would imply
$$k(a^1, a^2) + k(a^2, a^1) = 0$$

a contradiction to assumption (4.94). Hence we may assume $\ell_2 = 3$ and prove as before
$$\Phi^3(x_0^1, y_0^1) = \phi^2(x_0^1, y_0^1) = \Phi^1(x_0^1, y_0^1).$$

Repeating the preceding calculations with the index 3 replacing 2 and so on, after finitely many steps an index $j \leq m$ for which case 1 holds is reached and the proof is complete. ◄

REMARK 4.29. The proof of Theorem 4.28 shows in fact that a comparison principle holds for (SQVI), namely, that $u^j \leq w^j$, $j = 1,\ldots,m$, if $u = (u^1,\ldots,u^m)$ and $w = (w^1,\ldots,w^m)$ are, respectively, a sub- and a supersolution in the viscosity sense. ◁

By Theorems 4.27, 4.28, and Proposition 4.25 the value function $v = (v_1,\ldots,v^m)$ is the unique viscosity solution of (SQVI). Let us proceed now to the synthesis of an optimal control. By analogy with the results of §4.2 and §4.3, for fixed $x \in \mathbb{R}^N$ and $j \in A$ we set

(4.106) $$\widehat{\vartheta}^0 = 0, \qquad \widehat{a}^0 = j$$

and for $i = 1, 2, \ldots$

(4.107) $$\widehat{\vartheta}^i = \begin{cases} \inf\{\, t > \widehat{\vartheta}^{i-1} : v^{i-1}(y_x(t)) = (M^{i-1}v)(y_x(t))\,\} \\ +\infty \quad \text{if the preceding set is empty} \end{cases}$$

(4.108) $$\widehat{a}^i = \begin{cases} \text{any } \ell \neq \widehat{a}^{i-1} \text{ such that } (M^{i-1}v)(y_x(\widehat{\vartheta}^i)) = v^\ell(y_x(\widehat{\vartheta}^i)) + k(\widehat{a}^{i-1},\ell) \\ \widehat{a}^{i-1} \quad \text{if } \widehat{\vartheta}^i = +\infty\,. \end{cases}$$

The first thing to check is that $\widehat{\vartheta}^i \to +\infty$ as $i \to +\infty$. Indeed, we claim that $\widehat{\vartheta}^i \geq \widehat{\vartheta}^{i-1} + \sigma$ for some $\sigma > 0$ and $i = 2, \ldots$. This is trivially true if $\widehat{\vartheta}^{i-1} = +\infty$ for some i. Otherwise, assume that, for any $\sigma > 0$, $\widehat{\vartheta}^i < \widehat{\vartheta}^{i-1} + \sigma$ for some i. Then, (4.107) and (4.108) imply the existence of some $\widehat{\vartheta}^i \leq t < \widehat{\vartheta}^{i-1} + \sigma$ such that

$$v^{i-1}(y_x(t)) = v^\ell(y_x(t)) + k(\widehat{a}^{i-1},\ell), \quad \text{for some } \ell \in A,\ \ell \neq \widehat{a}^{i-1},$$

$$v^{i-2}(y_x(\widehat{\vartheta}^{i-1})) = v^{i-1}(y_x(\widehat{\vartheta}^{i-1})) + k(\widehat{a}^{i-2},\widehat{a}^{i-1}) \leq v^\ell(y_x(\widehat{\vartheta}^{i-1})) + k(\widehat{a}^{i-2},\ell)\,.$$

From assumption (4.94) it follows

$$0 < k(\widehat{a}^{i-2},\widehat{a}^{i-1}) + k(\widehat{a}^{i-1},\ell) - k(\widehat{a}^{i-2},\ell) \leq 2\omega(\sigma)$$

where ω is a modulus of continuity for $v(y_x(\,\cdot\,))$. This gives a contradiction and proves the claim.

Hence $\widehat{\alpha} = (\{\widehat{\vartheta}^i\},\{\widehat{a}^i\}) \in \mathcal{P}^j$ and it is not hard to show, following the lines of the proof of Theorem 4.12, that

$$J(x,\widehat{\alpha}) = v^j(x)\,.$$

Exercises

4.1. Consider the infinite horizon problem of §2.1 with control set $A = [0,1]$. Show that its value function v satisfies $v(x) \leq v(x,a)$ for all $a \in A$, where $v(x,a)$ is defined by (4.6).

4.2. For J and v as in (4.42) and (4.43) show that for any $\varepsilon > 0$ there exists $\vartheta_z^\varepsilon \geq 0$ such that

$$v(x) - v(z) \leq J(x,\vartheta_z^\varepsilon) - J(z,\vartheta_z^\varepsilon) + \varepsilon\,.$$

Use this to prove Proposition 4.8.

4.3. Set
$$\overline{u}(x) = \int_0^{+\infty} \ell(y_x(t)) e^{-\lambda t} dt .$$
Use the comparison results to prove that
$$v(x) \leq \min\{\overline{u}(x); \psi(x)\}, \qquad \forall x \in \mathbb{R}^N,$$
where v is defined by (4.43).

4.4. Observe that for $\alpha, \beta \in \mathbb{R}$ the relation $\max\{\alpha, \beta\} = 0$ is equivalent to $\alpha \leq 0$, $\beta \leq 0$, $\alpha\beta = 0$. Use this to prove that (VIO) and the complementarity system (4.50) are equivalent in the viscosity sense.

4.5. Give a direct proof of Theorem 4.11. Show, in particular, that if u_1, u_2 are, respectively, a sub- and a supersolution of (VIO), then $u_1 \leq u_2$.

4.6. Consider the optimal stopping time problem with $f(x, a) = f(x)$, $\ell(x, a) = \ell(x)$. Prove that
$$\lambda v - f \cdot Dv = \ell$$
in the open set $C = \{x \in \mathbb{R}^N : u(x) < \psi(x)\}$. Discuss the validity of the Lewy-Stampacchia type inequality
$$\min\{\lambda\psi - f \cdot D\psi; \ell\} \leq \lambda v - f \cdot Dv \leq \ell \qquad \text{in } \mathbb{R}^N .$$

4.7. In the framework of Exercise 4.6 assume that $\inf_{\mathbb{R}^N} \psi(x) > \lambda^{-1} \sup_{\mathbb{R}^N} \ell(x)$. Check that in this case $v(x) = \int_0^{+\infty} \ell(y_x(t)) e^{-\lambda t} dt$ and that $\vartheta^*(x) = +\infty$ for all x. Prove that in the case $\psi(x) \equiv \psi_0$ with $\psi_0 \leq \lambda^{-1} \inf_{\mathbb{R}^N} \ell(x)$ the value function is $v(x) \equiv \psi_0$ and that $\vartheta^*(x) = 0$ for all x.

4.8. For fixed $\beta = (\alpha, \{\vartheta^i\}, \{\xi^i\})$ as in (4.62), denote by y_0 the solution of
$$y_0'(t) = f(y_0(t), \alpha(t)), \qquad \text{for } t > 0,$$
$$y_0(0) = x .$$
Then recursively define y_i ($i = 0, 1, 2, \ldots$) by
$$y_i'(t) = f(y_i(t), \alpha(t)) \qquad \text{for } t > \vartheta^i,$$
$$y_i(t) = y_{i-1}(t) + \chi_{\{\vartheta^i = t\}} \xi^i \qquad \text{for } t \leq \vartheta^i$$
(here, $\chi_{\{\vartheta^i = t\}} = 1$ if $\vartheta^i = t$, $\chi_{\{\vartheta^i = t\}} = 0$ otherwise). Check that
$$y_i(t) = y_n(t) \qquad \forall t \in [0, \vartheta^i], \ \forall n \geq i$$
and define, consequently, for $t \geq 0$
$$y_x(t, \beta) := \lim_{i \to +\infty} y_i(t) .$$
Prove that $y_x(t, \beta)$ is a solution of (4.61). Prove also that
$$|y_x(t, \beta) - y_z(t, \beta)| \leq e^{Lt} |x - z|$$
for any $x, z \in \mathbb{R}^N$.

4. HJB INEQUALITIES

4.9. For $\mathcal{M}$ as in (4.70) show that
$$\|\mathcal{M}w - \mathcal{M}u\|_\infty \leq \|w - u\|_\infty \quad \forall w, u \in B(\mathbb{R}^N).$$

Use this and the inequality
$$(Tw)(x) - (Tu)(x) \leq e^{-\lambda \vartheta_\varepsilon}[(\mathcal{M}w)(y_x(\vartheta_\varepsilon, \beta)) - (\mathcal{M}u)(y_x(\vartheta_\varepsilon, \beta))] + \varepsilon$$

(which holds for any $\varepsilon > 0$ for suitable $\vartheta_\varepsilon \geq 0$) to conclude that
$$\|Tw - Tu\|_\infty \leq \|w - u\|_\infty \quad \forall u, w \in B(\mathbb{R}^N).$$

4.10. Let u_n be viscosity solutions of
$$\max\{\lambda u_n(x) + H_n(x, Du_n(x)); u_n(x) - (\mathcal{M}u_n)(x)\} = 0$$

in $\mathbb{R}^N$. Assume that $H_n \to H$, $u_n \to u$ locally uniformly and also that
$$\|(u_n - u)^-\|_\infty \longrightarrow 0 \quad \text{as } n \to +\infty.$$

Prove that

(i) if $u(x_0) < (\mathcal{M}u)(x_0)$ at some x_0 then, for small $r > 0$,
$$\lambda u_n + H_n(x, Du) = 0 \quad \text{in } B(x_0, r)$$

[Hint: use the condition on $(u_n - u)^-$ and $\mathcal{M}(u_n + \|(u_n - u)^-\|_\infty) = \mathcal{M}u_n + \|(u_n - u)^-\|_\infty \geq \mathcal{M}u$];

(ii) deduce from (i) and stability results that u is a supersolution of (QVI).

Finally, use the stability properties to show that u is a subsolution of (QVI).

4.11. Prove that the sequence $\{\widehat{\vartheta}^i\}$ defined by (4.88) satisfies the feasibility condition $\lim_{i \to +\infty} \widehat{\vartheta}^i = +\infty$ as required in (4.62).

4.12. Let $u = (u^1, \ldots, u^m)$ be a viscosity solution of (SQVI). Mimic the first part of the proof of Theorem 4.12 to show that
$$u^j(x) \leq (\mathcal{M}^j u)(x) \quad \forall x \in \mathbb{R}^N, \ \forall j \in \{1, \ldots, m\}.$$

4.13. Generalize the theory in §4.4 to the case of a state-dependent switching cost $k(x, a^j, a^\ell)$.

4.14. For Φ^j as in the proof of Theorem 4.28, use the inequality
$$\Phi^1(x_0^1, y_0^1) \geq \Phi^j(x, x), \quad \forall x \in \mathbb{R}^N, \ j \in \{1, \ldots, m\},$$

to deduce that if $u^1 = w^1$ then
$$u^j(x) - w^j(x) \leq o(1) \quad \text{as } \varepsilon \to 0^+$$

for all $j \in \{2, \ldots, m\}$.

5. Appendix:
Some results on ordinary differential equations

We begin with a result that is used many times throughout the book.

LEMMA 5.1 (GRONWALL INEQUALITY). *If $w \in L^1([t_0, t_1])$ and $h \in L^\infty([t_0, t_1])$ satisfy, for some $L \geq 0$,*

(5.1) $$w(t) \leq h(t) + L \int_{t_0}^{t} w(s) \, ds, \qquad \text{for a.e. } t \in [t_0, t_1],$$

then

(5.2) $$w(t) \leq h(t) + L e^{Lt} \int_{t_0}^{t} h(s) e^{-Ls} \, ds, \qquad \text{for a.e. } t \in [t_0, t_1].$$

If, in addition, h is nondecreasing, then

(5.3) $$w(t) \leq h(t) e^{L(t-t_0)}, \qquad \text{for a.e. } t \in [t_0, t_1].$$

PROOF. Define

$$z(t) := e^{-L(t-t_0)} \int_{t_0}^{t} w(s) \, ds$$

and note that z is absolutely continuous in $[t_0, t_1]$. After multiplying (5.1) by $e^{-L(t-t_0)}$ we get

$$z'(t) \leq h(t) e^{-L(t-t_0)}, \qquad \text{for a.e. } t \in [t_0, t_1].$$

We integrate both sides and multiply by $e^{L(t-t_0)}$ to obtain

$$\int_{t_0}^{t} w(s) \, ds \leq e^{Lt} \int_{t_0}^{t} h(s) e^{-Ls} \, ds, \qquad \text{for a.e. } t \in [t_0, t_1].$$

This inequality and (5.1) give (5.2).

If h is nondecreasing the integral in (5.2) is estimated by $h(t) \int_{t_0}^{t} e^{-Ls} \, ds$, and then (5.3) follows from a simple computation. ◂

Next we prove the existence of solutions of the integral equation

(5.4) $$y(t) = x + \int_{t_0}^{t} f(y(s), \alpha(s)) \, ds$$

for short time and for a given measurable control function α under the assumptions of Section 1. To do this we need two classical results of functional analysis. We recall that a set Y of functions $I \to \mathbb{R}^N$, $I \subseteq \mathbb{R}$, is *uniformly bounded* if there is a constant C such that $|G(t)| \leq C$ for all $t \in I$ and $G \in Y$, and it is called *equicontinuous* if for all $\varepsilon > 0$ there exists $\delta > 0$ such that $|G(t) - G(s)| \leq \varepsilon$ for all $t, s \in I$, $|t - s| \leq \delta$, and for all $G \in Y$.

THEOREM 5.2 (ASCOLI-ARZELÀ). *If $I \subseteq \mathbb{R}$ is compact and Y is a uniformly bounded and equicontinuous set of functions $I \to \mathbb{R}^N$, then any sequence in Y has a uniformly convergent subsequence.*

For the proof see, for instance, [Ru86].

THEOREM 5.3 (SCHAUDER FIXED POINT THEOREM). *Let X be a closed and convex subset of a Banach space. If $F : X \to X$ is continuous and $f(X)$ has compact closure, then there exists $x \in X$ such that $F(x) = x$.*

For the proof see, for instance, [GT83].

THEOREM 5.4 (LOCAL EXISTENCE OF TRAJECTORIES). *Assume (A_0) and (A_1), fix $x \in \mathbb{R}^N$ and set $M = M_x := \sup\{|f(z,a)| : |z - x| \leq 1, \ a \in A\}$. Then for any $t_0 \in \mathbb{R}$ and $\alpha \in \mathcal{A}$ there exists a lipschitzean solution y of (5.4) defined on $[t_0, t_0 + 1/M]$, and it satisfies*

(5.5) $\qquad |y(t) - x| \leq M(t - t_0) \qquad$ *for all t.*

PROOF. Set $t_1 = t_0 + 1/M$ and

$$X := \{g : [t_0, t_1] \to \overline{B}(x, 1) \text{ continuous}, \ g(t_0) = x\}.$$

For any $g \in X$ we can define

$$Fg(t) := x + \int_{t_0}^{t} f(g(s), \alpha(s)) \, ds, \qquad t_0 \leq t \leq t_1,$$

because the function to be integrated is bounded by M and it is measurable. Moreover,

$$|Fg(t) - x| \leq M(t_1 - t_0) = 1 \qquad \text{for all } t,$$

thus $F : X \to X$. We claim that F is continuous with respect to the uniform convergence. In fact, if $g_n, g \in X$, then for all $t \in [t_0, t_1]$

$$|Fg_n(t) - Fg(t)| \leq \int_{t_0}^{t_1} |f(g_n(s), \alpha(s)) - f(g(s), \alpha(s))| \, ds,$$

and the right-hand side tends to 0 as $n \to \infty$ by Lebesgue dominated convergence theorem if $g_n \to g$. Thus $Fg_n \to Fg$ uniformly.

Next we observe that the image $F(X)$ is equicontinuous because

(5.6) $\qquad |Fg(t) - Fg(\tau)| = \left| \int_{\tau}^{t} f(g(s), \alpha(s)) \, ds \right| \leq M |t - \tau|$

for all $g \in X$, $t, \tau \in [t_0, t_1]$. Then $F(X)$ has compact closure by the Ascoli-Arzelà Theorem 5.2.

Since F satisfies the assumptions of Schauder's Theorem 5.3, it has a fixed point y which is a solution of (5.4) in $[t_0, t_1]$. The estimate (5.6) gives the Lipschitz continuity of g with Lipschitz constant M, so (5.5) holds as well. ◂

Now we can prove the existence of solutions of the integral equation (5.4) for all times, their uniqueness and continuous dependence upon the initial state x.

THEOREM 5.5 (GLOBAL EXISTENCE OF TRAJECTORIES). *Assume* (A_0), (A_1), *and* (A_3). *Then for any* $t_0 \in \mathbb{R}$, $x \in \mathbb{R}^N$ *and* $\alpha \in \mathcal{A}$ *there is a unique solution* $y_x : [t_0, +\infty[\to \mathbb{R}^N$ *of* (5.4), *and it satisfies*

(5.7) $\qquad |y_x(t)| \leq (|x| + \sqrt{2K(t-t_0)}) e^{K(t-t_0)} \qquad$ *for all* $t > t_0$,

where $K := L + \sup_{a \in A} |f(0, a)|$. *Moreover, if* y_z *is the solution satisfying the initial condition* $y_z(t_0) = z$, *then*

(5.8) $\qquad |y_x(t) - y_z(t)| \leq e^{L(t-t_0)} |x - z|, \qquad$ *for all* $t \geq t_0$.

PROOF. We first prove (5.8) which implies the uniqueness of the solution of (5.4) by taking $x = z$. We compute

$$\frac{d}{dt} |y_x(t) - y_z(t)|^2 = 2[y_x(t) - y_z(t)] \cdot [f(y_x(t), a(t)) - f(y_z(t), a(t))]$$
$$\leq 2L |y_x(t) - y_z(t)|^2 \qquad \text{for a.e. } t \geq t_0,$$

where the inequality follows from (A_3). Since $\varphi(t) = |y_x(t) - y_z(t)|^2$ is absolutely continuous, by integrating both sides from t_0 to t we get

$$\varphi(t) \leq \varphi(t_0) + 2L \int_{t_0}^{t} \varphi(s) \, ds,$$

and thus by Gronwall inequality

$$\varphi(t) \leq |x - z|^2 e^{2L(t-t_0)},$$

which gives (5.8) for all t such that both solutions y_x and y_z exist.

Next we show the estimate (5.7). Note that (A_3) implies

$$[f(y, a) - f(0, a)] \cdot y \leq L |y|^2$$

and therefore

(5.9) $\qquad f(y, a) \cdot y \leq \sup_{a \in A} |f(0, a)| \, |y| + L |y|^2 \leq K(1 + |y|^2).$

Then for any control $\alpha \in \mathcal{A}$ the solution of (5.4) satisfies

$$\frac{d}{dt} |y(t)|^2 = 2y(t) \cdot f(y(t), \alpha(t)) \leq 2K(1 + |y(t)|^2),$$

thus integrating from t_0 to s

$$|y(s)|^2 \leq |x|^2 + 2K(s - t_0) + 2K \int_{t_0}^{s} |y(t)|^2 \, dt.$$

5. APPENDIX: SOME RESULTS ON ODES

Now Gronwall inequality gives for $s \leq t$

$$|y(s)|^2 \leq (|x|^2 + 2K(s-t_0)) e^{2K(s-t_0)},$$

which implies (5.7) for all t such that the solution exists.

To show the global existence, let us consider the supremum $\bar{t}$ of the times t_1 such that there exists a solution of (5.4) defined on $[t_0, t_1]$. Let us assume by contradiction that $\bar{t} < +\infty$. If a solution exists in $[t_0, \bar{t}]$ we have a contradiction because the solution can be continued on the right of $\bar{t}$ by the local existence Theorem 5.4. So we are left with the case that the maximal interval of existence is $[t_0, \bar{t}[$. We claim that $\lim_{t \to \bar{t}^-} y(t) = \bar{y} \in \mathbb{R}^N$. In fact we have shown that

$$|y(t)| \leq \left(|x| + \sqrt{2K(\bar{t}-t_0)}\right) e^{2K(\bar{t}-t_0)} =: C$$

for $t \in [t_0, \bar{t}[$. Then we set $M' := \sup\{f(y,a) : |y| \leq C, \ a \in A\}$ ($M' \leq +\infty$ by (A_1)), and get from the equation (5.4), for $s, t < \bar{t}$,

$$|y(t) - y(s)| \leq \int_s^t |f(y(\tau), a(\tau))| \, d\tau \leq M'|t-s|,$$

which implies the existence of the desired limit. Now it is easy to see that the extension of $y(\cdot)$ obtained by setting $y(\bar{t}) = \bar{y}$ is a solution of (5.4) in $[t_0, \bar{t}]$, which completes the proof. ◀

REMARK 5.6. Sharper estimates than (5.7) can be given. We can write, for all t such that $y(t) \neq 0$,

$$|y(t)| \frac{d}{dt}|y(t)| = \frac{1}{2}\frac{d}{dt}|y(t)|^2 = y(t) \cdot f(y(t), \alpha(t))$$
$$\leq C|y(t)| + L|y(t)|^2$$

by (5.9), where $C = \sup_{a \in A}|f(0,a)|$. We divide by $|y(t)|$, integrate in time and apply Gronwall inequality (5.2) to get, with some easy computations,

$$(5.10) \qquad |y_x(t)| \leq |x|e^{Lt} + \frac{C}{L}(e^{Lt} - 1).$$

Similarly, from

$$|y(t) - x|\frac{d}{dt}|y(t) - x| = (y(t) - x) \cdot f(y(t), a) \leq L|y(t) - x|^2 + (y(t) - x) \cdot f(x, a),$$

we get

$$(5.11) \qquad |y_x(t) - x| \leq \frac{1}{L} \sup_{a \in A}|f(x,a)|(e^{Lt} - 1). \ \triangleleft$$

REMARK 5.7. Most conclusions of Theorem 5.5 remain true if we weaken (A$_3$) to its local version

(5.12) $$(f(x,a) - f(y,a)) \cdot (x - y) \leq L_R |x - y|^2$$

for all $x, y \in B(0, R)$, $R > 0$, $a \in A$, plus the global bound

(5.13) $$f(y, a) \cdot y \leq K(1 + |y|^2)$$

for all $y \in \mathbb{R}^N$, $a \in A$. In fact (A$_0$), (A$_1$), and (5.12) are enough to get local existence and uniqueness of the trajectory by the same proof, and (5.13) is all we used to get the estimate (5.7) and the global existence. Of course (5.8) does not hold any more, but it is easy to show that for all $T > t_0$ there is C_T such that

(5.14) $$|y_x(t) - y_z(t)| \leq e^{C_T(t-t_0)} |x - z|, \qquad \text{for all } t \in [t_0, T]. \quad \triangleleft$$

We end this appendix with the classical theorem of differentiability of solutions of ordinary differential equations with respect to the initial data. We need a version that allows the vector field to be merely measurable with respect to the time variable. Its proof can be found, for instance, in [Ku86] and [Bre93].

We recall that the fundamental matrix $M(s,t)$ associated with the linear system

(5.15) $$\xi'(t) = A(t)\xi(t), \qquad t \in [t_0, t_1]$$

is the unique solution of the equation

$$M(s,t) = I + \int_t^s A(\tau) M(\tau, t) \, d\tau, \qquad s, t \in [t_0, t_1],$$

where $t \to A(t)$ is a bounded and measurable map from $[t_0, t_1]$ into the set of $N \times N$ matrices, and I is the identity $N \times N$ matrix. We recall also that the i-th column m_i of $M(\,\cdot\,, t_0)$ (i.e., $m_i(s) = M(s, t_0) e_i$ where e_i has the i-th component equal to 1 and the other components null) is the solution of (5.15) with initial condition $\xi(t_0) = e_i$, that is, it solves

$$m_i(s) = e_i + \int_{t_0}^s A(\tau) m_i(\tau) \, d\tau, \qquad s \in [t_0, t_1].$$

We consider the ordinary differential equation

(5.16) $$\begin{cases} y'(t) = F(y(t), t), & t \in \,]t_0, t_1[, \\ y(t_0) = x. \end{cases}$$

We assume it has a solution in the usual integral sense and denote it by $S(t, t_0, x) = y(t)$. We assume $F : \mathbb{R}^N \times [t_0, t_1] \to \mathbb{R}^N$ is bounded on any compact set and

- for any x the function $t \mapsto F(x, t)$ is measurable;

- for any t the function $x \mapsto F(x, t)$ is continuously differentiable; moreover, its Jacobian matrix $D_x F$ is bounded on $\mathcal{K} \times [t_0, t_1]$ for every compact $\mathcal{K} \subseteq \mathbb{R}^N$.

THEOREM 5.8. *Under the previous assumptions let* $\widehat{y}(\,\cdot\,) = S(\,\cdot\,,t_0,x_0)$ *be the solution of* (5.16) *with initial point* $x = x_0$. *Then, for any* $t \in [t_0,t_1]$, *the map* $x \mapsto S(t,t_0,x)$ *is continuously differentiable in a neighborhood of* x_0. *Moreover its Jacobian matrix at the point* x_0 *is*

$$D_x S(t,t_0,x_0) = M(t,t_0),$$

where $M(\,\cdot\,,\,\cdot\,)$ *is the fundamental matrix of the linear equation*

$$\xi'(t) = D_x F(\widehat{y}(t),t)\xi(t) \,.$$

Note that this result gives the differentiability of the trajectories of the system (S) (i.e., the solutions of (5.4)) with respect to the initial position for a fixed control function $\alpha \in \mathcal{A}$, that is, the differentiability of $x \mapsto y_x(t,\alpha)$, under the assumptions (A$_0$)–(A$_3$) if in addition $x \mapsto f(x,a)$ is continuously differentiable for all $a \in A$ with bounded Jacobian matrix over compact sets (i.e., $\omega_f(r,R) = L_R r$ in (A$_2$)).

6. Bibliographical notes

The classical theory and the history of Dynamic Programming were outlined in Chapter I.

The results of Sections 2.1 and 3.1 are essentially due to P.L. Lions [L82], see also [L83b, L85a, E83]. We have weakened some assumptions, in particular we have replaced the usual Lipschitz continuity of the dynamics (1.1) with condition (A$_3$). Under these assumptions the comparison Theorems 2.12 and 3.7 are proved by the techniques of Ishii [I84, I86, I88b], see also [BS94]. In connection with Remark 2.15 we recall that in some control problems a change in the set of admissible controls may lead to unexpected changes of the value, the so-called Lavrentiev phenomenon, see [Ce83, Zo92] and the references therein.

Verification theorems such as those of §2.2 are trivial applications of comparison results for viscosity sub- and supersolutions; they were explicitly observed in [B92, BSt93], with verification functions not necessarily continuous, see Chapter V. Some results based on different tools of nonsmooth analysis for locally Lipschitz verification functions can be found in Clarke's books [Cla83, Cla89] and in [VW90a]; for the classical result in the case of piecewise smooth value function we refer to [FR75].

The main result of §2.3, Theorem 2.32, is due to P.L. Lions and Souganidis [LS85]. Corollary 2.28 was proved by Mirica [Mi88, Mi90b] (see also [Mi92b]), and the Backward Dynamic Programming Principle was first explicitly observed by Soravia [Sor93b]. The name bilateral solution comes from [Ha91]. For other approaches to sub- and superoptimality principles see [Sw96a] and [Sor96b].

Subsection 2.4 mostly follows P.L. Lions and Souganidis [LS85, LS86], but they deal with the case of Lipschitz sub- and supersolutions. To treat merely continuous functions we need the generalized Dini (contingent) derivatives of Definition 2.36, which can be used even for semicontinuous functions. The classical Dini derivatives were introduced in [D1877], and the oldest paper that we know where the generalized version appears is [Yo68]. J.P. Aubin [Au81] derived the upper (respectively, lower) contingent derivative as the natural concept of derivative associated with the Bouligand-Severi contingent cone to the epigraph (respectively, hypograph) of the function. These tools of nonsmooth and set-valued analysis have a number of applications, see, for example, [AC84, AF90, Au91] and the references therein. The use of the Dini derivative for Hamilton-Jacobi equations goes back to Subbotin [Su80, Su84] for Lipschitz functions, and the generalized version appears in papers of his school [AT87, RS88] and of Frankowska [Fr87, Fr89b]. The equivalence between Dini-minimax and viscosity solutions was proved by many authors in the Lipschitz case by various methods under different assumptions [EI84, LS85, ST86]. In the general case it was observed in [Fr89b] that Dini solutions are viscosity solutions (Lemma 2.37 here), and Subbotin proved the equivalence [Su93b] by means of the property of the subdifferential described in Remark 2.44, see also [ClaL94] for another proof of this rather deep property. The main reference on the theory of minimax solutions is the monograph [Su95], we give some more historical information in Chapter VIII in connection with the theory of differential games. The regularity results Propositions 2.45 and 2.47 have been improved by Hijab [Hi91] and Lions and Souganidis [LS93] to show that for smooth Hamiltonian and discount rate λ large enough, the solution of the HJB equation becomes smooth and it solves the equation in the classical sense.

Subsection 2.5 originates from the papers of Frankowska [Fr89b] and Berkovitz [Be89], but it contains several additional results and our presentation is different (see [B95] for a more concise account). Some related results are in [CF91, RV91, Z93, Te95] and in [ZYL97] for the stochastic case; see also [Mi93b] where different methods are used. The existence of optimal open-loop controls, which is the main assumption for Theorem 2.61 on the existence of optimal multivalued feedbacks, was studied in depth by many authors under a number of different assumptions; see, for example, [Ce83, Bat78, BPi95] and the references therein. For a discussion of the classical problem of the synthesis of optimal single-valued feedbacks and some recent advances towards its solution we refer, respectively, to [FR75] and to [Mi90a, Sus90, PSus96].

Subsection 3.2 is taken from [I84]. Even if the results of §3.2 allow the cost functional and the value function to be unbounded without restrictions on their growth, they do not cover the case of a linear system with unbounded control set; in particular, they exclude the classical LQ problem (linear system and quadratic cost, see, e.g., [FR75, Son90, BDDM93]). Incidentally, this is essentially the only problem where the classical DP method works, because one finds a smooth (quadratic) solution of the HJB equation by means of a Riccati equation, and the optimal feedback obtained from the argmin of the Hamiltonian turns out to be linear. The characterization of the value function as the unique locally Lipschitz viscosity solution for some class of nonlinear non-quadratic problems containing the LQ problem can be found in [Ben88], in [CDaP90] for infinite dimensional systems, and in [BDL96, I97] where comparison theorems are proved. Similar results for the LQ problem with infinite horizon require more assumptions, because the HJB equation for this problem has two quadratic solutions; see [CDaP89, DL96]. The extensions to time-dependent Hamiltonians outlined in Exercises 3.5, 3.6, 3.7 can be carried further to treat the case of a merely measurable dependence on the time variable in the system f and therefore in H. This requires a generalization in the notion of viscosity solution, see [I85a, LP87, BJ87].

Semiconcavity results such as Theorem 3.26 in §3.3, and more general, can be proved either by working on the HJB equation, as in §4.2 in Chapter II and in [Kr60, Kr64, L82], or by working on the definition of value, as in [Kr67, CS87, CF91, Sin95]; our proof here follows the second strategy, but is different from the previous ones. The regularity Theorem 3.32 is based on the work of Subbotin and Subbotina [SS83]; see also [Ja77, Sua89]. Theorem 3.30 can be found in the book of Warga [Wa72] (see also [Ja77]) and Example 3.34 is in [Ja77]. The characterization of the differentiability of the value function in terms of uniqueness of the optimal trajectory is in Fleming [F69] for problems in the Calculus of Variations, where sharper smoothness results and more references can be found; we refer to Cannarsa and Frankowska [CF91] for some results on control problems.

The classical reference for the Pontryagin Maximum Principle is the pioneering monograph [PBGM62], but some version of the PMP can be found in every book on optimal control and we do not attempt to survey the many different proofs available in the literature. The connection between the PMP and the Dynamic Programming method in the case of smooth value was pointed out in Chapter 1 of [PBGM62]; see also [FR75]. The version of the PMP in §3.4, Theorem 3.42, in addition to the classical statement that (3.79) is a necessary condition of optimality, provides the connection (3.81) between the costate and the value, as well as the sufficiency of these two conditions for optimality. This formulation is due to Subbotina [Sua89], and our proof 1 follows hers with some variants, whereas proof 3 is based on Zhou [Z90]. On the other hand, the first proof of the PMP by means of viscosity solutions is due to Barron and Jensen [BJ86]. Other results on the connection between the PMP and DP can be found in [Fr89b, CF91, Z93, Te95] in the framework of viscosity solutions and in [ClaV87, VW90b, Mi92a] where different tools of nonsmooth analysis are used. Remark 3.49 suggests that the results of §3.4 are related to the classical method of characteristics (see §III.5.5). For some recent developments on generalizations of this method for HJB equations see [Mi90a, Mi93a, Sua91, Su95] and the references therein.

Our formulation of the monotone control problem in §4.1 is essentially the infinite horizon version of the evolution problem considered in [Bn85], see also [BJ80]. However, the proof of the comparison Theorem 4.6 is different from the one in [Bn85] since no use is made of smoothing techniques.

The optimal stopping problem in §4.2 can be seen as the prototype of optimal control problems whose value is characterized by a variational inequality. In the particular case where no control acts on the dynamics, this problem was analyzed by Dynamic Programming methods before the introduction of viscosity solutions (see, e.g., [Me80a, CDM81, Me82]). The synthesis procedure in this case,

Theorem 4.12, is based on the knowledge of the free boundary, a typical feature of variational and quasivariational inequalities. HJ variational inequalities also arise in some questions in singular control, see [FH94]. A general reference for this type of problems in the control of nondegenerate diffusion processes is the book [BL82].

The material in §4.3 is taken from [Ba85a, Ba85b]. The impulse control problem was considered also in a more general setting in [Mos76, Me80b]. Impulsive problems of different nature are considered in [BJM93] and [MR96a]. The problem of minimizing the maximum cost gives rise to a quasivariational inequality which was studied by the viscosity solution method in [BI89] and applied to the relaxation problem in [BJ95]. Let us also mention that some connections between the impulse and the monotone control problem with controls of bounded total variation are pointed out in [BJ80].

The optimal switching problem in §4.4 is perhaps the simplest problem in optimal control giving rise to a system of quasivariational inequalities. It was analyzed first by viscosity solutions in [CDE84], which we follow in our presentation; see also [E83].

The results of §4 were extended in various ways. We refer, for example, to [Y89] dealing with switching costs and usual as well as impulse controls, and [Le87] where the dynamics is piecewise deterministic; for the theory of systems of HJB equations see also [EL91, I92b].

Let us mention, finally, that many results of this chapter were extended to the optimal control of piecewise deterministic processes and/or jump-Markov processes: in these cases the first order HJB equation involves some integrodifferential terms. This was done in the above-cited [Le87] and also [Le89] (for impulse control), [S88] and [Say91] for the finite and the infinite horizon problem, respectively. A more recent paper on integrodifferential equations is [AlT96].

CHAPTER IV

Optimal control problems with continuous value functions: restricted state space

In this chapter we continue the study of optimal control problems with continuous value functions and consider cost functionals involving the exit time from a given domain, in particular time-optimal control, and infinite horizon problems with constraints on the state variables. The continuity of the value function for these problems is not as easy as in the previous chapter. For time-optimal control this is essentially the problem of small-time local controllability. We give the proof of just a few simple results on this topic, and state without proof several others. For each problem we characterize the value function as the unique viscosity solution of the appropriate Hamilton-Jacobi-Bellman equation and boundary conditions. We do not give all the applications of this theory, as verification functions and conditions of optimality: most of them can be obtained by the arguments of Chapter III and are left as exercises for the reader.

The system (S) and hypotheses (A_0)–(A_3) are the same as in Section 1 of Chapter III. We recall the basic estimates on the trajectories:

(0.1) $\quad |y_x(t,\alpha) - x| \leq M_x t \qquad$ for all $\alpha \in \mathcal{A}$ and $t \in [0, 1/M_x]$,

(0.2) $\quad |y_x(t,\alpha)| \leq (|x| + \sqrt{2Kt})e^{Kt} \qquad$ for all $\alpha \in \mathcal{A}$ and $t > 0$,

(0.3) $\quad |y_x(t,\alpha) - y_z(t,\alpha)| \leq e^{Lt}|x - z| \qquad$ for all $\alpha \in \mathcal{A}$ and $t > 0$.

1. Small-time controllability and minimal time functions

Throughout the first four sections of this chapter we are given a *target set* $\mathcal{T} \subset \mathbb{R}^N$ satisfying

$(A_\mathcal{T}) \qquad\qquad \mathcal{T}$ is closed with compact boundary $\partial\mathcal{T}$.

We study problems with initial state x in $\mathcal{T}^c := \mathbb{R}^N \setminus \mathcal{T}$, whose dynamics is stopped and the payoff computed when the system reaches either $\mathcal{T}$ or its interior int $\mathcal{T}$.

Therefore we define, for $\alpha \in \mathcal{A}$, the *exit time* from $\mathcal{T}^c$, or *entry time* in $\mathcal{T}$,

$$t_x(\alpha) := \begin{cases} +\infty & \text{if } \{t : y_x(t, \alpha) \in \mathcal{T}\} = \emptyset, \\ \min\{t : y_x(t, \alpha) \in \mathcal{T}\} & \text{otherwise,} \end{cases}$$

(where it is easy to show that the min exists), and the *exit time from* $\overline{\mathcal{T}^c}$, or *entry time in* int $\mathcal{T}$,

$$\widehat{t}_x(\alpha) := \begin{cases} +\infty & \text{if } \{t : y_x(t, \alpha) \in \text{int } \mathcal{T}\} = \emptyset, \\ \inf\{t : y_x(t, \alpha) \in \text{int } \mathcal{T}\} & \text{otherwise}. \end{cases}$$

In this section we discuss some continuity properties of the *minimal time function*

$$T(x) := \inf_{\alpha \in \mathcal{A}} t_x(\alpha), \qquad x \in \mathbb{R}^N,$$

and the *minimal interior-time function*

$$\widehat{T}(x) := \inf_{\alpha \in \mathcal{A}} \widehat{t}_x(\alpha), \qquad x \in \mathbb{R}^N.$$

Here we are extending the use of the term minimal time function to general targets. This name is commonly used in the literature for the case $\mathcal{T} = \{0\}$, and in this special case $\widehat{T} \equiv +\infty$, because the interior of the target, int $\mathcal{T}$, is empty. Note that the term minimal may seem inappropriate if a minimizing control does not exist. However, by standard results, a minimizing control for the first problem exists in $\mathcal{A}$ if $f(x, A)$ is convex, and it always exists among relaxed controls $\mathcal{A}^r$.

We introduce the notations

$$\mathcal{R}(t) := \{x \in \mathbb{R}^N : T(x) < t\}, \qquad t > 0,$$
$$\mathcal{R} := \bigcup_{t>0} \mathcal{R}(t) = \{x \in \mathbb{R}^N : T(x) < +\infty\},$$

and $\widehat{\mathcal{R}}(t)$ and $\widehat{\mathcal{R}}$ are defined by replacing T with $\widehat{T}$. The letter $\mathcal{R}$ stands for *reachable*: $\mathcal{R}(t)$ is the set of points reachable from the target in time less then t by the backward system $y' = -f(y, \alpha)$, as well as the set of starting points from which the system can reach the target in time less than t.

DEFINITION 1.1. The system (f, A) is *small-time controllable* on $\mathcal{T}$ (briefly STC$\mathcal{T}$) if $\mathcal{T} \subseteq \text{int } \mathcal{R}(t)$ for all $t > 0$. If $\mathcal{T} = \{0\}$ this property is called *small-time local controllability* (STLC). The system is *small-time controllable on* int $\mathcal{T}$, briefly STC$\overset{\circ}{\mathcal{T}}$, if $\mathcal{T} \subseteq \text{int } \widehat{\mathcal{R}}(t)$ for all $t > 0$. ◁

Note that STLC is equivalent to the continuity of T in 0 if $\mathcal{T} = \{0\}$, since $T(0) = 0$ by definition. The next proposition extends this observation to the general case. Throughout the section we use the notation

$$d(x) := \text{dist}(x, \mathcal{T}), \qquad x \in \mathbb{R}^N.$$

PROPOSITION 1.2. *Assume* (A_0), (A_1), (A_3), *and* (A_T). *Then the following statements are equivalent:*

(i) *the system is* STCT;

(ii) T *is continuous in* x *for all* $x \in \partial T$;

(iii) *there exists* $\delta > 0$ *and* $\omega_T : [0, \delta] \to [0, +\infty[$ *such that* $\lim_{s \to 0^+} \omega_T(s) = 0$ *and* $T(x) \leq \omega_T(d(x))$ *for all* $x \in B(T, \delta)$.[1]

PROOF. (i) $\Rightarrow$ (ii) For fixed $\varepsilon > 0$ and $x_0 \in \partial T$, since $T \subseteq \text{int}\, \mathcal{R}(\varepsilon)$ there exists δ such that $|x - x_0| < \delta$ implies $x \in \mathcal{R}(\varepsilon)$, that is, $T(x) < \varepsilon$. This gives the conclusion because $T(x) \geq 0$ and $T(x_0) = 0$.

(ii) $\Rightarrow$ (iii) Define $\omega(s) := \sup\{T(x) : x \text{ such that } d(x) \leq s\}$, for $s \geq 0$. We have to show that $\lim_{s \to 0^+} \omega(s) = 0$. We fix $\varepsilon > 0$ and for any $x_0 \in \partial T$ let $\delta(x_0) > 0$ be such that $T(x) < \varepsilon$ for all $x \in B(x_0, \delta(x_0))$. Since ∂T is compact we can select $x_i \in \partial T$, $i = 1, \ldots, n$, such that $\bigcup_{i=1}^n B(x_i, \frac{1}{2}\delta(x_i)) \supseteq \partial T$. Now set $\bar{s} := \min_{i=1,\ldots,n} \delta(x_i)/2$ and observe that $d(x) < \bar{s}$ implies $x \in X := \bigcup_{i=1}^n B(x_i, \delta(x_i))$. Therefore, by the construction of X, $T(x) < \varepsilon$ if $d(x) < \bar{s}$, and then $\omega(s) \leq \varepsilon$ for $s < \bar{s}$, which proves our claim.

(iii) $\Rightarrow$ (i) Fix $t > 0$ and let $s > 0$ be such that $\omega_T(s) < t$. Then $T(x) < t$ if $d(x) \leq \min\{s, \delta\}$, and so $\mathcal{R}(t) \supseteq B(T, \min\{s, \delta\})$, which immediately gives STCT. ◀

REMARK 1.3. Under the assumptions of Proposition 1.2, $T(x) > 0$ if and only if $x \notin T$. In fact, by (0.1), for any $\alpha \in \mathcal{A}$

$$d(x) \leq |y_x(t_x(\alpha), \alpha) - x| \leq M_x t_x(\alpha),$$

and $d(x) > 0$ for $x \notin T$ because T is closed. ◁

REMARK 1.4. Proposition 1.2, with STCT replaced by STC$\overset{\circ}{T}$, holds for $\widehat{T}$ under the additional assumption that T is the closure of its interior, $T = \overline{\text{int}\, T}$. In fact, the proofs that (iii) $\Rightarrow$ (i) $\Rightarrow$ (ii) keep holding essentially unchanged, and the proof that (ii) $\Rightarrow$ (iii) holds because any $x \in \partial T$ is the limit of a sequence in int T, thus the continuity of $\widehat{T}$ at such a point x implies $\widehat{T}(x) = 0$ because $\widehat{T} \equiv 0$ in int T.

Note that the condition $T = \overline{\text{int}\, T}$ is necessary for STC$\overset{\circ}{T}$. In fact if the distance c of x from $\overline{\text{int}\, T}$ is positive and less than 1, by estimate (0.1) $\hat{t}_x(\alpha) \geq c/M_x$ for all $\alpha \in \mathcal{A}$, and then $x \notin \widehat{\mathcal{R}}(t)$ for any $t \leq c/M_x$. Now it is easy to remove the restriction $c \leq 1$ and see that no set strictly larger than $\overline{\text{int}\, T}$ can be contained in $\widehat{\mathcal{R}}(t)$ for all $t > 0$. ◁

REMARK 1.5. It is clear that STC$\overset{\circ}{T}$ implies STCT. The reverse implication does not hold in general, even if $T = \overline{\text{int}\, T}$, as we see in the next example. In $\mathbb{R}^2$ take $T = \{(x_1, x_2) : |x_1| \leq |x_2| \text{ or } |x_1| \geq 1\}$ and the system $f((x_1, x_2), +) = (1, 0)$, $f((x_1, x_2), -) = (-1, 0)$, $A = \{+, -\}$ (we advise the reader to draw a picture). It is easy to compute explicitly T and $\widehat{T}$ and see that $\widehat{T}$ is discontinuous on

---
[1] $B(X, \delta) := \{x \in \mathbb{R}^N : \text{dist}(x, X) < \delta\}$.

$X := \{ (x_1, 0) : |x_1| < 1/2 \}$, in particular $\widehat{T}(0,0) = 1$ while $\lim_{\substack{x \to 0 \\ x \notin X}} \widehat{T} = 0$. Note also that T is lipschitzean and $\widehat{T} = T$ off X. In Chapter V we see that if $\partial \mathcal{T}$ is locally the graph of a Lipschitz function (a property failing only at the origin in the example), then STC$\overset{\circ}{\mathcal{T}}$ and STC$\mathcal{T}$ are equivalent. ◁

The next result lists some consequences of STC$\mathcal{T}$, which could be summarized by saying that T is a continuous function of $\mathbb{R}^N$ into $[0, +\infty]$.

PROPOSITION 1.6. *Assume* (A_0), (A_1), (A_3), $(A_\mathcal{T})$, *and* STC$\mathcal{T}$. *Then*

(i) $\mathcal{R}$ *is open;*

(ii) T *is continuous in* $\mathcal{R}$*;*

(iii) $\lim_{x \to x_0} T(x) = +\infty$ *for any* $x_0 \in \partial \mathcal{R}$.

PROOF. Fix $\varepsilon > 0$ and $\overline{x} \in \mathcal{R}$ and choose $\overline{\alpha} \in \mathcal{A}$ such that $\overline{t} := t_{\overline{x}}(\overline{\alpha}) \leq T(\overline{x}) + \varepsilon$. By (0.3), if $|x - \overline{x}| \leq \delta \, e^{-L\overline{t}}$, then

$$(1.1) \qquad |y_x(\overline{t}, \overline{\alpha}) - y_{\overline{x}}(\overline{t}, \overline{\alpha})| \leq |x - \overline{x}| e^{L\overline{t}} \leq \delta,$$

and thus $d(y_x(\overline{t}, \overline{\alpha})) \leq \delta$ because $y_{\overline{x}}(\overline{t}, \overline{\alpha}) \in \mathcal{T}$, where δ is as in property (iii) of Proposition 1.2. Then

$$(1.2) \qquad T(x) \leq \overline{t} + T(y_x(\overline{t})) \leq \overline{t} + \omega_T(d(y_x(\overline{t}))),$$

where the first inequality is a half Dynamic Programming Principle whose easy proof is left to the reader, and the second inequality follows from (iii) of Proposition 1.2. By (1.2), $x \in \mathcal{R}$, which proves (i). Moreover, the choice of $\overline{\alpha}$, (1.1), and (1.2) give

$$T(x) - T(\overline{x}) \leq \omega_T\bigl(|x - \overline{x}| e^{L(T(\overline{x}) + \varepsilon)}\bigr) + \varepsilon,$$

and by letting $\varepsilon \to 0^+$ and exchanging the roles of x and $\overline{x}$ we get

$$(1.3) \qquad |T(x) - T(\overline{x})| \leq \omega_T\bigl(|x - \overline{x}| e^{L(T(\overline{x}) \vee T(x))}\bigr),$$

which proves (ii).

To prove (iii) we fix $x_0 \in \partial \mathcal{R}$ and we suppose by contradiction there exist a constant M and a sequence $x_n \in \mathcal{R}$ such that $|x_n - x_0| \leq 1/n$ and $T(x_n) \leq M$ for all n. We choose $\overline{n}$ such that $e^{L(M+1)}/\overline{n} < \delta$ and $\overline{\alpha}$ such that $\overline{t} := t_{x_{\overline{n}}}(\overline{\alpha}) < T(x_{\overline{n}}) + 1 \leq M + 1$. Then, by (0.3),

$$|y_{x_0}(\overline{t}, \overline{\alpha}) - y_{x_{\overline{n}}}(\overline{t}, \overline{\alpha})| \leq |x_0 - x_{\overline{n}}| e^{L\overline{t}} < \delta,$$

which implies $d(y_{x_0}(\overline{t}, \overline{\alpha})) < \delta$ and thus $T(x_0) \leq \overline{t} + \omega_T(\delta)$ by Proposition 1.2 (iii). This gives $x_0 \in \mathcal{R}$, a contradiction to (i). ◀

REMARK 1.7. Inequality (1.3) allows us to estimate the modulus of continuity of T by the uniform modulus at the points of $\partial \mathcal{T}$, ω_T, on sets where T is bounded. For instance, $\omega_T(s) = C s^\gamma$ for some $C > 0$, $0 < \gamma \leq 1$, implies the Hölder continuity of exponent γ (Lipschitz continuity if $\gamma = 1$) of T on compact subsets of $\mathcal{R}$. ◁

REMARK 1.8. Proposition 1.6 holds for $\widehat{T}$ and $\widehat{\mathcal{R}}$ if STC$\mathcal{T}$ is replaced by STC$\overset{\circ}{\mathcal{T}}$ and $\mathcal{T} = \overline{\text{int}\,\mathcal{T}}$. The proof needs just a bit more care because $\widehat{t}_x$ is defined as an inf instead of a min. ◁

Next we give a brief account of the theory of controllability. First note that it is very easy to give examples where STC$\mathcal{T}$ fails for any compact $\mathcal{T}$ because the system cannot go to sufficiently many directions (take, for instance, A a singleton and f constant ...). In Chapter V we study the time-optimal control of non-controllable systems.

We begin by listing without proof some controllability conditions in the case $\mathcal{T} = \{0\}$, which is the most deeply studied in the literature. We give in each case the corresponding regularity of the minimal time function.

THEOREM 1.9. *Assume* $\mathcal{T} = \{0\}$ *and the system is linear, i.e.,*

$$f(y, a) = My + Qa, \qquad A = [-1, 1]^m,$$

where M is an $N \times N$ matrix and Q an $N \times m$ matrix. Then STC$\mathcal{T}$ holds if and only if the $N \times (Nm)$ controllability matrix

$$[Q \quad MQ \quad M^2Q \quad \ldots \quad M^{N-1}Q]$$

has full rank N. Moreover, T is locally Hölder continuous in $\mathcal{R}$ of exponent $1/(j+1)$, where j is the least integer such that

$$\operatorname{rank}[Q \quad MQ \quad \ldots \quad M^jQ] = N.$$

THEOREM 1.10. *Assume* $\mathcal{T} = \{0\}$ *and consider the linear system $f(y, a) = My + a$. Assume $A \subseteq \mathbb{R}^N$ is a compact convex neighborhood of the origin, and ∂A is a differentiable hypersurface (i.e., for each point of ∂A there is a unique supporting hyperplane for A). Then $T \in C^1(\mathcal{R} \smallsetminus \{0\})$.*

THEOREM 1.11. *Assume* $\mathcal{T} = \{0\}$, $A = [-1, 1]^m$, $f \in C^1$, *and* $f(0, 0) = 0$. *If the linearized system $My + Qa$, where*

$$M_{ij} = \frac{\partial f_i}{\partial y_j}(0, 0), \qquad Q_{ik} = \frac{\partial f_i}{\partial a_k}(0, 0),$$

is STC$\mathcal{T}$, then f is STC$\mathcal{T}$. If, moreover, $f \in C^\infty$, then T is locally Hölder continuous in $\mathcal{R}$ of exponent $1/(2j + 1 + \varepsilon)$ for any $\varepsilon > 0$, where j is defined as in Theorem 1.9.

THEOREM 1.12. *Assume* (A_0)–(A_3) *and* $\mathcal{T} = \{0\}$. *If*

(1.4) $$\inf_{a \in A} f(0, a) \cdot \gamma < 0 \quad \text{for any unit vector } \gamma,$$

then STC$\mathcal{T}$ holds. Moreover, (1.4) is necessary and sufficient for the local Lipschitz continuity of T in $\mathcal{R}$.

The proof of the sufficiency of (1.4) for STC$\mathcal{T}$ and the Lipschitz continuity of T are postponed to Section 4.

PROOF OF THE NECESSITY OF (1.4) FOR THE LIPSCHITZ CONTINUITY OF T.
We assume by contradiction that there exists γ, $|\gamma| = 1$, such that

(1.5) $$f(0,a) \cdot \gamma \geq 0 \qquad \text{for all } a \in \mathcal{A}.$$

Let $M := \sup\{|f(z,a)| : |z| \leq 2\}$, and note that for any $|x| \leq 1$, $\alpha \in \mathcal{A}$, and $t \leq 1/M$,

$$|y_x(t)| \leq 2,$$
$$|y_x(t)| \leq \left| x + \int_0^t f(x,\alpha(s))\,ds \right| + t\omega_f(Mt, 2),$$

where we have used (A$_2$) and (0.1). Then, by (A$_2$) again, for any $\alpha \in \mathcal{A}$,

(1.6) $$|f(y_x(t),\alpha) - f(0,\alpha)| \leq \omega_f(|x| + Mt + o(t), 2) \qquad \text{as } t \to 0^+.$$

Now pick $\overline{\alpha}$ such that $t_x(\overline{\alpha}) \leq T(x) + |x|$; then

(1.7) $$x = -\int_0^{t_x(\overline{\alpha})} f(y_x(t), \overline{\alpha}(t))\,dt,$$

and, by the Lipschitz continuity of T,

(1.8) $$t_x(\overline{\alpha}) \leq (c+1)|x|.$$

Plug (1.6) into (1.7) and use (1.8) to get

(1.9) $$x = -\int_0^{t_x(\overline{\alpha})} f(0, \overline{\alpha}(t))\,dt + t_x(\overline{\alpha})\,o(1) \qquad \text{as } |x| \to 0.$$

Next we choose $x = |x|\gamma$, take the scalar product of both sides of (1.9) with γ and use (1.5) and (1.8) to get

$$|x| \leq (c+1)|x|\,o(1) \qquad \text{as } |x| \to 0,$$

which gives the desired contradiction. ◂

REMARK 1.13. Condition (1.4) is called the *positive basis condition* (see Exercise 1.1 for the motivation of the name) and it is equivalent to the existence of $r > 0$ such that $\overline{\text{co}}\,f(0, A) \supseteq B(0, r)$. ◁

PROPOSITION 1.14. *Assume* (A$_0$)–(A$_3$) *and* $\mathcal{T} = \{0\}$. *Then a necessary condition for STC$\mathcal{T}$ is*

(1.10) $$\inf_{a \in A} f(0,a) \cdot \gamma \leq 0 \quad \text{for any unit vector } \gamma.$$

PROOF. The argument is essentially the same as in the previous proof. By contradiction we assume the existence of γ and $\varepsilon > 0$ such that

(1.11) $$f(0, a) \cdot \gamma \geq \varepsilon \quad \text{for all } a \in A.$$

By assumption T is continuous, thus $T(x) \leq \omega_T(|x|)$ by Proposition 1.2, and we can replace (1.8) by

$$t_x(\overline{\alpha}) = o(1) \quad \text{as } |x| \to 0.$$

This is enough to get (1.9) from (1.6) and (1.7). By using (1.11) in (1.9) we get

$$|x| \leq t_x(\overline{\alpha})(-\varepsilon + o(1)) \quad \text{as } |x| \to 0$$

and the right-hand side can be made negative for $x \neq 0$ small enough, a contradiction. ◄

Before stating next result we recall the definition of the Lie bracket, or Jacobi bracket, of two vector fields $g_1, g_2 \in C^1$:

$$[g_1, g_2](x) = Dg_2(x)g_1(x) - Dg_1(x)g_2(x),$$

where $Dg_i(x)$ is the Jacobian matrix of g_i at the point x. We say (by induction) that a vector field F is generated by the vector fields Q_i, $i = 1, \ldots, m$, by 0 bracket operations if it is a linear combination of the Q_i, and by k bracket operations if F is a linear combination of brackets $[G_1, G_2]$ with G_1, G_2 generated by $k - 1$ bracket operations.

THEOREM 1.15. *Assume* $\mathcal{T} = \{0\}$ *and the system is symmetric, i.e.,*

$$f(y, a) = Q(y)a = \sum_{i=1}^n a_i Q_i(y), \quad A = [-1, 1]^m,$$

where Q is an $N \times m$ matrix valued C^∞ function and the vector fields Q_i are its columns. If there exist N vector fields $F_1, \ldots, F_N$ generated by Q_i, $i = 1, \ldots, m$, by bracket operations and such that

$$\text{span}\{F_1(0), \ldots, F_N(0)\} = \mathbb{R}^N,$$

then STC$\mathcal{T}$ holds. Moreover, T is locally Hölder continuous with exponent $1/(j+1)$ where j is the minimum number of bracket operations necessary to generate all of the fields $F_1, \ldots, F_N$.

Now we prove some results for the case of a "smooth target". The smoothness of $\mathcal{T}$ we need is expressed in terms of the distance function from $\mathcal{T}$, $d(\,\cdot\,)$, as follows:

(1.12) $\quad d(\,\cdot\,)$ can be redefined in int $\mathcal{T}$ as a C^1 function on $B(\partial \mathcal{T}, c)$ for some $c > 0$.

This holds for instance if $T = \overline{\text{int}\, T}$ (which is a necessary condition for (1.12)) and ∂T is a C^2 manifold: it is enough to redefine $d(x)$ in int T as $-\,\text{dist}(x, \partial T)$ (this new function $d(\,\cdot\,)$ is the signed distance from ∂T). See also Remarks 1.20 and 1.21 for some generalizations. Note that under assumption (1.12) there exists an exterior normal $n(x) = Dd(x)$ for all $x \in \partial T$.

The first sufficient condition for controllability we consider is a very natural and simple one, though rather restrictive:

(1.13) $$\inf_{a \in A} f(x,a) \cdot n(x) < 0 \qquad \text{for all } x \in \partial T,$$

that is, at each point of the boundary of T the system can choose a direction pointing to the interior of T. Note the similarity between (1.13) and (1.4).

THEOREM 1.16. *Assume* (A_0)–(A_3), (A_T), *and* (1.12). *Then* (1.13) *implies* STCT *and* STC$\overset{\circ}{T}$. *Moreover*, (1.13) *is necessary and sufficient for the local Lipschitz continuity of* T *in* $\mathcal{R}$.

PROOF. We first show that there exist constants $C, \delta > 0$ such that

(1.14) $$T(x) \leq \widehat{T}(x) \leq C\, d(x) \qquad \text{for all } x \in B(T, \delta) \smallsetminus \text{int}\, T .$$

This implies STCT, STC$\overset{\circ}{T}$, and the local Lipschitz continuity of T by Proposition 1.2 (with $\omega_T(s) = Cs$) and Remarks 1.4, 1.7 and 1.8.

We fix $x_0 \in \partial T$ and use (1.12) and (1.13) to choose $a_0 \in A$, $\xi_0 > 0$, $0 < r_0 \leq C$ such that

(1.15) $$f(x, a_0) \cdot Dd(x) < -\xi_0 \qquad \text{for all } x \in B(x_0, r_0) .$$

Now let $\alpha_0(s) = a_0$ for all s and $y(s) := y_x(s, \alpha_0)$. In the following we restrict s to be less than 1, so $y(s)$ remains in a bounded set independent of the choice of x in $B(\partial T, C)$, by (0.2) and (A_T). Thus we have

(1.16) $$|y(s) - x| \leq Ms \qquad \text{for all } x \in B(\partial T, C),\text{ for } 0 \leq s \leq 1 .$$

By the Taylor expansion for $d(\,\cdot\,)$ and (1.16) we get

(1.17) $$\begin{aligned} d(y(s)) &= d(x) + \int_0^s f(y(s), a_0)\, ds \cdot Dd(x) + o(Ms) \\ &= d(x) + s\, f(x, a_0) \cdot Dd(x) + o(c_1 s) \\ &< d(x) - \frac{1}{2}\xi_0 s \qquad \text{for } 0 < s \leq \bar{s}, \end{aligned}$$

where in the second equality we have used (A_2) and c_1 is independent of x, while in the inequality we have used (1.15) and $0 < \bar{s} \leq 1$ is independent of x. Now we set $s_x := 2d(x)/\xi_0$, we note that $s_x \leq \bar{s}$ if $d(x) \leq \bar{s}\xi_0/2$, and then apply (1.17) to get $d(y(s_x)) < 0$; this implies that $y(s_x) \in \text{int}\, T$, thus

$$T(x) \leq \widehat{T}(x) < s_x = C\, d(x), \qquad \text{for all } x \in B(x_0, R_0) \smallsetminus \text{int}\, T,$$

where $C = 2/\xi_0$ and $R_0 = \min\{r_0, \bar{s}\xi_0/2\}$. The compactness of $\partial\mathcal{T}$ allows us to complete the proof of (1.14) by choosing a finite number of balls $B(x_i, R_i/2)$ covering $\partial\mathcal{T}$ and setting $\delta = \min_{i=0,\ldots,n} R_i/2$, as in the proof of Proposition 1.2.

The proof that (1.13) is necessary for the Lipschitz continuity is similar to the corresponding proof in Theorem 1.12. We assume by contradiction there exists x_0 where (1.13) is violated and we choose new coordinates such that $x_0 = 0$, so we have (1.5) with $\gamma = n(x_0)$. As before we seek a contradiction by looking at a sequence of points $x_n = \gamma/n$. Let α_n be such that $t_{x_n}(\alpha_n) \leq T(x_n) + |x_n|$ so that

$$(1.18) \qquad x_n + \int_0^{t_{x_n}(\alpha_n)} f(y_{x_n}(t), \alpha_n(t))\, dt =: y_n \in \partial\mathcal{T} .$$

We use (1.6) and (1.8), take the scalar product with γ and use (1.5) to get

$$(1.19) \qquad \frac{1}{n} = x_n \cdot \gamma \leq y_n \cdot \gamma + o(1/n) .$$

Now we estimate

$$|y_n| \leq |y_n - x_n| + |x_n| \leq M t_{x_n}(\alpha_n) + \frac{1}{n} \leq (MC + M + 1)\frac{1}{n},$$

where M is a local bound on $|f|$ and C a local Lipschitz constant for T. Then $y_n \cdot \gamma = o(1/n)$ because $y_n \in \partial\mathcal{T}$ and $\gamma = n(0)$ is orthogonal to $\partial\mathcal{T}$ at 0. Thus (1.19) gives a contradiction. ◀

The next result is the counterpart of Proposition 1.14 for smooth targets. Its proof is very similar and it is left to the reader.

PROPOSITION 1.17. *Assume* (A_0)–(A_3), $(A_\mathcal{T})$, *and* (1.12). *Then a necessary condition for* STC$\mathcal{T}$ *is*

$$(1.20) \qquad \inf_{a \in A} f(x, a) \cdot n(x) \leq 0 \quad \text{for all } x \in \partial\mathcal{T} .$$

Now we give another sufficient condition for controllability which is less restrictive than (1.13). This is similar to the hypotheses of Theorem 1.15, in particular we assume the system is symmetric.

THEOREM 1.18. *Assume* (A_0)–(A_3), $(A_\mathcal{T})$, *and* (1.12). *Assume that, in a neighborhood of any point* $x \in \partial\mathcal{T}$ *where* (1.13) *fails, the system is symmetric and there is a vector field* F *generated by* Q_i, $i = 1, \ldots, m$, *by bracket operations such that*

$$(1.21) \qquad F(x) \cdot n(x) \neq 0 .$$

Then STC$\mathcal{T}$ *and* STC$\overset{\circ}{\mathcal{T}}$ *hold. If, moreover, the number of bracket operations necessary to generate* F *can be bounded by* j *for all* $x \in \partial\mathcal{T}$, *then* T *is locally Hölder continuous in* $\mathcal{R}$ *with exponent* $1/(j+1)$.

For the proof we need the following result of geometric control theory.

LEMMA 1.19. *If the system is symmetric and F is generated by j bracket operations by Q_i, $i = 1, \ldots, m$, then there is $C > 0$ and, for each $s > 0$, a piecewise constant control $\alpha \in \mathcal{P}$ such that*

(1.22) $$y_x(s, \alpha) = x + C F(x) s^{j+1} + o(s^{j+1}) \qquad \text{as } s \to 0^+ .$$

PROOF OF THE LEMMA IN THE CASE $j = 1$. It is not restrictive to assume $F = [Q_i, Q_k]$. Denote $a^{(i)} \in A$ the vector whose components are all 0 except the i-th. Then set

$$\alpha(t) = \begin{cases} a^{(i)} & 0 \leq t \leq s/4, \\ a^{(k)} & s/4 < t \leq s/2, \\ -a^{(i)} & s/2 < t \leq 3s/4, \\ -a^{(k)} & 3s/4 < t \leq s . \end{cases}$$

A straightforward but tedious computation using the Taylor expansions of the trajectory gives

$$y_x(s, \alpha) = x + \frac{s^2}{16} [Q_i, Q_k](x) + o(s^2) \qquad \text{as } s \to 0^+ . \blacktriangleleft$$

PROOF OF THEOREM 1.18. With the lemma available the proof is very similar to that of Theorem 1.16. Given $x_0 \in \partial \mathcal{T}$, by (1.21) and the symmetry of the system, we can find $\xi_0 > 0$, $0 < r_0 \leq c$ such that

$$F(x) \cdot Dd(x) < -\xi_0 \qquad \text{for all } x \in B(x_0, r_0) .$$

Then we combine the Taylor expansion of $d(\cdot)$ with (1.22) to get

$$d(y_x(s, a)) = d(x) + C s^{j+1} F(x) \cdot Dd(x) + o(s^{j+1})$$
$$< d(x) - \frac{C\xi_0}{2} s^{j+1} \qquad \text{for } 0 < s \leq \bar{s},$$

which is the counterpart of (1.17). Now we choose

$$s_x := \left(\frac{2d(x)}{C\xi_0} \right)^{\frac{1}{j+1}}$$

and get

$$T(x) \leq \widehat{T}(x) < s_x = C' d^{1/(j+1)}(x) .$$

The conclusion is now achieved as in Theorem 1.16; we leave the details to the reader. $\blacktriangleleft$

REMARK 1.20. Theorems 1.16 and 1.18 can be extended to targets with piecewise smooth boundary in the following sense: $\mathcal{T} = \{x : g_i(x) \leq 0 \text{ for all } i = 1, \ldots, M\}$, where $g_i \in C^1(\mathbb{R}^N)$ and $|Dg_i(x)| > 0$ for all x such that $g_i(x) = 0$. It is enough to modify (1.13) and (1.21) by replacing $n(x)$ with $Dg_i(x)$ for all the "active constraints", that is, for all i such that $g_i(x) = 0$. The proof is essentially the same with d replaced by g_i for all i active at the given point $x_0 \in \partial \mathcal{T}$. In Section 4 we extend Theorem 1.16 to targets whose boundary has no smoothness properties. $\triangleleft$

1. SMALL-TIME CONTROLLABILITY

REMARK 1.21. Most of the preceding controllability results hold if measurable controls are replaced by piecewise constant controls and $T, \widehat{T}$ are replaced by $T^\sharp := \inf_{\alpha \in \mathcal{P}} t_x(\alpha)$, $\widehat{T}^\sharp := \inf_{\alpha \in \mathcal{P}} \widehat{t}_x(\alpha)$. This is the case, for instance, of Theorems 1.9, 1.12, 1.15, 1.16, and 1.18. ◁

The most restrictive assumption of this section is (A_3). In fact the minimum time problem makes perfect sense without knowing a priori that the trajectories of the system exist for a given time, so (A_3) can be replaced by

$$(1.23) \qquad (f(x,a) - f(y,a)) \cdot (x - y) \leq L_R |x - y|^2$$

for all $a \in A$, $|x|, |y| \leq R$ (which is satisfied, for instance, by any f of class C^1 if A is compact) because (1.23) is enough to ensure local existence and uniqueness of the trajectories (see Remark 5.7 in the appendix of Chapter III). Most of the results of this section are of local nature and keep holding under this assumption. However, in this case STC$\mathcal{T}$ does not imply the continuity of T in all of $\mathcal{R}$ (cfr. Proposition 1.6 (ii)) as shown by the example in Exercise 1.3. Nonetheless the continuity of T propagates in the so-called *escape controllability domain* which we define next. We restrict ourselves for simplicity to the case

$$(1.24) \qquad f \in C^1(\mathbb{R}^N \times \mathbb{R}^M), \qquad A \subseteq \mathbb{R}^M \text{ compact}, \qquad \mathcal{T} = \{0\} \,.$$

Let

$$\mathcal{R}(\leq t) := \{\, x \in \mathbb{R}^N : t_x(\alpha) \leq t \text{ for some } \alpha \in \mathcal{A} \,\} \,.$$

DEFINITION 1.22. The *escape time* is

$$T_0 := \sup\{\, t : \mathcal{R}(\leq t) \text{ is bounded}\, \} \,.$$

The escape controllability domain is

$$\mathcal{R}_0 := \mathcal{R}(T_0) = \{\, x : T(x) < T_0 \,\} \,.$$

It is easy to prove that

$$T_0 = \begin{cases} +\infty & \text{if } \mathcal{R} \text{ is bounded,} \\ \liminf_{|x| \to \infty} T(x) & \text{otherwise}\,. \end{cases} \quad \triangleleft$$

The next result, whose proof we omit, is an analogue of Proposition 1.6.

THEOREM 1.23. *Assume* (1.24) *and* STC$\mathcal{T}$. *Then*

(i) $\mathcal{R}_0$ *is open;*

(ii) T *is continuous in* $\mathcal{R}_0$;

(iii) $\lim_{\mathcal{R}_0 \ni x \to x_0} T(x) = T_0$ *for any* $x_0 \in \partial \mathcal{R}_0$;

(iv) $\lim_{\substack{|x| \to \infty \\ x \in \mathcal{R}_0}} T(x) = T_0$.

See Exercises 1.3, 1.4, 1.5, and Example 2.10 for some explicit examples of $\mathcal{R}_0$.

Exercises

1.1. A set of vectors $q_i \in \mathbb{R}^N$, $i = 1, \ldots, m$, is called a *positive basis* if for any $p \in \mathbb{R}^N$ there are $\lambda_i \geq 0$, $i = 1, \ldots, m$, such that $p = \sum_{i=1}^m \lambda_i q_i$. Prove that condition (1.4) holds if and only if $f(0, A)$ contains a positive basis. Prove that another equivalent condition is $\overline{co} f(0, A) \supseteq B(0, r)$ for some $r > 0$. (See [Pe68].)

1.2. Assume (A_0)–(A_3), $\mathcal{T}$ closed and unbounded, (1.12) and (1.13). Prove that T and $\widehat{T}$ are locally Lipschitz continuous, $\mathcal{R}$ is open, and $T(x) \to +\infty$ as $x \to x_0 \in \partial \mathcal{R}$.

1.3. Consider the system in $\mathbb{R}^2$

$$y' = \sum_{i=1}^{5} a_i f_i(y), \qquad a \in A,$$

where

$$f_1(y) = (-y_2(1 - y_1^2), y_1(1 + y_2^2)),$$
$$f_2(y) = (-1, 0),$$
$$f_3(y) = (-(y_2 - 8)(y_1 + 1/2), 0),$$
$$f_4(y) = (0, 1),$$
$$f_5(y) = (0, -1),$$

and A is the convex closure of the canonical basis of $\mathbb{R}^5$. Take $\mathcal{T} = \{0\}$. Prove that

(i) the trajectories corresponding to the control $(1, 0, 0, 0, 0)$ contained in the strip $S = \{(x_1, x_2) : |x_1| < 1\}$ are all the ellipses $x_1^2 + c x_2^2 = 1 - c$ for $c \in \,]0, 1[$, and the time taken to turn around each ellipse is 2π;

(ii) $S \subseteq \mathcal{R}$ $(\leq 2\pi + 1)$ (then the escape time $T_0 \leq 2\pi + 1$);

(iii) if $y_{(-1,0)}(t, \alpha) \in S^- := \{(x_1, x_2) : x_2 \leq 8\}$ for $t \leq t_0$, then $y_{(0,1)}(t, \alpha) \notin S$ for $t \leq t_0$;

(iv) if $y_{(-1,0)}(t, \alpha) \in S^+ := \{(x_1, x_2) : x_2 \geq 8\}$, then $t \geq 8$;

(v) $t_{(-1,0)}(\alpha) \geq 8$ for all $\alpha \in \mathcal{A}$;

(vi) $\lim_{S \ni x \to (-1,0)} T(x) \leq 2\pi + 1 < 8 \leq T(-1, 0)$ (then T is discontinuous at $(-1, 0)$).

1.4. Assume (A_0), (A_3), $\mathcal{T} = \{0\}$ and

$$|f(x, a)| \leq k(1 + |x|) \qquad \text{for all } x \in \mathbb{R}^N, a \in A.$$

Prove that $\lim_{|x| \to \infty} T(x) = +\infty$ and then the escape time $T_0 = +\infty$ and $\mathcal{R}_0 = \mathcal{R}$.

1.5. Let $N = 1$, $h(x) \geq 0$ be a smooth function such that $h(x) \equiv 0$ for $x \leq 0$ and $h(x) \equiv 1$ for $x \geq 1$. Consider the system $f(x, +) = -x^2 h(x)$, $f(x, -) = 1$, $A = \{+, -\}$, and the target $\mathcal{T} = \{1\}$. Prove that

(i) $T(x) = (x-1)/x$ for $x \geq 1$, $T(x) = 1 - x$ for $x \leq 1$, so that $T_0 = 1$, $\mathcal{R}_0 =]0, +\infty[$, $\mathcal{R} = \mathbb{R}$;

(ii) the system satisfies (A_0)–(A_3).

2. HJB equations and boundary value problems for the minimal time function: basic theory

In this section we formulate the Dynamic Programming Principle for the minimum time problem and derive a boundary value problem for the Hamilton-Jacobi-Bellman equation associated with the minimal time function

$$T(x) := \inf_{\alpha \in \mathcal{A}} t_x(\alpha),$$

where $t_x(\alpha)$ is the first time the system hits a given closed target $\mathcal{T}$. This boundary value problem has infinite data on $\partial \mathcal{R}$, where $\mathcal{R}$ is the set of starting points from which the system can reach the target. We also introduce also a rescaling of T which solves another boundary value problem where the (unknown) set $\mathcal{R}$ does not appear.

We use the notation

$$\chi_X(x) = \begin{cases} 1 & \text{if } x \in X, \\ 0 & \text{if } x \notin X. \end{cases}$$

PROPOSITION 2.1 (DYNAMIC PROGRAMMING PRINCIPLE). *Assume* (A_0), (A_1), (A_3). *Then, for all* $x \in \mathcal{R}$,

$$(2.1) \quad T(x) = \inf_{\alpha \in \mathcal{A}} \left\{ t \wedge t_x(\alpha) + \chi_{\{t \leq t_x(\alpha)\}} T(y_x(t, \alpha)) \right\}, \quad \text{for all } t \geq 0$$

and

$$(2.2) \quad T(x) = \inf_{\alpha \in \mathcal{A}} \left\{ t + T(y_x(t, \alpha)) \right\}, \quad \text{for all } t \in [0, T(x)].$$

PROOF. Let B and C denote, respectively, the right-hand sides of (2.1) and (2.2). Note that (2.1) reduces to the definition of $T(x)$ for $t > T(x)$. For $t \leq T(x)$ we have $B \leq C$ because the quantity in brackets in (2.1) is less than or equal to the quantity in brackets in (2.2). To show that $T(x) \geq C$ we observe that for all $\alpha \in \mathcal{A}$ and $t \leq T(x) \leq t_x(\alpha)$, by setting $\widetilde{\alpha}(s) = \alpha(t + s)$, we have

$$t_x(\alpha) = t + t_{y_x(t,\alpha)}(\widetilde{\alpha}) \geq t + T(y_x(t, \alpha)) \geq C,$$

and we reach the conclusion by taking the inf over $\mathcal{A}$.

To show that $T(x) \leq B$, we fix $\alpha \in \mathcal{A}$, $t \leq T(x)$, set $z := y_x(t, \alpha)$, and fix $\varepsilon > 0$ and $\alpha^1 \in \mathcal{A}$ such that $T(z) \geq t_z(\alpha^1) - \varepsilon$. We define

$$\overline{\alpha} = \begin{cases} \alpha(s) & s \leq t, \\ \alpha^1(s - t) & s > t. \end{cases}$$

Since $t \leq t_x(\overline{\alpha})$ we have

$$T(x) \leq t_x(\overline{\alpha}) = t + t_{y_x(t,\alpha)}(\alpha^1) \leq t + T(y_x(t,\alpha)) + \varepsilon$$
$$= t \wedge t_x(\alpha) + \chi_{\{t \leq t_x(\alpha)\}} T(y_x(y,\alpha)) + \varepsilon$$

because $t \leq t_x(\alpha)$. We conclude the proof by taking the inf over $\mathcal{A}$ and then letting ε go to 0. ◂

REMARK 2.2. The Dynamic Programming Principle for the minimal interior time function $\widehat{T}$ corresponding to the target int $\mathcal{T}$ is given by the same formulae (2.1) and (2.2), where T is replaced by $\widehat{T}$ and t_x by $\widehat{t}_x$. The proof is exactly the same provided one checks that for $t \leq \widehat{t}_x(\alpha)$

$$\widehat{t}_x(\alpha) = t + \widehat{t}_{y_x(t,\alpha)}(\alpha(\cdot + t)) \,.$$

We also note that in (2.1) we can replace $\chi_{\{t \leq t_x(\alpha)\}}$ with $\chi_{\{t < t_x(\alpha)\}}$ because for $t = t_x(\alpha)$ $y_x(t,\alpha) \in \mathcal{T}$ and $T \equiv 0$ on $\mathcal{T}$. The same statement holds for $\widehat{T}$ and $\widehat{t}_x$, even if $\widehat{T}$ may be positive on $\partial \mathcal{T}$, because $\widehat{T}(y_x(\widehat{t}(\alpha),\alpha)) = 0$ for all x and α, as the reader can easily check. ◁

PROPOSITION 2.3. *Assume* (A_0)–(A_3), $(A_\mathcal{T})$, $\mathcal{R} \setminus \mathcal{T}$ *open, and* $T \in C(\mathcal{R} \setminus \mathcal{T})$ (*e.g.*, STC$\mathcal{T}$ *holds*). *Then* T *is a viscosity solution of*

(2.3) $$\sup_{a \in A} \{-f(x,a) \cdot DT\} - 1 = 0 \qquad \text{in } \mathcal{R} \setminus \mathcal{T} \,.$$

PROOF. Let $\varphi \in C^1(\mathcal{R} \setminus \mathcal{T})$ and let x be a local maximum point of $T - \varphi$. Fix $a \in A$ and let $y_x(t)$ be the trajectory corresponding to the constant control function $\alpha(t) \equiv a$. Then for $t > 0$ small enough

$$\varphi(x) - \varphi(y_x(t)) \leq T(x) - T(y_x(t)),$$

and by the DPP (2.2) (note that $T(x) > 0$ by Remark 1.3)

$$\frac{\varphi(x) - \varphi(y_x(t))}{t} \leq 1 \,.$$

By letting $t \to 0^+$ we get

$$-D\varphi(x) \cdot f(x,a) \leq 1 \,.$$

We can take now the sup over A in the left-hand side, by the arbitrariness of a, and conclude that T is a subsolution of (2.3).

Next assume x is a local minimum point of $T - \varphi$. By (0.1) there is t_0 such that

$$\varphi(x) - \varphi(y_x(t,\alpha)) \geq T(x) - T(y_x(t,\alpha)), \qquad \text{for all } t \in [0,t_0], \alpha \in \mathcal{A} \,.$$

For each $\varepsilon > 0$ and $t \in \,]0, T(x)]$, by the DPP (2.2), there is $\overline{\alpha} \in \mathcal{A}$ and corresponding trajectory $\overline{y}$ such that

$$T(x) \geq t + T(\overline{y}_x(t)) - \varepsilon t,$$

and then
$$\frac{\varphi(x) - \varphi(\overline{y}_x(t))}{t} \geq 1 - \varepsilon.$$

As in the proof of Proposition III.2.8, (A_2) and (0.1) imply
$$\varphi(x) - \varphi(\overline{y}_x(t)) = -\int_0^t D\varphi(x) \cdot f(x, \overline{\alpha}(s))\, ds + o(t), \qquad \text{as } t \to 0^+,$$

and thus
$$\sup_{a \in A} \{ -D\varphi(x) \cdot f(x, a) \} \geq -\frac{1}{t} \int_0^t D\varphi(x) \cdot f(x, \overline{\alpha}(s))\, ds$$
$$\geq 1 - \varepsilon + o(1) \qquad \text{as } t \to 0^+.$$

After letting t and ε go to 0 we obtain that T is a viscosity supersolution of (2.3). ◄

REMARK 2.4. Under the assumptions (A_0)–(A_3), (A_T), $\widehat{\mathcal{R}} \setminus \mathcal{T}$ open and $\widehat{T} \in C(\widehat{\mathcal{R}} \setminus \mathcal{T})$, we have that $\widehat{T}$ is a viscosity solution of (2.3) in $\widehat{\mathcal{R}} \setminus \mathcal{T}$, and the proof is the same of Proposition 2.3 by means of Remark 2.2. ◁

EXAMPLE 2.5. If $f(x, a) = a$, A is the closed unit ball, then it is easy to prove that $T(x) = \text{dist}(x, \mathcal{T})$ and in this case the HJB equation (2.3) becomes the eikonal equation $|DT| - 1 = 0$ in $\mathcal{T}^c$ (cfr. Corollary II.2.16). ◁

Now we want to characterize T as the unique solution of the HJB equation (2.3) with suitable boundary conditions, namely,

(2.4)
$$T = 0 \qquad \text{on } \partial \mathcal{T},$$
$$T(x) \longrightarrow +\infty \qquad \text{as } x \to x_0 \in \partial \mathcal{R}.$$

Let us first give a simple result where the set $\mathcal{R}$ is supposed to be known. This is reasonable because the determination of $\mathcal{R}$ is a controllability problem: one has to find all the points from which there is a trajectory reaching $\mathcal{T}$. This can be considered as a preliminary investigation before looking for time-optimal trajectories. However, this is not completely satisfactory because $\mathcal{R}$ is not a datum of the problem. Later in this section we make a sort of rescaling of the cost functional leading to another boundary value problem whose solution simultaneously gives both T and $\mathcal{R}$. In Section 4 we also show that $\partial \mathcal{R}$ is a free boundary for the problem (2.3)–(2.4).

THEOREM 2.6. *Assume (A_0)–(A_3), (A_T), and STCT. Then T is the unique viscosity solution of (2.3) continuous in $\mathcal{R}$, bounded below and satisfying (2.4).*

PROOF. Let u_1 and u_2 be, respectively, a subsolution and a supersolution of (2.3). Define

(2.5)
$$v_i(x) = 1 - e^{-u_i(x)}, \qquad i = 1, 2.$$

By Proposition II.2.5 v_1 and v_2 are, respectively, a sub- and a supersolution of

(2.6) $$v + \sup_{a \in A}\{-f(x,a) \cdot Dv\} - 1 = 0 \quad \text{in } \mathcal{R} \setminus \mathcal{T}.$$

Moreover, v_i are bounded if u_i are bounded below and, by the boundary condition (2.4) on $\partial \mathcal{R}$, v_i can be extended in a unique way to $v_i \in BC(\overline{\mathcal{R} \setminus \mathcal{T}})$ satisfying the boundary conditions

(2.7) $$\begin{aligned} v_i &= 0 \quad \text{on } \partial \mathcal{T}, \\ v_i &= 1 \quad \text{on } \partial \mathcal{R}. \end{aligned}$$

Then we can apply the comparison results of Chapter III, more precisely Remark III.2.14, to get that $v_1 \le v_2$ and therefore $u_1 \le u_2$. We obtain the conclusion by exchanging the roles of u_1 and u_2. ◄

EXAMPLE 2.7. In this example we apply Theorem 2.6 to compute the minimal time function for the rocket railroad car. The system is $\ddot{z} = a \in [-1,1] = A$, and the target is to reach 0 with 0 speed. Then $N = 2$, $f((y_1, y_2), a) = (y_2, a)$, $\mathcal{T} = \{(0,0)\}$ and the HJB equation (2.3) is

(2.8) $$|T_{x_2}| - x_2 T_{x_1} - 1 = 0.$$

The trajectories corresponding to the constant controls -1 and $+1$ are, respectively, the parabolas $y_1 = -\frac{1}{2}y_2^2 + \text{constant}$ and $y_1 = \frac{1}{2}y_2^2 + \text{constant}$. The initial points from where the system can reach $\mathcal{T}$ with no switchings are

$$\Gamma_- := \{(x_1, x_2) : x_1 = -\tfrac{1}{2}x_2^2,\ x_2 \ge 0\},$$
$$\Gamma_+ := \{(x_1, x_2) : x_1 = \tfrac{1}{2}x_2^2,\ x_2 \le 0\},$$

and to minimize time we must choose $a = -1$ on Γ_- and $a = +1$ on Γ_+. Define

$$R_- := \{(x_1, x_2) : x_1 > -\tfrac{1}{2}x_2|x_2|\},$$
$$R_+ := \{(x_1, x_2) : x_1 < -\tfrac{1}{2}x_2|x_2|\}.$$

It is easy to see that starting in R_- (respectively, in R_+) and using the control $a = -1$ (respectively, $a = +1$) the system reaches Γ_+ (respectively, Γ_-) in finite time. Then $\mathcal{R} = \mathbb{R}^2$ and a single switching is enough to reach $\mathcal{T}$. We claim that the feedback map (see §III.2.2)

$$\Phi(x) = \begin{cases} -1 & \text{in } R_- \cup \Gamma_-, \\ +1 & \text{in } R_+ \cup \Gamma_+, \end{cases}$$

is optimal (we advise the reader to draw a picture of the trajectories corresponding to Φ). In fact it is clearly admissible, and a straightforward computation shows that the time taken by the system to reach $\mathcal{T}$ by using Φ is

$$u(x) = \begin{cases} x_2 + 2\sqrt{\tfrac{1}{2}x_2^2 + x_1}, & x \in \overline{R_-}, \\ -x_2 + 2\sqrt{\tfrac{1}{2}x_2^2 - x_1}, & x \in \overline{R_+}. \end{cases}$$

2. HJB EQUATIONS FOR THE MINIMAL TIME FUNCTION

By construction, $u \geq T$. Next we show that u is a subsolution of (2.3) in $\mathcal{T}^c$, so that $u \leq T$ by the proof of Theorem 2.6 and then $T = u$ (actually it is not hard to check directly that u is a supersolution, see Exercise 2.4). It is easy to check that u is a classical solution for $x \notin \Gamma := \Gamma_- \cup \Gamma_+$. As in Proposition II.2.9 we compute the superdifferential of u at $x \in \Gamma$ by decomposing its elements in the components tangential to Γ, $P_T Du(x)$, and orthogonal to Γ. The normal unit vector to Γ pointing inside R_+ is

$$n(x) := \varrho(-1, -|x_2|), \qquad \varrho := (1 + x_2^2)^{-1/2}.$$

If we denote $u_- := u_{|\bar{R}_-}$, $u_+ := u_{|\bar{R}_+}$ we have that

$$Du_-(x) \cdot n(x) = \begin{cases} -\infty & \text{for } x \in \Gamma_-, \\ \varrho/x_2 & \text{for } x \in \Gamma_+, \end{cases}$$

$$Du_+(x) \cdot n(x) = \begin{cases} \varrho/x_2 & \text{for } x \in \Gamma_-, \\ +\infty & \text{for } x \in \Gamma_+, \end{cases}$$

Then $D^+u(x) = \emptyset$ and so u is a viscosity subsolution of (2.8). ◁

The exponential transformation (2.5), also named the *Kružkov transform*, simultaneously reached two goals in the proof of Theorem 2.6: transforming the original HJB equation (2.3) into the PDE (2.6) with the good structure to apply the comparison results, and changing the singular boundary condition on $\partial \mathcal{R}$ (2.4) into a standard one (2.7). Next we push the consequences of this transformation further by observing that it amounts to considering the following payoff

$$J(x, \alpha) := \int_0^{t_x(\alpha)} e^{-t}\, dt = 1 - e^{-t_x(\alpha)},$$

which is a rescaled and discounted version of the original one $t_x(\alpha)$. The value function associated with the new payoff is

$$(2.9) \qquad v(x) := \inf_{\alpha \in \mathcal{A}} J(x, \alpha) = \begin{cases} 1 - e^{-T(x)} & \text{if } x \in \mathcal{R}, \\ 1 & \text{if } x \notin \mathcal{R}, \end{cases}$$

which satisfies $0 \leq v(x) \leq 1$ for all $x \in \mathbb{R}^N$. Note that, once v is known, both T and $\mathcal{R}$ are immediately recovered by the formulae

$$T(x) = -\log(1 - v(x)), \qquad \mathcal{R} = \{x : v(x) < 1\}.$$

The HJB equation for v is given by the following

LEMMA 2.8. *Assume* (A$_0$)–(A$_3$), (A$_T$), *and* $v \in BC(\mathbb{R}^N)$. *Then v is a viscosity solution of*

$$(2.10) \qquad v + \sup_{a \in A}\{-f(x, a) \cdot Dv\} - 1 = 0 \qquad \text{in } \mathcal{T}^c := \mathbb{R}^N \setminus \mathcal{T}.$$

PROOF 1. The Dynamic Programming Principle for this problem is

$$v(x) = \inf_{\alpha \in \mathcal{A}} \left\{ \int_0^{t \wedge t_x(\alpha)} e^{-s} ds + \chi_{\{t < t_x(\alpha)\}} e^{-t} v(y_x(t, \alpha)) \right\}$$

for all $t \geq 0$. The proof is essentially the same as those of Propositions III.2.5 and 2.1. The proof of (2.10) is then the same as those of Propositions III.2.8 and 2.3. ◀

PROOF 2. In the interior of $\mathcal{R}^c$ $v \equiv 1$, so v is a classical solution of (2.10). Moreover, v is a solution in $\mathcal{R} \smallsetminus \mathcal{T}$ by Propositions 2.3 and II.2.5. Finally, let $x \in \partial \mathcal{R}$ and observe that $y_x(t, \alpha) \in \mathcal{R}^c$ for all α and all t, so $v(x) - v(y_x(t)) \equiv 0$ for all t and α, in particular for $\alpha(\,\cdot\,) \equiv a$. If $\phi \in C^1$ and x is a maximum point for $v - \phi$ we get $\phi(x) - \phi(y_x(t)) \leq 0$, then $-D\phi \cdot f(x, a) \leq 0$ for all $a \in A$, and this says that v is a subsolution of (2.10) because $v(x) = 1$. Similarly, if x is minimum point for $v - \phi$ we have $\phi(x) - \phi(y_x(t)) \geq 0$, $-D\phi \cdot f(x, a) \geq 0$ for all $a \in A$, and then v is a supersolution of (2.10). ◀

As in Chapter III we want to compare v with any subsolution u_1 and supersolution u_2 in $BC(\mathbb{R}^N)$ of the boundary value problem

(2.11) $$\begin{cases} u + \sup_{a \in A}\{ -f(x, a) \cdot Du \} - 1 = 0 & \text{in } \mathcal{T}^c, \\ u = 0 & \text{on } \partial \mathcal{T}. \end{cases}$$

By this we mean that u_1, u_2 are, respectively, a sub- and a supersolution of the HJB equation such that $u_1 \leq 0$ on $\partial \mathcal{T}$, $u_2 \geq 0$ on $\partial \mathcal{T}$. We say that v is a *complete solution* of (2.11) in $BC(\mathbb{R}^N)$ if it is the maximal subsolution and the minimal supersolution in $BC(\mathbb{R}^N)$ (in particular, it is the unique solution in this set).

THEOREM 2.9. *Assume* (A$_0$)–(A$_3$), (A$_\mathcal{T}$), *and* STC$\mathcal{T}$. *Then v is the complete solution of* (2.11) *in* $BC(\mathbb{R}^N)$.

PROOF. By (2.9) and Proposition 1.6, $v \in BC(\mathbb{R}^N)$. By Lemma 2.8 v is a solution of (2.11). The Hamiltonian in (2.11) is a special case of (III.2.9) ($\ell \equiv 1$) and we get the conclusion by Remark III.2.14. ◀

As in Chapter III we can draw several consequences from this theorem: verification results such as Corollary III.2.18, where sub- and supersolutions of (2.11) are, respectively, verification and falsification functions; the equality $v = v^\sharp$ ($\sharp$ indicates the value function associated with piecewise constant controls) if the system is also small-time controllable to $\mathcal{T}$ by means of piecewise constant controls (see Remark 1.21), and then $\mathcal{T} = \mathcal{T}^\sharp$ and $\mathcal{R} = \mathcal{R}^\sharp$; the equality $v = \widehat{v}$ under the additional assumption STC$\overset{\circ}{\mathcal{T}}$, and then $\mathcal{T} = \widehat{\mathcal{T}}$ and $\mathcal{R} = \widehat{\mathcal{R}}$; relaxation and perturbation results. We leave the precise statements as exercises for the reader. Many of these results can be improved by the theory of discontinuous viscosity solutions we present in next chapter and we give some explicit applications there. For instance, in §V.4.2 we show that, for targets with Lipschitz boundary, the property

2. HJB EQUATIONS FOR THE MINIMAL TIME FUNCTION

of small-time controllability STC$\mathcal{T}$ is enough to prove that $T = T^\sharp$ and $T = \widehat{T}$, and therefore it is equivalent to STC$\overset{\circ}{\mathcal{T}}$ for such targets (cfr. Remark 1.5).

Let us make some comments on the assumptions of Theorems 2.6 and 2.9. The boundedness of $\partial \mathcal{T}$ in (A$_\mathcal{T}$) was assumed to simplify the study of STC$\mathcal{T}$ in Section 1 and it plays no role in the comparison theory. Indeed Theorems 2.6 and 2.9 hold for unbounded $\mathcal{T}$, with STC$\mathcal{T}$ replaced by the assumption that $\mathcal{R}$ is open and T is continuous in $\mathcal{R}$. The uniqueness assertion in Theorem 2.6 is false if extended to solutions unbounded below. Here is the counterexample: $f(x,a) = a$, A is the closed unit ball in $\mathbb{R}^N$, $\mathcal{T} = \{0\}$; the boundary value problem

$$\begin{cases} |Du| - 1 = 0 & \text{in } \mathbb{R}^N \setminus \{0\} \\ u(0) = 0 \end{cases}$$

has the classical solutions $T(x) = |x|$ and $-|x|$. Sections 3.2 and 5.1 of Chapter V are devoted to extending the theory to non-controllable systems, that is, to the case when STC$\mathcal{T}$ fails.

We have seen at the end of Section 1 that if we relax (A$_3$) to a local condition such as (1.23) or (1.24) and STC$\mathcal{T}$ holds, the minimum time function T is not necessarily continuous in all $\mathcal{R}$, but it is continuous in the escape controllability domain $\mathcal{R}_0$. By Theorem 1.23, under the assumptions (1.24) and STC$\mathcal{T}$, T is a viscosity solution of the boundary value problem

(2.12) $$\begin{cases} \max_{a \in A}\{-f(x,a) \cdot DT\} - 1 = 0 & \text{in } \mathcal{R}_0 \setminus \{0\}, \\ T(0) = 0, \quad \lim_{\mathcal{R} \ni x \to x_0} T(x) = T_0 & \text{if } x_0 \in \partial \mathcal{R}_0, \end{cases}$$

where T_0 is the escape time. If (A$_3$) holds, it is easy to see that T is the unique solution continuous and bounded below of (2.12) (note that it may be $\mathcal{R}_0 \subset \mathcal{R}$ even if (A$_3$) holds, see Exercise 1.5) but this is not the most interesting case. If we drop (A$_3$), uniqueness may fail as the following example shows.

EXAMPLE 2.10. Let $N = 1$, $f(x,a) = x^2 a$, $A = \{-1, +1\}$, $\mathcal{T} = \{1\}$. It is easy to check that $\mathcal{R} =]0, +\infty[$, $T(x) = 1/x - 1$ for $x \leq 1$, $T(x) = 1 - 1/x$ for $x > 1$, and so $T_0 = 1$ and $\mathcal{R}_0 =]1/2, +\infty[$. A second solution, continuous and bounded below, of

(2.13) $$\begin{cases} x^2 |u'| = 1, & \text{in } \mathcal{R}_0 \setminus \{1\}, \\ u(1) = 0, \quad u(1/2) = 1, \end{cases}$$

is $u(x) = 1/x - 1$. Note that both solutions are classical. ◁

However, Theorem 1.23 also gives the following information on the behavior of T at infinity:

(2.14) $$\lim_{\substack{|x| \to \infty \\ x \in \mathcal{R}_0}} T(x) = T_0.$$

The last result of this section says that if we add this boundary condition at infinity to the boundary value problem (2.12), then T is the unique solution.

THEOREM 2.11. *Assume (1.24) and STCT. Then T is the unique solution of (2.12) and (2.14) that is continuous in $\mathcal{R}_0$.*

PROOF. Let $u \in C(\mathcal{R}_0)$ be another solution of (2.12). Extend T and u by continuity to $\overline{\mathcal{R}}_0$ and take their exponential transformations

$$v := 1 - e^{-T}, \qquad w := 1 - e^{-u}.$$

By Proposition II.2.5 they solve equation (2.6) in $\mathcal{R}_0 \smallsetminus \mathcal{T}$. We apply the comparison principle of Exercise II.3.6 to the bounded open set $\Omega_r := \{\, x \in \mathcal{R}_0 : 0 < |x| < r \,\}$ and get

$$(2.15) \qquad \max_{\overline{\Omega}_r}(v-w)^+ = \max_{\partial \Omega_r}(v-w)^+ = \max_{\{x \in \mathcal{R}_0 : |x| = r\}}(v-w)^+ .$$

Now we let $r \to \infty$ so that the sets Ω_r invade $\mathcal{R}_0$. Since T and u satisfy (2.14), the right-hand side of (2.15) tends to 0, and we obtain $v \leq w$ in $\mathcal{R}_0$. The proof is completed by reversing the roles of v and w. ◀

The main drawback of this result is that it characterizes T only once $\mathcal{R}_0$ is known, even if it is not a datum of the problem. In Section 4 we give a uniqueness result where $\mathcal{R}_0$ appears as an unknown, so that $\partial \mathcal{R}_0$ can be considered as a free boundary.

The last remark is that one might suspect that T can be characterized as the unique solution of (2.12) by adding the natural sign condition, $T(x) > 0$ for $x \neq 0$, instead of (2.14). The next example shows that this is not the case.

EXAMPLE 2.12. Consider the system and target of Example 2.10 and define

$$u_\lambda(x) = \begin{cases} T(x), & \text{if } x \leq 1, \\ \lambda - |T(x) - \lambda|, & \text{if } x > 1 . \end{cases}$$

It is easy to check that, for all $\lambda > 0$, u_λ is a solution of (2.13), and, for $1/2 \leq \lambda < 1$, $u_\lambda(x) > 0$ for $x \neq 1$ and $u_\lambda(x) < T(x)$ for $x > 1/(1-\lambda)$. ◁

Exercises

2.1. In the assumptions of Proposition 2.1 let a function $S : \mathbb{R}^N \to [0, +\infty]$ satisfy the DPP at a given point $x \in \mathcal{R}$, that is

$$S(x) = \inf_\alpha \{ t \wedge t_x(\alpha) + \chi_{\{t \leq t_x(\alpha)\}} S(y_x(t, \alpha)) \}, \qquad \text{for all } t \geq 0 .$$

Prove that $S(x) = T(x)$.

2.2. Assume (A_0)–(A_3), $\mathcal{R}$ open, and T continuous in $\mathcal{R}$. Show that T is a solution of (2.3) in int $\mathcal{T}$ and that at any point of $\partial \mathcal{T}$ it satisfies the definition of subsolution and does not satisfy the definition of supersolution. [Hint: look at $D^+T(x)$ and $D^-T(x)$ by using Remark 1.3.]

2.3. In the assumptions of Remark 2.4 let $x \in \partial \mathcal{T}$ a point where $\widehat{T}(x) > 0$ but $\widehat{T}_{|\mathcal{R} \smallsetminus \text{int } \mathcal{T}}$ is continuous. Show that if $\phi \in C^1(\mathbb{R}^N)$ and x is a maximum point for $(u - \phi)_{|\mathcal{R} \smallsetminus \text{int } \mathcal{T}}$ then $\sup_a \{-f(x,a) \cdot D\phi(x)\} \leq 1$. This remark suggests an interpretation of the Dirichlet boundary condition in a weak sense which is studied in Section 4 of Chapter V.

2.4. In the setting of Example 2.7 show that

$$D^- u(x) = \{ P_T Du(x) + \lambda u(x) : \lambda \in \Lambda(x) \},$$

$$\Lambda(x) := \begin{cases}]-\infty, \varrho/x_2] & \text{if } x \in \Gamma_-, \\ [\varrho/x_2, +\infty[& \text{if } x \in \Gamma_+, \end{cases}$$

$$P_T Du(x) = \begin{cases} \varrho^2(-x_2, 1) & \text{if } x \in \Gamma_-, \\ \varrho^2(-x_2, -1) & \text{if } x \in \Gamma_+, \end{cases}$$

and check that u is a supersolution of (2.8).

2.5. Assume (A_0)–(A_3) and let T and T_p be, respectively, the minimal time functions associated with the targets $\{0\}$ and $\{p\}$, for a given $p \neq 0$. Assume small-time controllability on both targets and the existence of trajectories joining 0 with p and p with 0, so that the reachable sets for the two targets coincide:

$$\mathcal{R} := \{x : T(x) < +\infty\} = \{x : T_p(x) < +\infty\}.$$

Does $u(x) := T_p(x) - T_p(0)$ give a counterexample to Theorem 2.6?

3. Problems with exit times and non-zero terminal cost

In this and the next section the cost to minimize is

$$J(x, \alpha) := \begin{cases} \int_0^{t_x(\alpha)} \ell(y_x(s), \alpha(s)) e^{-\lambda s} \, ds + e^{-\lambda t_x(\alpha)} g(y_x(t_x(\alpha))), & \text{if } t_x(\alpha) < +\infty, \\ \int_0^\infty \ell(y_x(s), \alpha(s)) e^{-\lambda s} \, ds, & \text{if } t_x(\alpha) = +\infty, \end{cases}$$

where $t_x(\alpha)$ is the first time the trajectory $y_x(\cdot, \alpha)$ hits a given target $\mathcal{T}$, as in Sections 1 and 2; the running cost ℓ and the constant interest rate λ satisfy

(3.1) $\begin{cases} \ell : \mathbb{R}^N \times A \to \mathbb{R} \text{ is continuous;} \\ \text{there are a modulus } \omega_\ell \text{ and a constant } M \text{ such that} \\ |\ell(x, a) - \ell(y, a)| \leq \omega_\ell(|x - y|) \text{ and } 1 \leq \ell(x, a) \leq M \text{ for} \\ \text{all } x, y \in \mathbb{R}^N, a \in A; \\ \lambda \geq 0; \end{cases}$

the terminal cost g satisfies

(3.2) $\qquad g \in C(\mathcal{T}) \text{ and } g(x) \geq 0 \text{ for all } x \in \mathcal{T}.$

First of all we remark that, if $\lambda > 0$, the assumptions $\ell \geq 1$ and $g \geq 0$ are not more restrictive than the boundedness of ℓ and g from below. In fact, if they are not satisfied we define
$$C := \max\{1 - \inf_{\mathbb{R}^N \times A} \ell, -\lambda \inf_{\partial \mathcal{T}} g\},$$
$\widetilde{\ell} = \ell + C$, $\widetilde{g} = g + C/\lambda$, $\widetilde{v}$ the corresponding value function, and it is straightforward to check that $\widetilde{\ell}$ and $\widetilde{g}$ satisfy (3.1) and (3.2) and that $\widetilde{v}(x) = v(x) + C$ for all x. In the case $\lambda = 0$ the same remark holds for g, but the condition $\ell \geq 1$ can only be replaced by $\inf_{\mathbb{R}^N \times A} \ell > 0$. In this case, on the other hand, the upper bounds on ℓ and g can be dropped in some results (see, e.g., Exercises 4.1 and 4.1). The hypotheses on the sign of ℓ and g are used only in some of the following results, they are irrelevant for others such as, for instance, the Dynamic Programming Principle. The value function of the problem is
$$v(x) := \inf_{\alpha \in \mathcal{A}} J(x, \alpha).$$

We also state some simple results as the problem of minimizing $\widehat{J}$, which is obtained by replacing t_x by $\widehat{t}_x$ in the definition of J, whose value function is denoted by $\widehat{v}$. However, the main results on this problem are postponed to Chapter V.

3.1. Compatible terminal cost and continuity of the value function

The study of the continuity of v here is harder than in Section 1 because of the non-zero terminal cost g. We first prove that continuity at all points of $\partial \mathcal{T}$ implies continuity everywhere, then we give necessary and/or sufficient conditions for upper and lower semicontinuity at points of $\partial \mathcal{T}$. The property of small-time controllability on $\mathcal{T}$, STC$\mathcal{T}$, still plays an important role, but, in contrast to the case $g = 0$, it is neither sufficient nor necessary for the continuity of v as we show in Examples 3.1 and 3.2. At the end of the section we give a necessary and sufficient condition for the continuity of v in terms of sub- and supersolutions of the HJB equation.

EXAMPLE 3.1 (STC$\mathcal{T}$ IS NOT SUFFICIENT FOR THE CONTINUITY OF v). Assume $N = 2$, $A = \{1, 2\}$, $f(x, 1) = (0, 1)$, $f(x, 2) = (-1, 0)$, $\ell \equiv 1$, $\lambda = 0$, $\mathcal{T} = \{x : x_1 \leq 0 \text{ and } x_2 \geq 0\}$,
$$g(x) = \begin{cases} 0 & \text{if } x_2 = 0, \\ -2x_2 & \text{if } x_1 = 0, \ 0 \leq x_2 \leq 1, \\ -2 & \text{if } x_1 = 0, \ x_2 \geq 1. \end{cases}$$

It is easy to see that STC$\mathcal{T}$ holds and $T(x) = x_1 \vee 0 - x_2 \wedge 0$. It is also clear that $v(x) = T(x) = -x_2$ for $x_1 \leq 0$. On the other hand, if $x_1 > 0$ and $x_2 \leq 0$ we use the control $\alpha(t) = 1$ for $0 \leq t \leq -x_2 + 1$, $\alpha(t) = 2$ for $t > -x_2 + 1$, so that $t_x(\alpha) = -x_2 + 1 + x_1$, and $y_x(t_x(\alpha)) = (0, 1)$. Then
$$v(x) \leq J(x, \alpha) = x_1 - x_2 - 1 \qquad \text{for } x_1 > 0, \ x_2 \leq 0,$$

and therefore v is discontinuous on the half-line $\{(0, x_2) : x_2 \leq 0\}$. It is not hard to modify this example so that $\partial \mathcal{T}$ is bounded and v has the same lime of discontinuities. ◁

EXAMPLE 3.2 (STC$\mathcal{T}$ IS NOT NECESSARY FOR THE CONTINUITY OF v). The most trivial example is $f(x, a) \equiv 0$; then for any $\mathcal{T}$ and g

$$v(x) = \inf_{a \in A} \frac{\ell(x, a)}{\lambda} =: \bar{\ell}(x)$$

for $x \in \mathcal{T}^c$, thus v is continuous if and only if $g = \bar{\ell}_{|\mathcal{T}}$.

A less degenerate example is, for $N = 2$, $f(x, a) \equiv (1, 0)$, $\ell \equiv 1$, $\lambda = 0$, $\mathcal{T} = \{x : |x| \geq 1 \text{ or } x_1 \geq 0\}$, $g(x) = 0$ if $x_1 = 0$. Then $v(x) = -x_1$ for $x \in \mathcal{T}^c$ and it is continuous if and only if $g(x) = -x_1$ for $|x| = 1$. ◁

PROPOSITION 3.3. *Assume* (A$_0$), (A$_1$), (A$_3$), (A$_\mathcal{T}$), (3.1), (3.2), *and v continuous at each point of $\partial \mathcal{T}$.*

(i) *If $\lambda > 0$, then $v \in BUC(\mathbb{R}^N)$;*

(ii) *if $\lambda = 0$, then $\mathcal{R} = \{x : v(x) < +\infty\}$, $\lim_{x \to x_0} v(x) = +\infty$ for any $x_0 \in \partial \mathcal{R}$, and $v \in BUC(\mathcal{K})$ for any closed set $\mathcal{K}$ where the minimum time function T is bounded.*

PROOF. First of all we observe that there exists a modulus ω such that

(3.3) $\qquad |g(x) - v(z)| \leq \omega(|x - z|) \qquad$ for all $x \in \partial \mathcal{T}$, $z \in \mathbb{R}^N$.

It is enough to define $\omega(s) := \sup\{|g(x) - v(z)| : |x - z| \leq s, \ x \in \partial \mathcal{T}, \ z \in \mathbb{R}^N\}$ and use the continuity of v on $\partial \mathcal{T}$, the boundedness of $\partial \mathcal{T}$, and a compactness argument as in the proof of Proposition 1.2.

To prove (i) we note that the boundedness of v is trivial, we fix $\varepsilon > 0$ and $x, z \in \overline{\mathcal{T}^c}$, we assume $v(x) \geq v(z)$ and pick $\overline{\alpha} \in \mathcal{A}$ such that $v(z) \geq J(z, \overline{\alpha}) - \varepsilon$. As in the proof of the continuity of the value function for the infinite horizon problem, Proposition III.2.1, we fix τ such that $\int_\tau^\infty 2M e^{-\lambda t} dt \leq \varepsilon$, where M is a bound on ℓ. If $t_z(\overline{\alpha}) > \tau$ the trajectory $y_z(t, \overline{\alpha})$ stays at a positive distance from $\mathcal{T}$ for $t \leq \tau$ and therefore so does the trajectory $y_x(t, \overline{\alpha})$ if x is sufficiently close to z, by (0.3). Thus also $t_x(\overline{\alpha}) > \tau$, and the estimate $v(x) - v(z) < 3\varepsilon$ for $|x - z|$ small enough is found exactly as in the proof of Proposition III.2.1.

If, instead, $t_z(\overline{\alpha}) \leq \tau$, we first consider the case that

$$\overline{s} := t_x(\overline{\alpha}) \leq t_z(\overline{\alpha}) =: \overline{t}.$$

We define $\widetilde{\alpha} = \overline{\alpha}(\cdot + \overline{s})$, $\overline{z} := y_z(\overline{s}, \overline{\alpha})$, $\overline{x} := y_x(\overline{s}, \overline{\alpha}) \in \partial \mathcal{T}$, and compute

$$v(z) \geq \int_0^{\overline{s}} \ell(y_z, \overline{\alpha})) e^{-\lambda s} ds + \int_{\overline{s}}^{\overline{t}} \ell(y_z, \overline{\alpha}) e^{-\lambda s} ds + g(y_z(\overline{t}, \overline{\alpha})) e^{-\lambda \overline{t}} - \varepsilon$$
$$= \int_0^{\overline{s}} \ell(y_z, \overline{\alpha}) e^{-\lambda s} ds + e^{-\lambda \overline{s}} J(\overline{z}, \widetilde{\alpha}) - \varepsilon,$$

so that

$$
\begin{aligned}
(3.4) \quad v(x) - v(z) &\leq \int_0^{\bar{s}} |\ell(y_x, \bar{\alpha}) - \ell(y_z, \bar{\alpha})| e^{-\lambda s}\, ds + e^{-\lambda \bar{s}}(g(\bar{x}) - v(\bar{z})) + \varepsilon \\
&\leq \tau \omega_\ell(e^{LT}|x-z|) + \omega(e^{LT}|x-z|) + \varepsilon,
\end{aligned}
$$

where in the last inequality we have used (0.3), (3.1) and $\bar{s} \leq \tau$. Since the right-hand side of (3.4) can be made $\leq 2\varepsilon$ for $|x-z|$ small enough we reach the conclusion in this case.

On the other hand, if $\bar{t} < \bar{s}$ we define $\bar{x} := y_x(\bar{t}, \bar{\alpha})$ and use a similar calculation (which in this case is just a piece of the DPP, see Proposition 3.9) to get

$$
v(x) - v(z) \leq \int_0^{\bar{t}} \ell(y_x, \bar{\alpha}) e^{-\lambda s}\, ds + v(\bar{x}) e^{-\lambda \bar{t}} - v(z),
$$

which is $\leq$ the right-hand side of (3.4) by the definition of $\bar{\alpha}$, (0.3), (3.3), and $\bar{t} \leq \tau$. This completes the proof of (i).

To prove (ii), we note that the obvious inequalities

$$
T(x) \leq v(x) \leq M\,T(x) + \max_{\partial T} g \qquad \text{for all } x \in \mathbb{R}^N
$$

give the first two statements. Now set

$$
C_1 := M \sup_{x \in \mathcal{K}} T(x) + \max_{x \in T} g(x)
$$

so that $v(x) \leq C_1$ for $x \in \mathcal{K}$. Next fix $\varepsilon \in \,]0,1]$, $x, z \in \overline{\mathcal{K} \smallsetminus T}$, and $\bar{\alpha} \in \mathcal{A}$ such that $v(z) \geq J(z, \bar{\alpha}) - \varepsilon$. Then, by the lower bounds on g and ℓ, we have

$$
t_z(\bar{\alpha}) \leq J(z, \bar{\alpha}) \leq C_1 + 1 =: \tau.
$$

From now on the proof is exactly the same as the second part of the proof of (i). ◂

PROPOSITION 3.4. *Assume* (A_0), (A_1), (A_3), (A_T), (3.1), (3.2), *and* STCT. *Then, for every* $z \in \partial T$, v *is upper semicontinuous at* z, *that is,*

$$
\limsup_{x \to z} v(x) \leq v(z) = g(z).
$$

PROOF. For $x \notin T$ and $\varepsilon > 0$ we pick $\bar{\alpha} \in \mathcal{A}$ such that $t_x(\bar{\alpha}) \leq T(x) + \varepsilon$. Then

$$
\begin{aligned}
v(x) - g(z) &\leq \int_0^{t_x(\bar{\alpha})} \ell(y_x, \bar{\alpha}) e^{-\lambda s}\, ds + g(y_x(t_x(\bar{\alpha}))) - g(z) \\
&\leq M(T(x) + \varepsilon) + \omega_g(|y_x(t_x(\bar{\alpha})) - x| + |x - z|)
\end{aligned}
$$

and the right-hand side of the last inequality tends to zero as $x \to z$ and $\varepsilon \to 0^+$, thanks to (0.1) and because $T(x) \to 0$ by STCT. ◂

3. PROBLEMS WITH NON-ZERO TERMINAL COST

DEFINITION 3.5. We say that the terminal cost g is *compatible* (with the continuity of v) if g has an extension to a neighborhood $B(T, \delta)$ of T, still named g, which is lower semicontinuous at points of ∂T and such that $g(x) \leq J(x, \alpha)$ for all $\alpha \in \mathcal{A}$ and $x \in B(T, \delta)$. ◁

Clearly, the compatibility of g is equivalent to the lower semicontinuity of v at all points of $z \in T$. In fact, it is equivalent to the condition $g \leq v$ in $B(T, \delta)$; thus, for any $z \in T$,

$$v(z) = g(z) \leq \liminf_{x \to z} g(x) \leq \liminf_{x \to z} v(x),$$

which proves the sufficiency, and the necessity is obvious. Therefore we have the following

THEOREM 3.6. *Assume* (A_0), (A_1), (A_3), (A_T), (3.1), (3.2), *and* STCT. *Then the compatibility of g is equivalent to* $v \in BUC(\mathbb{R}^N)$ *if* $\lambda > 0$, *and to* $v \in C(\mathcal{R})$ *if* $\lambda = 0$.

PROOF. If $\lambda > 0$ it is enough to put together Propositions 3.3 and 3.4. If $\lambda = 0$ we have to recall that STCT implies $\mathcal{R}$ open and T continuous in $\mathcal{R}$, by Proposition 1.6, so that Proposition 3.3 (ii) applies to some neighborhood of each point of $\mathcal{R}$. ◂

In the case $\lambda = 0$ the compatibility condition can be rewritten in a different form by introducing the function

$$L(x, z) := \inf \left\{ \int_0^{t_x(\alpha)} \ell(y_x, \alpha) \, ds : \alpha \in \mathcal{A} \text{ such that } y_x(t_x(\alpha)) = z \right\} \leq +\infty,$$

for $x \notin T$, $z \in \partial T$, which represents the minimal cost to join x and z staying out of T. It is easy to prove that

$$(3.5) \qquad v(x) = \inf_{z \in \partial T} \{ L(x, z) + g(z) \},$$

and then the compatibility condition $g \leq v$ is equivalent to

$$(3.6) \qquad g(x) - g(z) \leq L(x, z) \qquad \text{for all } x \in B(T, \delta) \smallsetminus T,\ z \in \partial T.$$

In this case we can give some more explicit sufficient conditions for the compatibility of g.

PROPOSITION 3.7. *Assume* (A_0), (A_1), (A_3), (A_T), (3.1), (3.2), *and* $\lambda = 0$. *If either g can be extended to $B(T, \delta)$ so that*

$$(3.7) \qquad |g(x) - g(z)| \leq L^{-1} \log\left(1 + \frac{L|x-z|}{\sup_{a \in A} |f(x,a)|}\right)$$

for all $x \in B(T, \delta) \smallsetminus T$, $z \in \partial T$, where L is the constant in (A_3), *or*

$$(3.8) \qquad |g(x) - g(z)| \leq \left(\sup_{T^c \times A} |f|\right)^{-1} |x - z| \qquad \text{for all } x, z \in \partial T,$$

then g is compatible.

Note that (3.8) is just an upper bound on the Lipschitz constant of g on ∂T, and therefore it is much easier to check, but it is very restrictive and it requires that g is a constant if f is unbounded. However, (3.8) is also necessary for compatibility in the special case $f(x,a) = a$, $A = B(0,1)$, $\ell \equiv 1$ and T^c convex, see Exercise 3.2.

PROOF OF PROPOSITION 3.7. Set $z = y_x(t_x(\alpha))$. By the estimate (III.5.11) in Remark III.5.6

$$|z - x| \leq \frac{1}{L} \sup_A |f(x, \cdot)| (e^{Lt_x(\alpha)} - 1),$$

and thus $t_x(\alpha) \geq$ the right-hand side of (3.7). Now fix $\varepsilon > 0$ and $\overline{\alpha} \in \mathcal{A}$ such that

$$L(x,z) \geq \int_0^{t_x(\overline{\alpha})} \ell(y_x, \overline{\alpha}) \, ds - \varepsilon \geq t_x(\overline{\alpha}) - \varepsilon \, .$$

Then, by the arbitrariness of ε, (3.7) implies (3.6). Moreover, it is clear that (3.7) implies the continuity of the extended g at points of ∂T.

Now we assume (3.8) and f bounded (if f is unbounded the conclusion is very easy). Then g is Lipschitz on ∂T and can be extended off T preserving the same Lipschitz constant. Moreover, $|x - z| \leq \sup |f| t_x(\overline{\alpha})$ (for $x \notin T$ and z as before) implies $L(x,z) \geq (\sup|f|)^{-1}|x - z|$ and therefore (3.6). ◀

Next we give a necessary and sufficient condition for the compatibility of g which looks like (3.6) but can be checked only at points x, z of ∂T, so that one does not have to look for extensions of g satisfying (3.6). To do this we first extend $L(\cdot, \cdot)$ as follows

$$L(x,z) := \inf \Big\{ \int_0^\tau \ell(y_x, \alpha) \, ds : \alpha \in \mathcal{A} \text{ such that } y_x(\tau) = z \text{ and } \\ y_x(t) \notin T \text{ for all } t \in \,]0, \tau[\, \Big\}$$

for $x, z \in \overline{T^c}$. The candidate compatibility condition is

(3.9) $\quad g(x) - g(z) \leq L(x,z) \quad$ for all $x, z \in \partial T$.

PROPOSITION 3.8. *Assume* (A_0), (A_1), (A_3), (A_T), (3.1), (3.2), *and* $\lambda = 0$. *Then*
(i) (3.9) *is necessary for the compatibility of* g;
(ii) (3.9) *is sufficient for the compatibility of* g *if in addition for all* $x_0 \in \partial T$ *either*

(3.10) $\quad \lim_{\substack{x \to x_0 \\ x \notin T}} L(x_0, x) = 0$

or the minimal-time function T *is continuous at* x_0 (*e.g.*, STCT *holds*) *and for some* $r > 0$

(3.11) $\quad L(z,x) \leq L(x,z) \quad$ for all $z \in \partial T$, $x \in T^c$, $x, z \in B(x_0, r)$.

Condition (3.10) is the small-time controllability on x_0 of the backward system $y' = -f(y,a)$ with the state-constraint that the trajectories stay in $\mathcal{T}^c$ until they reach x_0. A simple sufficient condition for it is that the boundary $\partial \mathcal{T}$ is smooth (e.g., a Lipschitz surface) and $f(x_0, A) \supseteq B(0, r)$ for some $r > 0$, see Exercise 3.3. Condition (3.11) is satisfied, for instance, by symmetric systems, that is, systems whose trajectories can be run backwards at the same speed, if $\ell(\cdot, \cdot)$ depends only on the state variable.

PROOF OF PROPOSITION 3.8. To prove (i), we fix $x_0 \in \partial \mathcal{T}$ and first observe that

(3.12) $$\liminf_{x \to x_0} L(x, z) \leq L(x_0, z) .$$

In fact, for any $\varepsilon > 0$ there is α_0 such that $y_{x_0}(t) \notin \mathcal{T}$ for $t \in {]0, \tau[}$, $y_{x_0}(\tau) = z$ and

$$\int_0^\tau \ell(y_{x_0}, \alpha_0) \, ds \leq L(x_0, z) + \varepsilon .$$

We take $x_n = y_{x_0}(1/n, \alpha_0)$, $\alpha_n(\cdot) = \alpha_0(\cdot + 1/n)$, so that $y_{x_n}(\tau - 1/n, \alpha_n) = z$ and therefore

$$L(x_n, z) \leq \int_0^{\tau - 1/n} \ell(y_{x_n}, \alpha_n) \, ds \longrightarrow \int_0^\tau \ell(y_{x_0}, \alpha_0) \, ds,$$

which proves (3.12) by the arbitrariness of ε. Now we recall that the compatibility of g gives

$$g(x_0) \leq \liminf_{x \to x_0} v(x),$$

and then (3.5) and (3.12) imply (since "lim inf inf $\leq$ inf lim inf")

$$g(x_0) \leq \inf_{z \in \partial \mathcal{T}} \{ L(x_0, z) + g(z) \} .$$

This gives (3.9) by the arbitrariness of x_0.

To prove (ii) we first observe that

(3.13) $\qquad L(x_0, z) \leq L(x_0, x) + L(x, z) \qquad$ for all $x_0, z \in \overline{\mathcal{T}^c}$, $x \notin \mathcal{T}$,

because the right-hand side is the minimal cost to join x_0 and z touching x. Then (3.9) gives

(3.14) $\qquad g(x_0) - g(z) - L(x, z) \leq L(x_0, z) - L(x, z) \leq L(x_0, x)$

for $x_0, z \in \partial \mathcal{T}$, $x \notin \mathcal{T}$. Since $v = g$ on $\partial \mathcal{T}$ and g is continuous it is enough to prove that $g(x_0) \leq \liminf_n v(x_n)$ for any sequence $x_n \to x_0$, $x_n \notin \mathcal{T}$ for all n. By (3.5) and (3.14) we get

$$g(x_0) - \liminf_n v(x_n) = \limsup_n \sup_{z \in \partial \mathcal{T}} \{ g(x_0) - g(z) - L(x_n, z) \}$$
$$\leq \limsup_n L(x_0, x_n) = 0,$$

where in the last equality we have used the assumption (3.10). To treat the case where (3.10) does not hold we fix $\varepsilon > 0$ and $\alpha_n \in \mathcal{A}$ such that
$$t_{x_n}(\alpha_n) \leq T(x_n) + \varepsilon,$$
and set $z_n := y_{x_n}(t_{x_n}(\alpha_n))$ so that, by (3.1) and the definition of L,

(3.15) $$L(x_n, z_n) \leq M(T(x_n) + \varepsilon).$$

By (3.2), (3.9) and (3.13) we get
$$\begin{aligned} g(x_0) - g(z) - L(x_n, z) &\leq \omega_g(|x_0 - z_n|) + g(z_n) - g(z) - L(x_n, z) \\ &\leq \omega_g(|x_0 - z_n|) + L(z_n, z) - L(x_n, z) \\ &\leq \omega_g(|x_0 - z_n|) + L(z_n, x_n) \quad \text{for all } z \in \partial \mathcal{T}, \end{aligned}$$
where ω_g is the modulus of continuity of g on $\partial \mathcal{T}$. Since (3.5) gives
$$g(x_0) - v(x_n) \leq \sup_{z \in \partial \mathcal{T}} \{g(x_0) - g(z) - L(x_n, z)\},$$
we can now obtain, also using (3.11) and (3.15),
$$g(x_0) - v(x_n) \leq \omega_g(|x_0 - z_n|) + M(T(x_n) + \varepsilon),$$
where the right-hand side tends to 0 as $n \to \infty$ and $\varepsilon \to 0^+$ because $T(x_n) \to T(x_0) = 0$ and then $z_n \to x_0$ as well, as it is easy to show by contradiction. ◀

3.2. The HJB equation and a superoptimality principle

Next we derive the HJB equation for the optimal control problem of this section, we consider the Dirichlet boundary value problem for this equation with the terminal cost g as boundary data, and compare the value function with sub- and supersolutions of this Dirichlet problem.

PROPOSITION 3.9 (DYNAMIC PROGRAMMING PRINCIPLE). *Assume* (A_0), (A_1), (A_3), $(A_\mathcal{T})$, (3.1), *and* (3.2). *Then*

(3.16) $$v(x) = \inf_{\alpha \in \mathcal{A}} \left\{ \int_0^{t \wedge t_x(\alpha)} \ell(y_x, \alpha) e^{-\lambda s} ds \right. \\ \left. + \chi_{\{t < t_x(\alpha)\}} v(y_x(t, \alpha)) e^{-\lambda t} + \chi_{\{t \geq t_x(\alpha)\}} g(y_x(t_x(\alpha))) e^{-\lambda t_x(\alpha)} \right\}$$

for all $x \in \mathbb{R}^N$ *if* $\lambda > 0$, *for all* $x \in \mathcal{R}$ *if* $\lambda = 0$, *and for all* $t \geq 0$.

The proof of Proposition 3.9 is essentially the same as the proofs of Propositions III2.5 and 2.1 and we omit it (see Exercise 3.5). Note that (3.16) can be rewritten as

(3.17) $$v(x) = \inf_{\alpha \in \mathcal{A}} \left\{ \int_0^{t \wedge t_x(\alpha)} \ell(y_x, \alpha) e^{-\lambda s} ds + v(y_x(t \wedge t_x(\alpha))) e^{-\lambda(t \wedge t_x(\alpha))} \right\}.$$

3. PROBLEMS WITH NON-ZERO TERMINAL COST

REMARK 3.10. The DPP for $\hat{v}$ is given by the same formula (3.16) where v is replaced by $\hat{v}$ and t_x by $\hat{t}_x$. ◁

The Hamiltonian in the HJB equation for the problem with exit time is the same as for the infinite horizon problem, that is,

$$(3.18) \qquad H(x,p) := \sup_{a \in A} \{ -f(x,a) \cdot p - \ell(x,a) \}, \qquad x, p \in \mathbb{R}^N .$$

PROPOSITION 3.11. *Assume* (A_0)–(A_3), $(A_\mathcal{T})$, (3.1), (3.2), $\lambda > 0$, *and* $v \in C(\mathcal{T}^c)$. *Then v is a viscosity solution of*

$$(3.19) \qquad \lambda v + H(x, Dv) = 0 \qquad \text{in } \mathcal{T}^c .$$

The proof of Proposition 3.11 is the same as that of Proposition III.2.8.

For the case $\lambda = 0$ we proceed as for the minimal time function by considering the nonlinear transformation of v

$$(3.20) \qquad V(x) := \begin{cases} 1 - e^{-v(x)} & \text{if } x \in \mathcal{R}, \\ 1 & \text{if } x \notin \mathcal{R}, \end{cases}$$

which is the value function of the problem whose cost functional is

$$I(x,\alpha) := \int_0^{t_x(\alpha)} \ell(y_x, \alpha) \exp\left(-\int_0^t \ell(y_x,\alpha) \, ds \right) dt$$
$$+ \left(1 - e^{-g(y_x(t_x(\alpha)))} \right) \exp\left(-\int_0^{t_x(\alpha)} \ell(y_x,\alpha) \, ds \right) .$$

This is a problem with nonconstant interest rate, see Exercise III.2.4. The Hamiltonian for this problem is

$$\mathcal{H}(x,u,p) := \sup_{a \in A} \{ -f(x,a) \cdot p - \ell(x,a) + (\ell(x,a) - 1)u \},$$

for $u \in \mathbb{R}$, $x,p \in \mathbb{R}^N$.

PROPOSITION 3.12. *Assume* (A_0)–(A_3), $(A_\mathcal{T})$, (3.1), (3.2), *and* $\lambda = 0$.
(i) *If $\mathcal{R}$ is open and $v \in C(\mathcal{R} \setminus \mathcal{T})$, then v is a viscosity solution of*

$$(3.21) \qquad H(x, Dv) = 0 \qquad \text{in } \mathcal{R} \setminus \mathcal{T};$$

(ii) *if $V \in C(\mathcal{T}^c)$, then V is a viscosity solution of*

$$(3.22) \qquad V + \mathcal{H}(x, V, DV) = 0 \qquad \text{in } \mathcal{T}^c .$$

The continuity assumptions on v and V in the last two propositions are satisfied if g is compatible and STC$\mathcal{T}$ holds, by Theorem 3.6. The proof of (i) is the same as that of Proposition III.2.8, the proof of (ii) is the same as that of Lemma 2.8. As usual we characterize the value function v as the maximal subsolution and the minimal supersolution, that is, the complete solution, in a suitable set of functions, of the HJB equation with the natural boundary conditions. By subsolution (respectively, supersolution) of the Dirichlet condition

$$(3.23) \qquad u = g \quad \text{on } \partial \mathcal{T},$$

we mean a function $\leq g$ (respectively, $\geq g$) at each point of $\partial \mathcal{T}$.

PROPOSITION 3.13. *Assume* (A_0)–(A_3), $(A_\mathcal{T})$, (3.1), (3.2), STC$\mathcal{T}$, *and g compatible. Then,*

(i) *if $\lambda > 0$, v is the complete solution in $BC(\overline{\mathcal{T}^c})$ of (3.19) with the boundary condition (3.23);*

(ii) *if $\lambda = 0$, v is the unique solution of (3.21) continuous in $\mathcal{R} \setminus \operatorname{int} \mathcal{T}$, bounded below, and satisfying (3.23) and $v(x) \to +\infty$ as $x \to x_0 \in \partial \mathcal{R}$;*

(iii) *if $\lambda = 0$, V defined by (3.20) is the complete solution in $BC(\overline{\mathcal{T}^c})$ of (3.22) with the boundary condition $V = 1 - e^{-g}$ on $\partial \mathcal{T}$.*

PROOF. (i) It is enough to put together Theorem 3.6, Proposition 3.11, and the comparison result in Remark III.2.14.
(ii) v is a solution of the HJB equation with the desired properties because $v \geq 0$ and by Theorem 3.6 and Proposition 3.11. Uniqueness follows from the proof of Theorem 2.6.
(iii) The comparison assertion for equation (3.22) is an easy extension of Theorem III.2.12; see Exercise III.2.5 and Remark III.2.14. ◀

Next we give two comparison results where no continuity assumptions on the value function v are made a priori. They have several consequences, in particular necessary and sufficient conditions for the continuity of v in terms of existence of suitable sub- and supersolutions of the HJB equation. The first one is an immediate consequence of the global principle of suboptimality, Theorem III.2.33.

COROLLARY 3.14. *Assume* (A_0)–(A_3), $(A_\mathcal{T})$, (3.1), *and* (3.2). *If $u \in C(\overline{\mathcal{T}^c})$ is a subsolution of (3.19) such that $u \leq g$ on $\partial \mathcal{T}$ and u is bounded above if $\lambda > 0$, then $u \leq v$ in $\mathcal{T}^c$. In particular, the existence of such a u, satisfying in addition $u = g$ on $\partial \mathcal{T}$, is sufficient for the compatibility of g, and it is also necessary if STC$\mathcal{T}$ holds.*

PROOF. As we noted in Remark III.2.34, the assumption $\lambda > 0$ in Theorem III.2.33 can be dropped with no changes in the proof. If $\alpha \in \mathcal{A}$ is such that $t_x(\alpha) < +\infty$ we take $s = 0$, $t = t_x(\alpha)$ in statement (ii) of Theorem III.2.33 and use $u \leq g$ on $\partial \mathcal{T}$ to get $u(x) \leq J(x, \alpha)$. If $t_x(\alpha) = +\infty$, then for $\lambda = 0$ we have $J(x, \alpha) = +\infty$, while for $\lambda > 0$ we let $t \to +\infty$ in statement (ii) of Theorem III.2.33 and use the boundedness of u from above to get $u(x) \leq J(x, \alpha)$. The other statements are obvious, by the definition of compatible cost g and Theorem 3.6. ◀

The next theorem is one of the main results of this section.

3. PROBLEMS WITH NON-ZERO TERMINAL COST

THEOREM 3.15. *Assume* (A_0)-(A_3), (A_T), (3.1), *and* (3.2). *If* $u \in C(\overline{T^c})$ *is a supersolution of* (3.19) *bounded from below, then*

$$(3.24) \qquad u(x) \geq \inf_{\alpha \in \mathcal{A}} G(x, \alpha)$$

for all $x \in \overline{T^c}$, *where*

$$G(x,\alpha) := \begin{cases} \int_0^{t_x(\alpha)} \ell(y_x, \alpha) e^{-\lambda s} ds + e^{-\lambda t_x(\alpha)} u(y_x(t_x(\alpha))), & \text{if } t_x(\alpha) < +\infty \\ \int_0^\infty \ell(y_x, \alpha) e^{-\lambda s} ds, & \text{if } t_x(\alpha) = +\infty . \end{cases}$$

If, in addition, $u \geq g$ on ∂T, then $u \geq v$ in $\overline{T^c}$.

REMARK 3.16. In the proof of the theorem we also show the *global superoptimality principle*

$$(3.25) \qquad u(x) \geq \inf_{\alpha \in \mathcal{A}} \left\{ \int_0^{t \wedge t_x(\alpha)} \ell(y_x, \alpha) e^{-\lambda s} ds + e^{-\lambda(t \wedge t_x(\alpha))} u(y_x(t \wedge t_x(\alpha))) \right\}$$

for all $x \in \overline{T^c}$ and $t \geq 0$, under the same assumptions, except the boundedness below of u which is useless here. ◁

PROOF OF THEOREM 3.15. For simplicity we first give the proof under the additional assumption that u is bounded above, because in this case we can use the argument in the proof of Theorem III.2.32 and Remark III.2.34. Then we give the proof in the general case.

We recall from Chapter III the formula (III.2.46) in Remark III.2.34:

$$(3.26) \qquad u(x) \geq \inf_\alpha I(x, t, \alpha) \quad \text{for all } x \in T^c,\ 0 \leq t \leq T_\delta(x),\ \delta > 0,$$

where

$$(3.27) \qquad I(x, t, \alpha) := \int_0^t \ell(y_x(s, \alpha), \alpha(s)) e^{-\lambda s} ds + u(y_x(t, \alpha)) e^{-\lambda t},$$

and

$$T_\delta(x) := \inf\{t : \text{dist}(y_x(t, \alpha), \partial \Omega) \leq \delta,\ \alpha \in \mathcal{A}\} .$$

We extend I to $t = +\infty$ by setting

$$I(x, +\infty, \alpha) := \int_0^\infty \ell(y_x(s, \alpha), \alpha(s)) e^{-\lambda s} ds .$$

For given $x \in T^c$ and $\varepsilon > 0$ we are going to construct $\overline{\alpha} \in \mathcal{A}$ such that

$$(3.28) \qquad u(x) \geq I(x, t_x(\overline{\alpha}), \overline{\alpha}) - \varepsilon,$$

which immediately gives the desired inequality
$$u(x) \geq \inf_\alpha I(x, t_x(\alpha), \alpha) = \inf_\alpha G(x, \alpha) .$$

We define $x_1 := x$, $\tau_1 := T_1(x_1)$ and use (3.26) to get $\alpha_1 \in \mathcal{A}$ such that
$$u(x_1) \geq I(x_1, \tau_1, \alpha_1) - \varepsilon/2 .$$

Note that $y_{x_1}(\tau_1, \alpha_1) \in T^c$. By induction we define
$$x_k := y_{x_{k-1}}(\tau_{k-1}, \alpha_{k-1}), \qquad \tau_k := T_{1/k}(x_k),$$
and since $x_k \in T^c$ we use (3.26) to get $\alpha_k \in \mathcal{A}$ such that

(3.29) $$u(x_k) \geq I(x_k, \tau_k, \alpha_k) - \varepsilon \, 2^{-k} .$$

We also set $\sigma_k := \tau_1 + \cdots + \tau_k$, $\overline{\sigma} = \lim_k \sigma_k$,
$$\overline{\alpha}(s) := \begin{cases} \alpha_1(s) & \text{if } 0 \leq s < \sigma_1, \\ \alpha_2(s - \sigma_1) & \text{if } \sigma_1 \leq s < \sigma_2, \\ \cdots & \\ \alpha_k(s - \sigma_{k-1}) & \text{if } \sigma_{k-1} \leq s < \sigma_k, \\ \cdots & \\ \overline{a} & \text{if } \overline{\sigma} \leq s, \end{cases}$$

where the last line makes sense when $\overline{\sigma} < +\infty$, and $\overline{a} \in A$ is arbitrary. By the definition of x_k and $\overline{\alpha}$,

(3.30) $$y_x(s, \overline{\alpha}) = y_{x_k}(s - \sigma_{k-1}, \alpha_k) \in T^c \qquad \text{for all } s < \overline{\sigma}$$

and
$$e^{-\lambda \sigma_{k-1}} \int_0^{\tau_k} \ell(y_{x_k}, \alpha_k) e^{-\lambda s} \, ds = \int_{\sigma_{k-1}}^{\sigma_k} \ell(y_x, \overline{\alpha}) e^{-\lambda s} \, ds .$$

Then we use (3.27) and (3.29) to get

(3.31) $$\begin{aligned} u(x) &\geq \int_0^{\tau_1} \ell(y_x, \overline{\alpha}) e^{-\lambda s} \, ds + e^{-\lambda \sigma_1} u(x_2) - \varepsilon/2 \\ &\geq \int_0^{\sigma_2} \ell(y_x, \overline{\alpha}) e^{-\lambda s} \, ds + e^{-\lambda \sigma_2} u(x_3) - \varepsilon \left(\frac{1}{2} + \frac{1}{4}\right) \\ &\geq \cdots \\ &\geq I(x, \sigma_k, \overline{\alpha}) - \varepsilon(1 - 2^{-k}) . \end{aligned}$$

If $\overline{\sigma} = +\infty$ we use the boundedness from below of u to pass to the limit as $k \to \infty$ in (3.31) and get (3.28), because $t_x(\overline{\alpha}) = +\infty$ by (3.30) (note that this case cannot occur if $\lambda = 0$ because $\ell \geq 1$ by (3.1)).

On the other hand, in the case $\bar{\sigma} < +\infty$ we claim that

(3.32) $$\lim_{t \nearrow \bar{\sigma}} y_x(t, \bar{\alpha}) = \bar{x} \in \partial T .$$

Then again we can let $k \to \infty$ in (3.31) and obtain (3.28) because $t_x(\bar{\alpha}) = \bar{\sigma}$ by (3.30). To prove the claim we first note that the limit in (3.32) exists because any trajectory of the system is locally Lipschitz. Moreover $\bar{x} \in \overline{T^c}$. Let us assume by contradiction that $\bar{x} \in T^c$. We take $\delta > 0$ such that $B := B(\bar{x}, 3\delta) \subseteq T^c$ and set $C := \sup_{B \times A} |f| < +\infty$ by (A$_1$). For any $z \in B(\bar{x}, \delta)$, any trajectory starting at z remains in $B(\bar{x}, 2\delta)$ at least for the time δC^{-1}, and then $T_\delta(z) \geq \delta C^{-1}$. Thus there is $\bar{k}$ such that

$$\tau_k \geq \delta C^{-1} \qquad \text{for all } k > \bar{k},$$

and therefore $\bar{\sigma} = +\infty$. This gives the desired contradiction and completes the proof if u is bounded above.

To prove the global superoptimality principle in Remark 3.16, we fix $x \in T^c$, $\varepsilon > 0$, $\bar{t} > 0$, and modify slightly the construction of $\bar{\alpha}$ to get

(3.33) $$u(x) \geq I(x, \bar{t} \wedge t_x(\bar{\alpha}), \bar{\alpha}) - \varepsilon .$$

Of course, if $\bar{t} \geq \bar{\sigma} = t_x(\bar{\alpha})$, there is nothing to do. If $\bar{t} < \bar{\sigma}$, let j be such that $\sigma_{j-1} < \bar{t} \leq \sigma_j$, $\bar{\tau}_j := \bar{t} - \sigma_{j-1} \leq \tau_j$ and $\bar{\alpha}_j \in \mathcal{A}$ such that

$$u(x_j) \geq I(x_j, \bar{\tau}_j, \bar{\alpha}_j) - \varepsilon\, 2^{-j} .$$

Now redefine $\bar{\alpha}$ as before for $s < \sigma_{j-1}$, $\bar{\alpha}(s) := \bar{\alpha}_j(s - \sigma_{j-1})$ for $\sigma_{j-1} \leq s \leq \bar{t}$, and extend it arbitrarily for $s > \bar{t}$. Then the calculation in (3.31) now gives (3.33).

The additional assumption that u is bounded above has been used only to get (3.26), so it is enough to refine the cutoff argument of Theorem III.2.32 to prove (3.26) in the general case. For $\varepsilon > 0$ we set

$$\Omega_\varepsilon := \{ x \in T^c : \text{dist}(x, \partial T) > \varepsilon \text{ and } |x| < 1/\varepsilon \}$$

and take $h_\varepsilon, \varphi_\varepsilon \in C^1(\mathbb{R}^N)$ such that

$0 \leq h_\varepsilon \leq 1$ in $\mathbb{R}^N$, $h_\varepsilon \equiv 1$ in Ω_ε, $h_\varepsilon \equiv 0$ in $(\Omega_{3\varepsilon/4})^c$,

$0 \leq \varphi_\varepsilon \leq 1$ in $\mathbb{R}^N$, $\varphi_\varepsilon \equiv 1$ in $\Omega_{\varepsilon/2}$, $\varphi_\varepsilon \equiv 0$ in $(\Omega_{\varepsilon/4})^c$.

Define

$$u_\varepsilon(x) := \begin{cases} (\varphi_\varepsilon u)(x) & \text{in } T^c, \\ 0 & \text{in } T . \end{cases}$$

By multiplying the HJB equation (3.19) by $\varphi_\varepsilon h_\varepsilon$ it is easy to see that u_ε is a supersolution of

(3.34) $$\lambda w + \sup_{a \in A} \{ -f_\varepsilon(x, a) \cdot Dw - \ell_\varepsilon(x, a) \} = 0 \qquad \text{in } \mathbb{R}^N,$$

where $f_\varepsilon := \varphi_\varepsilon h_\varepsilon f$ and $\ell_\varepsilon := \varphi_\varepsilon h_\varepsilon \ell + (1 - h_\varepsilon)\lambda u_\varepsilon$. Now we proceed as in the proof of Theorem III.2.32: we consider the solution $w_\varepsilon(x,t)$ of the evolution equation associated with (3.34) (see (III.2.41)) with initial data $w_\varepsilon(x,0) = u_\varepsilon(x)$, and we observe that $u_\varepsilon \in BC(\mathbb{R}^N)$ is a (stationary) supersolution of this initial value problem and that the assumption of the comparison Theorem III.3.7 and Remark III.3.8 hold so that

(3.35) $$u(x) \geq u_\varepsilon(x) \geq w_\varepsilon(x,t) \quad \text{for } x \in \mathbb{R}^N, t \geq 0, \varepsilon > 0,$$

and

(3.36) $$w_\varepsilon(x,t) = \inf_\alpha \left\{ \int_0^t \ell_\varepsilon(y_x^{(\varepsilon)}, \alpha) e^{-\lambda s} ds + u_\varepsilon(y_x^{(\varepsilon)}(t)) e^{-\lambda t} \right\},$$

where $y_x^{(\varepsilon)}$ denotes the trajectories of the system associated with f_ε. For $x \in \mathcal{T}^c$, $\varepsilon > 0$, and $t < T_\varepsilon(x)$, $y_x^{(\varepsilon)}(s, \alpha) = y_x(s, \alpha) \in \Omega_\varepsilon$ for any α and all $s \leq t$. Since $\ell_\varepsilon = \ell$ and $u_\varepsilon = u$ in Ω_ε we combine (3.35) and (3.36) to get (3.26). ◀

A straightforward consequence of Corollary 3.14 and Theorem 3.15 is a representation formula for any bounded and continuous solution of the Dirichlet boundary value problem

(3.37) $$\begin{cases} \lambda u + H(x, Du) = 0 & \text{in } \mathcal{T}^c, \\ u = g, \end{cases}$$

for the Hamiltonian given by (3.18). As a uniqueness result it improves Proposition 3.13 considerably because the hypotheses of compatibility of g and STC$\mathcal{T}$ are dropped.

COROLLARY 3.17. *Assume* (A_0)-(A_3), $(A_\mathcal{T})$, (3.1), (3.2), *and* $\lambda > 0$. *If* $u \in BC(\overline{\mathcal{T}^c})$ *is a solution of* (3.37), *then*

$$u(x) = v(x) := \inf_\alpha \left\{ \int_0^{t_x} \ell(y_x, \alpha) e^{-\lambda s} ds + g(y_x(t_x)) e^{-\lambda t_x} \right\}$$

for all x.

REMARK 3.18. Under the additional assumption STC$\mathcal{T}$ there is an easy proof of Corollary 3.17 that avoids the (hard) Theorem 3.15. By Corollary 3.14, $u \leq v$, so $g = u_{|\partial \mathcal{T}}$ is compatible by definition and we conclude by Proposition 3.13. ◁

We end this section with a necessary and sufficient condition for the continuity of v which is an immediate consequence of Corollary 3.14, Theorem 3.15, and Proposition 3.3.

COROLLARY 3.19. *Under the assumptions of Corollary* 3.17, *the value function* v *is continuous in* $\overline{\mathcal{T}^c}$ *if and only if there exist* $u, w \in C(\overline{\mathcal{T}^c})$, u *a subsolution bounded above of the HJB equation* (3.19), w *a supersolution bounded below of* (3.19), $u = w = g$ *on* $\partial \mathcal{T}$. *Moreover, under this condition,* v *is the complete solution in* $BC(\overline{\mathcal{T}^c})$ *of* (3.37).

The statements of the last corollary also hold in the case $\lambda = 0$ without the assumption that the subsolution u is bounded above. However, the result is applicable only in the case that $\mathcal{R} = \mathbb{R}^N$. In the next section we give more precise results for the case $\lambda = 0$.

Exercises

3.1. Compute $\widehat{T}$ and $\widehat{v}$ in Example (3.1).

3.2. Let $f(x,a) = a$, $A = B(0,1)$, $h = 1$, T^c convex. Prove that
$$|g(x) - g(z)| \leq |x - z| \quad \text{for all } x, z \in \partial T$$
is necessary and sufficient for the compatibility of g.

3.3. Let $x_0 \in \partial T$. Assume there is $R > 0$, $G : B(x_0, R) \to X \subseteq \mathbb{R}^N$ Lipschitz continuous with a Lipschitz inverse, where X is a neighborhood of $G(x_0) = 0$, $G(T \cap B(x_0, R)) = \{x \in X : x_1 \leq 0\}$, where x_1 is the first component of x, and $G(\partial T \cap B(x_0, r)) = \{x \in X : x_1 = 0\}$ (G "straightens the boundary" near x_0). Assume there is $r > 0$ such that $f(x_0, A) \supseteq B(0, r)$, Show that $x \mapsto L(x_0, x)$ and $x \mapsto L(x, x_0)$ are Lipschitz continuous in a neighborhood of x_0.

3.4. Under the assumptions of Exercise 3.3 show that
$$L(x, z) = \widehat{L}(x, z) := \inf\left\{ \int_0^\tau \ell(y_x, \alpha)\, ds : \alpha \in \mathcal{A} \text{ such that } y_x(\tau) = z \text{ and } y_x(t) \in \overline{T^c} \text{ for all } t \right\}$$
for all $x, z \in \overline{T^c}$.

3.5. Prove Propositions 3.9 and 3.11 and Remark 3.10. Search for weaker assumptions on the data such that (3.16) remains true.

3.6. Let
$$\widehat{\mathcal{U}} := \{x \in \partial T : x = y_z(\widehat{t}_z(\alpha), \alpha) \text{ for some } z \in \mathbb{R}^N, \alpha \in \mathcal{A}\}.$$
Assume $\widehat{v}$ is continuous at each point of $\overline{\widehat{\mathcal{U}}}$. Prove that

(i) if $\lambda > 0$, then $\widehat{v} \in BUC(\mathbb{R}^N)$;

(ii) if $\lambda = 0$, then $\widehat{v} \in BUC(\mathcal{K})$ for any closed set $\mathcal{K}$ where $\widehat{T}$ is bounded.

4. Free boundaries and local comparison results for undiscounted problems with exit times

In this section we improve the basic theory for the HJB equations of problems with exit times developed in Sections 2 and 3, in the special case of null interest rate λ, so that the value function of the previous sections becomes
$$v(x) := \inf_{\alpha \in \mathcal{A}} \left\{ \int_0^{t_x(\alpha)} \ell(y_x(s), \alpha(s))\, ds + g(y_x(t_x(\alpha))) \right\},$$

for all x such that $t_x(\alpha) < +\infty$ for some α, $v(x) = +\infty$ otherwise. We first state the main results and then comment on them. Let us recall that the Hamiltonian is

(4.1) $$H(x,p) := \sup_{a \in A}\left\{-f(x,a) \cdot p - \ell(x,a)\right\}, \qquad x, p \in \mathbb{R}^N.$$

THEOREM 4.1. *Assume* (A_0)–(A_3), $(A_\mathcal{T})$, (3.1), (3.2). *Let* $\Omega \subseteq \mathbb{R}^N$ *be an open set*, $w_0 \in \mathbb{R} \cup \{+\infty\}$, $w \in C(\Omega)$ *be bounded below and satisfying*

$$\begin{cases} H(x, Dw) \geq 0 & \text{in } \Omega \setminus \mathcal{T}, \\ w \geq g & \text{on } \Omega \cap \mathcal{T}, \end{cases}$$

and

(4.2) $$\lim_{x \to x_0} w(x) = w_0 \qquad \text{for all } x_0 \in \partial\Omega,$$

(4.3) $$w(x) < w_0 \qquad \text{for all } x \in \Omega.$$

Then $v(x) \leq w(x)$ *for all* $x \in \Omega$.

THEOREM 4.2. *Under the assumptions of Theorem 4.1 let* Ω *be an open set such that* $\mathcal{T} \subseteq \Omega$, $w_0 \in \mathbb{R} \cup \{+\infty\}$, $w \in C(\Omega)$ *satisfy*

$$\begin{cases} H(x, Dw) \leq 0 & \text{in } \Omega \setminus \mathcal{T}, \\ w \leq g & \text{on } \mathcal{T}, \end{cases}$$

(4.2), *and* (4.3). *Then* $\Omega \supseteq \{x : v(x) < w_0\}$ *and* $v(x) \geq w(x)$ *for all* $x \in \Omega$.

These comparison results are different from usual because we are comparing the value function v with a super- or subsolution w on an open set Ω with no informations on the behavior of v on $\partial\Omega$, or on the continuity of v. Instead, we know that Ω is a sublevel set for w. An immediate consequence is the following uniqueness result, which is the counterpart of Corollary 3.17 for $\lambda = 0$.

COROLLARY 4.3. *Assume* (A_0)–(A_3), $(A_\mathcal{T})$, (3.1), (3.2), *and* $w_0 \in \mathbb{R} \cup \{+\infty\}$. *If there is a pair* (Ω, w) *such that* $\Omega \supseteq \mathcal{T}$ *is open*, $w \in C(\Omega)$ *is bounded below and satisfies*

(4.4) $$\begin{cases} H(x, Dw) = 0 & \text{in } \Omega \setminus \mathcal{T}, \\ w = g & \text{on } \mathcal{T}, \end{cases}$$

(4.2), *and* (4.3), *then* $\Omega = \{x : v(x) < w_0\}$ *and* $w = v$ *in* Ω.

Note that in the special case of time-optimal control, $\ell \equiv 1$, $g \equiv 0$, the conclusion of Corollary 4.3 is that $w = T$ in Ω and

$$\Omega = \begin{cases} \mathcal{R}(w_0) & \text{if } w_0 < +\infty, \\ \mathcal{R} & \text{if } w_0 = +\infty, \end{cases}$$

where T is the minimal time function and $\mathcal{R}(t), \mathcal{R}$ are the "reachable sets" defined in Section 1. Therefore Corollary 4.3 strengthens considerably the previous uniqueness results for (4.4), namely Theorem 2.6 and Proposition 3.13 (ii). In fact, for $w_0 = +\infty$, (4.2) and (4.3) reduce to

$$\lim_{x \to x_0} w(x) = +\infty \quad \text{for all } x_0 \in \partial\Omega, \tag{4.5}$$

and we see that the correct notion of solution of the boundary value problem (4.4), (4.5) is the pair (Ω, w), not just the function w. Moreover, if such a solution pair exists it must be $(\mathcal{R}, v)$, without assuming a priori that v is continuous. Thus $\partial\mathcal{R}$ is a *free boundary* for this problem, that is, a piece of the boundary which is not given but is part of the solution to be found.

Next we give the proofs of Theorem 4.1 and 4.2, then we turn to other applications and related results.

PROOF OF THEOREM 4.1. *Case* 1: $w_0 < +\infty$.

By (4.2) we can extend w continuously to $\overline{\Omega}$ by setting

$$w(x) := w_0 \quad \text{for all } x \in \partial\Omega. \tag{4.6}$$

Let $\tau_x(\alpha)$ be the exit time of the system from the open set $\Omega \smallsetminus \mathcal{T}$. By Theorem 3.15 we have

$$w(x) \geq \inf_{\alpha} \left\{ \int_0^{\tau_x(\alpha)} \ell(y_x, \alpha) \, ds + w(y_x(\tau_x(\alpha))) \right\}, \tag{4.7}$$

so if we prove that the infimum on the right-hand side does not change when we restrict ourselves to the controls α such that $\tau_x(\alpha) = t_x(\alpha)$, we immediately reach the conclusion because $w \geq g$ on $\partial\mathcal{T}$. Let us consider a control such that $\tau_x := \tau_x(\alpha) < +\infty$ and $y_x(\tau_x) \in \partial\Omega$. Since $\ell \geq 1$ we get from (4.6)

$$\int_0^{\tau_x} \ell(y_x, \alpha) \, ds + w(y_x(\tau_x)) \geq \tau_x + w_0 > w_0,$$

while $w(x) < w_0$ for $x \in \Omega$ by (4.3), so such a control is irrelevant for the calculation of the inf in (4.7).

Case 2: $w_0 = +\infty$. We recall the exponential (Kružkov) transformation

$$\Psi(t) := 1 - e^{-t}, \quad t \in \mathbb{R}, \tag{4.8}$$

and compare $W(x) := \Psi(w(x))$ and $V(x) := \Psi(v(x))$ after extending W continuously to $\overline{\Omega}$ by setting

$$W(x) := 1 \quad \text{for all } x \in \partial\Omega, \tag{4.9}$$

which is possible by (4.2). Since W is a supersolution of the HJB equation associated with V, the rest of the proof is similar to that of Case 1. We detail only the case $\ell \equiv 1$. (For the general case, involving a problem with nonconstant interest rate, see Exercise 4.1.) By Proposition II.2.5, W satisfies

$$W + \sup_{a \in A}\{-f(x,a) \cdot DW\} - 1 \geq 0 \quad \text{in } \Omega \setminus \mathcal{T},$$

so by Theorem 3.15, for all $x \in \Omega \setminus \mathcal{T}$,

$$W(x) \geq \inf_\alpha \left\{ \int_0^{\tau_x} e^{-s}\,ds + e^{-\tau_x} W(y_x(\tau_x)) \right\}.$$

If $y_x(\tau_x) \in \partial\Omega$, by (4.9) the term in brackets is 1, while $W(x) < 1$ for $x \in \Omega$, so the infimum does not change if we restrict ourselves to controls such that $\tau_x = t_x$, as in the proof of Case 1. Since $w \geq g$ on $\partial\mathcal{T}$ we then get

$$W(x) \geq \inf_\alpha \left\{ \int_0^{t_x} e^{-s}\,ds + e^{-t_x}\Psi(g(y_x(t_x))) \right\}$$
$$= \inf_\alpha \left\{ 1 - e^{-t_x} e^{-g(y_x(t_x))} \right\} = \Psi(v(x)) = V(x). \blacktriangleleft$$

PROOF OF THEOREM 4.2. We prove only the case $w_0 < +\infty$, since the case $w_0 = +\infty$ is obtained in a similar way after taking the exponential transforms (4.8) of w and v as in the previous proof: we leave this part as an exercise for the reader.

By (4.2) we can extend w continuously to $\overline{\Omega}$ by setting

(4.10) $$w(x) := w_0 \quad \text{for all } x \in \partial\Omega.$$

Let $\tau_x(\alpha)$ be the exit time of the system from the open set $\Omega \setminus \mathcal{T}$. Corollary 3.14 yields, for all $x \in \Omega$

(4.11) $$w(x) \leq \inf_\alpha \left\{ \int_0^{\tau_x} \ell(y_x, \alpha)\,ds + w(y_x(\tau_x)) \right\}.$$

To prove the first statement we assume by contradiction there is $x \notin \Omega$ such that $v(x) < w_0$, and fix $\overline{\alpha} \in \mathcal{A}$ such that

(4.12) $$\int_0^{t_x(\overline{\alpha})} \ell(y_x, \overline{\alpha})\,ds + g(y_x(t_x(\overline{\alpha}))) = w_0 - \delta.$$

for some $\delta > 0$. From $\ell \geq 1$ we get $t_x(\overline{\alpha}) < +\infty$. Since $\mathcal{T} \subseteq \Omega$ is closed and Ω is open, there is $t^* \in [0, t_x(\overline{\alpha})[$ such that $z := y_x(t^*, \overline{\alpha}) \in \partial\Omega$ and $y_x(t, \overline{\alpha}) \in \Omega$ for all $t \in]t^*, t_x(\overline{\alpha})]$. Therefore, $z_n := y_x(t^* + 1/n, \overline{\alpha}) \in \Omega$ satisfies $\tau_{z_n}(\alpha_n) = t_{z_n}(\alpha_n)$ for $\alpha_n(\cdot) := \overline{\alpha}(\cdot + t^* + 1/n)$. For these points we have, by (4.11),

$$w(z_n) \leq \int_0^{t_{z_n}(\alpha_n)} \ell(y_{z_n}, \alpha_n)\,ds + w(y_x(t_x(\overline{\alpha}), \overline{\alpha})).$$

Since $\ell \geq 0$, $w \leq g$, and by construction of z_n, α_n, we get from (4.12)

$$w(z_n) \leq w_0 - \delta.$$

But $w(z) = w_0$ by (4.10) and we get a contradiction by letting $n \to \infty$ because $w(z_n) \to w(z)$. Therefore $\Omega \supseteq \{x : v(x) < w_0\}$.

In view of (4.11) and $w \leq g$ on $\partial \mathcal{T}$, we can prove that $w \leq v$ in Ω by showing that the infimum in the definition of v does not change if it is restricted to controls α such that $\tau_x(\alpha) = t_x(\alpha)$. To this end we consider $\alpha \in \mathcal{A}$ such that

$$\int_0^{t_x(\alpha)} \ell(y_x, \alpha)\, ds + g(y_x(t_x(\alpha))) \leq v(x) + \varepsilon$$

for some $\varepsilon > 0$ to be chosen, and $\tau_x(\alpha) < t_x(\alpha)$. Then the integral in the left-hand side can be split as $\int_0^{\tau_x} + \int_{\tau_x}^{t_x}$ and we rewrite

$$\int_{\tau_x(\alpha)}^{t_x(\alpha)} \ell(y_x, \alpha)\, ds = \int_0^{t_z(\overline{\alpha})} \ell(y_z, \overline{\alpha})\, ds,$$

where $z := y_x(\tau_x(\alpha)) \in \partial \Omega$ and $\overline{\alpha}(\,\cdot\,) := \alpha(\,\cdot\, + \tau_x(\alpha))$, to get

$$v(x) + \varepsilon \geq \int_0^{\tau_x} \ell(y_x, \alpha)\, ds + v(z) \geq w_0,$$

where the last inequality follows from $\{x : v(x) < w_0\} \subseteq \Omega$ and $\ell \geq 0$. Now we observe that if $v(x) \geq w_0$ we get $v(x) \geq w(x)$ from (4.3). In the opposite case we choose $\varepsilon < w_0 - v(x)$, and we reach a contradiction if there exists a control as above. Then $v(x) \geq w(x)$ for all $x \in \Omega$. ◀

Next we consider the minimum time problem under merely local assumption on the system f, namely,

(4.13) $\qquad (f(x,a) - f(y,a)) \cdot (x - y) \leq L_R |x - y|^2$

for all $a \in A$, $R > 0$, $|x|, |y| \leq R$, instead of (A3). At the end of Section 2 we considered a boundary value problem for the minimal time function T in the escape controllability domain $\mathcal{R}_0$, and we proved the uniqueness of the solution under suitable assumptions, see Theorem 2.11. The next result is an improvement in the spirit of the previous Corollary 4.3: we show that $\mathcal{R}_0$ can be considered as part of the solution to be found, instead of a datum, so that $\partial \mathcal{R}_0$ is a free boundary.

THEOREM 4.4. *Assume* (A0)–(A2), (4.13), *and* (A$_\mathcal{T}$). *If there is a triple* (Ω, w_0, w) *such that* $\Omega \supseteq \mathcal{T}$ *is open and unbounded*, $w_0 \in \mathbb{R} \cup \{+\infty\}$, $w \in C(\Omega)$ *is bounded below and satisfies*

$$\begin{cases} \sup_{a \in A}\{-f(x,a) \cdot Dw\} - 1 = 0 & \text{in } \Omega \smallsetminus \mathcal{T}, \\ w = 0 & \text{in } \mathcal{T}, \\ w(x) \to w_0 & \text{as } x \to x_0 \in \partial\Omega \text{ and as } |x| \to \infty, \\ w(x) < w_0 & \text{for all } x \in \Omega, \end{cases}$$

then $\Omega = \mathcal{R}_0$, $w_0 = T_0$ *(the escape time) and* $w = T$ *in* Ω.

PROOF. Since $w < w_0$, $\lim_{|x|\to\infty} w(x) = w_0$ if and only if the sets
$$\Omega_t := \{\, x \in \Omega : w(x) < t \,\}$$
are bounded for all $t < w_0$. Therefore, once we have fixed such a t, we can extend the restriction of the system f to $\Omega_t \times A$ to a new system $f^* : \mathbb{R}^N \times A \to \mathbb{R}^N$ satisfying the global condition (A3) (for instance, $f^*(x,a) := f(x,a)h(x)$ with $h \in C^1(\mathbb{R}^N)$, $0 \le h \le 1$, $|Dh| \le 1$, $h \equiv 1$ in Ω_t and $h \equiv 0$ off some bounded set). Let $y_x^*(\,\cdot\,)$ and T^* be, respectively, the trajectories and the minimal time function corresponding to f^*, and
$$\mathcal{R}^*(t) := \{\, x \in \mathbb{R}^N : T^*(x) < t \,\}.$$
We can apply Corollary 4.3 in Ω_t and get $\Omega_t = \mathcal{R}^*$ and $w = T^*$ in Ω_t.

We are now going to prove $T^*(x) = T(x)$ for all $x \in \Omega_t$. Since $T^*(x) < t$, the only relevant controls $\alpha \in \mathcal{A}$ are those such that $t_x^*(\alpha) < t$, and for these controls $y_x^*(s) \in \mathcal{R}^*(t)$ for all $0 \le s \le t_x^*(\alpha)$ because $t_x^*(\alpha) = s + t_{y_x^*(s)}^*(\alpha(\,\cdot\,+s))$. Since these trajectories remain in $\mathcal{R}^*(t) = \Omega_t$, they are also trajectories for the original system f, so $T(x) \le T^*(x)$.

To prove the converse inequality, we argue by contradiction. If $T(x) < T^*(x)$ for some $x \in \mathcal{R}^*(t)$, let $\overline{\alpha} \in \mathcal{A}$ be such that $t_x(\overline{\alpha}) < T^*(x)$, so that the corresponding trajectory $y(\,\cdot\,) = y_x(\,\cdot\,,\overline{\alpha})$ exits $\mathcal{R}^*(t)$. Since $\mathcal{T} \subseteq \mathcal{R}^*(t)$ is closed and $\mathcal{R}^*(t)$ is open, there is $s' > 0$ such that $z := y(s') \in \partial\mathcal{R}^*(t)$ and $y(s) \in \mathcal{R}^*(t)$ for all $s \in \,]s', t_x(\overline{\alpha})] \ne \emptyset$. Then
$$t > T^*(x) > t_x(\overline{\alpha}) > t_z(\overline{\alpha}(\,\cdot\,+s')) = t_z^*(\overline{\alpha}(\,\cdot\,+s')) \ge T^*(z),$$
so $z \in \mathcal{R}^*(t)$, a contradiction to $z \in \partial\mathcal{R}^*(t)$, because $\mathcal{R}^*(t)$ is open. Therefore $T^* \equiv T$ in Ω_t.

Now we can conclude that $w = T$ in Ω because $\Omega = \cup_{0 \le t < w_0} \Omega_t$.

To prove the other statements let us first observe that $\mathcal{R}(w_0) \supseteq \Omega$ and Ω unbounded give $w_0 \ge T_0$. Next we show that $\mathcal{R}(t) = \Omega_t$ for all $t < w_0$, so that $w_0 = T_0$ because Ω_t are bounded for all t, and $\mathcal{R}_0 = \Omega$ because $\mathcal{R}_0 = \cup_{0 \le t < T_0} \mathcal{R}(t)$ by definition.

Since $T = w$ in Ω, $\mathcal{R}(t) \supseteq \Omega_t$ and a point $x \in \mathcal{R}(t) \smallsetminus \Omega_t$ should be out of Ω. If $T(x) < t$, there is a trajectory joining x with the target and lying in $\mathcal{R}(t)$. This is impossible for $x \notin \Omega$, because any trajectory reaching the target goes through Ω_s for all $0 \le s < w_0$. Then $\mathcal{R}(t) = \Omega_t$ for all $t < w_0$, and the proof is complete. ◂

Next we present a *local comparison result* whose main difference with respect to Theorems 4.1 and 4.2 is that we make less assumptions on the sub- and supersolution w and we can compare w with the value function v only an a subset of its domain. For simplicity, we restrict ourselves to the case $g \equiv 0$ and look at sub- and supersolutions of

(4.14) $$\begin{cases} H(x, Dw) = 0 & \text{in } B(x_0, r) \smallsetminus \mathcal{T}, \\ w = 0 & \text{on } \mathcal{T}, \end{cases}$$

where $x_0 \in \partial\mathcal{T}$ and H is given by (4.1). See Exercise 4.2 for extensions.

4. FREE BOUNDARIES AND LOCAL COMPARISON RESULTS

THEOREM 4.5. *Assume* (A_0)–(A_3), (A_T), (3.1), $x_0 \in \partial T$, $r > 0$, $w \in C(\bar{B}(x_0, r))$. *If w is a subsolution of* (4.14) *(respectively, a supersolution such that $w(x_0) = 0$ and $w(x) \geq 0$ for $|x - x_0| = r$), then there exists $s > 0$ such that $w(x) \leq v(x)$ (respectively, $\geq$) for all $x \in B(x_0, s)$.*

PROOF. For mere notational simplicity we assume $\ell \equiv 1$ and $x_0 = 0$. We denote with $\tau_x = \tau_x(\alpha)$ the exit time of the system from the open set $B(0, r) \smallsetminus T$.

We begin with the case of the supersolution. By Theorem 3.15

$$(4.15) \qquad w(x) \geq \inf_\alpha \{\, \tau_x(\alpha) + w(y_x(\tau_x(\alpha))) \,\} =: u(x)$$

and we want to show that, for $|x|$ small enough, the infimum does not change if it is restricted to controls such that $\tau_x(\alpha) = t_x(\alpha)$. Then $u(x) \geq T(x)$ because $w \geq 0$ on ∂T. Denote

$$\|f\| := \sup_{\bar{B}(0,r) \times A} |f|$$

and note that, for any control α and for $|x| \leq r$, $|y_x(t, \alpha) - x| \leq \|f\| t$ as long as the trajectory remains in $\bar{B}(0, r)$. Observe that the existence of the supersolution w implies $\|f\| > 0$ because $H(x, p) \equiv -1$ if $f \equiv 0$. If $|x| \leq s$ and $\tau_x(\alpha) \neq t_x(\alpha)$, then $|y_x(\tau_x(\alpha))| = r$ and we can estimate

$$(4.16) \qquad \tau_x(\alpha) \geq \frac{r - s}{\|f\|}\,.$$

Now we use $w(x_0) = 0$ and the continuity of w to choose $s > 0$ small enough so that

$$(4.17) \qquad \sup_{\bar{B}(x_0, s)} w \leq \frac{r - s}{2\|f\|}\,.$$

We also choose $0 < \delta < \dfrac{r - s}{2\|f\|}$, and take any $\alpha \in \mathcal{A}$ such that

$$\tau_x(\alpha) + w(y_x(\tau_x(\alpha))) \leq u(x) + \delta\,.$$

If $|x| \leq s$ and $\tau_x(\alpha) \neq t_x(\alpha)$, then (4.15), (4.16) and $w(z) \geq 0$ for $|z| = r$ imply

$$w(x) \geq \frac{r - s}{\|f\|} - \delta > \frac{r - s}{2\|f\|},$$

a contradiction to (4.17). Therefore $u(x) \geq T(x)$ for all $x \in \bar{B}(0, s)$, which is the desired conclusion in the case where w is a supersolution.

In the case where w is a subsolution we choose s satisfying (4.17) as before and assume by contradiction $w(x) > T(x)$ for some $x \in B(0, s)$ (in this case we can assume $\|f\| > 0$ because $T \equiv \infty$ in $B(0, r) \smallsetminus T$ if $\|f\| = 0$). Then there is $\bar{\alpha} \in \mathcal{A}$ such that

$$(4.18) \qquad t_x(\bar{\alpha}) < w(x)\,.$$

By Corollary 3.14

$$w(x) \leq \inf_\alpha \{\, \tau_x(\alpha) + w(y_x(\tau_x(\alpha))) \,\}$$
$$\leq \inf\{\, t_x(\alpha) : \alpha \in \mathcal{A}, \ t_x(\alpha) = \tau_x(\alpha) \,\},$$

where we used $w \leq 0$ on $\partial \mathcal{T}$ in the second inequality. Then, by (4.18), $\tau_x(\overline{\alpha}) < t_x(\overline{\alpha})$ and we can use (4.16) to get

$$\frac{r-s}{\|f\|} < w(x),$$

which gives a contradiction to (4.17). Therefore $w \leq T$ in $B(0,s)$. ◄

Now we turn to some applications of the previous theory. As we explained in §2.2 of Chapter III, a comparison result for the HJB equation can usually be translated into a verification theorem, where subsolutions are verification functions and supersolutions are falsification functions. For instance, if Ω and w are as in Theorem 4.2, and

$$J(x, \overline{\alpha}) := \int_0^{t_x(\overline{\alpha})} \ell(y_x, \overline{\alpha})\, ds + g(y_x(t_x(\overline{\alpha}))) \leq w(x)$$

for some $x \in \Omega$, then $\overline{\alpha}$ is an optimal control for x. This sufficient condition of optimality is also necessary if the value function v is continuous, by taking $w = v$ and $\Omega = \{\, z : v(z) < w_0 \,\}$ with $w_0 > v(x)$. Similarly, for Ω, w as in Theorem 4.1, a control $\overline{\alpha}$ is not optimal for $x \in \Omega$ if $J(x, \overline{\alpha}) > w(x)$, and the condition of non-optimality is necessary as well if v is continuous.

The second application concerns the continuity of the value function v. A first simple consequence of Theorem 4.2 is that the terminal cost g is compatible (see Definition 3.5) if, in addition to the hypotheses of the theorem, $w = g$ on $\partial \mathcal{T}$. Therefore in this case v is continuous in $\mathcal{R}$ if the system is small-time controllable on $\mathcal{T}$ (see Definition 1.1) by Theorem 3.6.

The next result characterizes the continuity of v even if STC$\mathcal{T}$ fails. It is a counterpart of Corollary 3.19 for $\lambda = 0$; note that here we can use local, instead of global, sub- and supersolutions. Its proof is trivial by Theorems 4.1 and 4.2 and Proposition 3.3.

COROLLARY 4.6. *Under the assumptions* (A$_0$)–(A$_3$), (A$_\mathcal{T}$), (3.1), *and* (3.2), $\mathcal{R}$ *is open and the value function v is continuous in $\mathcal{R}$ if and only if there exist Ω, Ω' open sets containing $\mathcal{T}$ and $w \in C(\Omega)$, $u \in C(\Omega')$, such that $H(x, Dw) \geq 0$ in $\Omega \setminus \mathcal{T}$, $H(x, Du) \leq 0$ in $\Omega' \setminus \mathcal{T}$, $w = u = g$ on $\mathcal{T}$, w satisfies (4.2), (4.3) for some $w_0 \leq +\infty$, and $u < u_0$, $\lim_{x \to x_0} u(x) = u_0$ for all $x_0 \in \partial \Omega'$, for some constant $u_0 \leq +\infty$.*

In the special case $g \equiv 0$ one obviously takes $u \equiv 0$ and gets a characterization of small-time controllability STC$\mathcal{T}$. A better result in this case is immediately obtained from Theorem 4.5, Proposition 1.2, and a standard compactness argument.

4. FREE BOUNDARIES AND LOCAL COMPARISON RESULTS

COROLLARY 4.7. *Assume* (A_0)–(A_3), *and* $(A_\mathcal{T})$. *Then the system* (f, A) *is* STC$\mathcal{T}$ *if and only if for all* $x_0 \in \partial \mathcal{T}$ *there exist* $r > 0$ *and* $w \in C(\overline{B}(x_0, r))$ *satisfying*

(4.19)
$$\begin{cases} \sup_{a \in A}\{ -f(x,a) \cdot Dw \} - 1 \geq 0 & in\ B(x_0, r) \smallsetminus \mathcal{T}, \\ w(x) \geq 0 & for\ x \in \partial(B(x_0, r) \smallsetminus \mathcal{T}), \\ w(x_0) = 0\ . \end{cases}$$

Since the result is of local nature we can actually replace (A_3) with (4.13).

Some explicit applications of Corollary 4.7 can be easily given. The first is in the case $\mathcal{T} = \{0\}$. Theorem 1.12 says that the minimal time function T is locally lipschitzean if and only if

(4.20)
$$\inf_{a \in A} f(0, a) \cdot \gamma < 0 \quad \text{for any unit vector } \gamma\ .$$

In Section 1 we proved the necessity part. Here is a short proof of the sufficiency part.

END OF THE PROOF OF THEOREM 1.12. By (4.20), (A_2), and a continuity argument, there are $r, \delta > 0$ such that

$$\inf_{a \in A} f(x, a) \cdot \gamma < -\delta \quad \text{for } |x| < r, |\gamma| = 1\ .$$

Then $w(x) := |x|/\delta$ is a classical solution of (4.19) with $x_0 = 0$, $\mathcal{T} = \{0\}$. By Corollary 4.7 the system is STC$\mathcal{T}$. Moreover, $T(x) \leq w(x)$ implies the local Lipschitz continuity of T by Remark 1.7. ◀

Now we extend this simple argument to an arbitrary closed and bounded target $\mathcal{T}$. We know from Theorem 1.16 that, for smooth targets, T is locally Lipschitz if and only if

$$\inf_{a \in A} f(x, a) \cdot n(x) < 0 \quad \text{for all } x \in \partial \mathcal{T},$$

where $n(x)$ is the exterior normal to $\mathcal{T}$. The next result uses a similar condition where $n(x)$ is replaced by the generalized exterior normals introduced in Chapter II, Definition II.2.17. It includes Theorems 1.12, 1.16, and the extension to piecewise smooth targets of Remark 1.21 as special cases. We recall that $N(x)$ denotes the set of all the exterior normals to $\mathcal{T}$ at x.

THEOREM 4.8. *Assume* (A_0)–(A_3), $(A_\mathcal{T})$, *and* $x_0 \in \partial \mathcal{T}$. *If there exist* $s, \delta > 0$ *such that*

(4.21)
$$\inf_{a \in A} f(x_0, a) \cdot \nu \leq -\delta$$

for any $\nu \in N(x)$ *with* $x \in \partial \mathcal{T} \cap B(x_0, s)$, *then the minimal time function* T *is locally lipschitzean in a neighborhood of* x_0.

PROOF. Define
$$F(x,p) := \inf_{a \in A} f(x,a) \cdot p$$

and observe that it is continuous in $\mathbb{R}^{2N}$ by (A$_2$). Let $\mathcal{N} := \{\nu \in N(x) : x \in \partial T \cap B(x_0, s)\}$. Since $|\nu| = 1$ for any normal, the function

$$G(x) := \max_{p \in \mathcal{N}} F(x,p)$$

exists and is continuous. Moreover, (4.21) implies $G(x_0) \leq -\delta$, so there is $r > 0$, $r \leq s/2$, such that

$$G(x) \leq -\frac{\delta}{2} \quad \text{for all } x \in B(x_0, r).$$

Then

(4.22) $\quad \sup_{a \in A} \{-f(x,a) \cdot \nu\} \geq \dfrac{\delta}{2} \quad \text{for all } \nu \in \mathcal{N}, x \in B(x_0, r).$

We claim that $w(x) := (2/\delta)\,\mathrm{dist}(x, T)$ satisfies (4.19). In fact, we have only to check the differential inequality. By Proposition II.2.14, either x has a unique projection onto T, $p(x)$, and then $D^- w(x) = \{(2/\delta)\nu^*\}$, where $\nu^* = \dfrac{x - p(x)}{|x - p(x)|}$ is a normal at $p(x)$, or $D^- w(x) = \emptyset$. In the latter case there is nothing to prove. In the former case, if $x \in B(x_0, r)$, then

$$|x_0 - p(x)| \leq |x_0 - x| + |x - p(x)| \leq 2r \leq s,$$

so $\nu^* \in \mathcal{N}$ and the claim follows from (4.22). Therefore Theorem 4.5 gives $T(x) \leq w(x)$ in some ball $B(x_0, r')$, $r' > 0$, and this gives the Lipschitz continuity of T in a smaller ball by the argument of the proof of Proposition 1.6. ◂

REMARK 4.9. It is easy to see that the main assumption of Theorem 4.8 is equivalent to the existence of $\sigma, \mu > 0$ such that

(4.23) $\quad \inf_{a \in A} f(x,a) \cdot \nu \leq -\mu \quad \text{for all } \nu \in N(x),$

for all $x \in \partial T \cap B(x_0, \sigma)$. By the methods of Section 1 it can also be proved that T is locally lipschitzean in all $\mathcal{R}$ if (4.23) holds for all $x \in \partial T$. Actually, the converse is also true: if T is lipschitzean in a neighborhood of T, then (4.23) holds for all $x \in \partial T$, see Exercise 4.5. ◁

Exercises

4.1. Complete the proofs of Theorems 4.1 and 4.2 in the case $w_0 = +\infty$. Show that both theorems hold if the following assumptions are dropped:

(i) the boundedness of T;

(ii) $g \geq 0$, provided g is bounded from below;

(iii) (3.1), provided $\ell : \mathbb{R}^N \times A \to \mathbb{R}$ is continuous, $\ell(x,a) \geq \ell_0 > 0$ for all x, a,

$$|\ell(x,a) - \ell(y,a)| \leq w(|x-y|, R), \qquad \text{for all } a \in A, \ |x|, |y| \leq R,$$

$R > 0$, for a suitable modulus w (and this follows from continuity if A is compact). The result can be checked in [Sor93a].

4.2. Extend Theorem 4.5 to the case of non-null continuous boundary data g, satisfying

$$g(x_0) < \frac{r}{\|f\|} + w_m, \quad g(x) \geq w_m \qquad \text{for all } x \in \partial T,$$

where $\|f\| := \sup_{\bar{B}(x_0,r) \times A} |f|$ and

$$w_m := \min_{\partial(B(x_0,r) \setminus T)} w \ .$$

Show also that Theorem 4.5 holds in the generality of Exercise 4.1. (See [BS91a].)

4.3. Suppose $T = \{0\}$, $\ell \equiv 1$, $g \equiv 0$, Ω open, $0 \in \Omega$ and let $w \in C(\Omega)$ solve $H(x, Dw) = 0$ in Ω and $w(x) > 0$ for $x \neq 0$, $w(0) = 0$. Use Theorems 4.1, 4.2, 4.5 and their proofs to find the largest set where we can ensure that $w = T$.

4.4. Assume (A_T) and let $\Omega \supseteq T$ be open and connected. Give an example of system f satisfying (A_0)- (A_3) such that $\mathcal{R} = \Omega$. (This example says that the free boundary $\partial \mathcal{R}$ does not have any regularity in general).

4.5. Assume f satisfies (A_0) with A compact and is lipschitzean with respect to the state variable, uniformly in the control variable. Let T be any closed set. Prove that the minimal time function T is locally lipschitzean in $\mathcal{R}$ if and only if for all $R > 0$ there exists $\mu_R > 0$ such that

$$\min_{a \in A} f(x,a) \cdot \nu \leq -\mu_R \qquad \forall \nu \in N(x), \ \forall x \in \partial T \cap B(0,R) \ .$$

5. Problems with state constraints

In this section we consider a problem where the trajectories of the system must verify a state-space constraint, that is, they have to remain in a given set for all time. More precisely, we are given an open set Ω, and for the initial position $x \in \overline{\Omega}$ we consider the following set of controls

$$\mathcal{A}_x := \{\alpha \in \mathcal{A} : y_x(t,\alpha) \in \overline{\Omega} \quad \text{for all } t \geq 0\} \ .$$

The problem is to minimize the infinite horizon cost

$$J(x,\alpha) := \int_0^\infty \ell(y_x(t), \alpha(t)) e^{-\lambda t} \, dt$$

over the set of admissible controls $\mathcal{A}_x$ which keep the state of the system in the closed set $\overline{\Omega}$ for all time. Therefore the value function is

$$v(x) := \inf_{\alpha \in \mathcal{A}_x} J(x, \alpha).$$

The basic assumption on ℓ and λ is (A$_4$), see Section 2.1 of Chapter III. We first study the continuity of v under the following additional assumptions

(5.1) $\quad\quad\quad\quad \partial\Omega$ is of class C^2 and compact;

(5.2) $\quad\quad\quad\quad K := \sup_{\mathbb{R}^N \times A} |f| < +\infty;$

(5.3) $\quad\quad\quad\quad \inf_{a \in A} f(x, a) \cdot n(x) < 0 \quad\quad$ for all $x \in \partial\Omega;$

where $n(x)$ is the exterior normal to Ω at x. Note that (5.3) means that at any point of the boundary of $\overline{\Omega}$ the system can choose a direction pointing inward Ω. It is rather intuitive that under this condition admissible controls exist, that is,

(5.4) $\quad\quad\quad\quad \mathcal{A}_x \neq \emptyset \quad\quad$ for all $x \in \overline{\Omega}$.

We first give a technical lemma, which in particular proves (5.4).

LEMMA 5.1. *Assume* (A$_0$)–(A$_4$) *with* $\omega_\ell(r) = L_\ell r$, *and* (5.1)–(5.3). *Then there exist* $t^*, C^* > 0$ *such that for any* $x \in \overline{\Omega}$ *and* $\alpha \in \mathcal{A}$ *there is* $\overline{\alpha} \in \mathcal{A}_x$ *satisfying*

(5.5) $\quad\quad\quad |J_{t^*}(x, \overline{\alpha}) - J_{t^*}(x, \alpha)| \leq C^* \sup_{0 \leq t \leq t^*} \mathrm{dist}(y_x(t, \alpha), \overline{\Omega}),$

where

(5.6) $\quad\quad\quad\quad J_{t^*}(x, \alpha) := \int_0^{t^*} \ell(y_x(t, \alpha), \alpha(t)) e^{-\lambda t} \, dt.$

PROOF. We define the *signed distance* from $\partial\Omega$

$$d(x) := \begin{cases} \mathrm{dist}(x, \partial\Omega), & \text{if } x \in \overline{\Omega}, \\ -\mathrm{dist}(x, \partial\Omega), & \text{if } x \notin \overline{\Omega}, \end{cases}$$

and recall that it is of class C^2 in a neighborhood of $\partial\Omega$ by (5.1). By (5.3), (A$_2$) and the compactness of $\partial\Omega$, there are $\delta, \zeta > 0$ such that, for any $x_0 \in \partial\Omega$ there is $a(x_0) \in A$ satisfying

(5.7) $\quad\quad\quad f(x, a(x_0)) \cdot Dd(x) > \zeta \quad\quad$ for all $x \in B(x_0, \delta)$.

Note also that by (5.2), for any $\alpha \in \mathcal{A}$

(5.8) $\quad\quad\quad\quad |y_x(t, \alpha) - x| \leq Kt \quad\quad$ for all $t > 0$.

5. PROBLEMS WITH STATE CONSTRAINTS

We want to determine two constants $t^*, k > 0$ such that for any $x \in \overline{\Omega}$ and $\alpha \in \mathcal{A}$, if we set

$$t_0 := t^* \wedge t_x(\alpha), \qquad z := y_x(t_0), \qquad \varepsilon := \sup_{0 \le t \le t^*} \mathrm{dist}(y_x(t, \alpha), \overline{\Omega}),$$

where $t_x(\alpha)$ is the exit time from Ω, and define

$$\begin{cases} \overline{\alpha} \equiv \alpha & \text{if } t_0 = t^*, \\ \overline{\alpha}(t) := \alpha(t)\chi_{[0,t_0[} + a(z)\chi_{[t_0,t_0+k\varepsilon]} + \alpha(t-k\varepsilon)\chi_{]t_0+k\varepsilon,\infty[} & \text{if } t_0 = t_x(\alpha), \end{cases}$$

then $y_x(t, \overline{\alpha}) \in \overline{\Omega}$ for all $t \in [0, t^*]$. This is proved by showing that $d(y_x(t, \overline{\alpha})) \ge 0$ for any such t. By repeating this construction on the intervals $[nt^*, (n+1)t^*]$ we obtain a control in $\mathcal{A}_x$.

We indicate for brevity

$$y(t) := y_x(t, \alpha), \qquad \overline{y}(t) = y_x(t, \overline{\alpha}) .$$

First observe that for $d(x) \ge \delta/2$ we can choose $t^* < \delta/2K$ and get from (5.8) $y(t) \in \overline{\Omega}$ for $t \in [0, t^*]$. So in this case $t_0 = t^*$ and $\overline{\alpha} = \alpha$. Now suppose $d(x) < \delta/2$ and $t_0 = t_x(\alpha) < t^* < \delta/2K$. Then $|\overline{y}(t) - x| < \delta/2$ by (5.8) for $t \le t^*$, and thus

$$|\overline{y}(t) - z| < \delta \qquad \text{for } t \in [0, t^*],$$

where $z = y(t_0) = \overline{y}(t_0) \in \partial\Omega$. Therefore (5.7) and the definition of $\overline{\alpha}$ give

$$f(\overline{y}(t), \overline{\alpha}(t)) \cdot Dd(\overline{y}(t)) > \zeta \qquad \text{for } t_0 \le t \le t^* \wedge (t_0 + k\varepsilon) .$$

We can use Taylor's formula for $d \circ \overline{y}$ centered at t_0 to get

$$d(\overline{y}(t)) > \zeta(t - t_0) \ge 0 \qquad \text{for } t_0 \le t \le t^* \wedge (t_0 + k\varepsilon)$$

and, in the case $t^* > t_0 + k\varepsilon$,

$$d(\overline{y}(t)) > \zeta k\varepsilon + I(t) \qquad \text{for } t \in [t_0 + k\varepsilon, t^*],$$

where

$$I(t) := \int_{t_0+k\varepsilon}^{t} Dd(\overline{y}(s)) \cdot f(\overline{y}(s), \overline{\alpha}(s)) \, ds .$$

We estimate the integrand function in $I(t)$ by adding and subtracting $Dd(\overline{y}(s)) \cdot f(y(s - k\varepsilon), \overline{\alpha}(s))$ and $Dd(y(s - k\varepsilon)) \cdot f(y(s - k\varepsilon), \overline{\alpha}(s))$:

$$Dd(\overline{y}(s)) \cdot f(\overline{y}(s), \overline{\alpha}(s)) \ge -(L + C)|\overline{y}(s) - y(s - k\varepsilon)| + (d \circ y)'(s - k\varepsilon),$$

where C is a Lipschitz constant for Dd in $B(z, \delta)$. Now we use Gronwall inequality to estimate

(5.9)
$$\begin{aligned} |\overline{y}(s) - y(s - k\varepsilon)| &\le |\overline{y}(t_0 + k\varepsilon) - z| e^{L(s - t_0 - k\varepsilon)} \\ &\le K k\varepsilon \, e^{L(s - t_0 - k\varepsilon)} \qquad \text{for } s > t_0 + k\varepsilon, \end{aligned}$$

where the last inequality follows from (5.8). Therefore we get, for a suitable constant C_1 and $t \in [t_0 + k\varepsilon, t^*]$,

$$d(\overline{y}(t)) \geq k\varepsilon[\zeta - C_1(e^{L(t-t_0-k\varepsilon)} - 1)] + d(y(t - k\varepsilon)).$$

Consequently, $d(\overline{y}(t)) \geq \dfrac{k\varepsilon\zeta}{2} - \varepsilon$ if we choose

$$t^* := \min\left\{\frac{\delta}{3K}, \frac{1}{L}\log\left(1 + \frac{\zeta}{2C_1}\right)\right\}$$

and use the definition of ε and d. Then we choose $k := 2/\zeta$ to obtain $d(\overline{y}(t)) \geq 0$ for all $t \in [0, t^*]$. This proves that $\overline{\alpha} \in \mathcal{A}_x$.

It remains to prove (5.5) and we do it in the case $t^* > t_0 + k\varepsilon$, since the other case is similar and easier. By (A_4) and the definition of $\overline{\alpha}$ we get

$$|J_{t^*}(x, \overline{\alpha}) - J_{t^*}(x, \alpha)|$$

$$\leq \int_{t_0}^{t^*} |\ell(\overline{y}, \overline{\alpha}) - \ell(y, \alpha)| e^{-\lambda t} dt$$

$$\leq \int_{t_0}^{t_0+k\varepsilon} |\ell(\overline{y}, \overline{\alpha})| dt + \int_{t_0+k\varepsilon}^{t^*} |\ell(\overline{y}(s), \overline{\alpha}(s)) - \ell(y(s - k\varepsilon), \overline{\alpha}(s))| ds$$

$$+ \int_{t^*-k\varepsilon}^{t^*} |\ell(y, \alpha)| dt$$

$$\leq Mk\varepsilon + t^* L_\ell \sup_{t_0+k\varepsilon \leq s \leq t^*} |\overline{y}(s) - y(s - k\varepsilon)| + Mk\varepsilon \leq k\varepsilon(2M + t^* L_\ell K e^{Lt^*}),$$

where we have used $\omega_\ell(r) = L_\ell r$ and (5.9). By the definition of ε the proof is now complete. ◀

THEOREM 5.2. *Assume* (A_0)–(A_4) *with* $\omega_\ell(r) = L_\ell r$, *and* (5.1)–(5.3). *Then the value function v is in* $BUC(\overline{\Omega})$.

PROOF. We assume for simplicity $\lambda = 1$. Let $x, z \in \overline{\Omega}$ with $|x - z| < r$. For any $\delta > 0$ there is $\alpha \in \mathcal{A}_z$ such that

(5.10) $$J_{t^*}(z, \alpha) + e^{-t^*} v(y_z(t^*, \alpha)) \leq v(z) + \delta,$$

where t^* is as in Lemma 5.1. Since $y_z(t, \alpha) \in \overline{\Omega}$, by the estimate (0.3)

(5.11) $$\sup_{0 \leq t \leq t^*} \text{dist}(y_x(t, \alpha), \overline{\Omega}) \leq r e^{Lt^*}.$$

Now let $\overline{\alpha} \in \mathcal{A}_x$ be as in Lemma 5.1 so that

(5.12) $$|J_{t^*}(x, \overline{\alpha}) - J_{t^*}(x, \alpha)| \leq C_0 r$$

for a suitable constant C_0. By the definition of $\overline{\alpha}$ and the estimates (5.9) and (5.11) there is also $C_1 > 0$ such that

$$|y_x(t, \overline{\alpha}) - y_x(t, \alpha)| \leq C_1 r \quad \text{for } t \in [0, t^*].$$

Then, by (0.3),

(5.13) $\quad |y_x(t,\overline{\alpha}) - y_z(t,\alpha)| \leq (C_1 + e^{Lt})r \quad$ for $t \in [0, t^*]$;

by (A_4) with $w_\ell(r) = L_\ell r$ and (5.12) there is $C > 0$ such that

(5.14) $\quad |J_{t^*}(x,\overline{\alpha}) - J_{t^*}(z,\alpha)| \leq Cr$.

Now let, for $r > 0$,

$$\omega(r) = \sup\{|v(x) - v(z)| : x, z \in \overline{\Omega},\ |x - z| < r\}.$$

Our goal is to show that $\omega \to 0$ as $r \to 0^+$, and for this it is enough to find $r_n \to 0^+$ such that $\omega(r_n) \to 0$ as $n \to \infty$. We use the Dynamic Programming Principle (see Proposition 5.5), (5.10), (5.13) and (5.14) to get

$$v(x) - v(z) \leq J_{t^*}(x,\overline{\alpha}) + e^{-t^*}v(y_x(t^*,\overline{\alpha})) - J_{t^*}(z,\alpha) - e^{-t^*}v(y_z(t^*,\alpha)) + \delta$$
$$\leq Cr + e^{-t^*}\omega(\widetilde{C}r) + \delta,$$

where $\widetilde{C} := C_1 + e^{Lt^*}$. Therefore

$$\omega(r) \leq Cr + e^{-t^*}\omega(\widetilde{C}r).$$

Now set $r_n := \widetilde{C}^{-n}$; since $\widetilde{C} > 1$, $r_n \to 0$ as $n \to \infty$. By an easy induction argument we obtain

$$\omega(r_n) \leq Cr_n \sum_{k=0}^{n-1} (\widetilde{C}e^{-t^*})^k + e^{-nt^*}\omega(1).$$

Now we assume, without loss of generality, $\widetilde{C}e^{-t^*} \neq 1$ to get

$$\omega(r_n) \leq C\frac{e^{-nt^*} - r_n}{\widetilde{C}e^{-t^*} - 1} + e^{-nt^*}\omega(1),$$

so that $\omega(r_n) \to 0$ as $n \to \infty$. ◂

An extension of Theorem 5.2 to the case of unbounded $\partial\Omega$ is in Exercise 5.1.

The assumption (5.3) in Theorem 5.2 looks rather strong and one may wonder whether it can be replaced by a weaker one, for instance by (5.4). The property (5.4) of existence of admissible controls is thoroughly studied in the literature under the name of *viability* of the set $\overline{\Omega}$ for the system (f, A).

The celebrated Viability Theorem gives a necessary and sufficient condition for a closed set to be viable for a differential inclusion with convex right-hand side, in particular, for a controlled system such that $f(x, A)$ is a convex set for all $x \in \mathbb{R}^N$. In our special case where $\overline{\Omega}$ has a smooth boundary, this condition is

(5.15) $\quad \inf_{a \in A} f(x, a) \cdot n(x) \leq 0 \quad$ for all $x \in \partial\Omega$.

The next example shows that the value function v can be discontinuous if (5.15) holds, even if the strict inequality in (5.3) fails at just one point. Therefore the viability property is not a sufficient condition for the continuity of v, though it is obviously necessary.

EXAMPLE 5.3. Let $\Omega = \{(x_1, x_2) \in \mathbb{R}^2 : x_2 < x_1^2\}$, $A = \{-1, 1\}$, $f(x, a) = (a, 0)$, for all $x \in \mathbb{R}^2$, $\lambda = 1$ and for all $a \in A$

$$\ell(x, a) = \begin{cases} 1, & x_1 < 0, \\ 1 - x_1 & 0 \leq x_1 \leq 1, \\ 0, & x_1 > 1. \end{cases}$$

It is easy to see that an optimal control is $\alpha \equiv -1$ if $x_1 < 0$ and $x_2 > 0$, $\alpha \equiv 1$ for all other points, and

$$v(x) = \begin{cases} 0, & x_1 > 1, x_2 \in \mathbb{R}, \\ 1, & x_1 < 0, x_2 > 0, \\ 1 - e^{x_1}(1 - e^{-1}), & x_1 < 0, x_2 \leq 0, \\ e^{x_1 - 1} - x_1, & 0 \leq x_1 \leq 1, x_2 \in \mathbb{R}. \end{cases}$$

Thus v is lower semicontinuous and discontinuous at all points $(x_1, 0)$ with $x_1 \leq 0$. Note that all the assumptions of Theorem 5.2 are satisfied except (5.3) at $x = (0, 0)$, where just (5.15) holds, and the boundedness of $\partial \Omega$, but it is easy to modify the example to one with the same features and bounded Ω. Note also that the value function does not change if we convexify the system by taking $A = [-1, 1]$. ◁

The next result is in the same spirit as Theorems 1.15 and 1.18 for time-optimal control. Near any point where (5.3) fails (i.e., the system can not choose a direction pointing inward Ω), we assume the system is symmetric and it generates a Lie bracket pointing inward Ω. The rather long and technical proof is omitted.

THEOREM 5.4. *Assume* (A_0), *A is a compact symmetric subset of $\mathbb{R}^m$,*

(5.16) $\qquad |f(x, a) - f(y, a)| \leq L|x - y| \qquad$ *for all x, y, a,*

(A_4) with $\omega_\ell(r) = L_\ell r$, (5.1), (5.2), and (5.15). Suppose that for any $x \in \partial \Omega$ such that $\min_{a \in A} f(x, a) \cdot n(x) = 0$, there is a neighborhood of x where the system is symmetric, that is, $f(y, a) = Q(y)a$ with the $N \times m$ matrix Q of class C^1, and there are two columns Q_i, Q_j of Q such that

$$[Q_i, Q_j](x) \cdot n(x) \neq 0.$$

Then the value function v is in $BUC(\Omega)$.

Note that the conclusion of the theorem is the uniform continuity of v in Ω, not in $\overline{\Omega}$, and it cannot be improved in view of the example in Exercise 5.3.

Next we turn to the study of the HJB equation for this problem. The first result is standard and we omit the proof (see Proposition III.2.5).

PROPOSITION 5.5 (DYNAMIC PROGRAMMING PRINCIPLE). *Assume* (A_0), (A_1), (A_3), *and* (A_4). *Then, for all $x \in \overline{\Omega}$ and $t > 0$,*

$$v(x) = \inf_{\alpha \in \mathcal{A}_x} \left\{ \int_0^t \ell(y_x(s), \alpha(s)) e^{-\lambda s}\, ds + v(y_x(t, \alpha)) e^{-\lambda t} \right\}.$$

5. PROBLEMS WITH STATE CONSTRAINTS

From the DPP one deduces by a standard argument that the value function v solves

(5.17) $$\lambda v + H(x, Dv) = 0$$

in Ω, where $H(x,p) = \sup_{a \in A}\{-f(x,a) \cdot p - \ell(x,a)\}$ (see Proposition III.2.8). However in this problem it is not clear what boundary condition v should satisfy. While in problems with exit times v equals the terminal cost on $\partial \Omega$ by definition, here we do not know it a priori on the boundary, so the Dirichlet problem is not appropriate.

To derive the correct boundary condition let us assume $\partial \Omega$ is C^1, v is C^1 in $\overline{\Omega}$, and there exists a continuous optimal feedback map Φ. Then v satisfies (5.17) in $\overline{\Omega}$ in the classical sense and

$$H(x, Dv(x)) = -f(x, \Phi(x)) \cdot Dv(x) - \ell(x, \Phi(x))\,.$$

The state constraint imposes $f(x, \Phi(x)) \cdot n(x) \leq 0$ for $x \in \partial \Omega$, so, for any $\beta \geq 0$,

(5.18) $$\begin{aligned} H(x, Dv(x)) &\leq -f(x, \Phi(x)) \cdot (Dv(x) + \beta n(x)) - \ell(x, \Phi(x)) \\ &\leq H(x, Dv(x) + \beta n(x)) \qquad \text{for } x \in \partial \Omega\,. \end{aligned}$$

Now note that for any smooth φ such that $v - \varphi$ has a minimum on $\overline{\Omega}$ at $x \in \partial \Omega$, by the Lagrange Multiplier Rule, $D\varphi(x) = Dv(x) + \beta n(x)$ for some $\beta \geq 0$. Therefore (5.18) and (5.17) give

(5.19) $$\lambda v(x) + H(x, D\varphi(x)) \geq 0\,.$$

This leads to the following

DEFINITION 5.6. Let $u \in C(\overline{\Omega})$. We say that

(i) u is a viscosity supersolution of (5.17) in $\overline{\Omega}$ if (5.19) holds for any $\varphi \in C^1(\mathbb{R}^N)$ such that $u - \varphi$ has a local minimum, relative to $\overline{\Omega}$, at $x \in \overline{\Omega}$;

(ii) u is a *constrained viscosity solution* of (5.17) in $\overline{\Omega}$ if it is a subsolution in Ω and a supersolution in $\overline{\Omega}$. ◁

The reader must be warned that this definition has to be handled with some care because some obvious properties of supersolutions in open sets do not hold for viscosity supersolutions in closed sets. Consider, for instance, the equation $|u'| = 1$. The classical solution $u(x) = x$ is *not* a viscosity supersolution in $\overline{\Omega} = [0,1]$. Indeed, if we take $\varphi \equiv 0$, $u - \varphi$ has a minimum at 0 but $|\varphi'(0)| \geq 1$ is false. Therefore

(a) a classical supersolution up to the boundary is not necessarily a viscosity supersolution in $\overline{\Omega}$,

(b) a solution (even classical) in an open set $\Omega' \supseteq \overline{\Omega}$ is not necessarily a viscosity supersolution in $\overline{\Omega}$.

Now we can give the good properties of constrained (viscosity) solutions.

THEOREM 5.7. *Assume* (A_0)–(A_4). *If the value function v is in $C(\overline{\Omega})$, then it is a constrained viscosity solution of the HJB equation* (5.17) *in* $\overline{\Omega}$.

PROOF. We only indicate the changes with respect to the proof of Proposition III.2.8. Assume $x \in \partial\Omega$ is a local minimum point of $v - \varphi$ relative to $\overline{\Omega}$, that is, for some $r > 0$

(5.20) $\qquad v(x) - v(z) \leq \varphi(x) - \varphi(z) \qquad$ for all $z \in B(x,r) \cap \overline{\Omega}$.

The difference with respect to the previous proof is the additional restriction $z \in \overline{\Omega}$ in (5.20). In the continuation of the proof (5.20) is used only with $z = y_x(t, \overline{\alpha})$, where $\overline{\alpha}$ is a control function suboptimal in the DPP. In particular, in the present case, $\overline{\alpha} \in \mathcal{A}_x$, thus $y_x(t, \overline{\alpha}) \in \overline{\Omega}$ and we can use (5.20) as in the previous proof. ◀

The comparison theorem for the current problem requires the following regularity assumption on the set Ω,

(5.21) $\qquad$ there are $\eta : \overline{\Omega} \to \mathbb{R}^N$ bounded and uniformly continuous and $c > 0$ such that $B(x + t\eta(x), ct) \subseteq \Omega$ for all $x \in \overline{\Omega}, 0 < t \leq c$.

Condition (5.21) has an easy geometric interpretation: for each $x \in \overline{\Omega}$ there is a truncated open cone with vertex in x contained in Ω, these cones depend continuously on x and have a uniform size. When applied to $x \in \partial\Omega$ this suggests that (5.21) is equivalent to saying that $\partial\Omega$ is locally the graph of Lipschitzean functions satisfying a uniform bound on their Lipschitz constants. This statement is made precise in Exercise 5.6. For instance, (5.21) is satisfied if $\partial\Omega$ is C^1 and bounded.

THEOREM 5.8. *Assume* (A_0), (5.2), (5.16), (A_4), *and* (5.21). *If $u_1 \in BUC(\overline{\Omega})$ is a subsolution of* (5.17) *in Ω and $u_2 \in BC(\overline{\Omega})$ is a supersolution of* (5.17) *in $\overline{\Omega}$, then $u_1 \leq u_2$.*

These two results allow us to characterize the value function in terms of solutions of the HJB equation as follows.

COROLLARY 5.9. *Under the assumptions of Theorem 5.8, if the value function v is in $BUC(\overline{\Omega})$, then it is the unique constrained viscosity solution of* (5.17) *in $BUC(\overline{\Omega})$. Moreover, it is the maximal subsolution in Ω belonging to $BUC(\overline{\Omega})$ and the minimal supersolution in $\overline{\Omega}$ belonging to $BC(\overline{\Omega})$.*

The proof of the Corollary is immediate because (5.2) (5.16) imply (A_1)–(A_3). Note that the value function has the regularity required in Corollary 5.9 under the assumptions of Theorem 5.2, but not under those of Theorem 5.4. For this case we need the following finer result whose proof involves the theory of discontinuous solutions which is developed in Chapter V.

THEOREM 5.10. *Under the assumptions of Theorem 5.8, if $v \in BUC(\Omega)$, then its unique continuous extension to $\overline{\Omega}$ is the unique constrained viscosity solution of* (5.17) *in $BUC(\overline{\Omega})$.*

5. PROBLEMS WITH STATE CONSTRAINTS

Before turning to the proof of Theorem 5.8 we recall the usual consequences of comparison and uniqueness results such as Corollary 5.9 and Theorem 5.10: relaxation and perturbation results, as well as verification theorems with subsolutions in Ω playing the role of verification functions and supersolutions in $\overline{\Omega}$ as falsification functions. The statements and proofs of these results are easy exercises, following the arguments of §2.2 in Chapter III.

PROOF OF THEOREM 5.8. The proof is based on the usual idea of looking at maximum points of an auxiliary function $\Phi(x, y)$. Now, however, we have no information on the subsolution u_1 at boundary points, so we have to avoid the possibility that the maximum $(\overline{x}, \overline{y})$ of Φ is such that $\overline{x} \in \partial\Omega$. This is achieved by adding an extra term to Φ involving the function η of assumption (5.21).

Assume by contradiction $m := \sup_{\overline{\Omega}}(u_1 - u_2) > 0$ and pick $z \in \overline{\Omega}$ such that

$$(5.22) \qquad u_1(z) - u_2(z) \geq m - \delta,$$

where $\delta > 0$ will be chosen small enough later. Define

$$\Phi(x, y) := u_1(x) - u_2(y) - \left|\frac{x-y}{\varepsilon} - \eta(z)\right|^2 - |y - z|^2,$$

where $\varepsilon > 0$ will be chosen small enough. First we are going to show that Φ achieves its maximum. To this end we choose $\varepsilon \leq c$, so that $z + \varepsilon\eta(z) \in \Omega$, and name ω_1 the modulus of continuity of u_1 to compute

$$(5.23) \qquad \Phi(z + \varepsilon\eta(z), z) = u_1(z + \varepsilon\eta(z)) - u_2(z) \geq m - \delta - \omega_1(C\varepsilon),$$

where $C := \sup_{\overline{\Omega}} |\eta|$ and we have used (5.22). On the other hand,

$$\Phi(x, y) \leq m + \omega_1(|x - y|) - \left|\frac{x-y}{\varepsilon} - \eta(z)\right|^2 - |y - z|^2.$$

If (x, y) is such that $\Phi(x, y) \geq \Phi(z + \varepsilon\eta(z), z)$ we have

$$(5.24) \qquad |y - z|^2 + \left|\frac{x-y}{\varepsilon} - \eta(z)\right|^2 \leq \delta + \omega_1(C\varepsilon) + \omega_1(|x-y|).$$

Since ω_1 is bounded, the points x, y satisfying this inequality are in $B(z, R)$ for a suitable R, so the search for the maximum of Φ can be restricted to a compact set. Let $(\overline{x}, \overline{y})$ be such a point:

$$\Phi(\overline{x}, \overline{y}) = \max_{\overline{\Omega} \times \overline{\Omega}} \Phi.$$

By (5.24) we get for some constant C_1

$$(5.25) \qquad |\overline{x} - \overline{y}| \leq C_1\varepsilon,$$

and plugging this into (5.24)

$$(5.26) \qquad |\overline{y} - z|^2 + \left|\frac{\overline{x}-\overline{y}}{\varepsilon} - \eta(z)\right|^2 \leq \delta + \omega_1(C\varepsilon) + \omega_1(C_1\varepsilon).$$

From this inequality we deduce that $\bar{x} \in \Omega$. In fact, there is $\varrho > 0$ such that
$$|\eta(x) - \eta(y)| < \frac{c}{2} \quad \text{if } |x - y| < \varrho;$$
also we can choose δ and ε such that the right-hand side of (5.26) is less than $\min\{\varrho^2, c^2/4\}$. Therefore (5.26) yields $|\bar{y} - z| \leq \varrho$ and, consequently,

(5.27) $$|\eta(\bar{y}) - \eta(z)| < \frac{c}{2};$$

by (5.26) we also have
$$|\bar{x} - \bar{y} - \varepsilon\eta(z)| \leq \varepsilon\frac{c}{2}.$$

This, combined with (5.27), implies
$$|\bar{x} - \bar{y} - \varepsilon\eta(\bar{y})| < \varepsilon c.$$

Thus $\bar{x} \in B(\bar{y} + \varepsilon\eta(\bar{y}), c\varepsilon) \subseteq \Omega$ by (5.21).

Now we can use the fact that u_1 is a subsolution of (5.17) in Ω and that $u_1 - \varphi$ has a maximum at $\bar{x}$, where $\varphi(x) := \left|\dfrac{x - \bar{y}}{\varepsilon} - \eta(z)\right|^2$ to get

(5.28) $$\lambda u_1(\bar{x}) + H\left(\bar{x}, \frac{2}{\varepsilon}\left(\frac{\bar{x} - \bar{y}}{\varepsilon} - \eta(z)\right)\right) \leq 0.$$

On the other hand, u_2 is a supersolution in $\overline{\Omega}$ and $u_2 - \psi$ has a minimum at $\bar{y}$, where
$$\psi(y) := -\left|\frac{\bar{x} - y}{\varepsilon} - \eta(z)\right|^2 - |y - z|^2,$$
so that
$$\lambda u_2(\bar{y}) + H\left(\bar{y}, \frac{2}{\varepsilon}\left(\frac{\bar{x} - \bar{y}}{\varepsilon} - \eta(z)\right) + 2(z - \bar{y})\right) \geq 0.$$

Next we subtract this inequality from (5.28) and estimate $u_1(\bar{x}) - u_2(\bar{y})$. To this end we observe that (5.2), (5.16), and (A$_4$) imply
$$H(x, p) - H(y, q) \leq K|p - q| + L|x - y|\,|p| + \omega_\ell(|x - y|),$$
see Lemma III.2.11. We assume for simplicity $\lambda = 1$ and get
$$u_1(\bar{x}) - u_2(\bar{y}) \leq 2K|z - \bar{y}| + \frac{2L}{\varepsilon}|\bar{x} - \bar{y}|\left|\frac{\bar{x} - \bar{y}}{\varepsilon} - \eta(z)\right| + \omega_\ell(|\bar{x} - \bar{y}|),$$
and observe that by (5.25) and (5.26) the right-hand side can be estimated by
$$\overline{C}(\sqrt{\delta} + o(1)) \quad \text{as } \varepsilon \to 0^+,$$
for a suitable constant $\overline{C}$. Now we use (5.23) to obtain
$$m - \delta - o(1) \leq \Phi(\bar{x}, \bar{y}) \leq u_1(\bar{x}) - u_2(\bar{y}) \leq \overline{C}(\sqrt{\delta} + o(1)) \quad \text{as } \varepsilon \to 0^+,$$
which gives a contradiction to $m > 0$ for ε, δ small enough. ◂

REMARK 5.11. Comparison results still hold between subsolutions in Ω and supersolutions in $\overline{\Omega}$ of general Hamilton-Jacobi equations of the form
$$u(x) + H(x, Du(x)) = 0 .$$
For example, comparison holds if Ω is bounded and H satisfies conditions (H$_1$) and (H$_3$) of §3 in Chapter II (see Exercise 5.7). Note that no comparison between sub- and supersolution on $\partial\Omega$ is required whereas (H$_3$) is needed (compare with Theorem II.3.1). ◁

Exercises

5.1. Extend Theorem 5.2 to the case that (5.1) and (5.3) are dropped and replaced by the following assumptions: the signed distance $d(\,\cdot\,)$ from $\partial\Omega$ is of class C^1 in $X := B(\partial\Omega, r)$ for some $r > 0$, and Dd is lipschitzean in X; there is a modulus $\tilde{\omega}$ such that $|f(x,a) - f(y,a)| \leq \tilde{\omega}(|x-y|)$ for all $x, y \in X$ and $a \in A$; there is $\zeta > 0$ such that $\inf_{a \in A} f(x,a) \cdot n(x) < -\zeta$ for all $x \in \partial\Omega$.

5.2. Prove that v is lipschitzean under the assumptions of Theorem 5.2 and for λ large enough. (Cfr. [LoT94].)

5.3. Consider $\Omega = \{(x_1, x_2) \in \mathbb{R}^2 : x_1^2 < x_2\}$, $f(x,a) = (2a_1, 2(x_1 + x_2)a_2)$, $A = \{(1,0), (-1,0), (0,1), (0,-1)\}$, $\lambda = 1$, $\ell(x,a) = \psi(x_2)$ where $\psi \in C^\infty([0, +\infty[)$ is nonincreasing, $\psi(r) \equiv 1$ if $0 \leq r \leq 1$, $\psi(r) \equiv 0$ if $r \geq 2$. Show that:

(i) the assumptions of Theorem 5.4 are satisfied;

(ii) v is discontinuous at $(0,0)$ (cfr. [Mo95]).

5.4. Prove that, under the assumptions of Theorem 5.4, the set of points $x \in \partial\Omega$ such that $\min_{a \in A} f(x,a) \cdot n(x) = 0$ has empty relative interior.

5.5. Assume H is differentiable with respect to p, $\partial\Omega$ is C^1 and $u \in C^1(\overline{\Omega})$ is a constrained viscosity solution of (5.17). Prove that

(i) u satisfies the nonlinear oblique derivative boundary condition
$$\frac{\partial H}{\partial p}(x, Du(x)) \cdot n(x) \geq 0, \qquad x \in \partial\Omega;$$

(ii) if in addition H is nondecreasing with respect to $|p|$, then u satisfies the Neumann boundary condition
$$\frac{\partial u}{\partial n}(x) \geq 0, \qquad x \in \partial\Omega .$$

5.6. Assume $\Omega \subseteq \mathbb{R}^N$ is open and $\partial\Omega$ is bounded. Prove that Ω satisfies (5.21) if and only if there exists $x_i \in \partial\Omega$, $i = 1, \ldots, m$, unit vectors η_i, Lipschitz continuous functions $f_i : \{v : v \cdot \eta_i = 0\} \to \mathbb{R}$, and $r > 0$ such that
$$\bigcup_{i=1,\ldots,m} B(x_i, r/2) \supseteq \partial\Omega$$
and $\Omega \cap B(x_i, 2r) = \{t\eta_i + v : v \cdot \eta_i = 0 \text{ and } t > f_i(v)\} \cap B(x_i, 2r)$ for $i = 1, \ldots, m$.

5.7. Assume Ω bounded and let $u_1, u_2 \in C(\overline{\Omega})$ be, respectively, a subsolution in Ω and a supersolution in $\overline{\Omega}$ of
$$u(x) + H(x, Du(x)) = 0 \, .$$
Prove that, under conditions (H$_1$) and (H$_3$) in §II.3, $u_1 \le u_2$ in $\overline{\Omega}$.

5.8. Assume Ω bounded and starshaped with respect to the origin. Assume also that, for some $k > 0$,
$$\text{dist}(x, \overline{\Omega}) \ge k\varepsilon, \qquad \forall x \in (1+\varepsilon)\partial\Omega, \, \forall \varepsilon > 0 \, .$$
Prove that if (H$_1$) of Chapter II holds then the same result as in Exercise 5.7 is true. [Hint: use the auxiliary function
$$\Phi(x, y) = (1+\varepsilon)u_1\left(\frac{x}{1+\varepsilon}\right) - u_2(y) - \frac{1}{2\varepsilon^2}|x-y|^2$$
and look for a maximum point $(\bar{x}, \bar{y})$ of Φ over $(1+\varepsilon)\overline{\Omega} \times \overline{\Omega}$ (cfr. [CDL90]).]

5.9. Consider equation (5.17) in a bounded set Ω with $\lambda = 0$. Assume the existence of $\underline{u}_0 \in C^1(\overline{\Omega})$ such that
$$H(x, D\underline{u}_0(x)) < 0, \qquad x \in \overline{\Omega} \, .$$
Prove that the comparison result in Exercise 5.7 still holds in this case. [Hint: mimic the proof of Theorem 5.9 in Chapter II.]

5.10. Assume (A$_0$)–(A$_4$) and let v be the value function of the infinite horizon problem without state constraints, that is, $v(x) = \inf_{\alpha \in \mathcal{A}} J(x, \alpha)$. Given $\Omega_0 \subseteq \mathbb{R}^N$, the *attainable set* from Ω_0 is the set
$$\Omega := \{z : z = y_x(t, \alpha) \text{ for some } x \in \Omega_0, \, t \ge 0, \, \alpha \in \mathcal{A}\} \, .$$
Prove that:

(i) v is a viscosity solution of (5.17) in $\overline{\Omega}$, that is, it is a supersolution in $\overline{\Omega}$ according to Definition 5.6 and for all $\varphi \in C^1(\mathbb{R}^N)$ such that $v - \varphi$ has maximum, relative to $\overline{\Omega}$, at $x \in \overline{\Omega}$ $\lambda v(x) + H(x, D\varphi(x)) \le 0$;

(ii) v is the unique bounded and continuous viscosity solution of (5.17) in $\overline{\Omega}$ (cfr. [IMZ91]).

6. Bibliographical notes

Time-optimal control is one of the most classical problems in control theory (it appears already in Carathéodory's book [Ca35]) and it has a large literature (see, e.g., [PBGM62, Lei66, LM67, HL69, FR75, Co85]). The property of small-time local controllability (STLC) is one of the main subjects of the so-called differential geometric control theory; introductory surveys are, for example, [Sus83, Bac86].

It has also different names, for instance it is called N-local controllability in the Russian literature.

The characterization of controllability for linear systems in Theorem 1.9 is classical (see, e.g., [MS82, Bi83, Co85]) and the Hölder continuity of the minimal time function T in this case can be found in [Liv80, Ra82, Gy84]. Theorem 1.10 can be found in [KP72, Bre80], see also the book [Co85] for more informations on controllability of linear systems, and [CSi95a] for semiconvexity and semiconcavity properties of T.

The controllability statement in Theorem 1.11 on linearized systems is again classical [MS82], and the estimate of the Hölder exponent of T was proved by Bianchini and Stefani [BiS90]. Theorem 1.12 is due to Petrov [Pe68, Pe70a], the proof we present here is new.

Theorem 1.15 is the prototype result in geometric theory of controllability, usually called Chow's theorem, see [Bac86] for its history. There is a large literature on small-time controllability for point-shaped targets and systems nonlinear in the state and affine in the control variables, see [Sus87, Kaw90, He91, BiS93] and the references therein. For the related Hölder continuity properties of T the reader should consult [Liv80, St91, BiS90] and the references therein.

There are not so many results for general targets. If the boundary is smooth, the sufficient condition for controllability (1.13) is very natural and it appears, for example, in [F61, Fri74] in the more general context of pursuit-evasion games. Theorem 1.16 and Remark 1.20 are taken from [BF90a], and the extension to general targets Theorem 4.8 can be found in [Sor93a], see also [Ve96]. Theorem 1.18 is taken from [BS91c] where necessary conditions for $1/2$-Hölder continuity are also studied. Lemma 1.19 is taken from [HH70] and it is implicit in Hörmander's work on hypoelliptic PDEs [Hor68], its proof for $j > 1$ can be deduced, for example, from Theorem 3.4 in [BiS93]. The Hölder continuity of T when the target is a lower-dimensional manifold is studied in [Sor92a]. For further regularity properties of the minimal time function and some applications we refer to [CSi95a, CSi95b, BLP97].

Definition 1.22, Theorem 1.23 and the example in Exercise 1.3 can be found in [BiS90], see also [Pe70b].

Section 2 is based on [B89]; observe that the example reported in Exercise 2.5 was proposed in [Ha91] as a possible counterexample to Theorem 2.6. Theorems 2.9 and 2.11 are taken, respectively, from [BF90a] and [BS92].

For linear systems with target the origin, the minimal time function is a bilateral solution of the HJB equation (see Remark 2.5 in Chapter VI and [Ha91]), and it also solves a different PDE derived by Hájek, see [Sta89]. For some earlier connections between the linear time-optimal problem and other notions of solution of the HJB equation we refer to [Co85, MP86].

Some of the results in Sections 1 and 2 are extended to systems with unbounded controls in [RS96].

The material of §3.1 is mostly new and it is based on unpublished joint work with Soravia. It extends the analysis in Chapter 5 of P.L. Lions' book [L82] where problems of the Calculus of Variations in smooth domains are treated. The compatibility conditions on the boundary data are formulated in [L82] by means of a comparison function different from $L(\cdot, \cdot)$, namely, the function $\widehat{L}(\cdot, \cdot)$ defined in Exercise 3.4, and they can be recovered from our results by Exercises 3.3 and 3.4. Some related results for controlled diffusions are in [L83a]. A different approach to the compatibility of the data in the Dirichlet problem for HJ equations is in [En86].

The main result in §3.2 is Theorem 3.15, whose proof follows Evans and Ishii [EI84], see also [LS85] (both papers also treat differential games). A uniqueness theorem such as Corollary 3.17 was proved by P.L. Lions [L83a] in the case of controlled degenerate diffusion processes as well.

Section 4 follows the papers of Bardi and Soravia [BS91a, Sor93a, BS92]. A complete proof of Corollary 4.3 in the case $w_0 = +\infty$ is in [BS91a] for the more general problem of differential games (see also [EJ89] for a similar result in some special cases); Theorems 4.1 and 4.2 are taken from [Sor93a] (where differential games are studied), Theorem 4.4 can be found in [BS92]. Theorem 4.5 is due to Evans and James [EJ89], see also [BS91a] for the case of non-zero terminal cost and [He88] for a connection of this result with the synthesis of optimal feedbacks.

Several results of Sections 1–4 were extended to the case of state-dependent constraints on the control in [Bo93], see also [K96a]. The problem of maximizing the blowup time of a controlled system gives rise to a free boundary problem similar to those of §4 which was studied in [BL96a].

Section 5 follows Soner [S86]. The proof of Lemma 5.1 using the signed distance from the boundary is new (see, e.g., the appendix to Chapter 14 in [GT83] for the smoothness of the distance function). The viability theorem can be found, for example, in the books [AC84, AF90, Au91]. Example 5.3 was shown to us by Soravia. Theorems 5.4 and 5.10 are due to Motta [Mo95]. Further properties of the value function of the infinite horizon problem with state constraints are studied in [Lo87, LoT94, ArL96], and the boundary condition is reformulated and improved in [IK96]. Finite horizon problems with state constraints, unbounded controls, and non-coercive cost were studied in [MR96b, MR96c].

The theory of constrained viscosity solutions was developed in several directions in [CDL90]. The condition satisfied at boundary points by constrained solutions is related to oblique derivative problems, as indicated by Exercise 5.5. Hamilton-Jacobi equations with Neumann boundary conditions were first studied by P.L. Lions [L85b, L86], linear oblique derivative problems in nonsmooth domains in [DI90, DI91], and fully nonlinear oblique derivative problems in [BaL91, I91].

CHAPTER V

Discontinuous viscosity solutions and applications

In this chapter we extend the theory of continuous viscosity solutions developed in Chapter II to include solutions that are not necessarily continuous. This has two motivations. The first is that many optimal control problems have a discontinuous value function and we want to extend to these problems the results of Chapters III and IV, in particular the characterization of the value function as the unique solution of the Hamilton-Jacobi-Bellman equation with suitable boundary conditions. The second motivation is more technical: viscosity solutions are stable with respect to certain relaxed semi-limits, that we call weak limits in the viscosity sense, which are semicontinuous sub- or supersolutions. These weak limits are used extensively in Chapters VI and VII to study the convergence of approximation schemes and several asymptotic limits, even for control problems where the value function is continuous.

In Section 1 we present a comparison result for semicontinuous sub- and supersolutions and prove the stability of viscosity solutions with respect to the weak limits. The reader interested in the approximation and asymptotic results of the following chapters should read this section.

Section 2 is devoted to Ishii's definition of a non-continuous, (i.e., not necessarily continuous) viscosity solution. We prove the existence of such solutions by means of Perron's method, and we remark that there is no uniqueness for the standard boundary value problems without some additional conditions.

In the remaining sections we describe three different ways to get uniqueness in classes of non-continuous functions. In Section 3 we reformulate Subbotin's generalized solutions in the viscosity language; we call them envelope solutions. In Section 4 we discuss a weak interpretation of Dirichlet boundary conditions which is related to the vanishing viscosity method. In Section 5 we describe the Barron-Jensen definition of discontinuous solution, that we call bilateral supersolutions. In all three sections the theory is applied to control problems involving the exit times of the system from a domain, in particular to the minimum time problem. In Section 5 we also consider the finite horizon problem with constraints on the endpoint of the trajectories.

1. Semicontinuous sub- and supersolutions, weak limits, and stability

We begin with some notation for the spaces of semicontinuous functions on a set $E \subseteq \mathbb{R}^N$:

$$USC(E) := \{\, u : E \to \mathbb{R} \text{ upper semicontinuous}\,\},$$
$$LSC(E) := \{\, u : E \to \mathbb{R} \text{ lower semicontinuous}\,\},$$
$$BUSC(E) := \{\, u \in USC(E) \text{ bounded}\,\},$$
$$BLSC(E) := \{\, u \in LSC(E) \text{ bounded}\,\}.$$

We recall that $u : E \to \mathbb{R}$ is upper (respectively, lower) semicontinuous if for any $x \in E$ and $\varepsilon > 0$ there is δ such that $f(y) < f(x) + \varepsilon$ (respectively, $f(y) > f(x) - \varepsilon$) for all $y \in E \cap B(x, \delta)$; Weierstrass' Theorem on the existence of maxima (respectively, minima) on compact sets holds for upper (respectively, lower) semicontinuous functions.

The definition of continuous viscosity sub- and supersolutions of a first order equation

$$(1.1) \qquad F(x, u, Dx) = 0 \quad \text{in } \Omega,$$

with $\Omega \subseteq \mathbb{R}^N$ open and $F : \Omega \times \mathbb{R} \times \mathbb{R}^N \to \mathbb{R}$ continuous, extends naturally to semicontinuous functions as follows.

DEFINITION 1.1. A function $u \in USC(\Omega)$ (respectively, $LSC(\Omega)$) is a *viscosity subsolution* (respectively, *supersolution*) of (1.1) if, for any $\varphi \in C^1(\Omega)$ and $x \in \Omega$ such that $u - \varphi$ has a local maximum (respectively, minimum) at x,

$$F(x, u(x), D\varphi(x)) \leq 0 \quad (\text{respectively}, \geq 0) . \quad \triangleleft$$

As for continuous semisolutions, there is an equivalent definition by means of the semidifferentials of u instead of test functions. In fact the definition of $D^+u(x)$ and $D^-u(x)$ makes sense for any real valued function, and $u \in USC(\Omega)$ (respectively, $LSC(\Omega)$) is a viscosity subsolution (respectively, supersolution) of (1.1) if and only if, for any $x \in \Omega$,

$$F(x, u, p) \leq 0 \quad (\text{respectively}, \geq 0)$$

for all $p \in D^+u(x)$ (respectively, $D^-u(x)$), as the reader can prove by the arguments of Chapter II, see Exercise 1.3.

Later in this section we show how the proofs of the comparison theorems of Chapters II and III can be modified to get some assertions for these larger classes of sub- and supersolutions. For future reference we make precise what we mean by *comparison principle* and give an example (whose proof is postponed to the end of the section).

DEFINITION 1.2. We say that the Dirichlet problem for equation (1.1) satisfies the comparison principle or, briefly, that the comparison principle holds for F, if for any $u_1 \in BUSC(\overline{\Omega})$ and $u_2 \in BLSC(\overline{\Omega})$ which are, respectively, a viscosity subsolution and supersolution of (1.1) such that $u_1 \leq u_2$ on $\partial\Omega$, we have $u_1 \leq u_2$ in Ω. ◁

THEOREM 1.3. *Assume $H : \Omega \times \mathbb{R}^N \to \mathbb{R}$ satisfies either the assumptions of Theorem II.3.5 or those of Theorem III.2.12. Then the Dirichlet problem for the equation*

(1.2) $$u + H(x, Du) = 0 \quad in\ \Omega$$

satisfies the comparison principle.

This result, whose proof is postponed to the end of this section, adds nothing to the problem of uniqueness of viscosity solutions, because a function which is simultaneously a subsolution and a supersolution according to Definition 1.1 is indeed continuous, and thus it is a solution in the same sense as in the previous chapters.

However, Theorem 1.3 has other important consequences. A first simple one is the strengthening of the notion of *complete solution* we used in Chapters II and III. Consider the boundary value problem of equation (1.2) with the boundary condition

(1.3) $$u = g \quad \text{on } \partial\Omega,$$

and call a *subsolution* (respectively, *supersolution*) *of the Dirichlet problem* (1.2)-(1.3) a subsolution of (1.2) in $BUSC(\overline{\Omega})$ (respectively $BLSC(\overline{\Omega})$) which is $\leq g$ (respectively, $\geq g$) on $\partial\Omega$. We call a complete solution of (1.2)-(1.3) a function which is at the same time the maximal subsolution and the minimal supersolution. Of course such a function is a solution in $BC(\overline{\Omega})$. Theorem 1.3 implies that the converse is also true: a solution of (1.2)-(1.3) in $BC(\overline{\Omega})$ is the complete solution.

In Chapters III and IV we have seen a variety of optimal control problems whose value function is the continuous solution of an initial-boundary value problem for a Hamilton-Jacobi-Bellman equation. For instance the value function of infinite horizon problems fits within Theorem 1.3 with $\Omega = \mathbb{R}^N$ (see III.2), and so does the value function of exit time problems with terminal cost g (see IV.3), under suitable assumptions. Theorem 1.3 and the fact that the value function is the complete solution give *verification theorems*, and then sufficient conditions of optimality, with merely upper semicontinuous (and bounded) verification functions, and they give *falsification theorems*, and then necessary conditions of optimality, with merely lower semicontinuous falsification functions. The precise statements of these results can be easily deduced from the arguments of §III.2.2.

A much more interesting and far-reaching application of Theorem 1.3 is the extension of the stability properties of viscosity solutions. Let us recall briefly what was proved on this matter in Chapter II. Let $F_\varepsilon : \Omega \times \mathbb{R} \times \mathbb{R}^N \to \mathbb{R}$, $0 < \varepsilon \leq 1$, be continuous and

(1.4) $$F_\varepsilon \rightrightarrows F \quad \text{on compacta of } \Omega \times \mathbb{R} \times \mathbb{R}^N \text{ as } \varepsilon \to 0^+,$$

($\Rightarrow$ means uniform convergence), $u_\varepsilon \in C(\Omega)$ be solutions of

(1.5) $$F_\varepsilon(x, u_\varepsilon, Du_\varepsilon) = 0 \quad \text{in } \Omega$$

and

(1.6) $$u_\varepsilon \rightrightarrows u \quad \text{on compacta of } \Omega \text{ as } \varepsilon \to 0^+.$$

Then the limit function u is a solution of the limit equation

(1.7) $$F(x, u, Du) = 0 \quad \text{in } \Omega,$$

see Proposition II.2.2.

If one wants to use this result to pass to the limit in the PDE (1.5), assuming (1.4) is known, the natural procedure is the following. First prove that u_ε satisfy on each compactum $\mathcal{K} \subseteq \Omega$ the uniform estimates

(1.8) $$\sup_{\mathcal{K}} |u_\varepsilon| \leq C_\mathcal{K}$$

(1.9) $$|u_\varepsilon(x) - u_\varepsilon(y)| \leq \omega_\mathcal{K}(|x - y|) \quad \text{for all } x, y \in \mathcal{K},$$

where the constant $C_\mathcal{K}$ and the modulus $\omega_\mathcal{K}$ do not depend on ε, then use the Ascoli-Arzelà compactness theorem and a diagonal procedure to extract a sequence $u_{\varepsilon_n} \rightrightarrows u$ on all compacta of Ω and so conclude that u solves (1.7). In the special case $\Omega = \mathbb{R}^N$ a uniqueness theorem for (1.7) would then ensure the convergence (1.6) for the whole net u_ε and not just for a subsequence; the same can be done when (1.7) is coupled with a boundary condition as (1.3), provided this boundary value problem satisfies a uniqueness theorem, $u_{\varepsilon|\partial\Omega} \to g$, and the estimates (1.8), (1.9) hold on each compactum $\mathcal{K} \subseteq \overline{\Omega}$.

In next results we want to dispense with the equicontinuity estimate (1.9).

DEFINITION 1.4. For the functions $u_\varepsilon : E \to \mathbb{R}$, $0 < \varepsilon \leq 1$, $E \subseteq \mathbb{R}^N$, the *lower weak limit* in E as $\varepsilon \to 0^+$ at the point $x \in E$ is

$$\underline{u}(x) = \liminf_{\varepsilon \to 0^+}{}_* u_\varepsilon(x) := \sup_{\delta > 0} \inf\{ u_\varepsilon(y) : y \in E, |x - y| < \delta \text{ and } 0 < \varepsilon < \delta \},$$

or, more concisely,

$$\underline{u}(x) = \liminf_{(y,\varepsilon) \to (x, 0^+)} u_\varepsilon(y);$$

the *upper weak limit* is

$$\overline{u}(x) = \limsup_{\varepsilon \to 0^+}{}^* u_\varepsilon(x) := \limsup_{(y,\varepsilon) \to (x, 0^+)} u_\varepsilon(y). \quad \triangleleft$$

If the functions u_ε are locally uniformly bounded, that is, (1.8) holds for each compactum $\mathcal{K}$, the weak limits are finite and define two functions $\underline{u}, \overline{u} : E \to \mathbb{R}$. Their main properties are described by the next two lemmata.

LEMMA 1.5. *The lower weak limit $\underline{u}$ is lower semicontinuous and the upper weak limit $\overline{u}$ is upper semicontinuous.*

PROOF. We only show that $\overline{u}$ is upper semicontinuous. We assume by contradiction $x_n \to x$ and $\overline{u}(x_n) \geq \overline{u}(x) + \delta$ for some $\delta > 0$ and all n large enough. By definition of $\overline{u}(x_n)$ there are $y_n, \varepsilon_n, \overline{n}$ such that $|y_n - x_n| < 1/n$, $\varepsilon_n \to 0^+$, and

$$u_{\varepsilon_n}(y_n) \geq \overline{u}(x_n) - \frac{\delta}{2} \geq \overline{u}(x) + \frac{\delta}{2} \quad \text{for } n > \overline{n}.$$

Then $\liminf_n u_{\varepsilon_n}(y_n) > \overline{u}(x)$, a contradiction to the definition of $\overline{u}(x)$ because $y_n \to x$. ◂

LEMMA 1.6. *Let $u_\varepsilon \in USC(E)$ (respectively, $LSC(E)$) be locally uniformly bounded, $\varphi \in C^1(E)$, $B := \overline{B}(x_0, r) \cap E$ be closed, and assume $x_0 \in B$ is a strict maximum (respectively, minimum) point for $\overline{u} - \varphi$ (respectively, $\underline{u} - \varphi$) on B. Then there exists a sequence $\{x_n\}$ in B and $\varepsilon_n \to 0^+$ such that x_n is a maximum (respectively, minimum) point for $u_{\varepsilon_n} - \varphi$ on B and*

$$\lim_n x_n = x_0, \qquad \lim_n u_{\varepsilon_n}(x_n) = \overline{u}(x_0) \quad (\textit{respectively, } \underline{u}(x_0)).$$

PROOF. We prove only the statement about the upper limit. Let $\varepsilon_n \to 0^+$ and $\{x^{(n)}\}$ be sequences such that

(1.10) $\qquad x^{(n)} \longrightarrow x_0, \quad u_{\varepsilon_n}(x^{(n)}) \longrightarrow \overline{u}(x_0) \qquad \text{as } n \to \infty.$

Let x_n be a maximum point of $u_{\varepsilon_n} - \varphi$ on B and extract subsequences, still denoted $\{x_n\}$ and $\{\varepsilon_n\}$, such that

(1.11) $\qquad\qquad x_n \longrightarrow \overline{x}, \quad u_{\varepsilon_n}(x_n) \longrightarrow s.$

Since $(u_{\varepsilon_n} - \varphi)(x_n) \geq (u_{\varepsilon_n} - \varphi)(x^{(n)})$, we obtain from (1.11), the definition of $\overline{u}$, and (1.10),

(1.12) $\qquad\qquad (\overline{u} - \varphi)(\overline{x}) \geq s - \varphi(\overline{x}) \geq (\overline{u} - \varphi)(x_0).$

Since x_0 is a strict maximum point for $\overline{u} - \varphi$ we get that $\overline{x} = x_0$. Then (1.12) also gives $s = \overline{u}(x_0)$. ◂

The next result states that viscosity sub- and supersolutions are stable with respect to upper and lower weak limits, respectively.

THEOREM 1.7. *Assume F_ε is continuous for any $\varepsilon \in \,]0,1]$, (1.4) holds, and let $u_\varepsilon : \Omega \to \mathbb{R}$ satisfy (1.8) for any compactum $\mathcal{K} \subseteq \Omega$ and all ε.*

(i) *If $u_\varepsilon \in USC(\Omega)$ is a subsolution of (1.5) for all ε, then the upper weak limit $\overline{u} \in USC(\Omega)$ is a subsolution of (1.7);*

(ii) *if $u_\varepsilon \in LSC(\Omega)$ is a supersolution of (1.5) for all ε, then the lower weak limit $\underline{u} \in LSC(\Omega)$ is a supersolution of (1.7).*

PROOF. By Lemma 1.5, $\overline{u}$ and $\underline{u}$ have the correct semicontinuity. The proof that they are, respectively, a subsolution and a supersolution follows from the definitions and a straightforward passage to the limit, just as in the proof of the previous stability result Proposition II.2.2, once we have replaced Lemma II.2.4 with Lemma 1.6. ◀

This result is particularly powerful when it can be used in conjunction with a comparison principle such as Theorem 1.3. In fact, we can prove the local convergence of the whole net u_ε under the same assumptions as before except for the equicontinuity estimate (1.9), which can be dropped altogether if $\Omega = \mathbb{R}^N$, and replaced by similar estimates just near the boundary in the general case.

COROLLARY 1.8. *Assume F_ε is continuous for any $\varepsilon \in \,]0,1]$, the comparison principle holds for F, and (1.4). Let $u_\varepsilon : \overline{\Omega} \to \mathbb{R}$ be continuous in Ω and satisfy*

$$F_\varepsilon(x, u_\varepsilon, Du_\varepsilon) = 0 \quad \text{in } \Omega,$$

and for some constant C

(1.13) $$\sup_{\overline{\Omega}} |u_\varepsilon| \leq C \quad \text{for all } \varepsilon > 0.$$

If $\Omega \neq \mathbb{R}^N$, assume in addition the existence of the limit

(1.14) $$\lim_{(y,\varepsilon) \to (x, 0^+)} u_\varepsilon(y) =: g(x) \quad \text{for any } x \in \partial\Omega.$$

Then $u_\varepsilon \rightrightarrows u$ on compacta of $\overline{\Omega}$ as $\varepsilon \to 0^+$, where u is the unique solution in $BC(\overline{\Omega})$ of

(1.15) $$\begin{cases} F(x, u, Du) = 0 & \text{in } \Omega, \\ u = g & \text{on } \partial\Omega. \end{cases}$$

PROOF. By definition $\underline{u} \leq \overline{u}$. By (1.13) $\underline{u}$ and $\overline{u}$ are bounded, they are semicontinuous in $\overline{\Omega}$ by Lemma 1.5, and (1.14) gives $\underline{u} = \overline{u} = g$ on $\partial\Omega$. Thus, by Theorem 1.7, the comparison principle gives $\overline{u} \leq \underline{u}$ and therefore

$$\underline{u} = \overline{u} =: u$$

is the solution of (1.15). The local uniform convergence of u_ε to u is the following elementary Lemma 1.9. ◀

LEMMA 1.9. *Assume that the functions u_ε as in Definition 1.4 satisfy (1.8) on a compact set $\mathcal{K}$ and*

(1.16) $$\underline{u} = \overline{u} =: u \quad \text{on } \mathcal{K}.$$

Then $u_\varepsilon \rightrightarrows u$ on $\mathcal{K}$ as $\varepsilon \to 0^+$.

PROOF. Since u is both upper and lower semicontinuous (by Lemma 1.5), it is continuous. Assume by contradiction there exist $\delta > 0$, $\varepsilon_n \to 0^+$, and $x_n \in \mathcal{K}$ such that either $u_{\varepsilon_n}(x_n) - u(x_n) > \delta$ or $u_{\varepsilon_n}(x_n) - u(x_n) < -\delta$. We may also assume $x_n \to x$ and use the continuity of u to get either $\overline{u}(x) - u(x) \geq \delta$ or $\underline{u}(x) - u(x) \leq -\delta$, which are both in contradiction to (1.16). ◀

As a first simple application of the new stability results to control theory, we improve Corollary III.2.22 on the continuous dependence of the value function v_h of the infinite horizon problem with respect to a small parameter $h \geq 0$ measuring the size of some perturbation of the data: the system's dynamics f_h, the control set A_h, the running cost ℓ_h and the interest rate λ_h.

As in §III.2.2 we assume $A_h \subseteq A \subseteq \mathbb{R}^M$ are compact for all h, the Hausdorff distance $d_H(A_h, A_0)$ tends to 0 as $h \to 0^+$, $f_h \to f$ and $\ell_h \to \ell$ uniformly on $\mathcal{K} \times A$ for each compact $\mathcal{K} \subseteq \mathbb{R}^N$, $\lim_{h \to 0} \lambda_h = \lambda_0 > 0$. We also assume the data satisfy for all $h \geq 0$ the structural assumptions (A0)–(A4) of Chapter III with the same A and a uniform bound M for the running costs ℓ_h, but now we allow the constant L measuring the regularity of f_h in (A3) and the modulus of continuity ω_ℓ of ℓ_h in (A4) to depend on h. In other words, we are now requiring the perturbations to be small just in the sup-norm without bounds on their oscillations. For instance,

$$\ell_h(x, a) = \ell_0(x, a) + h\varphi(x/h^\gamma, a)$$

with φ lipschitzean and bounded is allowed for all constants γ, but it satisfies the assumptions of §III.2.2 only if $\gamma \leq 1$ (or if φ is constant with respect to x ...).

THEOREM 1.10. *Under the preceding assumptions, v_h converges to v_0 as $h \to 0^+$, uniformly on compact sets of $\mathbb{R}^N$.*

PROOF. We know from §III.2 that v_h satisfies a HJB equation of the form

$$\lambda_h v_h + H_h(x, Dv_h) = 0 \quad \text{in } \mathbb{R}^N,$$

where $H_h \rightrightarrows H_0$ on compact sets of $\mathbb{R}^{2N}$ by the proof of Corollary III.2.22. The limit equation

$$\lambda_0 u + H_0(x, Du) = 0 \quad \text{in } \mathbb{R}^N$$

satisfies the comparison principle by Theorem 1.3, since $\lambda_0 > 0$. Moreover,

$$\sup_{\mathbb{R}^N} |v_h| \leq \frac{M}{\lambda_h} \quad \text{for all } h,$$

and we have a uniform bound because $\lambda_h \to \lambda_0 > 0$. Thus we can apply Corollary 1.8 and get the conclusion. ◀

REMARK 1.11. When Corollary 1.8 is applied to problems with $\Omega \neq \mathbb{R}^N$ it may not be easy to prove the convergence of u_ε near the boundary, as required by (1.14). In Section 4 we show that this can be avoided if one knows that the limit problem (1.15) has a solution $u \in BC(\overline{\Omega})$. This is done by combining the weak limit technique with a weak notion of boundary condition, the Dirichlet boundary condition in the viscosity sense. ◁

Next we show how we can adapt the proofs of the comparison theorems in Chapters II and III to semicontinuous semisolutions. We first consider the simpler case that Ω is bounded where the technicalities are kept to a minimum.

PROOF OF THEOREM 1.3 FOR Ω BOUNDED. We are going to indicate the changes with respect to the proof of Theorem II.3.1. The continuity of u_1 and u_2 is used three times in that proof. The first time it is used to produce a maximum point $(x_\varepsilon, y_\varepsilon)$ for the auxiliary function

$$\Phi_\varepsilon(x, y) = u_1(x) - u_2(y) - \frac{|x - y|^2}{2\varepsilon},$$

and this still works if Φ_ε is upper semicontinuous, as it is under the semicontinuity assumptions on u_1 and u_2.

It is used for the second time to deduce

(1.17) $$\frac{|x_\varepsilon - y_\varepsilon|^2}{2\varepsilon} \longrightarrow 0 \quad \text{as } \varepsilon \to 0^+$$

from

(1.18) $$|x_\varepsilon - y_\varepsilon| \longrightarrow 0 \quad \text{as } \varepsilon \to 0^+.$$

Now we argue as follows. We observe that $u_1 - u_2 \in USC(\overline{\Omega})$ and

$$S := \max_{\overline{\Omega}}(u_1 - u_2) \leq \Phi_\varepsilon(x_\varepsilon, y_\varepsilon) \leq u_1(x_\varepsilon) - u_2(y_\varepsilon)$$

immediately gives (1.17) if we prove that

(1.19) $$u_1(x_\varepsilon) - u_2(y_\varepsilon) \longrightarrow S \quad \text{as } \varepsilon \to 0^+.$$

For this purpose, we assume by contradiction that (1.19) does not hold, so by the compactness of $\overline{\Omega}$ there exists $\varepsilon_k \to 0^+$, $x_{\varepsilon_k} \to x^*$, $y_{\varepsilon_k} \to y^*$ such that

$$\ell := \lim_k (u_1(x_{\varepsilon_k}) - u_2(y_{\varepsilon_k})) > S.$$

But (1.18) implies $x^* = y^*$, and the upper semicontinuity of $(x, y) \mapsto u_1(x) - u_2(y)$ gives $\ell \leq u_1(x^*) - u_2(x^*) \leq S$, a contradiction.

The continuity of u_1 and u_2 is used for the last time to prove that

$$\ell' := \limsup_k \Phi_{\varepsilon_k}(x_{\varepsilon_k}, y_{\varepsilon_k}) \leq 0$$

for every sequence $\varepsilon_k \to 0^+$ such that $(x_{\varepsilon_k}, y_{\varepsilon_k}) \in \partial(\Omega \times \Omega)$. To obtain the same conclusion we assume by contradiction that $\ell' > 0$ for some $\varepsilon_k \to 0^+$. By the compactness of $\overline{\Omega}$ and (1.18) we can assume $x_{\varepsilon_k} \to \overline{x}$, $y_{\varepsilon_k} \to \overline{x}$. Then we use the upper semicontinuity of $(x, y) \mapsto u_1(x) - u_2(y)$ to get

$$\ell' \leq \limsup_k (u_1(x_{\varepsilon_k}) - u_2(y_{\varepsilon_k})) \leq u_1(\overline{x}) - u_2(\overline{x}) \leq 0,$$

where the last inequality holds because $\overline{x} \in \partial\Omega$. We have reached a contradiction and completed the proof. ◀

PROOF OF THEOREM 1.3 FOR Ω UNBOUNDED. We outline the variants to the proof of Theorem III.2.12 (see also Remark III.2.14); the case of a Hamiltonian as in Theorem II.3.5 is left as an exercise for the reader. In this proof the auxiliary function is

$$\Phi(x,y) := u_1(x) - u_2(y) - \frac{|x-y|^2}{2\varepsilon} - \beta(\langle x \rangle^m + \langle y \rangle^m),$$

where $\langle x \rangle = (1+|x|^2)^{1/2}$, $m = \min\{1, 1/K\}$, and $K > 0$ is a constant appearing in the assumptions on the Hamiltonian.

We assume by contradiction that $u_1 - u_2$ is larger than $\delta > 0$ somewhere, and determine $\beta > 0$ so that $\sup \Phi > \delta/2$. Since Φ is u.s.c., and negative for large $|x|$ and $|y|$, it achieves the maximum at some point $(\bar{x}, \bar{y}) = (x_\varepsilon, y_\varepsilon)$. Thus the goal is to get a contradiction to

$$(1.20) \qquad \Phi(x_\varepsilon, y_\varepsilon) \geq \frac{\delta}{2} > 0$$

by taking $\varepsilon > 0$ very small.

Since m and β have been fixed, it is easy to see that (1.20) and the boundedness of u_1, u_2 force x_ε and y_ε to remain in a compact set $\mathcal{K}$ for all $\varepsilon > 0$ (cfr. Remark III.2.13). The boundedness of u_1, u_2 implies as before

$$(1.21) \qquad |x_\varepsilon - y_\varepsilon| \longrightarrow 0 \quad \text{as } \varepsilon \to 0^+,$$

and we want to deduce that

$$(1.22) \qquad \frac{|x_\varepsilon - y_\varepsilon|^2}{2\varepsilon} \longrightarrow 0 \quad \text{as } \varepsilon \to 0^+.$$

We follow as much as possible the proof of the previous case where Ω was bounded by setting

$$S := \max_{x \in \mathcal{K}}[(u_1 - u_2)(x) - 2\beta \langle x \rangle^m]$$

and observing that

$$S \leq \Phi(x_\varepsilon, y_\varepsilon) \leq u_1(x_\varepsilon) - u_2(y_\varepsilon) - \beta(\langle x_\varepsilon \rangle^m + \langle y_\varepsilon \rangle^m) =: \psi(\varepsilon).$$

By the definition of Φ we get (1.22) if we show that $\psi(\varepsilon) \to S$ as $\varepsilon \to 0^+$, and to prove this we assume by contradiction the existence of $\varepsilon_k \to 0^+$ such that $\lim_k \psi(\varepsilon_k) > S$. By the compactness of $\mathcal{K}$ and (1.21) we can assume $x_{\varepsilon_k} \to x^*$ and $y_{\varepsilon_k} \to x^*$, thus the semicontinuity of u_1 and u_2 gives $\lim_k \psi(\varepsilon_k) \leq u_1(x^*) - u_2(x^*) - 2\beta \langle x^* \rangle^m \leq S$, which is the desired contradiction.

If $x_{\varepsilon_k}, y_{\varepsilon_k} \in \Omega$ for all k of some sequence $\varepsilon_k \to 0^+$, we conclude as in the proof of Theorem III.2.12. Otherwise we use the compactness of $\mathcal{K}$ and (1.21) to select $\varepsilon_k \to 0^+$ such that $x_{\varepsilon_k} \to x^*$, $y_{\varepsilon_k} \to x^*$ and observe that $x^* \in \partial\Omega$ because $(x_\varepsilon, y_\varepsilon) \in \partial(\Omega \times \Omega)$ for all ε small. By the semicontinuity of u_1 and u_2 we get

$$\limsup_k \Phi(x_{\varepsilon_k}, y_{\varepsilon_k}) \leq \limsup_k (u_1(x_{\varepsilon_k}) - u_2(y_{\varepsilon_k})) \leq u_1(x^*) - u_2(x^*) \leq 0,$$

where the last inequality comes from the boundary condition $u_1 \leq u_2$ on $\partial\Omega$. We have got a contradiction to (1.20) which completes the proof. ◂

Also the comparison theorems on initial value problems for evolutive HJB equations, Theorems II.3.7, III.3.7, and III.3.15, are still valid if the assumption of continuity of the semisolutions to compare is dropped and we suppose just the upper semicontinuity of the subsolution and the lower semicontinuity of the supersolution. The proofs are obtained by the arguments given above and we leave them to the reader, as well as some extensions: see Exercises 1.5, 1.6, 1.7.

The reader might now believe in the validity of a metatheorem saying that every comparison result for continuous viscosity semisolutions extends to semicontinuous ones. This is essentially true when the boundary conditions are achieved in the classical pointwise sense, but it can be false if the boundary conditions are understood in some weak sense. For instance, the comparison result for constrained viscosity solutions (Theorem IV.5.8) does not hold for a subsolution in Ω, $u_1 \in BUSC(\overline{\Omega})$, and a supersolution in $\overline{\Omega}$, $u_2 \in BLSC(\overline{\Omega})$, of $u + H(x, Du) = 0$. (The definition of u.s.c. supersolution in $\overline{\Omega}$ is the natural one: just add $\varphi \in C^1(\overline{\Omega})$ and $x \in \overline{\Omega}$ in Definition 1.1.) The counterexample is provided by Example IV.5.3. In fact it is not hard to check (either directly or by a Dynamic Programming argument) that the value functions $v_{\overline{\Omega}}$ and v_Ω for the problem with state constraints respectively in $\overline{\Omega}$ and Ω are, respectively, a l.s.c. supersolution in $\overline{\Omega}$ and an u.s.c. subsolution in Ω (both bounded), but $v_{\overline{\Omega}} < v_\Omega$ on the half-line of discontinuity.

In Section 3 we study the possibility of comparing semisolutions satisfying weak boundary conditions. For the constrained boundary value problem we prove that Theorem IV.5.8 holds for semicontinuous semisolutions, provided the subsolution is continuous at each point of the boundary of Ω.

Exercises

1.1. Let $u : \Omega \to \mathbb{R}$ be an arbitrary function. Prove that if $D^+u(x) \neq \emptyset$ (respectively, $D^-u(x) \neq \emptyset$), then u is u.s.c. at x (respectively, u is l.s.c. at x). Compute the sub- and superdifferentials of the characteristic functions of $\mathbb{R}_+$ and of $\mathbb{Q}$.

1.2. Discuss the validity of Lemma II.1.8 on the properties of D^+u and D^-u without the continuity assumptions on u.

1.3. Prove the following extensions of Lemma II.1.7. Assume $u \in USC(\Omega)$ (respectively, $u \in LSC(\Omega)$). Then $p \in D^+u(x)$ (respectively, $p \in D^-u(x)$) if and only if there exists $\varphi \in C^1(\Omega)$ such that $D\varphi(x) = p$ and $u - \varphi$ has a local maximum (respectively, minimum) at x.

1.4 (LIPSCHITZ CONTINUITY OF SUBSOLUTIONS FOR COERCIVE HAMILTONIANS). Assume that for some constants C and K

$$F(x, r, p) > 0 \qquad \forall x \in \Omega, \ r \in [-K, K], \ |p| \geq C.$$

Prove that a subsolution $u \in USC(\Omega)$ of (1.1), such that $|u| \leq K$, satisfies $|u(x) - u(y)| \leq C|x - y|$ for all x, y such that the line segment joining them is contained in Ω. In particular u is locally lipschitzean (cfr. Proposition II.4.1).

1.5 (COMPARISON PRINCIPLE FOR CAUCHY-DIRICHLET PROBLEMS). Assume $T \in \,]0, +\infty]$, $\lambda \geq 0$, $\Omega \subseteq \mathbb{R}^N$ is open, $H(x,t,p)$ is continuous on $\mathbb{R}^{2N+1}$ and satisfies the structure condition (2.18) of Chapter III for each fixed t with the same constants L, K and modulus ω_ℓ. Let $u_1 \in BUSC(\overline{\Omega} \times [0,T[)$ and $u_2 \in BLSC(\overline{\Omega} \times [0,T[)$ be, respectively, a sub- and supersolution of

$$u_t + \lambda u + H(x, t, D_x u) = 0 \quad \text{in } \Omega \times \,]0, T[$$

such that $u_1 \leq u_2$ on $\overline{\Omega} \times \{0\}$ and on $\partial\Omega \times [0, T[$. Prove that $u_1 \leq u_2$ on $\overline{\Omega} \times [0, T[$.

[Hints: one has to modify appropriately the proof of Theorem III.3.7 using the tricks of the proof of Theorem 1.3. There are two additional difficulties. The first is how to compare at $t = T$. We suggest to replace u_1 by $u_1 - \varepsilon/(T-t)$ and then let $\varepsilon \to 0^+$. Alternatively, one can extend u_1, u_2 to $t = T$ by taking their semicontinuous envelopes and then prove an analogue of Lemma II.2.10 for these extensions (we warn the reader that this lemma does not hold for arbitrary semicontinuous extensions to $\overline{\Omega} \times [0, T]$). The second difficulty is the dependence of H on t. This can be overcome by replacing the term $(|x-y|^2 + |t-s|^2)/2\varepsilon$ with $|x-y|^2/2\varepsilon + |t-s|^2/2\eta$ in the definition of the auxiliary function Φ, and letting η go to 0^+ before ε.] ◁

1.6 (LOCAL COMPARISON FOR INITIAL VALUE PROBLEMS). Prove that Theorem III.3.12 remains valid if the continuity assumption on u_1, u_2 is dropped and replaced by $u_1 \in BUSC(\overline{C})$ and $u_2 \in BLSC(\overline{C})$.

1.7 (COMPARISON OF LOCALLY BOUNDED SOLUTIONS TO INITIAL VALUE PROBLEMS). Prove that Theorem III.3.15 remains valid if the continuity assumption on u_1, u_2 is dropped and replaced by $u_1 \in USC(\mathbb{R}^N \times [0,T])$, $u_2 \in LSC(\mathbb{R}^N \times [0,T])$, $u_1 - u_2$ bounded on $\mathcal{K} \times [0,T]$ for any compact $\mathcal{K}$. ◁

1.8. Prove the following converse of Lemma 1.9: if u_ε converge uniformly on a compact $\mathcal{K}$ to a continuous function u as $\varepsilon \to 0^+$, then

$$\liminf_{\varepsilon \to 0^+}{}_* u_\varepsilon = \limsup_{\varepsilon \to 0^+}{}^* u_\varepsilon = u .$$

1.9. Let $I \subseteq \mathbb{R}$ be an open interval. Prove that

(i) $u \in USC(I)$ is nondecreasing if and only if it is a supersolution of $u' = 0$ in I;

(ii) $u \in LSC(I)$ is nondecreasing if and only if it is a subsolution of $-u' = 0$ in I (cfr. Lemma II.5.15).

2. Non-continuous solutions

2.1. Definitions, basic properties, and examples

In this subsection we extend the notion of viscosity solution to functions that are not necessarily continuous. Note that the extension of the definition of sub- and supersolutions to semicontinuous functions in Section 1 does not change the concept of solution, because a function that is simultaneously a sub- and a supersolution according to Definition 1.1 is automatically continuous. Therefore we need an additional tool, which is the notion of *upper* and *lower semicontinuous envelope* of a function $u : X \to \overline{\mathbb{R}} = [-\infty, +\infty]$, $X \subseteq \mathbb{R}^N$. These new functions are, respectively,

$$u^*(x) := \limsup_{y \to x} u(y) := \lim_{r \to 0^+} \sup\{\, u(y) : y \in X,\ |y - x| \leq r\,\},$$

$$u_*(x) := \liminf_{y \to x} u(y) := \lim_{r \to 0^+} \inf\{\, u(y) : y \in X,\ |y - x| \leq r\,\}.$$

The next proposition collects the properties of the semicontinuous envelopes.

PROPOSITION 2.1.

(i) $u_* \leq u \leq u^*$;

(ii) $u_* = -(-u)^*$;

if u is locally bounded, then $u^*, u_* : X \to \mathbb{R}$ satisfy

(iii) $\begin{cases} u^*(x) = \inf\{\, v(x) : v \in C(X),\ v \geq u\,\}, \\ u_*(x) = \sup\{\, v(x) : v \in C(X),\ v \leq u\,\}; \end{cases}$

(iv) $\begin{cases} u^*(x) = \min\{\, v(x) : v \in USC(X),\ v \geq u\,\}, \\ u_*(x) = \max\{\, v(x) : v \in LSC(X),\ v \leq u\,\}; \end{cases}$

(v) *u is u.s.c. at x_0 if and only if $u(x_0) = u^*(x_0)$, u is l.s.c. at x_0 if and only if $u(x_0) = u_*(x_0)$.*

PROOF. (i) and (ii) are trivial. The most important statement of this proposition is that $u^* \in USC(X)$ and $u_* \in LSC(X)$. This can be deduced from (iii), and then (iv) follows immediately from the characterization of upper semicontinuity of v at x by means of the equality $v(x) = \limsup_{y \to x} v(y)$.

Instead of giving the proof of (iii), which is left as an exercise for the reader, we prove directly that u^* is u.s.c. By definition, for any $x \in X$ and $\varepsilon > 0$, there is $r > 0$ such that

$$u^*(x) + \varepsilon > \sup\{\, u(y) : |y - x| < r\,\} \geq \sup\{\, u(y) : |y - z| < r - |x - z|\,\} \geq u^*(z)$$

for $z \in X$ such that $|z - x| < r$. Then $\limsup_{z \to x} u^*(z) \leq u^*(x)$, and so u^* is u.s.c. at x. Finally (v) follows easily from the definitions. ◀

Now we can give the definition of *non-continuous viscosity solution* for the equation

(2.1) $$F(x, u, Du) = 0 \quad \text{in } \Omega$$

with $\Omega \subseteq \mathbb{R}^N$ open and $F : \Omega \times \mathbb{R} \times \mathbb{R}^N \to \mathbb{R}$ continuous.

DEFINITION 2.2. A locally bounded function $u : \Omega \to \mathbb{R}$ is a non-continuous viscosity solution of (2.1) if u^* is a subsolution of (2.1) and u_* is a supersolution of (2.1) according to Definition 1.1. ◁

This definition makes sense because $u^* \in USC(\Omega)$ and $u_* \in LSC(\Omega)$ by Proposition 2.1 (iv). Note that here the word non-continuous means "not necessarily continuous". In fact, if $u \in C(\Omega)$, then $u = u^* = u_*$ and the definition coincides with the one used in the previous chapters. From now on the word solution means non-continuous viscosity solution according to Definition 2.2.

It is interesting to note that, for Hamiltonians satisfying the structural assumptions of any comparison theorem, a solution of (2.1) may be a discontinuous function only if it is discontinuous at some boundary point. This fact was proved directly for the value functions of optimal control problems with exit times in Proposition IV.3.3, but it is indeed a general property of Hamilton-Jacobi equations, as the next result shows.

THEOREM 2.3. *Assume the Dirichlet problem for (2.1) satisfies a comparison principle (Definition 1.2). If $u : \overline{\Omega} \to \mathbb{R}$ is a bounded solution of (2.1) which is continuous at all $x \in \partial\Omega$, then $u \in C(\overline{\Omega})$. In particular, for any $g \in C(\partial\Omega)$, there is at most one such solution satisfying $u = g$ on $\partial\Omega$.*

PROOF. By definition, $u^* \in BUSC(\overline{\Omega})$ is a subsolution of (2.1) and $u_* \in BLSC(\overline{\Omega})$ is a supersolution. By Proposition 2.1 (v), the continuity of u at boundary points gives
$$u^*(x) = u(x) = u_*(x) \quad \text{for all } x \in \partial\Omega .$$
Then, by the comparison principle,
$$u^* \leq u_* \quad \text{in } \overline{\Omega} .$$
Since the opposite inequality is trivial by definition, we obtain $u^* = u_*$ and thus u is continuous by Proposition 2.1 (i) and (v). ◄

Next we give some simple examples showing that the notion of non-continuous solution is rather weak. The first one shows that nowhere continuous functions can be solutions in that sense of very simple equations such as

(2.2) $$u' = 0 \quad \text{in } \mathbb{R} .$$

EXAMPLE 2.4. Let $u(x) \equiv 1$ if $x \in \mathbb{Q}$, $u(x) \equiv 0$ if $x \in \mathbb{R} \setminus \mathbb{Q}$, that is, $u = \chi_\mathbb{Q}$. Then $u^* \equiv 1$ and $u_* \equiv 0$, so u is a solution of (2.2). (See Exercise 2.9 for some properties of the solutions of (2.2).) ◁

The second example shows that a boundary value problem with a unique continuous solution may have infinitely many discontinuous solutions.

EXAMPLE 2.5. The Dirichlet problem

(2.3) $$u' = 1 \quad \text{in }]0,1[, \quad u(0) = 1, \; u(1) = 2,$$

has the continuous solution $u(x) = x + 1$, which is unique by Theorem II.5.9. On the other hand, $w(x) = x + \chi_\mathbb{Q}(x)$ is also a solution because $w^*(x) = x + 1$

and $w_*(x) = x$ are both classical solutions of the equations and the boundary conditions are satisfied. Similarly, for any dense subset X of $[0,1]$ such that $0, 1 \in X$, $x + \chi_X(x)$ is a solution of (2.3). However, it is enough to require the continuity of the solutions at the boundary points 0 and 1 to have uniqueness as it is easy to prove by the comparison theorems for semicontinuous semisolutions, Theorem 1.3, and the change of variables $v := 1 - e^{-h}$ (Kružkov transform, see §IV.2). ◁

Next we show that discontinuous value functions of optimal control problems satisfy the associated HJB equation in the sense of Definition 2.2. We have seen in Chapter IV examples of discontinuous value functions for problems with exit times or state constraints. Here we do not refer to a particular control problem because it is enough to assume a local optimality principle. In fact, in Chapter IV we proved the Dynamic Programming Principle without assuming the continuity of the value functions (Propositions IV.2.1 and IV.3.9).

We assume that the system with dynamics f and control constraint A satisfies the basic assumptions (A_0)–(A_3) of Chapter III, the running cost ℓ satisfies (A_4) in Chapter III, and $\lambda \geq 0$. We give the result for the stationary equation

$$(2.4) \qquad \lambda u + H(x, Du) = 0 \quad \text{in } \Omega,$$

with

$$(2.5) \qquad H(x,p) := \sup_{a \in A} \{ -f(x,a) \cdot p - \ell(x,a) \},$$

and $\Omega \subseteq \mathbb{R}^N$ open. A similar result holds for evolutive equations, see Exercise 2.1. We recall that $\mathcal{A}$ denotes the set of measurable control functions, and $y_x(\cdot, \alpha)$ is the trajectory of the system starting at x and corresponding to $\alpha \in \mathcal{A}$.

THEOREM 2.6. *Under the preceding hypotheses, assume u is locally bounded in Ω and that for all x there exists $\tau > 0$ such that, for all $0 < t < \tau$,*

$$(2.6) \qquad u(x) = \inf_{\alpha \in \mathcal{A}} \left\{ \int_0^t \ell(y_x, \alpha) e^{-\lambda s} ds + u(y_x(t, \alpha)) e^{-\lambda t} \right\}.$$

Then u is a solution of (2.4).

PROOF. As in the proof of Proposition III.2.8, the local suboptimality principle (i.e., the inequality $\leq$ in (2.6)) implies that u^* is a subsolution, while the local superoptimality principle (i.e., the inequality $\geq$ in (2.6)) gives that u_* is a supersolution.

To prove that u_* is a supersolution we take $\varphi \in C^1(\Omega)$ and $z \in \Omega$ such that $u_*(z) = \varphi(z)$ and $u_*(x) \geq \varphi(x)$ for all x in a neighborhood of z. We assume by contradiction $\lambda \varphi(z) + H(z, D\varphi(z)) < 0$. Then, for some $\varepsilon > 0$,

$$(2.7) \qquad \lambda \varphi(x) + H(x, D\varphi(x)) \leq -\varepsilon \quad \text{for all } x \in B(z, \varepsilon) \subseteq \Omega.$$

By the assumptions on f and A, there is $t \in \,]0, \tau[$ such that $y_x(s, \alpha) \in B(z, \varepsilon)$ for all $x \in B(z, \varepsilon/2)$, $0 < s \leq t$, and all $\alpha \in \mathcal{A}$. We fix such a t and set

$$\delta := \begin{cases} \varepsilon(1 - e^{-\lambda t})/2\lambda & \text{if } \lambda > 0, \\ \varepsilon t/2 & \text{if } \lambda = 0. \end{cases}$$

By the inequality $\geq$ in (2.6), for any x there is $\alpha \in \mathcal{A}$ such that

$$u(x) > -\delta + \int_0^t \ell(y_x, \alpha) \, e^{-\lambda s} \, ds + u(y_x(t, \alpha)) \, e^{-\lambda t}.$$

Since $u \geq u_* \geq \varphi$ and

$$\frac{d}{ds} \varphi(y_x(s)) \, e^{-\lambda s} = e^{-\lambda s}[-\lambda \varphi(y_x) + D\varphi(y_x) \cdot y_x']$$

a.e., we get, for $x \in B(z, \varepsilon/2)$,

$$u(x) - \varphi(x) > -\delta + \int_0^t e^{-\lambda s}[\ell(y_x, \alpha) - \lambda \varphi(y_x) + D\varphi(y_x) \cdot f(y_x, \alpha)](s) \, ds$$

(2.8)
$$\geq -\delta - \int_0^t e^{-\lambda s}[\lambda \varphi(y_x) + H(y_x, D\varphi(y_x))](s) \, ds$$

$$\geq -\delta + \int_0^t \varepsilon e^{-\lambda s} \, ds = \delta,$$

where in the last inequality we used (2.7) and the choice of t. Then

$$u_*(z) = \liminf_{x \to z} u(x) \geq \varphi(z) + \delta,$$

a contradiction to the choice of φ because $\delta > 0$.

The proof that u^* is a subsolution is similar. We fix $z \in \Omega$ and $\varphi \in C^1(\Omega)$ such that $u^*(z) = \varphi(z)$ and $u^*(x) \leq \varphi(x)$ for all x near z. We assume by contradiction that $\lambda \varphi(z) + H(z, D\varphi(z)) > 0$ and fix $\varepsilon > 0$, $\bar{a} \in A$ such that

(2.9) $\quad \lambda \varphi(x) - f(x, \bar{a}) \cdot D\varphi(x) - \ell(x, \bar{a}) \geq \varepsilon \quad \text{for all } x \in B(z, \varepsilon) \subseteq \Omega.$

We choose t small as before and apply the inequality $\leq$ in (2.6) to the constant control $\alpha(t) = \bar{a}$ to get

$$u(x) \leq \int_0^t \ell(y_x, \bar{a}) \, e^{-\lambda s} \, ds + u(y_x(t, \bar{a})) \, e^{-\lambda t}.$$

Now we use the inequalities $u \leq u^* \leq \varphi$ to perform the same calculation as in (2.8) and get, by (2.9),

$$u(x) - \varphi(x) \leq -2\delta < 0 \quad \text{for all } x \in B(z, \varepsilon/2).$$

By taking the lim sup as $x \to z$ we obtain $u^*(z) < \varphi(z)$, a contradiction to the choice of φ. ◄

Now we can give some interesting examples of discontinuous value functions solving the HJB equation. If we are given a closed target $\mathcal{T}$ with compact boundary, we recall that $t_x(\alpha)$ indicates the first time the trajectory associated with $\alpha \in \mathcal{A}$ and starting at x hits $\mathcal{T}$, that

$$T(x) := \inf_{\alpha \in \mathcal{A}} t_x(\alpha)$$

is the minimum time function, and

(2.10) $\quad v(x) := \inf_{\alpha \in \mathcal{A}} \int_0^{t_x(\alpha)} e^{-t}\,dt = \begin{cases} 1 - e^{-T(x)} & \text{if } T(x) < +\infty, \\ 1 & \text{if } T(x) = +\infty, \end{cases}$

is the Kružkov transform of T. In Section IV.1 we proved that v is continuous in $\mathcal{T}^c$ if and only if the system is small-time controllable on $\mathcal{T}$ (STC$\mathcal{T}$), and in this case v is the complete solution in $BC(\overline{\mathcal{T}^c})$ of

(2.11) $\quad \begin{cases} u + \sup_{a \in A}\{-f(x,a) \cdot Du\} - 1 = 0 & \text{in } \mathcal{T}^c, \\ u = 0 & \text{on } \partial\mathcal{T}. \end{cases}$

If STC$\mathcal{T}$ fails, by Theorem 2.6 v is still a solution in the sense of Definition 2.2. Thus we have the following necessary condition of optimality for control laws also in this case, as in §III.2.2.

COROLLARY 2.7. *A control law* $\mathbb{A}(x) = \alpha_x \in \mathcal{A}$ *defined on an open set* $\Omega \subseteq \mathbb{R}^N$ *is time-optimal, that is,* $T(x) = t_x(\alpha_x)$ *for all* $x \in \Omega$, *only if* $u(x) := 1 - e^{-t_x(\alpha_x)}$ *is a viscosity solution in* Ω *of the HJB equation in* (2.11).

PROOF. If $\mathbb{A}$ is optimal, then $u = v$ in Ω and it is enough to apply the DPP Proposition IV.3.9 and Theorem 2.6. ◂

If STC$\mathcal{T}$ fails, however, the problem of uniqueness and comparison for (2.11), related to sufficient conditions of optimality for control laws, is much harder. This is because v is discontinuous at all points of $\partial\mathcal{T}$ where STC$\mathcal{T}$ fails, so the boundary condition $v = 0$ at such points gives very poor information. In fact it is easy to construct a counterexample to the uniqueness for (2.11) similar to Example 2.5, even in cases where there is a complete continuous solution, see Exercise 2.2.

In Sections 3, 4, and 5 we see some appropriate interpretations of the boundary conditions which give uniqueness of solution of (2.11) in classes of discontinuous functions. To further motivate our interest in discontinuous value functions we give some explicit examples of time-optimal control problems missing STC$\mathcal{T}$. The first is one of the most classical problems in optimal control theory.

EXAMPLE 2.8 (ZERMELO NAVIGATION PROBLEM). Consider a boat moving with velocity of constant magnitude, which we normalize to 1, relative to a stream of constant velocity. We want to reach in minimum time a given compact target $\mathcal{T}$.

We choose the axes in $\mathbb{R}^2$ so that the stream velocity is $(\sigma, 0)$ with $\sigma > 0$. Then the system is

$$y_1' = \sigma + a_1,$$
$$y_2' = a_2,$$

with $a_1^2 + a_2^2 = 1$. It is easy to see that the system is small-time controllable on any target T if $\sigma < 1$, so we consider only the case $\sigma \geq 1$. Let us first take $T = \{0\}$. It is easy to see by an elementary geometric argument that the reachable set $\mathcal{R} := \{x : T(x) < +\infty\}$ is

$$\mathcal{R} = \{x : x_1 < 0 \text{ or } x_1 = x_2 = 0\} \qquad \text{if } \sigma = 1,$$
$$\mathcal{R} = \{x : x_1 \leq 0, \ |x_2| \leq -x_1(\sigma^2 - 1)^{-1/2}\} \quad \text{if } \sigma > 1.$$

Thus in both cases T is not contained in the interior of $\mathcal{R}$ and v is discontinuous at least in 0. Actually, it is not hard to compute T by standard control-theoretic methods and get

$$T(x) = \begin{cases} -\dfrac{|x|^2}{2x_1} & \text{if } \sigma = 1, \\ \dfrac{-x_1\sigma - [x_2^2(1-\sigma^2) + x_1^2]^{1/2}}{\sigma^2 - 1} & \text{if } \sigma > 1, \end{cases}$$

see Exercise 2.3. Then, for $\sigma = 1$, v is discontinuous only at 0, whereas for $\sigma > 1$ it is discontinuous at each point of $\partial\mathcal{R}$ (in fact at such points $T(x) = -x_1\sigma/(\sigma^2-1)$).

If the target is $T = \overline{B}(0,1)$, by similar arguments we compute for $\sigma > 1$

$$\mathcal{R} = T \bigcup \{x : x_1 \leq 1/\sigma, \ |x_2| \leq (\sigma^2-1)/\sigma + (1/\sigma - x_1)(\sigma^2-1)^{-1/2}\},$$

and again v jumps at all points of $\partial\mathcal{R}$. ◁

EXAMPLE 2.9. Consider the system

$$y_1' = -ay_2 + 1 - |a|,$$
$$y_2' = ay_1,$$

with $a \in \{-1, 0, 1\}$, and $T = \{0\}$. The only way to reach T by a piecewise smooth control is by first reaching the half-axis $S := \{(x_1, 0) : x_1 < 0\}$ and then using the control $a = 0$. Thus one guesses that the optimal control is to use $a = 1$ in the half-plane where $x_2 > 0$ and $a = -1$ in the half-plane where $x_2 < 0$. An easy computation shows that the time needed to reach S by such controls does not depend on $|x|$, and that the payoff associated with the candidate optimal control α is $t_x(\alpha) = |x| + \arccos(-x_1/|x|)$. It is not hard to prove that this is, in fact, the minimum time function (at least if we restrict ourselves to piecewise smooth controls). Note that in this problem $\mathcal{R} = \mathbb{R}^2$, the value function is discontinuous at 0, since it has a jump of height π along the x_1 axis at the origin, and it is continuous elsewhere. ◁

EXAMPLE 2.10. Consider the system of Example 2.9 with $a \in \{0,1\}$ and $\mathcal{T} = \{(1,0)\}$. Now one can reach $\mathcal{T}$ either from the half-axis $\{(x_1, 0) : x_1 < 1\}$, by using $a = 0$ as before, or from the arc $\{x_1 = (1 - x_2^2)^{1/2} : -1 \leq x_2 < 0\}$ by choosing $a = 1$. Again $\mathcal{R} = \mathbb{R}^2$. By analyzing the possible trajectories one realizes that the minimum time function is discontinuous at all points $(x_1, 0)$ with $0 \leq x_1 \leq 1$. In fact, for $x_1 \leq 1$ it is easy to see that

$$T(x_1, 0) = 1 - x_1,$$

while for $x_2 > 0$ the best behavior is first using $a = 1$ until one reaches $x_2 = 0$, so

$$T(x_1, x_2) = |x| + 1 + \arccos(-x_1/|x|). \quad \triangleleft$$

The points of discontinuity of the value function in Example 2.8 are two half-lines and they are parts of $\partial \mathcal{R}$, in Example 2.10 they are a bounded arc in the interior of $\mathcal{R}$.

In any case, a set of discontinuities of T that is not contained in $\mathcal{T}$ is called a *barrier*, because no optimal trajectory can cross it. An example of a time-optimal problem whose target is a half-space instead of a single point and having a barrier is in Exercise 2.6.

2.2. Existence of solutions by Perron's method

We begin with some properties of envelopes of viscosity sub- or supersolutions that generalize Proposition II.2.1.

PROPOSITION 2.11.

(a) *Let $\mathcal{S}$ be a set of functions such that w^* is subsolution of (2.1) for all $w \in \mathcal{S}$, and define*

$$u(x) := \sup_{w \in \mathcal{S}} w(x), \qquad x \in \Omega.$$

If u is locally bounded, then u^* is a subsolution of (2.1).

(b) *Let $\mathcal{Z}$ be a set of functions such that w_* is supersolution of (2.1) for all $w \in \mathcal{Z}$, and define*

$$u(x) := \inf_{w \in \mathcal{Z}} w(x), \qquad x \in \Omega.$$

If u is locally bounded, then u_* is a supersolution of (2.1).

PROOF. We only prove (a). Indeed (b) can be obtained by a similar argument or using the fact that, for a subsolution w of (2.1), $v = -w$ is a supersolution of

$$-F(x, -v, -Dv) = 0 \quad \text{in } \Omega$$

and vice versa. We fix $y \in \Omega$, $\varphi \in C^1(\Omega)$ such that $u^* - \varphi$ has a local maximum at y and $u^*(y) = \varphi(y)$. It is not restrictive to assume, for some $r > 0$,

$$(u^* - \varphi)(x) \leq -|x - y|^2 \qquad \text{for all } x \in \overline{B}(y, r)$$

2. NON-CONTINUOUS SOLUTIONS

by adding $|x-y|^2$ to φ. By definition of u^*, there is a sequence $x_n \to y$ such that

$$(u-\varphi)(x_n) \geq -\frac{1}{n} \qquad \text{for all } n.$$

By definition of u there is a sequence $\{v_n\}$ with $v_n \in \mathcal{S}$ such that

$$u(x_n) - \frac{1}{n} < v_n(x_n) \qquad \text{for all } n.$$

Then $(v_n - \varphi)(x_n) > -2/n$ for all n and

$$(v_n - \varphi)(x) \leq -|x-y|^2 \qquad \text{for } x \in \bar{B}(y,r),$$

and thus the same inequalities hold with v_n replaced by v_n^*. Therefore a maximum point y_n of $v_n^* - \varphi$ in $\bar{B}(y,r)$ satisfies, for n large enough,

$$-\frac{2}{n} < (v_n^* - \varphi)(y_n) \leq -|y_n - y|^2,$$

and letting $n \to +\infty$ we get

$$y_n \longrightarrow y \qquad \text{and} \qquad v_n^*(y_n) \longrightarrow \varphi(y) = u^*(y).$$

Since v_n^* is a subsolution of (2.1) we have $F(y_n, v_n^*(y_n), D\varphi(y_n)) \leq 0$, and letting $n \to \infty$

$$F(y, u^*(y), D\varphi(y)) \leq 0. \quad \blacktriangleleft$$

The following technical lemma, sometimes called the *Bump Lemma*, will allow us to prove that in some cases the sup of subsolution or the inf of supersolution are indeed solutions of the equations. We give it in a form that makes it usable in different situations.

LEMMA 2.12. *Let u be a function such that u^* is subsolution of (2.1) and u_* fails to be a supersolution at some point y. Then, for any $\delta > 0$ there is $\gamma > 0$ such that, for all $r < \gamma$, there exists a function w with the following properties: w^* is a subsolution of (2.1), $w \geq u$,*

(2.12) $$\sup_\Omega (w - u) > 0$$

and, for all $x \in \Omega$,

(2.13) $\qquad w(x) = u(x) \qquad\qquad\qquad\quad\;$ *if* $|x - y| \geq r$,
(2.14) $\qquad w(x) \leq u(x) \vee (u_*(y) + \delta) \quad$ *if* $|x - y| < r$.

PROOF. The argument is essentially the same as in the proof of Proposition II.2.1(c). It is not restrictive to assume $y = 0$. Since u_* fails to be a supersolution at this point, there exists $\varphi \in C^1(\Omega)$ such that

$$\varphi(0) = u_*(0), \qquad \varphi(x) \leq u_*(x) \text{ for small } |x|, \qquad F(0, u_*(0), D\varphi(0)) < 0.$$

Define, for $\varepsilon > 0$,

$$v(x) := \varphi(x) + \varepsilon - |x|^2,$$

and observe that, by continuity, v is a classical subsolution of $F(x, v, Dv) = 0$ in $B(0,r)$ for sufficiently small $\varepsilon, r > 0$. We choose r small enough to have, in addition,

(2.15) $$v(x) \leq u_*(0) + \frac{\delta}{2} + \varepsilon \quad \text{for } |x| < r.$$

Note that $(v - u)(x) \leq (v - u_*)(x) \leq (v - \varphi)(x) \leq 0$ for $|x| \geq \varepsilon^{1/2}$, and thus

(2.16) $$v(x) \leq u(x) \quad \text{for } |x| \geq r/2$$

if we choose $\varepsilon < r^2/4$. Moreover, if $x_n \to 0$ is such that $u(x_n) \to u_*(0)$, we have $\lim_n (v - u)(x_n) = \varepsilon > 0$, so, for all $r > 0$,

(2.17) $$\sup_{B(0,r)} (v - u) > 0.$$

Now we define

$$w(x) := \begin{cases} u(x) \vee v(x) & \text{if } |x| < r, \\ u(x) & \text{otherwise,} \end{cases}$$

and claim that w has the desired properties. In fact, $w(x) = u(x)$ for $|x| \geq r/2$ by (2.16). Then (2.13) holds and w^* is a subsolution of (2.1), because it coincides with u^* for $|x| > r/2$, while for $|x| < r$ we can apply Proposition 2.11. Moreover, (2.12) follows from (2.17), and (2.14) follows from (2.15) if we choose $\varepsilon \leq \delta/2$. ◄

REMARK 2.13. Of course there is an analogue of Lemma 2.12 for the case when u_* is a supersolution and u^* fails to be a subsolution. ◁

The first consequence of the last two results is a very general existence theorem for equation (2.1). This is the adaptation to first order equations of the classical *Perron's Method* for the Laplace equation.

THEOREM 2.14. *Assume there exists a subsolution u_1 and a supersolution u_2 of (2.1) such that $u_1 \leq u_2$. Then the functions*

(2.18) $$U(x) := \sup\{ w(x) : u_1 \leq w \leq u_2, \ w^* \text{ subsolution of } (2.1) \},$$
(2.19) $$W(x) := \inf\{ w(x) : u_1 \leq w \leq u_2, \ w_* \text{ supersolution of } (2.1) \},$$

are (non-continuous viscosity) solutions of (2.1).

PROOF. By Proposition 2.11 (a), U^* is a subsolution of (2.1). To prove that U_* is a supersolution we fix $y \in \Omega$ and first consider the case that $U_*(y) = u_2(y)$.

Let $\varphi \in C^1(\Omega)$ be such that $U_* - \varphi$ has a local minimum at y and $U_*(y) = \varphi(y)$. Then, in a neighborhood of y,

$$(u_2 - \varphi)(x) \geq (U_* - \varphi)(x) \geq 0 = (u_2 - \varphi)(y),$$

so $u_2 - \varphi$ has a local minimum at y. Since u_2 is a supersolution we get

$$F(y, u_2(y), D\varphi(y)) \geq 0,$$

which completes the proof in the case $U_*(y) = u_2(y)$.

The other possible case is

(2.20) $$U_*(y) + \delta \leq u_2(y) - \delta$$

for some $\delta > 0$. If we assume by contradiction that U_* fails to be a supersolution at y, then by Lemma 2.12 there is a subsolution w^* such that $w > U$ at some point of Ω, and $w \geq U \geq u_1$ everywhere. Moreover, since u_2 is lower semicontinuous, we can choose $r > 0$ small enough such that

$$u_2(y) - \delta \leq u_2(x) \qquad \text{for all } x \in \Omega, |x - y| < r.$$

Then we use (2.20), the properties (2.13) and (2.14) of w, and $U \leq u_2$, to get $w \leq u_2$. Then, by the definition of U, $w \leq U$ in Ω, a contradiction to the fact that $w > U$ at some point of Ω. Thus U is a solution of (2.1). The proof that W is a solution as well is similar, by means of Proposition 2.11 (b) and Remark 2.13. ◂

It is easy to deduce from Theorem 2.14 existence results for initial or boundary value problems associated with (2.1). In fact, if we assume in addition that on $\Gamma \subseteq \partial\Omega$

(2.21) $$\lim_{x \to y} u_2(x) = \lim_{x \to y} u_1(x) = g(y) \qquad \text{for all } y \in \Gamma,$$

then the solutions U and W of (2.1) defined by (2.18) and (2.19) have unique extensions U, W to $\Omega \cup \Gamma$, continuous at each point of Γ, satisfying the boundary condition

$$U = W = g \qquad \text{on } \Gamma.$$

In particular, we have the following result on the Dirichlet problem for (2.1).

COROLLARY 2.15. *Assume there exist a subsolution u_1 and a supersolution u_2 of (2.1), both bounded and satisfying (2.21) with $\Gamma = \partial\Omega$ and $g \in C(\partial\Omega)$. Suppose the Dirichlet problem for (2.1) satisfies a comparison principle. Then there exists a (unique) solution $u \in BC(\overline{\Omega})$ of (2.1) satisfying the boundary condition $u = g$ on $\partial\Omega$.*

PROOF. We extend u_1 and u_2 by setting $u_1 = u_2 = g$ on $\partial\Omega$. The continuity of g and (2.21) imply $u_1 \in BUSC(\overline{\Omega})$ and $u_2 \in BLSC(\overline{\Omega})$, thus $u_1 \leq u_2$ by the comparison principle. Then we can apply Theorem 2.14 to get a solution of (2.1) which can be extended continuously to $\partial\Omega$ by (2.21). The conclusion follows from Theorem 2.3. ◂

Next we give a stability result for discontinuous viscosity solutions with respect to pointwise convergence of monotone sequences. It is reminiscent of Harnack's convergence theorem for harmonic functions.

PROPOSITION 2.16.

(i) If $u_n \in USC(\Omega)$ (respectively, $u_n \in LSC(\Omega)$), $n \in \mathbb{N}$, are subsolutions (respectively, supersolutions) of (2.1) such that $u_n(x) \searrow u(x)$ (respectively, $u_n(x) \nearrow u(x)$) where u is locally bounded, then u is a subsolution (respectively, supersolution) of (2.1).

(ii) If, in addition, u_n are solutions of (2.1), then u is a solution of (2.1).

PROOF. First note that (ii) follows immediately from (i) and Proposition 2.11. We prove (i) for subsolutions. Since $u(x) = \inf u_n(x)$ for all x and $u_n \in USC(\Omega)$, we have $u \in USC(\Omega)$ (this follows immediately from the fact that u is u.s.c. if and only if $\{x : u(x) < t\}$ is open for all $t \in \mathbb{R}$). Let $\varphi \in C^1(\Omega)$ be such that $u - \varphi$ has a local max at y, $u(y) = \varphi(y)$. We can assume

$$(u - \varphi)(x) \leq -|x - y|^2 \quad \text{for } x \in \Omega, |x - y| \leq r,$$

by adding $|x - y|^2$ to φ. Consider the functions

$$w_n(x) := (u_n(x) - \varphi(x) + |x - y|^2)^+.$$

They are u.s.c. functions such that $w_n(x) \to 0^+$ on $\bar{B}(y, r)$. We can apply Dini's lemma: if a sequence of u.s.c. w_n is such that $w_n(x) \searrow w(x)$ for all x in a compact set and w is continuous, then the convergence is uniform. Then, after relabeling if necessary, we may assume that, for all n,

$$u_n(x) - \varphi(x) \leq \frac{1}{n} - |x - y|^2 \quad \text{for } x \in \Omega, |x - y| \leq r.$$

Let y_n be a maximum point of $u_n - \varphi$ in $\bar{B}(y, r)$. Then,

$$0 = (u - \varphi)(y) \leq (u_n - \varphi)(y) \leq (u_n - \varphi)(y_n) \leq \frac{1}{n} - |y_n - y|^2.$$

Hence $y_n \to y$ and $u_n(y_n) \to u(y)$. Since u_n is a subsolution of (2.1)

$$F(y_n, u_n(y_n), D\varphi(y_n)) \leq 0,$$

so we let $n \to \infty$ and get

$$F(y, u(y), D\varphi(y)) \leq 0. \quad \blacktriangleleft$$

REMARK 2.17. Another proof of Proposition 2.16 is easily obtained by combining Theorem 1.7 with next elementary lemma, whose proof is left to the reader. This lemma is of independent interest because it relates pointwise monotone limits with the weak limits of Definition 1.4. ◁

2. NON-CONTINUOUS SOLUTIONS

LEMMA 2.18. *Assume the functions $u_\varepsilon : E \to \mathbb{R}$, $0 < \varepsilon \le 1$, $E \subseteq \mathbb{R}^N$, are upper semicontinuous, $u_\varepsilon \ge u_{\varepsilon'}$ if $\varepsilon \ge \varepsilon'$, and set*

$$w(x) = \inf_{\varepsilon > 0} u_\varepsilon(x).$$

Then

$$w = w^* = \limsup{}^*_{\varepsilon \to 0^+} u_\varepsilon, \qquad w_* = \liminf{}_*{}_{\varepsilon \to 0^+} u_\varepsilon.$$

Exercises

2.1 (THEOREM 2.4 FOR EVOLUTIVE HJB EQUATIONS). Assume u is locally bounded in $\Omega \times {]0,T[}$ and for all (x,t) there exists $\tau > 0$ such that, for all $0 < \vartheta < \tau$,

$$u(x,t) = \inf_{\alpha \in \mathcal{A}} \left\{ \int_0^\vartheta \ell(y_x, \alpha) e^{-\lambda s}\, ds + u(y_x(\vartheta), t - \vartheta) e^{-\lambda \vartheta} \right\}.$$

Prove that u is a solution of

$$u_t + \lambda u + H(x, D_x u) = 0 \qquad \text{in } \Omega \times {]0,T[},$$

where H is given by (2.5).

2.2. Take $N = 1$, $A = [-1,1]$, $f(x,a) = a$, $\mathcal{T} = {]-\infty, -1]} \cup {[1, +\infty[}$. Check that in this case (2.11) is

$$u + |u'| - 1 = 0 \quad \text{in } {]-1, 1[}, \qquad u(-1) = u(1) = 0,$$

and find the unique solution of this problem in $C([0,1])$. Show that there are infinitely many discontinuous solutions.

2.3. Compute the optimal controls and the value function for the Zermelo Navigation Problem, Example 2.8, for $\mathcal{T} = \{0\}$ and all $\sigma \ge 0$. [Hint: use the Pontryagin Maximum Principle and the existence of optimal controls; check the result in Section 1.17 of [Lei66].]

2.4. Prove that the minimum time function of Example 2.9 is C^1 except on the x_1 axis and locally Hölder continuous of exponent $1/2$ in $\mathbb{R}^2 \setminus \mathcal{T}$.

2.5. Guess an explicit formula for the minimum time function of Example 2.10, by analyzing the trajectory associated with piecewise smooth controls.

2.6. Consider the system $y_1' = y_2$, $y_2' = a$, $-1 \le a \le 1$ and the target $\mathcal{T} = \{ x : x_1 \le x_2 \}$. Prove that the minimum time function is discontinuous at the points of $\partial \mathcal{T}$ such that $x_1 = x_2 \ge 1$ and on the barrier $\{ x : x_1 = 1/2 + x_2^2/2,\ 0 \le x_2 \le 1 \}$, and continuous elsewhere. [Hint: use the Pontryagin Maximum Principle and the existence of optimal controls; check the result in Section 5.5.3 of [BaO82].]

2.7. Check directly that the discontinuous value functions found in Exercise 2.3, Example 2.9, and Exercises 2.5 and 2.6 are viscosity solutions of the corresponding HJB equations.

2.8. Give another proof of the sufficiency statement in Petrov's Theorem IV.1.12 by using the comparison principle for non-continuous semisolutions, Theorem 1.3. [Hint: show that $u(x) = 1 - e^{C|x|}$ for $|x| \leq R$, $u(x) = 1$ elsewhere, is a supersolution of (2.11) for a suitable choice of C and R.]

2.9. Let $u : \mathbb{R} \to \mathbb{R}$ locally bounded. Prove that

(i) if $u' = 0$ in $\mathbb{R}$ (in the sense of Definition 2.2), then there exists $a \in [-\infty, +\infty]$ such that u is discontinuous at all $x \in \,]-\infty, a[$ and u is constant in $]a, +\infty[$;

(ii) if $-u' = 0$ in $\mathbb{R}$, then there is $a \in [-\infty, +\infty]$ such that u is discontinuous at all $x \in \,]a, +\infty[$ and u is constant in $]-\infty, a[$;

(iii) if $u' = 0$ and $-u' = 0$ in an open interval I, then u is either constant in I or discontinuous at each point of I.

3. Envelope solutions of Dirichlet problems

In this section we consider the Dirichlet boundary value problem

(3.1) $$\begin{cases} F(x, u, Du) = 0 & \text{in } \Omega, \\ u = g & \text{on } \partial\Omega, \end{cases}$$

with $\Omega \subseteq \mathbb{R}^N$ open, $F : \Omega \times \mathbb{R} \times \mathbb{R}^N \to \mathbb{R}$ continuous and $g : \partial\Omega \to \mathbb{R}$. Motivated by the theory of optimal control problems with exit times in Chapter IV and the further discussion in Section 2, we restrict ourselves to the case that (3.1) satisfies the comparison principle (see Definition 1.2), so that

(3.2) if u_1 and u_2 are, respectively, a sub- and a supersolution of (3.1), then $u_1 \leq u_2$.

We recall that u is a subsolution (respectively, supersolution) of (3.1) if $u \in BUSC(\overline{\Omega})$ (respectively, $BLSC(\overline{\Omega})$) is a viscosity subsolution (respectively, supersolution) of the differential equation $F = 0$ in Ω, which is $\leq g$ (respectively, $\geq g$) on $\partial\Omega$.

Our model problem is $F(x, u, p) = u + H(x, p)$, with H satisfying the structural assumptions quoted in Theorem 1.3. This is the case for the HJB equations arising in control problems with payoff involving the exit time of the system from the domain Ω, see Chapter IV. We know from Theorem 2.6 that the value function of such a problem is a (non-continuous viscosity) solution of the corresponding HJB equation. It also satisfies the boundary condition in (3.1), where g is the terminal cost to be paid when the system hits the boundary $\partial\Omega$. However, this is not very meaningful if it is discontinuous at boundary points, and in fact we saw in Example 2.5 that there are infinitely many such solutions in a very simple problem satisfying the comparison principle. On the other hand, if we require continuity at the boundary, that is, $u_* = u^* = g$ on $\partial\Omega$, then any solution is automatically continuous in $\overline{\Omega}$ by Theorem 2.3, and this rules out many interesting problems. In fact, the value functions in Examples 2.8–2.10 are discontinuous at some point of

$\partial\Omega$, and indeed we know from Chapter IV that the value function v of time-optimal problems ($g \equiv 0$) for systems which are not small-time controllable on the target satisfies $v^* > 0$ at some boundary point, while in problems with incompatible terminal cost g we have $v_* < g$.

Therefore we seek a notion of generalized solution of (3.1), and in particular of generalized Dirichlet boundary condition, which singles out the value function even when it is discontinuous. There are several ways to do this, and no one is better than the others in all situations. Sections 4 and 5 are devoted to this problem by different methods. The approach we describe in the present section is global, in the sense that we define a generalized solution of the whole boundary problem (3.1) and not of the Dirichlet boundary condition by itself.

The first subsection is devoted to the general theory, and the second deals with the application to the minimum time problem.

3.1. Existence and uniqueness of e-solutions

We denote

$$\mathcal{S} := \{\text{subsolutions of } (3.1)\}, \qquad \mathcal{Z} := \{\text{supersolutions of } (3.1)\}.$$

DEFINITION 3.1. Let $u : \overline{\Omega} \to \mathbb{R}$ be locally bounded.

(i) u is an *envelope viscosity subsolution* of (3.1), briefly, *e-subsolution*, if there exists $\mathcal{S}(u) \subseteq \mathcal{S}$, $\mathcal{S}(u) \neq \emptyset$, such that

$$u(x) = \sup_{w \in \mathcal{S}(u)} w(x), \qquad x \in \overline{\Omega};$$

(ii) u is an *e-supersolution* of (3.1) if there exists $\mathcal{Z}(u) \subseteq \mathcal{Z}$, $\mathcal{Z}(u) \neq \emptyset$, such that

$$u(x) = \inf_{w \in \mathcal{Z}(u)} w(x), \qquad x \in \overline{\Omega};$$

(iii) u is an *e-solution* of (3.1) if it is e-subsolution and e-supersolution. ◁

A trivial example of an e-subsolution of (3.1) is a subsolution u, by taking $\mathcal{S}(u) = \{u\}$. Note that, if u is also a supersolution of (3.1), we can take $\mathcal{S}(u) = \mathcal{S}$ as well (and $\mathcal{Z}(u) = \mathcal{Z}$), because in this case u is indeed the complete solution of (3.1) by the comparison principle (3.2), that is,

$$u = \max_{w \in \mathcal{S}} w = \min_{w \in \mathcal{Z}} w.$$

At this point it is not apparent whether a discontinuous value function of a time-optimal problem is an e-solution of the corresponding boundary value problem. This is a nontrivial result that we postpone to the next subsection.

At the moment the main motivation of the definition of e-solution comes from Perron's method of Section 2, since this method allows us to construct solutions of the equation

(3.3) $$F(x, u, Du) = 0 \quad \text{in } \Omega$$

as envelopes of certain sets of sub- or supersolutions. As a first application of the results of §2.2 we get some additional consistency properties of e-semisolutions with viscosity semisolutions.

PROPOSITION 3.2.
(i) *If u is an e-subsolution of (3.1), then u^* is a subsolution of (3.3) and $u(x) \leq g(x)$ for all $x \in \partial\Omega$;*
(ii) *if u is an e-supersolution of (3.1), then u_* is a supersolution of (3.3) and $u(x) \geq g(x)$ for all $x \in \partial\Omega$;*
(iii) *if u is an e-solution of (3.1), then u is a (non-continuous viscosity) solution of (3.3) and $u = g$ on $\partial\Omega$.*

PROOF. To prove (i) we observe that u^* is a subsolution of (3.3) by Proposition 2.11 (a). If $x \in \partial\Omega$, then $w(x) \leq g(x)$ for all $w \in \mathcal{S}$ and so $u(x) \leq g(x)$. The proof of (ii) is analogous and (iii) is a trivial consequence of Definition 2.2. ◀

It is easy to see that the converse of each statement in Proposition 3.2 is false, see Exercise 3.1.

Comparison and uniqueness for e-solutions are immediately obtained by the very definition and the comparison principle.

PROPOSITION 3.3. *Assume (3.2). Then*
(i) *any e-subsolution u and e-supersolution U of (3.1) satisfy $u \leq U$ in $\overline{\Omega}$;*
(ii) *if there exists an e-solution v of (3.1), then it is unique and*

$$v = \sup_{w \in \mathcal{S}} w = \inf_{W \in \mathcal{Z}} W .$$

PROOF. For any $w \in \mathcal{S}(u)$ and $W \in \mathcal{Z}(U)$, $w \leq W$ by (3.2), so that

$$u = \sup_{w \in \mathcal{S}(u)} w \leq \inf_{W \in \mathcal{Z}(U)} W = U$$

and (i) is proved; (ii) is an easy consequence of (i). ◀

The main result of this subsection is the following existence theorem for e-solutions.

THEOREM 3.4. *Assume (3.2) and $\mathcal{S} \neq \emptyset$, $\mathcal{Z} \neq \emptyset$.*
(i) *If there exists $\underline{w} \in \mathcal{S}$ continuous at each point of $\partial\Omega$ and such that $\underline{w} = g$ on $\partial\Omega$, then*

(3.4) $$U(x) := \sup_{w \in \mathcal{S}} w(x) = \min_{W \in \mathcal{Z}} W(x)$$

is the unique e-solution of (3.1), and there exists a sequence $w_n \in \mathcal{S}$ such that $w_n \nearrow U$.

(ii) *If there exists $\widetilde{w} \in \mathcal{Z}$ continuous at each point of $\partial\Omega$ and such that $\widetilde{w} = g$ on $\partial\Omega$, then*

(3.5) $$u(x) := \inf_{W \in \mathcal{Z}} W(x) = \max_{w \in \mathcal{S}} w(x)$$

is the unique e-solution of (3.1), *and there exists a sequence $W_n \in \mathcal{Z}$ such that $W_n \searrow u$.*

Before proving the theorem we make some comments on its assumptions and conclusions. The assumption $\mathcal{S} \neq \emptyset$, $\mathcal{Z} \neq \emptyset$ is of course necessary for the existence of e-solutions. An easy sufficient condition is

$$F(x, M, 0) \geq 0, \quad F(x, -M, 0) \leq 0 \quad \text{for all } x \in \Omega,$$

for some large constant M. This is satisfied in the HJB equation

(3.6) $$\lambda u + \sup_{a \in A} \{ -f(x, a) \cdot Du - \ell(x, a) \} = 0,$$

if ℓ is bounded, by $M = \lambda^{-1} \sup_{\mathbb{R}^N \times A} |\ell|$. The assumption on the existence of a continuous sub- or supersolution is not easy to verify for general boundary data g, but it is easily obtained in the homogeneous case $g \equiv 0$. In fact, $\underline{w} \equiv 0$ is a subsolution if $F(x, 0, 0) \leq 0$, and this is indeed the case for time-optimal control, whose equation is (3.6) with $\ell \equiv 1$. See also Remark 3.5 for a weaker assumption.

The characterization (3.4) or (3.5) for the e-solution is optimal. In fact, when a continuous solution exists it is

$$\max_{w \in \mathcal{S}} w(x) = \min_{W \in \mathcal{Z}} W(x),$$

but this formula implies continuity (because $\mathcal{S} \cap \mathcal{Z} \subseteq C(\overline{\Omega})$), so it is false in general. In (3.4) we weaken it just to allow $U \notin \mathcal{S}$, and in (3.5) we allow $u \notin \mathcal{Z}$. However, $U \in \mathcal{Z}$ and $u \in \mathcal{S}$ so, in particular, U is lower semicontinuous and bounded.

Next we turn to the proof, which is a variant of Perron's method.

PROOF OF THEOREM 3.4 (i). The function U defined by (3.4) is below any $W \in \mathcal{Z}$, by the comparison principle (3.2), so it is bounded and therefore it is an e-subsolution by its very definition. We get all the conclusions if we prove that $U \in \mathcal{Z}$, and we do it by constructing $V \in \mathcal{Z}$ such that $U \geq V$, so that $U = V$ by Proposition 3.3 (i).

We define, for $\varepsilon > 0$ and $x \in \overline{\Omega}$,

$$v_\varepsilon(x) := \sup\{ w(x) : w \in \mathcal{S}, \ w(x) = \underline{w}(x) \text{ if } \operatorname{dist}(x, \partial\Omega) < \varepsilon \},$$

and observe that this function is bounded. Moreover, $v_\varepsilon \geq v_\delta$ if $\varepsilon < \delta$, so we can define

$$V(x) := \lim_n (v_{1/n})_*(x).$$

Note that $U \geq v_\varepsilon \geq (v_\varepsilon)_*$ by definition, so $U \geq V$. We want to prove that $V \in \mathcal{Z}$. For this purpose we define

$$\Omega_\varepsilon := \{\, x \in \Omega : \operatorname{dist}(x, \partial\Omega) > \varepsilon \,\},$$

and claim that $(v_\varepsilon)_*$ is a supersolution of $F = 0$ in Ω_ε. Then V is also supersolution in Ω_ε by Proposition 2.16 (recall that $(v_{1/n})_* \nearrow V$), thus it is a supersolution of $F = 0$ in Ω. Moreover, for $x \in \partial\Omega$, $(v_\varepsilon)_*(x) = (v_\varepsilon)^*(x) = g(x)$ for all ε because $v_\varepsilon \equiv \underset{\sim}{w}$ in a neighborhood of x, and $\underset{\sim}{w}$ is continuous at x with $\underset{\sim}{w} = g(x)$, so we conclude that $V \in \mathcal{Z}$.

To prove the claim, we first observe that $(v_\varepsilon)^*$ is a subsolution of (3.3) by Proposition 2.11. We assume by contradiction that $(v_\varepsilon)_*$ fails to be a supersolution at $y \in \Omega_\varepsilon$. We can apply Lemma 2.12 to get, for any small $r > 0$, a subsolution w^* of (3.3) such that

(3.7) $$\sup_{\Omega}(w - v_\varepsilon) > 0,$$

and

$$w(x) = v_\varepsilon(x) \qquad \text{for } |x - y| \geq r \,.$$

We choose $r \leq \operatorname{dist}(y, \partial\Omega) - \varepsilon$. This gives $w(x) = \underset{\sim}{w}(x)$ if $\operatorname{dist}(x, \partial\Omega) < \varepsilon$. Since $\underset{\sim}{w} \in USC(\overline{\Omega})$ we get $w^*(x) = \underset{\sim}{w}(x)$ if $\operatorname{dist}(x, \partial\Omega) < \varepsilon$, and so $w^* \in \mathcal{S}$. Therefore, by definition of v_ε, we have $w^* \leq v_\varepsilon$. This is a contradiction to (3.7), which proves the claim.

To complete the proof we define $w_n := (v_{1/n})^*$, and observe that this is a nondecreasing sequence in $\mathcal{S}$ whose pointwise limit is $\geq V$ by the definition of V. On the other hand $w_n \leq U$ by the definition of U, and we have shown that $U = V$, so $w_n \nearrow U$. ◂

REMARK 3.5. We can prove (3.4) and the fact that U is the e-solution of (3.1) under the weaker assumptions that $\mathcal{S}$ and $\mathcal{Z}$ are nonempty and for all $z \in \partial\Omega$ there exists $w_z \in \mathcal{S}$ such that $\lim_{x \to z} w_z(x) = g(z)$. The proof is indeed simpler than that of Theorem 3.4. In fact, by the definition of U, $w_z \leq U$, so $U_*(z) \geq g(z)$ for all $z \in \partial\Omega$. By Theorem 2.14, U_* is a supersolution of (3.3), thus $U_* \in \mathcal{Z}$. This implies $U_* = U \in \mathcal{Z}$, which completes the proof. ◁

The next result gives a characterization of the e-solution in terms of pointwise limits of sequences in $\mathcal{S}$ and $\mathcal{Z}$.

PROPOSITION 3.6. *Under the assumptions of Theorem 3.4, a function v is the e-solution of (3.1) if and only if there exist sequences $w_n \in \mathcal{S}$, $W_n \in \mathcal{Z}$, such that $w_n(x) \to v(x)$ and $W_n(x) \to v(x)$ for all $x \in \overline{\Omega}$.*

PROOF. The "only if" part follows immediately from Theorem 3.4 and in fact one of the two sequences can be chosen constant and the other monotone. To prove the converse we observe that there cannot be any point y where $v(y) < \inf_{W \in \mathcal{Z}} W(y)$ or $v(y) > \sup_{w \in \mathcal{S}} w(y)$. By Theorem 3.4 the e-solution is $\inf_{W \in \mathcal{Z}} W = \sup_{w \in \mathcal{S}} w$, so v must coincide with it. ◂

3.2. Time-optimal problems lacking controllability

The main examples of discontinuous solutions of an HJB equation in Section 2 were the minimum time function T and its discounted version (or Kružkov transform) v, see (2.10) and Chapter IV. The Dirichlet boundary value problem associated with v is

(3.8) $$\begin{cases} u + H(x, Du) = 0 & \text{in } \mathcal{T}^c, \\ u = 0 & \text{on } \partial \mathcal{T}, \end{cases}$$

with

(3.9) $$H(x, p) := \sup_{a \in A}\{-f(x, a) \cdot p\} - 1,$$

where f is the dynamics of the controlled system and $\mathcal{T}$ the target. We assume $f : \mathbb{R}^N \times A \to \mathbb{R}^N$ continuous, $\mathcal{T}$ closed and nonempty, and, for simplicity,

(3.10) $$|f(x, a) - f(y, a)| \leq L |x - y| \quad \text{for all } x, y, a,$$

A and $\partial \mathcal{T}$ compact. Exercise 3.3 gives some extensions, and in Section 5 we study by different methods the case of nonconstant running and terminal costs (see Corollary 5.10). The main result of this subsection is the following.

THEOREM 3.7. *Under the previous assumptions v_* is the e-solution of* (3.8).

Before turning to the proof we make some comments. The Dirichlet problem (3.8) satisfies the comparison principle (3.2) by Theorem 1.3, and $u \equiv 0$ is a continuous subsolution of (3.8), so the assumptions of Theorem 3.4(i) are satisfied. Therefore the e-solution is a supersolution of (3.8) (in particular, it is l.s.c.), so v_* is the natural candidate. In fact it is a supersolution of the PDE by Theorem 2.6, and $v_* \geq 0$ on $\partial \mathcal{T}$ because $v \geq 0$ everywhere by definition. If, in particular, the system is convex (i.e., $f(x, A)$ is a convex set for all $x \in \mathbb{R}^N$), then T is lower semicontinuous (see, e.g., [HL69]) and so $v = v_*$.

What is left to prove is that v_* is an e-subsolution, that is, it can be approximated from below by subsolutions of (3.8). We construct approximations having a control-theoretic interpretation: roughly speaking, we increase the directions available to the system so that it becomes controllable and the value function is Lipschitz continuous.

The first preliminary result says that the minimum time function does not change if we add the null vector field to the system.

PROPOSITION 3.8. *Let $\widetilde{A} = A \cup \{\widetilde{a}\}$, $\widetilde{a} \notin A$, $\widetilde{f}(x, a) = f(x, a)$ if $a \in A$, $\widetilde{f}(x, \widetilde{a}) = 0$, for all x. Then the minimum time function $\widetilde{T}$ associated with the system $(\widetilde{f}, \widetilde{A})$ coincides with the minimum time function T associated with (f, A).*

SKETCH OF THE PROOF. The result is rather intuitive. Of course $\widetilde{T} \leq T$. Fix x such that $\widetilde{T}(x) < +\infty$, $\varepsilon > 0$, and $\alpha \in \widetilde{\mathcal{A}}$ such that $t_x(\alpha) \leq \widetilde{T}(x) + \varepsilon$. We are going to construct a control $\beta \in \mathcal{A}$ which, roughly speaking, skips all the times t such that $\alpha(t) = \tilde{a}$. Since at such times the velocity of the system is null, we expect to reach the target in a time not larger than $t_x(\alpha)$. We define

$$J := \{t \geq 0 : \alpha(t) = \tilde{a}\}, \qquad J^c := [0, \infty[\setminus J,$$

$$\tau(t) := t - \text{meas}([0,t] \cap J) = \int_0^t \chi_{J^c}(s)\,ds, \quad \tau^{-1}(s) := \min\{t \geq 0 : \tau(t) = s\},$$

$$\beta(t) := \alpha(\tau^{-1}(t)).$$

We claim that $y_x(\tau(t), \beta) = y_x(t, \alpha)$ for all $t \geq 0$. Then $T(x) \leq t_x(\beta) \leq \tau(t_x(\alpha)) \leq t_x(\alpha) \leq \tilde{T}(x) + \varepsilon$ and we conclude because ε is arbitrary. The proof of the claim is left to the reader (see Exercise 3.2). ◀

We approximate the system (f, A) with the controllable system

(3.11)
$$\begin{cases} y' = \tilde{f}(y, \alpha) + h\beta, \\ y(0) = x, \end{cases}$$

where $h > 0$. The control functions are $(\alpha, \beta) \in \tilde{\mathcal{A}} \times \mathcal{B}$, where $\mathcal{B} := \{\beta : [0, \infty[\to \bar{B}(0,1)$ measurable$\}$, and the trajectories are denoted by $y_x^h(\cdot, \alpha, \beta)$. The value functions are

$$T_h(x) := \inf_{\tilde{\mathcal{A}} \times \mathcal{B}} t_x^h, \qquad v_h(x) := \inf_{\tilde{\mathcal{A}} \times \mathcal{B}} \int_0^{t_x^h} e^{-t}\,dt,$$

where $t_x^h = t_x^h(\alpha, \beta)$ is the entry time in $\mathcal{T}$ of the trajectory of (3.11).

PROPOSITION 3.9. *For all $h > 0$, $v_h \in BC(\mathbb{R}^N)$ is a subsolution of (3.8) such that $v_h \equiv 0$ on $\mathcal{T}$.*

PROOF. Since the system (3.11) can go in any direction, T_h is locally Lipschitz continuous by Theorem IV.4.8 (in special cases simpler controllability results suffice: Theorem IV.1.12 if $\mathcal{T} = \{0\}$, Theorem IV.1.16 if the target is smooth). Thus v_h is a continuous solution of

$$u + \tilde{H}(x, Du) + h|Du| = 0 \quad \text{in } \mathcal{T}^c,$$

where $\tilde{H}(x,p) = H(x,p) \vee (-1)$ is the Hamiltonian associated with the system $(\tilde{f}, \tilde{A})$ as in (3.9). Therefore v_h is a subsolution of (3.8). ◀

The proof of Theorem 3.7 follows immediately from the following result, which is of independent interest.

THEOREM 3.10. *Under the assumptions of Theorem 3.7*

$$v_*(x) = \sup_{h>0} v_h(x) \qquad \text{for all } x \in \mathbb{R}^N.$$

3. ENVELOPE SOLUTIONS OF DIRICHLET PROBLEMS

PROOF. By Proposition 3.8 we can assume without loss of generality that there is $\tilde{a} \in A$ such that $f(x, \tilde{a}) = 0$ for all x, and drop the $\sim$ on f and A. Since $v_h \leq v$ for all h and v_h is continuous, we have $\sup_h v_h \leq v_*$. Let us assume by contradiction there exists $\overline{x}$ such that

(3.12) $$\sup_h v_h(\overline{x}) < v_*(\overline{x}) .$$

Then there exists $\varepsilon, h^* > 0$ and $(\alpha_h, \beta_h) \in \mathcal{A} \times \mathcal{B}$ such that

(3.13) $$1 - e^{-t_{\overline{x}}^h(\alpha_h, \beta_h)} \leq v_h(\overline{x}) + \varepsilon \leq v_*(\overline{x}) - \varepsilon \qquad \text{for } 0 < h \leq h^* .$$

We simplify the notations as follows:
$$\tau_h := t_{\overline{x}}^h(\alpha_h, \beta_h), \qquad y^h(t) := y_{\overline{x}}^h(t, \alpha_h, \beta_h) .$$

If $v_h(\overline{x}) \to 1$ as $h \to 0^+$ we get a contradiction to (3.12) because $v_* \leq 1$. Otherwise, there exists $\delta > 0$ such that $v_h(\overline{x}) < 1 - \delta$ for any $h > 0$, so $\tau := \sup_{h>0} \tau_h < +\infty$ and there exists $\xi_h \in \partial \mathcal{T}$ such that $y^h(\tau_h) = \xi_h$. By the compactness of $\partial \mathcal{T}$ we may assume $\xi_h \to \xi_0 \in \partial \mathcal{T}$ as $h \to 0^+$. Now we solve the system (3.11) backward starting at ξ_0, that is, we consider the solution $y_h(\cdot)$ of

$$\begin{cases} y' = f(y, \alpha_h) + h\beta_h, \\ y(\tau_h) = \xi_0, \end{cases}$$

and the solution $\widetilde{y}_h(\cdot)$ of

$$\begin{cases} y' = f(y, \alpha_h), \\ y(\tau_h) = \xi_0 . \end{cases}$$

For $0 \leq t \leq \tau_h$ we get from (3.10)

$$|y_h(t) - \widetilde{y}_h(t)| \leq \int_t^{\tau_h} L |y_h(s) - \widetilde{y}_h(s)| \, ds + h\tau,$$

hence, by Gronwall's inequality, $|y_h(t) - \widetilde{y}_h(t)| \leq h\tau e^{L(\tau_h - t)}$. Then we set $C_1 := \tau e^{L\tau}$ and get

(3.14) $$|y_h(0) - \widetilde{y}_h(0)| \leq C_1 h .$$

On the other hand, since y_h and y^h are the trajectory of system (3.11) passing at time τ_h at, respectively, ξ_0 and ξ_h, a standard estimate gives $|y_h(t) - y^h(t)| \leq |\xi_0 - \xi_h| e^{L\tau}$. Then, for $C_2 := e^{L\tau}$, we have

$$|y_h(0) - \overline{x}| = |y_h(0) - y^h(0)| \leq C_2 |\xi_0 - \xi_h| .$$

Putting this estimate together with (3.14) we get

(3.15) $$\widetilde{y}_h(0) \longrightarrow \overline{x} \qquad \text{as } h \to 0^+ .$$

By definition of $\widetilde{y}_h$ we have $s_h := t_{\widetilde{y}_h(0)}(\alpha_h) \leq \tau_h = t_{\overline{x}}^h(\alpha_h, \beta_h)$; hence, by (3.13),

$$v_*(\overline{x}) - v(\widetilde{y}_h(0)) \geq \varepsilon + 1 - e^{-\tau_h} - (1 - e^{-s_h}) \geq \varepsilon .$$

Then, by (3.15), $v_*(\overline{x}) \geq v_*(\overline{x}) + \varepsilon$, a contradiction. ◀

REMARK 3.11. It is easy to give verification theorems for time-optimal controls similar to Corollaries III.2.18–2.19 by taking e-subsolutions of (3.8) as verification functions and e-supersolutions as falsification functions. ◁

Exercises

3.1. Find a counterexample to the converse of each statement of Proposition 3.2. [Hint: use Example 2.5.]

3.2. Complete the proof of Proposition 3.8. (This is essentially an exercise in measure theory. The result can be checked on [BSt93].)

3.3. Extend Theorem 3.7 to the case where the Hamiltonian is

$$H(x,p) = \sup_{a}\{-f(x,a)\cdot p - \ell(x,a)\},$$

where $\ell \geq 1$ is a running cost satisfying the hypotheses of (A4) in Chapter III, to the case of $\partial \mathcal{T}$ unbounded, and to systems (f, A) satisfying the assumptions (A_0)–(A_3) of Chapter III.

3.4. Consider the necessary condition of time-optimality for control laws given in Corollary 2.7.

(i) Find a stronger necessary condition which is also sufficient;

(ii) state precisely the verification theorems alluded to in Remark 3.11.

4. Boundary conditions in the viscosity sense

In this section we continue the study of the Dirichlet boundary value problem

(4.1) $$\begin{cases} F(x, u, Du) = 0 & \text{in } \Omega, \\ u = g & \text{on } \partial\Omega, \end{cases}$$

in cases where a solution satisfying the boundary condition in the classical sense (i.e., $u^* = u_* = g$ on $\partial\Omega$) may not exist. While in Section 3 we gave a global notion of generalized solution of (4.1), here we give a local definition of non-continuous viscosity solution of the boundary condition, and use the definition of solution given in Section 2 for the PDE. In subsection 4.1 we show that these definitions have good stability properties and apply to value functions via standard Dynamic Programming arguments. In subsection 4.2 we give some comparison and uniqueness results when $\partial\Omega$ is a Lipschitz hypersurface and give some applications. The first is a further study of the value functions of control problems involving exit times from Ω. The second is a convergence theorem that completes the stability results of §4.1. In subsection 4.3 we study the non-continuous complete solution of the boundary value problem for time-optimal control.

Throughout the section we assume $\Omega \subseteq \mathbb{R}^N$ open, $F : \overline{\Omega} \times \mathbb{R} \times \mathbb{R}^N \to \mathbb{R}$ continuous, $g : \partial\Omega \to \mathbb{R}$ continuous.

4. BOUNDARY CONDITIONS IN THE VISCOSITY SENSE

4.1. Motivations and basic properties

We begin with an example of an optimal control problem which provides a first motivation to the definition of the Dirichlet boundary condition in the viscosity sense. Consider a system with dynamics f and control constraint A satisfying the basic assumptions of (A_0)–(A_3) of Chapter III. Let $\mathcal{T}$ be a closed target with nonempty interior int $\mathcal{T}$ and denote by $\widehat{t}_x(\alpha)$ the entry time in int $\mathcal{T}$ of the trajectory $y_x(\cdot)$ associated with the control function $\alpha \in \mathcal{A}$ as in §1 of Chapter IV. Consider the value function

(4.2) $$\widehat{v}(x) := \inf_{\alpha \in \mathcal{A}} \{ \widehat{t}_x(\alpha) + g(y_x(\widehat{t}_x(\alpha))) \}$$

and assume it is continuous and finite everywhere. In §3 of Chapter IV we gave the Dynamic Programming Principle for $\widehat{v}$ which can be written as

(4.3) $$\widehat{v}(x) = \inf_{\alpha \in \mathcal{A}} \{ t + \widehat{v}(y_x(t, \alpha)) \} \qquad \text{if } 0 < t \leq \widehat{t}_x(\alpha).$$

By a standard argument we deduce from it the HJB equation in the viscosity sense

(4.4) $$H(x, D\widehat{v}) = 0 \qquad \text{in } \Omega := \mathcal{T}^c,$$

where

$$H(x, p) := \sup_{a \in A} \{ -f(x, a) \cdot p \} - 1.$$

We claim that $\widehat{v}$ satisfies

(4.5) $$(\widehat{v} - g) \vee H(x, D\widehat{v}) \geq 0 \qquad \text{on } \partial \Omega,$$

in the viscosity sense: see the definition of constrained viscosity solution in §5 of Chapter IV. In fact, if Γ is the part of $\partial\Omega$ where $\widehat{v} < g$, it is not profitable to enter int $\mathcal{T}$ through Γ and pay the terminal cost g. Therefore, the control problem with the state constraint that forbids trajectories to enter int $\mathcal{T}$ through Γ has the same value function. Then $\widehat{v}$ satisfies the boundary condition $H(x, D\widehat{v}) \geq 0$ of state constrained problems on Γ. This proves (4.5).

Next we claim that $\widehat{v}$ also satisfies

(4.6) $$(\widehat{v} - g) \wedge H(x, D\widehat{v}) \leq 0 \qquad \text{on } \partial\Omega,$$

in the viscosity sense. To see this, let $x \in \partial\Omega$ be such that $\widehat{v}(x) > g(x)$. Then the system cannot enter int $\mathcal{T}$ at x, that is, $\widehat{t}_x(\alpha) > 0$ and $y_x(t, \alpha) \in \overline{\Omega}$ for all $\alpha \in \mathcal{A}$ and $t > 0$ small. Thus for a C^1 function φ such that $\widehat{v} - \varphi_{|\overline{\Omega}}$ has a max at x we have, for small t,

$$(\widehat{v} - \varphi)(y_x(t, \alpha)) \leq (\widehat{v} - \varphi)(x).$$

Now the DPP (4.3) gives

$$\varphi(x) - \varphi(y_x(t, \alpha)) \leq t \qquad \text{for } 0 < t \leq \widehat{t}_x(\alpha),$$

and by the usual argument (divide by $t > 0$ and let $t \to 0^+$) we get $H(x, D\varphi(x)) \leq 0$, which gives $H(x, D\widehat{v}) \leq 0$ on $\partial\Omega$ in the viscosity sense and proves (4.6).

In conclusion, $\widehat{v}$ satisfies the boundary value problem (4.4) (4.5) (4.6), which we summarize with the following notation

$$\begin{cases} H(x, Du) = 0 & \text{in } \Omega, \\ u = g \text{ or } H(x, Du) = 0 & \text{on } \partial\Omega. \end{cases}$$

Motivated by this example we give the following

DEFINITION 4.1. A function $u \in USC(\overline{\Omega})$ (respectively, $LSC(\overline{\Omega})$) is a (*viscosity*) *subsolution* (respectively, *supersolution*) of

(4.7) $\qquad\qquad u = g \quad \text{or} \quad F(x, u, Du) = 0 \quad \text{on } \partial\Omega$

if, for any $\varphi \in C^1(\overline{\Omega})$ and $x \in \partial\Omega$ such that $u - \varphi$ has a local maximum (respectively, minimum) at x,

$$(u - g)(x) \leq 0 \quad (\text{resp.}, \geq 0) \qquad \text{or} \qquad F(x, u(x), D\varphi(x)) \leq 0 \quad (\text{resp.}, \geq 0).$$

We also express this property by saying that subsolutions (respectively, supersolutions) of (4.7) satisfy

$$(u - g) \wedge F(x, u, Du) \leq 0 \qquad \text{on } \partial\Omega$$

(respectively,

$$(u - g) \vee F(x, u, Du) \geq 0 \qquad \text{on } \partial\Omega)$$

in the viscosity sense.

We say that u is a *v-subsolution* (respectively, *v-supersolution*) of the Dirichlet problem (4.1) if it is a (viscosity) subsolution (respectively, supersolution) of

(4.8) $\qquad\begin{cases} F(x, u, Du) = 0 & \text{in } \Omega, \\ u = g \text{ or } F(x, u, Du) = 0 & \text{on } \partial\Omega. \end{cases}$

Finally, a locally bounded function $u : \overline{\Omega} \to \mathbb{R}$ is a (*non-continuous viscosity*) *solution* of (4.7) (respectively, (4.8)) if u^* is a subsolution and u_* is a supersolution of (4.7) (respectively, (4.8)). ◁

REMARK 4.2. Definition 4.1 can be reformulated more concisely. Define $G : \overline{\Omega} \times \mathbb{R} \times \mathbb{R}^N \to \mathbb{R}$ by setting

$$G(x, u, p) = \begin{cases} F(x, u, p) & \text{if } x \in \Omega, \\ (u - g)(x) & \text{if } x \in \partial\Omega. \end{cases}$$

Note that $G_* = G^* = F$ in Ω, while $G_*(x, u, p) = (u - g(x)) \wedge F(x, u, p)$ and $G^*(x, u, p) = (u - g(x)) \vee F(x, u, p)$ on $\partial\Omega$. Then Definition 4.1 is equivalent to the following definition of viscosity subsolution (respectively, supersolution) of

(4.9) $\qquad\qquad G(x, u, Du) = 0 \qquad \text{in } \overline{\Omega}.$

A function $u \in USC(\overline{\Omega})$ is a subsolution (respectively, $u \in LSC(\overline{\Omega})$ is a supersolution) if for any $\varphi \in C^1(\overline{\Omega})$ such that $u - \varphi$ has a local maximum (respectively, minimum) at x

$$G_*(x, u(x), D\varphi(x)) \leq 0 \quad \text{(respectively, } G^*(x, u(x), D\varphi(x)) \geq 0\text{)}.$$

Therefore interpreting the boundary condition in the viscosity sense is a special case of defining viscosity solutions of discontinuous equations in non-open sets, see Exercise 4.1. ◁

Note that subsolutions (respectively, supersolutions) of (4.1) as defined and used in Sections 1 and 3 are obviously v-subsolutions (respectively, v-supersolutions) but the converse is not true in general. For instance, the value function $\widehat{v}$ of the previous motivating example is a v-subsolution but not a subsolution on a part of $\partial\mathcal{T}$ where the system is not small-time locally controllable. A more explicit example is the following.

EXAMPLE 4.3. Consider the Dirichlet problem

(4.10) $$\begin{cases} u' - 1 = 0 & \text{in }]0, 1[, \\ u(0) = 0, \quad u(1) = c\,. \end{cases}$$

This is a special case of the motivating example by taking $f(x) \equiv -1$ and A a singleton. Clearly, $\widehat{v}(x) = x$ for all $c \in \mathbb{R}$, so $\widehat{v}$ is not a subsolution of (4.10) in the sense of §1 if $c < 1$, and it is not a supersolution if $c > 1$. It is easy to check directly that $\widehat{v}$ is a v-solution of (4.10). ◁

The notion of v-subsolution (respectively, v-supersolution) of (4.1) is also weaker than that of e-subsolution (respectively, e-supersolution) given in §3, as the following proposition and example show.

PROPOSITION 4.4. *If u is an e-subsolution (respectively, e-supersolution) of (4.1), then u^* is a v-subsolution (respectively, u_* is a v-supersolution) of (4.1), that is, u^* is a viscosity subsolution (respectively, u_* is a supersolution) of (4.8).*

PROOF. By Proposition 3.2 we must check only the boundary condition. By definition, $u(x) = \sup_{w \in \mathcal{S}(u)} w(x)$, where each $w \in \mathcal{S}(u)$ is a subsolution of the PDE in (4.1) and $w \leq g$ on $\partial\Omega$. The proof is a variant of that of Proposition 2.11. We fix $y \in \partial\Omega$ such that $u^*(y) > g(y)$ and take $\varphi \in C^1(\overline{\Omega})$ such that $u^*(y) = \varphi(y)$ and, without loss of generality,

$$(u^* - \varphi)(x) \leq -|x - y|^2 \quad \text{for } x \in \overline{\Omega} \cap \overline{B}(y, r)\,.$$

We must prove that

(4.11) $$F(y, u^*(y), D\varphi(y)) \leq 0\,.$$

Select a sequence $x_n \to y$ such that

$$(u - \varphi)(x_n) \geq -\frac{1}{n} \quad \text{for all } n,$$

and $v_n \in \mathcal{S}(u)$ such that
$$u(x_n) - \frac{1}{n} < v_n(x_n) \quad \text{for all } n.$$

Then a maximum point y_n of $v_n - \varphi$ in $\overline{\Omega} \cap \bar{B}(y,r)$ satisfies, for n large,
$$-\frac{2}{n} < (v_n - \varphi)(y_n) \leq -|y_n - y|^2,$$

which gives
$$y_n \longrightarrow y \quad \text{and} \quad v_n(y_n) \longrightarrow \varphi(y) = u^*(y) > g(y).$$

Thus $y_n \notin \partial\Omega$ for n large enough, because $y_n \in \partial\Omega$ would imply $v_n(y_n) \leq g(y_n)$ and therefore a contradiction to the continuity of g by letting $n \to +\infty$. Then
$$F(y_n, v_n(y_n), D\varphi(y_n)) \leq 0$$

and we get (4.11) in the limit. ◀

EXAMPLE 4.5. Consider the Dirichlet problem

(4.12)
$$\begin{cases} u + |Du| - 1 = 0 & \text{in } \mathbb{R}^N \setminus \{0\}, \\ u(0) = 0. \end{cases}$$

The complete solution in $BC(\mathbb{R}^N)$ of this problem is $v(x) = 1 - e^{-|x|}$, thus it is also the e-solution of the problem. Any constant function $w \equiv c \leq 1$ is a v-subsolution of (4.12) but it is not an e-subsolution because it is above v in some neighborhood of 0. Note that $w \equiv 1$ is indeed a v-solution of (4.12) and $w > v$ everywhere. ◁

The key point of this example is that $\overline{\Omega} = \mathbb{R}^N$, that is, in terms of the target for time-optimal control, int $\mathcal{T} = \emptyset$. See Exercise 4.2 for a generalization. An example where int $\overline{\Omega} = \Omega$, that is, $\overline{\text{int }\mathcal{T}} = \mathcal{T}$, will be given soon (see Example 4.15).

The next property of v-sub- and v-supersolutions has a proof very similar to Propositions 2.11 and 4.4 and we leave it to the reader.

PROPOSITION 4.6. *Let $\mathcal{F}$ be a set of functions such that w^* is a subsolution (respectively, w_* is a supersolution) of (4.8) for all $w \in \mathcal{F}$. Set*
$$u(x) := \sup_{w \in \mathcal{F}} w(x) \quad (\text{respectively, inf}) \quad \text{for } x \in \overline{\Omega},$$

and assume u is locally bounded. Then u^* is a subsolution (respectively, u_* is a supersolution) of (4.8).

Next we show the very good stability properties of v-solutions. This is one of their main features, since the weak limits of essentially any perturbation of (4.1) are v-sub- or v-supersolutions of (4.1). In the next subsection we give comparison

4. BOUNDARY CONDITIONS IN THE VISCOSITY SENSE

theorems for v-semisolutions that allow us to prove, under some additional assumptions, the coincidence of the weak limits and therefore the uniform convergence of the solutions of the following three perturbed problems (see Corollary 4.27).

We begin with regular perturbations, as in Section 1. Assume $F_\varepsilon : \overline{\Omega} \times \mathbb{R} \times \mathbb{R}^N \to \mathbb{R}$ and $g_\varepsilon : \partial\Omega \to \mathbb{R}$ are continuous, $0 < \varepsilon \le 1$, and

(4.13) $$\begin{cases} F_\varepsilon \rightrightarrows F & \text{on compacta of } \overline{\Omega} \times \mathbb{R} \times \mathbb{R}^N, \\ g_\varepsilon \rightrightarrows g & \text{on compacta of } \partial\Omega, \text{ as } \varepsilon \to 0^+. \end{cases}$$

We consider the boundary value problem

(P_ε) $$\begin{cases} F_\varepsilon(x, u, Du) = 0 & \text{in } \Omega, \\ u = g_\varepsilon \text{ or } F_\varepsilon(x, u, Du) = 0 & \text{on } \partial\Omega. \end{cases}$$

For the function $u_\varepsilon : \overline{\Omega} \to \mathbb{R}$ assume for each compactum $\mathcal{K} \subseteq \overline{\Omega}$

(4.14) $$\sup_{\mathcal{K}} |u_\varepsilon| \le C_\mathcal{K} \quad \text{for } 0 < \varepsilon \le 1,$$

so that the weak limits

$$\underline{u} = \liminf_{\varepsilon \to 0^+} {}_* u_\varepsilon, \quad \overline{u} = \limsup_{\varepsilon \to 0^+} {}^* u_\varepsilon$$

are locally bounded in $\overline{\Omega}$ (see Definition 1.4).

PROPOSITION 4.7. *Assume (4.13). If u_ε satisfying (4.14) are such that u_ε^* is a subsolution (respectively, $u_{\varepsilon*}$ is a supersolution) of (P_ε) for $\varepsilon > 0$, then the upper weak limit $\overline{u}$ is a subsolution of (4.8) (respectively, the lower weak limit $\underline{u}$ is a supersolution of (4.8)).*

This result is better understood if we look at a simple example where $u_\varepsilon = g$ on $\partial\Omega$ in the classical sense for all $\varepsilon > 0$.

EXAMPLE 4.8. Consider the Dirichlet problem

$$\begin{cases} \max\{u', -\varepsilon u'\} - 1 = 0 & \text{in }]0, 1[, \\ u(0) = u(1) = 0. \end{cases}$$

This is the HJB equation for a time-optimal control problem whose unique viscosity solution in $BC([0, 1])$ is $u_\varepsilon(x) = x \wedge (1-x)/\varepsilon$. Then $\overline{u}(x) = x$, so $\overline{u}$ satisfies the boundary condition at $x = 1$ only in the viscosity sense. Note that $\overline{u}$ is indeed a continuous v-solution of the limit problem

(4.15) $$\begin{cases} u' - 1 = 0 & \text{in }]0, 1[, \\ u = 0 & \text{on } \{0, 1\}. \end{cases}$$

On the other hand, $\underline{u}(x) = \overline{u}(x)$ for $x < 1$, $\underline{u}(1) = 0$, so $\underline{u}$ is the pointwise limit and the supremum of u_ε as $\varepsilon \to 0^+$. Since u_ε is a subsolution of $u' - 1 = 0$, $\underline{u}$ is the e-solution of (4.15), as well as a v-solution. ◁

Before giving the proof of Proposition 4.7 we extend a bit Lemma 1.6. Since the proof requires minor modifications we leave it to the reader.

LEMMA 4.9. *Let $u_\varepsilon : \overline{\Omega} \to \mathbb{R}$ satisfy (4.14) for any compactum $\mathcal{K} \subseteq \overline{\Omega}$, $\varphi \in C^1(\overline{\Omega})$, and assume $\overline{u} - \varphi$ attains its maximum in $\overline{\Omega} \cap \overline{B}(x_0, r)$ only at x_0. Then there exist sequences $\varepsilon_n \to 0^+$ and x_n such that x_n is a maximum point for $u^*_{\varepsilon_n} - \varphi$ in $\overline{\Omega} \cap \overline{B}(x_0, r)$ and*

$$x_n \longrightarrow x_0 \quad \text{and} \quad u^*_{\varepsilon_n}(x_n) \longrightarrow \overline{u}(x_0) \quad \text{as } n \to \infty.$$

PROOF OF PROPOSITION 4.7. The proof is a variant of that of Theorem 1.7. Let us check the boundary condition. Let $\varphi \in C^1(\overline{\Omega})$ be such that $x_0 \in \partial\Omega$ is the unique maximum of $\overline{u} - \varphi$ in $\overline{\Omega} \cap \overline{B}(x_0, r)$ and consider the sequences ε_n, x_n produced by Lemma 4.9. If there are subsequences, still named ε_n, x_n, such that $x_n \in \partial\Omega$ and $u^*_{\varepsilon_n}(x_n) \leq g_{\varepsilon_n}(x_n)$, then $\overline{u}(x_0) \leq g(x_0)$ because $g_\varepsilon \rightrightarrows g$. For all other cases

$$F_{\varepsilon_n}(x_n, u^*_{\varepsilon_n}(x_n), D\varphi(x_n)) \leq 0$$

because u^*_ε is subsolution of (P_ε), and we conclude by (4.13). ◂

The next example concerns singular perturbations of the PDE via vanishing viscosity. Suppose $u_\varepsilon \in C^2(\Omega) \cap C(\overline{\Omega})$ are solutions of the elliptic boundary value problem

(4.16) $$\begin{cases} -\varepsilon a_{ij}(x) u_{x_i x_j} + F(x, u, Du) = 0 & \text{in } \Omega, \\ u = g & \text{on } \partial\Omega, \end{cases}$$

satisfying the uniform bound (4.14) in any compact $\mathcal{K} \subseteq \overline{\Omega}$. Here $\varepsilon > 0$, the coefficients a_{ij} are bounded in Ω; the matrix $(a_{ij}(x))$ is nonnegative definite for all $x \in \Omega$, we have used the summation convention, and $u_{x_i x_j} = \partial^2 u / \partial x_i \partial x_j$. The most classical example is $a_{ij} u_{x_i x_j} = \Delta u$, the Laplacian of u, and the term $-\varepsilon \Delta u$ is called artificial viscosity. For the existence of such solutions we refer, for instance, to [GT83, Kry87].

PROPOSITION 4.10. *Under the previous assumptions, the upper weak limit $\overline{u}$ of the solutions of (4.16) is a subsolution of (4.8) and the lower weak limit $\underline{u}$ is a supersolution of (4.8).*

PROOF. Let $\varphi \in C^2(\overline{\Omega})$ be such that x_0 is the unique maximum point of $\overline{u} - \varphi$ in $\overline{\Omega} \cap \overline{B}(x_0, r)$ and consider the points x_n and the sequence ε_n produced by Lemma 4.9. If $x_n \in \partial\Omega$ on a subsequence, then $u_{\varepsilon_n}(x_n) = g(x_n)$, and we get $\overline{u}(x_0) = g(x_0)$. Otherwise, for all small ε, $x_n \in \Omega$ is an interior maximum of $u_{\varepsilon_n} - \varphi$. Then $Du_{\varepsilon_n}(x_n) = D\varphi(x_n)$ and the Hessian matrix $D^2(u_{\varepsilon_n} - \varphi)(x_n)$ is nonpositive definite. Since $a_{ij}(x)$ is nonnegative definite, by standard linear algebra

$$-a_{ij}(x_n)(u_{\varepsilon_n} - \varphi)_{x_i x_j}(x_n) \geq 0,$$

4. BOUNDARY CONDITIONS IN THE VISCOSITY SENSE

and from the PDE in (4.16) we get

$$-\varepsilon a_{ij}(x_n)\varphi_{x_i x_j}(x_n) + F(x_n, u_{\varepsilon_n}(x_n), D\varphi(x_n)) \leq 0.$$

Finally, we let $\varepsilon_n \to 0^+$ and use the boundedness of a_{ij} to obtain

$$F(x_0, \overline{u}(x_0), D\varphi(x_0)) \leq 0. \quad \blacktriangleleft$$

Proposition 4.10 gives a motivation for naming "viscosity sense" the boundary condition defined in Definition 4.1. In fact, this boundary condition is satisfied by the limit of the vanishing viscosity approximation, even when a boundary layer occurs. We give a simple one-dimensional example to better illustrate this phenomenon.

EXAMPLE 4.11. Consider the problem

$$\begin{cases} -\varepsilon u'' - u' + u - 1 = 0 & \text{in }]0,1[, \\ u(0) = u(1) = 0. \end{cases}$$

The solution of this problem is

$$u_\varepsilon(x) = 1 + (e^\lambda - e^\mu)^{-1}[e^{\lambda x + \mu} - e^{\mu x + \lambda} - e^{\lambda x} + e^{\mu x}],$$

where

$$\lambda = \lambda(\varepsilon) = \frac{-1 + \sqrt{1 + 4\varepsilon}}{2\varepsilon}, \qquad \mu = \mu(\varepsilon) = \frac{-1 - \sqrt{1 + 4\varepsilon}}{2\varepsilon}.$$

As $\varepsilon \to 0^+$, $\lambda \to 1$ and $\mu \to -\infty$, so $\overline{u}(x) = 1 - e^{x-1}$ satisfies the boundary condition at $x = 0$ only in the viscosity sense. Note that $\underline{u}(x) = \overline{u}(x)$ for $x > 0$, $\underline{u}(0) = 0$, so both $\overline{u}$ and $\underline{u}$ are solutions of the limit problem (4.8) with $F(x, u, p) = -p + u - 1$, $g \equiv 0$, $\Omega =]0,1[$. The *boundary layer* is the small interval on the right of 0 where u_ε is not close to $\overline{u}$ even for $\varepsilon > 0$ small, because it is "pulled down" by the boundary condition. $\triangleleft$

The last example of stability property in this section concerns perturbations of the domain. Assume $\Omega_\varepsilon \subseteq \Omega$ are open, $0 < \varepsilon \leq 1$, and their complements $\mathcal{T}_\varepsilon := \Omega_\varepsilon^c$ are bounded and converge to $\mathcal{T} := \Omega^c$ in the Hausdorff metrics (see §2.2 in Chapter III), that is,

$$\lim_{\varepsilon \to 0^+} d_H(\mathcal{T}_\varepsilon, \mathcal{T}) = 0.$$

Assume $g \in C(\overline{\Omega})$, and let u_ε be solutions of

(4.17)
$$\begin{cases} F(x, u, Du) = 0 & \text{in } \Omega_\varepsilon, \\ u = g \text{ or } F(x, u, Du) = 0 & \text{on } \partial\Omega_\varepsilon, \end{cases}$$

for $\varepsilon > 0$. Extend u_ε to $\overline{\Omega}$ by setting $u_\varepsilon := g$ off Ω_ε and assume u_ε are uniformly bounded for $\varepsilon > 0$ on every compact $\mathcal{K} \subseteq \overline{\Omega}$, that is, (4.14) holds.

PROPOSITION 4.12. *Under the previous assumptions, the upper weak limit $\overline{u}$ of the solutions of* (4.17) *is a subsolution of* (4.8), *and the lower weak limit $\underline{u}$ is a supersolution of* (4.8).

The proof of this result is a variant of previous ones, and we leave it as an exercise for the reader.

We end this subsection with the proof that the value functions v and $\widehat{v}$ of the control problems studied in §IV.3 are solutions of the boundary value problem

(4.18) $$\begin{cases} \lambda u + H(x, Du) = 0 & \text{in } \Omega, \\ u = g \quad \text{or} \quad \lambda u + H(x, Du) = 0 & \text{on } \partial\Omega, \end{cases}$$

where

(4.19) $$H(x, p) := \sup_{a \in A} \{ -f(x, a) \cdot p - \ell(x, a) \},$$

even if they are not continuous. The payoffs of the optimal control problems under consideration are

$$J(x, \alpha) := \int_0^{t_x(\alpha)} \ell(y_x(s), \alpha(s)) e^{-\lambda s} \, ds + e^{-\lambda t_x(\alpha)} g(y_x(t_x(\alpha))),$$

where $t_x(\alpha)$ is the first time the trajectory $y_x(\,\cdot\,, \alpha)$ hits the target $\mathcal{T} := \Omega^c$, and $\widehat{J}(x, \alpha)$ is obtained by replacing $t_x(\alpha)$ with the entry time in int $\mathcal{T}$, $\widehat{t}_x(\alpha)$, in the definition of J. We recall that $\mathcal{A} := \{\, \alpha : [0, \infty[\to A \text{ measurable} \,\}$.

THEOREM 4.13. *Assume the hypotheses* (A_0)-(A_4) *of Chapter* III *and* $g \in BC(\partial\Omega)$. *Then the value functions*

$$v(x) = \inf_{\alpha \in \mathcal{A}} J(x, \alpha), \qquad \widehat{v}(x) = \inf_{\alpha \in \mathcal{A}} \widehat{J}(x, \alpha),$$

are both viscosity solutions of (4.18) *with H defined by* (4.19).

PROOF. Since $\lambda > 0$ we can assume without loss of generality $\lambda = 1$. Under the current assumptions v and $\widehat{v}$ are bounded and satisfy the Dynamic Programming Principle, Proposition IV.3.9. Then they satisfy the PDE in (4.18) by Theorem 2.6. We prove that v is a supersolution of the boundary condition in (4.18), all the other proofs being similar (and sometimes easier). This proof is a refinement of that of Theorem 2.6.

We take $z \in \partial\Omega$ and $\varphi \in C^1(\overline{\Omega})$ such that $v_*(z) = \varphi(z)$ and $v_*(x) \geq \varphi(x)$ for all x in a neighborhood of z. We assume by contradiction $\varphi(z) + H(z, D\varphi(z)) < 0$ and $\varphi(z) < g(z)$. Then, for some $\varepsilon > 0$,

(4.20) $\qquad \varphi(x) + H(x, D\varphi(x)) \leq -\varepsilon \qquad$ for $x \in B(z, \varepsilon) \cap \overline{\Omega}$,
(4.21) $\qquad \qquad \quad (\varphi - g)(x) \leq -\varepsilon \qquad$ for $x \in B(z, \varepsilon) \cap \partial\Omega$.

Now we fix $t > 0$ such that $y_x(s, \alpha) \in B(z, \varepsilon)$ for all $x \in B(z, \varepsilon/2)$, $0 < s \leq t$ and $\alpha \in \mathcal{A}$, and set $\delta := \varepsilon(1 - e^{-t})/2$. By the DPP, Proposition IV.3.9, for any $x \in B(z, \varepsilon/2)$ there is $\alpha \in \mathcal{A}$ such that

(4.22) $v(x) > \begin{cases} -\delta + \int_0^t \ell(y_x, \alpha) e^{-s}\, ds + v(y_x(t))\, e^{-t} & \text{if } t < t_x(\alpha), \\ -\delta + \int_0^{t_x(\alpha)} \ell(y_x, \alpha) e^{-s}\, ds + g(y_x(t_x(\alpha)))\, e^{-t_x(\alpha)} & \text{if } t \geq t_x(\alpha). \end{cases}$

In the first case, $t < t_x(\alpha)$, we use $v(y_x) \geq v_*(y_x) \geq \varphi(y_x)$ into the right-hand side and repeat the calculation (2.8) in the proof of Theorem 2.6 (based on (4.20) and the choice of t and δ) to get

(4.23) $\qquad\qquad\qquad v(x) - \varphi(x) > \delta\,.$

In the second case, $t \geq t_x(\alpha)$, we use (4.21) and the choice of t to observe that

$$g(y_x(t_x)) \geq \varphi(y_x(t_x)) + \varepsilon,$$

and plug this into the right-hand side of (4.22). Then we can repeat the calculation (2.8) in the proof of Theorem 2.6 to get

$$v(x) - \varphi(x) > -\delta + \int_0^{t_x} \varepsilon e^{-s}\, ds + \varepsilon e^{-t_x} = -\delta + \varepsilon > \delta\,.$$

Therefore (4.23) holds for all $x \in B(z, \varepsilon/2)$, so $v_*(z) \geq \varphi(z) + \delta > \varphi(z)$, a contradiction to the choice of φ. ◀

REMARK 4.14. The argument of the proof of Theorem 4.13 is local and uses only the local boundedness of v and the Dynamic Programming Principle. Therefore it applies to other situations, for instance, to the case $\lambda = 0$. ◁

EXAMPLE 4.15. Consider as in Remark IV.1.5,

$$\mathcal{T} = \{\,(x_1, x_2) \in \mathbb{R}^2 : |x_1| \leq |x_2| \text{ or } |x_1| \geq 1\,\}$$

and the system $f(x, 1) = (1, 0)$, $f(x, 2) = (-1, 0)$, $A = \{1, 2\}$. It is easy to see that the minimal time function is $T(x) = (|x_1| - |x_2|) \wedge (1 - |x_1|)$ and the minimal interior-time function is

$$\widehat{T}(x) = T(x) \quad \text{if } x_2 \neq 0, \qquad \widehat{T}(x_1, 0) = 1 - |x_1|\,.$$

Their associated Kružkov transforms, $v = 1 - e^{-T}$, $\widehat{v} = 1 - e^{-\widehat{T}}$, both satisfy Theorem 4.13. Note that v is the complete solution in $BC(\mathbb{R}^N)$, and therefore the e-solution, of

(4.24) $\qquad \begin{cases} u + |u_{x_1}| - 1 = 0 & \text{in } \Omega = \mathcal{T}^c, \\ u = 0 & \text{on } \partial\Omega, \end{cases}$

while $\widehat{v}$ is a discontinuous v-solution of this problem. ◁

4.2. Comparison results and applications to exit-time problems and stability

In this subsection we prove some comparison and uniqueness results for the boundary value problem

(4.25) $$\begin{cases} u + H(x, Du) = 0 & \text{in } \Omega, \\ u = g \text{ or } u + H(x, Du) = 0 & \text{on } \partial\Omega, \end{cases}$$

where $\Omega \subseteq \mathbb{R}^N$ is open, $H : \overline{\Omega} \times \mathbb{R}^N \to \mathbb{R}$ is continuous, and $g \in C(\partial\Omega)$, under some structural assumptions on H, some regularity conditions on $\partial\Omega$, and some continuity assumptions at $\partial\Omega$ for the semisolutions to compare. We show on examples the optimality of most of these conditions. Then we apply these results to minimal time and minimal interior-time functions T and $\widehat{T}$ and to the value functions v and $\widehat{v}$ of problems with terminal cost. No uniqueness theorem in this subsection allows the solution to be discontinuous: such results are postponed to the next subsection. Finally we prove a convergence theorem that completes the stability results in §4.1.

We begin with a result for Ω bounded where the assumptions and proof are simpler. Here are the hypotheses on H:

(4.26) $$|H(x,p) - H(y,p)| \leq L|p||x-y| + \omega_1(|x-y|),$$

where ω_1 is a modulus, and

(4.27) $$|H(x,p) - H(x,q)| \leq M|p-q|,$$

for all $x, y \in \overline{\Omega}$, $p, q \in \mathbb{R}^N$. They are satisfied by the Hamiltonian of control theory

(4.28) $$H(x,p) = \sup_{a \in A}\{-f(x,a) \cdot p - \ell(x,a)\},$$

where Ω is bounded, if, for instance, A is compact, f and ℓ are continuous, and

(4.29) $$|f(x,a) - f(y,a)| \leq L|x-y|, \quad |\ell(x,a) - \ell(y,a)| \leq \omega_1(|x-y|),$$

for all $x, y \in \overline{\Omega}$, $a \in A$.

The assumption on $\partial\Omega$ is to be a Lipschitz surface, as in §IV.5, or, more precisely,

(4.30) there are $c > 0$ and $\eta : \overline{\Omega} \to \mathbb{R}^N$ continuous such that $B(x + t\eta(x), ct) \subseteq \Omega$ for all $x \in \overline{\Omega}$, $0 < t \leq c$.

THEOREM 4.16. *Assume* (4.26), (4.27), (4.30) *and* Ω *bounded. Suppose* $u_1 \in USC(\overline{\Omega})$ *and* $u_2 \in LSC(\overline{\Omega})$ *are, respectively, a subsolution and a supersolution of*

(4.31) $$u + H(x, Du) = 0 \quad \text{in } \Omega,$$

u_1 *is continuous at each point of* $\partial\Omega$ *and* u_2 *satisfies*

(4.32) $$(u_2 - u_1) \vee (u_2 + H(x, Du_2)) \geq 0 \quad \text{on } \partial\Omega$$

4. BOUNDARY CONDITIONS IN THE VISCOSITY SENSE

in the viscosity sense (Definition 4.1). Then $u_1 \leq u_2$ in $\overline{\Omega}$. The same conclusion holds if u_2 instead of u_1 is assumed continuous at the points of $\partial\Omega$ and u_1 satisfies

$$(u_1 - u_2) \wedge (u_1 + H(x, Du_1)) \leq 0 \quad \text{on } \partial\Omega.$$

PROOF. We consider first the case $m := \max_{\partial\Omega}(u_1 - u_2) < \max_{\overline{\Omega}}(u_1 - u_2)$. If $m \leq 0$ we conclude immediately by Theorem 1.3. If $m > 0$, then $w = u_2 + m$ has the same properties of u_2, and, in addition, $u_1 \leq w$ on $\partial\Omega$. Then $u_1 \leq w$ in $\overline{\Omega}$ by Theorem 1.3, so $u_1 - u_2 \leq m$, a contradiction.

The other case is $\max_{\partial\Omega}(u_1 - u_2) = \max_{\overline{\Omega}}(u_1 - u_2)$. We assume by contradiction

$$(4.33) \qquad (u_1 - u_2)(z) = \max_{\overline{\Omega}}(u_1 - u_2) > 0,$$

for some $z \in \partial\Omega$. Define, for $0 < \varepsilon < c$ (see (4.30)),

$$\Phi(x, y) := u_1(x) - u_2(y) - \left|\frac{x-y}{\varepsilon} - \eta(z)\right|^2 - |y - z|^2,$$

and observe that Φ is u.s.c. in $\overline{\Omega} \times \overline{\Omega}$. Let $(x_\varepsilon, y_\varepsilon)$ be a point such that

$$\Phi(x_\varepsilon, y_\varepsilon) = \max_{\overline{\Omega} \times \overline{\Omega}} \Phi.$$

By (4.30) $z + \varepsilon\eta(z) \in \Omega$, so $\Phi(z + \varepsilon\eta(z), z) \leq \Phi(x_\varepsilon, y_\varepsilon)$ and we obtain, using the continuity of u_1 at z,

$$(4.34) \quad \begin{aligned} \left|\frac{x_\varepsilon - y_\varepsilon}{\varepsilon} - \eta(z)\right|^2 + |y_\varepsilon - z|^2 &\leq u_1(x_\varepsilon) - u_2(y_\varepsilon) - u_1(z + \varepsilon\eta(z)) + u_2(z) \\ &\leq u_1(x_\varepsilon) - u_2(y_\varepsilon) - (u_1 - u_2)(z) + \omega(\varepsilon) \end{aligned}$$

where ω is a modulus. Since u_1 and $-u_2$ are bounded above, we get

$$(4.35) \qquad \left|\frac{x_\varepsilon - y_\varepsilon}{\varepsilon}\right|^2 \leq C \quad \text{for all } \varepsilon.$$

Then $x_\varepsilon - y_\varepsilon \to 0$ as $\varepsilon \to 0^+$, hence $\limsup_{\varepsilon \to 0^+}(u_1(x_\varepsilon) - u_2(y_\varepsilon)) \leq \max_{\overline{\Omega}}(u_1 - u_2)$ by the compactness of $\overline{\Omega}$ and the upper semicontinuity of $u_1 - u_2$. Now we use (4.33) and (4.34) to get

$$(4.36) \qquad \frac{x_\varepsilon - y_\varepsilon}{\varepsilon} \longrightarrow \eta(z) \quad \text{and} \quad y_\varepsilon \longrightarrow z \quad \text{as } \varepsilon \to 0^+.$$

This implies

$$x_\varepsilon = y_\varepsilon + \varepsilon\eta(z) + o(\varepsilon) = y_\varepsilon + \varepsilon\eta(y_\varepsilon) + o(\varepsilon) \quad \text{as } \varepsilon \to 0^+,$$

so we can use (4.30), for ε small enough, to get $x_\varepsilon \in \Omega$. Then we can exploit the fact that u_1 is a subsolution of (4.31) and obtain

$$(4.37) \qquad u_1(x_\varepsilon) + H\left(x_\varepsilon, \frac{2}{\varepsilon}\left(\frac{x_\varepsilon - y_\varepsilon}{\varepsilon} - \eta(z)\right)\right) \leq 0.$$

On the other hand, (4.34) also implies
$$u_2(y_\varepsilon) \leq u_2(z) + u_1(x_\varepsilon) - u_1(z) + \omega(\varepsilon),$$
so we use $\lim_{\varepsilon \to 0^+} x_\varepsilon = z$ and the semicontinuity of u_1 to get
$$(4.38) \qquad \limsup_{\varepsilon \to 0^+} u_2(y_\varepsilon) \leq u_2(z).$$

Thus, by (4.36), the continuity of u_1 at z, and (4.33) we obtain $(u_2 - u_1)(y_\varepsilon) < 0$ for all ε small, and we can exploit either the PDE (4.31) or the boundary condition (4.32) to get
$$(4.39) \qquad u_2(y_\varepsilon) + H\left(y_\varepsilon, \frac{2}{\varepsilon}\left(\frac{x_\varepsilon - y_\varepsilon}{\varepsilon} - \eta(z)\right) + 2(z - y_\varepsilon)\right) \geq 0.$$

Combining this with (4.37) and using (4.26) and (4.27) we obtain
$$u_1(x_\varepsilon) - u_2(y_\varepsilon) \leq 2M|z - y_\varepsilon| + 2L\frac{|x_\varepsilon - y_\varepsilon|}{\varepsilon}\left|\frac{x_\varepsilon - y_\varepsilon}{\varepsilon} - \eta(z)\right| + \omega_1(|x_\varepsilon - y_\varepsilon|)$$
and the right-hand side tends to 0 as $\varepsilon \to 0^+$ by (4.35) and (4.36). Since (4.34) gives
$$(u_1 - u_2)(z) \leq u_1(x_\varepsilon) - u_2(y_\varepsilon) + \omega(\varepsilon),$$
for small ε, we get a contradiction to (4.33).

The second statement of the theorem is obtained by replacing the term $|y - z|^2$ in the auxiliary function Φ with $|x - z|^2$ and repeating the previous argument with the roles of the variables x and y and of the functions u_1 and u_2 exchanged. ◂

Note that Theorem 4.16 implies a comparison result between a subsolution of (4.31) (continuous at $\partial\Omega$) and a constrained supersolution, that is, a supersolution of $u + H(x, Du) = 0$ in $\overline{\Omega}$, which extends the theory of §IV.5. In fact, the boundary condition in this case is stronger than (4.32). The next result is a small variant of the previous one.

THEOREM 4.17. *Assume* (4.26), (4.27), (4.30) *and* Ω *bounded. If* u_1 *and* u_2 *are, respectively, a subsolution and a supersolution of* (4.25), *and they are continuous at each point of* $\partial\Omega$, *then* $u_1 \leq u_2$ *in* $\overline{\Omega}$.

PROOF. If $\max_{\partial\Omega}(u_1 - u_2) < \max_{\overline{\Omega}}(u_1 - u_2)$ we argue as in Theorem 4.16. In the opposite case we assume by contradiction (4.33), as before. We have to deal with two cases: (i) $u_2(z) < g(z)$, and (ii) $u_1(z) > g(z)$. In case (i) we proceed as before until (4.37). Next we want to get (4.39) to reach a contradiction as before. If $y_\varepsilon \in \Omega$, we have (4.39) from the PDE (4.31) and the definition of y_ε. If $y_\varepsilon \in \partial\Omega$, we use (4.38), (4.36), and the continuity of g to get
$$\limsup_{\varepsilon \to 0^+} (u_2 - g)(y_\varepsilon) = (u_2 - g)(z) < 0.$$

Thus, for ε small enough, the boundary condition (4.32) gives (4.39).

Case (ii) is treated similarly, as for the second statement of Theorem 4.16. ◂

Before turning to the case of unbounded Ω and to uniqueness results we discuss the optimality of the assumptions of the two comparison theorems. The new hypotheses with respect to the comparison principle for the standard Dirichlet problem (see Chapters II and III and §1) are (i) the regularity of $\partial\Omega$, (4.30); and (ii) the continuity of the semisolutions to compare. The next examples show that neither of these assumptions can be removed.

EXAMPLE 4.18. Let $H(x,p) = |p| - 1$, $\Omega = B(0,1) \setminus \{0\}$, $g(0) = 0$, $g(x) = 1$ for $|x| = 1$. Here $v(x) = 1 - e^{-|x|}$ is a Lipschitz continuous solution of (4.25) such that $v = g$ on $\partial\Omega$. Any $w \equiv c \leq 1$ is a subsolution of (4.25), and if $c > 0$ it violates the conclusions of Theorems 4.16 and 4.17. Note that $\widehat{v}(x) \equiv 1$ is another continuous solution of (4.25) (cfr. Example 4.5 and Exercise 4.2). Even if (4.30) is violated only at 0, the domain is rather bad because int $\overline{\Omega} \supsetneq \Omega$. The domain is more regular in Example 4.15 because int $\overline{\Omega} = \Omega$ and (4.30) does not hold just at 0; in such an example, however, the semisolutions violating the conclusions of Theorems 4.16 and 4.17 are not both continuous. ◁

It is easy to give examples where the conclusions of the comparison theorems are false because the semisolutions to compare are discontinuous at the boundary, even for smooth domains. In fact, it is enough that small-time controllability on $\partial\Omega$ be violated and we have $T_*(x) < T^*(x)$ at some point $x \in \partial\Omega$, because the continuity of the minimal time function T at each point of the boundary of the target is equivalent to small-time controllability on the target, see Proposition IV.1.2. Then v_* and v^* ($v = 1 - e^{-T}$) do not satisfy $v_* \geq v^*$, even though they are a super- and subsolution of (4.25) with $g \equiv 0$, H given by (4.28) with $\ell \equiv 1$, by Theorem 4.13. If Ω is bounded and nonconvex, one can easily choose a system with a single, constant direction and get the example (here the PDE is linear with constant coefficients). We propose an explicit example where the supersolution fails to be continuous at just one point of $\partial\Omega$, but the conclusion of Theorem 4.16 is false.

EXAMPLE 4.19. Take the system of Example 4.15, $g \equiv 0$, and

$$\Omega = \{(x_1, x_2) \in \mathbb{R}^2 : |x_1| < 1, \ |x_1| - 1 < x_2 < |x_1|\}.$$

The minimal time function T is as in Example 4.15 for $x_2 \geq 0$, and

$$T(x) = 1 - |x_1| + x_2 \quad \text{for } x_2 < 0.$$

Thus it is continuous in $\overline{\Omega}$ except on $X := \{(x_1, 0) : -1/2 < x_1 < 1/2\}$. We take $v := 1 - e^{-T}$. By Theorem 4.13 $v = v_*$ is a v-supersolution of (4.24) and $\widehat{v} = v^*$ is a v-subsolution, and both are v-solutions of (4.24). For every $x \in X$, $v_*(x) < v^*(x)$, so the conclusions of Theorems 4.16 and 4.17 do not hold. Here the domain Ω satisfies the regularity assumption (4.30), and it is easy to see that the corners of Ω can be smoothed to make $\partial\Omega$ a C^∞ curve while keeping the set of discontinuities of v unchanged. ◁

If Ω is unbounded, we give the assumptions for the comparison theorem only for the Bellman's Hamiltonian (4.28). For the general structural hypotheses see

Exercise 4.7. We do not give the proof since it is rather long and technical. It can be obtained by the methods of proof of Theorem 4.16 and of §III.2.1.

THEOREM 4.20. *The comparison statements in Theorems 4.16 and 4.17 remain valid if Ω is unbounded, provided that*

(i) *(4.30) holds with η bounded and uniformly continuous;*

(ii) *u_1 and u_2 are bounded;*

(iii) *H is given by (4.28) with $f : \overline{\Omega} \times A \to \mathbb{R}^N$ and $\ell : \overline{\Omega} \times A \to \mathbb{R}^N$ continuous and satisfying (4.29) for all $x, y \in \overline{\Omega}$, $a \in A$, A compact, f bounded in $\partial\Omega \times A$.*

Note that the assumptions in (iii) imply (4.26). However (iii) does not imply (4.27) unless f is bounded, and this is a restrictive assumption we want to avoid.

Next we turn to the uniqueness theorem for the boundary value problem (4.25). Denote

$$B(\overline{\Omega}) := \{\, u : \overline{\Omega} \to \mathbb{R} \text{ bounded}\,\},$$
$$BC_\partial(\overline{\Omega}) := \{\, u \in B(\overline{\Omega}) : u \text{ continuous at each point of } \partial\Omega\,\}.$$

THEOREM 4.21. *Assume the hypotheses (i) and (iii) of Theorem 4.20 and suppose there exists a solution $u \in BC_\partial(\overline{\Omega})$ of (4.25). Then*

(a) *$u \in C(\overline{\Omega})$ and it is the unique solution, the maximal subsolution and the minimal supersolution of (4.25) in $BC_\partial(\overline{\Omega})$;*

(b) *if, in addition, $u = g$ on $\partial\Omega$, then it is the unique solution, maximal subsolution and minimal supersolution of (4.25) in $B(\overline{\Omega})$, and $u = v = \widehat{v}$ in $\overline{\Omega}$, where v and $\widehat{v}$ are the value functions defined in Theorem 4.13 with $\lambda = 1$.*

PROOF. Since $u_*, u^* \in BC_\partial(\overline{\Omega})$ are, respectively, a super- and a subsolution of (4.25), Theorems 4.17 and 4.20 imply $u_* \geq u^*$, thus u is continuous. Then by the same theorems u is the maximal subsolution and minimal supersolution in $BC_\partial(\overline{\Omega})$. If w is another solution in $BC_\partial(\overline{\Omega})$, then $w^* \leq u \leq w_*$ and so $u = w$. This proves (a). We prove (b) in the same way by replacing the comparison statement in Theorem 4.17 with that of Theorem 4.16. The value functions v and $\widehat{v}$ are bounded solutions by Theorem 4.13, thus they coincide with u. ◄

The statement $u = v$ in (b) had already been proved under weaker assumptions in Chapter IV. On the other hand, the statement $v = \widehat{v}$ was not studied in Chapter IV and it deserves a more explicit statement.

COROLLARY 4.22. *Assume hypotheses (i) and (iii) of Theorem 4.20, $g \in BC(\partial\Omega)$, and let $v, \widehat{v}$ be the value functions defined in Theorem 4.13 with $\lambda = 1$. If v is continuous at each point of $\partial\Omega$, then $v = \widehat{v}$. This occurs in particular if either*

(j) *the system (f, A) is small-time locally controllable on $\mathcal{T} := \Omega^c$ (STCT, Definition IV.1.1) and the terminal data g are compatible (Definition IV.3.5),*

or

4. BOUNDARY CONDITIONS IN THE VISCOSITY SENSE

(jj) *there exist $\underline{u}, \widetilde{u} \in BC_\partial(\overline{\Omega})$ respectively a sub- and supersolution of (4.25) such that $\underline{u} = \widetilde{u} = g$ on $\partial\Omega$.*

PROOF. The first statement follows immediately from Theorems 4.13 and 4.21. The sufficiency of (j) for the continuity of v follows from Theorem IV.3.6 (this theorem is stated under the simplifying assumption that $\partial\mathcal{T}$ is bounded which can be easily dropped). The sufficiency of (jj) is easily proved by the comparison theorems 4.16 and 4.20 or by Perron's method (Corollary 2.15). ◀

The following control-theoretic consequence of the previous theory was announced in Chapter IV.

COROLLARY 4.23. *Assume the hypotheses of Corollary 4.22 and $\partial\Omega$ bounded. Then system (f, A) is small-time locally controllable on $\mathcal{T} := \Omega^c$ if and only if it is small-time locally controllable on* int $\mathcal{T}$ (STC$\overset{\circ}{\mathcal{T}}$, *Definition IV.1.1*).

Here the assumption that $\partial\Omega$ be bounded can be dropped; it is enough to extend a bit two results of §IV.1, where this assumption was made only to simplify some statements and proofs. On the other hand, we observed in Remark IV.1.4 that some regularity on $\partial\mathcal{T}$ such as (4.30) is necessary.

PROOF OF COROLLARY 4.23. STC$\overset{\circ}{\mathcal{T}}$ implies STC$\mathcal{T}$ by definition. If we assume STC$\mathcal{T}$ we have the continuity of the minimal-time function T by Proposition IV.1.6. Thus $v = 1 - e^{-T}$ is continuous, and therefore $\widehat{T}$ is continuous by Corollary 4.22. Since (4.30) implies $\mathcal{T} = \overline{\text{int}\,\mathcal{T}}$ (see Exercise 4.8) we can use Remark IV.1.4 and get STC$\overset{\circ}{\mathcal{T}}$. ◀

We have not yet described in some generality the situation where there exists a unique continuous solution of (4.25) which does *not* satisfy $u = g$ on $\partial\Omega$, so that $u \neq v$, even though this was what happened in the simple one-dimensional Examples 4.3 and 4.11. The next result describes a situation where $\widehat{v}$, instead of v, is the unique continuous solution of (4.25) and it does not satisfy $\widehat{v} = g$ on $\partial\Omega$, as in the Examples 4.3 and 4.11 (other examples are in Exercise 4.13).

We define the *usable part* of $\partial\Omega$ as the set of terminal points of trajectories of the system, that is,

$$\mathcal{U} := \{\, z \in \partial\Omega : z = y_x(t_x(\alpha)) \text{ for some } x \in \Omega,\, \alpha \in \mathcal{A} \,\}.$$

The following theorem deals with time-optimal control and says, roughly speaking, that $\widehat{v}$ is continuous if there is a target $\widetilde{\mathcal{T}}$ containing $\mathcal{U}$, smaller than the true target $\mathcal{T} = \Omega^c$, and an extension of the system inside $\mathcal{T}$, such that STC$\widetilde{\mathcal{T}}$ holds even if STC$\mathcal{T}$ does not.

THEOREM 4.24. *Assume the hypotheses of Corollary 4.22 and $\ell \equiv 1$, $g \equiv 0$. Suppose there exist an open set $\widetilde{\Omega} \supseteq \Omega$ and a system $\widetilde{f} : \overline{\widetilde{\Omega}} \times A \to \mathbb{R}^N$ such that $\widetilde{f} = f$ on $\Omega \times A$, satisfying the structural assumptions (A_0)–(A_3) of Chapter III and such that $(\widetilde{f}, A)$ is small-time controllable on $\widetilde{\mathcal{T}} := \widetilde{\Omega}^c$. Assume also $\widetilde{\mathcal{T}} \supseteq \mathcal{U}$ and $\partial\widetilde{\mathcal{T}}$ bounded. Then*

(i) $\widehat{v} \in C(\overline{\Omega})$ is the solution of (4.25);

(ii) $\widehat{v}(x) = v(x)$ for all $x \in \Omega$;

(iii) for $x \in \partial\Omega$, $\widehat{v}(x) = v(x) = 0$ if and only if $x \in \overline{\mathcal{U}}$.

PROOF. Let $\widetilde{T}$ be the minimal time function for $\widetilde{\mathcal{T}}$ and $\widetilde{v} = 1 - e^{-\widetilde{T}}$. By the controllability assumption on $(\widetilde{f}, A)$ we have $\widetilde{v} \in BC(\overline{\overline{\Omega}})$.

To prove that

(4.40) $$\widehat{v}^* \leq \widetilde{v} \quad \text{in } \overline{\Omega},$$

we can use Theorems 4.16 and 4.20. In fact $\widetilde{v} \in BC(\overline{\Omega})$ is a supersolution of the HJB equation in Ω, and for $x \in \partial\Omega$ such that $\widehat{v}^*(x) > \widetilde{v}(x)$ we have $\widehat{v}^* + H(x, D\widehat{v}^*) \leq 0$ at x in the viscosity sense, because $\widetilde{v}(x) \geq 0$ and $\widehat{v}^*$ is a subsolution of (4.25) with $g \equiv 0$. Then we can apply the comparison results and get (4.40).

Next we claim that

(4.41) $$\widetilde{v} \leq \widehat{v} \quad \text{in } \overline{\Omega}.$$

This gives $\widetilde{v} \leq \widehat{v}_*$ and then $\widetilde{v} = \widehat{v}$ in $\overline{\Omega}$ by (4.40). To prove (4.41) we need the following representation formula for $\widetilde{v}$:

(4.42) $$\widetilde{v}(x) = \inf_{\alpha \in \mathcal{A}} \left\{ \int_0^{t_x(\alpha)} e^{-s}\, ds + \widetilde{v}(y_x(t_x(\alpha))) e^{-t_x(\alpha)} \right\},$$

which holds by Corollary IV.3.17. We assume by contradiction that $\widehat{v}(x) < \widetilde{v}(x)$. Then there is $\overline{\alpha} \in \mathcal{A}$ such that

$$1 - e^{-\widehat{t}_x(\overline{\alpha})} < \widetilde{v}(x),$$

and $\widehat{t}_x(\overline{\alpha}) < +\infty$ because $\widetilde{v} \leq 1$. Thus we use (4.42) and $\widehat{t}_x \geq t_x$ to get

$$\widetilde{v}(y_x(t_x(\overline{\alpha}))) > 0.$$

This implies $y_x(t_x(\overline{\alpha})) \notin \widetilde{\mathcal{T}}$, a contradiction to $\mathcal{U} \subseteq \widetilde{\mathcal{T}}$. This proves the claim and therefore $\widehat{v} \in C(\overline{\Omega})$. In a similar way we obtain $\widetilde{v} \leq v$ in Ω, so $\widetilde{v} = v$ in Ω because $v \leq \widehat{v}$ by definition.

To prove (iii), we first observe that $\widetilde{v} = 0$ on $\mathcal{U} \subseteq \widetilde{\mathcal{T}}$, so $\widehat{v} = \widetilde{v} = 0$ on $\overline{\mathcal{U}}$ by continuity. On the other hand, if $x \notin \overline{\mathcal{U}}$ let $r := \text{dist}(x, \overline{\mathcal{U}}) > 0$. For any $z \in B(x, r/2)$ and $\alpha \in \mathcal{A}$, we have $|z - y_z(t_z(\alpha))| > r/2$ by definition of $\mathcal{U}$, so that, for some $c > 0$, $v(z) \geq c$ for all such z. Then $\limsup_{z \to x} v(z) > 0 = v(x)$. ◀

For the case of general running and terminal costs we give without proof two results under controllability conditions. In the first one we make an explicit assumption on the directions of the vector fields at $\partial\Omega$, in the second one we assume essentially a uniform small-time local controllability on each point of Ω (cfr. Exercise 4.15).

THEOREM 4.25. *Assume the hypotheses of Corollary 4.22 and $\partial\Omega$ a compact manifold of class C^2. Suppose for all $x \in \partial\Omega$*

$$\inf_{a \in A} f(x,a) \cdot n(x) < 0, \qquad \sup_{a \in A} f(x,a) \cdot n(x) > 0,$$

where $n(x)$ is the interior normal to $\partial\Omega$. Then $\widehat{v} \in C(\overline{\Omega})$ is the solution of (4.25).

THEOREM 4.26. *Assume the hypotheses of Corollary 4.22 and that there exist constants $C, r > 0$ and a modulus ω such that, for all $x, z \in \Omega$, $|x - z| < r$, there is $\alpha \in \mathcal{A}$ satisfying $y_x(t, \alpha) = z$ for some $t \leq \omega(|x - z|) \wedge t_x(\alpha)$. Then $\widehat{v} \in C(\overline{\Omega})$ is the solution of (4.25) and $\widehat{v}(x) = v(x)$ for all $x \in \Omega$.*

We end this subsection with a convergence theorem that completes the stability results on the three perturbation problems studied in the previous subsection: regular perturbations (Proposition 4.7), singular perturbations (Proposition 4.10) and perturbations of the domain (Proposition 4.12). We summarize all the results in a single statement because the proof is exactly the same in the three cases.

COROLLARY 4.27. *Assume either (4.26), (4.27), (4.30) and Ω bounded or the hypotheses (i) and (iii) of Theorem 4.20. Suppose there exists a solution $u \in BC(\overline{\Omega})$ of (4.25) such that $u = g$ on $\partial\Omega$. Set $F(x, r, p) = r + H(x, p)$ and assume either the hypotheses of Proposition 4.7, or those of Proposition 4.10, or those of Proposition 4.12. Then $u_\varepsilon \rightrightarrows u$ on compacta of $\overline{\Omega}$.*

PROOF. The three propositions imply in each case that the upper and lower weak limits, $\overline{u}$ and $\underline{u}$, are, respectively, a subsolution and a supersolution of (4.25). By the comparison theorems 4.16 and 4.20 we get

$$\overline{u} \leq u \leq \underline{u}.$$

Since $\underline{u} \leq \overline{u}$ by definition, we obtain $\overline{u} = \underline{u} = u$, and this implies the uniform convergence of u_ε to u on compacta by Lemma 1.9. ◀

The reader should compare this result with Corollary 1.8, see Remark 1.11. Corollary 4.27 can be applied to prove the stability of the value function of problems with restricted state space (either involving exit times or with state constraints) under perturbations of the data, as we did in Theorem 1.10 for problems in all $\mathbb{R}^N$, see Exercise 4.16.

The most restrictive assumption of Corollary 4.27 is the existence of a continuous solution of the limit problem. This will be dropped in the next subsection for the problem of time-optimal control.

4.3. Uniqueness and complete solution for time-optimal control

In this subsection we give some uniqueness and stability results for the boundary value problem (4.25) that allow the solution to be discontinuous. We limit ourselves to the time-optimal control problem and consider the discounted value functions

$$(4.43) \qquad v(x) = \inf_{\alpha \in \mathcal{A}} \int_0^{t_x(\alpha)} e^{-s}\, ds, \qquad \widehat{v}(x) = \inf_{\alpha \in \mathcal{A}} \int_0^{\widehat{t}_x(\alpha)} e^{-s}\, ds,$$

where t_x and $\hat{t}_x$ are the first entry time of the system in $\mathcal{T}$ and int $\mathcal{T}$, respectively. We assume the same hypotheses as in §3.2, namely, the function f defining the system is continuous in all variables and lipschitzean in the state variables, the control set A is compact, the target $\mathcal{T}$ is closed and nonempty and its boundary $\partial\mathcal{T}$ is bounded (see Exercise 4.17 for some extensions).

We know that both v and $\hat{v}$ solve the boundary value problem

(4.44) $\quad\begin{cases} u + H(x, Du) = 0 & \text{in } \mathcal{T}^c =: \Omega, \\ u_* \geq 0 & \text{on } \partial\mathcal{T}, \\ u^* \wedge (u + H(x, Du)) \leq 0 & \text{on } \partial\mathcal{T} \end{cases}$

with H given by (3.9), where the HJB equation in $\mathcal{T}^c$ is in the sense of noncontinuous viscosity solutions, Definition 2.2, u_* and u^* are the semicontinuous envelopes, and the second boundary condition on $\partial\mathcal{T}$ is in the viscosity sense, Definition 4.1. This follows from Theorem 4.13 and from the fact that $v, \hat{v} \geq 0$, so that we do not need to consider supersolutions in the viscosity sense at the boundary. We call a supersolution of (4.44) a supersolution $\tilde{u} \in BLSC(\overline{\Omega})$ of the HJB equation satisfying the first boundary condition $\tilde{u} \geq 0$, we call a subsolution of (4.44) a subsolution $\underline{u} \in BUSC(\overline{\Omega})$ of the HJB equation satisfying the second boundary condition $\underline{u} \wedge (\underline{u}+H) \leq 0$, and a solution of (4.44) is a function u such that u_* is a supersolution and u^* is a subsolution.

Recall that in Chapters III and IV we called complete solution of a boundary value problem a function that is the minimal supersolution and the maximal subsolution. The natural extension of this notion to non-continuous solutions is the following.

DEFINITION 4.28. *A locally bounded function w is a complete solution of (4.44) if w_* is the minimal supersolution and w^* is the maximal subsolution; that is, if w is a solution of (4.44) and any subsolution $\underline{u}$ and supersolution $\tilde{u}$ of (4.44) satisfy*

$$w_* \leq \tilde{u}, \qquad \underline{u} \leq w^*. \quad \triangleleft$$

The main result of this subsection is the following.

THEOREM 4.29. *Under the previous assumptions $\hat{v}$ is the maximal subsolution of (4.44). If, in addition, $\mathcal{T}$ is the closure of its interior,*

(4.45) $$\mathcal{T} = \overline{\text{int }\mathcal{T}},$$

then $\hat{v}$ is a complete solution.

Before giving the proof we describe several consequences.

COROLLARY 4.30. *Under the previous assumptions and (4.45)*

(i) *$\hat{v}$ is the unique upper semicontinuous complete solution of (4.44);*

(ii) *for any solution u of (4.44),*

$$\hat{v}_* \leq u_* \leq u \leq u^* \leq \hat{v},$$

so that $u(x) = v(x) = \hat{v}(x)$ at each point x where $\hat{v}$ is continuous;

(iii) $\widehat{v}_* = v_*$ is the unique lower semicontinuous solution of (4.44); moreover, it has a meager[1] set of points of discontinuity and a dense set of points of continuity.

PROOF. (i) $\widehat{v}$ is upper semicontinuous because it is a subsolution by Theorem 4.29, and there is at most one u.s.c complete solution because it must be the maximal subsolution.

(ii) is a trivial consequence of the definition of complete solution.

(iii) The equality $\widehat{v}_* = v_*$ is essentially the second statement of Theorem 4.29 because v_* is the minimal supersolution of (4.44) by Theorem 3.7. To prove uniqueness, let $w = w_*$ be a solution of (4.44). By definition of complete solution

$$\widehat{v}_* \leq w \quad \text{and} \quad w^* \leq \widehat{v}^*.$$

The second inequality, w l.s.c. and $\widehat{v}$ u.s.c., imply

$$w \leq (w^*)_* \leq \widehat{v}_*,$$

which gives $w = \widehat{v}_*$.

To prove the regularity result, set $u = \widehat{v}_* = v_*$ and consider the sets

$$C_n := \{\, x : u^*(x) - u(x) < 1/n \,\}, \quad n \in \mathbb{N},$$

which are open because $u^* - u$ is u.s.c. Of course u is continuous at x if and only if $x \in C_n$ for all n. We are going to prove that each C_n is dense and we observe first that

(4.46) $$(u^*)_* = u.$$

In fact the inequality $\geq$ follows from the lower semicontinuity of u, and the opposite one from taking the l.s.c. envelope of both sides of the inequality $u^* \leq \widehat{v}^* = \widehat{v}$. If C_n were not dense, for some x_0 it would be $u^*(x) \geq u(x) + 1/n$ for all x near x_0. Thus $(u^*)_*(x_0) \geq u(x_0) + 1/n$, because u is l.s.c., and we have reached a contradiction to (4.46). The intersection of all C_n is dense by Baire's theorem (see, e.g., [Ru86]). ◀

Note that a complete solution is unique if and only if it is continuous, because otherwise there are infinitely many functions with the same semicontinuous envelopes.

We remark that under the assumptions of Theorem 4.29 a lower semicontinuous complete solution of (4.44) may not exist. In fact, the only possible candidate is $\widehat{v}_* = v_*$, and in Example 4.15 we have a target satisfying (4.45), a linear system, and a continuous v such that $v(x) < \widehat{v}(x)$ on a line segment. This example also says that v may not be a complete solution of (4.44), even in a case where it is the complete solution of the usual Dirichlet problem for the HJB equation.

---

[1] A set is meager (or Baire's first category) if it is a countable union of nowhere dense sets (i.e., sets with empty interior), or, equivalently, if its complement is a countable intersection of dense open sets.

Note also that Corollary 4.30 (ii) says that $\hat{v}$ is the *unique solution in the sense of the graph*, that is, the graph of any solution u of (4.44) is contained in the completed graph of $\hat{v}$

$$\text{c-graph } \hat{v} := \{(x, r) \in \overline{\Omega} \times \mathbb{R} : \hat{v}_*(x) \le r \le \hat{v}^*(x)\}.$$

Theorem 4.29 is particularly fit for the study of the stability of the value functions with respect to perturbations. In fact, if u_ε are solutions of any of the perturbed problems studied in §4.1, in the special case $F(x, r, p) = r + H(x, p)$ and boundary data $g \equiv 0$, then the upper weak limit $\overline{u}$ is a subsolution of (4.44) (see Propositions 4.7, 4.10, 4.12), while the lower weak limit $\underline{u}$ is a supersolution of (4.44) provided $u^\varepsilon \ge 0$ in a neighborhood of the boundary. Then

$$\hat{v}_* \le \underline{u} \le \overline{u} \le \hat{v}$$

by Theorem 4.29, so the weak limits coincide at any point where $\hat{v}$ is continuous, and therefore $u_\varepsilon \rightrightarrows \hat{v}$ on compact sets where $\hat{v}$ is continuous. This is the natural extension of Corollaries 1.8 and 4.27 to the case that the value function of the limit problem is discontinuous.

Next we give two explicit examples of applications. The first is the analogue for time-optimal control of Theorem 1.10 for the infinite horizon problem. For each $\varepsilon \ge 0$ we are given a system with continuous dynamics $f_\varepsilon : \mathbb{R}^N \times A \to \mathbb{R}^N$, with compact control set $A_\varepsilon \subseteq A$ and satisfying

$$|f_\varepsilon(x, a) - f_\varepsilon(y, a)| \le L_\varepsilon |x - y| \quad \text{for all } x, y \in \mathbb{R}^N, a \in A_\varepsilon.$$

We assume $f_\varepsilon \rightrightarrows f_0$ on $\mathcal{K} \times A$ for each compact $\mathcal{K} \subseteq \mathbb{R}^N$ and that the Hausdorff distance $d_H(A_\varepsilon, A_0) \to 0$ as $\varepsilon \to 0^+$. For each ε we are given a compact target $\mathcal{T}_\varepsilon \supseteq \mathcal{T}$ and assume $d_H(\mathcal{T}_\varepsilon, \mathcal{T}_0) \to 0$ as $\varepsilon \to 0^+$. Let $v_\varepsilon, \hat{v}_\varepsilon$ be the discounted value functions of the time-optimal control problems corresponding to $\varepsilon \ge 0$, that is, they are given by (4.43) with $\mathcal{A}$ replaced by $\mathcal{A}_\varepsilon$ and the entry times $t_x, \hat{t}_x$ corresponding to the system f_ε and the target $\mathcal{T}_\varepsilon$.

THEOREM 4.31. *Assume the preceding hypotheses and let $\mathcal{T}_0$ be the closure of its interior. Then v_ε and $\hat{v}_\varepsilon$ converge uniformly to $v_0 = \hat{v}_0$ on every compact set where $\hat{v}_0$ is continuous.*

PROOF. We know that v_ε and $\hat{v}_\varepsilon$ satisfy for all $\varepsilon \ge 0$ problem (4.44) with H replaced by the Hamiltonian H_ε associated with the system $(f_\varepsilon, A_\varepsilon)$ and $\mathcal{T}$ replaced by $\mathcal{T}_\varepsilon$. It is easy to see, as in the proof of Corollary III.2.22, that $H_\varepsilon \rightrightarrows H_0$ on compacta of $\mathbb{R}^{2N}$. By the arguments of the proofs of Propositions 4.7 and 4.12 and the obvious inequalities $0 \le v_\varepsilon \le 1$, the weak limits

$$\underline{u} = \liminf_{\varepsilon \to 0^+} {}_* v_\varepsilon, \qquad \overline{u} = \limsup_{\varepsilon \to 0^+}{}^* v_\varepsilon$$

(see Definition 1.4) are, respectively, a supersolution and a subsolution of the limit problem (4.44) with $H = H_0$ and $\mathcal{T} = \mathcal{T}_0$. By Theorem 4.29 $\hat{v}_0$ is a complete solution of this problem, so the general argument explained above gives the uniform convergence of v_ε to $\hat{v}_0$ on compacta where $\hat{v}_0$ is continuous, and on these sets $v_0 = \hat{v}_0$ by Corollary 4.30. The proof for $\hat{v}_\varepsilon$ is obtained in the same way by taking the corresponding weak limit. ◀

The second application is to the vanishing viscosity approximation. Suppose $u_\varepsilon \in C^2(\Omega) \cap C(\overline{\Omega})$ are solutions of

(4.47) $$\begin{cases} -\varepsilon \Delta u + u + H(x, Du) = 0 & \text{in } \Omega = \mathcal{T}^c, \\ u = 0 & \text{on } \partial \Omega, \end{cases}$$

for $\varepsilon > 0$, and assume Ω is bounded. For the existence of such solutions, for example, in the case Ω has smooth boundary, we refer to [GT83, Kry87].

COROLLARY 4.32. *Assume the hypotheses of Theorem 4.29, including (4.45). Then $u_\varepsilon \rightrightarrows \hat{v}$ as $\varepsilon \to 0^+$ on compact sets where $\hat{v}$ is continuous.*

PROOF. We claim that $0 \leq u_\varepsilon \leq 1$ for all ε. By contradiction assume u_ε is negative somewhere. By the compactness of $\overline{\Omega}$ and the boundary condition there is $x_0 \in \Omega$ where u_ε attains a negative minimum. Since $Du_\varepsilon(x_0) = 0$, $\Delta u_\varepsilon(x_0) \geq 0$ and $H(x_0, 0) = -1$, we have

$$-\varepsilon \Delta u_\varepsilon(x_0) + u_\varepsilon(x_0) + H(x_0, Du_\varepsilon(x_0)) < -1,$$

a contradiction to the PDE in (4.47).

Next assume by contradiction u_ε is larger than 1 somewhere. Then there is a maximum point $x_1 \in \Omega$ where $u_\varepsilon(x_1) > 1$, so

$$-\varepsilon \Delta u_\varepsilon(x_1) + u_\varepsilon(x_1) + H(x_1, Du_\varepsilon(x_1)) \geq u_\varepsilon(x_1) - 1 > 0,$$

a contradiction again.

Now we can apply Proposition 4.10 to get that the weak limits $\underline{u}$ and $\overline{u}$ are, respectively, a supersolution and a subsolution of (4.44). By Theorem 4.29 and the general stability argument explained above we can conclude. ◀

Now we turn to the proof of Theorem 4.29, which is a simple consequence of the following two propositions.

PROPOSITION 4.33. *The value function $\hat{v}$ is the maximal subsolution of (4.44).*

PROOF. We assume int $\mathcal{T} \neq \emptyset$ because the case int $\mathcal{T} = \emptyset$ is much easier and we leave it as an exercise for the reader. Let w be a subsolution of (4.44) and let us first observe that it is not restrictive to assume $w \geq 0$. In fact, if this is not the case, we can replace w by $w^+ = w \vee 0$ which is a viscosity subsolution of (4.44) by Proposition 4.6, since it is u.s.c. and w and 0 are subsolutions.

Define for $h > 0$

$$\mathcal{T}_h := \{ x \in \mathcal{T} : \text{dist}(x, \partial \mathcal{T}) \geq h \} \quad \text{and} \quad v^h(x) := \inf_{\alpha \in \mathcal{A}} \int_0^{t_x^h(\alpha)} e^{-s} \, ds,$$

where t_x^h is the first entry time of the system in T_h. By Theorem 2.6 $(v^h)_*$ is a supersolution of the Dirichlet problem

(4.48) $$\begin{cases} u + H(x, Du) = 0 & \text{in } \mathbb{R}^N \setminus T_h, \\ u = 0 & \text{on } \partial T_h. \end{cases}$$

We extend w in int T by setting $w \equiv 0$ and note that $w \in USC(\mathbb{R}^N)$. We claim that w is a subsolution of (4.48) and that

(4.49) $$\widehat{v}(x) = \inf_{h>0} v^h(x) \quad \text{for all } x \in \mathbb{R}^N.$$

Then we get $w \leq \widehat{v}$ because the comparison principle, Theorem 1.3, gives $w \leq (v^h)_*$ in $\mathbb{R}^N$ for all h. Since $\widehat{v}^*$ is a subsolution of (4.44) by Theorem 4.13, we conclude that $\widehat{v} = \widehat{v}^*$ is the maximal subsolution.

To prove the first claim we first observe that w is trivially a subsolution of the boundary condition in (4.48) and of the HJB equation in T^c and in int $T \setminus T_h$. Next let $x_0 \in \partial T$ be a maximum point of $w - \varphi$ with $\varphi \in C^1(\mathbb{R}^N \setminus T_h)$. There are two possible cases. The first is that $w(x_0) > 0$, and we get the conclusion because w satisfies the second boundary condition in (4.44), hence

(4.50) $$w(x_0) + H(x_0, D\varphi(x_0)) \leq 0.$$

The second case is that $w(x_0) = 0$. Since we can assume $\varphi(x_0) = w(x_0) = 0$ and $\varphi(x) \geq w(x) \geq 0$ near x_0, we have $D\varphi(x_0) = 0$ and then (4.50) is satisfied.

It remains to prove (4.49). Note that by definition $\widehat{v} \leq v^h$ for all h. Assume by contradiction there is $\overline{x}$ such that $\widehat{v}(\overline{x}) < \inf_{h>0} v^h(\overline{x})$. Then there is $\varepsilon > 0$ such that

(4.51) $$\widehat{v}(\overline{x}) + 2\varepsilon < v^h(\overline{x}) \quad \text{for all } h > 0.$$

In particular, $\widehat{v}(\overline{x}) + 2\varepsilon < 1$; therefore, given $\overline{\alpha} \in \mathcal{A}$ such that

(4.52) $$1 - e^{-t_{\overline{x}}(\overline{\alpha})} \leq \widehat{v}(\overline{x}) + \varepsilon,$$

there is $\widetilde{t} \geq 0$ such that $z := y_{\overline{x}}(\widetilde{t}, \overline{\alpha}) \in \text{int } T$ and

(4.53) $$1 - e^{-\widetilde{t}} \leq 1 - e^{-t_{\overline{x}}(\overline{\alpha})} + \varepsilon.$$

By taking $h \leq \text{dist}(z, \partial T)$ we have $z \in T_h$, hence $v^h(\overline{x}) \leq 1 - e^{-\widetilde{t}}$ and by (4.52) and (4.53) we get a contradiction to (4.51). ◄

PROPOSITION 4.34. *If T is the closure of its interior, then*

$$\widehat{v}_*(x) = v_*(x) \quad \text{for all } x \in \mathbb{R}^N.$$

PROOF. Since $v \leq \hat{v}$ we have $v_* \leq \hat{v}_*$. We are going to prove that $\hat{v}_* \leq v$, which immediately gives the conclusion. Fix $x \in \mathbb{R}^N$ and $\alpha \in \mathcal{A}$ such that $t_x(\alpha) < +\infty$ (if such an α does not exist then $v(x) = 1$ and the desired inequality is verified). Set $\tau := t_x(\alpha)$, $\bar{x} := y_x(\tau, \alpha) \in \mathcal{T}$, and choose a sequence $\bar{x}_n \in \text{int}\,\mathcal{T}$ such that $\bar{x}_n \to \bar{x}$. Let $z_n(\,\cdot\,)$ be the solution of

$$\begin{cases} z' = f(z, \alpha) & 0 \leq t < \tau, \\ z(\tau) = \bar{x}_n. \end{cases}$$

Then, for all $0 \leq t \leq \tau$, $|z_n(t) - y_x(t,\alpha)| \leq |\bar{x}_n - \bar{x}| + \int_t^\tau L\,|z_n(s) - y_x(s,\alpha)|\,ds$, and by Gronwall's inequality

$$|z_n(t) - y_x(t,\alpha)| \leq |\bar{x}_n - \bar{x}|\,e^{L\tau}.$$

By plugging $t = 0$ into this inequality we get that $x_n := z_n(0)$ converge to x as $n \to \infty$. Moreover, $\hat{t}_{x_n}(\alpha) \leq \tau = t_x(\alpha)$ for all n, so

$$\hat{v}_*(x) \leq \liminf_{y \to x}(1 - e^{-\hat{t}_y(\alpha)}) \leq \liminf_n (1 - e^{-\hat{t}_{x_n}(\alpha)}) \leq 1 - e^{-t_x(\alpha)}.$$

Since α is arbitrary we have proved that $\hat{v}_*(x) \leq v(x)$. ◀

PROOF OF THEOREM 4.29. The first statement of the theorem is Proposition 4.33. In order to show that $\hat{v}$ is a complete solution it remains to prove that $\hat{v}_*$ is the minimal supersolution of (4.44), and this follows from Proposition 4.34 and Theorem 3.7. ◀

Exercises

4.1. Motivated by Remark 4.2, we define a viscosity subsolution (respectively, supersolution) of the equation $G(x, u, Du) = 0$ in a set $X \subseteq \mathbb{R}^N$, with $G : X \times \mathbb{R} \times \mathbb{R}^N \to [-\infty, +\infty]$ not necessarily continuous, a locally bounded function u such that $G_*(x, u(x), D\varphi(x)) \leq 0$ (respectively, $G^*(x, u(x), D\varphi(x)) \geq 0$) for every $\varphi \in C^1(X)$ and x local maximum (respectively, minimum) point for $u - \varphi$. Consider the infinite horizon problem with $f(x, a) = a$, $A = \mathbb{R}^N$, $\ell(x,a) = \ell(x)$ with $\ell \in BC(\mathbb{R}^N)$, and $\lambda = 1$. Compute the Hamiltonian H associated with this problem and the value function v, and observe that H is discontinuous and v is smooth. Prove that v is a viscosity solution of the HJB equation $u + H(x, Du) = 0$ in $\mathbb{R}^N$ in the sense just defined; however, if ℓ is not a constant, $v(x) + H(x, Dv(x)) < 0$ at some point x, so v does not solve the HJB equation in the classical sense.

4.2. Consider $\mathcal{T}$ closed, $\text{int}\,\mathcal{T} = \emptyset$, $d(x) = \text{dist}(x, \mathcal{T})$, $\mathcal{O}$ open, $\mathcal{O} \supseteq \mathcal{T}$,

$$k := \inf\{\,d(x) : x \in \partial\mathcal{O}\,\} > 0.$$

Consider the problem

$$\begin{cases} u + |Du| - 1 = 0 & \text{in } \Omega = \mathcal{O} \setminus \mathcal{T}, \\ u = 0 \quad \text{or} \quad u + |Du| - 1 = 0 & \text{on } \partial\mathcal{T}, \\ u = d & \text{on } \partial\mathcal{O}. \end{cases}$$

(This is a problem with mixed boundary conditions, but it is obvious how to interpret them.) Find a lipschitzean solution of this problem satisfying $u = 0$ on ∂T. Prove that any function $w \equiv c \leq k \wedge 1$ is a subsolution of the problem.

4.3. Prove that the function $\underline{u}$ in Example 4.11 is the e-solution of $-u' + u - 1 = 0$ in $]0, 1[$, $u(0) = u(1) = 0$.

4.4. Under the hypotheses (A_0)–(A_4) of Chapter III consider

$$J(x, \alpha, \tau) = \int_0^\tau e^{-\lambda s} \ell(y_x, \alpha) \, ds + e^{-\lambda \tau} g(y_x(\tau, \alpha))$$

and the value function $w(x) = \inf\{ J(x, \alpha, \tau) : \alpha \in \mathcal{A}, \, t_x(\alpha) \leq \tau \leq \hat{t}_x(\alpha) \}$. Prove that w is a viscosity solution of (4.18), (4.19).

4.5. If we drop the continuity assumption on the boundary data g, Definition 4.1 can be modified by replacing g with g^* in the definition of a subsolution and with g_* in the definition of a supersolution. Prove that Propositions 4.4 and 4.6 remain valid in this case. Check that this definition is consistent with that of Remark 4.2.

4.6. Use Theorems 4.16 and 4.20 to give verification results for infinite horizon problems with state constraints using semicontinuous verification and falsification functions.

4.7. (a) Assume A is compact, $f : \overline{\Omega} \times A \to \mathbb{R}^N$ is continuous, it satisfies

$$(f(x, a) - f(y, a)) \cdot (x - y) \leq L |x - y|^2$$

for all x, y, a and (4.29) for all a and $x, y \in B(\partial\Omega, \delta)$, $\delta > 0$, and it is bounded in $\partial\Omega \times A$. Assume ℓ is continuous and satisfies (4.29) for all x, y, a. Show that H defined by (4.28) satisfies (4.26), (4.27) for all p, q and $x, y \in B(\partial\Omega, \delta/2)$, and the inequality (2.7) in Chapter III for all $x, y \in \overline{\Omega}$.
(b) Prove Theorem 4.20 for H satisfying the structural conditions in (a). (This proof is rather long and delicate, the reader can check the result in [BS94].)

4.8. Prove that (4.30) implies $\text{int } \overline{\Omega} = \Omega$.

4.9. Use Theorem 4.21 to give verification theorems for the problems of minimizing J and $\hat{J}$.

4.10. Let $\mathcal{P}$ be the set of piecewise constant controls as in Chapter III. Let $v^\sharp(x) := \inf_{\alpha \in \mathcal{P}} J(x, \alpha)$. Assume the hypotheses of Corollary 4.22 and v continuous. Prove that $v^\sharp = v$.

4.11. Assume the hypotheses (i) and (iii) of Theorem 4.20 and $g \equiv 0$. Suppose there exists $\overline{v} \in C(\overline{\Omega})$ such that $\overline{v}(x) = v(x)$ for all $x \in \Omega$. Prove that $\overline{v} = \hat{v}$. (Check the result in [BS94].)

4.12. Assume the hypotheses of Corollary 4.22 and STC$\mathcal{T}$. Suppose there exists $\overline{v} \in C(\overline{\Omega})$ such that $\overline{v}(x) = v(x)$ for all $x \in \Omega$. Prove that $\overline{v} = \hat{v}$.

4. BOUNDARY CONDITIONS IN THE VISCOSITY SENSE   341

4.13. Let $f(x,a) = f(x) = (1,0)$ for all $x \in \mathbb{R}^2$, $\mathcal{T} = ([0,1]^2)^c$, $\ell \equiv 1$, $g \equiv 0$.

(i) Compute v and $\widehat{v}$;

(ii) show that the assumptions of Theorem 4.24 hold;

(iii) do the same for $\mathcal{T} = B(0,1)^c$.

4.14. Prove Theorem 4.26. [Hint: use Exercise 4.12.]

4.15. Give some explicit conditions on the system (f, A) ensuring the assumption of Theorem 4.26. [Hint: use the results on small-time local controllability of Chapter IV, e.g., Theorems IV.1.11, IV.1.12, and IV.1.15.]

4.16. Consider control systems on $\mathbb{R}^N$ with dynamics f_h and control set A_h, $0 \leq h \leq 1$, target sets $\mathcal{T}_h$, and value functions

$$v_h(x) = \inf_{\alpha \in \mathcal{A}_h} \left\{ \int_0^{t_x} \ell_h(y_x^{(h)}, \alpha) e^{-\lambda_h s} \, ds + e^{-\lambda_h t_x} g_h(y_x(t_x(\alpha))) \right\},$$

where $y_x^{(h)}(\cdot)$ is the trajectory of the system corresponding to α and $t_x = t_x^{(h)}(\alpha)$ is its first entry time in $\mathcal{T}_h$. Assume $f_h, A_h, \ell_h, \lambda_h$ are continuous with respect to h as in Theorem 1.10, $g_h \in C(\mathbb{R}^N)$ converge uniformly to g_0, and $\mathcal{T}_h \supseteq \mathcal{T}_0$ are compact and $d_H(\mathcal{T}_h, \mathcal{T}_0) \to 0$ as $h \to 0^+$, where d_H is the Hausdorff distance. Assume $\mathcal{T}_0^c$ satisfies (4.30),

(4.54) $|f_0(x,a) - f_0(y,a)| \leq L|x - y|$ for all $x, y \in B(\partial \mathcal{T}, \delta)$, $a \in A$,

for some $\delta > 0$, and $v_0 \in BC(\overline{\mathcal{T}_0^c})$. Prove that $v_h \rightrightarrows v_0$ on compacta of $\mathbb{R}^N$.

[Hint: if (4.54) holds for all $x, y \in \mathbb{R}^N$ it is enough to check the hypotheses of Corollary 4.27, otherwise one has to use Exercise 4.7 instead of Theorem 4.20.]

4.17. Extend Theorem 4.29 to systems (f, A) satisfying the assumptions (A_0)–(A_3) of Chapter III and to the case of unbounded $\partial \mathcal{T}$.

4.18. Extend Corollary 4.32 to the case that Ω is unbounded and u_ε is a classical solution of

$$\begin{cases} -a_{ij}^{(\varepsilon)} u_{x_i x_j} + u + H(x, Du) = 0 & \text{in } \Omega, \\ u = g_\varepsilon & \text{on } \partial \Omega, \\ 0 \leq \liminf_{|x| \to \infty} u(x) \leq \limsup_{|x| \to \infty} u(x) \leq 1, \end{cases}$$

where the matrices $(a_{ij}^{(\varepsilon)}(x))$ are positive semidefinite for all $x \in \Omega$ and $\varepsilon > 0$, $a_{ij}^{(\varepsilon)} \rightrightarrows 0$ on compacta, and $g_\varepsilon \rightrightarrows 0$ on compacta of $\partial \Omega$, $0 \leq g_\varepsilon \leq 1$.

4.19. Consider the targets $\mathcal{T}_h$ introduced in the proof of Proposition 4.33. Find systems (f_h, A_h) such that the corresponding value functions v^h are Lipschitz continuous and satisfy (4.49).

4.20. Under the assumptions of Corollary 4.30 prove that

$$v_*(x) = \liminf_{\substack{y \to x \\ y \neq x}} v(y)$$

for all x.

5. Bilateral supersolutions

In Chapter II, §5.1, we saw that if the Hamiltonian H is convex with respect to the p variables, then a continuous function u is a viscosity solution of

$$\lambda u + H(x, Du) = 0 \quad \text{in } \Omega, \tag{5.1}$$

where $\Omega \subseteq \mathbb{R}^N$ is open, if and only if it is a bilateral supersolution, that is,

$$\lambda u(x) + H(x, p) = 0 \quad \text{for all } x \in \Omega \text{ and } p \in D^-u(x). \tag{5.2}$$

In Chapter III, §2.3, we also saw that this formulation is natural for value functions in view of the Backward Dynamic Programming Principle.

In this section we exploit the fact that the notion of bilateral supersolution extends naturally to lower semicontinuous functions without taking semicontinuous envelopes, and prove uniqueness results in this framework with appropriate boundary conditions.

DEFINITION 5.1. A function $u \in LSC(\Omega)$ is a *bilateral (non-continuous viscosity) supersolution* of (5.1) if (5.2) holds, or, equivalently, for any $\varphi \in C^1(\Omega)$ and $x \in \Omega$ such that $u - \varphi$ has a local minimum at x,

$$\lambda u(x) + H(x, D\varphi(x)) = 0,$$

that is, equivalently, if u is a supersolution of both (5.1) and

$$-\lambda u - H(x, Du) = 0 \quad \text{in } \Omega \tag{5.3}$$

(Definition 1.1). ◁

In the first subsection we consider the Dirichlet boundary value problem, with lower semicontinuous boundary data satisfying a sort of compatibility condition. This theory can be compared with those of §§3.2 and 4.3 on time-optimal control. Here we consider nonconstant running and terminal costs and the target $\mathcal{T} = \Omega^c$ is an arbitrary closed set as in §3.2, without the assumption $\mathcal{T} = \overline{\text{int }\mathcal{T}}$ of §4.3. In the second subsection we consider the initial value problem for an evolutive HJB equation with lower semicontinuous initial data. This corresponds to finite horizon problems with lower semicontinuous terminal cost, and this generality includes problems with a constraint on the endpoint of the trajectories by assigning a large cost to trajectories violating the constraint. We also prove a result on the penalization of the constraint. This allows to approximate the constrained problem with unconstrained ones fitting into the theory of necessary and sufficient conditions of optimality given in §3.4 of Chapter III and satisfying the Extended Pontryagin Maximum Principle.

5.1. Problems with exit times and general targets

We consider the boundary value problem

$$\begin{cases} u + H(x, Du) = 0 & \text{in } \Omega, \\ u = g & \text{on } \partial\Omega, \end{cases} \tag{5.4}$$

with the Hamiltonian of the form

(5.5) $$H(x,p) := \sup_{a \in A} \{ -f(x,a) \cdot p - \ell(x,a) \},$$

associated with the problem of minimizing the cost functional

$$J(x, \alpha) := \begin{cases} \int_0^{t_x(\alpha)} \ell(y_x, \alpha) e^{-t} dt + e^{-t_x(\alpha)} g(y_x(t_x(\alpha))) & \text{if } t_x(\alpha) < +\infty, \\ \int_0^{+\infty} \ell(y_x, \alpha) e^{-t} dt & \text{if } t_x(\alpha) = +\infty, \end{cases}$$

over $\mathcal{A} := \{\alpha : [0, +\infty[\to A, \ \alpha \text{ measurable}\}$, where $t_x(\alpha)$ is the first exit time from the open set Ω of the trajectory $y_x(\cdot, \alpha)$ of the system

(S) $$\begin{cases} y' = f(y, \alpha), \\ y(0) = x. \end{cases}$$

We assume that

(5.6) $$\begin{cases} f : \mathbb{R}^N \times A \longrightarrow \mathbb{R}^N \text{ is continuous,} \\ |f(x,a) - f(z,a)| \leq L|x - z|, \\ |f(x,a)| \leq k(1 + |x|), \quad \text{for all } x, z, a; \end{cases}$$

(5.7) $$\begin{cases} \ell : \mathbb{R}^N \times A \longrightarrow \mathbb{R} \text{ is continuous,} \\ |\ell(x,a) - \ell(z,a)| \leq L|x - z|, \\ |\ell(x,a)| \leq M, \quad \text{for all } x, z, a; \end{cases}$$

(5.8) $$\begin{cases} g \in BLSC(\mathbb{R}^N) \text{ satisfies a global principle of suboptimality,} \\ \text{i.e., for all } x \in \mathbb{R}^N \text{ and } t > 0, \\ g(x) \leq \inf_{\alpha \in \mathcal{A}} \left\{ \int_0^t \ell(y_x, \alpha) e^{-s} ds + e^{-t} g(y_x(t, \alpha)) \right\} \end{cases}$$

This assumption should be compared with the compatibility condition on the terminal cost of Definition IV.3.5. A trivial example satisfying (5.8) for any system (f, A) and for $\ell \geq 0$ is $g \equiv 0$. In particular, the discounted version of the minimum time problem, that is, $\ell \equiv 1$ and $g \equiv 0$ (see Section IV.2), satisfies (5.7) and (5.8).

We interpret the Dirichlet boundary value problem (5.4) in the following sense.

DEFINITION 5.2. Given $g \in LSC(\mathbb{R}^N)$, a function $u \in LSC(\mathbb{R}^N)$ is a bilateral supersolution of (5.4) if it is a bilateral supersolution of the HJB equation in Ω, $u(x) = g(x)$ for all $x \in \Omega^c$, and it is a supersolution of

(5.9) $$-u - H(x, Du) \geq 0 \quad \text{in } \mathbb{R}^N. \quad \triangleleft$$

If g satisfies (5.8) the definition can be rephrased as follows. A function $u \in LSC(\overline{\Omega})$ which is a bilateral supersolution of the HJB equation in Ω, and such that

$u = g$ on $\partial\Omega$, can be extended to a bilateral supersolution of (5.4) by defining $u = g$ on $(\overline{\Omega})^c$ if and only if this extension satisfies

(5.10) $\qquad u + H(x,p) \leq 0 \qquad$ for all $x \in \partial\Omega$ and $p \in D^-u(x)$.

In fact the requirement at points $x \notin \overline{\Omega}$ is automatically fulfilled by assumption (5.8) (see Exercise 5.1).

PROPOSITION 5.3. *Assume* (5.6), (5.7). (5.8) *and consider the value function*

$$v(x) := \inf_{\alpha \in \mathcal{A}} J(x, \alpha).$$

Then its lower semicontinuous envelope v_* is a bilateral supersolution of (5.4).

PROOF. We know from Theorem 2.6 that v_* is a supersolution of the equation. The following Backward Dynamic Programming Principle implies that v_* is a supersolution of (5.9) by the arguments of the proofs of Corollary III.2.28 and Theorem 2.6. Finally, $v_* \leq v = g$ on Ω^c by definition, and $v_* \geq g$ on $\mathbb{R}^N$ because $v \geq g$ by (5.8) and g is l.s.c. ◀

LEMMA 5.4 (BACKWARD DPP). *Under the assumptions of Proposition 5.3,*

$$v(x) \geq v(y_x(-t, \alpha))e^t - \int_0^t \ell(y_x(-s), \alpha(-s))\, e^s\, ds$$

for all $x \in \mathbb{R}^N$, $t > 0$ and $\alpha : \,]-\infty, 0] \to A$ measurable.

PROOF. We claim that for all $x \in \mathbb{R}^N$, $t > 0$, $\alpha \in \mathcal{A}$

(5.11) $\qquad v(x) \leq \int_0^t \ell(y_x, \alpha)\, e^{-s}\, ds + e^{-t} v(y_x(t, \alpha))$.

In fact the usual DPP gives

(5.12) $\qquad v(x) \leq \int_0^{t \wedge t_x(\alpha)} \ell(y_x, \alpha)\, e^{-s}\, ds + e^{-t \wedge t_x(\alpha)} v(z)$,

where $z = y_x(t \wedge t_x(\alpha), \alpha)$. This proves the claim if $t \leq t_x(\alpha)$.

If $t > t_x(\alpha)$, we use that $v(z) = g(z)$, the principle of suboptimality in (5.8) at the point z, and the inequality $g \leq v$ (obvious from (5.8)) to get (5.11). Now the conclusion is reached exactly as in the proof of the Backward DPP Proposition III.2.25. ◀

The main result of this section is the following.

THEOREM 5.5. *Assume* (5.6), (5.7), *and* (5.8). *Then v_* is the minimal bounded supersolution of* (5.4) *and the maximal bounded supersolution of* (5.9) *such that $u \leq g$ on Ω^c. In particular, v_* is the unique bounded bilateral supersolution of* (5.4).

5. BILATERAL SUPERSOLUTIONS

The first statement of the theorem is already known from Section 3 in the case of time-optimal control, $\ell \equiv 1$, $g \equiv 0$, because v_* is the e-solution of (5.4), but the proof we give here is independent and completely different.

In the proof we need a one-parameter family of inf-convolution type approximations of a lower semicontinuous function.

LEMMA 5.6. *Let* $u \in BLSC(\mathbb{R}^N)$, $\varepsilon, C > 0$, *and define for* $(x,t) \in \mathbb{R}^{N+1}$

$$(5.13) \qquad u_\varepsilon(x,t) := \inf_{y \in \mathbb{R}^N} \{ u(y) + e^{-Ct}|x-y|^2/\varepsilon^2 \}.$$

Then

(i) u_ε *is locally Lipschitz continuous;*

(ii) *for all* $(x,t) \in \mathbb{R}^{N+1}$, $u_\varepsilon(x,t) \nearrow u(x)$ *as* $\varepsilon \to 0^+$;

(iii) $M_\varepsilon(x,t) := \arg\min_{y \in \mathbb{R}^N} \{ u(y) + e^{-Ct}|x-y|^2/\varepsilon^2 \} \neq \emptyset$ *for all* x, ε;

(iv) *for any* $y_\varepsilon \in M_\varepsilon(x,t)$

$$(5.14) \qquad |x - y_\varepsilon| \leq \varepsilon\, e^{Ct/2}(2\|u\|_\infty)^{1/2};$$

(v) $D^- u_\varepsilon(x,t) = \emptyset$ *if* $M_\varepsilon(x,t)$ *is not a singleton; if* $M_\varepsilon(x,t) = \{y_\varepsilon\}$, *then*

$$D^- u_\varepsilon(x,t) = \{ (p_\varepsilon, -C\, e^{-Ct}|x - y_\varepsilon|^2/\varepsilon^2) \}$$

with $p_\varepsilon := 2 e^{-Ct}(x - y_\varepsilon)/\varepsilon^2$, *and in this case* $p_\varepsilon \in D^- u(y_\varepsilon)$.

PROOF. (i) By the argument of Lemma II.4.11 (a), $u_\varepsilon(\,\cdot\,,t)$ is semiconcave, and the local Lipschitz continuity with respect to t is easily checked directly.

(ii) The proof is essentially the same as that of Lemma II.4.11 (b).

(iii–iv) As in the proof of Lemma II.4.12 it is easy to see that the search for the inf in (5.13) can be restricted to points $y = y_\varepsilon$ satisfying (5.14), so the minimum is attained by Weierstrass' theorem.

(v) We write $u_\varepsilon(x,t) = \min_{y \in B} g(x,t,y)$, with $B = \bar{B}(x, \varepsilon e^{Ct/2}(2\|u\|_\infty)^{1/2})$ and $g(x,t,y) := u(y) + e^{-Ct}|x-y|^2/\varepsilon^2$, and apply the results on marginal functions of Chapter II, in particular Proposition II.2.13. The required assumptions are satisfied because

$$D_x g(x,t,y) = 2e^{-Ct}(x-y)/\varepsilon^2, \qquad D_t g(x,t,y) = -Ce^{-Ct}|x-y|^2/\varepsilon^2,$$

and a straightforward calculation gives

$$g(x+h, t+s, y) - g(x,t,y) - D_x g(x,t,y) \cdot h - D_t g(x,t,y) s$$
$$= \frac{e^{-Ct}}{\varepsilon^2}\left(|h|^2 + (e^{-Cs} - 1)|x + h - y|^2 + Cs|x-y|^2\right) \leq C_1(|h|^2 + s^2)$$

for all h and s small and a constant C_1 independent of $y \in B$, so g is differentiable with respect to (x,t) uniformly in y. This gives the formula for $D^- u_\varepsilon(x,t)$. The last statement is obtained easily as in the proof of Lemma II.4.12 (iii). ◀

PROOF OF THEOREM 5.5. We show that if $u \in BLSC(\mathbb{R}^N)$, $w \in BLSC(\overline{\Omega})$ are, respectively, a supersolution of (5.9) and of

(5.15) $$w + H(x, Dw) \geq 0 \quad \text{in } \Omega,$$

such that $u \leq w$ on $\partial \Omega$, then $u \leq w$ in Ω. All the conclusions then follow from Proposition 5.3, since v is bounded.

First note that the Hamiltonian satisfies

(5.16) $$|H(x,p) - H(y,p)| \leq L|x-y|(1+|p|).$$

Consider the approximation u_ε of u defined by (5.13) with a constant C to be chosen appropriately. We claim that u_ε is a subsolution of

(5.17) $$U_t + U + H(x, D_x U) \leq Lm\varepsilon e^{Ct/2} \quad \text{in } \mathbb{R}^{N+1},$$

where $m := (2\|u\|_\infty)^{1/2}$. Since u_ε is locally lipschitzean and $H(x, \cdot)$ is convex, by Proposition II.5.1 it is enough to check (5.17) at all points of differentiability of u_ε. At such a point (x, t), Du_ε is the single element of $D^- u_\varepsilon$ computed in Lemma 5.6 (v) and

$$u_\varepsilon(x, t) = u(y_\varepsilon) + e^{-Ct} \frac{|x - y_\varepsilon|^2}{\varepsilon^2}.$$

By (5.9)

(5.18) $$u(y_\varepsilon) + H(y_\varepsilon, p_\varepsilon) \leq 0$$

because $p_\varepsilon \in D^- u(y_\varepsilon)$ (see Lemma 5.6 (v)). Then the left-hand side of (5.17) for $U = u_\varepsilon$ is

$$(-C+1) e^{-Ct} \frac{|x - y_\varepsilon|^2}{\varepsilon^2} + u(y_\varepsilon) + H(x, p_\varepsilon)$$

$$\leq (-C + 1 + 2L) e^{-Ct} \frac{|x - y_\varepsilon|^2}{\varepsilon^2} + L|x - y_\varepsilon| \leq Lm\varepsilon \, e^{Ct/2},$$

where in the first inequality we have used (5.16), (5.18), and to get the last inequality we have chosen $C = 2L + 1$ and used (5.14). This proves the claim.

Next define $W_\varepsilon(x,t) := w(x) + \frac{2Lm\varepsilon}{C} e^{Ct/2}$, which is a supersolution of

$$U_t + U + H(x, D_x U) \geq Lm\varepsilon \, e^{Ct/2} \quad \text{in } \Omega \times \mathbb{R}$$

by (5.15). Choose t_ε large enough to have $W_\varepsilon(x, t_\varepsilon) \geq \|u\|_\infty \geq u_\varepsilon(x, t_\varepsilon)$ (this is possible if $m > 0$; if $m = 0$, then $u \equiv 0$ and the conclusion is reached by a standard comparison theorem for (5.4)). Then $u_\varepsilon \leq W_\varepsilon$ on $\partial(\Omega \times \,]-\infty, t_\varepsilon[)$ because for $x \in \partial \Omega$

$$u_\varepsilon(x,t) \leq u(x) \leq w(x) \leq W_\varepsilon(x,t)$$

for all $t \in \mathbb{R}$. Since u_ε and W_ε are bounded and they are, respectively, a subsolution and a supersolution of the same Hamilton-Jacobi equation in $\Omega \times \,]-\infty, t_\varepsilon[$, we can apply the comparison Theorem 1.3 and get $u_\varepsilon \leq W_\varepsilon$ for $x \in \Omega$ and $t < t_\varepsilon$. Now we fix any such t and let $\varepsilon \to 0^+$ to get $u \leq w$ by Lemma 5.6 (ii). ◀

5. BILATERAL SUPERSOLUTIONS

REMARK 5.7. In the next subsection we show that, for the Cauchy problem, the unusual boundary condition (5.10) is equivalent to a more classical-looking one, see Definition 5.14. For the Dirichlet problem one might hope to replace (5.10) with the condition

(5.19) $$\liminf_{\mathcal{R} \ni y \to x} u(y) = g(x) \quad \text{for all } x \in \partial\Omega,$$

where $\mathcal{R} \subseteq \Omega$ is the set of starting points from which the system can reach $\partial\Omega$. However, the following example shows that this boundary condition does not ensure uniqueness. Consider the equation $u + |\partial u/\partial x_1| = 1$ in $\Omega = \mathbb{R}^N \smallsetminus \{0\}$ and the terminal cost $g \equiv 0$. Then $v(x) = 1 - \exp(-|x_1|)$ on the x_1 axis, and $v \equiv 1$ elsewhere (so $v = v_*$). A different bilateral solution u of this HJB equation satisfying (5.19) is obtained by setting $u = v$ except at the points of the x_1 axis where $x_1 < 0$, and $u = 1$ at such points. ◁

REMARK 5.8. It is easy to modify the proof of Theorem 5.5 to get the following comparison principle. If (5.6) and (5.7) hold, $\delta > 0$, and $u \in BLSC(B(\overline{\Omega}, \delta))$, $w \in BLSC(\overline{\Omega})$ are supersolutions, respectively, of $-u - H(x, Du) \geq 0$ in $B(\overline{\Omega}, \delta)$ and of (5.15), such that $u \leq w$ on $\partial\Omega$, then $u \leq w$ in Ω. In fact, it is enough to modify the approximating u_ε by taking the infimum over $B(\overline{\Omega}, \delta)$ instead of $\mathbb{R}^N$ in their definition (5.13), then observe that $u_\varepsilon \nearrow u(x)$ as $\varepsilon \to 0^+$ for all $(x, t) \in \overline{\Omega} \times \mathbb{R}$ and check that u_ε satisfies the inequality (5.17) in $\Omega \times \mathbb{R}$ for all $\varepsilon < e^{-Ct/2}\delta/m$. ◁

REMARK 5.9. If the running cost ℓ does not depend on a, A is compact and the system is convex (i.e., $f(x, A)$ is convex for all x, as it happens, for instance, with relaxed controls, see §III.2.2), then $v = v_*$, see Exercises 5.9, 5.10. Therefore we can use Theorem 5.5 to prove verification theorems such as Corollaries 2.18 and 2.19 of Chapter III, by taking bounded supersolutions of (5.4) as falsification functions and bounded supersolutions u of (5.9), $u \leq g$ on $\partial\Omega$, as verification functions. ◁

Next we discuss the connections among the bilateral supersolutions and the other notions of non-continuous solution introduced in this chapter. We begin with the envelope solutions of Section 2.

COROLLARY 5.10. *Assume* (5.6), (5.7), (5.8) *and g continuous at each point of $\partial\Omega$. Then v_* is the e-solution of* (5.4).

PROOF. By (5.8) and Theorem 2.6 g^* is a subsolution of the HJB equation in (5.4). By the continuity of g on $\partial\Omega$, g^* is a subsolution of the Dirichlet problem (5.4) which is continuous at the points of $\partial\Omega$, so the assumptions of Theorem 3.4 (i) are verified. Then the minimal bounded supersolution of (5.4) is the unique e-solution of (5.4) and we conclude by Theorem 5.5. ◂

We omit the proof of the next result, which compares supersolutions of (5.9) with subsolutions of the HJB equation and with v-subsolutions of the boundary conditions defined in Section 4.

THEOREM 5.11. *Assume* (5.5), (5.6), (5.7), $\lambda \geq 0$, *and* $\Omega \subseteq \mathbb{R}^N$ *open. Then*
(i) *for any* $u \in LSC(\Omega)$ *supersolution of* (5.3), u^* *is a subsolution of* (5.1);
(ii) *for any* $u \in USC(\Omega)$ *subsolution of* (5.1), u_* *is a supersolution of* (5.3);
(iii) *for any solution of* (5.1) (*Definition* 2.2) *such that* $(u^*)_* = u_*$, u_* *is a bilateral supersolution of* (5.1).

Assume, in addition, that g is continuous at each point of $\partial\Omega$. *Then*
(iv) *for any* $u \in LSC(\mathbb{R}^N)$ *supersolution of* (5.9), u^* *is a v-subsolution of* (5.4) (*see Definition* 4.1);
(v) *a bilateral supersolution of* (5.4) *is a v-solution of* (5.4).

Note that (iii) and (v) are immediate consequences of (ii) and (iv), respectively. Some hints for the proofs of the other statements are given in Exercises 5.5 and 5.6. Since the bilateral supersolution of (5.4) is unique, while v-solutions are not unique, the implications in (iv) and (v) cannot be inverted. Here is an explicit example, which is essentially Example 4.18.

EXAMPLE 5.12. The bilateral supersolution of
$$\begin{cases} u + |u'| = 1 & \text{in } \mathbb{R} \setminus \{0\}, \\ u(0) = 0, \end{cases}$$
is $1 - e^{-|x|}$, whereas $w(x) = 1$ for $x \neq 0$, $w(0) = 0$, is a lower semicontinuous v-solution, but it does not satisfy $-u - |u'| + 1 \geq 0$ at 0. ◁

We end this subsection with a stability result whose easy proof is left to the reader.

PROPOSITION 5.13. *Assume* (5.6), (5.7), (5.8), *and let H_n be a sequence of continuous Hamiltonians converging to H uniformly on compacta of* $\mathbb{R}^{2N}$. *If u_n is a bilateral supersolution of* (5.4) *with H replaced by H_n, the sequence u_n is locally uniformly bounded and in a neighborhood of* $\partial\Omega$ $u_n \geq g$ *for all n, then the lower weak limit* $\liminf_n {}_* u_n(x)$ *is the bilateral supersolution of the limit problem* (5.4).

5.2. Finite horizon problems with constraints on the endpoint of the trajectories

In this subsection we consider the initial value problem

(5.20) $$\begin{cases} u_t + \lambda u + H(x, Du) = 0 & \text{in } \mathbb{R}^N \times \,]0, T[, \\ u(x, 0) = g(x) & \text{in } \mathbb{R}^N, \end{cases}$$

where $u_t = \partial u / \partial t$, $Du = D_x u$, $\lambda \geq 0$, $T \in \,]0, +\infty]$, the Hamiltonian is given by (5.5) and g is lower semicontinuous. The associated optimal control problem is the minimization of the finite horizon cost functional

(5.21) $$J(x, t, \alpha) := \int_0^t \ell(y_x, \alpha) e^{-\lambda s} \, ds + e^{-\lambda t} g(y_x(t, \alpha))$$

5. BILATERAL SUPERSOLUTIONS

over $\mathcal{A}$, with the same notations of the previous subsection.

In §3 of Chapter III we studied this problem for continuous terminal cost g, whereas here we assume g merely lower semicontinuous. This allows us to consider problems with the constraint that the endpoint of the trajectory $y_x(t, \alpha)$ belongs to a given closed target set $\mathcal{T}$: given a terminal cost defined on $\mathcal{T}$, $\widetilde{g} \in BLSC(\mathcal{T})$, we want to minimize (5.21), with g replaced by $\widetilde{g}$, over controls $\alpha \in \mathcal{A}$ such that

(5.22) $$y_x(t, \alpha) \in \mathcal{T}.$$

This constrained minimization problem can be rewritten as a free minimization problem, provided we are interested only in horizons $t \leq T$ for a given $T < +\infty$. In fact we define

(5.23)
$$g(x) := \begin{cases} \widetilde{g}(x) & \text{if } x \in \mathcal{T}, \\ e^{\lambda T} C & \text{if } x \notin \mathcal{T}, \end{cases}$$

$$C := \begin{cases} \sup_{\mathcal{T}} |\widetilde{g}| + 2T \sup |\ell| + 1 & \text{if } \lambda = 0, \\ \sup_{\mathcal{T}} |\widetilde{g}| + \dfrac{1 - e^{-\lambda T}}{\lambda} 2 \sup |\ell| + 1 & \text{if } \lambda > 0, \end{cases}$$

and consider the problem of minimizing (5.21) with such g over all $\mathcal{A}$. Note that $g \in BLSC(\mathbb{R}^N)$, because $\mathcal{T}$ is closed, and $g \leq C$. Clearly the value function of this problem

$$v(x, t) := \inf_{\alpha \in \mathcal{A}} J(x, t, \alpha)$$

coincides with the value function of the constrained problem for all (x, t) such that there exists an admissible control, that is, a control α satisfying the terminal constraint (5.22). On the other hand, it is easy to see that $v(x, t) > C'$ if and only if an admissible control does not exist, where

$$C' = \begin{cases} \sup_{\mathcal{T}} |\widetilde{g}| + T \sup |\ell| & \text{if } \lambda = 0, \\ \sup_{\mathcal{T}} |\widetilde{g}| + \dfrac{1 - e^{-\lambda T}}{\lambda} \sup |\ell| & \text{if } \lambda > 0. \end{cases}$$

In other words, the big constants C and C' replace $+\infty$ when the terminal constraint cannot be fulfilled by any control α.

In view of the poor regularity of the initial data g we do not expect solutions of (5.20) to be better than lower semicontinuous. Since the HJB equation in (5.20) can be rewritten in the form (5.1), the definition 5.1 of bilateral supersolution applies. The initial condition will be interpreted as follows.

DEFINITION 5.14. We say that $u \in LSC(\mathbb{R}^N \times [0, T[)$ is a bilateral supersolution of (5.20) if it is a bilateral supersolution of the HJB equation in $\mathbb{R}^N \times]0, T[$, it coincides with g at $t = 0$, and for any $x \in \mathbb{R}^N$ there exist $x_n \to x$, $t_n \to 0^+$, $t_n > 0$ for all n, such that $g(x) = \lim_n u(x_n, t_n)$; in other words,

(5.24) $$g(x) = u(x, 0) = (u_{|\mathbb{R}^N \times]0,T[})_*(x, 0). \quad \triangleleft$$

We assume for simplicity that $\ell \equiv 0$ so that the Hamiltonian becomes

(5.25) $$H(x,p) := \sup_{a \in A} \{-f(x,a) \cdot p\}.$$

We know from §3 of Chapter III that we can always reduce to this case if the running cost ℓ satisfies (5.7) and $\lambda = 0$. For the general case see Exercise 5.11.

Under this assumption, if we find the value function

(5.26) $$v_\lambda(x,t) := \inf_{\alpha \in \mathcal{A}} e^{-\lambda t} g(y_x(t,\alpha))$$

for a given value of λ, we can easily get it for all λ because

$$v_\lambda(x,t) = e^{-\lambda t} v_0(x,t).$$

Moreover, we can consider a very general g satisfying merely

(5.27) $\qquad g : \mathbb{R}^N \longrightarrow \mathbb{R} \cup \{-\infty, +\infty\} \quad$ lower semicontinuous,

and transform the problem into an equivalent one with terminal cost in $BLSC(\mathbb{R}^N)$. It is enough to take a strictly increasing function ψ mapping the extended real line into a compact interval, for instance,

(5.28) $\qquad \psi(r) = \dfrac{2}{\pi} \arctan r, \quad \psi(-\infty) = -1, \quad \psi(+\infty) = 1.$

If we take $\psi \circ g \in BLSC(\mathbb{R}^N)$ as a new terminal cost and, for instance, $\lambda = 0$, the new value function is

$$V(x,t) := \inf_{\alpha \in \mathcal{A}} \psi(g(y_x(t,\alpha))) = \psi(v(x,t)),$$

where $v(x,t) = \inf_{\alpha \in \mathcal{A}} g(y_x(t,\alpha))$ is the original value function. Thus, once we know V, we can immediately recover v by taking ψ^{-1}:

$$v(x,t) = \begin{cases} -\infty & \text{if } V(x,t) = -1, \\ \tan(\tfrac{1}{2}\pi V(x,t)) & \text{if } |V(x,t)| \neq 1, \\ +\infty & \text{if } V(x,t) = 1. \end{cases}$$

Therefore the following theory is developed just for a bounded terminal cost, without loss of generality. Note that with a g as general as in (5.27) we can cover the case of the constraint (5.22) on the endpoint of the trajectories even for an unbounded $\tilde{g} \in LSC(\mathcal{T})$, by extending it to $g = +\infty$ on $\mathcal{T}^c$.

PROPOSITION 5.15. *Assume* (5.6), (5.25), $\lambda \geq 0$, $g \in LSC(\mathbb{R}^N)$, *and consider the value function* $v = v_\lambda$ *given by* (5.26). *Then its lower semicontinuous envelope* v_* *is a bilateral supersolution of* (5.20) *with* $T = +\infty$.

5. BILATERAL SUPERSOLUTIONS

PROOF. By standard arguments (i.e., the proofs of Proposition III.3.5 and Theorem 2.6) v_* is a supersolution of the HJB equation. The Backward Dynamic Programming Principle is

$$v(x,t) \geq e^{\lambda t} v(y_x(-\tau, \alpha), t + \tau)$$

for all $x \in \mathbb{R}^N$, $t, \tau > 0$, $\alpha :]-\infty, 0] \to A$ measurable, see Proposition III.3.20. Its proof under the current assumptions is easily obtained as in Chapter III. We deduce from it that v_* is a supersolution of

$$-u_t - \lambda u - H(x, Du) \geq 0 \quad \text{in } \mathbb{R}^N \times]0, +\infty[$$

by standard arguments.

It remains to check the initial condition. By definition $v(x, 0) = g(x)$, so

(5.29) $$v_*(x, 0) \leq g(x).$$

We claim that for any $x_n \to x$, $t_n \to 0^+$,

(5.30) $$\liminf_n v(x_n, t_n) \geq g(x),$$

and so the equality holds in (5.29). Moreover, it is easy to construct a sequence with $t_n > 0$ for all n such that the equality holds in (5.30). It is enough to fix $z \in \mathbb{R}^N$, $\alpha \in \mathcal{A}$ and $t > 0$ such that $y_z(t, \alpha) = x$ and choose $x_n := y_z(t - 1/n, \alpha)$. Then there is a trajectory starting from x_n and reaching x in time $1/n$, thus $v(x_n, 1/n) \leq g(x)e^{-\lambda/n}$ and by (5.30) we get $\lim_n v(x_n, 1/n) = g(x)$, which completes the proof of (5.24).

To prove the preceding claim we choose $\alpha_n \in \mathcal{A}$ such that

(5.31) $$v(x_n, t_n) \geq e^{-\lambda t_n} g(y_{x_n}(t_n, \alpha_n)) - \frac{1}{n}.$$

By the assumption (5.6) and the standard estimates on the trajectories, there is a constant c such that

$$|y_{x_n}(t_n, \alpha_n) - x_n| \leq c t_n \quad \text{for all } n$$

and so $y_{x_n}(t_n, \alpha_n) \to x$ as $n \to \infty$. Therefore, by taking the liminf of both sides of (5.31) and using the semicontinuity of g, we obtain (5.30). ◀

The next result is the main one of this subsection. Some of its assumptions are weakened in a subsequent remark.

THEOREM 5.16. *Assume* (5.6), (5.25), $\lambda \geq 0$, $T \in]0, +\infty]$, *and* $g \in BLSC(\mathbb{R}^N)$. *Then* v_* *is the minimal bounded supersolution of* (5.20) *and the maximal supersolution* $u \in BLSC(\mathbb{R}^N \times [0, T[)$ *of*

(5.32) $$-u_t - \lambda u - H(x, Du) \geq 0$$

in $\mathbb{R}^N \times]0, T[$, *such that* $u(\cdot, 0) \leq g$ *and*

(5.33) $$u = (u_{|\mathbb{R}^N \times]0, T[})_*.$$

In particular, v_* *is the unique bounded bilateral supersolution of* (5.20).

We remark that the regularity assumption on u at time $t = 0$ (5.33) cannot be dropped. A trivial counterexample when $\lambda = 0$ is $w := v_* + 1$ if $t > 0$, $w := g$ if $t = 0$.

PROOF OF THEOREM 5.16. We split the proof in two steps. The goal of Step 1 is to extend a solution of (5.32) in $\mathbb{R}^N \times]0,T[$ satisfying (5.33) to a solution of (5.32) in $\mathbb{R}^N \times]-1,T[$. Then what remains to prove is a comparison statement very similar to Theorem 5.5, which is tackled in Step 2.

Step 1. The argument of this step is an extension to semicontinuous functions of Lemma II.2.10.

Let $u \in BLSC(\mathbb{R}^N \times [0,T[)$ satisfy (5.33). Define

$$U(x,t) := \begin{cases} u(x,t) & \text{if } t \geq 0, \\ e^{-\lambda t} \sup_{x \in \mathbb{R}^N} u(x,0) & \text{if } t < 0, \end{cases}$$

and note that $U \in BLSC(\mathbb{R}^N \times]-1,T[)$. Clearly U solves (5.32) for $t < 0$. To prove that U solves (5.32) at $t = 0$ take $\varphi \in C^1(\mathbb{R}^{N+1})$ such that $U - \varphi$ has a strict minimum in $\bar{B}(\bar{x},r) \times [-r,r]$ at $(\bar{x},0)$. By (5.33) there are sequences $x_n \to \bar{x}$, $t_n \to 0$, $t_n > 0$ for all n such that $u(x_n,t_n) \to u(\bar{x},0)$. Define for $t > 0$

$$\psi_n(x,t) = u(x,t) - (\varphi(x,t) - t_n^2/t),$$

and denote by (z_n,s_n) a minimum point of ψ_n in $\bar{B}(\bar{x},r) \times]0,r]$. We claim that there is a subsequence, which we do not relabel, such that $(z_n,s_n) \to (\bar{x},0)$ and $u(z_n,s_n) \to u(\bar{x},0) = U(\bar{x},0)$. Then we can use (5.32) for $t > 0$ to get

$$-\varphi_t - \frac{t_n^2}{s_n^2} - \lambda u - H(z_n,D\varphi) \geq 0,$$

where all the functions are evaluated at (z_n,s_n). Now we can erase the negative term $-t_n^2/s_n^2$ and let $n \to +\infty$ to obtain

$$-\varphi_t(\bar{x},0) - \lambda U(\bar{x},0) - H(\bar{x},D\varphi(\bar{x},0)) \geq 0,$$

which is the desired inequality. To prove the claim assume $(z_n,s_n) \to (\tilde{x},\tilde{t})$ and $\psi_n(z_n,s_n) \to b$. Since $\psi_n \geq u - \varphi$,

(5.34) $$b \geq \liminf_n (u-\varphi)(z_n,s_n) \geq (u-\varphi)(\tilde{x},\tilde{t}).$$

On the other hand, we can pass to the limit in

$$\psi_n(z_n,s_n) \leq \psi_n(x_n,t_n) = (u-\varphi)(x_n,t_n) + t_n$$

to get

(5.35) $$b \leq (u-\varphi)(\bar{x},0).$$

Since $(\bar{x},0)$ is a strict minimum point of $u-\varphi$ in $\bar{B}(\bar{x},r)\times[0,r]$, (5.34) and (5.35) give $(\tilde{x},\tilde{t})=(\bar{x},0)$ and $b=(u-\varphi)(\bar{x},0)$. Finally

$$(u-\varphi)(\bar{x},0) \le \liminf_n (u-\varphi)(z_n,s_n)$$

and the first inequality in (5.34) imply that for a subsequence $(u-\varphi)(z_n,s_n) \to b$, and therefore $u(z_n,s_n) \to u(\bar{x},0)$, which completes the proof of the claim.

Step 2. Here we show that if $u \in BLSC(\mathbb{R}^N \times\,]-1,T[)$ solves (5.32) in $\mathbb{R}^N \times\,]-1,T[$, $w \in BLSC(\mathbb{R}^N \times [0,T[)$ is a supersolution of

(5.36) $$w_t + \lambda w + H(x,Dw) \ge 0 \quad \text{in } \mathbb{R}^N \times\,]0,T[$$

and $u(\,\cdot\,,0) \le w(\,\cdot\,,0)$, then $u \le w$ in $\mathbb{R}^N \times [0,T[$. All the conclusions then follow from Step 1 and Proposition 5.15.

The case $T=+\infty$ is covered by the comparison principle in Remark 5.8. This is clear if $\lambda > 0$, while for $\lambda = 0$ we replace u and w with $e^{-t}u$ and $e^{-t}w$, respectively, which satisfy (5.32) and (5.36), respectively, with $\lambda = 1$, by Proposition II.2.7 (which is easily seen to work under the current assumptions).

Now consider the case $T < +\infty$ and restrict ourselves to $\lambda = 0$. For the general case it is enough to replace u and w with $e^{\lambda t}u$ and $e^{\lambda t}w$, respectively. The proof is very similar to that of Theorem 5.5, and actually simpler. The main difference is the replacement of the approximation of Lemma 5.6 with the following one

(5.37) $$u_\varepsilon(x,t) := \inf_{\substack{y \in \mathbb{R}^N \\ -1 < s < T}} \left\{ u(y,s) + e^{-ct}\frac{|x-y|^2}{\varepsilon^2} + \frac{|t-s|^2}{\varepsilon^2} \right\},$$

where we choose $c = 2L$. By the arguments of Lemma 5.6 and Lemma II.4.12 it is easy to see that u_ε is locally lipschitzean, $u_\varepsilon(x,t) \nearrow u(x,t)$ as $\varepsilon \to 0^+$ for all $x \in \mathbb{R}^N$, $0 \le t < T$, and for $\varepsilon < [(T-t) \wedge 1]/m$, $m = (2\|u\|_\infty)^{1/2}$, the inf in (5.37) is attained at some $(y_\varepsilon, s_\varepsilon)$. Moreover, at any point of differentiability of u_ε,

$$Du_\varepsilon(x,t) = \frac{2e^{-ct}(x-y_\varepsilon)}{\varepsilon^2}, \quad \frac{\partial u_\varepsilon}{\partial t}(x,t) = 2\frac{t-s_\varepsilon}{\varepsilon^2} - \frac{ce^{-ct}|x-y_\varepsilon|^2}{\varepsilon^2},$$

and $(Du_\varepsilon(x,t), 2(t-s_\varepsilon)/\varepsilon^2) \in D^-_{(y,s)} u(y_\varepsilon, s_\varepsilon)$.

Then we use (5.32), the inequality $|H(x,p) - H(y,p)| \le L|x-y||p|$, and the choice $c = 2L$ to get

(5.38) $$\frac{\partial u_\varepsilon}{\partial t} + H(x, Du_\varepsilon) \le 0 \quad \text{in } \mathbb{R}^N \times\,]0, T-\varepsilon m[$$

at every point where u_ε is differentiable. Thus (5.38) holds in the viscosity sense as well, by the convexity of $H(x,\,\cdot\,)$ and Proposition II.5.1. Since for all x

$$u_\varepsilon(x,0) \le u(x,0) \le w(x,0),$$

we can use the comparison principle for initial value problems (see Theorem III.3.7 and Exercise 1.5) to get $u_\varepsilon(x,t) \le w(x,t)$ in $\mathbb{R}^N \times [0, T-\varepsilon m[$, and we conclude by letting $\varepsilon \to 0^+$. ◂

REMARK 5.17. By a classical compactness result for relaxed controls the bilateral supersolution of (5.20) v_* is, in fact, the value function v^r corresponding to relaxed controls $\mathcal{A}^r$. The compactness theorem states that, if $A \subseteq \mathbb{R}^M$ is compact, then for any sequence $\mu_n \in \mathcal{A}^r$ there is a subsequence, which we do not relabel, such that μ_n converge weak star on $[0,T]$ to $\mu \in \mathcal{A}^r$ and the corresponding trajectories $y_{x_n}(\,\cdot\,,\mu_n)$ converge uniformly to $y_x(\,\cdot\,,\mu)$ if $x_n \to x$ (see §2.2 of Chapter III for definition and notations and [Wa72] for more details and the proof). In order to show that $v_* = v^r$ it is enough to show that v^r is lower semicontinuous, by the uniqueness Theorem 5.16 and the fact that relaxed controls give rise to the same Hamiltonian H (see §III.2.2). Take $(x_n,t_n) \to (x,t)$ and $\mu_n \in \mathcal{A}^r$ such that $v^r(x_n,t_n) = g(y_{x_n}(t_n,\mu_n))e^{-\lambda t_n}$ (such optimal controls exist by the previous compactness theorem, but ε-optimal controls would do the job anyway). By extracting a subsequence as before and using the semicontinuity of g we get

$$\liminf_n v^r(x_n,t_n) \geq g(y_x(t,\mu))e^{-\lambda t} \geq v^r(x,t),$$

which completes the proof. It can also be shown that for any convex system (i.e., such that $f(x,A)$ is convex for any x) the value function is lower semicontinuous, thus it is the bilateral solution of (5.20) (see Exercise 5.8). ◁

REMARK 5.18. By Theorem 5.16 and Remark 5.17 we can get verification theorems as in §III.2.2, see Remark 5.9 and Exercise 5.14. ◁

REMARK 5.19. Theorem 5.16 admits an extension to the case of a locally lipschitzean system f. The different proof also allows us to compare v_* with locally bounded supersolutions of (5.20) and (5.32). Here is a more precise statement.

Assume $g \in LSC(\mathbb{R}^N)$ is locally bounded, f satisfies (5.6) with the global Lipschitz condition replaced by the following: for all $R > 0$ there is L_R such that for all $x,y \in B(0,R)$ and $a \in A$,

$$|f(x,a) - f(y,a)| \leq L_R|x-y|,$$

and (5.25), $\lambda = 0$, $T \in \,]0,+\infty[$. Then v_* is the unique locally bounded bilateral supersolution of (5.20). This follows from a comparison principle on the cones

$$\mathcal{C}_{x_0} = \{\,(x,t) : 0 < t < T,\ |x - x_0| \leq c(T-t)\,\}$$

for Hamiltonians such that $|H(x,p) - H(x,q)| \leq c|p-q|$ as in the proof of Theorem III.3.15. Here is the local comparison principle: if u is a BLSC supersolution of (5.32) in a neighborhood of $\mathcal{C}_{x_0}$ and $w \in BLSC(\overline{\mathcal{C}}_{x_0})$ is supersolution of $w_t + H(x,Dw) \geq 0$ in $\mathcal{C}_{x_0}$ such that $u(\,\cdot\,,0) \leq w(\,\cdot\,,0)$, then $u \leq w$ in $\mathcal{C}_{x_0}$. This can be proved as in Step 2 of the proof of Theorem 5.16 by means of the local comparison principle between sub- and supersolutions, see Theorem III.3.12 and Exercise 1.6. We leave the details to the interested reader. ◁

Next we give a stability result and discuss some applications.

THEOREM 5.20. *Under the hypotheses of Theorem 5.16 assume* $g_n \in BLSC(\mathbb{R}^N)$, $g_n \nearrow g$ *as* $n \to +\infty$. *Let* u_n *and* u *be, respectively, the bounded bilateral solution of* (5.20) *with initial data* g_n *and* g. *Then* $u_n \nearrow u$ *as* $n \to +\infty$.

PROOF. Extend u and all u_n for $t < 0$ by setting them equal to $e^{-\lambda t} \sup g$. Then all u_n solve (5.32) in $\mathbb{R}^N \times]-\infty, T[$ by Step 1 of the proof of Theorem 5.16. This theorem implies that u_n is a nondecreasing sequence, so we can define $w := \sup_n u_n$. By Lemma 2.18, w is the lower weak limit of $\{u_n\}$, thus $w \in BLSC(\mathbb{R}^N \times]-1, T[)$ is a supersolution of both (5.32) in $\mathbb{R}^N \times]-1, T[$ and (5.36) by the stability result Theorem 1.7. Moreover, $w(x, 0) = g(x)$ by definition. By Step 2 of the proof of Theorem 5.16 there is at most one solution to this problem, and by Step 1 the bilateral supersolution of (5.20) is such a solution. Then $w = u$ and the proof is complete. ◄

A first consequence of this stability theorem is that we can approximate the value function of a finite horizon problem with discontinuous terminal cost g by much smoother value functions corresponding to more regular costs g_n. If we define g_n as the inf-convolution of $g \in BLSC(\mathbb{R}^N)$ corresponding to the parameter $\varepsilon = \varepsilon_n$, with $\varepsilon_n \to 0^+$, we have that $g_n \nearrow g$ and each g_n is semiconcave in $\mathbb{R}^N$ by the proof of Lemma II.4.11. The value function

$$v_n(x, t) = \inf_{\alpha \in \mathcal{A}} e^{-\lambda t} g_n(y_x(t, \alpha))$$

is locally lipschitzean and it is the bilateral supersolution of (5.20) with g replaced by g_n, by Theorem 5.16. By Theorem 5.20 $v_n \nearrow v_*$ as $n \to +\infty$ where v is the value function with cost g, (5.26).

This fact is useful because we can apply to the approximating problems the theory on necessary and sufficient conditions of optimality developed in §3.4 of Chapter III: see Theorem III.3.37 on necessary conditions where the continuity of v_n is enough, and Theorem III.3.38 on sufficient conditions where the Lipschitz continuity of v_n is needed. If, in addition, the system f satisfies

(5.39) $\qquad |f(x - h, a) - 2f(x, a) + f(x + h, a)| \leq M |h|^2$

(e.g., if f is C^2 with respect to x with bounded second derivatives), then each v_n is locally semiconcave by Theorem III.3.26. Therefore we can apply to the approximating problems the necessary and sufficient conditions of optimality in the form of the Minimum Principle of Theorem III.3.39. If instead of (5.39) we assume that $f(\cdot, a)$ is C^1 for all a, then the Extended Pontryagin Maximum Principle, Theorem III.3.44, applies to the approximating problems.

The finite horizon problem with the constraint (5.22) on the endpoint of the trajectories can be approximated in a more explicit way by penalizing smoothly the violation of the constraint. We assume the terminal cost $\widetilde{g}$ is the restriction to the target $\mathcal{T}$ of a semiconcave function $\widetilde{g} : \mathbb{R}^N \to \mathbb{R}$. Then we approximate the discontinuous terminal cost g defined by (5.23) with

(5.40) $\qquad g_n(x) := \min\{\widetilde{g}(x) + n \operatorname{dist}^2(x, \mathcal{T}); e^{\lambda T} C\}.$

Clearly $g_n \nearrow g$, and it is not hard to show that all g_n are semiconcave. Therefore all the previous remarks on the regularity of v_n and on the applicability of the theory of Chapter III still hold.

Exercises

5.1. Prove that a function g satisfying (5.8) is a supersolution of (5.9).

5.2. Prove Theorem 5.5 under the assumptions (A_0)–(A_3) of Chapter III on the system (f, A) instead of (5.6).

5.3. Extend Theorem 5.5 to the equation $H(x, u, Du) = 0$ with

$$H(x, u, p) := \sup_{a \in A}\{-f(x, a) \cdot p - \ell(x, a) + k(x, a)u\},$$

where f satisfies (5.6) (or the assumptions of Exercise 5.2), ℓ satisfies (5.7) except the condition of boundedness which can be replaced by $|\ell(x, a)| \leq M(k(x, a) + 1)$ for all x, a, and $k : \mathbb{R}^N \times A \to \mathbb{R}$ is continuous, bounded above in $\mathcal{K} \times A$ for all compact $\mathcal{K}$, bounded below by a constant $k_0 > 0$, and lipschitzean in x uniformly with respect to a.

5.4. Prove that the uniqueness statement of Theorem 5.5 is false if the equation $u + H(x, Du) = 0$ is replaced by $H(x, Du) = 0$. [Hint: use the example in Remark II.5.11.]

5.5. Prove the statements (i) and (iv) of Theorem 5.11. [Hint: consider the approximation of Lemma 5.6 and use Lemma 2.18. The result can be checked in [Sor93b].]

5.6. Prove statement (ii) of Theorem 5.11. [Hint: use the approximation $u^{(\varepsilon)}(x) = \sup_{y \in \Omega}\{u(y) - |x - y|^\gamma/\varepsilon^\gamma\}$ for a suitable choice of $\gamma > 0$. The result can be checked in [Ba93].]

5.7. State and prove some result on the stability of the bilateral supersolution of (5.4) with respect to vanishing viscosity approximations and to perturbations of the domain, such as Propositions 4.10 and 4.12 for v-solutions.

5.8. Prove that the value function of the finite horizon problem is lower semicontinuous if the running cost ℓ does not depend on a, A is compact and $f(x, A)$ is convex for all x. [Hint: show that $f(x, A) = f^r(x, A^r)$, where "r" indicates relaxed controls, and use a suitable selection lemma, such as Lemma B in the Appendix of [FR75], see also the proof of Corollary 1.4 in Chapter VI.]

5.9. Assume A is compact and ℓ does not depend on a. Prove that the relaxed value function for the problem with exit times of §5.1 is lower semicontinuous. [Hint: see the proof in Remark 5.17.]

5.10. Prove the first statement of Remark 5.9. [Hint: use Exercise 5.9 and the hint to Exercise 5.8.]

5.11. Extend Theorem 5.16 and Remark 5.19 to the case of Hamiltonian H given by (5.5) instead of (5.25) with ℓ satisfying (5.7). [Hint: in Step 1 extend u for $t < 0$ as Ge^{-kt} for suitable large constants G and k. In Step 2 add to w a term of the form $\varepsilon k e^{ht}$ for suitable constants k and h.]

5.12. Give a direct proof that $u \in BLSC(\Omega \times [0,T[)$, $\Omega \subseteq \mathbb{R}^N$ open, solution of (5.32) in $\Omega \times]0,T[$ can be extended to a solution on $\Omega \times]-\infty,T[$ if and only if $u(x,0) = (u_{|\mathbb{R}^N \times]0,T[})_*(x,0)$ for all x such that $D^-_{(x,t)}u(x,0) \neq \emptyset$, or for all x such that $D^-_x u(x,0) \neq \emptyset$.

5.13. Assume (5.25), $\lambda = 0$, $T \in]0,+\infty[$, $g \in LSC(\mathbb{R}^N)$, f satisfies (5.6) with global Lipschitz continuity replaced by local, as in Remark 5.19. Let $u \in LSC(\mathbb{R}^N \times [0,T[)$ be a supersolution of (5.20) (respectively, of (5.32)) such that $u(\,\cdot\,,0) \leq g$ and (5.33) holds). Prove that $u \geq v_*$ (respectively, $\leq$). [Hint: use the change of variables (5.28).]

5.14. Give the verification theorems for the finite horizon problem of §5.2 corresponding to each comparison result of this subsection: Theorem 5.16, Remark 5.19 and Exercises 5.11 and 5.13.

5.15. Extend Theorem 5.20 to bilateral solutions u_n of (5.20) with H replaced by H_n satisfying the assumptions of Theorem 5.16 and $H_n \searrow H$.

5.16. Under the assumption of Theorem 5.16 let $g_n \in BLSC(\mathbb{R}^N)$ be uniformly bounded, $H_n \in C(\mathbb{R}^{2N})$, $H_n \rightrightarrows H$ on compacta. Let u_n be bilateral supersolutions of (5.20) with H,g replaced by H_n, g_n and set $u = \liminf_n {}_*u_n$. Prove that u is the bilateral supersolution of (5.20) with $g := u(\,\cdot\,,0)$.

5.17. Extend the results of §5.2 to systems $f(t,y,a)$ depending continuously on t.

6. Bibliographical notes

The possibility of extending the definition of viscosity sub- and supersolutions and the comparison principles to semicontinuous functions was remarked in [CEL84, CIL87, I87a], see in particular [I88b].

The "half-relaxed" semilimits that we simply call weak limits in Definition 1.4 are a special case of the general notion of Γ-limit in the Calculus of Variations (see, e.g., [DM93]). They were introduced in the theory of viscosity solutions by Barles and Perthame who proved the stability results Theorem 1.7 [BaP87] and Corollary 1.8 [BaP88]; this method is very powerful, as we show in Chapters VI and VII, and it is sometimes called the Barles-Perthame procedure.

The definition of non-continuous viscosity solution 2.2 is due to Ishii [I87a], as well as the connection with optimal control, Theorem 2.6 [I89]. The Zermelo Navigation Problem, one of the most classical problems in Optimal Control, was studied by Carathéodory [Ca35] and many other authors in the '30s (see the references in [BSt93]) and can be found in the books [Lei66, Ce83]. Example 2.10 is taken from [Sus89b].

Section 2.2 follows Ishii [I87a], Proposition 2.16 appears in [I88b]. Perron's method is a very versatile tool in the existence theory of viscosity solutions; it was used in [CDL90] for constrained solutions and extended to weakly coupled systems of HJ equations [EL91], integrodifferential equations [LeY91], as well as second order degenerate elliptic equations, see [CIL92] and the references therein.

Section 3.1 gives a viscosity solutions approach to a result obtained by Subbotin [Su93a, Su95] in the framework of minimax solutions, see also [RS88]. The definition of envelope solution is new and it is inspired by the classical Wiener's notion of generalized solution of the Dirichlet problem for the Laplace equation in nonsmooth domains [W25]; Subbotin's original definition of (generalized) minimax solution is essentially the equivalent property of Proposition 3.6. Theorem 3.4 was extended to second order degenerate elliptic equations in [BB95] where the stability properties of e-solutions are also studied.

Section 3.2 follows [BSt93].

The boundary conditions in the viscosity sense of §4.1 appear in [BaP87] and are systematically studied by Ishii [I89] and Barles and Perthame [BaP88]. Proposition 4.12 and Theorem 4.13 are taken, respectively, from [Sa90] and [I89]. The definition of solution for discontinuous Hamiltonians in Exercise 4.1 is also in [I89], while the example is taken from [Ba90a]; in [Ba90a] it is also proved that for a large class of unbounded control problems with discontinuous Hamiltonian the value function is the maximal subsolution.

Section 4.2 follows [I89] in the case of bounded domains, Theorem 4.16, and [BS94] for unbounded domains, Theorem 4.20. The definition of usable part of the boundary comes from the theory of differential games [Is65]; Theorem 4.24 is taken from [BS94]. The proof of Theorem 4.25 can be found in [CGS91].

Section 4.3 follows [BSt93]; statement (iii) of Corollary 4.30 is due to Barles and Perthame [BaP87] and it is the first uniqueness theorem for discontinuous viscosity solutions in the literature.

Related papers are [BaP90, T92].

The theory of bilateral supersolutions in §5 originates from the work of Barron and Jensen [BJ90] as revisited by Barles [Ba93]. Section 5.1 follows Soravia [Sor93b] and extends a bit his results because we do not assume the continuity of the boundary data. Discontinuous data that do not satisfy (5.8) were considered in [Bla97]. Some boundary conditions different from (5.10) but equivalent to it were studied recently in [WZ97] and [KN97]. The proof of Theorem 5.11 is in [Ba93].

Section 5.2 gives a new proof of the main result of [BJ90, BJ91].

The methods of this section were applied to equations with singular coefficients in [K96b, CamS95].

A different approach to control problems with semicontinuous value function was pursued by Frankowska using contingent derivatives and viability theory, see [Fr89b, Fr93]; a uniqueness theorem for the HJB equation and connections with the viscosity solutions theory of §5 are in [Fr93] (see also [FPR94]). The same approach was used in [CQS94b] for time-optimal control with state constraints.

Further recent results on discontinuous viscosity solutions are in [BL96b, BL97, Sor96b, BG97].

CHAPTER VI

Approximation and perturbation problems

In this chapter we consider some approximation and perturbation problems for Hamilton-Jacobi equations. The results are presented for the model case

(HJB) $\quad \lambda u(x) + \sup_{a \in A}\{-f(x,a) \cdot Du(x) - \ell(x,a)\} = 0 \quad \text{in } \mathbb{R}^N,$

although many of them hold for more general Hamiltonians and boundary value problems, as indicated in the remarks. Singularly perturbed versions of (HJB) motivated by discrete time and stochastic control are analyzed in §1 and §3, respectively. On the other hand, in §2 we consider regular perturbations of (HJB).

The methods of viscosity solutions and, in particular, the *weak limits* technique described in Chapter V, §1 (see Definition V.1.4, Theorem V.1.7 and Corollary V.1.8), prove to be effective in providing simple proofs of general convergence and stability results in the uniform norm. Moreover, estimates on the rate of convergence are established via appropriate implementations of the basic ideas appearing in the proofs of comparison theorems for viscosity solutions.

1. Semidiscrete approximation and ε-optimal feedbacks

In this section we develop a theme already touched upon in Chapter I, §7, namely, the convergence as $h \to 0^+$ of the solutions of the functional equation

(HJB)$_h \quad u_h(x) + \sup_{a \in A}\{-(1-\lambda h)u_h(x + h\,f(x,a)) - h\,\ell(x,a)\} = 0, \quad x \in \mathbb{R}^N$

to the viscosity solution of

(HJB) $\quad \lambda u(x) + \sup_{a \in A}\{-f(x,a) \cdot Du(x) - \ell(x,a)\} = 0, \quad x \in \mathbb{R}^N,$

and its application to the approximate synthesis of optimal feedback maps for the infinite horizon problem.

The existence, uniqueness and regularity theory for (HJB)$_h$, as well as its relationship with discrete time dynamic programming, are developed in the appendix of this chapter, Section 4. Equation (HJB)$_h$ can be interpreted as an approximation of (HJB). Indeed, if $u \in C^1(\mathbb{R}^N)$, then its directional derivative in the direction $f(x,a)$ satisfies

$$-h f(x,a) \cdot Du(x) = u(x) - u(x + h f(x,a)) + o(h)$$

as $h \to 0^+$. Hence, if u is a classical solution of (HJB), then

$$\sup_{a \in A}\{ \lambda h\, u(x) + u(x) - u(x + h f(x,a)) - h\, \ell(x,a) \} = o(h)$$

for all $x \in \mathbb{R}^N$. So, if we compare the preceding with (HJB)$_h$, it is reasonable to guess that u_h will be close to u for small $h > 0$.

The section is divided into three parts. In 1.1 we prove (Theorem 1.1) the convergence of u_h to u under a set of assumptions which guarantee that the unique viscosity solution of (HJB) is the value function of the infinite horizon problem (see Chapter III). This result has some relevant applications to the synthesis of ε-optimal controls and to the existence of optimal controls for the infinite horizon problem (see Theorem 1.3 and Corollary 1.4). In 1.2 and 1.3 some quantitative results on the rate of convergence are established, showing that $\sup_{\mathbb{R}^N} |u(x) - u_h(x)|$ behaves, for small h, as h^σ for some $\sigma \in]0,1]$, depending on the regularity of u and u_h.

The theory of this section provides the theoretical basis to some computational techniques explained in greater detail in Appendix A.

The following conditions on the data are assumed throughout the present section

(1.1) $f : \mathbb{R}^N \times A \to \mathbb{R}^N$, $\ell : \mathbb{R}^N \times A \to \mathbb{R}$ continuous, A compact,

(1.2) $|f(x,a) - f(y,a)| \leq L|x - y|$,

(1.3) $|\ell(x,a) - \ell(y,a)| \leq L_1|x - y|$, $|\ell(x,a)| \leq M$,

for $x, y \in \mathbb{R}^N$, $a \in A$, and

(1.4) $\lambda > 0$.

Note that (1.1) and (1.2) imply

(1.5) $|f(x,a)| \leq K(1 + |x|)$.

Under these assumptions, equation (HJB) has a unique viscosity solution in $BUC(\mathbb{R}^N)$ (see Chapter III), whereas (HJB)$_h$ has a unique solution in $BUC(\mathbb{R}^N)$ for each $h \in]0, 1/\lambda[$ (see the appendix of this chapter, Section 4).

1.1. Approximation of the value function and construction of optimal controls

The next result shows that the viscosity solution u of (HJB) (that is, the value of the infinite horizon problem) can be uniformly approximated by the solution of (HJB)$_h$, for small h. The proof employs the weak limits technique of Chapter V.

1. SEMIDISCRETE APPROXIMATION

THEOREM 1.1. *Let us assume* (1.1)-(1.4). *Then*

$$\sup_{\mathcal{K}} |u_h(x) - u(x)| \longrightarrow 0 \quad \text{as} \quad h \to {}^+,$$

for every compact $\mathcal{K} \subseteq \mathbb{R}^N$, *where* $u_h, u \in BUC(\mathbb{R}^N)$ *are, respectively, the solution of* $(HJB)_h$ *and the viscosity solution of* (HJB).

PROOF. It is easy to check that M/λ and $-M/\lambda$, with M as in (1.3), are super- and subsolutions of $(HJB)_h$, for any h. Hence, by the results of §4,

$$\sup_{\mathbb{R}^N} |u_h(x)| \leq M/\lambda.$$

Let us define then functions $\underline{u}, \overline{u}$ by

$$\underline{u}(x) = \liminf_{(y,h) \to (x, 0^+)} u_h(y), \quad \overline{u}(x) = \limsup_{(y,h) \to (x, 0^+)} u_h(y).$$

By Lemma V.1.5, $\underline{u}$ and $\overline{u}$ are, respectively, lower and upper semicontinuous and satisfy

(1.6) $$\underline{u} \leq \overline{u} \quad \text{in } \mathbb{R}^N.$$

Let us assume temporarily that $\underline{u}$ is a viscosity supersolution and $\overline{u}$ is a viscosity subsolution of (HJB) in the generalized sense of Definition V.1.1. Then, by the comparison principle, Theorem V.1.3, we obtain

$$\overline{u} \leq \underline{u} \quad \text{in } \mathbb{R}^N.$$

Taking (1.6) into account, then $\underline{u} \equiv \overline{u}$. Therefore, $u := \overline{u} \equiv \underline{u}$ turns out to be a continuous viscosity solution of (HJB) and also by Lemma V.1.9,

$$u_h \longrightarrow u \quad \text{locally uniformly in } \mathbb{R}^N \text{ as } h \to 0^+,$$

which proves the theorem.

Let us show then that $\underline{u}$ is a viscosity supersolution of (HJB). Let x_1 be a strict minimum point for $\underline{u} - \varphi$ in $\overline{B} = \overline{B}(x_1, r)$, $\varphi \in C^1(\mathbb{R}^N)$. By Lemma V.1.6, there exist $x_n \in \overline{B}$ and $h_n \to 0^+$ such that

(1.7) $$(u_{h_n} - \varphi)(x_n) = \min_{\overline{B}}(u_{h_n} - \varphi), \quad x_n \longrightarrow x_1, \quad u_{h_n}(x_n) \longrightarrow \underline{u}(x_1).$$

Since u_h solves $(HJB)_h$ we obtain

$$u_{h_n}(x_n) - (1 - \lambda h_n)u_{h_n}(x_n + h_n f(x_n, a_n)) - h_n \ell(x_n, a_n) = 0,$$

for some $a_n = a_n(x_n) \in A$. Add and subtract $\lambda h_n u_{h_n}(x_n)$ and use the first relation in (1.7) to get

$$(1 - \lambda h_n)[\varphi(x_n) - \varphi(x_n + h_n f(x_n, a_n))] + \lambda h_n u_{h_n}(x_n) - h_n \ell(x_n, a_n) \geq 0,$$

for sufficiently small h_n. Divide now by h_n, take a subsequence n_k such that $a_{n_k} \to \bar{a} \in A$ and let $k \to +\infty$ to obtain by (1.7)

$$-D\varphi(x_1) \cdot f(x_1, \bar{a}) + \lambda \underline{u}(x_1) - \ell(x_1, \bar{a}) \geq 0 .$$

This implies that $\underline{u}$ is a viscosity supersolution of (HJB).

Now let x_0 be a strict maximum point for $\bar{u} - \varphi$ on $\bar{B} := \bar{B}(x_0, r)$. By Lemma V.1.6 again, there exist $x_n \in \bar{B}$, $h_n \to 0^+$, such that

(1.8) $\quad (u_{h_n} - \varphi)(x_n) = \max_{\bar{B}}(u_{h_n} - \varphi), \quad x_n \longrightarrow x_0, \quad u_{h_n}(x_n) \longrightarrow \bar{u}(x_0) .$

From (HJB)$_h$ we obtain the inequality

$$u_{h_n}(x_n) - (1 - \lambda h_n) u_{h_n}(x_n + h_n f(x_n, a)) - h_n \ell(x_n, a) \leq 0,$$

for all $a \in A$. Then, by (1.8),

$$(1 - \lambda h_n)[\varphi(x_n) - \varphi(x_n + h_n f(x_n, a))] + \lambda h_n u_{h_n}(x_n) - h_n \ell(x_n, a) \leq 0$$

for all $a \in A$. The same argument as above shows then that $\bar{u}$ is a viscosity subsolution of (HJB). ◂

We proceed now to explain how the previous results can be applied to the construction of an approximate optimal control for the infinite horizon problem. An important tool is the following representation formula for u_h:

(1.9) $\quad u_h(x) = \inf_{\alpha \in \mathcal{A}_h} J_h(x, \alpha), \quad x \in \mathbb{R}^N .$

Here

$$\mathcal{A}_h := \{ \alpha \in \mathcal{A} : \alpha(s) \equiv \alpha(kh), \ s \in [kh, (k+1)h[, \ k \in \mathbb{N} \},$$

$$J_h(x, \alpha) := h \sum_{k=0}^{\infty} \ell(y_h(k), \alpha(kh))(1 - \lambda h)^k,$$

where $y_h(k)$ is recursively defined by

$$y_h(0) = x, \quad y_h(k+1) = y_h(k) + h f(y_h(k), \alpha(kh)) .$$

The proof of (1.9) can be found in §4. Note that this discrete time system is the Euler approximation scheme with step h associated with the continuous time system (S) of Chapter III, while J_h is a sort of trapezoidal approximation of the integral functional

$$J(x, \alpha) = \int_0^\infty \ell(y_x(t), \alpha(t)) e^{-\lambda t} dt .$$

The algorithm for the construction of an approximated optimal control is as follows: let u_h be the solution of (HJB)$_h$ and, for each fixed x, pick $a_h^*(x) \in A$ such that

(1.10) $\quad a_h^*(x) \in \arg\max_{a \in A} \{ -(1 - \lambda h)u_h(x + h f(x,a)) - h \ell(x,a) \}$.

Hence $a_h^*(x)$ satisfies

$$u_h(x) - (1 - \lambda h)u_h(x + h f(x, a_h^*(x)) - h \ell(x, a_h^*(x))) = 0$$

for all $x \in \mathbb{R}^N$. Define then $y_h^*(k)$ by

(1.11) $\quad y_h^*(0) = x, \quad y_h^*(k+1) = y_h^*(k) + h f(y_h^*(k), a_h^*(y_h^*(k)))$

for $k = 0, 1, 2, \ldots$, and $\alpha_h^* \in \mathcal{A}_h$ by

(1.12) $\quad\quad\quad\quad \alpha_h^*(s) := a_h^*(y_h^*([s/h]))$

($[s/h]$ is the largest integer less than or equal to s/h). From (HJB)$_h$, (1.9) and (1.10) it follows that

$$u_h(x) = J_h(x, \alpha_h^*) = \inf_{\alpha \in \mathcal{A}_h} J_h(x, \alpha).$$

The simple proof of this fact is detailed in §4. Therefore, in the framework of Theorem 1.1, we have

$$J_h(x, \alpha_h^*) \longrightarrow \inf_{\alpha \in \mathcal{A}} J(x, \alpha),$$

uniformly as $h \to 0^+$. The preceding discussion says that α_h^* can be viewed as an approximate (piecewise constant) optimal control for the infinite horizon problem and that its construction is reduced to the solution of (HJB)$_h$. The computational aspects are discussed in Appendix A.

We turn our attention now to the behavior of the piecewise constant controls α_h^* and the corresponding y_h^* as $h \to 0^+$. The next lemma and Theorem 1.1 imply that $\{\alpha_h^*\}$ is a minimizing sequence for the infinite horizon problem, that is,

(1.13) $\quad\quad\quad\quad J(x, \alpha_h^*) \longrightarrow \inf_{\alpha \in \mathcal{A}} J(x, \alpha)$.

LEMMA 1.2. *Assume* (1.1)–(1.4). *Then*

$$\lim_{h \to 0^+} \sup_{\alpha \in \mathcal{A}_h} |J(x,\alpha) - J_h(x,\alpha)| = 0.$$

If, moreover, either $\lambda > L$ and f is bounded or $\lambda > L + K$, where K is the constant in (1.5), *then*

$$|J(x,\alpha) - J_h(x,\alpha)| \leq Ch \quad \text{for all } \alpha \in \mathcal{A}_h,\ h \in\,]0, 1/\lambda[$$

for some $C > 0$.

PROOF. For fixed $\alpha \in \mathcal{A}_h$ we set $y(s) = y_x(s, \alpha)$ and $\tilde{y}(s) = y_h([s/h])$, where $y_h(k)$ is the solution of the discrete time system starting at x, defined above. Observe that $\tilde{y}$ can be expressed as

$$\tilde{y}(s) = x + \int_0^{[s/h]h} f(\tilde{y}(\tau), \alpha(\tau)) \, d\tau, \qquad s \geq 0.$$

Then

$$|y(s) - \tilde{y}(s)|$$
$$\leq \int_0^{[s/h]h} |f(y(\tau), \alpha(\tau)) - f(\tilde{y}(\tau), \alpha(\tau))| \, d\tau + \int_{[s/h]h}^s |f(y(\tau), \alpha(\tau))| \, d\tau.$$

Hence, using (1.2) and (1.5),

$$|y(s) - \tilde{y}(s)| \leq L \int_0^s |y(\tau) - \tilde{y}(\tau)| \, d\tau + K \int_{[s/h]h}^h (1 + |y(\tau)|) \, d\tau.$$

By estimate (III.1.4) the above gives

$$|y(s) - \tilde{y}(s)| \leq L \int_0^s |y(\tau) - \tilde{y}(\tau)| \, d\tau + K(1 + (|x| + \sqrt{2Ks})e^{Ks})h$$

(where we used the inequalities $[s/h]h \leq s \leq ([s/h] + 1)h$). Hence, by Gronwall's Lemma,

(1.14) $$|y(s) - \tilde{y}(s)| \leq Kh(1 + (|x| + \sqrt{2Ks})e^{Ks})e^{Ls}.$$

Now a simple computation shows that

(1.15) $$|J(x, \alpha) - J_h(x, \alpha)| \leq X_1 + X_2 + X_3, \qquad \forall \alpha \in \mathcal{A}_h,$$

with

$$X_1 = \left| h \sum_{k=0}^{[T/h]-1} \ell(\tilde{y}(kh), \alpha(kh))\beta^k - \int_0^T \ell(y(s), \alpha(s)) e^{-\lambda s} \, ds \right|,$$

$$X_2 = \left| h \sum_{k=[T/h]}^{\infty} M\beta^k \right|, \qquad X_3 = \int_T^{\infty} M e^{-\lambda s} \, ds.$$

Here M is given by (1.3), we set $\beta = 1 - \lambda h$, and $T > 0$ is arbitrary.

Let us proceed now to estimate the X_i's ($i = 1, 2, 3$). It is straightforward to check that

$$X_2 + X_3 \leq Mh \frac{(1 - \lambda h)^{[T/h]}}{\lambda h} + M \frac{e^{-\lambda T}}{\lambda}.$$

Since $[T/h] \log(1 - \lambda h) \to -\lambda T$ as $h \to 0^+$, we conclude that for any $\varepsilon > 0$ there exist $\overline{h} > 0$, $\overline{T} > 0$ such that

$$X_2 + X_3 \leq \varepsilon, \qquad \text{for all } 0 < h < \overline{h}, T > \overline{T}.$$

1. SEMIDISCRETE APPROXIMATION

Observe next that

$$X_1 \leq \int_0^{[T/h]h} \left|\ell(\widetilde{y}(s),\alpha(s))\beta^{[s/h]} - \ell(y(s),\alpha(s))e^{-\lambda s}\right|ds + \int_{[T/h]h}^T M e^{-\lambda s}\,ds$$

so that

(1.16) $$X_1 \leq X_4 + X_5 + Mh$$

with

$$X_4 = \int_0^T |\ell(\widetilde{y}(s),\alpha(s)) - \ell(y(s),\alpha(s))|\,e^{-\lambda s}\,ds,$$

$$X_5 = \int_0^T |\ell(\widetilde{y}(s),\alpha(s))|\,|e^{-\lambda s} - e^{-\lambda \vartheta [s/h]h}|\,ds\,.$$

Here

$$\vartheta = -\frac{\log(1-\lambda h)}{\lambda h}\,.$$

Now, by (1.3) and (1.14)

$$X_4 \leq \int_0^T L_1 |\widetilde{y}(s) - y(s)|\,e^{-\lambda s}\,ds \leq L_1 K(1+Y)h\,e^{LT} \leq \varepsilon$$

for sufficiently small h.

The estimate for X_5 is obtained by an application of the mean value theorem to the function $s \mapsto e^{-\lambda s}$. Since $[s/h]h \leq s \leq ([s/h]+1)h$ this gives

$$X_5 \leq TM\lambda|s - \vartheta[s/h]h| \leq TM\lambda((1-\vartheta)T + \vartheta h)\,.$$

Now $\vartheta \to 1$ as $h \to 0^+$ and consequently $X_5 \leq \varepsilon$ for sufficiently small h. The first statement now follows from (1.15), (1.16).

In order to prove the remaining part of the statement, observe that

$$|J(x,\alpha) - J_h(x,\alpha)| \leq X + Y \qquad \forall \alpha \in \mathcal{A}_h,$$

where

$$X = \int_0^{+\infty} |\ell(y(s),\alpha(s)) - \ell(\widetilde{y}(s),\alpha(s))|\,e^{-\lambda s}\,ds,$$

$$Y = \int_0^{+\infty} |\ell(\widetilde{y}(s),\alpha(s))|\,\left|e^{-\lambda s} - e^{-\vartheta\lambda[s/h]h}\right|ds\,.$$

Now (1.3) and (1.14) yield

$$X \leq KL_1 h \int_0^{+\infty} (1 + (|x| + \sqrt{2Ks})\,e^{Ks})e^{(L-\lambda)s}\,ds\,.$$

If $\lambda > K + L$ one easily obtains from the preceding that

$$X \leq Ch \quad \text{for some } C > 0.$$

In order to estimate Y let us observe that by (1.3) and the mean value theorem

$$Y \leq M \int_0^{+\infty} \max\left\{e^{-\lambda s}; e^{-\lambda \vartheta [s/h]h}\right\} |\lambda s - \lambda \vartheta [s/h]h|\, ds.$$

Since $\vartheta > 1$ and $s \leq ([s/h] + 1)h$ we have

$$\max\left\{e^{-\lambda s}; e^{-\lambda \vartheta [s/h]h}\right\} \leq e^{\vartheta \lambda h} e^{-\lambda s}.$$

This, together with the inequality $|s - \vartheta[s/h]h| \leq (\vartheta - 1 + h)(s + 1)$, gives

$$Y \leq M\lambda(\vartheta - 1 + h)e^{\vartheta \lambda h} \int_0^{+\infty} (s+1)e^{-\lambda s}\, ds.$$

From the above and the fact that $\lim_{h \to 0^+} (\vartheta - 1)/h = \lambda/2$ one concludes that

$$Y \leq Ch \quad \text{for some } C > 0.$$

The preceding and the similar estimate previously obtained for X complete the proof in the case $\lambda > K + L$.

Under the stronger assumption

$$M_f := \sup |f| < +\infty$$

the estimate

$$|y(s) - \widetilde{y}(s)| \leq M_f h e^{LS}$$

holds instead of (1.14). In this case the term X is bounded above by

$$L_1 M_f h \int_0^{+\infty} e^{(L-\lambda)s}\, ds$$

and a fortiori, if $\lambda > L$, by $L_1 M_f h/(\lambda - L)$. ◀

It is well known that minimizing sequences of optimal control problems may not have a limit in any classical sense due to a highly oscillatory behavior. Therefore it seems natural to approach the question of convergence of α_h^* as $h \to 0^+$ in the framework of relaxed controls. We refer to III.2.2 for the notations employed hereafter.

The next result affirms, roughly speaking, that the discrete time optimal control problem $\inf_{\alpha \in A_h} J_h(x, \alpha)$ converges, in an appropriately weak sense, to the relaxation of the infinite horizon problem $\inf_{\alpha \in A} J(x, \alpha)$. More precisely, if we denote by y_h^* the solution of

(1.17)
$$\dot{y}(t) = f(y(t), \alpha_h^*(t)),$$
$$y(0) = x,$$

with α_h^* given by (1.12), the theorem says that the pair (α_h^*, y_h^*) converges to an optimal relaxed pair (μ^*, y^*) for the infinite horizon problem.

THEOREM 1.3. *Let us assume* (1.1)–(1.4). *Let* $\alpha_h^* \in \mathcal{A}_h$ *be the control defined by* (1.10), (1.11), *and* (1.12), *where* $u_h \in BUC(\mathbb{R}^N)$ *is the solution of* (HJB)$_h$, *and let* y_h^* *be the solution of* (1.17). *Then for any* $x \in \mathbb{R}^N$ *there exist* $\mu^* \in \mathcal{A}^r$, $y^* : [0, +\infty[\to \mathbb{R}^N$ *and a sequence* $h_n \to 0$ *as* $n \to +\infty$ *such that, for any* $T > 0$,

$$\text{(1.18)} \qquad \int_0^T \varphi(s, \alpha_{h_n}^*(s)) \, ds \longrightarrow \int_0^T \int_A \varphi(s, a) \, d\mu^*(s) \, ds$$

for all $\varphi \in L^1(0, T; C(A))$ (*i.e., for all* $\varphi : [0, T] \to C(A)$ *Lebesgue integrable*), *and*

$$\text{(1.19)} \qquad y_{h_n}^* \longrightarrow y^* \qquad \text{uniformly in } [0, T] \, .$$

In addition,

$$\text{(1.20)} \qquad y^*(t) = x + \int_0^t \int_A f(y^*(s), a) \, d\mu^*(s) \, ds \qquad \text{for } t \geq 0,$$

$$\text{(1.21)} \qquad \lim_{n \to +\infty} J_{h_n}(x, \alpha_{h_n}^*) = \lim_{n \to +\infty} J(x, \alpha_{h_n}^*) = J^r(x, \mu^*) = \inf_{\alpha \in \mathcal{A}} J(x, \alpha) \, .$$

PROOF. The crucial point is that for any $T > 0$ the restrictions to $[0, T]$ of measures $\mu \in \mathcal{A}^r$ form a (convex) sequentially compact subset of $L^\infty(0, T; \mathcal{A}^r)$ equipped with the weak-* topology (see [Wa72]). This means that we can select $h_n \to 0$ as $n \to +\infty$ and $\mu^* \in \mathcal{A}^r$ in such a way that (1.18) holds true.

By definition $y_{h_n}^*$ satisfies

$$\text{(1.22)} \qquad y_{h_n}^*(t) = x + \int_0^t f(y_{h_n}^*(s), \alpha_{h_n}^*(s)) \, ds, \qquad t \geq 0 \, .$$

From (1.2) it follows that

$$|y_{h_n}^*(t) - x| \leq L \int_0^t |y_{h_n}^*(s) - x| \, ds + \int_0^t |f(x, \alpha_n^*(s))| \, ds$$

and, consequently,

$$|y_{h_n}^*(t) - x| \leq L \int_0^t |y_{h_n}^*(s) - x| \, ds + T \sup_{a \in A} |f(x, a)|,$$

for $t \in [0, T]$. By Gronwall's inequality, $\{y_{h_n}^*\}$ is a uniformly bounded and equicontinuous sequence of functions on $[0, T]$. Therefore (1.19) holds, modulo the extraction of a subsequence.

To prove (1.20), observe that

$$\text{(1.23)} \qquad \int_0^t [f(y_{h_n}^*(s), \alpha_{h_n}^*(s)) - f(y^*(s), \alpha_{h_n}^*(s))] \, ds \longrightarrow 0$$

as $n \to +\infty$, as a consequence of (1.2) and (1.19). The choice $\varphi(s, a) := f(y^*(s), a)$ in (1.18) yields

$$\int_0^t f(y^*(s), \alpha_{h_n}^*(s)) \, ds \longrightarrow \int_0^t \int_A f(y^*(s), a) \, d\mu^*(s) \, ds \, .$$

Combining the above with (1.23) gives (1.20). A similar argument shows that

$$\text{(1.24)} \qquad \int_0^t \ell(y_{h_n}^*(s), \alpha_{h_n}^*(s)) \, e^{-\lambda s} \, ds \longrightarrow \int_0^t \int_A \ell(y^*(s), a) \, e^{-\lambda s} \, d\mu^*(s) \, ds,$$

uniformly in $[0, T]$.

To prove (1.21) observe first

$$|J(x, \alpha_{h_n}^*) - J^r(x, \mu^*)| \le X_n + Y_n$$

with

$$X_n = \int_0^T \left| \ell(y_{h_n}^*(s), \alpha_{h_n}^*(s)) - \int_A \ell(y^*(s), a) \, d\mu^*(s) \right| e^{-\lambda s} \, ds,$$

$$Y_n = \int_T^{+\infty} \left| \ell(y_{h_n}^*(s), \alpha_{h_n}^*(s)) - \int_A \ell(y^*(s), a) \, d\mu^*(s) \right| e^{-\lambda s} \, ds \, .$$

By (1.24), $X_n \to 0$ as $n \to +\infty$; on the other hand,

$$|Y_n| \le 2M \int_T^{+\infty} e^{-\lambda s} \, ds,$$

as follows from (1.3). Consequently

$$\lim_{n \to +\infty} J(x, \alpha_{h_n}^*) = J^r(x, \mu^*) \, .$$

By Lemma 1.2, $\lim_{n \to +\infty} J(x, \alpha_{h_n}^*) = \lim_{n \to +\infty} J_{h_n}(x, \alpha_{h_n}^*)$. From Theorem 1.1 we know that $\lim_{n \to +\infty} J_{h_n}(x, \alpha_{h_n}^*) = \inf_{\alpha \in \mathcal{A}} J(x, \alpha)$, so the proof is complete. ◀

As an application of Theorem 1.3 we have an existenceexistence of an optimal control result of an optimal control for the infinite horizon problem.

COROLLARY 1.4. *Assume (1.1)–(1.4) and*

$$\text{(1.25)} \qquad f(x, A) \times \ell(x, A) \text{ is convex for all } x \in \mathbb{R}^N \, .$$

Then there exists $\alpha^* \in \mathcal{A}$ *such that*

$$J(x, \alpha^*) = \inf_{\alpha \in \mathcal{A}} J(x, \alpha) \, .$$

PROOF. Let y^*, μ^* be as in Theorem 1.3. We observe first that (1.1) and (1.25) imply that $f(x, A) \times \ell(x, A)$ is convex and compact for each $x \in \mathbb{R}^N$. On the other hand,

$$f^r(x, A^r) \times \ell^r(x, A^r) = \overline{\text{co}}(f(x, A) \times \ell(x, a))$$

for all $x \in \mathbb{R}^N$ (see Exercise III.2.10) so this set coincides with $f(x, A) \times \ell(x, A)$. By definition, y^* and μ^* satisfy, for all $s \ge 0$,

$$\int_A f(y^*(s), a) \, d\mu^*(s) \in f^r(y^*(s), A^r),$$

$$\int_A \ell(y^*(s), a) \, d\mu^*(s) \in \ell^r(y^*(s), A^r) \, .$$

This means that for any $s \geq 0$ there exists $a^*(s) \in A$ such that

$$f(y^*(s), a^*(s)) = \int_A f(y^*(s), a) \, d\mu^*(s),$$
$$\ell(y^*(s), a^*(s)) = \int_A \ell(y^*(s), a) \, d\mu^*(s).$$

By general measurable selection results (see, e.g., Appendix B in [FR75]) we can assume that $s \mapsto a^*(s)$ is a measurable function α^*. By (1.20)

$$y^*(t) = x + \int_0^t f(y^*(s), \alpha^*(s)) \, ds,$$

and therefore

$$J(x, \alpha^*) = \int_0^\infty e^{-\lambda s} \int_A \ell(y^*(s), a) \, d\mu^*(s) \, ds = J^r(x, \mu^*).$$

Then (1.21) of Theorem 1.3 gives the conclusion. ◀

1.2. A first result on the rate of convergence

We have seen in Theorem 1.1 that $\sup_{\mathcal{K}} |u_h - u| \to 0$ as $h \to 0^+$ for every compact $\mathcal{K} \subseteq \mathbb{R}^N$. The next result shows, under an extra regularity assumption on u, that the convergence is uniform over all $\mathbb{R}^N$ and the rate of convergence can be estimated.

THEOREM 1.5. *Let us assume* (1.1)–(1.4) *and*

(1.26) $\qquad |f(x,a)| \leq M \qquad \text{for } (x,a) \in \mathbb{R}^N \times A.$

Assume also

(1.27) $\qquad u \in C^{0,\gamma}(\mathbb{R}^N) \qquad \text{for some } \gamma \in \,]0,1]\,.$

Then

$$\sup_{\mathbb{R}^N} |u_h(x) - u(x)| \leq Ch^{\gamma/2} \qquad \text{for all } h \in \,]0, 1/\lambda[,$$

for some constant $C > 0$ depending on the various bounds on the data but not on h.

PROOF. Consider for $\varepsilon, \delta > 0$ the auxiliary function

$$\Psi(x,y) = u(x) - u_h(y) - \frac{|x-y|^2}{2\varepsilon} - \delta(\mu(x) + \mu(y)),$$

where $\mu \in C^1(\mathbb{R}^N)$ is chosen to satisfy

(1.28) $\qquad \mu \geq 0, \quad \mu(0) = 0, \quad |D\mu| \leq 1, \quad \lim_{|x| \to +\infty} \mu(x) = +\infty,$

(see Exercise 1.2). It is not hard to check that the boundedness of u, u_h implies the existence of some $(\bar{x}, \bar{y})$ (depending on ε, δ) such that

$$\Psi(\bar{x}, \bar{y}) \geq \Psi(x, y) \quad \text{for all } x, y \in \mathbb{R}^N .$$

The inequality $\Psi(\bar{x}, \bar{y}) \geq \Psi(0,0)$ immediately gives

(1.29) $$\frac{|\bar{x} - \bar{y}|^2}{2\varepsilon} \leq C,$$

with $C = 2(\sup_{\mathbb{R}^N} |u(x)| + \sup_{\mathbb{R}^N} |u_h(x)|)$. Now use $\Psi(\bar{x}, \bar{y}) \geq \Psi(\bar{y}, \bar{y})$ to get, on the account of (1.27), (1.28),

(1.30) $$\frac{|\bar{x} - \bar{y}|^2}{2\varepsilon} \leq u(\bar{x}) - u(\bar{y}) + \delta(\mu(\bar{y}) - \mu(\bar{x})) \leq C |\bar{x} - \bar{y}|^\gamma + \delta |\bar{x} - \bar{y}| .$$

From (1.29) it follows that $|\bar{x} - \bar{y}| \leq \sqrt{2\varepsilon C} < 1$ for small ε. Hence (1.30) gives the estimate

(1.31) $$|\bar{x} - \bar{y}| \leq C \varepsilon^{1/(2-\gamma)} .$$

From equation (HJB)$_h$ we have

(1.32) $$u_h(\bar{y}) - (1 - \lambda h)u_h(\bar{y} + h f(\bar{y}, a^*)) - h \ell(\bar{y}, a^*) = 0$$

for some $a^* = a^*(\bar{y}) \in A$. On the other hand, since

$$x \longmapsto u(x) - \left[u_h(\bar{y}) + \frac{|x - \bar{y}|^2}{2\varepsilon} + \delta(\mu(x) + \mu(\bar{y})) \right]$$

has a maximum at $\bar{x}$, from (HJB) we obtain

(1.33) $$\lambda u(\bar{x}) \leq f(\bar{x}, a^*) \cdot \left(\frac{\bar{x} - \bar{y}}{\varepsilon} + \delta D\mu(\bar{x}) \right) + \ell(\bar{x}, a^*) .$$

The inequality $\Psi(\bar{x}, \bar{y}) \geq \Psi(\bar{x}, \bar{y} + h f(\bar{y}, a^*))$ gives

$$u_h(\bar{y} + h f(\bar{y}, a^*)) \geq u_h(\bar{y}) + \frac{|\bar{x} - \bar{y}|^2 - |\bar{x} - \bar{y} - h f(\bar{y}, a^*)|^2}{2\varepsilon} + \delta X_1,$$

with

$$X_1 = \mu(\bar{y}) - \mu(\bar{y} + h f(\bar{y}, a^*)) .$$

Consequently,

$$u_h(\bar{y} + h f(\bar{y}, a^*)) \geq u_h(\bar{y}) + \frac{2h f(\bar{y}, a^*) \cdot (\bar{x} - \bar{y}) - h^2 |f(\bar{y}, a^*)|^2}{2\varepsilon} + \delta X_1 .$$

Multiply the above inequality by $1 - \lambda h$ and substitute the result into (1.32) to obtain

$$u_h(\bar{y}) \geq (1 - \lambda h) \left[u_h(\bar{y}) + \frac{2h f(\bar{y}, a^*) \cdot (\bar{x} - \bar{y}) - h^2 |f(\bar{y}, a^*)|^2}{2\varepsilon} + \delta X_1 \right] + h \ell(\bar{y}, a^*) .$$

Hence, dividing by h,

$$-\lambda u_h(\overline{y}) \leq -(1-\lambda h)\left[f(\overline{y},a^*)\cdot\frac{\overline{x}-\overline{y}}{\varepsilon} - \frac{h}{2\varepsilon}|f(\overline{y},a^*)|^2 + \delta\frac{X_1}{h}\right] - \ell(\overline{y},a^*).$$

Now, add this to (1.33) to get, using (1.3),

(1.34) $\qquad \lambda(u(\overline{x}) - u_h(\overline{y})) \leq X_2 + X_3 + X_4 + X_5 + M\,|\overline{x}-\overline{y}|,$

where

$$X_2 = (f(\overline{x},a^*) - f(\overline{y},a^*))\cdot\frac{\overline{x}-\overline{y}}{\varepsilon}, \qquad X_3 = \lambda h\,f(\overline{y},a^*)\cdot\frac{\overline{x}-\overline{y}}{\varepsilon},$$
$$X_4 = \frac{h}{2\varepsilon}|f(\overline{y},a^*)|^2, \qquad X_5 = \delta\,f(\overline{x},a^*)\cdot D\mu(\overline{x}) + \lambda\delta X_1.$$

Let us proceed then to estimate the various terms on the right-hand side of (1.34). We have

$$|X_2| \leq L\frac{|\overline{x}-\overline{y}|^2}{\varepsilon}, \qquad |X_3| \leq \lambda Mh\frac{|\overline{x}-\overline{y}|}{\varepsilon},$$
$$|X_4| \leq \frac{h}{2\varepsilon}M^2, \qquad |X_5| \leq 2\delta CM.$$

Therefore, using (1.31) and choosing $\varepsilon = h^{(2-\gamma)/2}$, we obtain

(1.35) $\qquad u(\overline{x}) - u_h(\overline{y}) \leq C(h^{\gamma/2} + h^{(\gamma+1)/2} + h^{\gamma/2} + \delta),$

for a suitable constant C. Hence, for small h,

$$u(\overline{x}) - u_h(\overline{y}) \leq C(h^{\gamma/2} + \delta).$$

Then the inequality $\Psi(\overline{x},\overline{y}) \geq \Psi(x,x)$ yields

$$u(x) - u_h(x) - 2\delta\,\mu(x) \leq u(\overline{x}) - u_h(\overline{y}) \leq C(h^{\gamma/2} + \delta),$$

for all $x \in \mathbb{R}^N$. Let $\delta \to 0$ in the above to conclude that

$$\sup_{\mathbb{R}^N}(u(x) - u_h(x)) \leq Ch^{\gamma/2}.$$

To prove the inequality $u_h(x) - u(x) \leq Ch^{\gamma/2}$ it is enough to interchange the roles of u and u_h in the auxiliary function Ψ and proceed as before. ◀

REMARK 1.6. From the theory of Chapter III we know that (1.27) holds with $\gamma = 1$ if $\lambda > L$, $\gamma = \lambda/L$ if $\lambda < L$ and $\gamma = 1-\delta$ for any $\delta > 0$ if $\lambda = L$. ◁

1.3. Improving the rate of convergence

The estimate of the rate of convergence given by Theorem 1.5 can be improved under a uniform semiconcavity assumption on the solutions of (HJB)$_h$. We have the following result.

THEOREM 1.7. *Let us assume (1.1)–(1.4), $\lambda > L$, and*

(1.36) $$|f(x,a)| \leq M \quad \text{for } (x,a) \in \mathbb{R}^N \times A.$$

Assume also that the solution u of (HJB) is Lipschitz continuous, and that the solution u_h of (HJB)$_h$ satisfies

(1.37) $$u_h(x+z) - 2u_h(x) + u_h(x-z) \leq C|z|^2 \quad \text{for all } x, z \in \mathbb{R}^N,$$

for some $C > 0$ independent of h. Then there exists $C' > 0$ independent of h such that

$$\sup_{\mathbb{R}^N} |u(x) - u_h(x)| \leq C'h \quad \text{for all } h \in \,]0, 1/\lambda[\,.$$

In the proof we need the following technical lemma which exploits a useful consequence of (1.37).

LEMMA 1.8. *Let u_h be the solution of (HJB)$_h$. If (1.37) holds, then for any $a \in A$*

(1.38) $u_h(\overline{x} + h f(\overline{x}, a))$
$$\leq u_h(\overline{x}) + h f(\overline{x}, a) \cdot \frac{\overline{x} - \overline{y}}{\varepsilon} + \frac{1}{2}Ch^2|f(\overline{x},a)|^2 + \delta h |f(\overline{x},a)|,$$

at any global maximum point $(\overline{x}, \overline{y})$ of

$$\Psi(x,y) := u_h(x) - u(y) - \frac{|x-y|^2}{2\varepsilon} - \delta(\mu(x) + \mu(y)),$$

where $\mu \in C^1(\mathbb{R}^N)$ satisfies (1.28).

PROOF. First use the inequality $\Psi(\overline{x}, \overline{y}) \geq \Psi(\overline{x} + x, \overline{y})$, with arbitrary $x \in \mathbb{R}^N$, to obtain

(1.39) $$u_h(\overline{x} + x) - u_h(\overline{x}) - x \cdot \frac{\overline{x} - \overline{y}}{\varepsilon} - \frac{|x|^2}{2\varepsilon} - \delta D\mu(\overline{x}) \cdot x \leq 0$$

for all $x \in \mathbb{R}^N$. Now consider the function

$$\eta(x) := u_h(\overline{x} + x) - u_h(\overline{x}) - \left(\frac{\overline{x} - \overline{y}}{\varepsilon} + \delta D\mu(\overline{x})\right) \cdot x.$$

Using assumption (1.37) and inequality (1.39) it is easy to check that

$$\begin{cases} \eta(x+z) - 2\eta(x) + \eta(x-z) = u_h(\overline{x}+x+z) - 2u_h(\overline{x}+x) + u_h(\overline{x}+x-z) \\ \qquad\qquad\qquad\qquad\qquad \leq C|z|^2, \\ \eta(0) = 0, \quad \limsup_{|x| \to 0} \frac{\eta(x)}{|x|} \leq 0. \end{cases}$$

1. SEMIDISCRETE APPROXIMATION

This implies

(1.40) $$\eta(x) \leq \frac{C}{2}|x|^2 \quad \text{for all } x \in \mathbb{R}^N,$$

see Exercise 1.3. Let us now evaluate η at $x = h f(\bar{x}, a)$ with arbitrary $a \in A$ to obtain using (1.40)

$$u_h(\bar{x} + h f(\bar{x}, a)) - u_h(\bar{x}) - \left(\frac{\bar{x} - \bar{y}}{\varepsilon} + \delta D\mu(\bar{x})\right) \cdot h f(\bar{x}, a) \leq \frac{C}{2} h^2 |f(\bar{x}, a)|.$$

Since $|D\mu| \leq 1$, from the above inequality the statement follows. ◀

PROOF OF THEOREM 1.7. Here we denote as C any constant independent of h. Let us show first that

(1.41) $$\sup_{\mathbb{R}^N}(u_h(x) - u(x)) \leq Ch.$$

For this purpose, if Ψ is as in the proof of Theorem 1.5 and $(\bar{x}, \bar{y})$ is a global maximum point for Ψ, then from (HJB) it follows that

(1.42) $$\lambda u(\bar{y}) - f(\bar{y}, a^*) \cdot \left(\frac{\bar{x} - \bar{y}}{\varepsilon} - \delta D\mu(\bar{y})\right) - \ell(\bar{y}, a^*) \geq 0,$$

for some $a^* = a^*(\bar{y}) \in A$. On the other hand, (HJB)$_h$ gives

$$u_h(\bar{x}) - (1 - \lambda h) u_h(\bar{x} + h f(\bar{x}, a^*)) - h \ell(\bar{x}, a^*) \leq 0.$$

By Lemma 1.8 and multiplication by $(1 - \lambda h)/h$, then

(1.43) $$\lambda u_h(\bar{x}) \leq (1 - \lambda h) f(\bar{x}, a^*) \cdot \frac{\bar{x} - \bar{y}}{\varepsilon} + \frac{C}{2}(1 - \lambda h) h |f(\bar{x}, a^*)|^2$$
$$+ (1 - \lambda h) \delta |f(\bar{x}, a^*)| + \ell(\bar{x}, a^*).$$

From now on the argument is exactly the same as in the proof of Theorem 1.5: subtracting (1.42) from (1.43) and estimating the various terms we obtain

(1.44) $$u_h(\bar{x}) - u(\bar{y}) \leq C\left[\frac{|\bar{x} - \bar{y}|^2}{\varepsilon} + h\frac{|\bar{x} - \bar{y}|}{\varepsilon} + |\bar{x} - \bar{y}| + h + \delta\right].$$

Now the Lipschitz continuity of u yields the estimate

$$|\bar{x} - \bar{y}| \leq C\varepsilon;$$

see (1.31) in the proof of Theorem 1.5. This, combined with (1.44) and the choice $\varepsilon = h$, proves (1.41).

The inequality

(1.45) $$\sup_{x \in \mathbb{R}^N}(u(x) - u_h(x)) \leq Ch$$

is a consequence of the representation of u as the value function of the infinite horizon problem

(1.46) $$u(x) = \inf_{\alpha \in \mathcal{A}} J(x, \alpha)$$

(see Chapter III), as well of the analogous one for u_h,

(1.47) $$u_h(x) = \inf_{\alpha \in \mathcal{A}_h} J_h(x, \alpha)$$

(the proof of (1.47) can be found in the appendix, §4, Theorem 4.2). Indeed, from (1.46), (1.47) we obtain

$$u(x) - u_h(x) \leq \inf_{\alpha \in \mathcal{A}_h} J(x, \alpha) - \inf_{\alpha \in \mathcal{A}_h} J_h(x, \alpha) \leq \sup_{\alpha \in \mathcal{A}_h} |J(x, \alpha) - J_h(x, \alpha)|.$$

The estimate (1.45) then follows by Lemma 1.2. ◄

REMARK 1.9. A sufficient condition for the Lipschitz continuity of u is $\lambda > L$ (see Proposition III.2.1). On the other hand, the uniform one-sided estimate (1.37) holds if the further conditions

$$\lambda > 2L$$

and

$$|f(x + z, a) - 2f(x, a) + f(x - z, a)| \leq C|z|^2,$$
$$\ell(x + z, a) - 2\ell(x, a) + \ell(x - z, a) \leq Q|z|^2,$$

hold for some $C, Q > 0$ and all $x, z \in \mathbb{R}^N$, $a \in A$.

To check this it is enough to apply Proposition 4.5 in the appendix with

$$\widehat{f} = hf, \quad \widehat{\ell} = h\ell, \quad \beta = 1 - \lambda h, \quad h \in \,]0, 1/\lambda[\,.$$

Observe for this purpose that the constants Q, L, L_1, and C in the estimate (4.10) there must be multiplied by h. Observe also that $\lambda > 2L$ implies $(1 - \lambda h) \cdot (1 + Lh)^2 < 1$ for $h \in \,]0, 1/\lambda[$, thus condition (4.9) holds.

Let us point out that under the above assumptions on λ, f, ℓ the result of Theorem 1.7 is optimal, as shown by the following example. Take $N = 1$, A a singleton and let $f(\,\cdot\,, a), \ell(\,\cdot\,, a)$ be C_0^∞ functions on $\mathbb{R}$ such that $f(x, a) = -x$, $\ell(x, a) = x$ for $x \in [0, 1]$. It is easy to check that

$$u_h(x) = \frac{x}{1 + \lambda - \lambda h}, \quad u(x) = \frac{x}{1 + \lambda}, \quad \text{for } x \in [0, 1].$$

This gives

$$\sup_{x \in [0,1]} |u(x) - u_h(x)| = \frac{\lambda h}{(1 + \lambda)^2 - \lambda h(1 + \lambda)}$$

and, consequently,

$$\sup_{x \in [0,1]} \frac{|u(x) - u_h(x)|}{h^{1+\alpha}}$$

diverges as $h \to 0^+$, for any $\alpha > 0$. ◁

Exercises

1.1. Prove that Theorem 1.5 implies the following:

$$\sup |u - u_h| \leq Ch^{1/2} \quad \text{if } \lambda > L,$$
$$\sup |u - u_h| \leq Ch^{\lambda/2L} \quad \text{if } \lambda < L.$$

1.2. Check that

$$\mu(x) = \begin{cases} -A|x|^3 + B|x|^2 + C|x| & \text{for } |x| < e, \\ \log |x| & \text{for } |x| \geq e, \end{cases}$$

satisfies conditions (1.28), (2.13), and (3.7) for a suitable choice of positive constants A, B, C.

1.3. Assume that $\eta : \mathbb{R}^N \to \mathbb{R}$ satisfies, for some $C > 0$,

$$\eta(x+z) - 2\eta(x) + \eta(x-z) \leq C|z|^{1+\tau}$$

for all $x, z \in \mathbb{R}^N$ and some $\tau \in\,]0, 1]$, as well as

$$\eta(0) = 0, \quad \limsup_{|x| \to 0} \frac{\eta(x)}{|x|} \leq 0.$$

Prove that

$$\eta\left(\frac{x}{2^{k+1}}\right) \geq \frac{\eta(x)}{2^k} - \frac{C}{2^k} \left|\frac{x}{2}\right|^{1+\tau} \frac{2^\tau - 2^{-\tau(k-1)}}{2^\tau - 1}$$

for any $x \neq 0$, $k = 0, 1, \ldots$. Deduce from this the inequality

$$\eta(x) \leq \frac{C}{2(2^\tau - 1)} |x|^{1+\tau}, \quad \forall x \in \mathbb{R}^N.$$

1.4. Assume that u_h satisfies

$$u_h(x+z) - 2u_h(x) + u_h(x-z) \leq C|z|^{1+\tau}, \tau \in\,]0,1]$$

for all $x, z \in \mathbb{R}^N$. Use Exercise 1.3 to prove that the estimate in Theorem 1.7 holds in the form

$$\sup |u - u_h| \leq Ch^\tau.$$

1.5. Prove the analogue of Theorem 1.1 for solutions u_h of

$$\sup_{a \in A}\{(1 + \lambda h)u_h(x) - u_h(x + hf(x,a)) - h\ell(x,a)\} = 0.$$

2. Regular perturbations

In this section we address the following question: how much is the value function of an optimal control problem modified by perturbations of the various data not affecting their regularity or their structural properties?

The approach adopted here is to look at the (HJB) equation satisfied by the value function and prove general convergence results for a class of such equations whose Hamiltonians satisfy the same structural conditions. We present a typical result in this direction, which applies to the infinite horizon problem (see Theorem 2.1) and a variant of it (Theorem 2.4) motivated by a regularization question in minimal time theory.

Let us consider then the equations

$$\text{(HJB)}_0 \qquad u_0(x) + \sup_{a \in A_0} \{ -f_0(x,a) \cdot Du_0(x) - \ell_0(x,a) \} = 0, \qquad x \in \mathbb{R}^N,$$

and

$$\text{(HJB)}_h \qquad u_h(x) + \sup_{a \in A_h} \{ -f_h(x,a) \cdot Du_h(x) - \ell_h(x,a) \} = 0, \qquad x \in \mathbb{R}^N.$$

We assume, for $h \in [0,1]$,

(2.1) $\qquad f_h : \mathbb{R}^N \times A \to \mathbb{R}^N, \quad \ell_h : \mathbb{R}^N \times A \to \mathbb{R}$ continuous,

(2.2) $\qquad |f_h(x,a) - f_h(y,a)| \leq L\,|x-y|,$

(2.3) $\qquad |\ell_h(x,a) - \ell_h(y,a)| \leq M\,|x-y|,$

(2.4) $\qquad |\ell_h(x,a)| \leq M,$

for $x, y \in \mathbb{R}^N$, $a \in A$, with constants L, M independent of $h \in [0,1]$ and

(2.5) $\qquad A_h \subseteq A, \quad A$ compact.

Since we are interested in the rate of convergence of u_h to u_0 as $h \to 0^+$, the perturbed data A_h, f_h, ℓ_h are assumed to be of order 1 in h. More precisely,

(2.6) $\quad \sup_{\mathbb{R}^N} \{ |f_h(x,a) - f_0(x,a)| + |\ell_h(x,a) - \ell_0(x,a)| \} \leq Ch \qquad$ for all $a \in A$,

(2.7) $\quad d_H(A_h, A_0) \leq Ch,$

for some $C > 0$. The notation d_H stands for the *Hausdorff distance* between the sets A_h and A_0, that is

$$d_H(A_h, A_0) = \max \Big\{ \max_{a \in A_h} d(a, A_0), \max_{a \in A_0} d(a, A_h) \Big\},$$

where d is the usual Euclidean distance. Let us recall that the uniform convergence of u_h to u_0 has been proved in III.2.2. In the next result, the rate of convergence is established under slightly more restrictive assumptions.

2. REGULAR PERTURBATIONS

THEOREM 2.1. *Let us assume* (2.1)–(2.7) *and*

(2.8) $$|f_h(x,a) - f_h(x,a')| + |\ell_h(x,a) - \ell_h(x,a')| \leq L|a - a'|$$

for all $a, a' \in A$, $h \in [0,1]$. *If* $u_0 \in C^{0,\gamma}(\mathbb{R}^N)$ *for some* $\gamma \in \,]0,1]$, *then*

$$\sup_{\mathbb{R}^N} |u_0(x) - u_h(x)| \leq Ch^\gamma \quad \text{for } h \in [0,1],$$

for some constant $C > 0$.

PROOF. Let us set for $h \in [0,1]$

$$H_h(x,p) = \sup_{a \in A_h} \{-f_h(x,a) \cdot p - \ell_h(x,a)\}.$$

The first step is to show that, for any $h \in [0,1]$,

(2.9) $$|H_h(x,p) - H_0(x,p)| \leq Ch(|p| + 1) \quad \text{for all } (x,p) \in \mathbb{R}^N \times \mathbb{R}^N.$$

For fixed (x,p) let $\bar{a} \in A_0$ be such that

$$H_0(x,p) = -f_0(x,\bar{a}) \cdot p - \ell_0(x,\bar{a}).$$

By (2.7) there exists $\tilde{a} \in A_h$ such that

(2.10) $$|\bar{a} - \tilde{a}| \leq Ch;$$

by definition of H_h, then

$$H_h(x,p) \geq -f_h(x,\tilde{a}) \cdot p - \ell_h(x,\tilde{a}),$$

so that

$$H_0(x,p) - H_h(x,p) \leq (f_h(x,\tilde{a}) - f_0(x,\bar{a})) \cdot p + (\ell_h(x,\tilde{a}) - \ell_0(x,\bar{a})).$$

It is now easy to show, using (2.10) and assumptions (2.6) and (2.8) that

$$H_0(x,p) - H_h(x,p) \leq Ch(|p| + 1),$$

for some C. The other half of inequality (2.9) can be established in a similar way.

Simple computations show that, under the assumptions made, the Hamiltonians H_h satisfy

(2.11) $$|H_h(x,p) - H_h(y,p)| \leq L|x - y|(1 + |p|)$$
(2.12) $$|H_h(x,p) - H_h(x,q)| \leq C|p - q|(1 + |x|),$$

for some C, for all $h \in [0,1]$ and x, y, p, q in $\mathbb{R}^N$. Let us consider now the auxiliary function

$$\Psi(x,y) = u_0(x) - u_h(y) - \frac{|x-y|^2}{2\varepsilon} - \delta(\mu(x) + \mu(y)),$$

with $\mu \in C^1(\mathbb{R}^N)$ such that

(2.13) $\quad\begin{cases} \mu \geq 0, \quad \mu(0) = 0, \quad \lim_{|x| \to +\infty} \mu(x) = +\infty, \\ |D\mu(x)| \leq 1, \quad |x||D\mu(x)| \leq 1 \end{cases}$

(see Exercise 1.2). The proof of Theorem 1.5 shows that

(2.14) $\quad\quad\quad\quad\quad\quad |\bar{x} - \bar{y}| \leq C\varepsilon^{1/(2-\gamma)},$

where $\bar{x}, \bar{y}$ is a global maximum point for Ψ on $\mathbb{R}^N \times \mathbb{R}^N$. Since u_0 is a viscosity subsolution of (HJB)$_0$ and u_h is a viscosity supersolution of (HJB)$_h$, we have

(2.15) $\quad\quad u_0(\bar{x}) + H_0\left(\bar{x}, \dfrac{\bar{x} - \bar{y}}{\varepsilon} + \delta D\mu(\bar{x})\right) \leq 0,$

(2.16) $\quad\quad u_h(\bar{y}) + H_h\left(\bar{y}, \dfrac{\bar{x} - \bar{y}}{\varepsilon} - \delta D\mu(\bar{y})\right) \geq 0.$

Therefore

(2.17) $\quad u_0(\bar{x}) - u_h(\bar{y}) \leq H_h\left(\bar{y}, \dfrac{\bar{x} - \bar{y}}{\varepsilon} - \delta D\mu(\bar{y})\right) - H_0\left(\bar{x}, \dfrac{\bar{x} - \bar{y}}{\varepsilon} + \delta D\mu(\bar{x})\right).$

The right-hand side of the preceding inequality can be written as $R_1 + R_2$ with

$$R_1 = H_h\left(\bar{y}, \dfrac{\bar{x} - \bar{y}}{\varepsilon} - \delta D\mu(\bar{y})\right) - H_h\left(\bar{x}, \dfrac{\bar{x} - \bar{y}}{\varepsilon} + \delta D\mu(\bar{x})\right),$$
$$R_2 = H_h\left(\bar{x}, \dfrac{\bar{x} - \bar{y}}{\varepsilon} + \delta D\mu(\bar{x})\right) - H_0\left(\bar{x}, \dfrac{\bar{x} - \bar{y}}{\varepsilon} + \delta D\mu(\bar{x})\right).$$

Now (2.9), (2.11), and (2.12) provide the estimates

$$|R_1| \leq L|\bar{x} - \bar{y}|\left(1 + \left|\dfrac{\bar{x} - \bar{y}}{\varepsilon} - \delta D\mu(\bar{y})\right|\right) + C\delta|D\mu(\bar{x}) + D\mu(\bar{y})|(1 + |\bar{x}|),$$
$$|R_2| \leq Ch\left(1 + \left|\dfrac{\bar{x} - \bar{y}}{\varepsilon} + \delta D\mu(\bar{x})\right|\right).$$

Using (2.13) and (2.14) we obtain, for some $C > 0$,

$$|R_1| \leq C\left(\varepsilon^{1/(2-\gamma)} + \varepsilon^{\gamma/(2-\gamma)} + \delta\varepsilon^{1/(2-\gamma)} + \delta\right)$$

(observe that $|D\mu(\bar{y})||\bar{x}| \leq |D\mu(\bar{y})|(|\bar{x} - \bar{y}| + |\bar{y}|)$),

$$|R_2| \leq C\left(h + h\varepsilon^{(\gamma-1)/(2-\gamma)} + \delta h\right).$$

Let us now choose $\varepsilon = h^{2-\gamma}$ in the above and plug the previous estimates into (2.17) to get

(2.18) $\quad\quad u_0(\bar{x}) - u_h(\bar{y}) \leq C(h^\gamma + h + \delta h + \delta) \leq C(h^\gamma + \delta h + \delta),$

for small h. Finally, the inequality $\Psi(\bar{x}, \bar{y}) \geq \Psi(x, x)$ and (2.18) yield

$$u_0(x) - u_h(x) - 2\delta\mu(x) \leq u_0(\bar{x}) - u_h(\bar{y}) \leq C(h^\gamma + \delta h + \delta).$$

Let $\delta \to 0$ in the above to obtain

$$u_0(x) - u_h(x) \leq Ch^\gamma \quad \text{for all } x \in \mathbb{R}^N.$$

The inequality $u_h(x) - u_0(x) \leq Ch^\gamma$ is established in a similar way. ◄

REMARK 2.2. A simple situation where the preceding result applies is as follows. Let A_0, f_0, ℓ_0 be given and define

$$A_h = A_0 \times h\bar{B}(0,1),$$
$$f_h(x,a,v) = f_0(x,a) + hv,$$
$$\ell_h(x,a,v) = \ell_0(x,a) + h|v|^2,$$

for any $a \in A_0$, $v \in \bar{B}(0,1)$ and $h \in [0,1]$. This amounts to a "fattening" of the original control set A_0, giving rise to a new control system (f_h, A_h) with better controllability properties (see Corollary III.2.23). ◁

REMARK 2.3. If the control system is unperturbed (i.e., $f_h \equiv f_0$, $A_h \equiv A_0$), then the estimate is

$$\sup_{\mathbb{R}^N} |u_0(x) - u_h(x)| \le Ch$$

(see Exercise 2.2). ◁

An important observation is that the special structure of the Hamiltonians (namely, the convexity in p) does not play any role. Actually, Theorem 2.1 remains true for general Hamiltonians satisfying (2.9), (2.11), and (2.12). This fact has application to differential games problems (see Chapter VIII).

Next we describe an application of Theorem 2.1 (rather, a slight variant of it) that provides the justification of a natural regularization procedure for the minimal time function. We confine ourselves to the case of linear control systems and target $\mathcal{T} = \{0\}$. With the notations of §1 in Chapter IV, let

$$T(x) = \inf_{\alpha \in \mathcal{A}} t_x(\alpha), \quad x \in \mathbb{R}^N,$$

where $t_x(\alpha)$ is the entry time in the target $\mathcal{T} = \{0\}$ for the control system

(F, A)
$$\begin{cases} \dot{y}(t) = Fy(t) + \alpha(t), \\ y(0) = x. \end{cases}$$

Here, F is a given $N \times N$ matrix,

$$\alpha \in \mathcal{A} = \{\alpha : [0, +\infty[\to A \subseteq \mathbb{R}^N, \ \alpha \text{ measurable}\}.$$

Let us consider, for $h \in]0,1]$, the perturbed control systems

(F, A_h)
$$\begin{cases} \dot{y}_h(t) = Fy_h(t) + \alpha_h(t), \\ y_h(0) = x, \end{cases}$$

with $\alpha_h \in \mathcal{A}_h = \{\alpha : [0, +\infty[\to A_h := A + h\bar{B}(0,1), \ \alpha \text{ measurable}\}$ and the corresponding minimal time function

$$T_h(x) = \inf_{\alpha \in \mathcal{A}_h} t_x(\alpha).$$

If we assume

(2.19) $\qquad\qquad A$ compact and convex, $0 \in A$,

then, for any $h > 0$, A_h is compact and strictly convex with smooth ∂A_h and $A_h \supseteq \bar{B}(0, h)$. Hence $T_h \in C^1(\mathcal{R}_h \setminus \{0\})$ by Theorem IV.1.10 and by Proposition IV.2.3 it is a classical solution of

$$\sup_{a \in A_h} \{-(Fx + a) \cdot DT_h(x)\} = 1 \quad \text{in } \mathcal{R}_h \setminus \{0\}.$$

We also know that $T \in C(\mathcal{R} \setminus \{0\})$ and is a viscosity solution of

$$\sup_{a \in A} \{-(Fx + a) \cdot DT(x)\} = 1 \quad \text{in } \mathcal{R} \setminus \{0\},$$

provided the system (F, A) is small-time locally controllable on the origin, see Definition IV.1.1 and Proposition IV.2.3.

The next result shows that if T is γ-Hölder continuous in a neighborhood of $\{0\}$, then $T \simeq T_h + Ch^\gamma$, locally in $\mathcal{R}$, for small $h > 0$. This fact has some useful consequence for the synthesis of time-optimal control for the system (F, A) (see Remark 2.5).

THEOREM 2.4. *Let us assume A compact and*

(2.20) $\qquad \exists M > 0,\ \gamma \in\ ]0, 1]$ *such that* $T(x) \leq M |x|^\gamma$, *for all x in a neighborhood of $\{0\}$.*

Then for every compact $K \subset \mathcal{R}$ there exists $C = C(K)$ such that

$$\sup_K |T(x) - T_h(x)| \leq Ch^\gamma.$$

PROOF. Since $A \subseteq A_h$ we have $T_h \leq T$, so that it is enough to prove

(2.21) $\qquad\qquad \sup_K (T(x) - T_h(x)) \leq Ch^\gamma.$

Let us define for $x \in \mathbb{R}^N$ new unknown functions

$$u_h(x) = \chi(T_h(x)), \qquad u(x) = \chi(T(x))$$

via the Kružkov transformation

$$\chi(r) = \begin{cases} \dfrac{1 - e^{-\lambda r}}{\lambda} & r < +\infty, \\ \dfrac{1}{\lambda} & r = +\infty, \end{cases}$$

with $\lambda > 0$ to be chosen later. By Propositions IV.1.2 and IV.1.6 u_h, u belong to $BC(\mathbb{R}^N)$ and satisfy, respectively,

(HJB)$_h$ $\qquad \lambda u_h(x) + \sup\limits_{a \in A_h} \{-(Fx + a) \cdot Du_h(x)\} = 1 \quad$ in $\mathbb{R}^N \setminus \{0\},$

(HJB) $\qquad \lambda u(x) + \sup\limits_{a \in A} \{-(Fx + a) \cdot Du(x)\} = 1 \quad$ in $\mathbb{R}^N \setminus \{0\},$

2. REGULAR PERTURBATIONS

in the viscosity sense. Let us observe that in the present case (i.e., $A_h = A + h\bar{B}(0,1)$, $f_h(x,a) = Fx + a$, for any $a \in A + \bar{B}(0,1)$, $\ell_h(x,a) \equiv 1$) the assumptions of Theorem 2.1 are trivially satisfied.

Now, from Propositions IV.1.2 and IV.1.6 and Remark IV.1.7, it follows that (F, A) satisfies (STLC) and $T \in C^{0,\gamma}(K)$ for all compact $K \subset \mathcal{R}$. Then it is easy to prove that $u = \chi(T) \in C^{0,\gamma}(\{x : T(x) \leq 2\})$. We claim that this implies $u \in C^{0,\gamma}(\mathbb{R}^N)$ if $\lambda > \gamma L$, where L is the Euclidean norm of the matrix F. To prove the claim we consider the set $\Omega := \{x \in \mathcal{R} : T(x) > 1\}$ and denote with ω and ω', respectively, the modulus of continuity of u in $\mathbb{R}^N$ and in Ω. Then

$$(2.22) \qquad \omega(r) \leq \omega'(r) + Cr^\gamma, \qquad r \geq 0.$$

We fix $x, z \in \Omega$ such that $u(z) \leq u(x)$, and $\varepsilon > 0$. By the Dynamic Programming Principle for the problem of minimizing $J(x, \alpha) := \int_0^{t_x(\alpha)} e^{-\lambda s}\, ds$, there exists $\alpha \in \mathcal{A}$ such that

$$u(z) \geq \int_0^1 e^{-\lambda s}\, ds + e^{-\lambda} u(y_z(1,\alpha)) - \varepsilon,$$

$$u(x) \leq \int_0^1 e^{-\lambda s}\, ds + e^{-\lambda} u(y_x(1,\alpha)).$$

Then

$$u(x) - u(z) \leq e^{-\lambda} \omega(|y_z(1,\alpha) - y_x(1,\alpha)|) + \varepsilon,$$

and by standard estimates on the trajectories of the system we get

$$\omega'(r) \leq e^{-\lambda}\omega(e^L r) + \varepsilon.$$

By (2.22) and the arbitrariness of ε we obtain

$$(2.23) \qquad \omega(r) \leq e^{-\lambda}\omega(e^L r) + Cr^\gamma.$$

Given $0 < r < 1$, we fix $n \in \mathbb{N}$ such that

$$e^{-(n+1)L} \leq r < e^{-nL}.$$

We use (2.23) $n+1$ times and compute

$$\omega(r) \leq \omega(e^{-nL}) \leq e^{-(n+1)\lambda}\omega(e^L) + C\sum_{j=1}^n e^{-jL\gamma + (j-n)\lambda}.$$

Since

$$\sum_{j=1}^n e^{-jL\gamma + (j-n)\lambda} = e^{-n\lambda}\sum_{j=1}^n e^{j(\lambda - L\gamma)}$$

$$= \left(e^{-L\gamma(n+1)+\lambda} - e^{(1-n)\lambda - L\gamma}\right)\left(e^{\lambda - L\gamma} - 1\right)^{-1},$$

by the choice $\lambda > \gamma L$ we obtain, for some constant C',

$$\omega(r) \leq C' e^{-(n+1)L\gamma} \leq C' r^\gamma,$$

which proves the claim.

Let us consider now the auxiliary function Ψ employed in the proof of Theorem 2.1. If $\sup_{\mathbb{R}^N \times \mathbb{R}^N} \Psi$ is attained at some $(\overline{x}, \overline{y}) \in (\mathbb{R}^N \setminus \{0\}) \times (\mathbb{R}^N \setminus \{0\})$, then

$$u(x) - u_h(x) \leq C' h^\gamma \qquad \text{for all } x \in \mathbb{R}^N$$

for some $C' > 0$. This follows by the same argument as in the proof of Theorem 2.1. If, on the contrary, $\overline{x} = 0$, we get from the inequality $\Psi(x,x) \leq \Psi(\overline{x}, \overline{y})$,

$$u(x) - u_h(x) - 2\delta\,\mu(x) \leq 0 \qquad \text{for all } x \in \mathbb{R}^N .$$

Hence, letting $\delta \to 0^+$,

$$u(x) - u_h(x) \leq 0 \qquad \text{for all } x \in \mathbb{R}^N .$$

Since $A_h \supseteq A$, we have $T_h \leq T$ and therefore $u_h \leq u$ because χ is increasing. Thus we conclude that

(2.24) $$0 \leq \sup_{\mathbb{R}^N}(u(x) - u_h(x)) \leq C' h^\gamma .$$

For any $x \in \mathcal{R}$ we have $T_h(x) \leq T(x) < +\infty$. The mean value property for χ, when applied with $r = T(x)$, $s = T_h(x)$ gives

$$u(x) - u_h(x) = e^{-\lambda \xi}(T(x) - T_h(x))$$

for some $\xi = \xi(x) \in (T_h(x), T(x))$. Consequently, (2.24) and the above yield

$$T(x) - T_h(x) \leq e^{\lambda T(x)}(u(x) - u_h(x)) \leq C' e^{\lambda T(x)} h^\gamma .$$

Inequality (2.21) now follows with $C = C' e^{\lambda \sup_K T}$. ◀

REMARK 2.5. Since $T_h \in C^1$ when (2.19) holds, the classical synthesis procedure applies to equation $(HJB)_h$. This allows us to construct some $\alpha_h^* \in \mathcal{A}_h$ such that $T_h(x) = t_x(\alpha_h^*)$. From Theorem 2.4 it follows that $T(x) \leq t_x(\alpha_h^*) + Ch^\gamma$.

Note also that T is a bilateral solution of the HJB equation (see Definition III.2.27), because it is a uniform limit of classical solutions. ◁

REMARK 2.6. Sufficient conditions for the validity of (2.20) are provided by the results of Chapter IV (see, for example, Theorem IV.1.9). ◁

Exercises

2.1. For $A = [-1,1]^N$, set $f_0(x,a) = \sigma(x)a$, $a \in A$, where $\sigma(x) = (\sigma_{ij}(x))$ is a semidefinite positive $N \times N$ matrix for each $x \in \mathbb{R}^N$. Set, for $h \in [0,1]$,

$$f_h(x,a) = (\sigma(x) + hI)a, \qquad \ell_h(x,a) \equiv 1 .$$

Assume that for some $L > 0$

$$|\sigma_{ij}(x) - \sigma_{ij}(y)| \leq L|x-y| \qquad \forall\, x,y \in \mathbb{R}^N .$$

Show that the assumptions of Theorem 2.1 are satisfied. Discuss the STC conditions of IV.1 for the dynamics f_h.

2.2. In the setting of Theorem 2.1 take $f_h \equiv f_0$, $A_h \equiv A_0$. Check that

$$|H_h(x,p) - H_0(x,p)| \leq Ch .$$

Show that, in this case,

$$\sup_{\mathbb{R}^N} |u_0(x) - u_h(x)| \leq Ch .$$

[Hint: Observe that the term $|R_2|$ in the proof is estimated by Ch; choose then $\varepsilon = h^{(2-\gamma)/\gamma}$.]

2.3. Consider continuous functions $H_h : \Omega \times \mathbb{R}^N \to \mathbb{R}$, $h \in [0,1]$, satisfying conditions (2.9), (2.11), and (2.12) for any $x, y \in \Omega$, a bounded set in $\mathbb{R}^N$. Let u_h, u_0 be, respectively, viscosity solutions of

$$u_h + H_h(x, Du_h) = 0 \quad \text{in } \Omega, \, h > 0,$$
$$u_0 + H_0(x, Du_0) = 0 \quad \text{in } \Omega,$$

with $u_h = u_0 = 0$ on $\partial\Omega$. Assume that $u \in C^{0,\gamma}(\overline{\Omega})$ for some $\gamma \in\,]0,1]$ and that

$$|u_h(x) - u_h(y)| \leq C|x-y|^\gamma, \qquad \text{for } x \in \Omega,\, y \in \partial\Omega .$$

Follow the lines of the proof of Theorem 2.1 to show that

$$\sup_{\overline{\Omega}} |u_h - u_0| \leq C'h^\gamma,$$

for some constant C' independent of h.

3. Stochastic control with small noise and vanishing viscosity

Here we return to the topic outlined in Chapter I, §8. In the spirit of the present chapter the aim is to prove the convergence and estimates for the vanishing viscosity approximation

$$(\text{VHJB})_\varepsilon \qquad -\varepsilon \Delta u_\varepsilon(x) + \lambda u_\varepsilon(x) + \sup_{a \in A}\{-f(x,a) \cdot Du_\varepsilon(x) - \ell(x,a)\} = 0$$

of equation

(HJB) $$\lambda u(x) + \sup_{a \in A}\{-f(x,a) \cdot Du(x) - \ell(x,a)\} = 0.$$

Despite the different nature of the perturbed equation, the results in this section exhibit a strong analogy with those of §1. For example, a common feature is the loss of one half in the rate of convergence (see Theorems 1.5 and 3.2), due to the singular character shared by the perturbed equations (HJB)$_h$ and (VHJB)$_\varepsilon$.

We consider here equations (HJB) and (VHJB)$_\varepsilon$ in $\mathbb{R}^N$ and assume that the Hamiltonian

$$H(x,p) = \sup_{a \in A}\{-f(x,a) \cdot p - \ell(x,a)\}$$

satisfies

(3.1) $\quad |H(x,p) - H(y,p)| \leq L|x-y|(1+|p|),$

(3.2) $\quad |H(x,p) - H(x,q)| \leq M|p-q|,$

(3.3) $\quad |H(x,0)| \leq M,$

for any x,y,p,q. These conditions are fulfilled, for example, if f and ℓ satisfy (1.1), (1.2), (1.3), and (1.26) in §1. It is well known from classical results on quasilinear elliptic equations (see, for example, [L82, GT83, Kry87]) that for any $\varepsilon > 0$ (VHJB)$_\varepsilon$ has a unique bounded solution $u_\varepsilon \in C^2(\mathbb{R}^N)$. Moreover, standard arguments in stochastic control theory show that

(3.4) $$u_\varepsilon(x) = \inf E_x \int_0^{+\infty} \ell(y_x^\varepsilon(t), \alpha(t)) e^{-\lambda t}\, dt,$$

where y_x^ε is the solution of the Ito's stochastic differential equation

$$\begin{cases} dy(t) = f(y(t), \alpha(t))\, dt + \sqrt{2\varepsilon}\, dw(t), \\ y(0) = x. \end{cases}$$

Here w is an N-dimensional standard Brownian motion, E_x denotes the expectation, and the infimum in (3.4) is taken on the class of progressively measurable processes with values in A (see [FR75, Kry80, FS93]). On the other hand, by the theory of Chapter III, the value function of the infinite horizon problem, namely,

(3.5) $$u(x) = \inf_{\alpha \in \mathcal{A}} \int_0^{+\infty} \ell(y_x(t), \alpha(t)) e^{-\lambda t}\, dt$$

is the unique viscosity solution of (HJB) in $BUC(\mathbb{R}^N)$.

THEOREM 3.1. *Let us assume* (3.1)–(3.3) *and* $\lambda > 0$. *Then*

$$\sup_{\mathcal{K}} |u_\varepsilon(x) - u(x)| \longrightarrow 0 \quad \text{as } \varepsilon \to 0^+,$$

for every compact $\mathcal{K} \subseteq \mathbb{R}^N$.

PROOF. It is straightforward to check that the constants $\pm M/\lambda$ with M as in (3.3) are, respectively, a sub- and a supersolution of $(\text{VHJB})_\varepsilon$, for any $\varepsilon > 0$. Through the maximum principle for elliptic equations we get then the uniform bound

$$\sup_{\mathbb{R}^N} |u_\varepsilon(x)| \leq M/\lambda.$$

Then set

$$\underline{u}(x) = \liminf_{(y,\varepsilon) \to (x,0^+)} u_\varepsilon(y),$$
$$\overline{u}(x) = \limsup_{(y,\varepsilon) \to (x,0^+)} u_\varepsilon(y), \qquad x \in \mathbb{R}^N.$$

Functions $\underline{u}, \overline{u}$ are, respectively, lower and upper semicontinuous. The claim is that $\underline{u}$ is a supersolution and $\overline{u}$ is a subsolution of (HJB) according to the definition in Chapter V, §1. To prove the claim, let $\varphi \in C^2(\mathbb{R}^N)$ and $\underline{x}$ be a strict minimum for $\underline{u} - \varphi$ in $\overline{B} = \overline{B}(\underline{x}, r)$. By Lemma V.1.6 there exist $x_n \to \underline{x}$ and $\varepsilon_n \to 0^+$ such that

$$(u_{\varepsilon_n} - \varphi)(x_n) = \min_{\overline{B}}(u_{\varepsilon_n} - \varphi), \qquad u_{\varepsilon_n}(x_n) \longrightarrow \underline{u}(\underline{x}).$$

From the minimum property above it follows that

$$-\Delta u_{\varepsilon_n}(x_n) \leq -\Delta \varphi(x_n);$$

hence $(\text{VHJB})_\varepsilon$ gives

$$-\varepsilon_n \Delta \varphi(x_n) + \lambda u_{\varepsilon_n}(x_n) + H(x_n, D\varphi(x_n)) \geq 0.$$

Now let $n \to \infty$ to obtain

$$\lambda \underline{u}(\underline{x}) + H(\underline{x}, D\varphi(\underline{x})) \geq 0.$$

The proof that $\overline{u}$ is a viscosity subsolution of (HJB) is completely similar. By the comparison principle, Theorem V.1.3, the conclusion follows. ◂

An extra regularity assumption on u allows us to get the uniform convergence over all $\mathbb{R}^N$ and to estimate the rate of convergence.

THEOREM 3.2. *Let us assume* (3.1)–(3.3), $\lambda > 0$ *and*

(3.6) $\qquad u \in C^{0,\gamma}(\mathbb{R}^N), \quad \text{for some } \gamma \in \,]0,1]\,.$

Then, for some $C > 0$,

$$\sup_{\mathbb{R}^N} |u_\varepsilon(x) - u(x)| \leq C\varepsilon^{\gamma/2} \qquad \text{as } \varepsilon \to 0^+.$$

PROOF. Consider
$$\Psi(x,y) = u_\varepsilon(x) - u(y) - \frac{|x-y|^2}{2\alpha} - \delta(\mu(x) + \mu(y))$$
with $\alpha, \delta > 0$ and $\mu \in C^2(\mathbb{R}^N)$ such that

(3.7) $\begin{cases} \mu \geq 0, \quad \mu(0) = 0, \quad \lim_{|x| \to +\infty} \mu(x) = +\infty, \\ |D\mu(x)| + |\Delta\mu(x)| \leq 1, \quad |x||D\mu(x)| \leq 1. \end{cases}$

If $(\overline{x}, \overline{y})$ is a global maximum for Ψ, then

(3.8) $\quad D_x\Psi(\overline{x},\overline{y}) = Du_\varepsilon(\overline{x}) - \frac{\overline{x}-\overline{y}}{\alpha} - \delta D\mu(\overline{x}) = 0,$

(3.9) $\quad \Delta_x\Psi(\overline{x},\overline{y}) = \Delta u_\varepsilon(\overline{x}) - \frac{N}{\alpha} - \delta \Delta\mu(\overline{x}) \leq 0.$

Hence (VHJB)$_\varepsilon$ yields

(3.10) $\quad \lambda u_\varepsilon(\overline{x}) + H\left(\overline{x}, \frac{\overline{x}-\overline{y}}{\alpha} + \delta D\mu(\overline{x})\right) \leq N\frac{\varepsilon}{\alpha} + \varepsilon\delta \Delta\mu(\overline{x}).$

On the other hand, $\overline{y}$ minimizes $-\Psi(\overline{x},y)$. From (HJB) it follows that
$$\lambda u(\overline{y}) + H\left(\overline{y}, \frac{\overline{x}-\overline{y}}{\alpha} - \delta D\mu(\overline{y})\right) \geq 0.$$
Therefore
$$\lambda(u_\varepsilon(\overline{x}) - u(\overline{y}))$$
$$\leq H\left(\overline{y}, \frac{\overline{x}-\overline{y}}{\alpha} - \delta D\mu(\overline{y})\right) - H\left(\overline{x}, \frac{\overline{x}-\overline{y}}{\alpha} + \delta D\mu(\overline{x})\right) + N\frac{\varepsilon}{\alpha} + \varepsilon\delta \Delta\mu(\overline{x}).$$
Using assumptions (3.1) and (3.2) exactly as in the proof of Theorem 2.1 we obtain, for some $C > 0$,

(3.11) $\quad u_\varepsilon(\overline{x}) - u(\overline{y}) \leq C(\alpha^{1/(2-\gamma)} + \alpha^{\gamma/(2-\gamma)} + \delta\alpha^{1/(2-\gamma)} + \delta) + \lambda\left(N\frac{\varepsilon}{\alpha} + \varepsilon\right)$

since $|\overline{x} - \overline{y}| \leq C\alpha^{1/(2-\gamma)}$.

The inequality $\Psi(\overline{x},\overline{y}) \geq \Psi(x,x)$ and (3.11) then give
$$u_\varepsilon(x) - u(x) - 2\delta\mu(x) \leq C\left(\alpha^{1/(2-\gamma)} + \alpha^{\gamma/(2-\gamma)} + \delta\alpha^{1/(2-\gamma)} + \delta + \frac{\varepsilon}{\alpha} + \varepsilon\right),$$
for any $x \in \mathbb{R}^N$. Let us choose now $\alpha = \varepsilon^{(2-\gamma)/2}$ in the preceding to obtain
$$u_\varepsilon(x) - u(x) - 2\delta\mu(x) \leq C\left(\varepsilon^{1/2} + \varepsilon^{\gamma/2} + \delta\varepsilon^{1/2} + \delta + \varepsilon^{\gamma/2} + \varepsilon\right)$$
for all $x \in \mathbb{R}^N$. If we send $\delta \to 0^+$ in the above we conclude that
$$\sup_{\mathbb{R}^N}(u_\varepsilon(x) - u(x)) \leq C\varepsilon^{\gamma/2}$$
for small $\varepsilon > 0$. The inequality $\sup_{\mathbb{R}^N}(u(x) - u_\varepsilon(x)) \leq C\varepsilon^{\gamma/2}$ is proved by exchanging the roles of u and u_ε in Ψ. ◂

REMARK 3.3. Let us point out that the particular structure of H (namely, its convexity in p) is irrelevant in the proofs of Theorems 3.1 and 3.2. Then the results hold for general H satisfying (3.1), (3.2), and (3.3). ◁

REMARK 3.4. Standard verification arguments in stochastic control show that any feedback map a_ε^* such that

$$a_\varepsilon^*(x) \in \arg\max_{a \in A}\{ -f(x,a) \cdot Du_\varepsilon(x) - \ell(x,a) \}$$

is optimal for problem (3.4). From this point of view one can see Theorems 3.1 and 3.2 as the basis for the justification of those numerical methods for deterministic control problems relying on the viscosity approximation. ◁

The final result of this section is analogous to Theorem 1.7. We assume that the viscosity solution u of (HJB) is Lipschitz continuous and

(3.12) $\quad \exists C > 0$ such that $\Delta u_\varepsilon(x) \leq C$ for all $x \in \mathbb{R}^N$ and $\varepsilon > 0$.

Condition (3.12) amounts to the uniform semiconcavity of u_ε (compare with (1.37)). The validity of (3.12) can be ensured under further assumptions on H. This can be proved by elliptic PDEs techniques (see, e.g., §§2.4 and 6.2 in [L82]) which are outside the scope of this book. Exercise 3.1 gives just a flavor of this type of result.

THEOREM 3.5. *Let us assume* (3.1)–(3.3), (3.12) *and*

(3.13) $\quad\quad\quad\quad\quad\quad\quad\quad u \in C^{0.1}(\mathbb{R}^N)\,.$

Then

$$\sup_{\mathbb{R}^N} |u_\varepsilon - u(x)| \leq C\varepsilon,$$

for some $C > 0$.

PROOF. Proceed as in the proof of Theorem 3.2 and observe that because of (3.8) and (3.12), from (VHJB)$_\varepsilon$ it follows that

(3.14) $\quad\quad \lambda u_\varepsilon(\overline{x}) + H\left(\overline{x}, \dfrac{\overline{x} - \overline{y}}{\alpha} + \delta D\mu(\overline{x})\right) = \varepsilon \Delta u_\varepsilon(\overline{x}) \leq C\varepsilon\,.$

As a consequence of this and (3.13), inequality (3.11) now becomes

$$u_\varepsilon(\overline{x}) - u(\overline{y}) \leq C(\alpha + \alpha + \delta\alpha + \delta) + \lambda C\varepsilon\,.$$

At this point choose $\alpha = \varepsilon$ and conclude as in the proof of Theorem 3.2. ◂

REMARK 3.6. For the Dirichlet boundary value problem associated with (HJB), some results similar to Theorem 3.1 on the convergence of the vanishing viscosity approximation have been proved in Chapter V, see Corollaries V.4.27 and V.4.32.
◁

Exercises

3.1. Assume $\lambda > 0$, $F \in C^2(\mathbb{R})$ convex, $\ell \in C^2(\mathbb{R})$ semiconcave. Suppose that the equations
$$-\varepsilon u'' + F(u') + \lambda u = \ell \quad \text{in } \mathbb{R}$$
have a solution $u = u_\varepsilon \in C^4(\mathbb{R}) \cap B(\mathbb{R})$ for each $\varepsilon > 0$. Differentiate the above to prove that $w = u''$ satisfies for some $C > 0$

(i) $\quad -\varepsilon w'' + F'(u')w' + \lambda w \leq C \quad \text{in } \mathbb{R}.$

Deduce from (i) that $\sup_\mathbb{R} u_\varepsilon'' \leq C$.

4. Appendix: Dynamic Programming for Discrete Time Systems

This appendix contains a few results on the *discrete time infinite horizon problem*

(4.1) $\quad \inf_{\alpha \in \mathcal{A}} J(x, \alpha), \quad J(x, \alpha) := \sum_{n=0}^{\infty} \ell(y_n, a_n)\beta^n.$

Here, y_n is given by the recursion

(S) $\quad \begin{cases} y_{n+1} = y_n + f(y_n, a_n), & n = 0, 1, 2, \ldots \\ y_0 = x \in \mathbb{R}^N \end{cases}$

where $a_n \in A$, a topological space, and $f : \mathbb{R}^N \times A \to \mathbb{R}^N$ is a given mapping. We denote by $\mathcal{A}$ the set of all sequences $\alpha = \{a_n\} \subseteq A$ and by $y_n(x, \alpha)$ the corresponding trajectory of (S).

Problem (4.1) can be seen as a discrete time version of the infinite horizon problem (see Chapter III). Some relevant properties of the value function are proved and the Bellman functional equation characterizing the value is established. A main feature of the discrete time case is that the synthesis of optimal feedback controls can be easily performed starting from the discrete Bellman equation.

Let us recall that we applied the results presented here in §1 of this chapter. More precisely, this was done in the case
$$f_h = hf, \quad \ell_h = h\ell, \quad \beta_h = 1 - \lambda h$$
where $h \in \,]0, 1/\lambda[$, $\lambda > 0$.

We assume that ℓ and β satisfy

(A$_1$) $\quad \beta \in \,]0,1[; \quad \exists M \geq 0 : |\ell(x,a)| \leq M, \quad \forall x \in \mathbb{R}^N, \forall a \in A.$

Consider now the value function
$$v(x) := \inf_{\alpha \in \mathcal{A}} J(x, \alpha).$$

The next result is the statement of the Dynamic Programming optimality Principle.

PROPOSITION 4.1. *Assume* (A_1). *Then v satisfies*
$$v(x) = \inf_{a \in A} \{ \beta v(x + f(x,a)) + \ell(x,a) \}, \qquad x \in \mathbb{R}^N.$$

PROOF. Fix any $x \in \mathbb{R}^N$ and any $\alpha = \{a_0, a_1, \dots\} \in \mathcal{A}$. It is easy to check that
$$J(x, \alpha) = \ell(x, a_0) + \beta J(x + f(x, a_0), \overline{\alpha})$$
with $\overline{\alpha} = \{a_1, a_2, \dots\}$. The above identity follows from the fact that
$$y_{n+1}(x, \alpha) = y_n(y_1(x, \alpha), \overline{\alpha}) = y_n(x + f(x, a_n), \overline{\alpha}).$$
By definition of v
$$J(x, \alpha) \geq \ell(x, a_0) + \beta v(x + f(x, a_0)).$$
Hence
$$v(x) \geq \inf_{a \in A} \{ \ell(x, a) + \beta v(x + f(x, a)) \}.$$
To prove the reverse inequality, set $z = x + f(x, a)$ with arbitrary fixed $a \in A$. By definition of $v(z)$, for any $\varepsilon > 0$ there exists $\alpha^\varepsilon = \{a_n^\varepsilon\} \in \mathcal{A}$ such that
$$(4.2) \qquad v(z) \geq J(z, \alpha^\varepsilon) - \varepsilon.$$
Observe now that
$$J(x, \hat{\alpha}) = \ell(x, a) + \beta J(x + f(x, a), \alpha)$$
where $\hat{\alpha} = \{a, a_0, a_1, \dots\}$, $a \in A$. This identity with $\alpha = \alpha^\varepsilon$ and (4.2) gives
$$v(x) \leq \ell(x, a) + \beta v(x + f(x, a)) + \beta \varepsilon.$$
Since $a \in A$ and $\varepsilon > 0$ were arbitrary,
$$v(x) \leq \inf_{a \in A} \{ \ell(x, a) + \beta v(x + f(x, a)) \}. \qquad \blacktriangleleft$$

Proposition 4.1 asserts that v satisfies the Dynamic Programming equation, or Bellman functional equation
$$(\text{DBE}) \qquad u(x) = \inf_{a \in A} \{ \beta u(x + f(x, a)) + \ell(x, a) \}, \qquad x \in \mathbb{R}^N,$$
a discrete version of the Hamilton-Jacobi-Bellman equation for the infinite horizon problem (see Chapter I, Chapter III §2, Chapter VI §1).

Denoting by G the nonlinear operator on the right-hand side of (DBE), any function u such that
$$u(x) \leq (Gu)(x) \qquad \forall x \in \mathbb{R}^N$$
is called a *subsolution* of (DBE). Similarly, u is a *supersolution* provided
$$u(x) \geq (Gu)(x) \qquad \forall x \in \mathbb{R}^N.$$

Under the assumptions made, v is in fact characterized as the unique solution of (DBE). Moreover, a comparison principle between sub- and supersolutions of (DBE) holds.

THEOREM 4.2. *Assume* (A$_1$). *Then* $v \in B(\mathbb{R}^N)$ *and v is the unique solution of* (DBE) *in* $B(\mathbb{R}^N)$. *Also, if* $u_1, u_2 \in B(\mathbb{R}^N)$ *are, respectively, a sub- and a supersolution of* (DBE) *then*

(4.3) $$u_1 \leq u_2 \quad \text{in } \mathbb{R}^N.$$

PROOF. Observe that from (A$_1$) and the definition of v

$$|v(x)| \leq \sum_{n=0}^{\infty} M\beta^n = \frac{M}{1-\beta} \quad \forall x \in \mathbb{R}^N,$$

so that $v \in B(\mathbb{R}^N)$. By Proposition 4.1 v is both a sub- and a supersolution of (DBE); from the comparison principle (4.3) it follows that v is the unique solution of (DBE) in $B(\mathbb{R}^N)$.

In order to prove (4.3), observe that by definition of supersolution for any $\varepsilon > 0$ and $x \in \mathbb{R}^N$ there exists $a^\varepsilon = a^\varepsilon(x) \in A$ such that

$$u_2(x) \geq \beta u_2(x + f(x, a^\varepsilon)) + \ell(x, a^\varepsilon) - \varepsilon.$$

On the other hand

$$u_1(x) \leq \beta u_1(x + f(x, a^\varepsilon)) + \ell(x, a^\varepsilon),$$

since u_1 is a subsolution. Hence

$$u_1(x) - u_2(x) \leq \beta u_1(x + f(x, a^\varepsilon)) - \beta u_2(x + f(x, a^\varepsilon)) + \varepsilon, \quad \forall x \in \mathbb{R}^N,$$

so that

(4.4) $$\sup_{\mathbb{R}^N}(u_1 - u_2) \leq \beta \sup_{\mathbb{R}^N}(u_1 - u_2) + \varepsilon.$$

Since $\beta \in {]}0, 1{[}$ this implies (4.3). ◂

REMARK 4.3. For $\beta = 1$ uniqueness does not hold. Actually, if $u \in B(\mathbb{R}^N)$ is a solution of (DBE) the same is true for $w(x) = u(x) + c$ for any constant c. ◁

In order to proceed in our program leading to the description of the synthesis procedure (see Theorem 4.6) we make some further assumptions on the data, namely

(A$_2$) $\quad \exists L \geq 0 : |f(x, a) - f(z, a)| \leq L|x - z|, \quad \forall x, z \in \mathbb{R}^N, \forall a \in A,$
(A$_3$) $\quad\quad\quad |\ell(x, a) - \ell(z, a)| \leq \omega(|x - z|), \quad \forall x, z \in \mathbb{R}^N, \forall a \in A,$

for some modulus ω. Observe that (A$_2$) implies

(4.5) $\quad |y_n(x, \alpha) - y_n(z, \alpha)| \leq (1 + L)^n |x - z|, \quad \forall \alpha \in \mathcal{A}, n = 0, 1, 2, \ldots$

Then we have

PROPOSITION 4.4. *Assume* (A_1), (A_2), *and* (A_3). *Then* $v \in BUC(\mathbb{R}^N)$. *Moreover, if* (A_3) *holds with* $\omega(r) = L_1 r$ *and* $0 < \beta < 1/(1+L)$, *then*

$$|v(x) - v(z)| \leq \frac{L_1}{1 - \beta(1+L)}|x - z|, \qquad \forall\, x, z \in \mathbb{R}^N. \tag{4.6}$$

PROOF. We have seen already that

$$|v(x)| \leq \frac{M}{1-\beta} \qquad \forall\, x \in \mathbb{R}^N,$$

i.e., $v \in B(\mathbb{R}^N)$. To prove the uniform continuity, observe that for fixed z in $\mathbb{R}^N$ and arbitrary $\varepsilon > 0$ there is $\alpha^\varepsilon = \{a_n^\varepsilon\} \in \mathcal{A}$ such that

$$v(z) \geq J(z, \alpha^\varepsilon) - \frac{\varepsilon}{3}.$$

Hence

$$v(x) - v(z) \leq J(x, \alpha^\varepsilon) - J(z, \alpha^\varepsilon) + \frac{\varepsilon}{3}, \qquad \forall\, x, z \text{ in } \mathbb{R}^N.$$

Consequently, by (A_1),

$$v(x) - v(z) \leq \sum_{n=0}^{k} |\ell(y_n(x, \alpha^\varepsilon), a_n^\varepsilon) - \ell(y_n(z, \alpha^\varepsilon), a_n^\varepsilon)|\beta^n + \sum_{n=k+1}^{\infty} 2M\beta^n + \frac{\varepsilon}{3}$$

for any $k \in \mathbb{N}$. Let us choose now k such that $\sum_{n=k+1}^{\infty} 2M\beta^n < \varepsilon/3$ and use (A_3) and (4.5) to get

$$v(x) - v(z) \leq \sum_{n=0}^{k} \omega((1+L)^n |x - z|)\beta^n + \frac{2\varepsilon}{3}.$$

Since $\omega(r) \to 0$ as $r \to 0$, it is easy to check that this sum can be made smaller than $\varepsilon/3$ for $|x - z|$ small enough. Therefore, v is uniformly continuous.

Let us proceed now to the proof of (4.6). Arguing as before, we obtain easily

$$v(x) - v(z) \leq \sum_{n=0}^{k} L_1(1+L)^n \beta^n |x - z| + \frac{2\varepsilon}{3}$$

for large enough k. Since $0 < \beta < 1/(1+L)$, we can let $k \to +\infty$ to obtain

$$v(x) - v(z) \leq \frac{L_1}{1 - \beta(1+L)}|x - z| + \frac{2\varepsilon}{3}.$$

Since ε was arbitrary this proves (4.6). ◀

The next proposition states sufficient conditions for the *semiconcavity* of v, (see Chapter II, §4.2, Chapter III, and §1.3 in this chapter).

PROPOSITION 4.5. *Assume* (A$_1$), (A$_2$), *and* (A$_3$) *with* $\omega(r) = L_1 r$. *Assume also*

(4.7) $\quad \exists C > 0: \quad |f(x+z,a) - 2f(x,a) + f(x-z,a)| \leq C|z|^2,$

(4.8) $\quad \exists Q > 0: \quad \ell(x+z,a) - 2\ell(x,a) + \ell(x-z,a) \leq Q|z|^2,$

for all $x, z \in \mathbb{R}^N$, $a \in A$, *and*

(4.9) $$0 < \beta < \frac{1}{(1+L)^2}.$$

Then

(4.10) $\quad v(x+z) - 2v(x) + v(x-z) \leq \left(Q + \frac{L_1 C}{L}\right) \frac{1}{1 - \beta(1+L)^2} |z|^2$

for all $x, z \in \mathbb{R}^N$.

PROOF. A simple consequence of the definition of v is that

$$v(x+z) - 2v(x) + v(x-z) \leq \sup_{\alpha \in \mathcal{A}}[J(x+z,\alpha) - 2J(x,\alpha) + J(x-z,\alpha)].$$

Observe that for fixed $\alpha = \{a_n\} \in \mathcal{A}$ we have

$$J(x+z,\alpha) - 2J(x,\alpha) + J(x-z,\alpha) = \sum_{n=0}^{\infty} A_n \beta^n,$$

where

$$A_n := \ell(y_n^x + (y_n^{x+z} - y_n^x), a_n) - 2\ell(y_n^x, a_n) + \ell(y_n^x - (y_n^{x+z} - y_n^x), a_n) \\ + \ell(y_n^{x-z}, a_n) - \ell(y_n^x - (y_n^{x+z} - y_n^x), a_n).$$

In the above we used the notation $y_n^\xi = y_n(\xi, \alpha)$. From assumptions (4.8) and (A$_3$) we obtain

(4.11) $\quad A_n \leq Q|y_n^{x+z} - y_n^x|^2 + L_1|y_n^{x+z} - 2y_n^x + y_n^{x-z}|.$

Let us use now (4.7) and (A$_2$) to estimate the quantity

$$\alpha_n := |y_n^{x+z} - 2y_n^x + y_n^{x-z}|.$$

By definition of y_n^ξ we have

$$\alpha_{n+1} \leq \alpha_n + |f(y_n^{x+z}, a_n) - 2f(y_n^x, a_n) + f(y_n^{x-z}, a_n)| \\ \leq \alpha_n + |f(y_n^x + (y_n^{x+z} - y_n^x), a_n) - 2f(y_n^x, a_n) + f(y_n^x - (y_n^{x+z} - y_n^x), a_n)| \\ + |f(y_n^{x-z}, a_n) - f(y_n^x - (y_n^{x+z} - y_n^x), a_n)|.$$

Therefore, using also estimate (4.5),

$$\alpha_{n+1} \leq \alpha_n + C|y_n^{x+z} - y_n^x|^2 + L\alpha_n \leq (1+L)\alpha_n + C(1+L)^{2n}|z|^2.$$

By recursion this implies

$$\alpha_n \leq C|z|^2 \sum_{k=1}^{2n-1}(1+L)^k = C|z|^2 \frac{(1+L)^{2n}-(1+L)}{L}, \quad n=1,2,\ldots$$

The previous inequality, together with (4.11), yields

$$A_n \leq \left(Q + \frac{L_1 C}{L}\right)(1+L)^{2n}|z|^2, \quad n=1,2,\ldots$$

Hence, using assumption (4.9),

$$v(x+z) - 2v(x) + v(x-z) \leq \sum_{n=0}^{\infty} A_n \beta^n \leq \left(Q + \frac{L_1 C}{L}\right)\frac{1}{1-\beta(1+L)^2}|z|^2. \quad \blacktriangleleft$$

In the next result we state some sufficient conditions for the existence of an optimal feedback control for the infinite horizon problem and give the procedure for its construction from the equation (DBE).

THEOREM 4.6. *Assume* (A_1), (A_2), *and* (A_3) *and*

(4.12) $a \mapsto \ell(x,a)$ *is lower semicontinuous for each fixed* x,

(4.13) $a \mapsto f(x,a)$ *is continuous for each fixed* x,

(4.14) A *is compact*.

Then the control $\alpha^* = \{a_n^*\} \in \mathcal{A}$ *defined by*

$$a_n^* \in \arg\min_{A}\left\{\ell(y_n^*,\cdot) + \beta v(y_n^* + f(y_n^*,\cdot))\right\} \quad (n = 0,1,\ldots),$$

where $y_0^* = x$ *and* $\{y_n^*\}$ *is the trajectory of* (S) *corresponding to* α^*, *is optimal for* x.

PROOF. Under the assumptions made, v is the unique solution of (DBE) in $BUC(\mathbb{R}^N)$ (see Theorem 4.2 and Proposition 4.4). By (4.12), (4.13) the function

$$a \longmapsto \beta v(x + f(x,a)) + \ell(x,a)$$

is lower semicontinuous for each fixed x. Since A is compact, for every $x \in \mathbb{R}^N$ there exists $a(x) \in A$ such that

$$\beta v(x + f(x,a(x))) + \ell(x,a(x)) = \inf_{a \in A}\left\{\beta v(x + f(x,a)) + \ell(x,a)\right\}.$$

On the account of (DBE) this means that the set

$$F(x) = \{a \in A : v(x) = \beta v(x + f(x,a)) + \ell(x,a)\}$$

is nonempty for all $x \in \mathbb{R}^N$. For fixed x, set $y_0^* = x$ and choose any $a_0^* \in F(x)$. Define then recursively the sequences $\{y_n^*\} \subset \mathbb{R}^N$ and $\alpha^* = \{a_n^*\}$ by setting

$$y_n^* = y_{n-1}^* + f(y_{n-1}^*, a_{n-1}^*), \quad a_n^* \in F(y_n^*).$$

for $n = 1, 2, \ldots$ By definition of F, y_{n+1}^*, y_n^*, a_n^* are related by

$$\beta^n v(y_n^*) - \beta^{n+1} v(y_{n+1}^*) = \beta^n \ell(y_n^*, a_n^*), \qquad n = 0, 1, \ldots .$$

Therefore

$$\sum_{n=0}^{\infty} \beta^n (v(y_n^*) - \beta v(y_{n+1}^*)) = \sum_{n=0}^{\infty} \ell(y_n^*, a_n^*) \beta^n .$$

It is straightforward to check that $\sum_{n=0}^{\infty} \beta^n (v(y_n^*) - \beta v(y_{n+1}^*)) = v(x)$; hence

$$v(x) = \sum_{n=0}^{\infty} \ell(y_n^*, a_n^*) \beta^n = J(x, \alpha^*)$$

and the proof is complete. ◂

Exercises

4.1. Consider the value function

$$v_h(x) = \inf_{\alpha \in \mathcal{A}_h} J_h(x, \alpha),$$

where J_h and $\mathcal{A}_h$ are defined in §1.1. Use Theorem 4.2, Propositions 4.4 and 4.5 to deduce that for $h \in \,]0, 1/\lambda[$ the following estimates hold:

(a) $\quad |v_h(x)| \le \dfrac{M}{\lambda}$ \hfill if $\lambda > 0$;

(b) $\quad |v_h(x) - v_h(z)| \le \dfrac{L_1}{\lambda - L} |x - z|$ \hfill if $\lambda > L$;

(c) $\quad v_h(x + z) - 2v_h(x) + v_h(x - z) \le \dfrac{LM + L_1 C}{L(\lambda - 2L)} |z|^2$ \hfill if $\lambda > 2L$

for all $x, z \in \mathbb{R}^N$.

4.2. Take $A = \mathbb{R}^N$, $f(x, a) = Mx + a$, $\ell(x, a) = |x|^2 + |a|^2$ where M is a $N \times N$ matrix. The DBE equation is

$$u(x) = \inf_{a \in \mathbb{R}^N} \left\{ \beta u((I + M)x + a) + |x|^2 + |a|^2 \right\} .$$

(i) Show that $u(x) := Qx \cdot x$ is a solution provided that Q satisfies the matrix Riccati equation

$$Q = I + \beta (I + M)^t (I + \beta Q)^{-1} Q (I + M) .$$

(ii) Compute Q in the case $N = 1$ and show that

$$u(x) = \left((1 + M)^2 + 1 - 1/\beta + \sqrt{((1 + M)^2 + 1 - 1/\beta)^2 + 4/\beta} \right) \frac{x^2}{2} .$$

4.3. For any control $\alpha = \{a_n\} \in \mathcal{A}$ and trajectory $y_n(x, \alpha)$ of system (S) define

$$n(x, \alpha) := \begin{cases} \min\{n \in \{0, 1, 2, \ldots\} : y_n(x, \alpha) \in \mathcal{T}\}, \\ +\infty \text{ if } y_n(x, \alpha) \notin \mathcal{T} \text{ for all } n, \end{cases}$$

where $\mathcal{T}$ is a given subset of $\mathbb{R}^N$. Consider the minimum time function

$$N(x) := \min_{\alpha \in \mathcal{A}} n(x, \alpha)$$

and the reachable set

$$\mathcal{R} := \{x \in \mathbb{R}^N : N(x) < +\infty\}.$$

Show that if A is compact, f continuous, and $\mathcal{T}$ closed, then N satisfies

$$N(x) = 1 + \min_{a \in A} N(x + f(x, a)), \quad \forall x \in \mathcal{T} \smallsetminus \mathcal{T}.$$

4.4. Consider the minimum time function N in the case $\mathcal{T} = \bar{B}(0, 1)$, $A = \bar{B}(0, R)$, $f(x, a) = a$. Show that the mapping $x \mapsto a^*(x) := -Rx/|x|$ is an optimal feedback map and compute $N(x)$.

5. Bibliographical notes

The problem of convergence of semidiscrete approximations of Hamilton-Jacobi-Bellman equations can be traced back to Bellman's own work [Bel57, BelD62]. Our presentation in §1 basically follows [CD83] and [CDI84] (see also [F61, F64, Cu69, M79, CDM81] for previous convergence results in particular cases, and [CL84, CDF89], where connections with different approximations procedures are discussed). Note, however, that the proof of convergence in Theorem 1.1 uses ideas of [BF90a], where the weak limits technique of Barles-Perthame [BaP87, BaP88] was first applied to the discretization of HJB equations. The estimates in Theorem 1.5 were proved in [CDI84, Sou85a]; the role of semiconcavity to improve estimates as well as the convergence of suboptimal feedbacks to optimal relaxed controls are results from [CDI84]. We refer to the papers [Kr66b, CL84, Sou85a, Sou85b, BaSo91] for general consistency and convergence results. Section 1 also includes a simple but apparently new constructive proof of existence of optimal controls for the infinite horizon problem, Corollary 1.4.

Semidiscrete approximation schemes of the type described in §1 have been applied to several optimal control problems; see, for example, the papers [BF90a, BF90b] for the minimum time, [LoT94, CFa96] for the case of state constraints, [Cam96] for $\mathcal{H}_\infty$ control, [CFa95b] for impulse problems, and [BS91b, BFS94, Sor97b, BFS97] for some differential games. Higher order schemes have been proposed in [FaF90, FaF94, GTi92]. Extensions to the case of discontinuous value functions can be found in [BFS94, BBF95]. Other related papers are [GR85, MeR86, Lo86, RV91, Ta94, AL94, DMG95, LS95], and [RT92] for problems in shape from shading, [LoS93] for some infinite dimensional case. Recently, approximation schemes based on the viability theory have been proposed in [CQS94a, CQS94b, CQS97].

Similar approximation schemes have been considered and their convergence proved in the stochastic case. Let us mention in this respect the books [Kus77, KD92, FS93] and the papers [Me89, CFa95a].

The material of §2 is from the papers [BSa91a, BSa92], where more general problems in optimal control and differential games are also considered. Theorem 2.4 refines a result in [BBi90]; see also the references therein. Related works are [Ja77, Pe79, Sa90]. For a different approach to well-posedness in optimal control see the recent book [DZ93] and, for perturbation results, [Ben88].

The vanishing viscosity approximation to Hamilton-Jacobi equations studied in §3 is very classical (see, e.g., [Kr64, Kr66a, F69]). Theorems 3.2 and 3.5 slightly improve some results in [L82] (see also [K91]). A more extensive treatment of this important topic, including in particular the connections with stochastic control, is outside the scope of the present book. The reader is referred to the book by Fleming and Soner [FS93] (see also [FR75, BL82]). The related topic of boundary conditions in the viscosity sense is treated in Chapter V. We refer to [FSo86] for asymptotic expansions, to [PS88] and [CLS89] for the vanishing viscosity, respectively, in the Neumann boundary value problem and in the Cauchy problem with infinite data, and to [LSV87, EM94, BM96] for approximation of Hamilton-Jacobi equations via nonlinear degenerate diffusion.

The use of viscosity solutions methods in large deviations problems was initiated by Evans and Ishii in [EI85] and surveys on this topic can be found in the books [E90, FS93, Ba94]; other results on the exponential decay of solutions in singular perturbation problems for elliptic equations are in [Kam84, Kam86, Kam88, B87, BP90, BP91, BaP88, BaP90, Ei87, Ei90, Ei93, IK91a].

CHAPTER VII

Asymptotic problems

In this chapter we consider several asymptotic problems in optimal control. Our approach is to pass to the limit as the relevant parameter goes to zero in the Hamilton-Jacobi-Bellman equation satisfied by the value function and characterize the limit value function as the viscosity solution of the limit equation.

A useful tool in this approach is the *weak limit technique* described in §V.1. The main advantage provided by this technique in the context of the present chapter lies in the fact that it requires only a priori bounds in the supremum norm.

Throughout the chapter we assume without explicit mention that conditions (A_0)-(A_4) of Chapter II hold. For simplicity, we assume also that the control set A is compact and that (A_4) is satisfied with $\omega_\ell(r) = Lr$, even though most of the results remain valid in greater generality.

1. Ergodic problems

In this section we investigate the limiting behavior of the value function v_λ of the infinite horizon problem as the discount rate $\lambda \to 0^+$. In §1.1 we consider the case of state constraints and show that, under a controllability assumption, λv_λ converges to some constant χ_0 that can be evaluated as the infimum of a long run averaged cost criterion (see Theorem 1.1 and Proposition 1.3). Section 1.2 deals with the unconstrained case, with special attention to a particular case, namely, that of the optimal stopping time. In this case a dissipativity condition on the dynamics allows us to obtain the same results as in the constrained case (Theorem 1.6). Moreover, an estimate on the rate of convergence is established (Proposition 1.9).

1.1. Vanishing discount in the state constrained problem

We consider here the limiting behavior as $\lambda \to 0^+$ of the value function v_λ of the infinite horizon problem with state constraints discussed in Chapter IV, §5. Let us recall for convenience that, for $\lambda > 0$, v_λ is defined by

$$v_\lambda(x) = \inf_{\alpha \in \mathcal{A}_x} \int_0^{+\infty} \ell(y_x(s,\alpha), \alpha(s)) e^{-\lambda s} \, ds, \qquad x \in \overline{\Omega},$$

where Ω is a bounded open set in $\mathbb{R}^N$ with smooth boundary and

$$\mathcal{A}_x = \{ \alpha \in \mathcal{A} : y_x(s,\alpha) \in \overline{\Omega}, \ \forall s \geq 0 \}.$$

Let us also recall that the following boundary condition on the dynamics

(1.1) $$\inf_{a \in A} f(x,a) \cdot n(x) < 0 \qquad \forall x \in \partial\Omega,$$

where $n(x)$ is the exterior normal to Ω at x, and the standard assumptions on ℓ, f, imply that $v_\lambda \in C(\overline{\Omega})$ and it is the unique constrained viscosity solution of the Bellman equation

(HJB)$_\lambda$ $$\lambda v_\lambda(x) + \sup_{a \in A} \{ -f(x,a) \cdot Dv_\lambda(x) - \ell(x,a) \} = 0 \qquad x \in \overline{\Omega}$$

(i.e., v_λ is a viscosity subsolution in Ω and a viscosity supersolution in $\overline{\Omega}$ of (HJB)$_\lambda$).

As a preliminary remark, let us observe that the Hamiltonian

$$F_\lambda(x,r,p) = \lambda r + \sup_{a \in A} \{ -f(x,a) \cdot p - \ell(x,a) \}$$

converges locally uniformly as $\lambda \to 0^+$ to

$$F_0(x,r,p) = \sup_{a \in A} \{ -f(x,a) \cdot p - \ell(x,a) \}.$$

Therefore, by Proposition II.2.2, one can pass to the limit in (HJB)$_\lambda$ if v_λ is known to converge locally uniformly in $\mathbb{R}^N$ as $\lambda \to 0^+$. However this cannot be true in general. Indeed, if ℓ is a constant, then $|v_\lambda| \equiv |\ell|/\lambda \to +\infty$ as $\lambda \to 0^+$.

Let us make the controllability assumption

(1.2) $$\exists r > 0 \text{ such that } B(0,1/r) \subseteq \overline{\text{co}} f(x,A), \text{ for all } x \in \overline{\Omega}.$$

If (1.2) holds, then

(1.3) $$|v_\lambda(x) - v_\lambda(z)| \leq Mr|x-z| \qquad \forall x, z \in \Omega,$$

where $M = \sup_{\mathbb{R}^N \times A} |\ell(x,a)|$, as it follows easily from Proposition III.2.3.

The uniform Lipschitz estimate (1.3) suggests the introduction of the function

$$w_\lambda(x) := v_\lambda(x) - v_\lambda(x_0),$$

where x_0 is arbitrarily fixed in $\overline{\Omega}$. It is easy to check that w_λ satisfies

(HJB)$'_\lambda$ $$\lambda w_\lambda(x) + \sup_{a \in A} \{ -f(x,a) \cdot Dw_\lambda(x) - \ell(x,a) \} + \lambda v_\lambda(x_0) = 0, \qquad x \in \overline{\Omega}$$

in the constrained viscosity sense. The asymptotic behavior of (HJB)$'_\lambda$ as $\lambda \to 0^+$ is described by the next result.

1. ERGODIC PROBLEMS

THEOREM 1.1. *Let us assume* (1.1) *and* (1.2). *Then there exists* $\chi_0 \in \mathbb{R}$ *such that*

(1.4) $$\lim_{\lambda \to 0^+} \lambda v_\lambda(x) = \chi_0, \quad \textit{uniformly in } \overline{\Omega} .$$

Moreover,

$$\lim_{n \to +\infty} w_{\lambda_n} = w_0, \quad \textit{uniformly in } \overline{\Omega},$$

for some $\lambda_n \to 0$, *and* w_0 *is a constrained viscosity solution of*

(HJB)$'_0$ $$\sup_{a \in A}\{-f(x,a) \cdot Dw_0 - \ell(x,a)\} + \chi_0 = 0 \quad \textit{in } \overline{\Omega}.$$

PROOF. From (1.3) it follows immediately that

(1.5) $$\begin{aligned} |\lambda v_\lambda(x) - \lambda v_\lambda(z)| &\le \lambda Mr|x-z|, \\ |w_\lambda(x)| &\le Mr|x-x_0|, \\ |w_\lambda(x) - w_\lambda(z)| &\le Mr|x-z|, \end{aligned}$$

for any x, z, x_0 in $\overline{\Omega}$. Also, the uniform bound

$$|\lambda v_\lambda(x)| \le \sup_{\mathbb{R}^N \times A} |\ell(x,a)|$$

is easily obtained either directly from the definition of v_λ or by comparison results in Chapter III. Therefore, by the Ascoli-Arzelà Theorem,

$$\lim_{n \to +\infty} \lambda_n v_{\lambda_n} \longrightarrow \chi_0 \in C(\overline{\Omega}), \quad \lim_{n \to +\infty} w_{\lambda_n} \longrightarrow w_0 \in C(\overline{\Omega}),$$

uniformly in $\overline{\Omega}$, for some sequence $\lambda_n \to 0$ as $n \to +\infty$. It is an immediate consequence of the first inequality in (1.5) that χ_0 is a constant, and consequently, $\lambda_n w_{\lambda_n}(x) \to 0$ uniformly in $\overline{\Omega}$. Let us check now that (χ_0, w_0) satisfy (HJB)$'_0$. By Proposition II.2.2 we can pass to the limit in (HJB)$'_{\lambda_n}$ as $n \to \infty$ to obtain

$$\sup_{a \in A}\{-f(x,a) \cdot Dw_0 - \ell(x,a)\} + \chi_0 = 0 \quad \textit{in } \Omega$$

in the viscosity sense. In order to check the supersolution condition on $\partial\Omega$, let $x \in \partial\Omega$ be a local strict minimum for $w_0 - \varphi$, $\varphi \in C^1(\overline{\Omega})$. Then by Lemma II.2.4 there exist local minimum points x_{λ_n} for $w_{\lambda_n} - \varphi$ such that

$$x_{\lambda_n} \longrightarrow x, \quad w_{\lambda_n}(x_{\lambda_n}) \longrightarrow w_0(x).$$

Since w_λ is a supersolution of (HJB)$'_\lambda$ on $\overline{\Omega}$,

$$\lambda_n w_{\lambda_n}(x_{\lambda_n}) + \sup_{a \in A}\{-f(x_{\lambda_n},a) \cdot D\varphi(x_{\lambda_n}) - \ell(x_{\lambda_n},a)\} + \lambda_n v_{\lambda_n}(x_0) \ge 0$$

and the desired statement follows by letting $n \to +\infty$ in the above.

It remains to be proved that χ_0 is uniquely defined by (HJB)$'_0$ and that the whole family $\{\lambda v_\lambda\}$ converges to χ_0. For this purpose, let $H(x,p) = \sup_{a \in A}\{-f(x,a) \cdot p - \ell(x,a)\}$ and consider the equation

$$(1.6) \qquad H(x, Du) + \alpha = 0 \quad \text{in } \overline{\Omega}.$$

The claim is that there is a unique $\alpha \in \mathbb{R}$ for which (1.6) has a constrained viscosity solution $u \in C(\overline{\Omega})$.

Indeed, suppose that u and $\tilde{u}$ are constrained viscosity solutions of (1.6) corresponding to α, $\tilde{\alpha}$ with $\tilde{\alpha} < \alpha$.

Fix $\delta > 0$ such that

$$-\tilde{\alpha} + \delta \tilde{u}(x) \geq -\alpha + \delta(u(x) + C), \qquad \forall x \in \overline{\Omega}$$

with $C > \sup_{\overline{\Omega}}|\tilde{u}(x) - u(x)|$. Observe now that $\tilde{u}$ is a constrained solution of

$$\delta \tilde{u} + H(x, D\tilde{u}) = -\tilde{\alpha} + \delta \tilde{u} \quad \text{in } \overline{\Omega}$$

while $u + C$ satisfies

$$\delta(u + C) + H(x, D(u + C)) = -\alpha + \delta(u + C).$$

By the comparison result for constrained solution (see §IV.5) we conclude

$$\tilde{u} \geq u + C,$$

a contradiction to the choice of C. Therefore $\tilde{\alpha} \geq \alpha$. Reversing the roles of α, $\tilde{\alpha}$ the claim is proved. Using this fact for equation (HJB)$'_0$, the proof is easily completed. ◀

REMARK 1.2. Uniqueness does not hold in general for equation (HJB)$'_0$. Indeed, it is clear that $u + C$ is a solution of (HJB)$'_0$ if u is such and C is any constant.

The next simple example shows that (HJB)$'_0$ can have a very large set of constrained viscosity solutions.

Let Ω be any smooth bounded open set in $\mathbb{R}^N$ and $A = \{a \in \mathbb{R}^N : |a| \leq 1\}$. For $f(x,a) \equiv -a$ and $\ell(x,a) \equiv |a|$ it is easy to compute

$$\sup_{a \in A}\{-f(x,a) \cdot p - \ell(x,a)\} = (|p| - 1)^+.$$

The unique constrained viscosity solution of (HJB)$_\lambda$, $\lambda > 0$, is $v_\lambda \equiv 0$ so that $\chi_0 = 0$ and $w_\lambda \equiv 0$ in this case. Since $H(x,p) \equiv (|p| - 1)^+ \geq 0$ for all p, any continuous function is a supersolution on $\overline{\Omega}$ of (HJB)$'_0$; on the other hand it is easy to check that any Lipschitz continuous u with $|Du(x)| \leq 1$ a.e. in Ω is a subsolution of (HJB)$'_0$. Because of the lack of uniqueness for (HJB)$'_0$ we do not know in general if the whole family $\{w_\lambda\}$ converges to w_0. ◁

The next result states that the constant χ_0 appearing in (HJB)$'_0$ can be interpreted in terms of a *long run averaged cost* criterion for the infinite horizon problem with state constraints.

1. ERGODIC PROBLEMS

PROPOSITION 1.3. *Under the assumptions of Theorem 1.1, the constant χ_0 is given by*

$$\chi_0 = \inf_{\alpha \in \mathcal{A}_x} \left\{ \liminf_{T \to +\infty} \frac{1}{T} \int_0^T \ell(y_x(s), \alpha(s)) \, ds \right\}.$$

PROOF. By the DPP (see Proposition IV.5.5),

$$(1.7) \quad \lambda(v_\lambda(x) - e^{-\lambda T} v_\lambda(y_x(t, \alpha))) \leq \lambda \int_0^T \ell(y_x(s), \alpha(s)) \, ds + \lambda \psi(T),$$

for any $x \in \overline{\Omega}$, $T > 0$, $\lambda > 0$ and $\alpha \in \mathcal{A}_x$, where

$$\psi(T) = \int_0^T (e^{-\lambda s} - 1) \ell(y_x(s), \alpha(s)) \, ds.$$

An immediate computation shows that

$$\lambda \psi(T) \leq M\lambda\left(\frac{1}{\lambda} - \frac{1}{\lambda} e^{-\lambda T} - T\right), \qquad M = \sup_{\mathbb{R}^N \times A} |\ell(x, a)|.$$

Let us fix $\delta > 0$. For λ, T such that $\lambda T = \delta$, (1.7) yields

$$\frac{\delta}{T}\left(v_{\delta/T}(x) - e^{-\delta} v_{\delta/T}(y_x(T, \alpha))\right) \leq \frac{\delta}{T} \int_0^T \ell(y_x(s), \alpha(s)) \, ds + M(1 - e^{-\delta} - \delta).$$

Thanks to the uniform convergence in (1.4) and the compactness of $\overline{\Omega}$ we obtain from the above, as $T \to +\infty$,

$$(1 - e^{-\delta})\chi_0 \leq \delta \liminf_{T \to +\infty} \frac{1}{T} \int_0^T \ell(y_x(s), \alpha(s)) \, ds + M(1 - e^{-\delta} - \delta).$$

The preceding yields, dividing by δ and letting $\delta \to 0^+$,

$$\chi_0 \leq \liminf_{T \to +\infty} \frac{1}{T} \int_0^T \ell(y_x(s), \alpha(s)) \, ds, \qquad \forall \alpha \in \mathcal{A}_x.$$

To prove the opposite inequality, fix again $\delta > 0$. By Proposition IV.5.5, there exists $\overline{\alpha} \in \mathcal{A}_x$ such that

$$\delta^2 + \lambda(v_\lambda(x) - e^{-\lambda T} v_\lambda(y_x(T, \overline{\alpha}))) \geq \lambda \int_0^T \ell(y_x(s), \overline{\alpha}(s)) \, ds + \lambda \psi(T),$$

and it is easy to see that $\overline{\alpha}$ can be chosen independent of T (see the proof of the DPP, Proposition III.2.5). For $\lambda T = \delta$ this gives

$$\delta^2 + \frac{\delta}{T}\left(v_{\delta/T}(x) - e^{-\delta} v_{\delta/T}(y_x(T, \overline{\alpha}))\right) \geq \frac{\delta}{T} \int_0^T \ell(y_x(s), \overline{\alpha}(s)) \, ds - M(1 - e^{-\delta} - \delta)$$

for any $T > 0$. Hence, by the same arguments as in the first part of the proof, as $T \to +\infty$ we obtain

$$\delta^2 + (1 - e^{-\delta})\chi_0 \geq \delta \liminf_{T \to +\infty} \frac{1}{T} \int_0^T \ell(y_x(s), \overline{\alpha}(s))\,ds - M(1 - e^{-\delta} - \delta).$$

Now take the inf over $\mathcal{A}_x$, divide by δ and let $\delta \to 0$ in the above, to get

$$\chi_0 \geq \inf_{\alpha \in \mathcal{A}_x} \left\{ \liminf_{T \to +\infty} \frac{1}{T} \int_0^T \ell(y_x(s), \alpha(s))\,ds \right\}. \quad \blacktriangleleft$$

REMARK 1.4. The constant χ_0 is also related to the asymptotic behavior as $T \to +\infty$ of the finite horizon problem with state constraint $\overline{\Omega}$, for any terminal cost g. The value function of this problem is

$$v(x, T) = \inf_{\alpha \in \mathcal{A}_x} \left\{ \int_0^T \ell(y_x(s, \alpha), \alpha(s))\,ds + g(y_x(T, \alpha)) \right\}$$

and it is not hard to show that $\chi_0 = \lim_{T \to +\infty} T^{-1} v(x, T)$ uniformly in $\overline{\Omega}$ (see Exercise 1.1). $\triangleleft$

1.2. Vanishing discount in the unconstrained case: optimal stopping

We consider now the same limit problem as in §1.1 in the unconstrained case $\Omega = \mathbb{R}^N$. In this setting a result of the type of Theorem 1.1 still holds true. Namely, if we consider

$$v_\lambda(x) = \inf_{\alpha \in \mathcal{A}} \int_0^{+\infty} \ell(y_x(s, \alpha), \alpha(s)) e^{-\lambda s}\,ds \qquad x \in \mathbb{R}^N,$$

the value function of the unconstrained infinite horizon problem, we have

THEOREM 1.5. *Let us assume* (1.2). *Then there exists a constant χ_0 such that*

$$\lambda v_\lambda(x) \longrightarrow \chi_0,$$

locally uniformly in $\mathbb{R}^N$ as $\lambda \to 0^+$. Moreover, for any fixed $x_0 \in \mathbb{R}^N$, $w_{\lambda_n}(x) = v_{\lambda_n}(x) - v_{\lambda_n}(x_0)$ converges locally uniformly in $\mathbb{R}^N$ for some $\lambda_n \to 0^+$ to $w_0 \in C(\mathbb{R}^N)$ such that

$$\sup_{a \in A}\{-f(x, a) \cdot Dw_0(x) - \ell(x, a)\} + \chi_0 = 0 \qquad \text{in } \mathbb{R}^N$$

in the viscosity sense.

The proof of the preceding result is essentially the same as that of Theorem 1.1.

A major difficulty arises, however, if one tries to establish an analogue of Proposition 1.3, because in this case the controlled trajectories are not necessarily bounded as $T \to +\infty$ and the limit function w_0 is only locally bounded in $\mathbb{R}^N$.

1. ERGODIC PROBLEMS

We describe in what follows some ergodic type results in the unconstrained case on a simplified model, namely, the optimal stopping problem (see §III.4.2). The value function is then

$$(1.8) \quad v_\lambda(x) = \inf_{\vartheta \geq 0}\left\{\int_0^\vartheta \ell_1(y_x(s))\, e^{-\lambda s}\, ds + \ell_0(y_x(\vartheta))\, e^{-\lambda \vartheta}\right\} = \inf_{\vartheta \geq 0} J_x^\lambda(\vartheta)$$

and is the unique viscosity solution of

$$(1.9) \quad \max\{\lambda v_\lambda(x) - f_1(x) \cdot Dv_\lambda(x) - \ell_1(x);\, v_\lambda(x) - \ell_0(x)\} = 0 \qquad x \in \mathbb{R}^N.$$

Let us observe first that in this case the system is governed by the discrete control set $A = \{0, 1\}$ with dynamics $f(x, 0) = 0$, $f(x, 1) = f_1(x)$. Hence $\overline{co}f(x, A)$ is the line segment joining 0 with $f_1(x)$ and the controllability assumption (1.2) does not hold. We replace it with a *dissipativity condition* on f_1, namely,

$$(1.10) \quad \exists F > 0 \text{ such that } (f_1(x) - f_1(z)) \cdot (x - z) \leq -F|x - z|^2 \qquad \forall x, z \in \mathbb{R}^N.$$

This condition implies, via Gronwall's inequality, that

$$(1.11) \quad |y_x(t) - y_z(t)| \leq e^{-Ft}|x - z| \qquad \forall x, z \in \mathbb{R}^N,\ \forall t > 0,$$

where $y_x(t)$ is the solution of

$$y'(t) = f_1(y(t)), \qquad y(0) = x.$$

Moreover, if f_1 is Lipschitz continuous, then

$$(1.12) \qquad \exists! x_0 \text{ such that } f_1(x_0) = 0$$

(this is a standard result in monotone operator theory, see Exercise 1.2). Properties (1.11) and (1.12) imply

$$(1.13) \quad |y_x(t) - x_0| \leq e^{-Ft}|x - x_0| \leq |x - x_0| \qquad \forall x \in \mathbb{R}^N,\ \forall t > 0,$$

or, in other words, that any closed ball $\overline{B}(x_0, R)$ is invariant for the flow defined by f_1.

The limiting behavior of (1.8) is described by the next result. We denote by w_λ the function

$$w_\lambda(x) = v_\lambda(x) - v_\lambda(x_0),$$

where x_0 is as in (1.12).

THEOREM 1.6. *Let us assume* (1.10). *Then*

$$\lim_{\lambda \to 0^+} \lambda v_\lambda(x) = \chi_0 = \min\{\ell_1(x_0); 0\}.$$

Moreover, if $\ell_1(x_0) \geq 0$, then at least a subsequence of $\{v_\lambda\}$ converges locally uniformly as $\lambda \to 0^+$ to a viscosity solution v_0 of

(1.14) $\quad \max\{-f_1(x) \cdot Dv_0(x) - \ell_1(x); v_0(x) - \ell_0(x)\} = 0 \quad x \in \mathbb{R}^N$.

If, on the contrary, $\ell_1(x_0) < 0$, then $\{w_\lambda\}$ converges locally uniformly as $\lambda \to 0^+$ to the unique viscosity solution w_0 of

(1.15) $\quad \begin{cases} -f_1(x) \cdot Dw_0(x) = \ell_1(x) - \ell_1(x_0) & x \in \mathbb{R}^N, \\ w_0(x_0) = 0. \end{cases}$

PROOF. Let

$$M = \max\left\{\frac{1}{\lambda} \sup_{\mathbb{R}^N} |\ell_1(x)|; \sup_{\mathbb{R}^N} |\ell_0(x)|\right\}.$$

It is easy to check that M is a supersolution of (1.9) whereas $-M$ is a subsolution of (1.9). Hence, by the comparison Theorem II.3.5,

$$|v_\lambda(x)| \leq M \quad \forall x \in \mathbb{R}^N,$$

so that

(1.16) $\quad |\lambda v_\lambda(x)| \leq \max\left\{\sup_{\mathbb{R}^N} |\ell_1(x)|; \lambda \sup_{\mathbb{R}^N} |\ell_0(x)|\right\} \quad \forall x \in \mathbb{R}^N$.

In order to obtain uniform Lipschitz estimates on $\{\lambda v_\lambda\}$ we apply Proposition II.4.2. For this purpose, let us observe that $u_\lambda = (\lambda + F)v_\lambda$ satisfies (see Exercise 1.7)

$$u_\lambda(x) + H_\lambda(x, Du_\lambda(x)) = 0 \quad x \in \mathbb{R}^N,$$

in the viscosity sense, where

$$H_\lambda(x, p) = \max\left\{-\frac{1}{\lambda} f_1(x) \cdot p - \frac{\lambda + F}{\lambda} \ell_1(x); -(\lambda + F)\ell_0(x)\right\}.$$

Also, for any $C > 0$, $x, z, p \in \mathbb{R}^N$,

(1.17) $H_\lambda(x, Cp) - H_\lambda(z, Cp)$
$\geq \min\left\{\frac{C}{\lambda}(f_1(z) - f_1(x)) \cdot p - \frac{\lambda + F}{\lambda}(\ell_1(x) - \ell_1(z)); (\lambda + F)(\ell_0(z) - \ell_0(x))\right\}$

(see again Exercise 1.7). For $p = \dfrac{x - z}{|x - z|}$, the above and (1.10) yield

$H_\lambda\left(x, C\dfrac{x-z}{|x-z|}\right) - H_\lambda\left(z, C\dfrac{x-z}{|x-z|}\right)$

$\geq \min\left\{\dfrac{CF}{\lambda} - \dfrac{\lambda + F}{\lambda} L_1; -(\lambda + F)L_0\right\}|x - z|$

where L_0, L_1 are Lipschitz constants for ℓ_0, ℓ_1. Choosing $C \geq \max\{(1+F)L_0; L_1\}$ we get

$$H_\lambda\left(x, C\frac{x-z}{|x-z|}\right) - H_\lambda\left(z, C\frac{x-z}{|x-z|}\right) \geq -C|x-z|, \quad \forall \lambda \in\,]0,1[\,.$$

Therefore, by Proposition II.4.2,

$$(\lambda + F)|v_\lambda(x) - v_\lambda(z)| \leq C|x-z| \quad \forall \lambda \in\,]0,1[\,.$$

This implies, of course,

(1.18) $$|v_\lambda(x) - v_\lambda(z)| \leq \frac{C}{F}|x-z| \quad \forall \lambda \in\,]0,1[$$

and

(1.19) $$|\lambda v_\lambda(x) - \lambda v_\lambda(z)| \leq \frac{\lambda C}{F}|x-z| \quad \forall \lambda \in\,]0,1[\,.$$

From (1.16) and (1.19) it is easy to deduce that a subsequence of λv_λ converges to some constant χ_0.

Let us now show that $\chi_0 = \min\{\ell_1(x_0); 0\}$. The representation formula (1.8) at the equilibrium point x_0 gives

$$v_\lambda(x_0) = \inf_{\vartheta \geq 0}\left\{\ell_1(x_0)\int_0^\vartheta e^{-\lambda s}\, ds + \ell_0(x_0)e^{-\lambda \vartheta}\right\}.$$

An elementary computation then yields

$$v_\lambda(x_0) = \begin{cases} \ell_0(x_0) & \text{if } \ell_1(x_0) - \lambda\ell_0(x_0) > 0, \\ \dfrac{\ell_1(x_0)}{\lambda} & \text{if } \ell_1(x_0) - \lambda\ell_0(x_0) \leq 0\,. \end{cases}$$

Hence, for sufficiently small $\lambda > 0$,

(1.20) $$\lambda v_\lambda(x_0) = \begin{cases} \lambda\ell_0(x_0) & \text{if } \ell_1(x_0) > 0, \\ \lambda\min\{\ell_0(x_0); 0\} & \text{if } \ell_1(x_0) = 0, \\ \ell_1(x_0) & \text{if } \ell_1(x_0) < 0, \end{cases}$$

and the formula for χ_0 is proved.

To prove the second part of the theorem, let us first treat the case $\ell_1(x_0) \geq 0$. From (1.18) and (1.20) it follows that

(1.21) $$|v_\lambda(x)| \leq |v_\lambda(x_0)| + \frac{C}{F}|x-x_0| \leq |\ell_0(x_0)| + \frac{C}{F}|x-x_0|\,.$$

This uniform estimate, together with (1.18), allows us to apply the Ascoli-Arzelà theorem and conclude that at least a subsequence of $\{v_\lambda\}$ converges uniformly to some v_0 as $\lambda \to 0^+$. The assertion that v_0 is a viscosity solution of (1.14) is then a direct consequence of Proposition II.2.2.

Now consider the case $\ell_1(x_0) < 0$; it is immediate to check that $w_\lambda = v_\lambda - v_\lambda(x_0)$ is a viscosity solution of

(1.22) $\max\{\lambda w_\lambda(x) - f_1(x) \cdot Dw_\lambda(x) - \ell_1(x) + \lambda v_\lambda(x_0);$
$$w_\lambda(x) - \ell_0(x) + v_\lambda(x_0)\} = 0 \qquad x \in \mathbb{R}^N .$$

As a consequence of (1.18), $\{v_\lambda\}$ is locally equibounded and equi-Lipschitz continuous. Let w_0 be any subsequential uniform limit of $\{w_\lambda\}$ as $\lambda \to 0^+$. In the present case (see (1.20)) $v_\lambda(x_0) = \lambda^{-1}\ell_1(x_0) \to -\infty$ as $\lambda \to 0^+$. Hence, for sufficiently small λ, the max in equation (1.22) is attained by the first member at any $x \in \mathbb{R}^N$, so that w_λ is a viscosity solution of the linear equation

$$\lambda w_\lambda(x) - f_1(x) \cdot Dw_\lambda(x) - \ell_1(x) + \lambda v_\lambda(x_0) = 0 \qquad x \in \mathbb{R}^N .$$

The conclusion that w_0 is a viscosity solution of (1.15) is again a consequence of Proposition II.2.2.

As for the uniqueness of the solution of (1.15), let us observe that

$$u_i(x) = \int_0^t (\ell_1(y_x(s)) - \ell_1(x_0))\,ds + u_i(y_x(t)) \qquad \forall t > 0, x \in \mathbb{R}^N$$

if u_i ($i = 1, 2$) are viscosity solutions of (1.15) (by Proposition II.5.18 and some calculations, or by Remark III.2.34). Hence

$$(u_1 - u_2)(x) = (u_1 - u_2)(y_x(t)) \qquad \forall t > 0, x \in \mathbb{R}^N .$$

By assumption (1.10) and its consequence (1.13), we can let $t \to +\infty$ and obtain

$$(u_1 - u_2)(x) = (u_1 - u_2)(x_0) = 0 \qquad \forall x \in \mathbb{R}^N,$$

and the proof is complete. ◂

REMARK 1.7. Equation (1.14) does not have a unique solution. As an example, suppose that the stopping cost ℓ_0 satisfies

$$-f_1(x) \cdot D\ell_0(x) - \ell_1(x) \leq 0 \qquad x \in \mathbb{R}^N$$

in the viscosity sense. Then $u \equiv \ell_0$ is a solution of (1.14), but u does not necessarily coincide with $v_0 = \lim_{\lambda \to 0^+} v_\lambda$. Actually, if $\ell_1(x_0) = 0$ and $\ell_0(x_0) > 0$, then (1.20) gives $v_\lambda(x_0) = 0$ for all λ, so that $0 = v_0(x_0) < u(x_0) = \ell_0(x_0)$. ◁

We conclude this section with some properties of the undiscounted value:

(1.23) $\quad v_0(x) := \inf_{\vartheta \geq 0}\left\{\int_0^\vartheta \ell_1(y_x(s))e^{-\lambda s}\,ds + \ell_0(y_x(\vartheta))\right\} = \inf_{\vartheta \geq 0} J_x^0(\vartheta)$

and an estimate on $v_\lambda - v_0$.

PROPOSITION 1.8. *Let us assume* (1.10) *and*

(1.24) $$\ell_1(x_0) > 0 .$$

Then for any $R > 0$ there exists $T > 0$ such that

(1.25) $$v_0(x) = \inf_{0 \leq \vartheta \leq T} J_x^0(\vartheta) \qquad \forall\, x \in \bar{B}(x_0, R) .$$

Moreover, for some $K > 0$,

$$|v_0(x) - v_0(z)| \leq K|x - z|, \qquad \forall\, x, z \in \bar{B}(x_0, R) .$$

PROOF. By the continuity of ℓ_1 and (1.24) we can choose $\delta > 0$ such that

$$\ell_1(y) > \frac{1}{2}\ell_1(x_0) \qquad \forall\, y \in \bar{B}(x_0, \delta) .$$

Thanks to (1.13), there exists $T_1 > 0$ independent on $x \in \bar{B}(x_0, R)$ such that

$$y_x(s) \in B(x_0, \delta), \qquad \forall\, s \geq T_1, \quad \forall\, x \in \bar{B}(x_0, R) .$$

For $\vartheta > T_1$ we have

$$J_x^0(\vartheta) = J_x^0(T_1) + \int_{T_1}^{\vartheta} \ell_1(y_x(s))\, ds + \ell_0(y_x(\vartheta)) - \ell_0(y_x(T_1))$$

so that

$$J_x^0(\vartheta) \geq J_x^0(T_1) + \frac{1}{2}\ell_1(x_0)(\vartheta - T_1) - 2M_0,$$

where $M_0 = \sup_{\bar{B}(x_0, R)} |\ell_0(x)|$. Hence, for $T = T_1 + \dfrac{4}{\ell_1(x_0)} \sup_{\bar{B}(x_0, R)} |\ell_0(x)|$,

$$\inf_{\vartheta \geq T} J_x^0(\vartheta) \geq J_x^0(T) .$$

On the other hand, since $T_1 < T$,

$$J_x^0(T_1) \geq \inf_{0 \leq \vartheta \leq T} J_x^0(\vartheta)$$

and (1.25) is proved. In order to prove the Lipschitz continuity of v_0, let x, $z \in \bar{B}(x_0, R)$. Then

$$J_x^0(\vartheta) - J_z^0(\vartheta) \leq L_1 \int_0^{\vartheta} |y_x(s) - y_z(s)|\, ds + L_0|y_x(\vartheta) - y_z(\vartheta)|;$$

where L_0, L_1 are Lipschitz constant for ℓ_0, ℓ_1. Hence, from (1.11),

$$J_x^0(\vartheta) - J_z^0(\vartheta) \leq (L_1 T + L_0)|x - z| \qquad \forall\, 0 \leq \vartheta \leq T$$

and, by (1.25), this implies

$$|v_0(x) - v_0(z)| \leq (L_1 T + L_0)|x - z|, \qquad \forall\, x, z \in \bar{B}(x_0, R) . \blacktriangleleft$$

PROPOSITION 1.9. *Let us assume* (1.10) *and* (1.24). *Then for any* $R > 0$ *there exists* $C_R > 0$ *such that*

(1.26) $$|v_\lambda(x) - v_0(x)| \leq C_R \lambda \qquad \forall x \in \bar{B}(x_0, R)$$

for sufficiently small $\lambda > 0$.

PROOF. Let us fix $R > 0$. For any $x \in \bar{B}(x_0, R)$ and $\varepsilon > 0$ let $\vartheta^* = \vartheta^*_{\lambda,x,\varepsilon}$ be such that

(1.27) $$J_x^\lambda(\vartheta^*) \leq v_\lambda(x) + \varepsilon .$$

We claim that

(1.28) $$\limsup_{\lambda \to 0^+} \vartheta^* \leq \frac{1}{\ell_1(x_0)}(2M_0 + \frac{L_1 R}{F} + \varepsilon) .$$

Indeed, (1.27) yields for $M_0 \geq \sup|\ell_0|$,

$$M_0 + \varepsilon \geq \int_0^{\vartheta^*} (\ell_1(y_x(s)) - \ell_1(x_0)) e^{-\lambda s} \, ds + \ell_1(x_0) \frac{1 - e^{-\lambda \vartheta^*}}{\lambda} - M_0 e^{-\lambda \vartheta^*} .$$

Hence, by (1.11),

$$M_0 + \varepsilon \geq -L_1 |x - x_0| \int_0^{\vartheta^*} e^{-(F+\lambda)s} \, ds + \ell_1(x_0) \frac{1 - e^{-\lambda \vartheta^*}}{\lambda} - M_0 e^{-\lambda \vartheta^*} ;$$

this implies

$$\ell_1(x_0) \frac{1 - e^{-\lambda \vartheta^*}}{\lambda} \leq \varepsilon + 2M_0 + \frac{L_1}{F + \lambda}|x - x_0| \leq \varepsilon + 2M_0 + \frac{L_1}{F} R$$

for all $x \in \bar{B}(x_0, R)$. Therefore

$$\vartheta^* \leq -\frac{1}{\lambda} \log\left(1 - \frac{\lambda}{\ell_1(x_0)}(\varepsilon + 2M_0 + \frac{L_1 R}{F})\right)$$

and the claim follows.

For any $x \in \bar{B}(x_0, R)$ and $\vartheta \geq 0$ we have

(1.29) $$|J_x^\lambda(\vartheta) - J_x^0(\vartheta)| \leq M_1 \int_0^\vartheta (1 - e^{-\lambda s}) \, ds + M_0(1 - e^{-\lambda \vartheta}) =: \varphi(\lambda, \vartheta) .$$

Thanks to (1.25), (1.28), for small λ we have

$$\sup_{\vartheta \geq 0} |J_x^\lambda(\vartheta) - J_x^0(\vartheta)| = \sup_{0 \leq \vartheta \leq T_M} |J_x^\lambda(\vartheta) - J_x^0(\vartheta)|$$

for some $T_M < +\infty$ independent of $x \in \bar{B}(x_0, R)$. Hence there exists $\vartheta_M \in [0, T_M]$ such that

$$|v_\lambda(x) - v_0(x)| \leq \sup_{\vartheta \geq 0} |J_x^\lambda(\vartheta) - J_x^0(\vartheta)| \leq \varphi(\lambda, \vartheta_M) .$$

Since

$$\varphi(\lambda, \vartheta_M) = M_1 \frac{\lambda \vartheta_M + e^{-\lambda \vartheta_M} - 1}{\lambda} + M_0(1 - e^{-\lambda \vartheta_M}),$$

(1.26) follows. ◂

Exercises

1.1. Consider the finite horizon problem
$$v(x,t) = \inf_{\alpha \in \mathcal{A}_x} \int_t^T \ell(y_x(s,\alpha), \alpha(s))\, ds + g(y_x(T,\alpha)),$$
where g is any given bounded continuous function on $\mathbb{R}^N$. Use the results of §III.3 (see, in particular, Remark III.3.10) to show that
$$\chi_0 = \lim_{T \to +\infty} \frac{1}{T} v(x,0), \quad \text{uniformly in } \overline{\Omega}.$$

1.2. Suppose that f_1 satisfies condition (1.10) and
$$|f_1(x) - f_1(z)| \leq L|x - z|, \quad \forall x, z \in \mathbb{R}^N.$$
Show that $T_\varrho(x) := x + \varrho f_1(x)$ ($\varrho > 0$) satisfies
$$|T_\varrho(x) - T_\varrho(z)|^2 \leq (1 - 2\varrho F + L^2 \varrho^2)|x - z|^2.$$
Deduce from this that the equation
$$f_1(x) = 0$$
has a unique solution.

1.3. For f_1 as in Exercise 1.2, let x_0 be the unique point with $f_1(x_0) = 0$. Show that
$$t_x \geq -\frac{1}{F} \log \frac{\delta}{R}, \quad \forall x \in \bar{B}(x_0, R),$$
where $t_x = \inf\{\, t \geq 0 : y_x(t) \in \bar{B}(x_0, \delta)\,\}$.

1.4. Let $u \in C(\overline{\Omega}) \cap \mathrm{Lip}(\Omega)$ be a solution of
$$\sup_{a \in A}\{\,-f(x,a) \cdot Du(x)\,\} = 0 \quad \text{in } \overline{\Omega}.$$

(i) Show that $z_\lambda := \frac{1}{\lambda} u$ is a solution of
$$\lambda z_\lambda + \sup_{a \in A}\{\,-f(x,a) \cdot Dz_\lambda(x) - u(x)\,\} = 0 \quad \text{in } \overline{\Omega}$$
and therefore
$$z_\lambda(x) = \inf_{\alpha \in \mathcal{A}_x} \int_0^{+\infty} u(y_x(s,\alpha)) e^{-\lambda s}\, ds;$$

(ii) show, under the assumptions of Theorem 1.1, that u is a constant.

1.5. Consider the Dirichlet problem

$$-f(x) \cdot Du(x) - \ell(x) = 0 \quad \text{in } \Omega,$$
$$u = 0 \quad \text{on } \Gamma_0,$$

where $\Gamma_0 := \{x \in \partial\Omega : t_x = 0\}$ and $t_x := \inf\{t \geq 0 : y_x(t) \notin \overline{\Omega}\}$. Let $S := \{x \in \Omega : t_x = +\infty\}$. Use the results of §II.5 to show that the preceding has at most one viscosity solution $u \in BC(\mathbb{R}^N)$ that satisfies the additional condition

$$\lim_{T \to +\infty} \frac{1}{T} \int_0^T u(y_x(s))\, ds = 0 \quad \text{uniformly in } S.$$

1.6. Consider $f \in C^1$ and suppose that $S := \{x \in \Omega : t_x = +\infty\}$ is measurable. Show that either S has zero measure or there exists $U \subseteq S$ with positive measure such that

$$\operatorname{div} f(x) \leq 0, \quad \forall x \in U,$$

where $\operatorname{div} f$ denotes the divergence of the vector field f. [Hint: set $U(t) = \{y \in \mathbb{R}^N : y = y_x(t), x \in U\}$ and use Liouville's volume formula $\operatorname{meas}(U(t)) = \int_U e^{t \operatorname{div} f(y_x(t))}\, dx$.]

1.7. Use the elementary facts

$$\max\{a; b\} = 0 \iff \max\{a; \gamma b\} = 0$$
$$\max\{\delta a + c; \delta a + d\} = \delta a + \max\{c; d\}$$
$$\max\{a; b\} - \max\{c; d\} \geq \min\{a - c; b - d\},$$

for $a, b, c, d \in \mathbb{R}$ and $\gamma, \delta > 0$, to complete the first part of the proof of Theorem 1.6.

1.8. Assume the existence of some $\varphi \in C^1(\overline{\Omega})$ such that

$$\begin{cases} \max\{-f_1(x) \cdot D\varphi(x) - \ell_1(x); \varphi(x) - \ell_0(x)\} < 0, \\ \varphi \leq v_0, \end{cases}$$

in $\overline{B}(x_0, R)$, where v_0 is given by (1.25). Prove under the standard assumption of f, ℓ_1, ℓ_0, that $v_0 \geq u$ for any viscosity solution u of

$$\begin{cases} \max\{-f_1(x) \cdot Du(x) - \ell_1(x); u(x) - \ell_0(x)\} \leq 0 & \text{in } \overline{B}(x_0, R), \\ u(x_0) \leq \ell_0(x_0). \end{cases}$$

[Hint: modify the proof of Theorem II.5.9 also taking into account Theorem IV.5.8.]

1.9. In the framework of Proposition 1.9 prove that if $\ell_1(x) \geq \ell_1(x_0)\ \forall x \in \mathbb{R}^N$, then the following variants of estimates (1.28) and (1.26) hold:

$$\vartheta^* \leq \frac{1}{\ell_1(x_0)}(2M_0 + \varepsilon), \quad \forall x \in \mathbb{R}^N,$$
$$|v_\lambda(x) - v_0(x)| \leq C\lambda, \quad \forall x \in \mathbb{R}^N.$$

2. Vanishing switching costs

In this section we analyze the limiting behavior of the value function $v_\varepsilon = (v_\varepsilon^1 \ldots v_\varepsilon^m)$ of the optimal switching problem when the positive switching costs $k_\varepsilon(i,j) \to 0$ as $\varepsilon \to 0^+$.

Let us recall from §III.4.4, to which we refer for notations, assumptions and basic results, that v_ε is defined in the case of discount rate $\lambda = 1$ as

$$v_\varepsilon^i(x) = \inf_{\alpha \in \mathcal{P}^i} \sum_{n=1}^\infty \left(\int_{\vartheta^{n-1}}^{\vartheta^n} \ell(y_x(s,\alpha), a^{n-1}) e^{-s} ds + k_\varepsilon(a^{n-1}, a^n) e^{-\vartheta^n} \right)$$
$$= \inf_{\alpha \in \mathcal{P}^i} J_\varepsilon(x, \alpha), \qquad \text{for } i \in A = \{1, \ldots, m\}.$$

and it is the unique solution of the system of quasivariational inequalities

(2.1) $\qquad \max\{ v_\varepsilon^i - f^i \cdot Dv_\varepsilon^i - \ell^i ; v_\varepsilon^i - \min_{j \neq i}(v_\varepsilon^j + k_\varepsilon(i,j))\} = 0 \qquad \text{in } \mathbb{R}^N.$

Setting formally $\varepsilon \to 0^+$ in (2.1) we find that the limit vector $\overline{v} = \lim_{\varepsilon \to 0^+} v_\varepsilon$, if any, should satisfy

(2.2) $\qquad \max\{ \overline{v}^i - f^i \cdot D\overline{v}^i - \ell^i; \overline{v}^i - \min_{j \neq i} \overline{v}^j \} = 0 \qquad \text{in } \mathbb{R}^N,$

for all $i \in \{1, \ldots, m\}$. The above yields, in particular, $\overline{v}^i \leq \min_{j \neq i} \overline{v}^j$ for all $i \in \{1, \ldots, m\}$ so that necessarily

$$\overline{v}^1 = \overline{v}^2 = \cdots = \overline{v}^m.$$

Denoting by v the common value of $\overline{v}^i$, $i \in \{1, \ldots, m\}$, from (2.2) we deduce that v should satisfy

$$\max_{i \in \{1,\ldots,m\}} \{ v - f^i \cdot Dv - \ell^i \} = 0 \qquad \text{in } \mathbb{R}^N.$$

This is the Hamilton-Jacobi-Bellman equation for the infinite horizon problem with control set $A = \{1, \ldots, m\}$. By the theory in §III.2, the guess is that

(2.3) $\qquad v(x) = \inf_{\alpha \in \mathcal{P}} \int_0^{+\infty} \ell(y_x(s, \alpha), \alpha(s)) e^{-s} ds.$

The preceding heuristics are confirmed by the next result.

THEOREM 2.1. *Let us assume*

(2.4) $\quad k_\varepsilon(i,j) > 0, \quad k_\varepsilon(i,i) = 0, \quad k_\varepsilon(i,j) < k_\varepsilon(i,\ell) + k_\varepsilon(\ell,j), \qquad i \neq j \neq \ell,$
$\qquad\qquad \lim_{\varepsilon \to 0^+} k_\varepsilon(i,j) = 0 \quad \forall i,j \in \{1, \ldots, m\}.$

Then

$$\lim_{\varepsilon \to 0^+} v_\varepsilon^i = v \qquad \forall i \in \{1, \ldots, m\},$$

locally uniformly in $\mathbb{R}^N$, *where* $v \in BUC(\mathbb{R}^N)$ *is given by* (2.3).

PROOF. The first step is to prove that

(2.5) $$\sup_{\mathbb{R}^N} |v_\varepsilon^i(x)| \leq M$$

for some constant M independent of ε, i. This follows from the fact that

$$M = \max_{1 \leq i \leq m} \sup_{\mathbb{R}^N} |\ell^i(x)|$$

satisfies

(2.6) $\max\{M - \ell^i(x); M - \min_{j \neq i}(M + k_\varepsilon(i,j))\} \geq M - \ell^i(x) \geq 0, \quad \forall x \in \mathbb{R}^N.$

On the other hand,

$$-M - \ell^i(x) \leq 0 \quad \forall x \in \mathbb{R}^N, \, i \in \{1, \ldots, m\}$$
$$-M - \min_{j \neq i}\{-M + k_\varepsilon(i,j)\} \leq 0,$$

so that

(2.7) $\max\{-M - \ell^i(x); -M - \min_{j \neq i}\{-M + k_\varepsilon(i,j)\}\} \leq 0, \quad \forall x \in \mathbb{R}^N.$

Inequalities (2.6) and (2.7) show that $M, -M$ are, respectively, super- and subsolutions of (2.1). Hence, by comparison, (2.5) follows.

Define then

$$\underline{u}^i(x) = \liminf_{(\varepsilon,y) \to (0^+,x)} v_\varepsilon^i(y), \qquad \overline{u}^i(x) = \limsup_{(\varepsilon,y) \to (0^+,y)} v_\varepsilon^i(y).$$

Let us observe at this point that for any $i \in \{1, \ldots, m\}$,

(2.8) $\quad v_\varepsilon^i(x) \leq v_\varepsilon^j(x) + k_\varepsilon(i,j) \quad \forall x \in \mathbb{R}^N, \, \forall j \neq i.$

Actually, if this were not true, then there would exist $x_0 \in \mathbb{R}^N$, $i_0 \neq j_0 \in \{1, \ldots, m\}$ such that

(2.9) $$v_\varepsilon^{i_0}(x) > v_\varepsilon^{j_0}(x) + k_\varepsilon(i_0, j_0)$$

for x in an open neighborhood B of x_0. It is not difficult to show the existence of $\varphi \in C^1(\mathbb{R}^N)$ such that $v_\varepsilon^{i_0} - \varphi$ has a local maximum at some $x_1 \in \overline{B}$. Since v_ε is a subsolution of (2.1), this implies

$$v_\varepsilon^{i_0}(x_1) \leq \min_{j \neq i_0}\{v_\varepsilon^j(x_1) + k_\varepsilon(i_0, j)\} \leq v_\varepsilon^{j_0}(x_1) + k_\varepsilon(i_0, j_0),$$

a contradiction to (2.9). From (2.8) and (2.4) it follows that

$$\underline{u}^1 \equiv \underline{u}^2 \equiv \cdots \equiv \underline{u}^m, \qquad \overline{u}^1 \equiv \overline{u}^2 \equiv \cdots \equiv \overline{u}^m.$$

Let us denote by $\underline{u}(x)$ and $\overline{u}(x)$, respectively, the common values of $\underline{u}^i(x)$ and $\overline{u}^i(x)$. By construction, $\underline{u}$ and $\overline{u}$ are, respectively, lower and upper semicontinuous and

(2.10) $$\underline{u} \leq \overline{u}.$$

The claim is that $\underline{u}, \overline{u}$ are, respectively, a super- and a subsolution of

(2.11) $$u(x) + \max_{i \in \{1,\ldots,m\}} \{-f^i(x) \cdot Du(x) - \ell^i(x)\} = 0 \qquad x \in \mathbb{R}^N,$$

in the viscosity sense (see Definition V.1.1).

Admitting this claim, the comparison Theorem V.1.3 and (2.10) imply $\underline{u} = \overline{u}$. Therefore, if u denotes the common value of $\underline{u}$ and $\overline{u}$, u is a bounded continuous viscosity solution of (2.11). Moreover, by the stability results of §V.1 (see Lemma 1.9),

$$v_\varepsilon^i \longrightarrow u$$

locally uniformly in $\mathbb{R}^N$, for all $i \in \{1, \ldots, m\}$. By the theory of Chapter III, §2, $u = v$ and the proof is complete.

In order to prove the claim, let $\varphi \in C^1(\mathbb{R}^N)$ and let $\overline{x}$ be a strict maximum point for $\overline{u} - \varphi$ in $\overline{B} = \overline{B}(\overline{x}, R)$. Then by Lemma V.1.6, there exist $\varepsilon_n \to 0^+$ and, for all i, a sequence $(x_n^i)_{n \in \mathbb{N}}$ in $\overline{B}$ such that $(v_{\varepsilon_n}^i - \varphi)(x_n^i) = \max_{\overline{B}}(v_{\varepsilon_n}^i - \varphi)$,

(2.12) $$x_n^i \longrightarrow \overline{x}, \quad v_{\varepsilon_n}^i(x_n^i) \longrightarrow \overline{u}^i(\overline{x}) = \overline{u}(\overline{x}), \qquad \text{as } n \to \infty.$$

Since $v_\varepsilon = (v_\varepsilon^1, \ldots, v_\varepsilon^m)$ is a viscosity subsolution of (2.1),

$$v_{\varepsilon_n}^i(x_n^i) - f^i(x_n^i) \cdot D\varphi(x_n^i) - \ell^i(x_n^i) \leq 0, \qquad \forall i \in \{1, \ldots, m\}.$$

A passage to the limit as $n \to \infty$ in the above proves, by (2.12), that $\overline{u}$ is a viscosity subsolution of (2.11).

On the other hand, let $\varphi \in C^1(\mathbb{R}^N)$ and $\underline{x}$ be a strict local minimum for $\underline{u} - \varphi$. As before, we can find $\varepsilon_n \to 0^+$ and $(x_n^i)_{n \in \mathbb{N}}$ in $\overline{B} = \overline{B}(\underline{x}, R)$ such that $(v_{\varepsilon_n}^i - \varphi)(x_n^i) = \min_{\overline{B}}(v_{\varepsilon_n}^i - \varphi)$,

(2.13) $$x_n^i \longrightarrow \underline{x}, \quad v_{\varepsilon_n}^i(x_n^i) \longrightarrow \underline{u}^i(\underline{x}) = \underline{u}(\underline{x}), \qquad \text{as } n \to \infty.$$

Now choose $i_n \in \{1, \ldots, m\}$ such that

$$(v_{\varepsilon_n}^{i_n} - \varphi)(x_n^{i_n}) = \min_{j \in \{1,\ldots,m\}} \min_{\overline{B}}(v_{\varepsilon_n}^j - \varphi),$$

and set $x_n := x_n^{i_n}$. Then, by (2.4),

(2.14) $$v_{\varepsilon_n}^{i_n}(x_n) \leq v_{\varepsilon_n}^j(x_n) < v_{\varepsilon_n}^j(x_n) + k_{\varepsilon_n}(i_n, j), \qquad \forall j \neq i_n.$$

Since $v_\varepsilon = (v_\varepsilon^1, \ldots, v_\varepsilon^m)$ is a viscosity supersolution of (2.1), and (2.14) holds, we have

$$v_{\varepsilon_n}^{i_n}(x_n) - f^{i_n}(x_n) \cdot D\varphi(x_n) - \ell^{i_n}(x_n) \geq 0.$$

Passing if necessary to a subsequence we may assume that $i_n \to \underline{i}$, and so $i_n = \underline{i}$ for all n large enough. Thus we can send $n \to \infty$ in the inequality above and use (2.13) to get

$$\underline{u}(\underline{x}) - f^{\underline{i}}(\underline{x}) \cdot D\varphi(\underline{x}) - \ell^{\underline{i}}(\underline{x}) \geq 0 \ .$$

Consequently, $\underline{u}$ is a viscosity supersolution of (2.11) and the proof is complete. ◀

Exercises

2.1. Generalize Theorem 2.1 to the case where the switching costs $k_\varepsilon(i,j)$ may depend on x.

2.2. Give a direct proof (i.e., without using the Hamilton-Jacobi-Bellman equations (2.1)) of Theorem 2.1.

2.3. Try to estimate the rate of convergence of v_ε^i to v in the case $k_\varepsilon(i,j) = k\varepsilon^\gamma$ for all i, j where k and γ are positive constants.

3. Penalization

The purpose of this section is to indicate how classical penalization methods can be easily implemented in the framework of Hamilton-Jacobi-Bellman equations through the technique of viscosity solutions.

Two examples are treated: penalization of unilateral constraints on the value function arising in optimal stopping problems and penalization of state constraints.

Let us mention here that also the limit problems considered in §4.2 can be regarded as examples of penalization techniques.

3.1. Penalization of stopping costs

Let us consider the value function of the optimal stopping problem with discount $\lambda = 1$ defined as

$$(3.1) \qquad v(x) = \inf_{\vartheta \geq 0} \left\{ \int_0^\vartheta \ell_1(y_x(s)) e^{-s} \, ds + \ell_0(y_x(\vartheta)) e^{-\vartheta} \right\} \ .$$

We know that in the framework of §III.4.2, $v \in BUC(\mathbb{R}^N)$ and is the unique solution of

$$(3.2) \qquad \max\{v(x) - f(x) \cdot Dv(x) - \ell_1(x); v(x) - \ell_0(x)\} = 0, \qquad x \in \mathbb{R}^N \ .$$

Let us associate with (3.2) the semilinear HJB equation

$$(3.3) \qquad u^\varepsilon(x) - f(x) \cdot Du^\varepsilon(x) - \ell_1(x) + \frac{1}{\varepsilon}(u^\varepsilon - \ell_0)^+(x) = 0, \qquad x \in \mathbb{R}^N,$$

with $\varepsilon > 0$. The simple observation that

$$\frac{1}{\varepsilon}(r - \ell_0(x))^+ = \sup_{a \in [0, 1/\varepsilon]} a(r - \ell_0(x)) \qquad \forall x \in \mathbb{R}^N, \ r \in \mathbb{R},$$

allows us to rewrite (3.3) as

$$\text{(3.4)} \qquad \sup_{a\in[0,1/\varepsilon]}\{(1+a)u^\varepsilon(x) - f(x)\cdot Du^\varepsilon(x) - (\ell_1(x) + a\ell_0(x))\} = 0,$$

$x \in \mathbb{R}^N$. The usual dynamic programming technique (see Chapter III) shows that

$$\text{(3.5)} \qquad v^\varepsilon(x) := \inf_{\alpha\in\mathcal{A}_\varepsilon} \int_0^{+\infty} (\ell_1(y_x(t)) + \alpha(t)\ell_0(y_x(t)))\, e^{-t-\int_0^t \alpha(s)\,ds}\, dt,$$

where

$$\mathcal{A}_\varepsilon = \{\alpha : [0,+\infty[\ \to [0,1/\varepsilon],\ \alpha \text{ measurable}\},$$

is a viscosity solution of (3.4) (indeed, its unique viscosity solution in view of the uniqueness results in §III.2). Hence the representation formula (3.5) shows that v^ε can be interpreted as the value of a discounted infinite horizon problem with the time varying discount factor

$$\int_0^t (1+\alpha(s))\,ds$$

and the running cost

$$\widehat{\ell}(x,a) = \ell_1(x) + a\ell_0(x), \qquad (x,a) \in \mathbb{R}^N \times [0,1/\varepsilon],$$

which incorporates the stopping cost ℓ_0, after multiplication by the penalization factor $a \in [0, 1/\varepsilon]$. Let us observe that the penalizing effect in (3.3) occurs only at those points x where $u^\varepsilon(x) > \ell_0(x)$ and becomes stronger and stronger as $\varepsilon \to 0^+$.

Concerning the behavior of v^ε as $\varepsilon \to 0^+$ we have the following quite natural result:

THEOREM 3.1. *Under the assumptions on f, ℓ, ℓ_0 as in §III.4.2, we have* $\lim_{\varepsilon\to 0^+} v^\varepsilon = v$, *locally uniformly in $\mathbb{R}^N$, with v given by (3.1).*

PROOF. It is a simple matter to check that the constant

$$M = \max\Big\{\sup_{\mathbb{R}^N} |\ell_0(x)|; \sup_{\mathbb{R}^N} |\ell_1(x)|\Big\}$$

is a supersolution of (3.3) for any $\varepsilon > 0$. Similarly, $-M$ is a subsolution of (3.3), for any $\varepsilon > 0$. Hence, by comparison (see Chapter III, §2)

$$-M \leq v^\varepsilon(x) \leq M \qquad \forall x \in \mathbb{R}^N,\ \varepsilon > 0.$$

Let us introduce then the weak limits $\underline{u}, \overline{u}$ studied in §V.1:

$$\underline{u}(x) = \liminf_{(\varepsilon,y)\to(0^+,x)} v^\varepsilon(y), \qquad \overline{u}(x) = \limsup_{(\varepsilon,y)\to(0^+,x)} v^\varepsilon(y).$$

By Lemma V.1.5, $\underline{u}, \overline{u}$ are, respectively, lower and upper semicontinuous and satisfy

(3.6) $$\underline{u} \leq \overline{u} \quad \text{in } \mathbb{R}^N .$$

Let us show that $\overline{u}$ is a subsolution of (3.2) in the generalized viscosity sense of Chapter V. For this purpose, let $\varphi \in C^1(\mathbb{R}^N)$ and $\overline{x}$ a strict maximum point for $\overline{u} - \varphi$ in $\overline{B} = \overline{B}(\overline{x}, r)$. Then by Lemma V.1.6 there exists $x^n \in \overline{B}$ and $\varepsilon_n \to 0^+$ such that

(3.7) $\quad (v^{\varepsilon_n} - \varphi)(x^n) = \max_{\overline{B}}(v^{\varepsilon_n} - \varphi), \quad x^n \longrightarrow \overline{x}, \quad v^{\varepsilon_n}(x^n) \longrightarrow \overline{u}(\overline{x}) .$

Since v^ε is a viscosity subsolution of (3.4),

$$\sup_{a \in [0, 1/\varepsilon_n]} \{(1+a)v^{\varepsilon_n} - f \cdot D\varphi - (\ell_1 + a\ell_0)\} \leq 0 \quad \text{at } x^n .$$

Let us choose $a = 0$ in the above to obtain

(3.8) $$v^{\varepsilon_n}(x^n) - f(x^n) \cdot D\varphi(x^n) - \ell_1(x^n) \leq 0 .$$

On the other hand, the choice $a = 1/\varepsilon_n$ gives

$$(1 + 1/\varepsilon_n)v^{\varepsilon_n}(x^n) - f(x^n) \cdot D\varphi(x^n) - \ell_1(x^n) - \frac{1}{\varepsilon_n}\ell_0(x^n) \leq 0,$$

yielding

(3.9) $$\varepsilon_n v^{\varepsilon_n}(x^n) + (v^{\varepsilon_n} - \ell_0)(x^n) \leq \varepsilon_n f(x^n) \cdot D\varphi(x^n) + \varepsilon_n \ell_1(x^n) .$$

Now let $n \to +\infty$ in (3.8) and (3.9) to obtain, thanks to (3.7),

$$\overline{u}(\overline{x}) - f(\overline{x}) \cdot D\varphi(\overline{x}) - \ell_1(\overline{x}) \leq 0, \quad (\overline{u} - \ell_0)(\overline{x}) \leq 0 .$$

The two inequalities show that $\overline{u}$ is an upper semicontinuous viscosity subsolution of (3.2).

We proceed now to prove that $\underline{u}$ is a lower semicontinuous viscosity supersolution of (3.2). Let then $\varphi \in C^1(\mathbb{R}^N)$ and $\underline{x}$ be a strict minimum point for $\underline{u} - \varphi$ in $\overline{B} = \overline{B}(\underline{x}, r)$. By Lemma V.1.6 again, there exist $x^n \in \overline{B}$ and $\varepsilon_n \to 0^+$ such that

(3.10) $\quad (v^{\varepsilon_n} - \varphi)(x^n) = \min_{\overline{B}}(v^{\varepsilon_n} - \varphi), \quad x^n \longrightarrow \underline{x}, \quad v^{\varepsilon_n}(x^n) \longrightarrow \underline{u}(\underline{x}) .$

If $\underline{u}(\underline{x}) \geq \ell_0(\underline{x})$, then trivially

$$\max\{\underline{u}(\underline{x}) - f(\underline{x}) \cdot D\varphi(\underline{x}) - \ell_1(\underline{x}); \underline{u}(\underline{x}) - \ell_0(\underline{x})\} \geq 0,$$

so that $\underline{u}$ is a viscosity supersolution of (3.2). If, on the contrary, $\underline{u}(\underline{x}) < \ell_0(\underline{x})$, then by (3.10) and the continuity of ℓ_0,

(3.11) $$\frac{1}{\varepsilon_n}(v^{\varepsilon_n} - \ell_0)^+(x^n) = 0,$$

3. PENALIZATION

for n large enough. Since v^ε is a viscosity supersolution of (3.3) we deduce

$$v^{\varepsilon_n}(x^n) - f(x^n) \cdot D\varphi(x^n) - \ell_1(x^n) \geq 0 .$$

Let $n \to +\infty$ in the preceding to obtain

$$\underline{u}(\underline{x}) - f(\underline{x}) \cdot D\varphi(\underline{x}) - \ell_1(\underline{x}) \geq 0$$

and, a fortiori,

$$\max\{\underline{u} - f \cdot D\varphi - \ell_1; \underline{u} - \ell_0\} \geq 0 \qquad \text{at } \underline{x} .$$

This shows that $\underline{u}$ is a lower semicontinuous viscosity supersolution of (3.2). At this point we observe that (3.2) is equivalent to

(3.12) $$\max_{a \in \{0,1\}} \{v(x) - f(x,a) \cdot Dv(x) - \ell(x,a)\} = 0 \qquad x \in \mathbb{R}^N,$$

with

$$f(x,0) = 0, \qquad \ell(x,0) = \ell_0(x),$$
$$f(x,1) = f(x), \qquad \ell(x,1) = \ell_1(x) .$$

Therefore we can apply the comparison principle in the discontinuous case (see Chapter V, §1) to conclude that

$$\overline{u} \leq \underline{u} \qquad \text{in } \mathbb{R}^N .$$

Taking (3.6) into account, $\overline{u} \equiv \underline{u}$. We have therefore proved that $u = \overline{u} = \underline{u}$ is a bounded continuous viscosity solution of (3.2). Since the viscosity solution of (3.2) is unique, u coincides with v given by (3.1); by Lemma V.1.9 the convergence of v^ε to v is locally uniform and the proof is complete. ◀

3.2. Penalization of state constraints

Let us consider now the infinite horizon problem with state constraints as described in §IV.5 (in the case of discount rate $\lambda = 1$). Its value function is defined for $x \in \overline{\Omega}$ as

(3.13) $$v_{\overline{\Omega}}(x) = \inf_{\alpha \in \mathcal{A}_x} \int_0^{+\infty} \ell(y_x(t,\alpha), \alpha(t)) e^{-t} \, dt = \inf_{\alpha \in \mathcal{A}_x} J(x,\alpha),$$

where Ω is an open bounded subset of $\mathbb{R}^N$ with sufficiently smooth boundary $\partial\Omega$ and

$$\mathcal{A}_x = \{\alpha \in \mathcal{A} : y_x(t,\alpha) \in \overline{\Omega}, \ \forall t > 0\} .$$

We know that under suitable assumptions on the data (see §IV.5), including

(3.14) $$\inf_{a \in A} f(x,a) \cdot n(x) < 0 \qquad \forall x \in \partial\Omega,$$

where $n(x)$ is the outward normal at $x \in \partial\Omega$, $v_{\overline{\Omega}} \in BUC(\overline{\Omega})$ and is characterized as the unique solution of

(3.15) $\quad u(x) + \sup_{a \in A}\{ -f(x,a) \cdot Du(x) - \ell(x,a) \} = 0, \qquad x \in \Omega,$

(3.16) $\quad u(x) + \sup_{a \in A}\{ -f(x,a) \cdot Du(x) - \ell(x,a) \} \geq 0, \qquad x \in \overline{\Omega},$

in the viscosity sense. A first obvious remark is that

$$v_{\overline{\Omega}}(x) \geq v(x) \qquad \forall x \in \overline{\Omega},$$

where v is the value function of the corresponding unconstrained problem; that is

$$v(x) = \inf_{\alpha \in \mathcal{A}} \int_0^{+\infty} \ell(y_x(t,\alpha), \alpha(t)) e^{-t} dt = \inf_{\alpha \in \mathcal{A}} J(x,\alpha) \,.$$

A rather natural question, both from a theoretical and numerical point of view, is to approximate $v_{\overline{\Omega}}$ by the value function v^ε of a suitable penalization of the corresponding unconstrained problem. One possibility is to look at

$$v^\varepsilon(x) = \inf_{\alpha \in \mathcal{A}}\left\{ J(x,\alpha) + \frac{1}{\varepsilon}\int_0^{+\infty} p(y_x(t,\alpha)) e^{-t} dt \right\} =: \inf_{\alpha \in \mathcal{A}} J^\varepsilon(x,\alpha), \qquad x \in \mathbb{R}^N.$$

Here ε is a positive parameter and p is a distance-like function vanishing on $\overline{\Omega}$, whose effect is to penalize the running cost of those trajectories that do not satisfy the state constraint. More precisely, we assume that p satisfies

(3.17) $\quad p \in BUC(\mathbb{R}^N), \qquad p \equiv 0 \text{ in } \overline{\Omega}, \qquad p > 0 \text{ in } \mathbb{R}^N \setminus \overline{\Omega}\,.$

We know then by the results in §III.2 that $v^\varepsilon \in BUC(\mathbb{R}^N)$ and is the unique viscosity solution of

(3.18) $\quad v^\varepsilon(x) + \sup_{a \in A}\{ -f(x,a) \cdot Dv^\varepsilon(x) - \ell(x,a) - \frac{1}{\varepsilon}p(x) \} = 0, \qquad x \in \mathbb{R}^N.$

The next result describes the asymptotics of v^ε as $\varepsilon \to 0^+$.

THEOREM 3.2. *Under the assumptions on f, ℓ as in §IV.5 and (3.17)*

(3.19) $\quad v(x) \leq v^{\varepsilon'}(x) \leq v^\varepsilon(x) \leq v_{\overline{\Omega}}(x) \qquad \forall x \in \overline{\Omega},\ 0 < \varepsilon < \varepsilon',$

(3.20) $\quad \lim_{\varepsilon \to 0^+} v^\varepsilon = v_{\overline{\Omega}}, \text{ uniformly in } \overline{\Omega}\,.$

PROOF. The inequalities (3.19) are the simple consequence of the definitions of $v, v^\varepsilon, v_{\overline{\Omega}}$. In particular, the inequality $v^\varepsilon \leq v_{\overline{\Omega}}$ in $\overline{\Omega}$ follows from the fact that $J^\varepsilon(x,\alpha) = J(x,\alpha)$, for any $x \in \overline{\Omega}$ and $\alpha \in \mathcal{A}_x$; hence

$$v^\varepsilon(x) = \min\left\{ \inf_{\alpha \in \mathcal{A}_x} J^\varepsilon(x,\alpha);\ \inf_{\alpha \in \mathcal{A} \setminus \mathcal{A}_x} J^\varepsilon(x,\alpha) \right\}$$
$$= \min\left\{ v_{\overline{\Omega}}(x);\ \inf_{\alpha \in \mathcal{A} \setminus \mathcal{A}_x} J^\varepsilon(x,\alpha) \right\} \leq v_{\overline{\Omega}}(x)\,.$$

3. PENALIZATION

In order to prove (3.20), let us introduce the weak limits

$$\underline{u}(x) = \liminf_{(\varepsilon,y)\to(0^+,x)} v^\varepsilon(y), \qquad \overline{u}(x) = \limsup_{(\varepsilon,y)\to(0^+,x)} v^\varepsilon(y),$$

as in §V.1. Functions $\underline{u}, \overline{u}$ are well-defined in $\overline{\Omega}$ on the basis of (3.19). By definition, $\underline{u}, \overline{u}$ are, respectively, lower and upper semicontinuous on $\overline{\Omega}$ and satisfy

$$\underline{u} \leq \overline{u}.$$

From the very definition of $\overline{u}$ we have

$$\overline{u} \leq v_{\overline{\Omega}}$$

(recall that (3.14) implies the continuity of $v_{\overline{\Omega}}$ on $\overline{\Omega}$). The plan of the proof is to show that $\underline{u}$ is a viscosity supersolution of (3.16). At this point the comparison Theorem V.4.16 shows that

$$\underline{u} \geq v_{\overline{\Omega}}.$$

Hence $\underline{u} = \overline{u} = v_{\overline{\Omega}}$ and v^ε converge uniformly to $v_{\overline{\Omega}}$ in $\overline{\Omega}$ as $\varepsilon \to 0^+$ (see Lemma V.1.9).

Let us prove then that $\underline{u}$ satisfies (3.16) in the viscosity sense. For this purpose, let $\underline{x}$ be a strict minimum of $\underline{u} - \varphi$ in $\overline{B}(\underline{x},r) \cap \overline{\Omega}$, $\varphi \in C^1(\overline{\Omega})$, $r > 0$. Extend φ as a C^1 function on $\mathbb{R}^N$, still denoted by φ, and set

$$\widetilde{\varphi}(x) = \varphi(x) - k|x - \underline{x}|^2$$

where $k > 0$ will be selected later. Let x^ε be a minimum point of $v^\varepsilon - \widetilde{\varphi}$ in $\overline{B}(\underline{x},r)$; we can extract a sequence $\varepsilon_n \to 0^+$ such that $x_n := x^{\varepsilon_n}$ converges to some x^* and $v^{\varepsilon_n}(x_n)$ converges to some γ. We claim now that $x_n \in B(\underline{x},r)$ for all n if k is chosen large enough and that $x^* \in \overline{\Omega}$. Indeed, $x_n \in \partial B(\underline{x},r)$ would imply by (3.19) that

$$(v^{\varepsilon_n} - \widetilde{\varphi})(x_n) \geq (v - \varphi)(x_n) + kr^2 \geq \min_{\overline{B}(\underline{x},r)}(v - \varphi) + kr^2;$$

on the other hand, using (3.19) again,

$$(v^{\varepsilon_n} - \widetilde{\varphi})(x_n) \leq (v^{\varepsilon_n} - \widetilde{\varphi})(\underline{x}) \leq (v_{\overline{\Omega}} - \varphi)(\overline{x}).$$

These two inequalities are contradictory for sufficiently large k. Therefore the first claim is proved and this implies that x_n is a local minimum point for $v^{\varepsilon_n} - \widetilde{\varphi}$. Since v^{ε_n} is a supersolution of (3.18), then

(3.21) $$v^{\varepsilon_n}(x_n) + H(x_n, D\widetilde{\varphi}(x_n)) \geq \frac{1}{\varepsilon_n} p(x_n)$$

with $H(x,q) = \sup_{a \in A}\{-f(x,a) \cdot q - \ell(x,a)\}$.

To prove the second claim, assume by contradiction that $x^* \notin \overline{\Omega}$. Hence there is $\delta > 0$ such that $p(x_n) \geq \delta$ for $n \geq n_\delta$, as a consequence of (3.17). This leads to a contradiction to the fact that the left-hand side of (3.21) is bounded uniformly with respect to n, as follows from (3.19).

Let us now show that $x^* = \underline{x}$ and $\gamma = \underline{u}(\underline{x})$. Indeed, by the definition of $\underline{u}$ there exists a sequence $z_n \to \underline{x}$ such that $v^{\varepsilon_n}(z_n) \to \underline{u}(\underline{x})$. From the choice of x_n

$$(v^{\varepsilon_n} - \widetilde{\varphi})(x_n) \leq (v^{\varepsilon_n} - \widetilde{\varphi})(z_n)$$

whence, by definition of γ, $(\underline{u} - \widetilde{\varphi})(x^*) \leq \gamma - \widetilde{\varphi}(x^*) \leq (\underline{u} - \widetilde{\varphi})(\underline{x})$. Since we know that $x^* \in \overline{\Omega}$ and $\underline{x}$ is a strict minimum for $\underline{u} - \widetilde{\varphi}$ in $\overline{B}(\underline{x}, r) \cap \overline{\Omega}$ as well, the above yields $x^* = \underline{x}$ and $\gamma = \underline{u}(\underline{x})$.

At this point we let $n \to +\infty$ in (3.21) to obtain

$$\underline{u}(\underline{x}) + H(\underline{x}, D\varphi(\underline{x})) \geq \limsup_{n \to \infty} \frac{1}{\varepsilon_n} p(x_n) \geq 0,$$

showing that $\underline{u}$ is a supersolution of (3.16). ◀

Exercises

3.1. Give a direct proof of the fact that v^ε defined by (3.5) converges to v given by (3.1) as $\varepsilon \to 0$.

3.2. Generalize the result of Theorem 3.1 to the case where

$$v(x) = \inf_{\alpha \in \mathcal{A}} \inf_{\vartheta \geq 0} \left\{ \int_0^\vartheta \ell_1(y_x(s,\alpha), \alpha(s)) e^{-s} \, ds + \ell_0(y_x(\vartheta, \alpha)) e^{-\vartheta} \right\}.$$

3.3. In the framework of Theorem 3.1, try to estimate in terms of ε $\sup_{\mathbb{R}^N} |v^\varepsilon(x) - v(x)|$ as $\varepsilon \to 0$.

3.4. Give a direct proof of Theorem 3.2.

4. Singular perturbation problems

We analyze here a quite general singular perturbation problem in optimal control and some related topics concerning the approximation of the infinite horizon problem and of the monotone control problem (see §III.2 and §III.4.1) with similar problems where feasible controls satisfy a bound on the time derivative.

4.1. The infinite horizon problem for systems with fast components

We consider here a system described by $N + M$ state variables whose evolution is governed by

$$(4.1) \qquad \begin{aligned} \dot{y}(t) &= f(y(t), \zeta(t), \alpha(t)) \\ \varepsilon \dot{\zeta}(t) &= g(y(t), \zeta(t), \beta(t)) \end{aligned} \qquad t > 0,$$

with initial conditions

(4.2) $$y(0) = x, \quad \zeta(0) = z.$$

Here, $f : \mathbb{R}^N \times \mathbb{R}^M \times A \to \mathbb{R}^N$, $g : \mathbb{R}^N \times \mathbb{R}^M \times B \to \mathbb{R}^M$, $\varepsilon > 0$ is a small parameter and A, B are given sets where the controls α, β take their values.

For small values of the parameter $\varepsilon > 0$ the y components of the state vary slowly in comparison with the fast variables ζ. This situation is common in many applications and it is a convenient approach to *reduced order modeling*. Indeed, setting $\varepsilon = 0$ reduces (4.1) to

(4.3) $$\dot{y}(t) = f(y(t), \zeta(t), \alpha(t)),$$
$$0 = g(y(t), \zeta(t), \beta(t)).$$

Problem (4.3) is no longer a system of ODE's: this explains the terminology *singular perturbation*. Observe that if the algebraic equation

(4.4) $$g(x, z, b) = 0$$

has a unique solution $\bar{z} = \bar{z}(x, b) \in \mathbb{R}^M$ for any fixed $x \in \mathbb{R}^N, b \in B$, then a substitution in (4.3) yields the reduced N-dimensional system

(4.5) $$\dot{y}(t) = f(y(t), \bar{z}(y(t), \beta(t)), \alpha(t)),$$
$$y(0) = x.$$

The optimal control problem that we consider in this section is to minimize the performance index

(4.6) $$J^\varepsilon(x, z, \alpha, \beta) = \int_0^{+\infty} \ell(y^\varepsilon_{x,z}(t), \zeta^\varepsilon_{x,z}(t), \alpha(t)) e^{-\lambda t} dt,$$

where $y^\varepsilon_{x,z}, \zeta^\varepsilon_{x,z}$ is the solution of (4.1), (4.2), over

$$(\mathcal{A} \times \mathcal{B})_{x,z,\varepsilon} = \{ (\alpha, \beta) : [0, +\infty[\to A \times B, \ \zeta^\varepsilon_{x,z}(t) \in \overline{\mathcal{O}}, \ \forall t > 0 \}.$$

Here, $\mathcal{O}$ is a connected subset of $\mathbb{R}^M$ with smooth boundary $\partial \mathcal{O}$ playing the role of a state constraint on the fast variable ζ. We also assume that $\mathcal{O}$ is open.

We assume the controllability condition on the fast variables ζ

(4.7) $$\bar{B}(0,1) \subseteq \overline{\text{co}}\, g(x, z, B) \quad \forall x \in \mathbb{R}^N, \ z \in \overline{\mathcal{O}},$$

as well as the uniform Lipschitz continuity assumption

(4.8) $$|f(x, z, a) - f(x', z', a)| + |g(x, z, b) - g(x', z', b)| \leq L(|x - x'| + |z - z'|),$$

for all $x, x' \in \mathbb{R}^N$, $z, z' \in \overline{\mathcal{O}}$, with L independent of the control settings $(a, b) \in A \times B$. We assume also that

(4.9) $$|\ell(x, z, a) - \ell(x', z', a)| \leq L(|x - x'| + |z - z'|),$$

for all x, x' in $\mathbb{R}^N$, $z, z' \in \overline{\mathcal{O}}$, with L independent of $a \in A$, and

(4.10) $\qquad 0 \leq \ell(x, z, a) \leq M \qquad \forall x \in \mathbb{R}^N, z \in \overline{\mathcal{O}}, a \in A.$

The question that we address here is the limiting behavior of the value function

(4.11) $\qquad v^\varepsilon(x, z) = \inf_{(\mathcal{A} \times \mathcal{B})_{x,z,\varepsilon}} J^\varepsilon(x, z, \alpha, \beta), \qquad (x, z) \in \mathbb{R}^N \times \overline{\mathcal{O}}$

as $\varepsilon \to 0^+$. Note that under this set of assumptions nothing can be said about the solvability of the algebraic equation (4.4). Nonetheless we are able to find a reduced N-dimensional system describing the limit problem. In fact, the controllability condition (4.7) allows the system to reach any point $z_1 \in \overline{\mathcal{O}}$ from any starting point z in a lap of time of order $|z - z_1|/\varepsilon$. Therefore the value function v^ε is expected to be less and less sensitive to z as ε decreases, and to be independent of z in the limit $\varepsilon \to 0^+$. Moreover, since the control β does not appear in the equations for the slow variables y, nor in the cost functional J^ε, we expect its role reduces to driving optimally the ζ variables within $\overline{\mathcal{O}}$. Since this can be done arbitrarily fast as $\varepsilon \to 0^+$, we expect β to disappear in the limit problem, and ζ to take the role of a control varying in $\overline{\mathcal{O}}$. The preceding heuristic is confirmed by the following result.

THEOREM 4.1. *Let us assume* (4.7), (4.8), (4.9), (4.10), *and* $\lambda > 0$. *Then*

$$\lim_{\varepsilon \to 0^+} v^\varepsilon(x, z) = v(x), \quad \text{locally uniformly in } \mathbb{R}^N \times \mathcal{O},$$

with v given by

(4.12) $\qquad v(x) = \inf_{\substack{\alpha \in \mathcal{A} \\ \zeta \in \mathcal{Z}}} \int_0^{+\infty} \ell(y_x(t), \zeta(t), \alpha(t)) e^{-\lambda t} dt \qquad x \in \mathbb{R}^N,$

where $y_x(t) = y_x(t, \zeta, \alpha)$ satisfies

(4.13) $\qquad \begin{aligned} \dot{y}_x(t) &= f(y_x(t), \zeta(t), \alpha(t)), \\ y_x(0) &= x, \end{aligned}$

with ζ in the set

$$\mathcal{Z} = \{\zeta : [0, +\infty[\to \overline{\mathcal{O}} : \zeta \text{ measurable}\}.$$

PROOF. Let us start by observing that the exterior normal vector to $\partial(\mathbb{R}^N \times \overline{\mathcal{O}})$ at (x, z) is given by $n(x, z) = (0_N, n(z))$ where 0_N is the null vector in $\mathbb{R}^N$ and $n(z)$ is the exterior normal to $\partial \mathcal{O}$ at z. Moreover, by (4.7) and Carathéodory's theorem on convex hulls, it is not hard to show that for all $(x, z) \in \partial(\mathbb{R}^N \times \overline{\mathcal{O}})$ there exists $\overline{b} = \overline{b}(x, z) \in B$ such that

$$g(x, z, \overline{b}) \cdot n(z) \leq -\frac{1}{M+1},$$

4. SINGULAR PERTURBATION PROBLEMS

where M is the dimension of the vector g. Hence

(4.14) $\quad \left(f(x,z,a), \dfrac{1}{\varepsilon} g(x,z,\bar{b})\right) \cdot n(x,z) = \dfrac{1}{\varepsilon} g(x,z,\bar{b}) \cdot n(z) \leq -\dfrac{1}{\varepsilon(M+1)}$

for all $(x,z) \in \partial(\mathbb{R}^N \times \overline{\mathcal{O}})$. Therefore, by the results on the infinite horizon problem with state constraints of Chapter IV, §5 (see also Exercise IV.5.1), the value function $v^\varepsilon \in BUC(\mathbb{R}^N \times \overline{\mathcal{O}})$ and satisfies

(4.15) $\quad \lambda v^\varepsilon + \sup\limits_{a \in A}\{-f(x,z,a) \cdot D_x v^\varepsilon - \ell(x,z,a)\} - \dfrac{1}{\varepsilon} \inf\limits_{b \in B} g(x,z,b) \cdot D_z v^\varepsilon = 0$

in $\mathbb{R}^N \times \mathcal{O}$, as well as

(4.16) $\quad \lambda v^\varepsilon + \sup\limits_{a \in A}\{-f(x,z,a) \cdot D_x v^\varepsilon - \ell(x,z,a)\} - \dfrac{1}{\varepsilon} \inf\limits_{b \in B} g(x,z,b) \cdot D_z v^\varepsilon \geq 0$

in $\mathbb{R}^N \times \partial\mathcal{O}$, in the viscosity sense. On the other hand, by the theory of Chapter III, $v \in BUC(\mathbb{R}^N)$ and is the unique viscosity solution of

(4.17) $\quad \lambda v(x) + \sup\limits_{(a,z) \in A \times \overline{\mathcal{O}}}\{-f(x,z,a) \cdot Dv(x) - \ell(x,z,a)\} = 0 \quad \text{in } \mathbb{R}^N.$

The uniform bound $|v^\varepsilon(x,z)| \leq M/\lambda$ for all $(x,z) \in \mathbb{R}^N \times \overline{\mathcal{O}}$ is an immediate consequence of assumption (4.10). We then define the weak limits

$$\underline{u}(x,z) = \liminf\limits_{(\varepsilon,y,\zeta) \to (0^+,x,z)} v^\varepsilon(y,\zeta)$$

$$\overline{u}(x,z) = \limsup\limits_{(\varepsilon,y,\zeta) \to (0^+,x,z)} v^\varepsilon(y,\zeta).$$

Functions $\underline{u}, \overline{u}$ are, respectively, lower and upper semicontinuous and satisfy

(4.18) $\quad \underline{u} \leq \overline{u} \quad \text{in } \mathbb{R}^N \times \overline{\mathcal{O}}.$

The plan of the remaining part of the proof is to show that

(4.19) $\quad v(x) \leq \underline{u}(x,z) \quad \forall (x,z) \in \mathbb{R}^N \times \overline{\mathcal{O}},$

(4.20) $\quad \overline{u}(x,z) = \overline{u}(x) \quad \forall (x,z) \in \mathbb{R}^N \times \mathcal{O},$

(4.21) $\quad \overline{u}(x) \leq v(x) \quad \forall x \in \mathbb{R}^N.$

From these and (4.18) the conclusion follows by Lemma V.1.9, as in the proofs of the various asymptotic results in this chapter.

In order to prove (4.19), let us observe that v satisfies in the viscosity sense

(4.22) $\quad \lambda v(x) + \sup\limits_{a \in A}\{-f(x,z,a) \cdot Dv(x) - \ell(x,z,a)\} \leq 0 \quad \text{in } \mathbb{R}^N,$

for all $z \in \mathcal{O}$. This is a straightforward consequence of (4.17). Also, since v does not depend on $z \in \overline{\mathcal{O}}$,

(4.23) $$-\frac{1}{\varepsilon} \inf_{b \in B} g(x, z, b) \cdot D_z v(x) = 0 \quad \text{in } \mathbb{R}^N \times \mathcal{O}$$

in the classical and, a fortiori, viscosity sense. Hence v is a viscosity subsolution of (4.15), so that by the comparison Theorem IV.5.8,

$$v(x) \le v^\varepsilon(x, z) \quad \forall (x, z) \in \mathbb{R}^N \times \overline{\mathcal{O}}, \ \forall \varepsilon > 0,$$

and (4.19) follows, by the continuity of v.

Let us proceed now to show that the $\overline{u}$ is a subsolution of

(4.24) $$|D_z \overline{u}(x, z)| = 0 \quad \text{in } \mathbb{R}^N \times \mathcal{O}$$

in the viscosity sense. For this purpose, let $\varphi \in C^1(\mathbb{R}^N \times \mathcal{O})$ and $(\overline{x}, \overline{z})$ be a strict maximum point for $\overline{u} - \varphi$ in $\overline{B}_r = \overline{B}((\overline{x}, \overline{z}), r)$. By Lemma V.1.6 there exist $\varepsilon_n \to 0^+$ and $(x^n, z^n) \in \overline{B}_r$ such that

$$(v^{\varepsilon_n} - \varphi)(x^n, z^n) = \max_{\overline{B}_r}(v^{\varepsilon_n} - \varphi),$$
$$(x^n, z^n) \longrightarrow (\overline{x}, \overline{z}), \quad v^{\varepsilon_n}(x^n, z^n) \longrightarrow \overline{u}(\overline{x}, \overline{z}).$$

Then, from (4.15)

(4.25) $\lambda v^{\varepsilon_n}(x^n, z^n) + \sup_{a \in A}\{-f(x^n, z^n, a) \cdot D_x \varphi(x^n, z^n) - \ell(x^n, z^n, a)\}$
$$\le \frac{1}{\varepsilon_n} \inf_{b \in B} g(x^n, z^n, b) \cdot D_z \varphi(x^n, z^n).$$

The right-hand side of the above is the same as

$$\frac{1}{\varepsilon_n} \inf_{\eta \in \overline{co}g(x^n, z^n, B)} \eta \cdot D_z \varphi(x^n, z^n);$$

by assumption (4.7) we obtain

$$\frac{1}{\varepsilon_n} \inf_{\eta \in \overline{co}g(x^n, z^n, B)} \eta \cdot D_z \varphi(x^n, z^n) \le \frac{1}{\varepsilon_n} \inf_{\eta \in \bar{B}(0,1)} \eta \cdot D_z \varphi(x^n, z^n)$$
$$= -\frac{1}{\varepsilon_n} |D_z \varphi(x^n, z^n)|.$$

The left-hand side of (4.25) is bounded uniformly in n; hence letting $n \to +\infty$ in (4.25) yields $|D_z \varphi(\overline{x}, \overline{z})| \le 0$, which proves that $\overline{u}$ is a subsolution of (4.24). This implies that $\overline{u}$ is locally Lipschitz continuous as a function of $z \in \mathcal{O}$ (see Chapter V, Exercise 1.4). This easily implies, using (4.24) again, that $\overline{u}$ is constant with respect to z, and we obtain (4.20).

The proof will be complete after we have established the validity of (4.21). To this end we show that $\overline{u} = \overline{u}(x)$ is a viscosity subsolution of (4.17). Indeed, if

$\varphi \in C^1(\mathbb{R}^N)$ and $\overline{x}$ is a strict local maximum point for $\overline{u} - \varphi$, then $(\overline{x}, \overline{z})$ is a strict maximum for $\overline{u} - \varphi - |z - \overline{z}|^2$ in $\overline{B}_r := \overline{B}((\overline{x}, \overline{z}), r)$, for any fixed $\overline{z} \in \mathcal{O}$. Hence, by Lemma V.1.6, there exist $(x^n, z^n) \to (\overline{x}, \overline{z})$ and $\varepsilon_n \to 0^+$ such that

$$(v^{\varepsilon_n} - \varphi)(x^n, z^n) = \max_{\overline{B}}(v^{\varepsilon_n} - \varphi), \qquad v^{\varepsilon_n}(x_n, z^n) \longrightarrow \overline{u}(\overline{x}) .$$

Since v^ε is a viscosity subsolution of (4.15) and $z^n \in \mathcal{O}$ for large n,

(4.26) $\lambda v^{\varepsilon_n}(x^n, z^n) + \sup_{a \in A}\{ -f(x^n, z^n, a) \cdot D\varphi(x^n) - \ell(x^n, z^n, a) \}$

$$\leq \frac{1}{\varepsilon_n} \inf_{b \in B} g(x^n, z^n, b) \cdot 2(z^n - \overline{z}) .$$

Now (4.7) gives

$$\inf_{b \in B} g(x^n, z^n, b) \cdot (z^n - \overline{z}) \leq \inf_{\eta \in \tilde{B}(0,1)} \eta \cdot (z^n - \overline{z}) \leq 0 .$$

Letting $n \to +\infty$ in (4.26) we conclude that

$$\lambda \overline{u}(\overline{x}) + \sup_{a \in A}\{ -f(\overline{x}, \overline{z}, a) \cdot D\varphi(\overline{x}) - \ell(\overline{x}, \overline{z}, a) \} \leq 0 \qquad \forall \overline{z} \in \mathcal{O} .$$

This shows, after taking the supremum over $\overline{\mathcal{O}}$, that $\overline{u}$ is a viscosity subsolution of (4.17). By the comparison Theorem V.1.3 we obtain (4.21) and the proof is complete. ◂

The result of Theorem 4.1 can be improved by showing that v^ε converges to v locally uniformly in $\mathbb{R}^N \times \overline{\mathcal{O}}$ (see Exercise 4.2). The same remark applies to Corollary 4.2.

To conclude this section we discuss the limiting behavior as $\varepsilon \to 0^+$ of the value functions of the Lipschitz control problem with prescribed Lipschitz constant $1/\varepsilon$. The Lipschitz control problem models the optimal regulation of those control systems that do not allow, for technological reasons, an instantaneous switching between different control settings. Its value function is

(4.27) $\quad v^\varepsilon(x, z) := \inf_{\zeta \in \mathcal{L}_z^\varepsilon} J(z, \zeta) := \inf_{\mathcal{L}_z^\varepsilon} \int_0^{+\infty} \ell(y_x(t, \zeta), \zeta(t)) e^{-\lambda t} dt,$

where $y_x(\,\cdot\,, \zeta)$ solves

(4.28) $\qquad\qquad\qquad \begin{aligned} \dot{y}(t) &= f(y(t), \zeta(t)), \\ y(0) &= x, \end{aligned}$

and

$$\mathcal{L}_z^\varepsilon = \{ \zeta : [0, +\infty[\to \overline{\mathcal{O}} : |\zeta(t) - \zeta(t')| \leq \varepsilon^{-1}|t - t'|, \ \zeta(0) = z \}$$

is the class of Lipschitz controls with prescribed initial value $z \in \overline{\mathcal{O}}$ and fixed Lipschitz constant $1/\varepsilon > 0$. The behavior of v^ε as $\varepsilon \to 0^+$ can be analyzed in the framework of Theorem 4.1. Indeed, a feasible $\zeta \in \mathcal{L}_z^\varepsilon$ can be interpreted as a fast state variable driven by the state equation

(4.29)
$$\dot{\zeta}(t) = \frac{1}{\varepsilon}\beta(t),$$
$$\zeta(0) = z \in \overline{\mathcal{O}},$$

with

$$\beta \in \mathcal{B}_{z,\varepsilon} = \{\beta : [0,+\infty[\to \overline{B}(0,1) : \zeta_z^\varepsilon(t) \in \overline{\mathcal{O}} \text{ for } t > 0\}.$$

COROLLARY 4.2. *Under the assumptions* (4.8), (4.9), (4.10), *and* $\lambda > 0$, v^ε *given by* (4.27) *converge locally uniformly in* $\mathbb{R}^N \times \mathcal{O}$ *as* $\varepsilon \to 0^+$ *to* v *given by*

$$v(x) := \inf_{\zeta \in \mathcal{Z}} J(x,\zeta)$$

where $\mathcal{Z} = \{\zeta : [0,+\infty[\to \overline{\mathcal{O}} : \zeta \text{ measurable}\}$.

PROOF. The statement follows from Theorem (4.1) when applied with A a singleton, $B = \overline{B}(0,1) \subset \mathbb{R}^M$, $f(x,z,a) = f(x,z)$, $g(x,z,b) = b$. Observe that with these choices condition (4.7) is trivially satisfied. ◂

From this result we obtain easily that the value function in the infinite horizon problem does not change if we restrict the set of admissible controls from merely measurable functions to Lipschitz continuous functions with given initial point, provided there is no bound on the Lipschitz constant. More precisely

(4.30) $\quad v(x) = \inf_{\zeta \in \mathcal{Z}_{\text{Lip}}} J(x,\zeta) = \inf\{J(x,\zeta) : \zeta \in \mathcal{Z}_{\text{Lip}}, \zeta(0) = z\}, \quad \forall z \in \overline{\mathcal{O}},$

where $\mathcal{Z}_{\text{Lip}} = \{\zeta : [0,+\infty[\to \overline{\mathcal{O}} : \zeta \text{ Lipschitz continuous}\}$.

4.2. Asymptotics for the monotone control problem

An approximation of the monotone control problem which is appropriate in the modeling of control systems that do not allow instantaneous switching between different control setting is

(4.31) $$v^\varepsilon(x,z) = \inf_{\zeta \in \mathcal{M}_z^\varepsilon} J(x,\zeta)$$

where J is as in (4.27) and

$$\mathcal{M}_z^\varepsilon = \{\zeta : [0,+\infty[\to \mathbb{R} : 0 \leq \zeta(t) - \zeta(t') \leq (t-t')/\varepsilon \text{ for } t \geq t', \zeta(0) = z\}$$

is the class of one-dimensional monotone nondecreasing controls with fixed upper bound $1/\varepsilon$ on the time derivative.

The preceding model is appropriate for systems whose evolution is driven by a non renewable resource (see Chapter III, §4.1). Much in the same way as for the case of Lipschitz controls (see the previous subsection) we interpret any $\zeta \in \mathcal{M}_z^\varepsilon$ as a "fast" variable driven by the dynamics

$$\dot\zeta(t) = \frac{1}{\varepsilon}\beta(t),$$
$$\zeta(0) = z \in \mathbb{R},$$

with $\beta \in \mathcal{B} = \{\,\beta : [0,+\infty[\to [0,1] : \beta \text{ measurable}\,\}$.

We again take A as a singleton, $f(x,z,a) = f(x,z)$, $g(x,z,b) = b$, but now B is the closed interval $[0,1]$ as dictated by the monotonicity condition. Hence, condition (4.7) does not hold. Consequently, the behavior of v^ε as $\varepsilon \to 0^+$ cannot be deduced from Theorem 4.1. However, a somewhat similar asymptotic result holds as indicated in the following. Heuristically, as $\varepsilon \to 0^+$ the upper bound on the time derivative of feasible controls becomes larger and larger so that one expects v^ε to converge to

$$(4.32) \qquad v(x,z) = \inf_{\zeta \in \mathcal{M}_z} J(x,\zeta)$$

with $\mathcal{M}_z = \{\,\zeta : [0,+\infty[\to \mathbb{R} : 0 \le \zeta(t) - \zeta(t') \text{ for } t \ge t',\ \zeta(0) = z\,\}$. Unlike the previous example, the limit function here depends on z, due to the unilateral constraint on ζ. As expected, this is reflected by a unilateral constraint on $\partial v/\partial z$, namely,

$$(4.33) \qquad -\frac{\partial v}{\partial z}(x,z) \le 0,$$

in the viscosity sense.

THEOREM 4.3. *Let us assume* (4.8), (4.9), (4.10) *and* $\lambda > 0$. *Then*

$$\lim_{\varepsilon \to 0^+} v^\varepsilon(x,z) = v(x,z) \quad \text{locally uniformly in } \mathbb{R}^N \times \mathbb{R}$$

where v is given by (4.32).

PROOF. Since v^ε is viewed as the value function of a standard infinite horizon problem with state variables (y,ζ), by the results of §III.2.1 $v^\varepsilon \in BUC(\mathbb{R}^N \times \mathbb{R})$ and satisfies equation (4.15) which in the present case (recall that $f(x,z,a) = f(x,z)$, $g(x,z,b) = b$ and $B = [0,1]$) takes the form

$$(4.34) \qquad \lambda v^\varepsilon(x,z) - f(x,z)\cdot D_x v^\varepsilon(x,z) - \ell(x,z) + \frac{1}{\varepsilon}\left(-\frac{\partial v^\varepsilon}{\partial z}(x,z)\right)^+ = 0$$

in $\mathbb{R}^N \times \mathbb{R}$. By the arguments of §III.4.1 we see that v satisfies the following Bellman equation

$$(4.35) \qquad \max\{\,\lambda v(x,z) - f(x,z)\cdot D_x v(x,z) - \ell(x,z);\, -\frac{\partial v}{\partial z}(x,z)\,\} = 0$$

in $\mathbb{R}^N \times \mathbb{R}$, in the viscosity sense. Observe that any viscosity subsolution of the above satisfies (4.33).

It is immediate to check that $M = \sup_{\mathbb{R}^N \times \mathbb{R}} |\ell(x,z)/\lambda|$ is a supersolution of (4.34), whereas $-M$ is a subsolution. Hence, by comparison, $|v^\varepsilon(x,z)| \leq M$ for any $(x,z) \in \mathbb{R}^N \times \mathbb{R}$ and $\varepsilon > 0$. We then define

$$\underline{u}(x,z) = \liminf_{(\varepsilon,y,\zeta) \to (0^+,x,z)} v^\varepsilon(y,\zeta), \qquad \overline{u}(x,z) = \limsup_{(\varepsilon,y,\zeta) \to (0^+,x,z)} v^\varepsilon(y,\zeta).$$

Functions $\underline{u}, \overline{u}$ are, respectively, lower and upper semicontinuous and satisfy

$$\underline{u} \leq \overline{u} \qquad \text{in } \mathbb{R}^N \times \mathbb{R}.$$

The claim now is that $\overline{u}$ is an upper semicontinuous viscosity subsolution of (4.35). For this purpose, let $\varphi \in C^1(\mathbb{R}^N \times \mathbb{R})$ and $(\overline{x}, \overline{z})$ be a strict maximum point for $\overline{u} - \varphi$ in $\overline{B}_r := \overline{B}(\overline{\xi}, r)$ where $\overline{\xi} = (\overline{x}, \overline{z})$. By Lemma V.1.6, there exist $\xi^n \in \overline{B}_r$ and $\varepsilon_n \to 0^+$ such that

(4.36)
$$(v^{\varepsilon_n} - \varphi)(\xi^n) = \max_{\overline{B}_r}(v^{\varepsilon_n} - \varphi),$$
$$\xi^n \longrightarrow \overline{\xi}, \qquad v^{\varepsilon_n}(\xi^n) \longrightarrow \overline{u}(\overline{\xi}).$$

Since v^ε is a viscosity subsolution of (4.34), we have

(4.37) $$\lambda v^{\varepsilon_n}(\xi^n) - f(\xi^n) \cdot D_x\varphi(\xi^n) - \ell(\xi^n) \leq 0,$$

(4.38) $$\left(-\frac{\partial \varphi}{\partial z}(\xi^n)\right)^+ \leq -\varepsilon_n(\lambda v^{\varepsilon_n}(\xi^n) - f(\xi^n) \cdot D_x\varphi(\xi^n) - \ell(\xi^n)).$$

Thanks to (4.36) and the uniform bound on v^ε, we can let $n \to +\infty$ in (4.37), (4.38) to obtain

$$\lambda \overline{u}(\overline{\xi}) - f(\overline{\xi}) \cdot D_x\varphi(\overline{\xi}) - \ell(\overline{\xi}) \leq 0, \qquad \left(-\frac{\partial \varphi}{\partial z}(\overline{\xi})\right)^+ \leq 0.$$

Hence $-\frac{\partial \varphi}{\partial z}(\overline{\xi}) \leq 0$. This implies that $\overline{u}$ is a viscosity subsolution of (4.35).

Let us proceed now to show that $\underline{u}$ is a viscosity supersolution of (4.35). If $\underline{\xi} = (\underline{x}, \underline{z})$ is a strict minimum point for $\underline{u} - \varphi$, $\varphi \in C^1(\mathbb{R}^N \times \mathbb{R})$, and $-\frac{\partial \varphi}{\partial z}(\underline{\xi}) \geq 0$, then the supersolution condition is trivially satisfied. If, on the contrary, $-\frac{\partial \varphi}{\partial z}(\underline{\xi}) < 0$, then by Lemma V.1.6 there exist $\xi^n \in \overline{B}_r := \overline{B}(\underline{\xi}, r)$ and $\varepsilon_n \to 0^+$ such that

(4.39)
$$(v^{\varepsilon_n} - \varphi)(\xi^n) = \min_{\overline{B}_r}(v^{\varepsilon_n} - \varphi),$$
$$\xi^n \longrightarrow \underline{\xi}, \qquad v^{\varepsilon_n}(\xi^n) \longrightarrow \underline{u}(\underline{\xi}).$$

By the continuity of $\partial\varphi/\partial z$ we have $\left(-\frac{\partial \varphi}{\partial z}(\xi^n)\right)^+ = 0$ for sufficiently large n, so that (4.34) yields

$$\lambda v^{\varepsilon_n}(\xi^n) - f(\xi^n) \cdot D_x\varphi(\xi^n) - \ell(\xi^n) \geq 0.$$

If we let $n \to +\infty$ in the preceding we obtain, by (4.39),

$$\lambda \underline{u}(\underline{\xi}) - f(\underline{\xi}) \cdot D_x\varphi(\underline{\xi}) - \ell(\underline{\xi}) \geq 0$$

and, a fortiori,

$$\max\{ \lambda \underline{u}(\underline{x}, \underline{z}) - f(\underline{x}, \underline{z}) \cdot D_x\varphi(\underline{x}, \underline{z}) - \ell(\underline{x}, \underline{z}); -\frac{\partial \varphi}{\partial z}(\underline{x}, \underline{z}) \} \geq 0 \ .$$

Now we claim that a comparison principle holds for semicontinuous sub- and supersolutions of (4.35): it can be obtained by combining the arguments of the proofs of Theorem III.4.6 and Theorem V.1.3 and we leave it as an exercise for the reader. Then $\underline{u} = \overline{u} = v$, and this implies the local uniform convergence of v^ε to v by Lemma V.1.9. ◀

Exercises

4.1. Assume that $B(0,1) \subseteq g(x, z, B)$, $\forall x \in \mathbb{R}^N$, $\forall z \in \overline{\mathcal{O}}$. Prove that for any $(x, z) \in \mathbb{R}^N \times \partial\Omega$ there exists $t^* > 0$ and $\gamma : [0, t^*] \to \mathbb{R}^N \times \overline{\Omega}$ such that:

γ is continuous, $\gamma(0) = (x, z)$,

$\gamma(t) \in \mathbb{R}^N \times \Omega$, $\forall t \in]0, t^*[$,

$v^\varepsilon(x, z) - v^\varepsilon(\gamma(t)) \leq 2Mt$, $\forall t \in [0, t^*], \forall 0 < \varepsilon \leq 1$.

4.2. Use Exercise 4.1 to show that v^ε converges pointwise to v in $\mathbb{R}^N \times \overline{\mathcal{O}}$. Conclude by a monotonicity argument that v^ε converges to v locally uniformly in $\mathbb{R}^N \times \overline{\mathcal{O}}$. This improves the result of Theorem 4.1.

4.3. Show that the convergence in Corollary 4.2 is monotone and prove (4.30).

5. Bibliographical notes

Our presentation of the vanishing discount problem in presence of state constraints in §1.1 is modeled on that in [CDL90] with a more explicit use of controllability conditions and with new proofs based on the weak limit technique described in Chapter V.

Recent results on the asymptotics as λ tends to zero for equation (HJB)$_\lambda$ with different boundary conditions (modeling, for example, the reflection of the state when it hits the boundary) can be found in [Ar95a, Ar95b].

The role of dissipativity conditions on the dynamics in the ergodic problem for the unconstrained optimal stopping time problem was pointed out in [CDMe86]. Our treatment in §1.2 is inspired by [CDMe88], with partly different proofs, and by some unpublished work of Capuzzo Dolcetta and H. Ishii on the estimate of the rate of convergence. It is worthwhile pointing out that similar ergodic problems arise in connection with $\mathcal{H}_\infty$ control (see for example [FM92b, FM95, Cam96]). Different results on the ergodic problem for the infinite horizon problems can be found in [Col89]; we refer also to the book [CH87] for some related topics.

Let us mention also that there is a wide literature on ergodic problems for second order Hamilton-Jacobi equations connected with optimal control of stochastic processes (see, for example, [Gi85, LP86, CDG86] and [Ro83] for a survey).

The convergence result for vanishing switching costs in §2 is taken from [CDE84] with a slightly different proof.

The penalization technique has been widely employed in the domain of elliptic and parabolic variational and quasivariational inequalities connected with free boundary problems, see [BL82] as a general reference on this topic. Subsection 3.1 can be seen as a variation of this theme in connection with variational inequalities for a first-order partial differential operator. Related results can be found in [CDMe88]. In §3.2 a different kind of problem is considered where the penalization term bears on the boundary condition. The asymptotic result there is taken from [CDL90] with a new proof using weak limits, as usual in this chapter.

Singular perturbation is an important topic in system theory and in optimal control, see, e.g., [Kok84, Ben88] and the references therein. The result on optimal control of systems with fast components in §4.1 seems to be new; our PDE treatment is inspired by [L85c] (see also [S93] for a similar approach to a stochastic problem). For some extensions see [Bag93, BBCD94, BagB96, An96]. The Lipschitz control problem that we study as a byproduct is also considered in [BEJ84] for differential games.

The asymptotic result for the monotone control problem in §4.2 is essentially taken from the original paper [Bn85] with a new proof.

To conclude, let us mention that many asymptotic problems of different nature have been analyzed by means of viscosity solutions. We refer, for example, to [LPV86, E89, E92, B90, Alv97b, HI97] for homogenization, to [Bn93] for the related problem of averaging, to [FrSo86, Di92] for blowup of solutions, to [L82, Ba85c, DR93] for large time behavior in evolutive HJ equations, and to [DIS90] for an asymptotic analysis of queuing systems. Viscosity solutions methods were applied to the "geometric optics" approximation of wavefronts propagation in reaction-diffusion equations and systems [ES89, BES90, MSou94], as well as to several other asymptotic problems related to the propagation of fronts (such as De Giorgi's definition of the motion by mean curvature via singular perturbations [DeG90, DeG92]); see, for instance, [ESS92, BSS93, SSou96] and the survey paper [Sou95].

CHAPTER VIII

Differential Games

In this chapter we consider two-person zero-sum differential games. Let us describe them. We are given a system

(0.1) $$\begin{cases} y'(t) = f(y(t), a(t), b(t)), & t > 0, \\ y(0) = x, \end{cases}$$

where

(0.2) $$\begin{cases} f : \mathbb{R}^N \times A \times B \to \mathbb{R}^N \text{ is continuous,} \\ A, B \text{ are compact metric spaces,} \end{cases}$$

and $b = b(\,\cdot\,) \in \mathcal{B} := \{\,\text{measurable functions } [0, +\infty[\to B\,\}$ and $a = a(\,\cdot\,) \in \mathcal{A}$ are, respectively, the control functions of the second and the first player. The solution of (0.1) is denoted by $y_x(\,\cdot\,; a, b)$. We are also given a cost functional $J(x, a(\,\cdot\,), b(\,\cdot\,))$, which the first player wants to minimize and the second player wants to maximize. In other words, $-J$ is the cost the second player has to pay, so the sum of the costs of the two players is null, which explains the name "zero-sum". Finally we are given an *information pattern* for the two players prescribing, roughly speaking, that each of them choose his or her own control at each instant of time without knowing the future choices of the opponent. This is made rigorous in Section 1 by introducing the notion of *nonanticipating strategy*, where one player knows the current and past choices of the control made by his opponent. In Section 3 we assume instead that one player knows just the current state and its past history, and model this information pattern by means of *feedback strategies* (strategies are denoted by Greek letters, α and β, that is why we have changed notation with respect to the previous chapters and we now denote control functions by a and b). These *dynamic* information patterns are more interesting for applications than the one where the first player chooses over $\mathcal{A}$ and the second player over $\mathcal{B}$, namely, the *static game* whose lower and upper value functions v_s and u_s are

(0.3) $$v_s(x) := \sup_{b \in \mathcal{B}} \inf_{a \in \mathcal{A}} J(x, a, b) \leq \inf_{a \in \mathcal{A}} \sup_{b \in \mathcal{B}} J(x, a, b) =: u_s(x).$$

Note that the lower value corresponds to a game where the second player makes his choice with the information of the whole future response of the first player to any control function $b \in \mathcal{B}$, and the roles are reversed in the case of the upper value. The static game is actually a two-person zero-sum game over the infinite dimensional sets $\mathcal{A}$ and $\mathcal{B}$ in the sense of classical game theory, but it is not fit to be analyzed by the Dynamic Programming method.

There are two main motivations to the theory of (two-person, zero-sum) differential games. The first is modeling a real conflict situation where two controllers have opposite interests, so that the loss of one coincides with the gain of the other. The most classical example is the *pursuit-evasion game* where the first $N/2$ state coordinates describe a moving object controlled by the first player, the other coordinates describe another object controlled by the second player, and the cost J is the *capture time*, that is, the first time when some prescribed coordinates of the first object get close enough to the corresponding coordinates of the second object.

The second motivation comes from optimal control problems with a single controller where, however, the system and/or the cost functional are affected by some unknown disturbance b. The traditional approach of stochastic control theory consists of choosing a statistical model for the disturbance and minimizing the expected value of the cost. In some cases this approach may not be appropriate because minimizing the expected value does not give any information on the dispersion of the realizations, and does not guarantee against some possible "catastrophic" events. Moreover, the statistical model may not be good enough. In these cases it may be useful to adopt a pessimistic, risk-averse approach and model the disturbance as the control of a second player willing to maximize the cost. This leads to a two-person zero-sum differential game, and it is equivalent to minimizing the cost in the case of the worst possible behavior of the disturbance. We refer the reader to Appendix B for the connections with robust optimal control and $\mathcal{H}_\infty$ control.

In this chapter we study only the infinite horizon differential game, where the cost functional is

$$(0.4) \qquad J(x, a, b) := \int_0^\infty \ell(y_x(t), a(t), b(t)) e^{-\lambda t}\, dt,$$

where $y_x(t) = y_x(t; a, b)$ is the solution of (0.1) corresponding to $a \in \mathcal{A}$ and $b \in \mathcal{B}$. The reason for this choice is that the infinite horizon is our model problem in Chapter I and it is the first one studied in Chapter III. However, all the other cost functionals considered in this book can be studied in the context of differential games by the same methods, as well as many approximation, perturbation, and asymptotic problems of Chapters VI and VII. In fact, the proofs of the main results of the book are mostly based on the HJB equations, and similar equations are satisfied in the viscosity sense by the value functions of differential games. As an important example we prove in Section 3 the convergence of approximations by discrete time games.

We assume throughout the chapter that the system satisfies, for some constant L,

$$(0.5) \qquad (f(x, a, b) - f(y, a, b)) \cdot (x - y) \leq L|x - y|^2$$

for all $x, y \in \mathbb{R}^N$, $a \in A$, $b \in B$. Under this assumption we have the usual estimates on the trajectories:

(0.6) $\quad |y_x(t, a, b) - y_z(t, a, b)| \leq e^{Lt}|x - z|, \quad\quad t > 0,$

(0.7) $\quad |y_x(t, a, b) - x| \leq M_x t, \quad\quad t \in [0, 1/M_x],$

(0.8) $\quad |y_x(t, a, b)| \leq (|x| + \sqrt{2Kt})e^{Kt},$

for all $a \in \mathcal{A}$, $b \in \mathcal{B}$, where

$$M_x := \max\{|f(z, a, b)| : |x - z| \leq 1, \ a \in A, b \in B\},$$
$$K := L + \max\{|f(0, a, b)| : a \in A, b \in B\}.$$

1. Dynamic Programming for lower and upper values

We begin with the definition of nonanticipating (or *causal*) strategy.

DEFINITION 1.1. A *strategy* for the first player is a map $\alpha : \mathcal{B} \to \mathcal{A}$; it is *nonanticipating*, if, for any $t > 0$ and $b, \widetilde{b} \in \mathcal{B}$, $b(s) = \widetilde{b}(s)$ for all $s \leq t$ implies $\alpha[b](s) = \alpha[\widetilde{b}](s)$ for all $s \leq t$.

We denote with Γ the set of nonanticipating strategies for the first player. Similarly, the set of nonanticipating strategies for the second player is

$$\Delta := \{\beta : \mathcal{A} \to \mathcal{B} : a(s) = \widetilde{a}(s)$$
$$\text{for all } s \leq t \text{ implies } \beta[a](s) = \beta[\widetilde{a}](s) \text{ for all } s \leq t\}. \quad \triangleleft$$

EXAMPLE 1.2. A trivial example of strategy $\alpha \in \Gamma$ is a constant one: for a fixed $\overline{a} \in \mathcal{A}$, $\alpha[b] \equiv \overline{a}$ for all $b \in \mathcal{B}$. Another example of nonanticipating strategy is $\alpha[b](s) = \Phi(b(s))$ where $\Phi : B \to A$ is a given map such that $\Phi(b(\,\cdot\,))$ is measurable for any $b \in \mathcal{B}$. $\quad \triangleleft$

EXAMPLE 1.3. A time-dependent feedback map $\Phi : \mathbb{R}^N \times [0, \infty[\to A$ is admissible for x if for every $b : [0, T[\to B$ measurable and $T \in]0, +\infty]$ there exists a unique solution of

$$\begin{cases} q'(t) = f(q(t), \Phi(q(t), t), b(t)), & 0 < t \leq T, \\ q(0) = x, \end{cases}$$

such that $t \mapsto \Phi(q(t), t)$ is measurable. We associate with Φ (and x) the strategy

$$\alpha[b](t) := \Phi(q(t), t),$$

which is nonanticipating and such that $y_x(\,\cdot\,, \alpha[b], b) = q(\,\cdot\,)$ for all $b \in \mathcal{B}$. $\quad \triangleleft$

EXAMPLE 1.4. One may want to apply a feedback map to some weighted average of the past history of the state instead of its current value. Then one supposes that for any measurable b and $T \in]0, +\infty]$ there exists a unique solution of

(1.1) $\quad \begin{cases} q'(t) = f(q(t), \Phi(I(t), t), b(t)), & 0 < t \leq T, \\ I(t) = \int_0^t q(s) \, d\mu_t, \\ q(0) = x, \end{cases}$

such that $t \mapsto \Phi(I(t), t)$ is measurable, where, for all t, μ_t is a given probability measure on $[0, t]$. Given Φ, μ_t, $t \geq 0$, and x, the strategy

$$\alpha[b](t) := \Phi(I(t), t)$$

is nonanticipating and such that $y_x(\,\cdot\,; \alpha[b], b) = q(\,\cdot\,)$ for all $b \in \mathcal{B}$. ◁

EXAMPLE 1.5. A strategy that is not nonanticipating is, for instance, $\alpha[b](s) := \Phi(b(s + \tau))$ for any fixed $\tau > 0$ and $\Phi : B \to A$ such that $\Phi(b(\,\cdot\, + \tau))$ is measurable for all $b \in \mathcal{B}$. ◁

Other examples are given in §3.1.

Now we can define the lower and upper values of a differential game.

DEFINITION 1.6. The *lower value* of a game with cost functional $J : \mathbb{R}^N \times \mathcal{A} \times \mathcal{B} \to \mathbb{R}$ is

$$v(x) := \inf_{\alpha \in \Gamma} \sup_{b \in \mathcal{B}} J(x, \alpha[b], b);$$

the *upper value* is

$$u(x) := \sup_{\beta \in \Delta} \inf_{a \in \mathcal{A}} J(x, a, \beta[a]) \,.$$

If $v(x) = u(x)$, we say that the game with initial point x has a value. ◁

Note that

(1.2) $\qquad v_s \leq v \leq u_s \qquad$ and $\qquad v_s \leq u \leq u_s,$

where u_s, v_s are the upper and lower values of the static game, see (0.3). In fact, for any $x \in \mathbb{R}^N$ and $\varepsilon > 0$ there is $\overline{\alpha} \in \Gamma$ such that

$$v(x) + \varepsilon \geq J(x, \overline{\alpha}[b], b) \geq \inf_{a \in \mathcal{A}} J(x, a, b) \qquad \text{for all } b \in \mathcal{B},$$

which proves the inequality $v_s \leq v$, while the inequality $v \leq u_s$ follows from Example 1.2, because Γ contains a copy of $\mathcal{A}$. The inequalities about u are proved similarly. Note also that the inequality

(1.3) $\qquad v(x) \leq u(x) \qquad$ for all x,

which would justify the terms "lower" and "upper", is not obvious at first glance. (Actually one might first guess the opposite inequality because $\inf \sup \geq \sup \inf$ if they are taken over the same sets; here, however, the inf in the definition of v is taken over strategies whereas in the definition of u it is taken over controls, and similarly the sup is taken over different sets in the two definitions, so the inequality $v \geq u$ is false in general, see Example 1.7.) We prove (1.3) for the infinite horizon problem in a rather indirect way, by using the Hamilton-Jacobi equations associated with the game. An intuitive reason why (1.3) is true is that, even if the players choose simultaneously with no knowledge of the future behavior of the opponent, in the definition of lower value the first player makes his choice of the strategy α with the information of the second player's optimal response to any α. Roughly speaking, at each instant of time the first player knows the choice the second one is making at the same time. This informational advantage goes to the second player in the definition of upper value, so (1.3) is plausible. The reader can be easily convinced of this argument by looking at Example 1.7, where actually (1.3) holds with the strict inequality.

The information pattern described is not always realistic for the applications because of the advantage given to the player using strategies. However, it is reasonable to believe that any more fair game has an outcome between $v(x)$ and $u(x)$. Indeed, this is the case for the game with feedback strategies that we study in Section 3 (see also Exercise 1.5). For these reasons it is interesting to give conditions ensuring the existence of a value, that is, the equality $v(x) = u(x)$.

From now on we study the lower and upper values for the infinite horizon problem, that is, with the cost functional given by (0.4), under the following assumptions on the running cost and the discount rate:

(1.4)
$$\begin{cases} \ell : \mathbb{R}^N \times A \times B \to \mathbb{R} \text{ is continuous,} \\ \text{there are a modulus } \omega_\ell \text{ and a constant } M \text{ such that} \\ |\ell(x,a,b) - \ell(y,a,b)| \leq \omega_\ell(|x-y|) \text{ and} \\ |\ell(x,a)| \leq M, \text{ for all } x, y \in \mathbb{R}^N, a \in A, b \in B; \\ \lambda > 0 \,. \end{cases}$$

Therefore in the rest of this chapter the lower and upper values are

$$v(x) := \inf_{\alpha \in \Gamma} \sup_{b \in \mathcal{B}} \int_0^\infty \ell(y_x(t), \alpha[b](t), b(t)) e^{-\lambda t} \, dt,$$

$$u(x) := \sup_{\beta \in \Delta} \inf_{a \in \mathcal{A}} \int_0^\infty \ell(y_x(t), a(t), \beta[a](t)) e^{-\lambda t} \, dt,$$

for $x \in \mathbb{R}^N$. First we give an example where $v(x) < u(x)$ for all x. (We refer to Exercise 2.1 for an example where $v_s(x) < v(x) = u(x) < u_s(x)$ for some x.)

EXAMPLE 1.7. Let $N = 1$, $\lambda = 1$, $f(x,a,b) = (a-b)^2$, $A = B = [-1,1]$, $\ell(x,a,b) = \ell(x) = \text{sgn}\, x \cdot (1 - e^{-|x|})$. Since the running cost ℓ is strictly increasing and $y' = f \geq 0$, the best move of the minimizing player "a" is to keep the system still, so his optimal strategy is

$$\alpha^*[b](t) = b(t) \qquad \text{for all } t \,.$$

With this choice $J(x, \alpha[b], b) = \ell(x)$ for all b, thus

$$v(x) = \ell(x) \qquad \text{for } x \in \mathbb{R} \,.$$

On the other hand the best choice of the maximizing player "b" is to make $(a-b)^2$ as large as possible, so his optimal strategy is

$$\beta^*[a](t) = \begin{cases} 1 & \text{if } a(t) \leq 0, \\ -1 & \text{if } a(t) > 0, \end{cases}$$

and the best choice of the first player under this strategy is $a \equiv 0$. Then we compute

$$J(x, \beta^*[0], 0) = \int_0^\infty \ell(x+t) e^{-t}\, dt = 1 - \frac{e^{-x}}{2} \qquad \text{for } x \geq 0,$$

and claim that

$$u(x) = 1 - \frac{e^{-x}}{2} = \ell(x) + \frac{e^{-x}}{2} > v(x), \qquad \text{for } x \geq 0\,.$$

A similar calculation can be made for $x < 0$. This claim can be checked by the verification theorems of the next section, for instance. ◁

As usual, we begin with a regularity result for the value functions.

PROPOSITION 1.8. *Assume (0.2), (0.5), and (1.4). Then v and u are in $BUC(\mathbb{R}^N)$. If, moreover, $\omega_\ell(r) = L_\ell r$ (i.e., ℓ is lipschitzean in y uniformly in a and b), then v and u are Hölder continuous with the following exponent γ:*

$$\gamma = \begin{cases} 1 & \text{if } \lambda > L, \\ \text{any } \gamma < 1 & \text{if } \lambda = L, \\ \lambda/L & \text{if } \lambda < L\,. \end{cases}$$

PROOF. The estimate $|v(x)| \leq M/\lambda$ is an immediate consequence of (1.4). We fix $x, z \in \mathbb{R}^N$, $\varepsilon > 0$ and first pick $\overline{\alpha} \in \Gamma$ such that

$$v(z) \geq \sup_{b \in \mathcal{B}} J(z, \overline{\alpha}[b], b) - \varepsilon/2,$$

then $\overline{b} \in \mathcal{B}$ such that

$$v(x) \leq \sup_{b \in \mathcal{B}} J(x, \overline{\alpha}[b], b) \leq J(x, \overline{\alpha}[\overline{b}], \overline{b}) + \varepsilon/2\,.$$

Thus

$$v(x) - v(z) \leq \int_0^\infty e^{-\lambda t} |\ell(y_x, \overline{\alpha}[\overline{b}], \overline{b}) - \ell(y_z, \overline{\alpha}[\overline{b}], \overline{b})|\, dt + \varepsilon,$$

and from now on the proof is exactly the same as that of Proposition 2.1 in Chapter III, in view of the assumption (1.4) and the estimate (0.6). The proof for the upper value is analogous. ◀

The next result is the cornerstone of this section.

THEOREM 1.9 (DYNAMIC PROGRAMMING PRINCIPLE). *Assume* (0.2), (0.5), *and* (1.4). *Then for all* $x \in \mathbb{R}^N$ *and* $t > 0$,

$$(1.5) \quad v(x) = \inf_{\alpha \in \Gamma} \sup_{b \in \mathcal{B}} \Big\{ \int_0^t \ell(y_x(s), \alpha[b](s), b(s)) e^{-\lambda s} \, ds + v(y_x(t; \alpha[b], b)) e^{-\lambda t} \Big\},$$

$$(1.6) \quad u(x) = \sup_{\beta \in \Delta} \inf_{a \in \mathcal{A}} \Big\{ \int_0^t \ell(y_x(s), a(s), \beta[a](s)) e^{-\lambda s} \, ds + u(y_x(t; a, \beta[a])) e^{-\lambda t} \Big\}.$$

PROOF. We give only the proof of (1.5), since the other is similar. Denote by $w(x)$ the right-hand side of (1.5). We fix $\varepsilon > 0$ and for any $z \in \mathbb{R}^N$ we pick $\alpha_z \in \Gamma$ such that

$$(1.7) \quad v(z) \geq \sup_{b \in \mathcal{B}} J(z, \alpha_z[b], b) - \varepsilon.$$

For simplicity we set $\lambda = 1$.

We first prove that $v(x) \leq w(x)$. We choose $\overline{\alpha} \in \Gamma$ such that

$$(1.8) \quad w(x) \geq \sup_{b \in \mathcal{B}} \Big\{ \int_0^t \ell(y_x, \overline{\alpha}[b], b) e^{-s} \, ds + v(y_x(t; \overline{\alpha}[b], b)) e^{-t} \Big\} - \varepsilon.$$

Now we define $\delta \in \Gamma$ as follows

$$\delta[b](s) := \begin{cases} \overline{\alpha}[b](s), & s \leq t, \\ \alpha_z[b(\cdot + t)](s - t), & s > t, \end{cases}$$

with $z := y_x(t, \overline{\alpha}[b], b)$. Note that

$$y_x(s + t; \delta[b], b) = y_z(s; \alpha_z[b(\cdot + t)], b(\cdot + t)) \quad \text{for } s > 0,$$

so by the change of variables $\tau = s + t$

$$J(z, \alpha_z[b(\cdot + t)], b(\cdot + t)) = \int_t^{+\infty} \ell(y_x(\tau), \delta[b](\tau), b(\tau)) e^{t-\tau} \, d\tau.$$

Then by (1.8) and (1.7) we obtain

$$w(x) \geq \sup_{b \in \mathcal{B}} \int_0^\infty \ell(y_x, \delta[b], b) e^{-s} \, ds - 2\varepsilon \geq v(x) - 2\varepsilon,$$

which gives the desired inequality because ε is arbitrary.

To prove the inequality $w(x) \leq v(x)$ we pick $b_1 \in \mathcal{B}$ such that

$$(1.9) \quad w(x) \leq \int_0^t \ell(y_x, \alpha_x[b_1], b_1) e^{-s} \, ds + v(y_x(t; \alpha_x[b_1], b_1)) e^{-t} + \varepsilon,$$

where α_x is defined as in (1.7). Now for each $b \in \mathcal{B}$ define $\widetilde{b} \in \mathcal{B}$ by

$$(1.10) \quad \widetilde{b}(s) := \begin{cases} b_1(s), & s \leq t, \\ b(s - t), & s > t, \end{cases}$$

and then define $\underline{\alpha} \in \Gamma$ by

(1.11) $$\underline{\alpha}[b](s) := \alpha_x[\widetilde{b}](s+t) .$$

Next set

(1.12) $$z := y_x(t; \alpha_x[b_1], b_1)$$

and choose $b_2 \in \mathcal{B}$ such that

(1.13) $$v(z) \leq J(z, \underline{\alpha}[b_2], b_2) + \varepsilon .$$

We claim that
$$w(x) \leq J(x, \alpha_x[\widetilde{b}_2], \widetilde{b}_2) + 2\varepsilon,$$

which gives, by (1.7), $w(x) \leq v(x) + 3\varepsilon$, and thus the conclusion by the arbitrariness of $\varepsilon > 0$. To prove the claim observe that, by (1.10) and (1.11),

$$y_x(\tau; \alpha_x[\widetilde{b}_2], \widetilde{b}_2) = \begin{cases} y_x(\tau, \alpha_x[b_1], b_1), & \tau \leq t, \\ y_z(\tau - t, \underline{\alpha}[b_2], b_2), & \tau > t, \end{cases}$$

so by the change of variables $\tau = s + t$

(1.14) $$J(z, \underline{\alpha}[b_2], b_2) = \int_t^{+\infty} \ell(y_x(\tau), \alpha_x[\widetilde{b}_2](\tau), \widetilde{b}_2(\tau)) e^{t-\tau} \, d\tau .$$

Now we use (1.9), (1.10), (1.12), (1.13) and (1.14) to get

$$w(x) \leq \int_0^\infty \ell(y_x(\tau), \alpha_x[\widetilde{b}_2](\tau), \widetilde{b}_2(\tau)) e^{-\tau} \, dt + 2\varepsilon,$$

which is the claim. ◀

We proceed as in the previous chapters and derive a partial differential equation for each value. For differential games the equations are called Hamilton-Jacobi-Isaacs (briefly, HJI) and they involve the following Hamiltonians

(1.15) $$H(x, p) := \min_{b \in B} \max_{a \in A} \{ -f(x, a, b) \cdot p - \ell(x, a, b) \},$$

(1.16) $$\widetilde{H}(x, p) := \max_{a \in A} \min_{b \in B} \{ -f(x, a, b) \cdot p - \ell(x, a, b) \} .$$

THEOREM 1.10. *Assume* (0.2), (0.5), *and* (1.4). *Then the lower value v is a viscosity solution of*

(1.17) $$\lambda v + H(x, Dv) = 0 \quad \text{in } \mathbb{R}^N,$$

and the upper value u is a viscosity solution of

(1.18) $$\lambda u + \widetilde{H}(x, Du) = 0 \quad \text{in } \mathbb{R}^N .$$

The equation in (1.17) is called the *lower HJI equation* and that in (1.18) is the *upper HJI equation*. We prove only that the lower value solves the lower HJI equation, the other proof being similar. For differential games, unlike the case of a single controller, proving that v is a subsolution is not easier than proving it is a supersolution. Indeed, for the first part we need the following

LEMMA 1.11. *Under the assumptions of Theorem* 1.10, *let* $x \in \mathbb{R}^N$ *and* $\phi \in C^1(\mathbb{R}^N)$ *be such that*

(1.19) $$\lambda \phi(x) + H(x, D\phi(x)) = \vartheta > 0 .$$

Then there exists $\alpha^* \in \Gamma$ *such that for all* $b \in \mathcal{B}$ *and* $t > 0$ *small enough*

(1.20) $$\int_0^t \{ \ell(y_x, \alpha^*[b], b) + f(y_x, \alpha^*[b], b) \cdot D\phi(y_x) - \lambda \phi(y_x) \} e^{-\lambda s} \, ds \leq -\frac{\vartheta t}{4}.$$

where $y_x = y_x(\,\cdot\,, \alpha^*[b], b)$.

PROOF. Set for simplicity $\lambda = 1$ and define

$$\Lambda(z, a, b) := \phi(z) - \ell(z, a, b) - f(z, a, b) \cdot D\phi(z) .$$

By the definition of H and (1.19)

$$\min_{b \in B} \max_{a \in A} \Lambda(x, a, b) = \vartheta,$$

so for any $b \in B$ there is $a = a(b) \in A$ such that

$$\Lambda(x, a, b) \geq \vartheta .$$

Since $\Lambda(x, a, \,\cdot\,)$ is uniformly continuous, we have in fact

$$\Lambda(x, a, \zeta) \geq \frac{3\vartheta}{4} \qquad \text{for all } \zeta \in B(b, r) \cap B$$

for some $r = r(b) > 0$. Because B is compact, there exist finitely many points $b_1, \ldots, b_n \in B$ and $r_1, \ldots, r_n > 0$ such that

$$B \subseteq \bigcup_{i=1}^n B(b_i, r_i),$$

and for $a_i = a(b_i)$

$$\Lambda(x, a_i, \zeta) \geq \frac{3\vartheta}{4} \qquad \text{for } \zeta \in B(b_i, r_i) \cap B .$$

Define $\Phi : B \to A$ by setting

$$\Phi(b) = a_k \qquad \text{if } b \in B(b_k, r_k) \setminus \bigcup_{i=1}^{k-1} B(b_i, r_i) .$$

It is easy to prove that, for any $b \in \mathcal{B}$, $\Phi(b(\,\cdot\,))$ is measurable, so we can define $\alpha^* \in \Gamma$ by setting
$$\alpha^*[b](s) := \Phi(b(s)),$$
see Example 1.2. By definition of Φ
$$\Lambda(x, \Phi(b), b) \geq \frac{3\vartheta}{4} \quad \text{for all } b \in B,$$
and by the continuity of Λ and estimate (0.7) there is $t > 0$ such that

(1.21) $$\Lambda(y_x(s), \alpha^*[b](s), b(s)) \geq \frac{\vartheta}{2} \quad \text{for } 0 \leq s \leq t$$

and for each $b \in \mathcal{B}$. Now we multiply both sides of (1.21) by e^{-s} and integrate from 0 to t to obtain (1.20) for t small enough. ◀

PROOF OF THEOREM 1.10. We set $\lambda = 1$ and we begin with the proof that v is a subsolution of (1.17). Let $\phi \in C^1(\mathbb{R}^N)$ and x be such that $v - \phi$ has a local maximum at x and $v(x) = \phi(x)$. We assume by contradiction that (1.19) holds and use Lemma 1.11 to find $\alpha^* \in \Gamma$ such that for all $b \in \mathcal{B}$ and all t small enough

(1.22) $$\int_0^t \ell(y_x, \alpha^*[b], b) e^{-s}\, ds + e^{-t}\phi(y_x(t)) - \phi(x) \leq -\frac{\vartheta t}{4},$$

where $y_x = y_x(\,\cdot\,, \alpha^*[b], b)$. Since $v - \phi$ has a local maximum at x and $v(x) = \phi(x)$, by (0.7) we have
$$e^{-t}\phi(y_x(t)) - \phi(x) \geq e^{-t}v(y_x(t)) - v(x)$$
for t small enough. By plugging this into (1.22) we obtain
$$\inf_{\alpha \in \Gamma} \sup_{b \in \mathcal{B}} \left\{ \int_0^t \ell(y_x, \alpha[b], b) e^{-s}\, ds + e^{-t}v(y_x(t, \alpha[b], b)) \right\} - v(x) \leq -\frac{\vartheta t}{4} < 0,$$
which is a contradiction to the inequality "$\leq$" in the DPP (1.5). Then v is a subsolution of (1.17).

Next we show that v is a supersolution of (1.17). Let $\phi \in C^1(\mathbb{R}^N)$, x be such that $v - \phi$ has a local minimum at x, and $v(x) = \phi(x)$. Assume by contradiction
$$\phi(x) + H(x, D\phi(x)) = -\vartheta < 0\,.$$
By the definition of H there exists $b^* \in B$ such that
$$\phi(x) - f(x, a, b^*) \cdot D\phi(x) - \ell(x, a, b^*) \leq -\vartheta$$
for all $a \in A$. For $t > 0$ small enough and any $\alpha \in \Gamma$
$$\phi(y_x(s)) - f(y_x(s), \alpha[b^*](s), b^*) \cdot D\phi(y_x(s)) - \ell(y_x(s), \alpha[b^*](s), b^*) \leq -\frac{\vartheta}{2}$$

for $0 \leq s \leq t$, where $y_x(s) = y_x(s; \alpha[b^*], b^*)$. We multiply by e^{-s} and integrate from 0 to t to get

(1.23) $$\phi(x) - e^{-t}\phi(y_x(t)) - \int_0^t \ell(y_x, \alpha[b^*], b^*) e^{-s} \, ds \leq -\frac{\vartheta t}{4}.$$

for t small enough. From

$$\phi(x) - e^{-t}\phi(y_x(t)) \geq v(x) - e^{-t}v(y_x(t))$$

we obtain

$$e^{-t}v(y_x(t)) + \int_0^t \ell(y_x, \alpha[b^*], b^*) e^{-s} \, ds \geq \frac{\vartheta t}{2} + v(x)$$

and thus

$$\inf_{\alpha \in \Gamma} \sup_{b \in \mathcal{B}} \left\{ \int_0^t \ell(y_x, \alpha[b], b) e^{-s} \, ds + v(y_x(t; \alpha[b], b)) e^{-t} \right\} > v(x).$$

This is a contradiction to the DPP (1.5) and completes the proof that v is a viscosity solution of (1.17). ◂

Exercises

1.1. Define $\Gamma^* := \{\alpha : \mathcal{B} \to \mathcal{A}\}$, the set of all strategies for the first player (not necessarily nonanticipating). Prove that

$$\inf_{\alpha \in \Gamma^*} \sup_{b \in \mathcal{B}} J(x, \alpha[b], b) = v_s(x),$$

where v_s is defined by (0.3).

1.2. Replace the assumption that A and B are compact with the following: A and B are separable metric spaces; for all bounded $X \subseteq \mathbb{R}^N$ f is bounded on $X \times A \times B$; $|f(x, a, b) - f(y, a, b)| \leq \omega_f(|x-y|, R)$ for all $x, y \in B(0, R)$, $a \in A$, $b \in B$ and $R > 0$, where ω_f is a modulus (cfr. (A$_1$) and (A$_2$) in Chapter III). Redefine H and $\widetilde{H}$ by replacing min max with inf sup in (1.15) and max min with sup inf in (1.16). Prove that all the results of this section remain true in this more general setting. [Hint: in the proof of Lemma 1.11 use Lindelöff's theorem, that is, any open cover of a separable metric space has a countable subcover.]

1.3. In analogy with the definitions of §2.2 of Chapter III we call presynthesis of strategies for the first player a map $\mathbf{a} : \mathbb{R}^N \supseteq \Omega \to \Gamma$ and set

$$J_\mathbf{a}(x) := \sup_{b \in \mathcal{B}} J(x, \mathbf{a}(x)[b], b).$$

We say $\mathbf{a}$ is optimal at x if $J_\mathbf{a}(x) = v(x)$. Prove that a necessary condition of optimality for $\mathbf{a}$ at all points of an open set Ω' is that $J_\mathbf{a}$ is a viscosity solution of the lower HJI equation in Ω'. Similarly, for a presynthesis $\mathbf{b} : \Omega \to \mathcal{B}$ for the second player we set $J_\mathbf{b}(x) := \inf_{\alpha \in \Gamma} J(x, \alpha[\mathbf{b}(x)], \mathbf{b}(x))$ and say it is optimal at x if $J_\mathbf{b}(x) = v(x)$. Find a necessary condition of optimality for $\mathbf{b}$ in an open set.

1.4. Compute $u(x)$, for $x < 0$, and v_s, u_s, H, $\widetilde{H}$ in Example 1.7.

1.5. A strategy α such that, for any $t > 0$ and $b, \widetilde{b} \in \mathcal{B}$, $b(s) = \widetilde{b}(s)$ for all $s < t$ implies $\alpha[b](s) = \alpha[\widetilde{b}](s)$ for all $s \leq t$, is called a *slightly delayed* strategy and it is clearly nonanticipating.

(i) Find which strategies in the Examples 1.2–1.4 are slightly delayed.

(ii) Define the lower and upper values v_d and u_d by replacing Γ and Δ in Definition 1.6 with the corresponding sets of slightly delayed strategies, and prove they satisfy the Dynamic Programming Principle.

(iii) Prove that v_d and u_d are supersolutions of (1.17) and subsolutions of (1.18).

(iv) Compute v_d and u_d in Example 1.7.

1.6. Adopt all the assumptions of Exercise 1.2, except the first, and replace it by: A, B topological spaces. Prove that u and v are supersolutions of the lower HJI equation (1.17) and subsolutions of the upper HJI equation (1.18).

1.7. Assume $w \in C(\mathbb{R}^N)$ satisfies for all $x \in \mathbb{R}^N$ and small $t > 0$,

$$w(x) \geq \inf_{\alpha \in \Gamma'} \sup_{b \in \mathcal{B}'} \left\{ \int_0^t \ell(y_x(s), \alpha[b](s), b(s)) e^{-\lambda s} \, ds + w(y_x(t; \alpha[b], b)) e^{-\lambda t} \right\}$$

(respectively, $\leq$), where $\mathcal{B}' \subseteq \mathcal{B}$ contains the constant control functions and $\Gamma' \subseteq \Gamma$ contains all the constant strategies of the type $\alpha[b](t) = \overline{a}$ for all t, for some $\overline{a} \in A$. Prove that w is a supersolution of the lower HJI equation (1.17) (respectively, subsolution of upper HJI equation (1.18)).

2. Existence of a value, relaxation, verification theorems

In this section we characterize the lower and upper values for the infinite horizon problem as the unique solution of the lower and the upper HJI equations, respectively. From this result it is easy to derive the announced inequality $v \leq u$, and the equality under the assumption $H = \widetilde{H}$, so that the game has a value. As a consequence, we show that any game has a value if both players use relaxed controls. Finally, we give verification theorems in the form of necessary and sufficient conditions of optimality for strategies and controls.

THEOREM 2.1. *Assume* (0.2), (0.5), *and* (1.4). *Then*

(i) *the lower value v is the complete solution (i.e., the minimal supersolution and maximal subsolution) of (1.17) in $BC(\mathbb{R}^N)$; in particular, it is the unique solution of (1.17) in $BC(\mathbb{R}^N)$;*

(ii) *the upper value u is the complete solution of (1.18) in $BC(\mathbb{R}^N)$.*

PROOF. By Proposition 1.8 and Theorem 1.10 it is enough to check the assumptions of a suitable comparison theorem for sub- and supersolutions of (1.17) and

(1.18). We want to apply Theorem 2.12 and Remark 2.13 of Chapter III, so we have to prove that H and $\widetilde{H}$ satisfy the inequality

(2.1) $H(y, \mu(x-y) - \tau y) - H(x, \mu(x-y) + \gamma x)$
$$\leq \mu L |x-y|^2 + \tau K(1+|y|^2) + \gamma K(1+|x|^2) + \omega(|x-y|)$$

for some constants L, K and a modulus ω, for all $x, y \in \mathbb{R}^N$, $\mu, \tau, \gamma > 0$.

We fix $b' \in B$ such that
$$H(x, p) \geq -f(x, a, b') \cdot p - \ell(x, a, b') \quad \text{for all } a \in A,$$

then we pick $a' \in A$ such that
$$H(y, q) \leq -f(y, a', b') \cdot q - \ell(y, a', b'),$$

where $p = \mu(x-y) + \gamma x$, $q = \mu(x-y) - \tau y$. The left-hand side of (2.1) is less than or equal to

$$\mu(f(x, a', b') - f(y, a', b')) \cdot (x-y) + \tau f(y, a', b') \cdot y$$
$$+ \gamma f(x, a', b') \cdot x + \ell(x, a', b') - \ell(y, a', b'),$$

which can be estimated by the right-hand side of (2.1) where L is the constant in (0.5), $\omega = \omega_\ell$, and

$$K := L + K', \quad K' := \max\{|f(0, a, b)| : a \in A, b \in B\}.$$

In fact, $(f(z, a, b) - f(0, a, b)) \cdot z \leq L|z|^2$ by (0.5), thus
$$f(z, a, b) \cdot z \leq L|z|^2 + K'|z| \leq K(1+|z|^2).$$

The proof of (2.1) with H replaced by $\widetilde{H}$ is similar. ◂

COROLLARY 2.2. *Assume* (0.2), (0.5), *and* (1.4). *Then*
$$v(x) \leq u(x) \quad \text{for all } x \in \mathbb{R}^N.$$

If, in addition,

(2.2) $$H(x, p) = \widetilde{H}(x, p) \quad \text{for all } x, p \in \mathbb{R}^N,$$

then $v \equiv u$ and the game has a value for all initial points.

PROOF. The Hamiltonians H and $\widetilde{H}$ defined by (1.15) and (1.16) are, respectively, $\min_B \max_A$ and $\max_A \min_B$ of the same function. Therefore
$$H(x, p) \geq \widetilde{H}(x, p) \quad \text{for all } x, p \in \mathbb{R}^N.$$

Thus v is a subsolution of the upper HJI equation (1.18) and u is a supersolution of the lower HJI equation (1.17). By either (i) or (ii) of Theorem 2.1 we conclude that $v \leq u$.

The second statement follows immediately from Theorem 2.1 because the two HJI equations (1.17) and (1.18) coincide if $H = \widetilde{H}$. ◂

The sufficient condition (2.2) for the existence of a value is called *Isaacs' condition* or *solvability of the small game*. It is easy to give conditions on the data A, B, f, and ℓ implying (2.2). In fact, if we set

$$F(a,b) := -f(x,a,b) \cdot p - \ell(x,a,b),$$

(2.2) says that F has the property

(2.3) $$\min_{b \in B} \max_{a \in A} F(a,b) = \max_{a \in A} \min_{b \in B} F(a,b).$$

The simplest case when this occurs is for any system and cost with the separation property

$$f(x,a,b) = f_1(x,a) + f_2(x,b), \qquad \ell(x,a,b) = \ell_1(x,a) + \ell_2(x,b),$$

which is often verified in the applications. More generally, (2.3) is equivalent to saying that the static (two-person, zero sum) game over the sets A and B with payoff F has a saddle. This is one of the basic problems in classical game theory and it has been studied in depth after the first pioneering result of von Neumann. The minimax theorem we report without proof is not the most general in the literature but it is sufficient for giving two more examples of games with the Isaacs' condition.

THEOREM 2.3. *Assume $F := A \times B \to \mathbb{R}$ is continuous, A, B are convex compact spaces, the sets $\{\, b \in B : F(\bar{a}, b) \geq t \,\}$ and $\{\, a \in A : F(a, \bar{b}) \leq t \,\}$ are convex for all $t \in \mathbb{R}, \bar{a} \in A, \bar{b} \in B$. Then (2.3) holds.*

EXAMPLE 2.4. Let the control sets be finite dimensional, $A \subseteq \mathbb{R}^m$, $B \subseteq \mathbb{R}^k$, the system be affine in the control variables, that is,

$$f(x,a,b) = f_1(x) + f_2(x)a + f_3(x)b,$$

where $f_2(x)$ is an $N \times m$ matrix and $f_3(x)$ is an $N \times k$ matrix, and the cost be bilinear in the control variables, that is,

$$\ell(x,a,b) = \ell_1(x) + a \cdot \ell_2(x)b,$$

where $\ell_2(x)$ is an $m \times k$ matrix. Then it is straightforward to check that Theorem 2.3 applies for all $x, p \in \mathbb{R}^N$, so that $H = \tilde{H}$ and the game has a value. More generally, the same holds if $\ell(x, \cdot, b)$ is convex for all x, b, and $\ell(x, a, \cdot)$ is concave for all x, a. ◁

★ ★ ★

Next we consider relaxed controls as we did in §2.2 of Chapter III for one-person problems. As we proved there, for those problems the value does not change using relaxed controls, and the advantage in using them is the existence of an optimal control which we proved in §1.1 of Chapter VI. For two-person games we show that the value always exists if we use relaxed controls for both players. In particular, if the original game does not have a value, then either the lower or the upper value must be different from the relaxed value.

2. EXISTENCE OF A VALUE, RELAXATION, VERIFICATION

We denote by A^r (respectively, B^r) the set of Radon probability measures on A (respectively, B) and define, for $x \in \mathbb{R}^N$, $m \in A^r$, $n \in B^r$,

$$f^r(x, m, n) := \int_B \int_A f(x, a, b)\, dm\, dn, \qquad \ell^r(x, m, n) := \int_B \int_A \ell(x, a, b)\, dm\, dn,$$

where the order of integration is irrelevant by Fubini's Theorem. Note that $f^r(x, \delta_a, \delta_b) = f(x, a, b)$, where δ_c is the Dirac measure concentrated at c. A similar formula holds for ℓ, so that f^r and ℓ^r can be viewed as extensions of f and ℓ, respectively. As in Section III.2.2, A^r and B^r are viewed as subsets of the dual spaces of $C(A)$ and $C(B)$, respectively, with the weak star topology.

LEMMA 2.5. *If A, B, f, ℓ satisfy* (0.2), (0.5), *and* (1.4), *then* A^r, B^r, f^r, ℓ^r *satisfy* (0.2), (0.5), *and* (1.4) *as well.*

PROOF. The compactness of A^r and B^r with the weak star topology is a classical result in functional analysis; see [Wa72]. The other properties are straightforward to check, as in the proof of Lemma III.2.20. ◂

By this lemma there exists a unique trajectory of the relaxed system, that is, a solution of

$$\begin{cases} y'(t) = f(y(t), \mu(t), \nu(t)), & t > 0, \\ y(0) = x, \end{cases}$$

for every $\mu \in \mathcal{A}^r$, $\nu \in \mathcal{B}^r := \{$ measurable functions $[0, +\infty[\to B^r \}$, and we denote it by $y_x(t) = y_x(t; \mu, \nu)$. Then we can extend the cost functional to relaxed controls:

$$J^r(x, \mu, \nu) := \int_0^\infty e^{-\lambda t} \ell^r(y_x(t), \mu(t), \nu(t))\, dt.$$

Now we can consider the lower and upper value functions corresponding to the following three games:

(i) only the first player can use relaxed controls, and the lower and upper value are denoted by v_A^r and u_A^r;

(ii) only the second player can use relaxed controls, and the values are v_B^r, u_B^r;

(iii) both players can use relaxed controls and the values are v^r, u^r.

The next result compares these six values and the original v and u. Note that some inequalities are obvious, for instance

(2.4) $$v \leq v_B^r, \qquad u_A^r \leq u.$$

THEOREM 2.6. *Assume* (0.2), (0.5), *and* (1.4). *Then*

(2.5) $$v = v_A^r, \qquad u = u_B^r,$$
(2.6) $$v_B^r = v^r = u^r = u_A^r.$$

In particular,

(a) *the game where both players can use relaxed controls has a value;*

(b) *if the original game has a value, then it coincides with the relaxed value v^r.*

PROOF. To prove the first formula in (2.5) we observe that for all $b \in B$

$$\max_{m \in A^r}\{-f^r(x,m,b) \cdot p - \ell^r(x,m,b)\} = \max_{a \in A}\{-f(x,a,b) \cdot p - \ell(x,a,b)\}$$

by the proof of Corollary III.2.21. Then

$$H_A^r(x,p) := \min_{b \in B} \max_{m \in A^r}\{-f^r(x,m,b) \cdot p - \ell^r(x,m,b)\} = H(x,p),$$

so both v and v_A^r are solutions of the lower HJI equation (1.17), by Theorem 1.10. Thus they coincide by the uniqueness Theorem 2.1.

The proof of the second formula in (2.5) is similar, by using the upper HJI equation (1.18).

Next we prove that the game where both players use relaxed controls has a value. By Corollary 2.2 it is enough to show that this game satisfies the Isaacs' condition, namely, that the following upper and lower Hamiltonians coincide,

$$H^r(x,p) := \min_{n \in B^r} \max_{m \in A^r} F(m,n), \quad \widetilde{H}^r(x,p) := \max_{m \in A^r} \min_{n \in B^r} F(m,n),$$

where $F(m,n) := -f^r(x,m,n) \cdot p - \ell^r(x,m,n)$. We claim that the assumptions of the minimax Theorem 2.3 hold true. In fact, F is continuous and A^r, B^r are compact by Lemma 2.5. Moreover, it is straightforward to check that A^r, B^r are convex and F is bilinear. By Theorem 2.3 we get $H^r = \widetilde{H}^r$ and then $v^r = u^r$.

The identity $v_B^r = v^r$ is obtained by applying the first formula in (2.5) to the game on A and B^r. Similarly $u_A^r = u^r$ follows from the second formula in (2.5) applied to the game on A^r and B.

The last conclusion (b) (i.e., $v = u$ implies $v = v^r$) follows immediately from (2.4) and (2.6). ◂

★ ★ ★

Before giving a verification theorem for the lower value we need some definitions.

DEFINITION 2.7. A strategy $\alpha^* \in \Gamma$ is optimal at the initial point x if

$$v(x) = \sup_{b \in \mathcal{B}} J(x, \alpha^*[b], b) . \quad \triangleleft$$

DEFINITION 2.8. A control $b^* \in \mathcal{B}$ is optimal at x if

$$v(x) = \inf_{\alpha \in \Gamma} J(x, \alpha[b^*], b^*) .$$

Note that if α^* and b^* are optimal at x, then

$$J(x, \alpha^*[b], b) \leq v(x) = J(x, \alpha^*[b^*], b^*) \leq J(x, \alpha[b^*], b^*)$$

for all $b \in \mathcal{B}$ and $\alpha \in \Gamma$. This means that the first (respectively, second) player cannot improve his guaranteed outcome, represented by $v(x)$, by a unilateral deviation from his optimal strategy (respectively, control). Thus the pair (α^*, b^*) is a *Nash equilibrium*, in the terminology of the general theory of noncooperative games. ◁

PROPOSITION 2.9. *Assume* (0.2), (0.5), *and* (1.4). *Then the following statements are equivalent:*

(i) $\alpha^* \in \Gamma$ *is optimal at* x_0;

(ii) *there exists* $u \in BUC(\mathbb{R}^N)$ *subsolution of the lower HJI equation* (1.17) *such that*

(2.7) $$\sup_{b \in \mathcal{B}} J(x_0, \alpha^*[b], b) \leq u(x_0);$$

(iii) *there exists* $u \in BUSC(\mathbb{R}^N)$ *subsolution of* (1.17) *satisfying* (2.7).

Also, the following statements are equivalent:

(j) $b^* \in \mathcal{B}$ *is optimal at* x_0;

(jj) *there exists* $w \in BUC(\mathbb{R}^N)$ *supersolution of* (1.17) *such that*

(2.8) $$w(x_0) \leq \inf_{\alpha \in \Gamma} J(x_0, \alpha[b^*], b^*);$$

(jjj) *there exists* $w \in BLSC(\mathbb{R}^N)$ *supersolution of* (1.17) *satisfying* (2.8).

PROOF. (i) implies (ii) by taking $u = v$, the lower value function. Obviously (ii) implies (iii). Now we assume (iii) and recall the comparison principle for semicontinuous sub- and supersolutions in Chapter V (see Definition V.1.2). By the proof of Theorem 2.1 the Hamiltonian H of the lower HJI equation has enough regularity to apply Theorem V.1.3 and get $u(x) \leq v(x)$ for all x. Thus by (2.7) and the definition of v we obtain that α^* is optimal at x_0.

The proof of the equivalence between (j), (jj) and (jjj) is analogous. ◂

We see from this result that viscosity subsolutions of the lower HJI equation can be used as verification functions for the optimality of strategies of the first player, whereas supersolutions are verification functions for controls of the second player. It is obvious how to define optimality for strategies $\beta \in \Delta$ of the second player and controls $a \in \mathcal{A}$ of the first. It is easy to give a verification theorem for them similar to Proposition 2.9, by using super- and subsolutions of the upper HJI equation (1.18) as verification functions.

Exercises

2.1. Consider the infinite horizon differential game with $N = 4$, $\lambda = 1$, $A = B = \{-1, 1\}$,

$$f(x) = (-ax_2, ax_1, -bx_4, bx_3),$$
$$\ell(x) = \arctan((y_1 - y_3)^2 + (y_2 - y_4)^2).$$

(i) Prove that $u(x) = v(x) = \ell(x)$ for all x;

(ii) find some x such that $v_s(x) < v(x) < u_s(x)$, where v_s and u_s are defined by (0.3).

2.2. Prove that a function $F : A \times B \to \mathbb{R}$ satisfies (2.3) if and only if it has a saddle, namely, a pair a^*, b^* such that

$$F(a, b^*) \leq F(a^*, b^*) \leq F(a^*, b)$$

for all $a \in A$, $b \in B$.

2.3. State and prove a verification theorem for strategies $\beta \in \Delta$ and controls $a \in \mathcal{A}$.

2.4. Assume that the static game over $\mathcal{A}$ and $\mathcal{B}$ with payoff $J(x, \cdot, \cdot)$ has a saddle (a^*, b^*). Prove that the strategy $\alpha^*[b] = a^*$, for all b, and the control b^* are optimal at x according to Definitions 2.7 and 2.8.

2.5. Compute all the relaxed upper and lower value functions for Example 1.7. Show in particular that $v^r = u$.

2.6. Consider the lower and upper values v_d, u_d defined in Exercise 1.5 using the slightly delayed strategies. Prove that

(i) $v \leq v_d \leq u$ and $v \leq u_d \leq u$;

(ii) $v = v_d = u_d$ if Isaacs' condition (2.2) holds.

2.7. Assume $\mathcal{A}' \subseteq \mathcal{A}$, $\mathcal{B}' \subseteq \mathcal{B}$ are closed with respect to translation in time ($a \in \mathcal{A}$, $\tau > 0$ implies $a(\cdot + \tau) \in \mathcal{A}'$) and concatenation ($a_1, a_2 \in \mathcal{A}'$, $\tau > 0$, $a(t) = a_1(t)$ for $t \leq \tau$, $a(t) = a_2(t - \tau)$ for $t > \tau$, implies $a \in \mathcal{A}'$), as, for instance, piecewise constant controls. Replace $\mathcal{A}$, $\mathcal{B}$ by $\mathcal{A}'$, $\mathcal{B}'$, respectively, in the definition of nonanticipating strategy and of lower and upper value, and call $v^\sharp$ and $u^\sharp$ the new values. Prove that $v^\sharp = v$ and $u^\sharp = u$.

3. Comparison with other information patterns and other notions of value

3.1. Feedback strategies

In Section 1 and 2 we used the notion of nonanticipating strategy which models a situation where a player knows the current and past choices of the control made by his or her opponent. This is not the case in many practical examples where a player knows at most the current state of the system, and perhaps can keep track of the past history of the state. To model this different information pattern we introduce the *feedback strategies*. We denote by $\mathcal{Y}$ the set of trajectories of the system, that is,

$$\mathcal{Y} := \{ y : [0, +\infty[\to \mathbb{R}^N :$$
$$\exists x \in \mathbb{R}^N, a \in \mathcal{A}, b \in \mathcal{B} \text{ such that } y(t) = y_x(t, a, b) \text{ for all } t \}.$$

3. OTHER INFORMATION PATTERNS

DEFINITION 3.1. A map $\zeta : \mathcal{Y} \to \mathcal{A}$ is a feedback (or closed-loop) strategy (with perfect memory and perfect state measurement) for the first player if it is nonanticipating (or causal) (i.e., for any $t > 0$ and $y, \widetilde{y} \in \mathcal{Y}$, $y(s) = \widetilde{y}(s)$ for all $s \leq t$ implies $\zeta[y](s) = \zeta[\widetilde{y}](s)$ for all $s \leq t$), and *playable*, that is, for any $x \in \mathbb{R}^N$, $b : [0, T[\to B$ measurable and $T \in \,]0, +\infty]$ there exists a unique solution of

(3.1)
$$\begin{cases} q'(t) = f(q(t), \zeta[q](t), b(t)), & 0 < t \leq T, \\ q(0) = x \,. \end{cases}$$

We denote this solution by $y_x(\,\cdot\,;\zeta, b)$ (with a slight abuse of notation) and the corresponding control function $\zeta[q]$ by $\alpha_x(\,\cdot\,;\zeta, b)$. A feedback strategy ξ for the second player is defined in the obvious symmetric way, the trajectory and the control function corresponding to ξ, x, and $a \in \mathcal{A}$ are denoted, respectively, by $y_x(\,\cdot\,;a, \xi)$ and $\beta_x(\,\cdot\,;a, \xi)$. We denote by $\mathcal{F}$ and $\mathcal{G}$, respectively, the set of feedback strategies for the first and the second player. ◁

The most trivial example of feedback strategy is a constant one: for a given $\overline{a} \in \mathcal{A}$, $\zeta[y] = \overline{a}$ for all $y \in \mathcal{Y}$. Here are some more interesting examples.

EXAMPLE 3.2. Let $\Phi : \mathbb{R}^N \times [0, \infty[\to A$ be a time-dependent feedback map, admissible for every $x \in \mathbb{R}^N$ as defined in Example 1.3. We can associate with Φ a feedback strategy by setting

$$\zeta[y](t) := \Phi(y(t), t),$$

provided $\zeta[y]$ is measurable for all $y \in \mathcal{Y}$. Note that this strategy is *memoryless* in the following sense: for any $t > 0$ and $y, \widetilde{y} \in \mathcal{Y}$, $y(s) = \widetilde{y}(s)$ for all $s \geq t$ implies $\zeta[y](s) = \zeta[\widetilde{y}](s)$ for all $s \geq t$. ◁

EXAMPLE 3.3. Let Φ be a feedback map acting on some average of the history of the state, as in Example 1.4. We can associate with Φ a feedback strategy by setting

$$\zeta[y](t) := \Phi\left(\int_0^t y(s)\, d\mu_t, t\right),$$

provided $\zeta[y]$ is measurable for all $y \in \mathcal{Y}$ and there is a unique solution of (1.1) for every $b \in \mathcal{B}$, $T \in \,]0, +\infty]$ and $x \in \mathbb{R}^N$. ◁

EXAMPLE 3.4. Given a map $\Phi : \mathbb{R}^N \to A$ and an increasing sequence of times $\{t_n\}$ such that $t_0 = 0$ and $\lim_n t_n = +\infty$, we define a feedback strategy as follows,

$$\zeta[y](t) = \Phi(y(t_n)) \quad \text{if } t_n \leq t < t_{n+1}\,.$$

We call such a strategy *sampled-data*, but it is also named K-strategy in Isaacs' terminology. Note that in this case the trajectory $y_x(\,\cdot\,;\zeta, b)$ exists unique for all $b \in \mathcal{B}$ and $x \in \mathbb{R}^N$, and it generates a piecewise constant control function $\alpha_x(\,\cdot\,;\zeta, b)$, with no assumption whatsoever on Φ. ◁

The next lemma gives the connection between feedback strategies and nonanticipating strategies in the sense of Definition 1.1.

LEMMA 3.5. *For any given $\zeta \in \mathcal{F}$ and $x \in \mathbb{R}^N$, the strategy $\alpha : \mathcal{B} \to \mathcal{A}$, $\alpha[b] = \alpha_x(\,\cdot\,;\zeta,b)$ is nonanticipating and $y_x(\,\cdot\,,\alpha[b],b) = y_x(\,\cdot\,;\zeta,b)$ for all $b \in \mathcal{B}$.*

PROOF. If $b, \tilde{b} \in \mathcal{B}$, $b(s) = \tilde{b}(s)$ for all $s < T$, then $y_x(s;\zeta,b) = y_x(s;\zeta,\tilde{b})$ for all $s \leq T$ because of the uniqueness requirement in Definition 3.1. Thus $\alpha_x(s;\zeta,b) = \alpha_x(s;\zeta,\tilde{b})$ for all $s \leq T$ because ζ is nonanticipating. Therefore α is slightly delayed (see Exercise 1.5) and, a fortiori, it is nonanticipating. ◀

Now we define the *feedback values* and compare them to the values of Definition 1.6.

DEFINITION 3.6. The *A-feedback value* of a game with cost functional $J : \mathbb{R}^N \times \mathcal{A} \times \mathcal{B} \to \mathbb{R}$ is

$$v_A(x) := \inf_{\zeta \in \mathcal{F}} \sup_{b \in \mathcal{B}} J(x, \alpha_x(\,\cdot\,;\zeta,b), b),$$

and the *B-feedback value* is

$$v_B(x) := \sup_{\xi \in \mathcal{G}} \inf_{a \in \mathcal{A}} J(x, a, \beta_x(\,\cdot\,;a,\xi)) .$$

If $v_A(x) = v_B(x)$ we say that the game with initial point x has a *feedback value*, which is denoted by $v_F(x)$. ◁

By Lemma 3.5 we immediately obtain

(3.2) $\qquad v(x) \leq v_A(x) \qquad$ and $\qquad v_B(x) \leq u(x) \qquad$ for all x,

where v and u are the usual lower and upper values of the game (Definition 1.6). Note that, in general, there are nonanticipating strategies that do not correspond to a feedback strategy, for instance, the second in Example 1.2. Therefore the inequalities in (3.2) might be strict at some point.

In the rest of this subsection we study the infinite horizon problem, that is, the cost functional is given by (0.4). The main result is that under Isaacs' condition the feedback value exists and coincides with the usual value. We introduce the notation

$$J_t(x;\zeta,b) := \int_0^t \ell(y_x(s;\zeta,b), \alpha_x(s;\zeta,b), b(s)) e^{-\lambda s} \, ds,$$

$$J_t(x;a,\xi) := \int_0^t \ell(y_x(s;a,\xi), a(s), \beta_x(s;a,\xi)) e^{-\lambda s} \, ds,$$

for $t \in [0,+\infty]$, $x \in \mathbb{R}^N$, $\zeta \in \mathcal{F}$, $b \in \mathcal{B}$, $\xi \in \mathcal{G}$, $a \in \mathcal{A}$. Thus in the rest of this subsection the A-feedback and B-feedback values are

$$v_A(x) := \inf_{\zeta \in \mathcal{F}} \sup_{b \in \mathcal{B}} J_\infty(x;\zeta,b),$$

$$v_B(x) := \sup_{\xi \in \mathcal{G}} \inf_{a \in \mathcal{A}} J_\infty(x;a,\xi) .$$

3. OTHER INFORMATION PATTERNS

PROPOSITION 3.7. *Assume* (0.2), (0.5), *and* (1.4). *Then* $v_A, v_B \in BUC(\mathbb{R}^N)$. *If, moreover,* $\omega_\ell(r) = L_\ell r$, *then* v_A *and* v_B *are Hölder continuous with the exponent* γ *given in Proposition* 1.8.

PROOF. It is essentially the same as that of Proposition 1.8. ◂

As in Section 1 the DPP is the cornerstone of our theory.

THEOREM 3.8 (DYNAMIC PROGRAMMING PRINCIPLE). *Assume* (0.2), (0.5), *and* (1.4). *Then for all* $x \in \mathbb{R}^N$ *and* $t > 0$

$$
(3.3) \qquad v_A(x) = \inf_{\zeta \in \mathcal{F}} \sup_{b \in \mathcal{B}} \{ J_t(x; \zeta, b) + v_A(y_x(t; \zeta, b)) e^{-\lambda t} \},
$$

$$
(3.4) \qquad v_B(x) = \sup_{\xi \in \mathcal{G}} \inf_{a \in \mathcal{A}} \{ J_t(x; a, \xi) + v_B(y_x(t; a, \xi)) e^{-\lambda t} \}.
$$

PROOF. We prove only (3.3) for $\lambda = 1$. We follow the scheme of the proof of Theorem 1.9 and fix $\varepsilon > 0$, $z \in \mathbb{R}^N$, and $\zeta_z \in \mathcal{F}$ such that

$$
(3.5) \qquad v_A(z) \geq \sup_{b \in \mathcal{B}} J_\infty(z; \zeta_z, b) - \varepsilon .
$$

We call $w(x)$ the right-hand side of (3.3) and begin with the proof that $v_A(x) \leq w(x)$. We choose $\overline{\zeta} \in \mathcal{F}$ such that

$$
(3.6) \qquad w(x) \geq \sup_{b \in \mathcal{B}} \{ J_t(x; \overline{\zeta}, b) + v_A(y_x(t; \overline{\zeta}, b)) e^{-t} \} - \varepsilon,
$$

and we define

$$
\sigma[y](s) := \begin{cases} \overline{\zeta}[y](s), & s \leq t, \\ \zeta_{y(t)}[y(\cdot + t)](s - t), & s > t, \end{cases}
$$

for any $y \in \mathcal{Y}$. We claim that $\sigma \in \mathcal{F}$. Clearly σ is nonanticipating, and it is not hard to check that the trajectory associated with σ exists unique and is given by

$$
y_x(s; \sigma, b) = \begin{cases} y_x(s; \overline{\zeta}, b), & s \leq t, \\ y_z(s - t; \zeta_z, b(\cdot + t)), & s > t, \end{cases}
$$

where

$$
z := y_x(t; \overline{\zeta}, b),
$$

for any $b \in \mathcal{B}$ and $x \in \mathbb{R}^N$. Proceeding as in the proof of Theorem 1.9 we deduce from (3.5) and (3.6)

$$
w(x) \geq \sup_{b \in \mathcal{B}} J_\infty(x; \sigma, b) - 2\varepsilon \geq v_A(x) - 2\varepsilon,
$$

which gives the desired inequality.

To prove the inequality $w(x) \le v_A(x)$ we pick $b_1 \in \mathcal{B}$ such that

(3.7) $$w(x) \le J_t(x;\zeta_x,b_1) + v_A(z)e^{-t} + \varepsilon,$$

where

$$z := y_x(t;\zeta_x,b_1).$$

Now for any $y \in \mathcal{Y}$ such that $y(0) = z$ we define $\widetilde{y} \in \mathcal{Y}$ by

$$\widetilde{y}(s) := \begin{cases} y_x(s;\zeta_x,b_1), & s \le t, \\ y(s-t), & s > t, \end{cases}$$

then we fix any $\bar{a} \in \mathcal{A}$ and define

$$\underset{\sim}{\zeta}[y](s) := \begin{cases} \zeta_x[\widetilde{y}](s+t), & \text{if } y(0) = z, \\ \bar{a}(s), & \text{otherwise}. \end{cases}$$

It is not hard to see that $\underset{\sim}{\zeta} \in \mathcal{F}$ and $y_p(s;\underset{\sim}{\zeta},b) = y_p(s;\bar{a},b)$ if $p \ne z$, and

(3.8) $$y_z(s;\underset{\sim}{\zeta},b) = y_x(t+s;\zeta_x,\widetilde{b}) \qquad \text{for } s \ge 0,$$

where for any $b \in \mathcal{B}$

$$\widetilde{b}(s) := \begin{cases} b_1(s), & s \le t, \\ b(s-t), & s > t. \end{cases}$$

The rest of the proof is essentially the same as that of Theorem 1.9: we choose $b_2 \in \mathcal{B}$ such that

$$v_A(z) \le J_\infty(z;\underset{\sim}{\zeta},b_2) + \varepsilon,$$

and use (3.7) and (3.8) to obtain, after some calculations,

$$w(x) \le J_\infty(x;\zeta_x,\widetilde{b}_2) + 2\varepsilon.$$

Then by (3.5)

$$w(x) \le v_A(x) + 3\varepsilon,$$

which gives the desired inequality by the arbitrariness of ε. ◀

REMARK 3.9. It would be desirable to be able to prove the DPP for the value function defined by means of some subset of feedback strategies, such as strategies generated by feedback maps (see Example 3.2), or sampled-data strategies (see Example 3.4). A set of strategies smaller than $\mathcal{F}$ for which the previous proof works is the set of feedback strategies with *finite memory*. These strategies are those $\zeta \in \mathcal{F}$ with the property that for all $y \in \mathcal{Y}$ there exists a set of "times to keep in memory" $S(\zeta, y) \subseteq [0, +\infty[$ such that for any $t > 0$ $S(\zeta, y) \cap [0, t]$ is finite, and

3. OTHER INFORMATION PATTERNS

$\widetilde{y} \in \mathcal{Y}$, $\widetilde{y}(s) = y(s)$ for all $s \geq t$ and all $s \in S(\zeta, y)$ implies $\zeta[\widetilde{y}](s) = \zeta[y](s)$ for all $s \geq t$. Clearly, a memoryless strategy has finite memory (take $S(\zeta, y) = \emptyset$ for any y). Thus the strategies of Example 3.2 have finite memory. Also the sampled-data strategies of Example 3.4 have finite memory, by taking $S(\zeta, y) = \{t_n : n \in \mathbb{N}\}$ for any $y \in \mathcal{Y}$. On the other hand, the strategies of Example 3.3 do not have finite memory if, for instance, the probability μ_t has a density for some t. ◁

COROLLARY 3.10. *Assume* (0.2), (0.5), *and* (1.4). *Then* v_A *and* v_B *are supersolutions of the lower* HJI *equation* (1.17) *and subsolutions of the upper* HJI *equation* (1.18).

PROOF. We set $\lambda = 1$ and we begin with the proof that v_A is a supersolution of (1.17), which is essentially the same as the proof that v is a supersolution in Theorem 1.10. In fact, after fixing $\phi \in C^1(\mathbb{R}^N)$ and a local minimum point x of $v_A - \phi$ such that $v_A(x) = \phi(x)$, we assume by contradiction

$$\phi(x) + H(x, D\phi(x)) = -\vartheta < 0,$$

and obtain from the definition of H the existence of $b^* \in B$ such that

$$\phi(x) - f(x, a, b^*) \cdot D\phi(x) - \ell(x, a, b^*) \leq -\vartheta \qquad \text{for all } a \in A.$$

Then for $t > 0$ small enough and any $\zeta \in \mathcal{F}$,

$$\phi(y_x(s)) - f(y_x(s), \alpha_x(s), b^*) \cdot D\phi(y_x(s)) - \ell(y_x(s), \alpha_x(s), b^*) \leq -\frac{\vartheta}{2}$$

for $0 \leq s \leq t$, where $y_x(s) := y_x(s; \zeta, b^*)$ and $\alpha_x(s) := \alpha_x(s; \zeta, b^*)$. Now we perform exactly the same calculations as in the proof of Theorem 1.10 and reach a contradiction to the inequality "$\geq$" in the DPP (3.3).

To prove that v_A is a subsolution of (1.18) we fix $\phi \in C^1(\mathbb{R}^N)$ and x such that $v_A - \phi$ has a local maximum at x and $v_A(x) = \phi(x)$. We assume by contradiction

$$\phi(x) + \widetilde{H}(x, D\phi(x)) = \vartheta > 0$$

and use the definition of $\widetilde{H}$, (1.16), to find $a^* \in A$ such that

$$\phi(x) - f(x, a^*, b) \cdot D\phi(x) - \ell(x, a^*, b) \geq \vartheta \qquad \text{for all } b \in B.$$

Now we define the constant strategy $\zeta^* \in \mathcal{F}$ as

$$\zeta^*[y](s) = a^* \qquad \text{for all } y \in \mathcal{Y} \text{ and } s \geq 0.$$

For $t > 0$ small enough and any $b \in \mathcal{B}$ we have

$$\phi(y_x(s)) - f(y_x(s), a^*, b(s)) \cdot D\phi(y_x(s)) - \ell(y_x(s), a^*, b(s)) \geq \frac{\vartheta}{2}$$

for $0 \leq s \leq t$, where $y_x(s) := y_x(s; \zeta^*, b)$. We multiply both sides by e^{-s} and integrate from 0 to t to obtain an inequality such as (1.20), with $\alpha^*[b]$ replaced by a^*. So we can repeat the first part of the proof of Theorem 1.10 and get a contradiction to the inequality "$\leq$" in the DPP (3.3). This completes the proof of the statements about v_A. The proofs concerning v_B are similar and we leave them to the reader. ◀

Here is the main result of this subsection.

THEOREM 3.11. *Assume (0.2), (0.5), and (1.4). Then*

(3.9) $$v \leq v_A \leq u, \qquad v \leq v_B \leq u.$$

In particular, if $H = \widetilde{H}$, then $v_A = v_B = v = u$ and the game has a feedback value for all initial points.

PROOF. Some of the inequalities in (3.9) are already known, see (3.2). All of them follow from the comparison theorem for HJI equations, Theorem 2.1, from Corollary 3.10 and the continuity of v_A and v_B, Proposition 3.7. If Isaacs' condition holds, then $v = u$ by Corollary 2.2, and (3.9) then implies $v_A = v = v_B$. ◄

REMARK 3.12. This result applies for instance to the game where both players use relaxed controls: by Theorem 2.6 this game has a value, thus it also has a feedback value and they coincide. ◁

REMARK 3.13. Theorem 3.11 remains valid if in the definition of v_A and v_B we use only the sets of feedback strategies with finite memory instead of the whole $\mathcal{F}$ and $\mathcal{G}$. In fact, by Remark 3.9 the DPP still holds in this case, and it is immediate to check that the proofs of Corollary 3.10 and Theorem 3.11 work as well. ◁

★ ★ ★

As in Section 2 we say that a strategy $\zeta^* \in \mathcal{F}$ (respectively, $\xi^* \in \mathcal{G}$) is optimal at the initial point x if

(3.10) $$v_A(x) = \sup_{b \in B} J_\infty(x; \zeta^*, b) \qquad (\text{resp.}, \ v_B(x) = \inf_{a \in A} J_\infty(x; a, \xi^*)),$$

and we can easily prove a verification theorem similar to Proposition 2.9(a) using subsolutions of the lower HJI equation as verification functions. If one has a pair of optimal strategies $\zeta^* \in \mathcal{F}, \xi^* \in \mathcal{G}$, a natural question to ask is whether they can be played one against the other and form a saddle point for a game with payoff J_∞ over some pair of sets of strategies $\mathcal{F}' \subseteq \mathcal{F}$ and $\mathcal{G}' \subseteq \mathcal{G}$. To make this precise we say that $\mathcal{F}'$ and $\mathcal{G}'$ are *compatible* if for any $x \in \mathbb{R}^N$, $\zeta \in \mathcal{F}'$, and $\xi \in \mathcal{G}'$, there exists a unique solution of

(3.11) $$\begin{cases} q' = f(q, \zeta[q], \xi[q]), & t > 0, \\ q(0) = x, \end{cases}$$

and define

$$\overline{J}(x; \zeta, \xi) := J(x, \zeta[q], \xi[q]).$$

Now we can give the answer to the previous question.

PROPOSITION 3.14. *Assume $x \in \mathbb{R}^N$ is a point where the game has a feedback value $v_F(x) := v_A(x) = v_B(x)$ and the strategies $\zeta^* \in \mathcal{F}$ and $\xi^* \in \mathcal{G}$ are optimal. If $\zeta^* \in \mathcal{F}' \subseteq \mathcal{F}$ and $\xi^* \in \mathcal{G}' \subseteq \mathcal{G}$ with $\mathcal{F}'$ and $\mathcal{G}'$ compatible, then*

$$(3.12) \quad v_F(x) = \overline{J}(x;\zeta^*,\xi^*) = \max_{\xi \in \mathcal{G}'} \min_{\zeta \in \mathcal{F}'} \overline{J}(x;\zeta,\xi)$$
$$= \min_{\zeta \in \mathcal{F}'} \max_{\xi \in \mathcal{G}'} \overline{J}(x;\zeta,\xi).$$

PROOF. The existence of the feedback value and the optimality of ζ^*, ξ^* give

$$J_\infty(x;\zeta^*,b) \leq J_\infty(x;a,\xi^*) \quad \text{for all } a \in \mathcal{A}, b \in \mathcal{B}.$$

Plug into this inequality first $b = \xi^*[q^*]$, then $a = \zeta^*[q^*]$, where q^* is the solution of (3.11) corresponding to ζ^* and ξ^*, and get

$$J_\infty(x;\zeta^*,b) \leq \overline{J}(x;\zeta^*,\xi^*) \leq J_\infty(x;a,\xi^*) \quad \text{for all } a \in \mathcal{A}, b \in \mathcal{B}.$$

This proves the first equality in (3.12). Now take any $\zeta \in \mathcal{F}'$ and plug $a = \zeta[\bar{q}]$ into the last inequality, where $\bar{q}$ is the solution of (3.11) corresponding to ζ and ξ^*, to obtain

$$\overline{J}(x;\zeta^*,\xi^*) \leq \overline{J}(x;\zeta,\xi^*) \quad \text{for all } \zeta \in \mathcal{F}'.$$

Similarly we get

$$\overline{J}(x;\zeta^*,\xi) \leq J(x;\zeta^*,\xi^*) \quad \text{for all } \xi \in \mathcal{G}',$$

and we reach the conclusion by an easy standard argument (see Exercise 2.2). ◄

An example of compatible sets $\mathcal{F}'$, $\mathcal{G}'$ is the sampled-data feedback strategies of Example 3.4. With this choice the term on the right-hand side of the last equality in (3.12) is the value function used in the pioneering book of Isaacs. Many authors define the value as the payoff $\overline{J}$ computed at a saddle point over a suitable pair of sets of compatible strategies. Any such definition is equivalent to the feedback value by Proposition 3.14 and the following converse.

PROPOSITION 3.15. *Assume $\mathcal{F}' \subseteq \mathcal{F}$ and $\mathcal{G}' \subseteq \mathcal{G}$ are compatible and contain the constant strategies. Suppose $\zeta^* \in \mathcal{F}'$ and $\xi^* \in \mathcal{G}'$ are such that for some $x \in \mathbb{R}^N$,*

$$(3.13) \quad \overline{J}(x;\zeta^*,\xi) \leq \overline{J}(x;\zeta^*,\xi^*) \leq \overline{J}(x;\zeta,\xi^*) \quad \text{for all } \zeta \in \mathcal{F}', \xi \in \mathcal{G}'.$$

If the feedback value exists at x, then (3.12) holds and ζ^*, ξ^* are optimal at x, that is, (3.10) holds.

PROOF. Since $\mathcal{F}'$ contains the constant strategies, (3.13) implies

$$\overline{J}(x;\zeta^*,\xi^*) \leq \inf_{a \in \mathcal{A}} J_\infty(x;a,\xi^*) \leq v_B(x),$$

and similarly

$$v_A(x) \leq \sup_{b \in \mathcal{B}} J_\infty(x;\zeta^*,b) \leq \overline{J}(x;\zeta^*,\xi^*).$$

Then all the conclusions follow from the existence of the feedback value, i.e., $v_A(x) = v_B(x) =: v_F(x)$. ◄

See Exercises 3.4 and 3.5 for extensions of Propositions 3.14 and 3.15 where the optimal strategies are replaced by optimizing sequences of strategies.

★ ★ ★

Some authors consider different notions of strategies that can be played one against the other by taking uniform limits of trajectories, usually called *motions of the game* generated by a pair of strategies. This is the case in the theory of *positional differential games* by Krasovskiĭ and Subbotin and in Berkovitz's theory. Following these approaches one can define a new lower value $\tilde{v}$ and a new upper value $\tilde{u}$ as, respectively, the sup inf and the inf sup over the same pair of the sets of strategies. Therefore one automatically has the inequality $\tilde{v} \leq \tilde{u}$, and if the new value exists (i.e., $\tilde{v} = \tilde{u}$), it makes sense looking for a saddle point and indeed the existence of saddle points can be proved under suitable assumptions. On the other hand, given a pair of strategies, the motion and payoff generated by them may not be unique. The description of these theories goes beyond the scope of this book. However, we explain briefly how they can be connected to the formulations of differential games presented in this chapter. This is done by proving that $\tilde{v}$ and $\tilde{u}$ satisfy the following weak super- and suboptimality principles, respectively,

$$(3.14) \qquad \tilde{v}(x) \geq \sup_{b \in B} \inf_{\mu \in \mathcal{A}^r} \left\{ \int_0^t \ell^r(y_x(s), \mu(s), b) e^{-\lambda s} \, ds + \tilde{v}(y_x(t; \mu, b)) e^{-\lambda t} \right\},$$

$$(3.15) \qquad \tilde{u}(x) \leq \inf_{a \in A} \sup_{\nu \in \mathcal{B}^r} \left\{ \int_0^t \ell^r(y_x(s), a, \nu(s)) e^{-\lambda s} \, ds + \tilde{u}(y_x(t; a, \nu)) e^{-\lambda t} \right\},$$

for all $0 < t < t_0$, where $\mathcal{A}^r$ and $\mathcal{B}^r$ denote the sets of relaxed control functions and ℓ^r the relaxed running cost. The next result says that any two functions with these properties coincide with the value of the game.

THEOREM 3.16. *Assume* (0.2), (0.5), *and* (1.4). *Let* $\tilde{v}, \tilde{u} \in BC(\mathbb{R}^N)$ *be such that* $\tilde{v} \leq \tilde{u}$ *and satisfy, respectively,* (3.14) *and* (3.15). *If the game has a value (i.e.,* $v = u$, *e.g., if* $H = \bar{H}$), *then* $\tilde{v} = \tilde{u} = v$.

PROOF. We claim that $\tilde{v}$ is a supersolution of the lower HJI equation (1.17) and $\tilde{u}$ is a subsolution of the upper HJI equation (1.18). Then by Theorem 2.1 $v \leq \tilde{v}$ and $\tilde{u} \leq u$ and we get the conclusion.

To prove the claim we first observe, as in the proof of Theorem 2.6, that the Hamiltonian H coincides with the Hamiltonian H_A^r of the lower HJI equation corresponding to the game where the first player uses relaxed controls and the second player standard controls. With this remark the proof of the claim is essentially the same as that of Corollary 3.10, with (3.14) and (3.15) replacing the DPP (3.3). ◀

REMARK 3.17. In the theory of positional differential games inequalities of the form of the weak super- and suboptimality principles (3.14) (3.15) are called, respectively, *stability* of $\tilde{v}$ (respectively, $\tilde{u}$) *with respect to the first* (respectively, second) *player*, briefly *a-stability* (respectively, *b-stability*). The motivation of this name can be explained by considering the corresponding inequalities for games with finite horizon and null running cost and interest rate, that is, with payoff

$$J(x,t,a,b) := g(y_x(t;a,b))$$

for a given $g \in C(\mathbb{R}^N)$. Then the weak superoptimality principle is

$$\tilde{v}(x,t) \geq \sup_{b \in B} \inf_{\mu \in \mathcal{A}^r} \tilde{v}(y_x(s;\mu,b), t-s)$$

for all $0 < s \leq t$, which is usually proved by considering any level $r \in \mathbb{R}$ and constructing for any $b \in B$ and (x,t) in the sublevel set of $\tilde{v}$,

$$S(r) := \{ (x,t) : \tilde{v}(x,t) \leq r \},$$

a $\mu \in \mathcal{A}^r$ such that $(y_x(s;\mu,b), t-s) \in S(r)$. This says that the first player can keep the system in the set $S(r)$ for any choice of a constant control of second player. In the terminology of Krasovskiĭ and Subbotin the set $S(r)$ is an *a-stable bridge*. Similarly, the weak suboptimality principle for $\tilde{u}$ is related to the b-stability of the superlevel sets of $\tilde{u}$. ◁

3.2. Approximation by discrete time games

In this subsection we consider some discrete time games associated with the infinite horizon differential game by a standard approximation scheme with step h. The main result is the convergence of the values of a majorant and a minorant discrete time game to the upper and lower value defined in §1 as h tends to 0. It is the extension to two-person games of Theorem 1.1 of Chapter VI. The interest of this result is threefold. As for problems with one player in §1 of Chapter VI, the discrete time games have an optimal strategy and an optimal control in feedback form that can be synthesized from the solution of the (discrete) Dynamic Programming equation, so we have approximated optimal feedbacks for the differential game. Moreover, the discrete time game is the first step of a fully discrete approximation scheme suitable for the numerical solution of the differential game: see Appendix A. Finally, we recall that limits of values of discrete time games were used for quite a while to define the value of a differential game, so the convergence result gives the equivalence of these definitions with the one of §1.

We denote by $A^{\mathbb{N}}$ and $B^{\mathbb{N}}$ the sets of sequences taking values, respectively, in A and B and consider for $\{a_n\} = a \in A^{\mathbb{N}}$, $\{b_n\} = b \in B^{\mathbb{N}}$, and $h > 0$ the discrete time system

$$\begin{cases} y(k+1) = y(k) + hf(y(k), a_k, b_k), & k \in \mathbb{N}, \\ y(0) = x, \end{cases}$$

whose trajectory is denoted by $y_h(k) = y_h(k; x, a, b)$.

The payoff functional is

$$J_h(x,a,b) := \sum_{k=0}^{\infty} h\ell(y_h(k), a_k, b_k)\sigma^k, \qquad \sigma := 1 - \lambda h.$$

The *fair discrete time game* has the following rules: at each step k the first player chooses a_k and the second player chooses b_k simultaneously; both know all previous moves and in particular the current state of the game $y_h(k)$. In the *minorant game* the rules are the same except that the first player knows b_k before choosing a_k, that is, the second player must commit himself first at each move. In the *majorant game* the advantage in the information pattern is reversed and goes to the second player because the first must commit himself first at each move. The value of the minorant game $v_h(x)$ can be formalized by defining a *nonanticipating discrete strategy* for the first player as a mapping $\alpha : B^{\mathbb{N}} \to A^{\mathbb{N}}$ such that, for any $k \in \mathbb{N}$, $b_j = \tilde{b}_j$ for all $j \leq k$ implies $\alpha[b]_j = \alpha[\tilde{b}]_j$ for all $j \leq k$, and setting

$$v_h(x) := \inf_{\alpha \in \Lambda} \sup_{b \in B^{\mathbb{N}}} J_h(x, \alpha[b], b),$$

where Λ denotes the set of nonanticipating discrete strategies for the first player. Similarly, the value of the majorant game $u_h(x)$ is defined as

$$u_h(x) := \sup_{\beta \in \Theta} \inf_{a \in A^{\mathbb{N}}} J_h(x, a, \beta[a]),$$

where Θ denotes the set of nonanticipating strategies for the second player. Clearly, any reasonable notion of value for the fair game should belong to the interval $[v_h(x), u_h(x)]$. Next we characterize v_h and u_h by means of the lower and upper *Dynamic Programming equations*

(3.16) $\qquad v_h(x) = \max_{b \in B} \min_{a \in A} \{ \sigma v_h(x + h f(x, a, b)) + h\ell(x, a, b) \},$

(3.17) $\qquad u_h(x) = \min_{a \in A} \max_{b \in B} \{ \sigma u_h(x + h f(x, a, b)) + h\ell(x, a, b) \},$

which are discrete versions of the lower and upper HJI equations (1.17) and (1.18).

THEOREM 3.18. *Assume* (0.2), (1.4), $h \in]0, 1/\lambda[$, *and that for some* L,

(3.18) $\quad |f(x, a, b) - f(y, a, b)| \leq L |x - y| \qquad$ *for all* $x, y \in \mathbb{R}^N$, $a \in A$, $b \in B$.

Then the value function v_h (respectively, u_h) of the minorant (respectively, majorant) game is the unique solution in $BUC(\mathbb{R}^N)$ of the lower (respectively, upper) DP equation (3.16) *(respectively,* (3.17)*).*

PROOF. We begin with the proof that $v_h \in BUC(\mathbb{R}^N)$. The estimate $|v_h(x)| \leq M/(1-\sigma)$ is an immediate consequence of (1.4). We fix $x, z \in \mathbb{R}^N$, $\varepsilon > 0$ and pick $\overline{\alpha} \in \Lambda$ such that

$$v_h(z) \geq \sup_{b \in B^{\mathbb{N}}} J_h(z, \overline{\alpha}[b], b) - \varepsilon/6,$$

then $\overline{b} \in B^{\mathbb{N}}$ such that

$$v_h(x) \leq J_h(x, \overline{\alpha}[\overline{b}], \overline{b}) + \varepsilon/6.$$

3. OTHER INFORMATION PATTERNS

Thus

$$v_h(x) - v_h(z) \leq J_h(x, \overline{\alpha}[\overline{b}], \overline{b}) - J_h(z, \overline{\alpha}[\overline{b}], \overline{b}) + \varepsilon/3,$$

and from now on the proof is exactly the same as that of Proposition VI.4.4.

Next we prove that v_h satisfies (3.16). Call $w(x)$ the right-hand side of (3.16). Observe that the min in (3.16) exists because $v_h, f,$ and ℓ are continuous and A is compact. Moreover, it is easy to deduce from (0.2) and (1.4) that the function $b \mapsto \min_{a \in A} \{ \ldots \}$ is continuous, so the max also exists because B is compact. We fix $\varepsilon > 0$ and for any $z \in \mathbb{R}^N$ we pick $\alpha_z \in \Lambda$ such that

(3.19) $$v_h(z) \geq \sup_{b \in B^{\mathbb{N}}} J_h(z, \alpha_z[b], b) - \varepsilon \, .$$

We first prove that $v_h(x) \leq w(x)$. By definition of w, for all $b \in B$ there exists $\overline{a}(b) \in A$ such that

(3.20) $$w(x) \geq \sigma v_h(x + hf(x, \overline{a}(b), b) + h\ell(x, \overline{a}(b), b)) \, .$$

We define $\delta \in \Lambda$ as follows:

$$(\delta[b])_0 := \overline{a}(b_0), \qquad (\delta[b])_k := (\alpha_z[\{b_{n+1}\}])_{k-1} \qquad \text{if } k \geq 1,$$

with

(3.21) $$z := x + hf(x, \overline{a}(b_0), b_0),$$

where $\{b_{n+1}\} \in B^{\mathbb{N}}$ denotes the sequence $n \mapsto b_{n+1}$. Note that, if we set $z(n) = y_h(n; z, \alpha_z[\{b_{n+1}\}], \{b_{n+1}\})$, we have

$$J_h(x, \delta[b], b) = h\ell(x, \overline{a}(b_0), b_0) + \sum_{n=1}^{\infty} h\ell(z(n-1), (\delta[b])_n, b_n)\sigma^n$$
$$= h\ell(x, \overline{a}(b_0), b_0) + \sigma J_h(z, \alpha_z[\{b_{n+1}\}], \{b_{n+1}\}).$$

Thus, by (3.19),

$$v_h(x) \leq \sup_{b \in B^{\mathbb{N}}} J_h(x, \delta[b], b) \leq \sup_{b_0 \in B} \{ h\ell(x, \overline{a}(b_0), b_0) + \sigma v_h(z) + \sigma \varepsilon \},$$

and we obtain $v_h(x) \leq w(x)$ by (3.20), (3.21) and the arbitrariness of ε.

To prove the inequality $w(x) \leq v_h(x)$ we pick $\overline{b} \in B$ such that

(3.22) $$w(x) = \min_{a \in A} \{ \sigma v_h(x + hf(x, a, \overline{b})) + h\ell(x, a, \overline{b}) \}$$

and define for any $b \in B^{\mathbb{N}}$

$$\widetilde{b}_k := \begin{cases} \overline{b}, & k = 0, \\ b_{k-1}, & k \geq 1 \, . \end{cases}$$

Observe that $(\alpha_x[\tilde{b}])_0$ depends only on $\bar{b}$ because the strategy α_x defined by (3.19) is nonanticipating, and set

(3.23) $$z := x + hf(x, \bar{a}, \bar{b}), \qquad \bar{a} := (\alpha_x[\tilde{b}])_0 \;.$$

Define also $\underset{\sim}{\alpha} \in \Lambda$ by

(3.24) $$(\underset{\sim}{\alpha}[b])_k := (\alpha_x[\tilde{b}])_{k+1},$$

and choose $b^* \in B^{\mathbb{N}}$ such that

(3.25) $$v_h(z) \leq J_h(z, \underset{\sim}{\alpha}[b^*], b^*) + \varepsilon \;.$$

We claim that

(3.26) $$w(x) \leq J_h(x, \alpha_x[\tilde{b^*}], \tilde{b^*}) + \varepsilon,$$

which gives, by (3.19)

$$w(x) \leq v_h(x) + 2\varepsilon,$$

and thus the conclusion by the arbitrariness of $\varepsilon > 0$. To prove the claim note that, by (3.22) and (3.23),

(3.27) $$w(x) \leq \sigma v_h(z) + h\ell(x, \bar{a}, \bar{b})$$

and that, by (3.23) and (3.24)

(3.28) $$\sigma J_h(z, \underset{\sim}{\alpha}[b^*], b^*) + h\ell(x, \bar{a}, \bar{b}) = J_h(x, \alpha_x[\tilde{b^*}], \tilde{b^*}) \;.$$

By putting together (3.27), (3.25) and (3.28) we get the claim (3.26). This completes the proof that v_h satisfies (3.16).

Next we prove that v_h is the unique solution of (3.16) by showing that two bounded functions u and w such that

(3.29) $$u(x) \leq \max_{b \in B} \min_{a \in A} \{ \sigma u(x + hf(x, a, b)) + h\ell(x, a, b) \}$$

(3.30) $$w(x) \geq \max_{b \in B} \min_{a \in A} \{ \sigma w(x + hf(x, a, b)) + h\ell(x, a, b) \}$$

satisfy $u \leq w$. In fact, by (3.29) there exists $\bar{b} \in B$ such that

$$u(x) \leq \sigma u(x + hf(x, a, \bar{b})) + h\ell(x, a, \bar{b}) \qquad \text{for all } a \in A,$$

and by (3.30) there is $\bar{a} \in A$ such that

$$w(x) \geq \sigma w(x + hf(x, \bar{a}, \bar{b})) + h\ell(x, \bar{a}, \bar{b}) \;.$$

Putting the last two inequalities together we get

$$u(x) - w(x) \leq \sigma \sup_{z \in \mathbb{R}^N} (u - w)(z) < +\infty \qquad \text{for all } x \in \mathbb{R}^N \;.$$

Since $\sigma < 1$ we conclude that $\sup_{\mathbb{R}^N}(u - w) \leq 0$. This completes the proof of all statements concerning v_h. The results about u_h can be proved in a similar way. ◂

3. OTHER INFORMATION PATTERNS

Next we give the main result of this subsection. It is an extension to HJI equations of Theorem 1.1 of Chapter VI.

THEOREM 3.19. *Assume (0.2), (1.4), and (3.18). Then the value function v_h (respectively, u_h) of the minorant (respectively, majorant) game converge uniformly on every compact set of $\mathbb{R}^N$ as $h \to 0^+$ to the lower (respectively, upper) value v (respectively, u) of the game (see Definition 1.6).*

REMARK 3.20. If $v_h(x)$ and $u_h(x)$ converge to the same limit $w(x)$ as $h \to 0^+$, then $w(x)$ is called the *Fleming's value* of the game. The last theorem implies that when the value defined in §1 exists (e.g., if the Isaacs' condition holds, by Corollary 2.2), then the Fleming's value also exists and the two values coincide. ◁

PROOF OF THEOREM 3.19. We prove only the statement about v_h and leave the part on u_h as an exercise for the reader. The proof is similar to that of Theorem VI.1.1. We observe that v_h, $h > 0$, are uniformly bounded and consider the weak limits

$$\underline{u}(x) := \liminf_{(y,h) \to (x,0^+)} v_h(y), \qquad \overline{u}(x) := \limsup_{(y,h) \to (x,0^+)} v_h(y).$$

We show that $\underline{u} \in BLSC(\mathbb{R}^N)$ is a supersolution and $\overline{u} \in BUSC(\mathbb{R}^N)$ is a subsolution of the lower HJI equation (1.17). Then the comparison principle Theorem V.1.3 (whose assumptions are verified by the proof of Theorem 2.1) implies that

$$\overline{u} \leq v \leq \underline{u},$$

and both inequalities are actually equalities because $\underline{u} \leq \overline{u}$ by definition. Thus v_h converge uniformly to v on compact sets by the properties of weak limits, see Lemma V.1.9.

We begin with the proof that $\underline{u}$ is a supersolution of (1.17). The semicontinuity of $\underline{u}$ follows from Lemma V.1.5. By (3.16) v_h satisfies

$$v_h(x) + \max_{a \in A} \{ -\sigma v_h(x + f(x,a,b)) - h\ell(x,a,b) \} \geq 0 \qquad \text{for all } b \in B,$$

so it is a supersolution of the discrete Bellman equation (HJB)$_h$ of Chapter VI for any fixed value of $b \in B$. Thus by the proof of Theorem VI.1.1 for any $\varphi \in C^1(\mathbb{R}^N)$ and x_1, a strict local minimum point of $\underline{u} - \varphi$, there exists $\overline{a} = \overline{a}(b)$ such that

$$\lambda \underline{u}(x_1) - D\varphi(x_1) \cdot f(x_1, \overline{a}, b) - \ell(x_1, \overline{a}, b) \geq 0.$$

Then

$$\lambda \underline{u}(x_1) + \min_{b \in B} \max_{a \in A} \{ -D\varphi(x_1) \cdot f(x_1, a, b) - \ell(x_1, a, b) \} \geq 0,$$

which is the desired inequality.

Next we consider $\overline{u}$ and a strict maximum point x_0 for $\overline{u} - \varphi$ on $S := \overline{B}(x_0, r)$. By Lemma V.1.6 there exist $x_n \in S$, $h_n \to 0^+$ such that

(3.31) $\quad (v_{h_n} - \varphi)(x_n) = \max_S (v_{h_n} - \varphi), \quad x_n \to x_0, \quad v_{h_n}(x_n) \to \overline{u}(x_0).$

By (3.16) there exist $b_n \in B$ such that

$$v_{h_n}(x_n) - \sigma v_{h_n}(x_n + h_n f(x_n, a, b_n)) - h_n \ell(x_n, a, b_n) \leq 0 \qquad \text{for all } a \in A.$$

We use the first relation in (3.31) and the definition of σ to get

$$(1 - \lambda h_n)[\varphi(x_n) - \varphi(x_n + h_n f(x_n, a, b_n))] + \lambda h_n v_{h_n}(x_n) - h_n \ell(x_n, a, b_n) \leq 0.$$

Now we extract a subsequence of b_n convergent to $\overline{b} \in B$, and without relabeling, we divide by h_n and pass to the limit in the last inequality to obtain, by (3.31),

$$-D\varphi(x_0) \cdot f(x_0, a, \overline{b}) + \lambda \overline{u}(x_0) - \ell(x_0, a, \overline{b}) \leq 0 \qquad \text{for all } a \in A.$$

Then,

$$\lambda \overline{u}(x_0) + \min_{b \in B} \max_{a \in A} \{ -D\varphi(x_0) \cdot f(x_0, a, b) - \ell(x_0, a, b) \} \leq 0,$$

which proves that $\overline{u}$ is a subsolution of (1.17). ◀

Now we explain how the preceding results can be applied to constructing approximated optimal strategies for the infinite horizon differential game. The next result shows how to synthesize an optimal strategy in feedback form for the minorant game from the solution of the lower DP equation (3.16). It is the extension to games of a standard result in the theory of optimal control for discrete time systems, see the Appendix to Chapter VI. Of course a similar result holds for the majorant game, by using the upper DP equation (3.17).

THEOREM 3.21. *Let* $h \in \,]0, 1/\lambda[$, $w \in BC(\mathbb{R}^N)$ *be a solution of* (3.16), *and* $F: \mathbb{R}^N \times B \to A$ *be such that*

$$F(y, b) \in \arg\min_{a \in A} \{ \sigma w(y + hf(y, a, b)) + h\ell(y, a, b) \}.$$

For any sequence $b \in B^{\mathbb{N}}$ *and any* $x \in \mathbb{R}^N$ *consider the sequence* $\{z_n[b]\}$ *in* $\mathbb{R}^N$ *defined by*

$$\begin{cases} z_{n+1} = z_n + hf(z_n, F(z_n, b_n), b_n), \\ z_0 = x, \end{cases}$$

and define $\alpha^*[b]_n := F(z_n, b_n)$. *Then* $\alpha^* \in \Lambda$ *is optimal at the initial point* x, *and*

(3.32) $\qquad v_h(x) = \max_{b \in B^{\mathbb{N}}} J_h(x, \alpha^*[b], b) = J_h(x, \alpha^*[b^*], b^*)$

for any $b^* \in B^{\mathbb{N}}$ *such that* $b_n^* \in G(z_n[b^*])$ *for all* n, *where*

$$G(y) := \arg\max_{b \in B} \min_{a \in A} \{ \sigma w(y + hf(y, a, b)) + h\ell(y, a, b) \}.$$

Note that the optimal strategy α^* for the minorant game, and the corresponding maximizing control b^*, can be computed if one knows a solution of (3.16), namely the value function v_h, and if one can solve for every $y \in \mathbb{R}^N$ the problem

$$\max_{b \in B} \min_{a \in A} \{ \sigma v_h(y + hf(y,a,b)) + h\ell(y,a,b) \},$$

which is the right-hand side of the lower DP equation (3.16).

PROOF OF THEOREM 3.21. For any $b \in B^{\mathbb{N}}$ by definition of $z_n[b]$ and (3.16),

$$w(z_n) \geq \sigma w(z_{n+1}) + h\ell(z_n, F(z_n, b_n), b_n)$$

and the equality holds if and only if $b_n \in G(z_n)$. Thus we get, by induction,

$$w(x) \geq \sigma^n w(z_n) + h \sum_{k=0}^{n-1} \sigma^k \ell(z_k, F(z_k, b_k), b_k),$$

and, letting $n \to \infty$,

$$w(x) \geq J_h(x, \alpha^*[b], b),$$

where the equality holds if $b_n \in G(z_n)$ for all n. ◀

There is a natural way to extend the optimal discrete time strategy $\alpha^* \in \Lambda$ and the control $b^* \in B^{\mathbb{N}}$ to continuous time, by defining $b_h^* \in \mathcal{B}$ as

(3.33) $$b_h^*(t) := b_{[t/h]}^*,$$

where $[\cdot]$ denotes the integer part of a real number, and $\alpha_h^* \in \Gamma$ as

(3.34) $$\alpha_h^*[b](t) := \alpha^*[\widetilde{b}]_{[t/h]},$$

where for any $b \in \mathcal{B}$ we set

$$\widetilde{b}_n := b(nh).$$

Note that $b_h^* \in \mathcal{B}_h$ where

$$\mathcal{B}_h := \{ b \in \mathcal{B} : b(s) = b(kh) \text{ for all } s \in [kh, (k+1)h[, k \in \mathbb{N} \},$$

and that $\alpha_h^*[b] \in \mathcal{A}_h$ for all $b \in \mathcal{B}$.

The next result says that in some sense α_h^* is an approximately optimal strategy for the infinite horizon game with continuous time.

COROLLARY 3.22. *Under the assumptions of Theorem 3.19 we have*

(3.35) $$\lim_{h \to 0^+} J(x, \alpha_h^*[b_h^*], b_h^*) = v(x),$$

and for any $\varepsilon > 0$

(3.36) $$\sup_{b \in \mathcal{B}_h} J(x, \alpha_h^*[b], b) \leq v(x) + \varepsilon$$

for all h sufficiently small.

PROOF. It is not hard to show, as in the proof of Lemma VI.1.2, that (0.2), (1.4), and (3.18) imply

(3.37) $\quad |J(x,a,b) - J_h(x,\widetilde{a},\widetilde{b})| \leq \omega(h) \quad$ for all $a \in \mathcal{A}_h$, $b \in \mathcal{B}_h$

where ω is a modulus. Then

$$|J(x,\alpha_h^*[b_h^*], b_h^*) - J_h(x,\alpha^*[b^*], b^*)| \leq \omega(h),$$

and we get the first conclusion (3.35) from (3.32) and Theorem 3.19. Now observe that (3.32) and (3.37) imply

$$\sup_{b \in \mathcal{B}_h} J(x,\alpha_h^*[b], b) \leq J(x,\alpha_h^*[b_h^*], b_h^*) + 2\omega(h).$$

Then we can use (3.35) to get the last conclusion (3.36). ◀

Under some additional assumptions we can refine Theorem 3.19 and Corollary 3.22 by giving an estimate of the error made if the value function v is replaced by v_h and if the approximately optimal strategy α_h^* is chosen by the first player.

THEOREM 3.23. *In addition to* (0.2), (1.4), *and* (3.18) *assume f is bounded, ℓ is lipschitzean in y uniformly in a and b (i.e., $\omega_\ell(r) = L_\ell r$), and let γ be the exponent of Hölder continuity of v (see Proposition 1.8). Then there is a constant C such that*

(3.38) $\quad |v_h(x) - v(x)| \leq Ch^{\gamma/2},$

(3.39) $\quad |J(x,\alpha_h^*[b_h^*], b_h^*) - v(x)| \leq Ch^{\gamma/2},$

(3.40) $\quad \left|\sup_{b \in \mathcal{B}_h} J(x,\alpha_h^*[b], b) - v(x)\right| \leq Ch^{\gamma/2},$

for all $x \in \mathbb{R}^N$ and $h \in \,]0, 1/\lambda[$, where α_h^* and b_h^* are defined by (3.34), (3.33), *and* (3.32).

SKETCH OF THE PROOF. The proof of (3.38) is very similar to that of Theorem VI.1.5 and it is left as an exercise for the reader (see Exercise 3.8). Under the additional assumption on ℓ, the modulus ω in (3.37) is $\omega(h) = Ch$. Then (3.39) and (3.40) are immediately obtained by the argument of the proof of Corollary 3.22. ◀

* * *

We end this section with a brief discussion of the approximation procedure which is the basis of Friedman's theory. A detailed description of this method goes beyond the scope of the book, but we explain how it is connected to the various formulations of differential games presented in this chapter. In Friedman's theory the differential game is approximated by a lower and an upper δ-game, where $\delta > 0$ is a small parameter. In these games the state is governed by the

3. OTHER INFORMATION PATTERNS

continuous-time system (0.1), but decisions are taken only at the discrete times δn, $n \in \mathbb{N}$. The information pattern is the following: in the lower δ-game at time δn the second player chooses his control in the interval of time $]\delta n, \delta(n+1)]$, and with this information available the first player makes his choice for the same time interval; in the upper δ-game the information advantage is reversed. This can be formalized by introducing the lower and upper δ-strategies and leads to the definition of a lower δ-value $v_\delta(x)$ and an upper δ-value $u_\delta(x)$, where x is the initial state of the system. It is rather intuitive that $v_\delta \leq v$ and $u \leq u_\delta$ for all $\delta > 0$, and that, setting $\delta_n = T/n$ for a fixed T, the sequence v_{δ_n} is nondecreasing and u_{δ_n} is nonincreasing. The limits of these sequences are, respectively, Friedman's lower and upper value, and if they coincide we have the existence of *Friedman's value*. To compare these values with the notions introduced previously in this chapter we need only to observe that the following form of the Dynamic Programming Principle holds for the lower and upper δ-game, respectively:

(3.41)
$$v_\delta(x) = \inf_{\alpha \in \Gamma_\delta} \sup_{b \in \mathcal{B}_\delta} \left\{ \int_0^\delta \ell(y_x(s), \alpha[b](s), b(s)) e^{-\lambda s} \, ds + v_\delta(y_x(\delta; \alpha[b], b)) e^{-\lambda \delta} \right\}.$$

(3.42)
$$u_\delta(x) = \sup_{\beta \in \Delta_\delta} \inf_{a \in \mathcal{A}_\delta} \left\{ \int_0^\delta \ell(y_x(s), a(s), \beta[a](s)) e^{-\lambda s} \, ds + u_\delta(y_x(\delta; a, \beta[a])) e^{-\lambda \delta} \right\}.$$

where

$$\mathcal{A}_\delta := \{ a : [0, \delta] \to A \text{ measurable} \}, \qquad \mathcal{B}_\delta := \{ b : [0, \delta] \to B \text{ measurable} \},$$
$$\Gamma_\delta := \{ \alpha : \mathcal{B}_\delta \to \mathcal{A}_\delta \}, \qquad \Delta_\delta := \{ \beta : \mathcal{A}_\delta \to \mathcal{B}_\delta \}.$$

See Exercise 3.9 for a simple definition of v_δ and u_δ with these properties.

THEOREM 3.24. *Assume (0.2), (0.5), and (1.4). For all small $\delta > 0$ let v_δ and u_δ be uniformly bounded functions satisfying, respectively, (3.41) and (3.42). Then v_δ and u_δ converge uniformly on every compact set of $\mathbb{R}^N$ as $\delta \to 0^+$, respectively, to the lower value v and the upper value u of the game, see Definition 1.6.*

SKETCH OF THE PROOF. To study the limit of v_δ we first show that (3.41) implies

(3.43) $$v_\delta(x) = \sup_{b \in \mathcal{B}} \inf_{a \in \mathcal{A}} \left\{ \int_0^\delta \ell(y_x(s), a(s), b(s)) e^{-\lambda s} \, ds + v_\delta(y_x(\delta; a, b)) e^{-\lambda \delta} \right\}.$$

Then we consider the weak limits

$$\underline{v}(x) := \liminf_{(y,\delta) \to (x, 0^+)} v_\delta(y), \qquad \overline{v}(x) := \limsup_{(y,\delta) \to (x, 0^+)} v_\delta(y),$$

and show that $\underline{v}$ is a supersolution and $\overline{v}$ a subsolution of the lower HJI equation (1.17). This is achieved by the arguments of the proof of Theorem 3.19, after replacing the discrete DP equation (3.16) with (3.43). The conclusion follows from the comparison principle Theorem 1.3 which gives $\overline{v} \leq v \leq \underline{v}$, and the inequality $\underline{v} \leq \overline{v}$ is trivial. ◀

Exercises

3.1. Let $\{\Phi_n\}$ be a sequence of maps $\mathbb{R}^N \to A$ and $\{t_n\}$ be an increasing sequence such that $t_0 = 0$ and $\lim_n t_n = +\infty$. Define for $y \in \mathcal{Y}$

$$\zeta[y](t) = \Phi_n(y(t_n)) \quad \text{if } t_n \leq t < t_{n+1}.$$

Prove that ζ is a feedback strategy with finite memory.

3.2. Prove that the DPP still holds if $\mathcal{F}$ and $\mathcal{G}$ are replaced by the feedback strategies with finite memory in the definition of v_A and v_B. [Hint: follow the proof of Theorem 3.8 and show that $S(\sigma, y) = \{t_i \in S(\bar{\zeta}, y) : t_i < t\} \cup \{t\} \cup \{t_i + t : t_i \in S(\zeta_{y(t)}, y(\cdot + t))\}$.]

3.3. State and prove a verification theorem for feedback strategies using subsolutions of the HJI equations as verification functions.

3.4. We say that a sequence of strategies $\{\zeta_n\}$ in $\mathcal{F}$ (respectively, $\{\xi_n\}$ in $\mathcal{G}$) is optimizing at the initial point x if

$$v_A(x) = \limsup_n \sup_{b \in \mathcal{B}} J_\infty(x; \zeta_n, b)$$

(respectively, $v_B(x) = \lim_n \inf_{a \in \mathcal{A}} J_\infty(x; a, \xi_n)$). Prove the following extension of Proposition 3.14. Assume x is a point where the game has feedback value and there are optimizing sequences $\{\zeta_n\}$ in $\mathcal{F}' \subseteq \mathcal{F}$ and $\{\xi_n\}$ in $\mathcal{G}' \subseteq \mathcal{G}$, with $\mathcal{F}'$ and $\mathcal{G}'$ compatible. Then

$$(3.44) \qquad v_F(x) = \lim_n \overline{J}(x; \zeta_n, \xi_n) = \sup_{\xi \in \mathcal{G}'} \inf_{\zeta \in \mathcal{F}'} \overline{J}(x; \zeta, \xi)$$

$$= \inf_{\zeta \in \mathcal{F}'} \sup_{\xi \in \mathcal{G}'} \overline{J}(x; \zeta, \xi).$$

3.5. Prove the following extension of Proposition 3.15. Assume the game has feedback value at x and that $\mathcal{F}'$, $\mathcal{G}'$ are compatible, contain the constant strategies, and satisfy

$$\sup_{\xi \in \mathcal{G}'} \inf_{\zeta \in \mathcal{F}'} \overline{J}(x; \zeta, \xi) = \inf_{\zeta \in \mathcal{F}'} \sup_{\xi \in \mathcal{G}'} \overline{J}(x; \zeta, \xi).$$

Then there exist optimizing sequences of strategies (see Exercise 3.4) $\{\zeta_n\}$ in $\mathcal{F}'$ and $\{\xi_n\}$ in $\mathcal{G}'$ such that (3.44) holds.

3.6. Given $g : \mathbb{R}^N \times B \to \mathbb{R}^P$, set $\mathcal{W} := \{g(y(\cdot), b(\cdot)) : y \in \mathcal{Y}, b \in \mathcal{B}\}$ and define an output state feedback strategy as a nonanticipating map $\sigma : \mathcal{W} \to \mathcal{A}$ such that for any $x \in \mathbb{R}^N$, $b : [0, T[\to B$ measurable and $T \in]0, +\infty]$ there exists a unique solution of

$$(3.45) \qquad \begin{cases} q'(t) = f(q(t), \sigma[w](t), b(t)), & 0 < t \leq T, \\ w(t) = g(q(t), b(t)), \\ q(0) = x. \end{cases}$$

3. OTHER INFORMATION PATTERNS

The variable w is called the output or observed state. We indicate with $\mathcal{I}$ the set of these strategies. Since for a general g the state q cannot be reconstructed from the output w, this information pattern is called imperfect state measurement and we have a game with partial observation. Define

$$W(x) := \inf_{\sigma \in \mathcal{I}} \sup_{b \in \mathcal{B}} J(x, \sigma[w], b),$$

where w comes from the solution of (3.45) for given, x, σ, and b. Prove that

(i) if $g(q,b) = b$, then $W = v$, the lower value of the game, and if $g(q,b) = q$, then $W = v_A$, the A-feedback value;

(ii) $\mathcal{I}$ contains the constant strategies;

(iii) for any $\sigma \in \mathcal{I}$ there exists $\alpha \in \Gamma$ such that, for any $b \in \mathcal{B}$, $\sigma[w] = \alpha[b]$, where w solves (3.45);

(iv) $v \leq W \leq u_s$, the upper value of the static game;

(v) if $g(q,b) = \widetilde{g}(q)$, then $v_A \leq W$;

(vi) if the Isaacs' condition holds and $u_s = u$, the upper value of the game, then $W = v_F$, the feedback value (results of this kind are sometimes called Certainty Equivalence Principle);

(vii) W is a supersolution of the lower HJI equation (1.17).

3.7. State and prove a result for the majorant game corresponding to Theorem 3.21 and a result for the upper value corresponding to Corollary 3.22.

3.8. Prove Theorem 3.23. [Hint: for the proof of (3.38) note that the proof of Theorem VI.1.5 has to be modified only at two points, namely, formulas (VI.1.32) and (VI.1.33), where one must choose first a suitable $b^* \in B$ and then a^*.]

3.9. Define the strategies of the first player anticipating the second player's moves for a time at most $\delta > 0$, as,

$$\Lambda := \big\{ \alpha : \mathcal{B} \to \mathcal{A} : \text{for any } n \in \mathbb{N} \text{ and } b, \widetilde{b} \in \mathcal{B},\ b(s) = \widetilde{b}(s) \text{ for all } s \leq n\delta \text{ implies}$$
$$\alpha[b](s) = \alpha[\widetilde{b}](s) \text{ for all } s \leq n\delta \big\},$$

and denote with Π the strategies of second player with the same property. Define

$$v_\delta(x) := \inf_{\alpha \in \Lambda} \sup_{b \in \mathcal{B}} J(x, \alpha[b], b),$$
$$u_\delta(x) := \sup_{\beta \in \Pi} \inf_{a \in \mathcal{A}} J(x, a, \beta[a]).$$

Prove that v_δ and u_δ satisfy, respectively, (3.41) and (3.42).

4. Bibliographical notes

The theory of (two-person, zero-sum) differential games started at the beginning of the '60s with the work of Isaacs [Is65] in the U.S.A. and of Pontryagin and his school [PBGM62, P66] in the Soviet Union.

The main motivation at that time was the study of military problems, and pursuit-evasion games were the most important examples. Isaacs based his work on the Dynamic Programming method. He analyzed many special cases of the PDE that is now called Hamilton-Jacobi-Isaacs, trying to solve them more or less explicitly and synthesizing optimal feedbacks from the solution. He began a study of singular surfaces (which are essentially the sets where the value function is discontinuous or not differentiable) which was continued mainly by Breakwell, Bernhard [Ber77, BrB90, Ber92], and Lewin [Lew94], and led to the explicit solution of some low-dimensional but highly nontrivial games. We refer to the recent book [Lew94] for a survey of this theory and to [Br89, Mel94] for some connections with viscosity solutions.

As for the second motivation for two-person, zero-sum differential games, that is, the worst case analysis of a controlled system with a disturbance, it is already present at the early stages of the theory [F61]. It has been made rigorous more recently by Barron and Jensen within the theory of risk-aversion [BJ89] (see also [BSa91b]) and by many authors in connection with $\mathcal{H}_\infty$ and robust control. For this topic we refer to the book of Basar and Bernhard [BaBe91] in the case of linear systems and to [FM92a, FM95, Jam92, Jam93, Sor96a, DME97] and Appendix B of this book for nonlinear systems and the use of viscosity solutions.

The notions of strategies and value of Sections 1 and 2 are due to Varaiya [Va67], Roxin [Ro69], and Elliott and Kalton [EK72, EK74], and Evans and Souganidis [ES84] first proved that such a value is a viscosity solution of the Hamilton-Jacobi-Isaacs equation for finite horizon problems. Theorems 1.10 and 2.1 are adaptations of their results to the infinite horizon problem. (Indeed, to our knowledge, the first connection between differential games and viscosity solutions was made by Souganidis in his Ph.D. thesis of 1983 for the Fleming and Friedman's values). The book [El87] contains a presentation of this material.

Similar results have been obtained for many different problems, indeed almost any cost functional considered in Chapters III and IV has been studied in the context of differential games by these methods. We mention [EI84] for problems with exit times in bounded domains, [L85b] for games with reflected processes, [BS89] for time-optimal control (i.e., pursuit-evasion games) and [BS91a, Sor93a] for more general problems with exit times and the study of free boundaries as in §IV.4 (Soravia applied some of these results to the study of the stability of dynamical systems with two competing controllers [Sor93a, Sor95] and to some models of front propagation [Sor94a]).

We refer to [Al91, K95, BKS96] for problems with state constraints, [Bn90] for games with maximum cost, [Y90, EL91, IK91b] for problems with switching cost and [LeY92] if the process is in addition piecewise deterministic, and [BJM93,

Y94, Ra95] for systems with impulses. Example 1.7 is adapted from [Be64]. The classical minimax Theorem 2.3 can be found in [AE84], for instance.

The notion of feedback strategy of §3.1 is taken from Soravia [Sor94b], see also [El77] and [BaBe91]. The results of this subsection are new; Propositions 3.14 and 3.15 are an attempt to make a connection with the Isaacs-Breakwell theory, inspired by Bernhard's lectures [Ber92].

The theory of positional differential games was developed in the '70s by N.N. Krasovskii and Subbotin [KS74]. The value of the game in this theory is characterized as the unique function that is stable with respect to both players (i.e., which satisfies weak super- and suboptimality principles of the form (3.14) and (3.15)), see Remark 3.17). Subbotin [Su80] characterized these properties in terms of inequalities for Dini directional derivatives, in the case of locally Lipschitz functions (see also [SS78] and [Su84]). The equivalence between functions satisfying these inequalities and viscosity sub- and supersolutions was proved in different ways, see [EI84, LS85, ST86]. Later this notion of solution was extended to general HJ equations and to merely continuous functions [Su91a, Su91b] by means of the generalized Dini derivatives, see Definitions 2.36 and 2.38 in Chapter III. The equivalence between minimax and viscosity solutions was proved in [Su93b], see also §2.4 in Chapter III and [ClaL94]. The theory was extended to discontinuous solutions [RS88, Su93a]. A comprehensive reference for the theory of minimax solutions is the book [Su95]. Some more recent papers on constructive and numerical methods within this theory are [Su97, CLS97, ChSu95, Ta94, Ta97, PBKTZ94]. For a survey of other developments of the theory of positional differential games we refer to the book of N.N. Krasovskii and A.N. Krasovskii [KK95].

The Berkovitz's theory starts with the paper [Be85] and combines features of the Isaacs, Friedman and Krasovskii-Subbotin approaches. The connection with minimax and viscosity solutions was established in [Be88], see also [Gh91] for games with state constraints. We refer to [Be94] for a survey of this theory.

Subsection 3.2 adapts to the infinite horizon problem the results obtained for pursuit-evasion games in [BS91b], see also [Sor92b]. Fleming introduced his notion of value in 1961 [F61] and proved its existence for finite horizon problems and for some special cases of pursuit-evasion games; see also [F64]. These were the first general and rigorous results about the existence of a value. The connection of this definition of value with viscosity solutions can be found in [Sou85b] for the case of finite horizon.

Friedman's theory was developed in the early '70s and is summarized in the books [Fri71, Fri74]. The connection with viscosity solutions was studied in [BEJ84] for the finite horizon problem, and in [Sor92b] for games with a target.

Many results of Chapters V and VI were extended to differential games. Indeed many classical examples of pursuit-evasion games have discontinuous value, and this was one of the main motivations for developing the theory of discontinuous viscosity solutions of Chapter V. The characterization of the value function as the unique e-solution of §V.3.2 holds for pursuit-evasion games where the pursuer uses relaxed controls [BFS97], see also [Su95, BBF95] for other notions of value.

A related result for problems with open targets has been proved in [BG97]. Some applications in §V.4.2 have been carried over to games in [BS94], whereas the uniqueness result of §V.4.3 is currently open for games. The theory of bilateral supersolutions described in §V.5 does not seem to extend to games because it uses heavily the convexity of the Hamiltonian.

The semidiscrete approximations of §3.2 and §VI.1.1 do not converge in general if the value function is discontinuous. In this case the scheme was modified in [BBF95] to get the convergence in a suitable sense. This paper also studies a fully discrete scheme, as in [BFS94] (see also Appendix A of the present book). Estimates of the rate of convergence such as in §VI.1.2 have been first studied for games by Souganidis [Sou85b], see also [Sou85a, AT87, Bot94] and [Sor97b]. Results similar to those of §VI.2 can be found in [BSa91a, BSa92] for games with targets. Differential games with Lipschitz controls are studied in [BEJ84], and the results extend Corollary VII.4.2.

Stochastic two-person zero-sum differential games have been studied in the framework of viscosity solutions by Fleming and Souganidis [FSo89], see also [BaSo91, Ka95, Sw96b]; for differential games involving infinite dimensional systems we refer to [KSS97] and the reference therein.

Finally we give some additional references. A book devoted to pursuit-evasion games is [Ha75]; the volume [P90] contains several contributions by authors of Pontryagin's school; some recent papers on numerical methods are [TiG94, Ti95, Pes94b]. The theory of viability has been applied to differential games by J.P. Aubin and his school [Au91, CQS94a]; let us mention [CQS97] for an algorithm to compute the possibly discontinuous value function of pursuit-evasion games with state constraints.

For the general theory of N-person differential games we refer to the books [BaO82] and [Kle93]. This theory has a wide range of applications from economics to environmental management, ecology, and biology, see [BaH94, Ol95] and the references therein.

APPENDIX A

Numerical Solution of Dynamic Programming Equations

by Maurizio Falcone

As shown in the book, the Dynamic Programming approach to the solution of deterministic optimal control problems is essentially based on the characterization of the value function in terms of a partial differential equation, the Hamilton-Jacobi-Bellman equation. In the case of deterministic optimal control problems, which we consider here, this is a nonlinear first order equation $H(x, u(x), Du(x)) = 0$ where H is convex with respect to Du. However, this approach is very flexible since a similar characterization can be obtained for the value function of stochastic optimal control problems (in which case the equation will be of the second order and/or an integro-differential equation depending on the stochastic process describing the dynamics) and of differential games (in which case we still have a first order equation but we lose the convexity with respect to Du).

One feature of this approach is particularly interesting in many applications: it permits computation of approximate optimal controls in feedback form and, as a consequence, approximate optimal trajectories.

Although these are very appealing properties if compared to the results given by the open-loop approach based on the Pontryagin Maximum Principle, the Dynamic Programming approach suffers the problem of the "rise of dimension". In fact, to compute the value function we need to solve a Hamilton-Jacobi-Bellman equation in the domain where the initial condition for the dynamics is taken. On the other hand, the necessary conditions of the Pontryagin Maximum Principle often correspond to the solution of a two-point boundary value problem for a system of ordinary differential equations involving the state and the adjoint variables. This simple observation implies that the number of computations needed by the Dynamic Programming approach can be huge when the state variable belongs to a space of high dimension ($\mathbb{R}^N$ with $N = 4, 6$ or more). This is one of the main motivations to develop efficient algorithms to solve equations of this type. Of course, when comparing the Dynamic Programming approach with the Pontryagin

Maximum Principle one has to remember that the former gives a global minimum while the latter gives only a necessary condition which is satisfied by local extrema.

The basic idea in the foreground of the numerical methods we present in this Appendix is to start the construction of the algorithm by a discretization of the original control problem and to use a discrete version of the Dynamic Programming Principle. Of course, this is not the only possible approach and some references to other methods will be given in the last section containing also some historical remarks.

The approximation scheme is obtained in two steps. First we look at the original control problem and make a time discretization with time step $h = \Delta t$. The value function v_h corresponding to this new discrete-time optimal control problem is a natural candidate to approximate v (see Chapter VI). A discrete Dynamic Programming Principle (see Chapter VI, Appendix) gives the characterization of v_h and, at the same time, a semi-discrete approximation scheme. The second step consists in a space discretization with space step $k = \Delta x$ and it results in a finite dimensional problem that we can finally solve.

The schemes obtained by this technique have a nice "built-in" control interpretation which is also useful to establish convergence results and a priori estimates in L^∞, both for the semi-discrete and the fully-discrete schemes. Moreover, the synthesis of approximate feedback controls is simple and effective.

The aim of this appendix is mainly expository, so we give here only some representative results. The interested reader should look at the papers mentioned in the references and in our last section for the most general version of the results. In Section 1 we give a short presentation of the results related to the approximation of the unconstrained infinite horizon problem that we will use as our model problem. In that section the main results related to our discretization technique will be presented in detail. Section 2 is devoted to the modifications needed to deal with problems with state constraints. Section 3 deals with "target" problems, such as the minimum time problem and pursuit-evasion games. In Section 4 we give some hints for the construction of the algorithms. Finally, Section 5 contains some commented references and historical remarks.

1. The infinite horizon problem

We consider the following system of controlled differential equations in $\mathbb{R}^N$

(1.1) $$\begin{cases} y'(t) = f(y(t), \alpha(t)) \\ y(0) = x \end{cases}$$

where $\alpha(t) \in \mathcal{A} \equiv L^\infty([0, +\infty[; A)$ and A is a compact subset of $\mathbb{R}^M$. We will refer to the set $\mathcal{A}$ as the *set of admissible controls*. The cost functional related to the infinite horizon problem is

(1.2) $$J_x(\alpha) \equiv \int_0^\infty \ell(y(s), \alpha(s)) e^{-\lambda s} \, ds \,.$$

1. THE INFINITE HORIZON PROBLEM

We will make the following assumptions on the data

(1.3) $\quad f : \mathbb{R}^N \times A \to \mathbb{R}^N$ and $\ell : \mathbb{R}^N \times A \to \mathbb{R}$ are continuous;

(1.4) $\quad |f(x_1,a) - f(x_2,a)| \leq L_f |x_1 - x_2|$ for any $a \in A$ and $\|f\|_\infty \leq M_f$;

(1.5) $\quad |\ell(x_1,a) - \ell(x_2,a)| \leq L_\ell |x_1 - x_2|$ for any $a \in A$ and $\|\ell\|_\infty \leq M_\ell$.

Assumptions (1.4) guarantee that there exists a unique solution trajectory y defined in $[0, +\infty[$ for any fixed control α. The value function v, defined for any initial state x, is

$$v(x) \equiv \inf_{\alpha \in \mathcal{A}} J_x(\alpha) .$$

1.1. The Dynamic Programming equation

Let us recall from Chapter III that the value function v of the *unconstrained* problem above is the unique viscosity solution of

(HJ) $\quad \lambda u(x) + \sup_{a \in A} \{-f(x,a) Du(x) - \ell(x,a)\} = 0, \quad$ for $x \in \mathbb{R}^N$.

Making a discretization in time of the original control problem, which consists in replacing the dynamics (1.1) by a one-step scheme (e.g., by the Euler method) and the cost functional (1.2) by its discretization by a quadrature formula (e.g., the rectangle rule), one can get a new control problem in discrete time. The value function v_h for this problem (as we have seen in Chapter VI) satisfies a discrete dynamic programming principle which gives the following approximation scheme,

(HJ)$_h$ $\quad u_h(x) = \min_{a \in A} \{(1 - \lambda h) u_h(x + h f(x,a)) + h \ell(x,a)\},$ for $x \in \mathbb{R}^N$.

Under our assumptions (1.3)–(1.5) and for $\lambda > L_f$, the family of functions v_h is equibounded (by M_f/λ) and equicontinuous

(1.6) $$|v_h(x) - v_h(y)| \leq \frac{L_\ell}{\lambda - L_f} |x - y| .$$

Then, by the Ascoli-Arzelà theorem we can pass to the limit and prove that it converges locally uniformly to v for h going to 0. Moreover, the following estimate holds true,

(1.7) $$\|v - v_h\|_\infty \leq C h^{1/2}$$

(see §VI.1). In order to compute an approximate value function and solve (HJ)$_h$ we have to make a further step: a discretization in space.

We start building a grid in the state space and, to simplify our presentation, we will assume that there exists a bounded polyhedron $\Omega \subset \mathbb{R}^N$ such that, for h sufficiently small

(1.8) $$x + h f(x,a) \in \overline{\Omega} \quad \forall x \in \overline{\Omega} \quad \forall a \in A$$

i.e., Ω contains the discrete controlled trajectories for all times $t \geq 0$. We will see in the next section how we can get rid of this assumption and treat different boundary conditions on $\partial \Omega$. We construct a regular triangulation of Ω made by a family of simplices S_j, such that $\overline{\Omega} = \bigcup_j S_j$, denoting by x_i, $i = 1, \ldots, L$, the nodes of the triangulation and by

$$k \equiv \max_j \mathrm{diam}(S_j)$$

the size of the mesh (here and in the sequel $\mathrm{diam}(B)$ will denote the diameter of the set B). We look for a solution of

$$(\mathrm{HJ})_h^k \qquad u(x_i) = \min_{a \in A}\{(1 - \lambda h)u(x_i + hf(x_i, a)) + h\ell(x_i, a)\}, \quad i = 1, \ldots, L$$

in the space of piecewise linear functions on Ω

(1.9) $\qquad W^k \equiv \{w : \Omega \to \mathbb{R} : w(\,\cdot\,) \in C(\Omega) \text{ and } Dw(x) = c_j \; \forall x \in S_j, \; \forall j\}$.

We will denote by $\|\cdot\|_\infty$ the max norm for vectors, i.e., $\|Z\|_\infty = \max_{i=1,\ldots,L} |Z_i|$.

THEOREM 1.1. *Assume (1.3)–(1.5) and (1.8). Then for any $h \in \,]0, 1/\lambda]$ there exists a unique solution v_h^k of $(\mathrm{HJ})_h^k$ in W^k.*

PROOF. By our assumption (1.8), starting from any x in Ω we will reach points which still belong to Ω. Then, for every $u \in W^k$ we have

$$u(x_i + hf(x_i, a)) = \sum_{j=1}^{L} \lambda_{ij}(a) u(x_j)$$

where $\lambda_{ij}(a)$ are the coefficients of the convex combination representing the point $x_i + hf(x_i, a)$, i.e.,

(1.10) $\qquad\qquad x_i + hf(x_i, a) = \sum_{j=1}^{L} \lambda_{ij}(a)\, x_j$

(1.11) $\qquad\qquad 0 \leq \lambda_{ij}(a) \leq 1 \text{ and } \sum_{j=1}^{L} \lambda_{ij}(a) = 1, \text{ for any } a \in A.$

Then $(\mathrm{HJ})_h^k$ is equivalent to the following fixed point problem in finite dimension

(1.12) $\qquad\qquad\qquad U = T(U)$

where the map $T : \mathbb{R}^L \to \mathbb{R}^L$ is defined componentwise as

(1.13) $\qquad\qquad (T(U))_i \equiv \min_{a \in A}[(1 - \lambda h)\Lambda(a)\, U + hF(a)]_i,$

$U_i \equiv u(x_i)$, $F_i(a) \equiv \ell(x_i, a)$ and $\Lambda(a)$ is the $L \times L$ matrix of the coefficients λ_{ij} satisfying (1.10)–(1.11) for $i, j = 1, \ldots, L$. Note that the matrix Λ satisfies the same properties of the transition matrix of a Markov chain process with L states. This can be interpreted as saying that the transition probability from the node x_i to the node x_j under the dynamics f is $\lambda_{ij}(a)$. This interpretation is particularly useful when one wants to extend the scheme to deal with diffusion control processes (see the Bibliographical notes for some references).

T is a contraction mapping. In fact, let $\bar{a}$ be a control giving the minimum in $T(V)_i$, we have

$$[T(U) - T(V)]_i \leq (1 - \lambda h)[\Lambda(\bar{a})(U - V)]_i \leq (1 - \lambda h)\max_{i,j}|\lambda_{ij}(a)|\, \|U - V\|_\infty$$
$$\leq (1 - \lambda h)\|U - V\|_\infty \,.$$

Switching the role of U and V, we can conclude that

(1.14) $$\|T(U) - T(V)\|_\infty \leq (1 - \lambda h)\|U - V\|_\infty$$

and the thesis follows for $h \in {]}0, 1/\lambda]$. ◂

REMARK 1.2. The existence of (at least) one control a^* giving the minimum in (1.13) relies on the continuity of the data and on the compactness of the set of controls A. Note also that the search for a *global minimum* over A is not an easy task. Let us define

$$L_i(a, U) \equiv (1 - \lambda h)\Lambda_i(a)U + hF_i(a) \,.$$

To compute the fixed point, we have to find the argument of the minimum for each component i, i.e., we have to solve the problems

$$\min_{a \in A} L_i(a, U) \,.$$

Note that $L_i(a, \cdot)$ is an affine function for every i and that when A is finite (e.g., $A \equiv \{a_1, \ldots, a_n\}$), the minimum can be obtained by direct comparison at each node. So one simple way to solve the problem is to replace A by a finite set of controls constructing a mesh over A. Depending on the structure of the control problem this may or may not affect the result. For example, when the optimal controls are bang-bang a careful discretization (essentially only of the boundary of A) can give very accurate results. ◁

Since T is a contraction mapping in $\mathbb{R}^N$, the sequence

(1.15) $$U^n = T(U^{n-1}), \qquad U^0 = Z$$

will converge to U^*, for any $Z \in \mathbb{R}^N$. Moreover, the following estimate holds true:

$$\|U_n - U^*\|_\infty \leq (1 - \lambda h)^n \|Z - U^*\|_\infty \,.$$

We want to give an estimate for the approximation error in the $L^\infty(\Omega)$ norm.

THEOREM 1.3. *Let v_h and v_h^k be the solutions of* (HJ)$_h$ *and* (HJ)$_h^k$. *Assume that* (1.4), (1.5) *and* (1.8) *hold true and that* $\lambda > L_f$, *then*

(1.16) $$\|v_h - v_h^k\|_\infty \leq \frac{L_\ell}{\lambda(\lambda - L_f)} \frac{k}{h}.$$

PROOF. For any $x \in \Omega$, we can write

$$|v_h(x) - v_h^k(x)| \leq \left|\sum_j \lambda_j (v_h(x) - v_h(x_j))\right| + \left|\sum_j \lambda_j v_h(x_j) - v_h^k(x_j)\right|.$$

where the λ_j are the coefficients of a convex combination. By the equations, we have

(1.17) $$\begin{aligned} v_h(x_j) - v_h^k(x_j) &\leq (1 - \lambda h)\left[v_h(x_j + hf(x_j, \bar{a})) - v_h^k(x_j + hf(x_j, \bar{a}))\right] \\ &\leq (1 - \lambda h)\|v_h - v_h^k\|_\infty \end{aligned}$$

where $\bar{a}$ is a control giving the minimum in (HJ)$_h^k$. This implies, as in Theorem 1.1, that

$$|v_h(x_j) - v_h^k(x_j)| \leq (1 - \lambda h)\|v_h - v_h^k\|_\infty.$$

By the Lipschitz continuity of v_h stated in (1.6), we get

(1.18) $$|v_h(x) - v_h(x_j)| \leq \frac{L_\ell}{\lambda - L_f} k.$$

By (1.17) and (1.18),

(1.19) $$\|v_h - v_h^k\|_\infty \leq (1 - \lambda h)\|v_h - v_h^k\|_\infty + \frac{L_\ell}{\lambda - L_f} k.$$

Then, we conclude

(1.20) $$\|v_h - v_h^k\|_\infty \leq \frac{L_\ell}{\lambda(\lambda - L_f)} \frac{k}{h}. \quad \blacktriangleleft$$

Coupling (1.20) with the estimate (1.7) we obtain the following global estimate for the numerical solution.

THEOREM 1.4. *Let v and v_h^k be the solutions of* (HJ)$_h$ *and* (HJ)$_h^k$. *Assume that* (1.3)-(1.5) *and* (1.8) *hold true and that* $\lambda > L_f$, *then*

(1.21) $$\|v - v_h^k\|_\infty \leq Ch^{1/2} + \frac{L_\ell}{\lambda(\lambda - L_f)} \frac{k}{h}.$$

The contraction mapping argument used in the proof of Theorem 1.1 does not provide a very efficient method to compute the solution since the contraction coefficient $(1 - \lambda h)$ tends to 1 as h tends to 0. This means that we would expect

1. THE INFINITE HORIZON PROBLEM

to have a very slow convergence whenever we try to improve the accuracy of our numerical solution by refining the step sizes.

To overcome this difficulty one can accelerate the convergence by exploiting the monotonicity of the map T that is, $T(U) \geq T(Z)$ if $U \geq Z$. Let us introduce the basic ideas of the *monotone acceleration procedure*. We define the set of subsolutions of (1.12) as

$$\mathcal{U} \equiv \{U \in \mathbb{R}^L : U \leq T(U)\} .$$

It is easy to prove that the set $\mathcal{U}$ is convex, closed. Moreover, the fixed point U^* of the map T is also the maximal element of $\mathcal{U}$, since it is the limit of all monotone nondecreasing sequences starting in $\mathcal{U}$. These properties are fundamental to show that starting from any $Z \in \mathcal{U}$ the sequence defined in (1.15) belongs to $\mathcal{U}$ for every n and converges monotonically (increasing) to U^*. We can accelerate the convergence using the contraction mapping iteration just to define a direction of displacement in $\mathcal{U}$. We will follow that direction up to the boundary of $\mathcal{U}$ before restarting the algorithm by the contraction mapping iteration. We have then the following

ALGORITHM
STEP 0: Take $U^0 \in \mathbb{R}^L$
 Set $n = 0$
STEP 1: Compute
$$U^{n+1/2} = T(U^n)$$

STEP 2: Compute
$$U^{n+1} = U^n + \overline{\mu}(U^{n+1/2} - U^n) \text{ where } \overline{\mu} \equiv \max\{\mu \in \mathbb{R}_+ : U^{n+1} \in \mathcal{U}\}$$

STEP 3: Stopping criterion: is $||U^{n+1} - U^n||_\infty < \varepsilon$?
 if YES then STOP else GOTO STEP 1

Note that if it is possible to derive an explicit optimal step μ to reach the boundary of the set of subsolutions this will produce a very fast algorithm. When this is not possible, the computation of μ will be obtained via an iterative one dimensional zero finding algorithm. In that case, the algorithm is usually faster if one spends more iterations in locating the point on the boundary before changing direction.

REMARK 1.5. When the set of controls is finite $A \equiv \{a_1, \ldots, a_m\}$ we can adopt a different method to find U^*. Instead of the contraction mapping argument coupled with the acceleration technique we can apply the *simplex method*. Note that $\mathcal{U}$ is identified by L linear constraints In fact, $\mathcal{U} = \{U \in \mathbb{R}^L : g_i(U, a) \leq 0$ for any $i = 1, \ldots, L$ and $a \in A\}$, where

$$g_i(U, a) = U - (1 - \lambda h)(\Lambda(a)U)_i - hF_i(a) .$$

Then, we look for a solution V^* of the problem

(1.22) $$\max_{V \in \mathcal{U}} \sum_{j=1}^{L} V_j .$$

It is evident by the above remarks that (1.22) can be solved by the simplex method or other methods of linear programming. Moreover, V^* and U^* must coincide as the following simple argument shows. As we have seen, the fixed point U^* is also the maximal element of the set of subsolutions $\mathcal{U}$, then by definition $U_i^* \geq V_i^*$ so that $\sum_{i=1}^{L} U_i^* \geq \sum_{i=1}^{L} V_i^*$; hence, by the definition of V^* and the fact that $U^* \in \mathcal{U}$,

$$(1.23) \qquad \sum_{i=1}^{L} U_i^* = \sum_{i}^{L} V_i^*$$

In fact, if there is an index i such that $V_i^* < U_i^*$, the maximality of U^* would imply

$$(1.24) \qquad \sum_{i}^{L} U_i^* > \sum_{i}^{L} V_i^*$$

which contradicts (1.23). ◁

1.2. Synthesis of feedback controls

We go back to one of the most important problems for real applications: the reconstruction of an approximate optimal feedback control. We will construct them by means of our numerical approximation of v.

Let us consider equation $(HJ)_h$ and define $a^* = a_h^*(x) \in A$ as the control setting such that

$$u_h(x) = (1 - \lambda h)u_h(x + hf(x, a^*(x))) + h\ell(x, a^*(x))$$

(if that control is not unique we can always define a selection). Let our discrete dynamics be obtained by an Euler discretization of (1.1)

$$\begin{cases} y_{m+1} = y_m + hf(y_m, a_m), & m \in \mathbb{N} \\ y_0 = x \end{cases}$$

and define a sequence of approximate feedback controls using our equation:

$$(1.25) \qquad a_m^* \equiv a_h^*(y_m) \qquad m = 0, 1, 2 \ldots$$

We define,

$$(1.26) \qquad v_h(x) = \inf_{\{a_m\}} J_x^h(\{a_m\})$$

where

$$(1.27) \qquad J_x^h(\{a_m\}) = \sum_{m=0}^{+\infty} h\ell(y_m, a_m)(1 - \lambda h)^m .$$

In the Appendix to Chapter VI it was proved that

$$v_h(x) = J_x^h(\{a_m^*\}) .$$

In Chapter VI the following results were established.

1. THE INFINITE HORIZON PROBLEM

THEOREM 1.6. *Let the assumptions (1.3)-(1.5) be satisfied. Let $\{a_m^*\}$ be the sequence defined in (1.25) and let*

$$\alpha_h^*(s) \equiv a^*(y_m) \quad \text{for any } s \in [mh, (m+1)h[\text{ and } m = 0, 1, 2, \ldots$$

Then

$$J_x^h(\alpha_h^*(\,\cdot\,)) \longrightarrow \inf_{\alpha(\cdot) \in \mathcal{A}} J_x(\alpha(\,\cdot\,)) \quad \text{and} \quad J_x(\alpha_h^*(\,\cdot\,)) \longrightarrow \inf_{\alpha(\cdot)} J_x(\alpha(\,\cdot\,))$$

locally uniformly in $\mathbb{R}^n$, as $h \to 0^+$.

Now the question arises: can the synthesis of feedback control based on the numerical solution of $(HJ)_j^k$ also produce an "almost optimal" control for the functional related to the continuous problem? The answer to this question is positive under additional assumptions. Let us define the functions $L^k : \Omega \times A \to \mathbb{R}$

$$L^k(x, a) \equiv (1 - \lambda h)v_h^k(x + hf(x, a)) + h\ell(x, a)$$

and $L : \Omega \times A \to \mathbb{R}$

$$L(x, a) \equiv (1 - \lambda h)v_h(x + hf(x, a)) + h\ell(x, a) \;.$$

Sometimes in the following we will use the short notation $w = v_h^k$ when we will not need to emphasize the dependence of the numerical solution from h and k. We can naturally associate with any $x \in \Omega$ a control $a_w(x) \in A$ such that

(1.28) $$L^k(x, a_w(x)) = \min_{a \in A} L^k(x, a) = v_h^k(x) \;.$$

That control of course could not be unique but we can always select one using a convex criterion (e.g., picking the control with minimum norm which minimizes $L^k(x, \,\cdot\,)$). In the same way we can select a (unique) control $a(x)$ such that

(1.29) $$L(x, a(x)) = \min_{a \in A} L(x, a) = v_h(x) \;.$$

Let us define the piecewise constant control

$$\alpha_w^*(s) \equiv a_w(y_m) \quad \text{for } s \in [mh, (m+1)h[$$

where $\{y_m\}$ is the solution trajectory of the discrete time system

$$y_{m+1} = y_m + hf(y_m, a_w(y_m)), \quad y_0 = x \;.$$

THEOREM 1.7. *Let the assumptions (1.3)-(1.5) be satisfied. Moreover, assume that for any x there exists a unique $a(x)$ minimizing L. Then, for any $x \in \Omega$*

$$J_x^h(\alpha_w^*(\,\cdot\,)) \longrightarrow \inf_{\alpha(\cdot)} J_x^h(\alpha(\,\cdot\,)) \quad \text{as } k \to 0^+ \;.$$

PROOF. The estimate (1.16) shows that L^k converges uniformly on $\Omega \times A$ to L for k tending to 0. It is clear that the feedback control defined in (1.28) for h and x fixed depends only on k so we will denote it here as $a_k(x)$. Since A is compact from any sequence $\{a_k(x)\}$ of controls giving the minimum in L^k we can always extract a subsequence $\{a_{n_k}\}$ converging to a limit, say $\{\bar{a}(x)\}$. We have then

$$L^{n_k}(x, a_x^{n_k}) - L(x, \bar{a}(x)) \longrightarrow 0$$

as $k \to 0$. Moreover, by the definition of $a_k(x)$ passing to the limit for k tending to 0 one gets

$$L(x, \bar{a}(x)) \leq L(x, a(x)) \, .$$

By the definition of $a(x)$ the reverse inequality is also true, so that

$$L(x, \bar{a}(x)) = L(x, a(x))$$

which implies $\bar{a}(x) = a(x)$. That argument can be repeated for any subsequence; then we conclude that $a_k(x)$ tends to $a(x)$ for k tending to 0. By definition,

$$h\ell(x, a_k(x)) = L^k(x, a_k(x)) - (1 - \lambda h) v_h^k(x + h f(x, a_k(x)))$$

for any $x \in \Omega$. Taking $x = y_m$ and multiplying by $(1 - \lambda h)^m$ one gets

$$h\ell(y_m, a_k(y_m))(1 - \lambda h)^m$$
$$= v_h^k(y_m)(1 - \lambda h)^m - (1 - \lambda h)^{m+1} v_h^k(y_m + h f(y_m, a_k(y_m))) \, .$$

Summing on m and passing to the limit for k tending to 0 we get the result. ◀

THEOREM 1.8. *Let the assumptions (1.3)-(1.5) be satisfied. Moreover, assume that for any x there exists a unique control $a(x)$ minimizing L. Then, for any $x \in \Omega$*

$$J_x(\alpha_w^*(\,\cdot\,)) \longrightarrow \inf_{\alpha(\,\cdot\,)} J_x(\alpha(\,\cdot\,))$$

for h, k and k/h tending to 0.

PROOF. The proof can be obtained by coupling the a priori error estimates with the traditional approximation of the integral by a series. We want to prove that

$$|J_x(\alpha_w^*(\,\cdot\,)) - v(x)| \text{ tends to } 0 \, .$$

We can write

$$|J_x(\alpha_w^*(\,\cdot\,)) - v(x)| \leq |J_x(\alpha_w^*(\,\cdot\,)) - w(x)| + |w(x) - v(x)| \, .$$

The last term is estimated by the result in Theorem 1.4 and tends to 0. The second term can be estimated by

$$(1.30) \quad |J_x(\alpha_w^*(\,\cdot\,)) - w(x)| \leq |J_x(\alpha_w^*(\,\cdot\,)) - J_x^h(\alpha_w^*(\,\cdot\,))| + |J_x^h(\alpha_w^*(\,\cdot\,)) - w(x)| \, .$$

Since $w = v_h^k$ the second term of the above sum converges to 0 by Theorem 1.7. Following the lines of Theorem 1.6 one can prove that the first term in the right-hand side of (1.30) tends to 0 for h and k going to 0 and this ends the proof. ◀

1.3. Numerical tests

We present some numerical results in order to show how our numerical approximation behaves for smooth and nonsmooth solutions. The first two tests are in $\mathbb{R}^1$, while the second is in $\mathbb{R}^2$.

TEST 1. We set $\Omega =]-1, 1[$,

$$A = [0, 1], \qquad \lambda = 1, \qquad f(x, a) = -xa .$$

We consider a cost function ℓ which is just Lipschitz continuous,

$$\ell(x, a) = x .$$

The viscosity solution in Ω is

$$v(x) = \begin{cases} x & \text{for } x < 0, \\ x/2 & \text{elsewhere} . \end{cases}$$

The optimal choice is to take $a^* = 0$ whenever $x < 0$ and to move on the left at maximum speed ($a^* = 1$) whenever $x \geq 0$. This causes a kink at $x = 0$. Figure 1 shows the exact solution (straight line) and the approximate solution (small circles). ◁

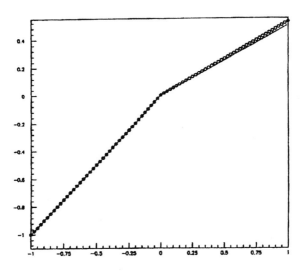

FIGURE 1: Value function ($h = 0.1$, $k = 0.1$).

TEST 2. We set $\Omega =]-1, 1[$,

$$A = [-1, 1], \qquad \lambda = 1, \qquad f(x, a) = a(1 - |x|) .$$

We consider a cost function ℓ which is just Lipschitz continuous,

$$\ell(x, y, a) = 3(1 - |x|) .$$

For numerical purposes we have extended the cost and the vector field outside Ω just setting them to 0.

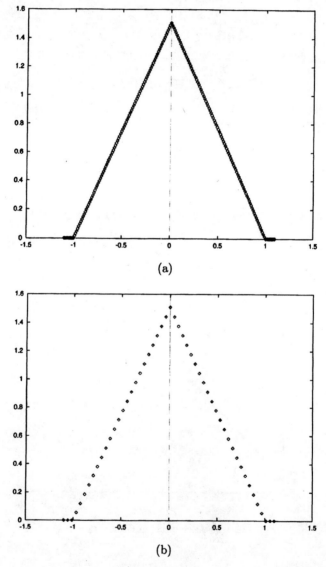

FIGURE 2: Value functions ($h = 0.01$), (a) Euler ($k = 0.011$), (b) Runge-Kutta ($k = 0.044$).

1. THE INFINITE HORIZON PROBLEM

In this test the optimal trajectories are quite simple. For $x > 0$ the optimal choice is $a^* = 1$ and for $x < 0$ the optimal choice is $a^* = -1$. By a direct computation on the optimal trajectories one can compute the value function. The viscosity solution in Ω is

$$v(x) = \begin{cases} \frac{3}{2}(x+1) & \text{for } x < 0, \\ \frac{3}{2}(1-x) & \text{elsewhere}. \end{cases}$$

The solution has one sharp kink at the point $x = 0$. It can be seen that the numerical solution is close to the exact solution also at that point for a discretization based on 201 grid points (Figure 2a). Note that at $x = 0$ one has two different optimal choices for the control $a^* = 1$ and $a^* = -1$ and this is the main cause for the jump in the derivative of the solution. Using a higher order method for the discretization of the system of differential equations and of the cost functional, the approximate solution can even be improved. In fact, coupling a Runge-Kutta scheme with a trapezoid rule produces an accurate approximation with just 51 grid points (see Figure 2b). Several other couplings between a one-step method for ordinary differential equations and a quadrature formula for the cost are possible (see the hints at the end of this Appendix). ◁

TEST 3. Let us consider an example in $\mathbb{R}^2$. We set $\Omega =]-1, 1[^2$,

$$A = \bar{B}(0,1) \subset \mathbb{R}^2, \qquad \lambda = 1$$
$$f_1(x, y, a) = ay, \qquad f_2(x, y, a) = 0.$$

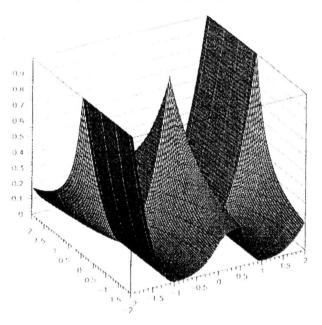

FIGURE 3: Value function ($h = 0.05$, $k = 0.025$).

We consider the cost function ℓ,

$$\ell(x,y,a) = (|x|-1)^2 \ .$$

In this example one can compute the exact solution. The numerical solution represented in Figure 3 has an L^∞ error of 0.025. Note that the algorithm gives an accurate approximate solution also at the origin where the exact solution has a cusp. ◁

2. Problems with state constraints

One major interest in view of real applications is to see how state constraints can be included in the model. In practical problems, the system has often to satisfy some restrictions (e.g., on the velocity or on the acceleration) which can be written as state constraints for the dynamics (1.1).

We will introduce such constraints in our infinite horizon model. Let Ω be an open bounded convex subset of $\mathbb{R}^N$ with regular boundary ($n(x)$ being its outward normal at the point $x \in \partial\Omega$). For any initial position $x \in \overline{\Omega}$, we require that the state remains in $\overline{\Omega}$ for all $t \geq 0$. As a consequence we will consider admissible with respect to the state constraint only the (open-loop) control functions such that the corresponding trajectory of (1.1) never leaves $\overline{\Omega}$. We will denote by $\mathcal{A}(x)$ such a subset of $\mathcal{A}$, i.e., for all $x \in \overline{\Omega}$ we define

$$\mathcal{A}(x) \equiv \{\alpha(\cdot) \in \mathcal{A} : y_x(t, \alpha(t)) \in \overline{\Omega}, \ \forall t \geq 0\},$$

where $y_x(t, \alpha(t))$ denotes the solution trajectory of (1.1) corresponding to α. The value function for the constrained problem is

(2.1) $$v(x) = \inf_{\alpha \in \mathcal{A}(x)} J_x(\alpha) \ .$$

By the theory of §IV.5, v is the unique constrained viscosity solution of

$$\lambda u(x) + \sup_{a \in A}\{-f(x,a) \cdot Du(x) - \ell(x,a)\} = 0 \ .$$

An approximation scheme for the constrained problem can be introduced adapting to this situation the scheme for unconstrained problems. In order to understand the scheme it is useful to note that at each internal point we can choose any control in A since, at least for a small time, we can move in any direction without leaving Ω. On the other hand, at each point on $\partial\Omega$ not all the controls in A are allowed since some of them correspond to directions pointing outward with respect to the constraint $\overline{\Omega}$. This means that the set of admissible controls will depend on x (in a rather irregular way if we do not make additional assumptions on the boundary of Ω), and the "right equation" for the value function should be:

$$\lambda u(x) + \sup_{a \in A(x)}\{-f(x,a) \cdot Du(x) - \ell(x,a)\} = 0$$

where

$$A(x) = \begin{cases} A & \text{for } x \in \Omega, \\ \{a \in A : f(x,a) \text{ points inward the constraint}\} & \text{for } x \in \partial\Omega. \end{cases}$$

If we assume that the set of constraints has a regular boundary so that the outward normal $n(x)$ is uniquely defined at each point of $\partial\Omega$, the above definition can be written as

$$A(x) = \{a \in A : f(x,a) \cdot n(x) < 0\}, \text{ for any } x \in \partial\Omega.$$

Note that the assumption (see §IV.5)

(2.2) $$\min_{a \in A} f(x,a) \cdot n(x) < 0, \quad \forall x \in \partial\Omega$$

guarantees that $A(x)$ is not empty for any $x \in \overline{\Omega}$.

Although this is not the easiest way to set the problem from the point of view of the theory of partial differential equations (the dependence of A on x will make it difficult to prove existence and uniqueness results), it can give a hint on the definition of constrained viscosity solution. Assume $x \in \Omega$, then $A(x) = A$ and we see that the equation is satisfied in the open set. Now take $x \in \partial\Omega$, then $A(x) \subset A$ and we have

$$\lambda v(x) + \sup_{a \in A}\{-f(x,a) \cdot Dv(x) - \ell(x,a)\}$$
$$\geq \lambda v(x) + \sup_{a \in A(x)}\{-f(x,a) \cdot Dv(x) - \ell(x,a)\} = 0$$

which means that the constrained viscosity solution is a solution in Ω and a supersolution on its boundary. In this way, the state constraints appear in the equation as a constraint on the set of admissible controls at each point.

From the point of view of the algorithm, we can still define a discrete time system and a discrete formula for the cost as in Section 2. We just have to define a multivalued map $A_h : \overline{\Omega} \to A$ which associates with each x the set of controls which are admissible for the discrete system, i.e.,

$$A_h(x) = \{a \in A : x + hf(x,a) \in \Omega\}, \text{ for every } x \in \overline{\Omega}.$$

At least when Ω is convex and when it has a reasonably smooth boundary it can be proved under assumption (2.2) that there exists a time step $\overline{h} > 0$ such that for any $h \leq \overline{h}$,

$$A_h(x) \neq \emptyset, \quad x \in \overline{\Omega}.$$

The approximation scheme which results from writing that discretization on the mesh is

$(\text{HJC})_h^k \quad u(x_i) = \inf_{a \in A_h(x_i)} \{(1 - \lambda h)u(x_i + hf(x_i,a)) + h\ell(x_i,a)\}, \quad i = 1,\ldots,L$

where x_i, $i = 1, \ldots, L$ is a node of a regular triangulation of Ω. Let W^k denote our usual space of piecewise linear functions defined in (1.9). The following convergence result holds true.

THEOREM 2.1. *Let (1.3), (1.4), (1.5), and (2.2) be satisfied. Moreover, let $\lambda > L_f$ and $h \in {]}0, 1/\lambda]$. Then there exists a unique solution v_h^k of $(HJC)_h^k$ in W^k. Moreover, v_h^k converges uniformly to v for h, k and k/h tending to 0.*

The proof that $(HJC)_h^k$ has a unique solution in W^k is based on a fixed point argument. In fact, defining $U_j = u(x_j)$ and $F_j(a) = \ell(x_j, a)$, it is easy to prove that $T : \mathbb{R}^N \to \mathbb{R}^N$

$$\left(T(U)\right)_i \equiv \min_{a \in A_h(x_i)} \left\{ (1 - \lambda h) \sum_{j=1}^N \lambda_{ij}(a) U_j + h F_i(a) \right\}, \quad i \in I$$

is a contraction mapping. Also in this case it is crucial to have an acceleration technique to speed up the convergence.

TEST 1. We set $\Omega = {]}0, 3{[}^2 \setminus (Q_1 \cup Q_2)$, where

$$Q_1 = (2, 1.5) + Q, \quad Q_2 = (1, 2) + Q \quad \text{and} \quad Q = [-0.2, 0.2]^2,$$
$$A = \bar{B}(0, 1) \subset \mathbb{R}^2, \quad \lambda = 1,$$
$$f(x_1, x_2, a) = a.$$

Let us consider the cost function ℓ,

$$\ell(x, a) = ((x_1 - 3)^2 + (x_2 - 3)^2)^{1/2} + 10(1 - x_2)^+$$

where $x \equiv (x_1, x_2) \in \bar{\Omega}$. This cost function is very high near the x_1 axis and it depends only on the distance from $(3, 3)$ for $x_2 > 1$. The dynamics is such that $A(x)$ is a proper subset of A on each point of the boundary. In the interior points the system changes instantaneously its direction so that it can avoid the obstacles Q_i, $i = 1, 2$, and runs at maximum speed toward $(3, 3)$, the absolute minimum point for the cost.

Note that in Figures 4, 5, and 6 the x_1-axis is on the left (i.e., it is the vertical axis in Figures 5 and 6) and the value on the obstacles has been set arbitrarily to zero (just to have a better graphical representation). In the computation the unit ball has been approximated by 9 points (8 on the boundary $\partial B(0, 1)$ and 1 at the origin). To compute the optimal trajectories starting from previous knowledge of the value function, we have used a one-step approximation scheme for the system of ordinary differential equations. The optimal control to be applied at each point of the trajectory is obtained by looking for a control which gives the minimum in the right-hand side of $(HJC)_h^k$. That search has been done over 37 controls (36 on the boundary and 1 at the origin). Figure 6 shows the approximate optimal trajectories obtained by that method. They start from several points on the boundary of Ω and point to the upper right corner, the point $(3, 3)$. Note that they avoid the obstacles and try to go around them in a very reasonable way. ◁

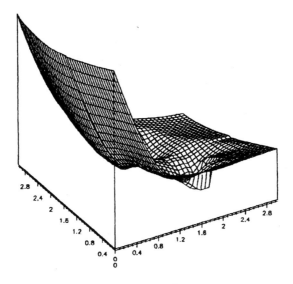

FIGURE 4: Value function ($h = 0.1$, $k = 0.187$).

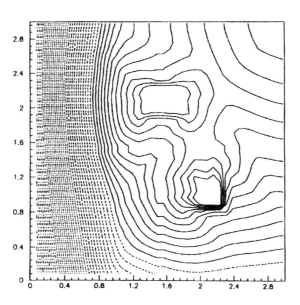

FIGURE 5: Level curves.

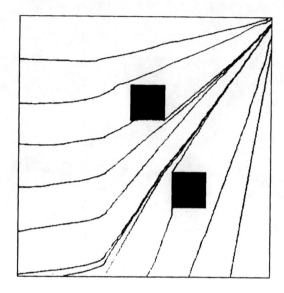

FIGURE 6: Computed optimal trajectories.

3. Minimum time problems and pursuit-evasion games

3.1. Time-optimal control

Another class of classical control problems is that of the so called "target problems". One of the most famous is the following minimum time problem.

Assume that the dynamics is given by (1.1) and that the assumptions (1.3)–(1.5) are satisfied. We are also given a closed target set $\mathcal{T} \subset \mathbb{R}^N$, and the problem consists of driving the system to the target in minimum time from its initial position x. It has been shown in Section 1 of Chapter IV that the the minimum time function

$$T(x) = \inf_{\alpha \in \mathcal{A}} \{t > 0 : y_x(t, \alpha) \in \mathcal{T}\}$$

is continuous over the reachable set $\mathcal{R}$ if a local controllability condition such as

$$\inf_{a \in A} f(x, a) \cdot n(x) < 0 \qquad \forall x \in \partial \mathcal{T}$$

is satisfied (here $n(x)$ is the exterior normal to $\mathcal{T}$ at x). Moreover, under that assumption T satisfies the boundary value problem

(3.1) $$\begin{cases} \max_{a \in A}\{-f(x, a) \cdot DT(x)\} = 1 & \text{in } \mathcal{R} \setminus \mathcal{T}, \\ T(x) = 0 & \text{in } \mathcal{T}, \\ \lim_{x \to x_0} T(x) = +\infty & \text{for any } x_0 \in \partial \mathcal{R}. \end{cases}$$

3. MINIMUM TIME PROBLEMS AND PURSUIT-EVASION GAMES

The following change of variable (see Chapter IV)

$$v(x) \equiv \begin{cases} 1 - e^{-T(x)} & \text{if } T(x) < +\infty \\ 1 & \text{if } T(x) = +\infty \end{cases}$$

is a rescaling of time which is useful to prove that T is the unique viscosity solution of the boundary value problem (3.1). In fact, one can prove that the exponential transform v is the unique viscosity solution of the boundary value problem

$$\begin{cases} v(x) + \sup_{a \in A}\{-f(x,a) \cdot Dv(x)\} = 1 & \text{in } \mathbb{R}^N \setminus \mathcal{T} \\ v(x) = 0 & \text{for } x \in \partial \mathcal{T}. \end{cases}$$

Note that by the same rescaling we have also eliminated the cumbersome boundary condition on $\partial \mathcal{R}$.

Applying the Euler discretization to the dynamics we can define a new discrete minimum time problem. T will be replaced by the quantity $hN_h(x)$ where $N_h(x)$ is the minimum number of steps necessary to reach the target. As in the continuous problem we can rescale the time, and we can prove by the methods of Chapter VI, Appendix, that the function

(3.2) $$v_h(x) \equiv \begin{cases} 1 - e^{-hN_h(x)} & \text{if } N_h(x) < +\infty \\ 1 & \text{if } N_h(x) = +\infty \end{cases}$$

is the unique bounded solution of

(3.3) $$\begin{cases} u_h(x) = \min_{a \in A}\{e^{-h}u_h(x + hf(x,a))\} + 1 - e^{-h}, & \text{in } \mathbb{R}^N \setminus \mathcal{T} \\ u_h(x) = 0 & \text{in } \mathcal{T}. \end{cases}$$

Naturally, to compute our solution we have to restrict the problem to a bounded set Q and to solve the equation (3.3) in $Q \setminus \mathcal{T}$. Note that this also means that we need to add an additional boundary condition on ∂Q. Since the set Q is arbitrary one can choose a cube in $\mathbb{R}^N$ containing the target set (but of course other choices are possible and can be even more efficient on specific problems). Working on a fixed grid on Q we are lead to the approximation scheme,

$$\begin{cases} u_h(x_i) = \min_{a \in A}\{e^{-h}u_h(x_i + hf(x,a))\} + 1 - e^{-h}, & \text{for } i \in I_{\text{in}} \\ u_h(x_i) = 0 & \text{for } i \in I_\mathcal{T} \\ u_h(x_i) = 1 & \text{for } i \in I_{\text{out}} \end{cases}$$

where

$$I_{\text{out}} \equiv \{i : x_i + hf(x_i, a) \notin Q \text{ for any } a\}$$
$$I_\mathcal{T} \equiv \{i : x_i \in \mathcal{T} \cap Q\}$$
$$I_{\text{in}} \equiv i \in I \setminus (I_{\text{out}} \cup I_\mathcal{T}).$$

Naturally, one would like to know if we can apply the same approximation technique to problems where the local controllability assumption is not satisfied and T is discontinuous. In fact, it can be proved that the same scheme can produce a reasonable approximation also in those cases. Let us introduce the time

$$t(\alpha) = t_x(\alpha) = \inf\{t \in [0, +\infty[\, : \, y_x(t, a(\,\cdot\,)) \in \text{int}(T)\},$$

and the function

$$\widehat{v}(x) = \inf_{\alpha(\cdot)} \{1 - e^{-t(\alpha)}\},$$

which corresponds to the exponential transform of the minimum time necessary to enter the interior of T and not just T. Under rather general hypotheses, it can be shown that our numerical approximation converges to $\widehat{v}$ uniformly on any compact set where $\widehat{v}$ is continuous.

TEST 1. We set $Q = \,]-2, 2[^2$, the dynamics is $\ddot{z} = a$, $A = [-1, 1]$ and $T = \{0\}$ or $T = \bar{B}(0, \varepsilon)$. When $T = \{0\}$ this is the well known example of the rocket railroad car, where the optimal control is bang-bang (see §IV.2). The optimal trajectories, corresponding to the feedback controls $a_* = 1$ and $a^* = -1$, are arcs of parabolas.

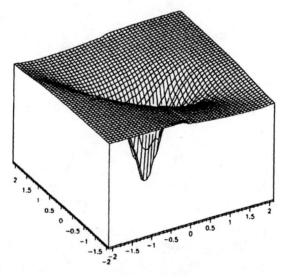

FIGURE 7: Value function ($h = 0.01$, $k = 0.15$).

The graph and the level curves of the computed value function are, respectively, in Figures 7 and 8; Figure 9 shows the optimal trajectories for two slightly different targets. Note that the approximation of the trajectories is also accurate for the pointwise target $T = \{0\}$ (Figure 9a), although this produces a degenerate boundary condition: one can easily recognize the families of parabolas corresponding to the two optimal choices of the control. ◁

3. MINIMUM TIME PROBLEMS AND PURSUIT-EVASION GAMES

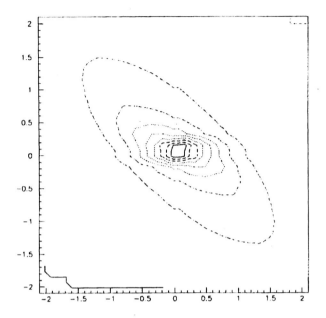

FIGURE 8: Level curves.

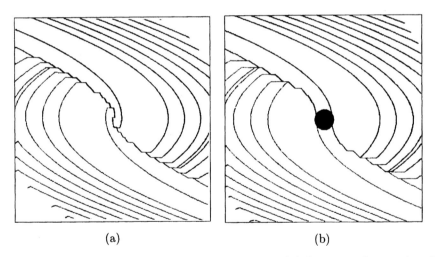

(a) (b)

FIGURE 9: Computed optimal trajectories: (a) $\mathcal{T} = \{0\}$ ($h = 0.01$, $k = 0.15$) and (b) $\mathcal{T} = \overline{B}(0,1)$.

TEST 2. We set $Q =]-2,2[^2$, the dynamics will be $\dot{x} = c(x)a$, $A = \overline{B}(0,1)$ and $\mathcal{T}$ will have different shapes. Let us consider first the simple case $c(x) = 1$. In that situation the value function coincides with the distance function from the target. Figure 10 shows the solution for T being a rhomb R, the level curves of the minimum time function are smooth as they should be. Figures 11 and 12 shows the minimum time function corresponding to a nonconvex target L (an L-shaped curve).

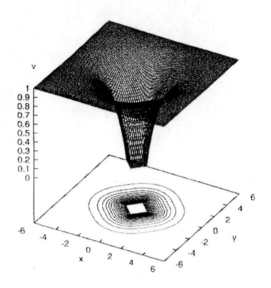

FIGURE 10: Minimum time function and level curves for $T = R$ and $c(x) \equiv 1$.

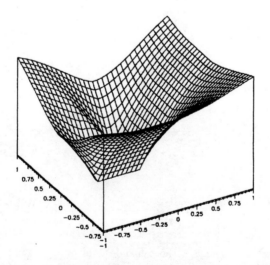

FIGURE 11: Minimum time function for $T = L$ and $c(x) \equiv 1$ ($h = 0.01$, $k = 0.25$).

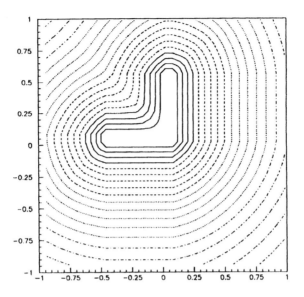

FIGURE 12: Level curves for $\mathcal{T} = \bar{B}(0,1)$ and $c(x) \equiv 1$.

Let us consider the same problem for $\mathcal{T} = \bar{B}(0,1)$ and for $c(x) = e^{x_1 x_2}$. Now the solution does not correspond to the distance function and becomes much more complicated (Figure 13). ◁

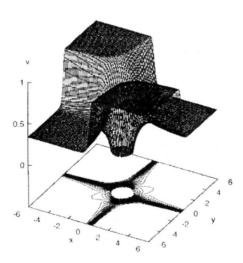

FIGURE 13: Value function and level curves for $\mathcal{T} = \bar{B}(0,1)$ and $c(x) \equiv e^{x_1 x_2}$.

3.2. Pursuit-evasion games

Another classical target problem is the so called "pursuit-evasion game". We are given a dynamical system controlled by two players

(3.4) $$\begin{cases} y'(t) = f(y, a(t), b(t)), & t > 0 \\ y(0) = x \end{cases}$$

where, $a(t) \in A$, $b(t) \in B$, A and B are given compact sets. We are also given a closed target set $\mathcal{T} \subset \mathbb{R}^N$, and we consider the generalized pursuit-evasion game where the first player "a" seeks to minimize the time $T(x)$ taken by the system to reach $\mathcal{T}$, while the second player "b" wants to maximize it. The function $T(x)$ represents the lower value function of this game in the sense of Varaiya, Roxin, Elliott and Kalton. As shown in Chapter VIII the Isaacs equation corresponding to the minimum capture time for the game is

$$\min_{b \in B} \max_{a \in A} \{-f(x, a, b) \cdot DT(x)\} = 1 \quad \text{in } \mathcal{R} \setminus \mathcal{T}$$

where $\mathcal{R}$ denotes the reachable set.

It was proved in Chapter VIII that the exponential transform of the minimum capture time v is a viscosity solution of the Isaacs equation

(3.5) $$v(x) + \min_{b \in B} \max_{a \in A} \{-f(x, a, b) \cdot Dv(x) - 1\} = 0 \quad \text{in } \mathbb{R}^N \setminus \mathcal{T} \equiv \mathcal{T}^c.$$

and that, if v is continuous, it is the unique bounded viscosity solution of (3.5) satisfying the natural Dirichlet boundary condition

$$v(x) = 0 \quad \text{for } x \in \partial \mathcal{T}.$$

By the standard discretization of the previous sections we obtain the following discrete Isaacs equation

(3.6) $$\begin{cases} v_h(x) = \beta \max_b \min_a v_h(x + hf(x, a, b)) + 1 - \beta & \text{in } \mathbb{R}^N \setminus \mathcal{T}, \\ v_h(x) = 0 & \text{on } \mathcal{T}, \end{cases}$$

where $\beta = e^{-h}$.

Again, to solve the problem numerically we have to restrict it in a bounded set Q. The choice of this set is arbitrary, the only request being that it should contain $\mathcal{T}$. For example, one can take a cube in $\mathbb{R}^N$ or the intersection of two sets of state constraints Q_1 and Q_2 (respectively, for the first and the second player). The boundary condition to be imposed on ∂Q will of course affect the solution (unless Q is invariant with respect to the controlled dynamics), and it can drastically change the result of the game in the interior of Q. Without entering into details on the advantages and disadvantages of other possibilities, let us take the same boundary condition we used in the minimum time problem. In the pursuit-evasion game the effect of this boundary condition is more tricky. Setting $v = 1$ outside

3. MINIMUM TIME PROBLEMS AND PURSUIT-EVASION GAMES

Q gives an advantage to the evader who can also win by leaving the domain of computation. However, it is sometimes better to know who will have an advantage in the "numerical game" since this will make it possible to interpret correctly the results. For example, with the above "boundary condition" we know that the numerical value function will be greater or equal to the real value function. This means that, if the evader is captured in our simulation, this will also be true in the real game; if not we cannot conclude anything on the real game. Let us consider a mesh over Q and denote by x_i its nodes, $i = 1, \ldots, L$. Let us introduce the following subsets of indices:

$$I_{\text{out}} \equiv \{i : \forall a \in A \ \exists \widehat{b} \in B \text{ such that } x_i + hf(x_i, a, \widehat{b}) \notin Q\}$$
$$I_T \equiv \{i : x_i \in T \cap Q\} \text{ and } I_{\text{in}} \equiv \{i : i \notin I_{\text{out}} \cup I_T\}.$$

Let us define the map $S : \mathbb{R}^L \to \mathbb{R}^L$ componentwise as follows:

$$S_i(V) \equiv \begin{cases} \beta \max_b \min_a P_i(a, b, V) + 1 - \beta & \text{if } i \in I_{\text{in}} \\ 1 & \text{if } i \in I_{\text{out}} \\ 0 & \text{if } i \in I_T \end{cases}$$

where $\beta \equiv e^{-h}$, $P_i(a, b, V) \equiv \sum_{j=1}^{L} \lambda_{ij}(a, b) V_j$ and

$$\sum_{j=1}^{L} \lambda_{ij}(a, b) x_j = x_i + hf(x_i, a, b), \qquad \lambda_{ij} \in [0, 1], \quad \sum_{j=1}^{L} \lambda_{ij} = 1.$$

Note that the definition of S_i for $i \in I_{\text{out}}$ should be interpreted as a discrete boundary condition. It is easy to prove the following

PROPOSITION 3.1.
(i) $S : [0, 1]^L \to [0, 1]^L$;
(ii) S *is monotone with respect to the partial order;*
(iii) S *is a contraction in* $[0, 1]^L$ *with respect to the norm* $\|V\|_\infty = \max_{i=1,\ldots,L} |V_i|$.

The unique fixed point of S is our approximate solution on the nodes of the grid belonging to Q. To have an approximate solution defined everywhere we can extend the triangulation over Q and define by linear interpolation an affine function on each simplex; that is, we define $w : Q \to [0, 1]$ as the unique function such that

(3.7)
$$\begin{cases} w(x) = \sum_j \lambda_j w(x_j) & \text{if } x = \sum_j \lambda_j x_j, \\ w(x_i) = \beta \max_b \min_a w(x_i + hf(x_i, a, b)) + 1 - \beta & \text{if } i \in I_{\text{in}} \\ w(x_i) = 1 & \text{if } i \in I_{\text{out}} \\ w(x_i) = 0 & \text{if } i \in I_T. \end{cases}$$

The condition $w(x_i) = 1$ off Q means that the first player loses the game if the system exits Q before reaching the target. This is a natural approximation for the case of the single player or for bounded targets, but it is not realistic for true pursuit-evasion games. It would be more realistic to consider a game with state constraints in Q in which each player loses the game if his state variables exit Q first. Under local controllability assumptions v is continuous and our approximation w will converge to v uniformly on any compact set of $\mathbb{R}^N$. The technique of the proof of that convergence result is based on weak limits (see Chapter V). As a final remark, we just observe that the convergence of the same scheme is also valid for games where v is discontinuous (i.e., in presence of barriers). In those cases, we expect the convergence to be uniform only in the regions where the solution is continuous.

3.3. Numerical tests

Let us analyze the numerical results related to two simple examples of 1-dimensional pursuit-evasion games for which the solution can be computed exactly. In both cases the set Q is invariant for the controlled dynamics so that the changes to the solution due to the artificial boundary condition do not appear. The first game has a continuous value function while the second has a slight modification in the dynamics which produces discontinuities (barriers).

TEST 1. We consider the following dynamics for the players

$$(3.8) \qquad \begin{aligned} x' &= a(x-1)(x+1)v_1 \\ y' &= b(y-1)(y+1)v_2 \end{aligned}$$

where x, y are the (1-dimensional) state variables respectively of the pursuer and the evader and v_1, v_2 are two positive real parameters representing their relative velocities. We choose $A = B \equiv [-1, 1]$.

We study this game in $Q \equiv [-1, 1]^2$ which is invariant with respect to the trajectories and we set $\mathcal{T} \equiv \{(x, y) : x = y\}$. Starting the game from (x_0, y_0), $x_0 < y_0$ the optimal strategies for the evader and the pursuer will be $b^* \equiv -1$ and $a^* \equiv -1$, since this corresponds for both to the motion toward 1 at the maximum speed.

We can solve explicitly (3.8) obtaining

$$x^*(t) = \frac{1 - k_1^2 e^{-2v_1 t}}{1 + k_1 e^{-2v_1 t}}, \qquad y^*(t) = \frac{1 - k_2^2 e^{-2v_2 t}}{1 + k_2 e^{-2v_2 t}},$$

where $k_1 \equiv (|x_0 - 1|/|x_0 + 1|)^{1/2}$ and $k_2 \equiv |y_0 - 1|/|y_0 + 1|^{1/2}$.

Setting $x^*(t) = y^*(t)$ we can compute the value function of the game. When $v_1 \leq v_2$ that equation implies $x_0 = y_0$; then the game has no solution and $T(x) = +\infty$ off the diagonal of Q except for $(x_0, y_0) \in \partial Q \cap (\{y = -1\} \cup \{y = 1\})$ where the game always has a solution.

For $v_1 > v_2$ the capture time is

$$t^* = \frac{1}{2(v_1 - v_2)} \ln (k_1/k_2)^2$$

and by the exponential transform we get the value

$$v(x_0, y_0) = 1 - \left(\frac{k_1}{k_2}\right)^{1/(v_2-v_1)}$$

valid on $\{(x,y) : x \leq y\} \cap Q$. This will also be the unique solution of the Isaacs equation in that subdomain.

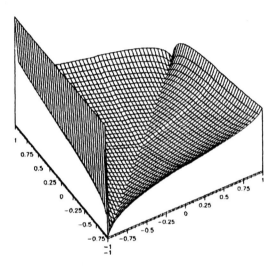

FIGURE 14: Value function.

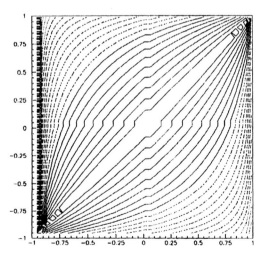

FIGURE 15: Level curves.

For $x_0 > y_0$, repeating the same argument we get

$$v(x_0, y_0) = 1 - \left(\frac{k_2}{k_1}\right)^{1/(v_2-v_1)}.$$

For $v_1 > v_2$, the exact solution is continuous and the approximation is accurate, in fact the maximum error on the nodes in $\mathrm{int}(Q)$ is 0.05962329. Figures 14 and 15 show the numerical results obtained by applying the algorithm on a grid of 1849 nodes ($k = 0.04$) and $h = 0.01$. ◁

TEST 2. Let the dynamics of the pursuer and the evader be given respectively by

$$\begin{cases} x' = ax(x-1)(x+1)v_1 \\ y' = by(y-1)(y+1)v_2 \end{cases}$$

where a, b, v_1, v_2 have the same meaning as in the previous test and $A = B \equiv [-1, 1]$. We study the game in the square $Q \equiv [-1, 1]^2$ (which is always invariant for the dynamics) and set $\mathcal{T} \equiv \{(x, y) : x = y\}$.

We divide the square into four regions by the lines $x = 0$ and $y = 0$. In $R_2 \equiv [-1, 0] \times [0, 1]$ and $R_4 = [0, 1] \times [-1, 0]$ the game has no solution and $T(x) \equiv +\infty$. In the remaining two squares R_1 and R_3, the game has a solution depending on the values of v_1 and v_2.

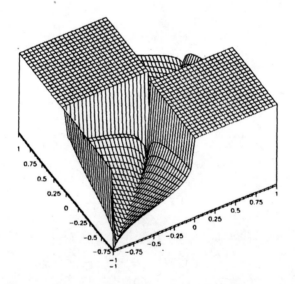

FIGURE 16: Value function.

As in Test 1 we can compute the exact solution, for example, for a starting point $(x_0, y_0) \in R_1$, $x_0 < y_0$, assuming $v_1 > v_2$. This solution is

$$v(x_0, y_0) = 1 - \left(\frac{k_1}{k_2}\right)^{1/(v_2-v_1)}$$

where
$$k_1 = \frac{\sqrt{|x_0^2 - 1|}}{x_0}, \qquad k_2 \equiv \frac{\sqrt{|y_0^2 - 1|}}{y_0}.$$

Figures 16 and 17 show the numerical results obtained by applying the algorithm on a grid of 1849 nodes ($k = 0.04$) and $h = 0.01$.

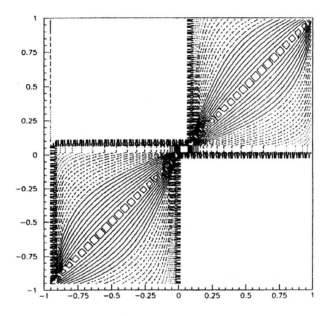

FIGURE 17: Level curves.

Note that the detection of R_2 and R_4 is quite accurate (they correspond to the region where the solution is flat since $v \equiv 1$ there). There is a little smoothing effect which rounds up the discontinuities of the solution, but in R_1 and R_3 the approximation is accurate when $v_1 > v_2$ and the solution is continuous. The maximum error on the nodes belonging to the interior of the regions $R_i, i = 1, \ldots, 4$, is 0.05754636. ◁

4. Some hints for the construction of the algorithms

The general algorithm to compute the value function is essentially based on a fixed point argument. Once we have an approximate value function we can use it to construct approximate feedbacks and approximate optimal trajectories at any point in the domain of computation. For example, dealing with the infinite horizon problem this can be simply obtained by taking

$$a^*(x) = \arg\min_{a \in A}\{(1 - \lambda h)v(x + hf(x, a)) + h\ell(x, a)\}$$

and in the pursuit-evasion game taking the couples $(a^*(x), b^*(x))$ which give the saddle value (maxmin) on the right-hand side of equation (3.6).

Note that the algorithm gives an approximate optimal control at each node of the grid without extra computations.

In this section we will give some hints and suggestions on different aspects connected to the construction of the algorithm. For a detailed list of references on the single topics we refer to the Bibliographical Notes.

Acceleration algorithm

The acceleration technique is crucial to get a fast solution with a reasonable accuracy. There are several ways to accelerate convergence. The first (described in Section 1) is based on the monotonicity properties of the operator T and requires use of the contraction mapping iteration only to recover a direction of displacement and reach the boundary of the set of subsolutions $\mathcal{U}$. Note that the boundary of $\mathcal{U}$ is where $U_i = (T(U))_i$ for at least one i; this implies that it can be reached either by an iterative root finding procedure (e.g., bisection) or by an explicit characterization of $\overline{\mu}$. Naturally, the second alternative is much more efficient (when feasible); the iterative procedure in fact requires several evaluations of the operator T to determine if we are still in or out of the set $\mathcal{U}$. Practical experience tells us that the algorithm works better when the accuracy in locating points on the boundary of $\mathcal{U}$ is high (although this will require more iterations in the root finding procedure).

A different acceleration technique which can be coupled with the previous one is based on the so called "iteration in policy space". We fix the controls at the nodes of the grid and compute a value function corresponding to them; then we fix the value and look for the corresponding optimal feedback controls on the grid. An intermediate choice in the algorithm is to compute the solution by the contraction mapping, keeping the approximate feedback controls fixed for a number of iterations (e.g., ten). This choice results in a considerable gain in CPU time.

Boundary conditions

We have seen how to treat Dirichlet boundary conditions and state constraints boundary conditions. A simple way to automatically enforce state constraints boundary conditions is to extend the solution outside the domain of computation Ω so that its value is greater than its maximum in Ω. Since the fixed point operator always looks for the minimum over the control set, this will imply that only the control satisfying the constraints will really play a role in the computation. This is exactly what we have done setting the solution equal to 1 outside Q in the minimum time problem (remember that in that problem $0 \leq v \leq 1$).

It is also possible to enforce Neumann type boundary conditions. Essentially this can be done by adding to the domain of computation a "frame" of width Δx so that one can compute a discrete normal (or directional) derivative.

Evolutive problems

In some problems (see §III.3), the value function is characterized in terms of an evolutive Hamilton-Jacobi equation and we have to deal with a Cauchy problem

$$\begin{cases} u_t + H(x,t,u,Du) = 0, \\ u(x,0) = u_0(x) . \end{cases}$$

The scheme corresponding to it is

$$\begin{cases} u^{n+1} + H^\Delta(x,t,u,Du) = 0, \\ u^0(x) = u_0(x), \end{cases}$$

where the discrete Hamiltonian H^Δ is computed using the space discretization discussed above. Note that the above mentioned evolutive problem can also be seen as a problem in $\mathbb{R}^{N+1}$ where the state is $z = (x,t)$, just adding to the dynamics one equation, namely $t' = 1$.

The finite horizon problem in $[0,T]$ leads to an evolutive Hamilton-Jacobi equation with a terminal condition. Problems of this kind can be solved backward or can be transformed to a Cauchy problem by a change of variable.

Local grid refinements

When we know (for example on the basis of an analysis of the characteristics) where are the switching curves for the optimal feedback controls, we can use an adapted grid to have a greater accuracy in the computations around those curves. A more sophisticated (and efficient) method consists of modifying the grid as far as the algorithm computes the solution on the basis of a posteriori local error estimators. This method can be particularly efficient, especially when the solution is not differentiable, and can produce accurate results also for meshes with a small number of nodes.

High-order methods

Although the results which have been presented here through some model problems are based on the simplest approximation schemes for the dynamics and the cost functional (we coupled the Euler method and the rectangle rule), one can improve the global efficiency of the algorithm by coupling different methods. In particular, a careful coupling of different higher order methods can result in very accurate approximations also on course grids. The coupling must satisfy some compatibility conditions and is effective as far as the methods used for both the dynamics and the costs produce local errors of the same order. For example, a good balance is obtained coupling the Heun method with the trapezoid rule. These methods are particularly effective in regions where the value function has some smoothness properties.

Synthesis of feedback controls and optimal trajectories

The main goal of our computations is to recover approximate controls and optimal trajectories. This task is particularly simple once we know an approximate solution w of the Hamilton-Jacobi equation. Knowing w we can compute the feedback corresponding to point $x \in Q$ by simply looking at the set of controls $A^*(x)$ giving the minimum (or the maxmin) in the fixed point operator. If that set is not a singleton we can introduce an additional criterion to make a selection. For example, when A is convex we can choose among them the control with minimum norm or the control which minimizes a strictly convex function. The unique feedback control can be used now to obtain an approximate optimal trajectory simply by applying a one step scheme to the dynamics, i.e.,

$$(4.1) \qquad \begin{cases} y_{n+1} = y_n + h\Phi(h, y_n, a^*(y_n)) \\ y_0 = x \end{cases}$$

where Φ is the numerical reconstruction of the vector field (i.e., the Henrici function of a one-step method). Moreover, along the trajectories we can introduce also a "dynamic selection criterion" which consists of keeping fixed the control as far as the control which was optimal in the last step is still optimal at the new point. This "inertia" has an important stabilization effect on the approximate optimal trajectories.

Domain decomposition and parallel algorithms

We can considerably enlarge the number of problems which can be solved by the Dynamic Programming approach using a domain decomposition strategy. As we mentioned in the introduction, Dynamic Programming has a major drawback in the "rise of dimension". In many real problems the number of variables is easily greater than three, and solving a partial differential equation in such a large dimension is a very complicated task, easily beyond the possibilities of the last generation computers. However, splitting a problem given over a domain Ω into a sequence of problems set in subdomains of manageable size Ω_r, $r = 1, \ldots, d$, is the basic idea of domain decomposition techniques and can be really helpful. Roughly speaking, instead of one problem in dimension N we will compute the solutions of d problems each of which has dimension N_r, $r = 1, \ldots, d$, where $N \approx N_1 + \cdots + N_d$. Of course, when splitting the problem, one has to decide which boundary conditions have to be enforced on the internal interfaces. As has been proved recently, there are transmission conditions which guarantee the continuity of the value function.

5. Bibliographical notes

As we mentioned, other approximation schemes can be adopted. Some of them go back to the Bellman school as [La67] or have appeared before the development of the theory of viscosity solutions. The interested reader should look at the book by Kushner [Kus77] (for stochastic control problems) or at the paper by Gonzalez-Rofman [GR85]. We should also mention that there is a large amount of

5. BIBLIOGRAPHICAL NOTES

literature dealing with numerical methods for optimal control problems based on the Pontryagin principle; we refer for that approach to [BH87, Pes94a] and references therein. More recently, the approach to control problems based on viability theory has also produced some algorithmic results as in [SP94] and [QSP95].

After the foundation of the theory of viscosity solution, the approximation of Hamilton-Jacobi equations was studied by Capuzzo-Dolcetta [CD83], Crandall and Lions [CL84], Souganidis [Sou85a], Capuzzo-Dolcetta and Ishii [CDI84], Falcone and co-authors [Fa87, Fa94, FGL94, BF90a, BF90b, FG94] (see also the survey paper [CDF89]), and Rouy [Rou92]. Theorem 1.3 and 1.4 are in [Fa87]. The assumption $\lambda > L_f$ may be restrictive in some cases and can be dropped by refining the proofs. The convergence of the scheme can be proved by the method of weak limits in the viscosity sense as in Chapter VI, §1.1, see [BF90a, BS91b] and [BFS94] while an estimate of the convergence rate has been obtained by the maximum principle argument of Chapter VI, §1.2, see [Sor97b].

High-order approximation schemes have been proposed recently either by adapting schemes used in the study of hyperbolic problems as in [OS91] or by applying Dynamic Programming arguments as in [FaF90, FaF94, FaF96]. More recently, a convergence result for some high-order schemes ("filtered" to preserve an upper bound on weak second order finite differences) has been obtained in [LS95] for Hamilton-Jacobi equations with convex Hamiltonians. Other approximation schemes are based on the discretization of the representation formulae for viscosity solutions of Hamilton-Jacobi equations. They use a fast Legendre-Fenchel transform and their application has been studied mainly for evolutive equations with convex Hamiltonians in [Bren89] and [Cor96]. A result on the convergence of approximation schemes in an abstract setting is contained in Barles and Souganidis [BaSo91].

The acceleration technique was introduced in [Fa87]. Another acceleration technique has been studied in [GS90]. The acceleration in policy space goes back to Howard [Ho60].

A dynamic grid refinement was introduced and implemented by Grüne [Gru97].

The convergence analysis and the a priori estimates for the schemes related to constrained problems are contained in [CFa96]. For a similar analysis related to the minimum time problem we refer to [BF90b] (see also [BSt93], [BBF95] and [BFS97] for a convergence result related to the discontinuous case). Further extensions to evolutive problems and impulsive control problems have been developed in [FG94, CFa95b] and [Ti91].

The numerical solution of the Isaacs equation was studied by Alziary [Al91] for the classical problem of a lion pursuing a slower antelope in a bounded and closed domain. Gonzalez and Tidball [GTi90, Ti91, TiG94, Ti95] worked on deterministic games with stopping times (see also the references therein). The convergence results for the generalized pursuit-evasion game are contained in [BFS94] (for the continuous case) and in [BBF95] (covering the discontinuous case). Let us also mention the paper [Sou85b] dealing with the convergence of approximation schemes for finite horizon games. The domain decomposition strategy with overlapping regions between the subdomains was studied in [FLS94]. An extension to the case without overlapping was considered in [CFLS94]. Sun [Sun93] studied similar techniques for stochastic control problems.

Finally let us mention optimal control problems for diffusion processes and the corresponding second order Hamilton-Jacobi-Bellman equations. Approximation schemes for those problems were studied in [Kus77, KD92, LMr80, Q80, FS93] and [CFa95a]. Specific applications of multilevel techniques for second order Hamilton-Jacobi equations were studied in [Hop86] and [Ak90]. More recently, in [CL96] new schemes for evolutive second order Hamilton-Jacobi equations of degenerate type (with applications to the motion by mean curvature) have been proposed.

Acknowledgments

I wish to thank A. Seghini, T. Giorgi, P. Lanucara, F. Camilli and all the students that have contributed to the development of the codes and to discussions of the theoretical and the algorithmic issues.

APPENDIX B

Nonlinear $\mathcal{H}_\infty$ control

by Pierpaolo Soravia

This appendix deals with a recent subject of systems theory which has been growing in popularity in recent years. We will present some results concerning the problem of *robust attenuation of disturbances* in controlled systems and the so called nonlinear $\mathcal{H}_\infty$ *control* problem. Our goal is to show how the theory of viscosity solutions can be applied to this problem and contributes to its understanding. The idea of the problem is to introduce a quantitative way to measure how disturbances affect a stable or asymptotically stable equilibrium of a controlled dynamical system. The point of view is entirely deterministic as opposed to the probabilistic one of stochastic control; the so called *worst case approach*, though historically important, is less developed and acknowledged than the probabilistic one, but has recently found a number of important applications in the disciplines that make use of control theory. Despite of the different nature, there is an interesting link between these two methods which is known as risk-sensitive control. This developing chapter of control theory goes however beyond the scope of this appendix. We set ourselves in the general framework of nonlinear systems, where the theory of viscosity solutions plays a crucial role, and refer to the literature for a more detailed discussion of specific results that can be proved for special subclasses of systems. Hence, we do not presume to give a complete treatment of the up-to-date theory of the subject here, but we intend to point out to what extent the theory of viscosity solutions is needed and helpful to approach the problem in great generality. We are concerned only with the state space formulation, while the original setting for linear systems was the frequency domain. We will not discuss the partial information case in great detail, as a fully satisfactory framework for this important side of the problem, for general nonlinear systems, is not yet developed.

In the following we indicate the system in the form

$$(0.1) \quad \begin{cases} \dot{y} = f(y, a, b), & y(0) = x \in \mathbb{R}^N, \\ z = \ell(y, a, b) \,. \end{cases}$$

Here as usual $y(\,\cdot\,) = y_x(\,\cdot\,;a,b)$ denotes the state, $a(\,\cdot\,)$ the control, $b(\,\cdot\,)$ the deterministic unknown disturbance on the system, or *input*, and $z(\,\cdot\,)$ the cost function. Later in the chapter we will add to the system (0.1) the equation

(0.2) $$w = g(y,a,b),$$

where $w(\,\cdot\,)$ is called the *observed output*.

Very special cases of the above are the following: linear systems

(0.3) $$\begin{cases} f(x,a,b) = Cx + Pa + Qb, \\ \ell(x,a,b) = |\bar{\ell}(x,a)|^p, \quad \bar{\ell}(x,a) = Dx + Ea; \end{cases}$$

and nonlinear-affine systems

(0.4) $$\begin{cases} f(x,a,b) = f_1(x) + f_2(x)a + f_3(x)b, \\ \ell(x,a,b) = |\bar{\ell}(x,a)|^p, \quad \bar{\ell}(x,a) = \ell_1(x) + \ell_2(x)a, \end{cases}$$

where C, P, Q, D, E are matrices and $f_1, f_2, f_3, \ell_1, \ell_2$ are suitably defined matrix valued functions.

In both examples the function $\bar{z}(\,\cdot\,) = \bar{\ell}(y(\,\cdot\,), a(\,\cdot\,))$ corresponding to a trajectory $y(\,\cdot\,;a,b)$ of (0.1) is called the *output* of the system. The real number $p \in [1, +\infty[$ that appears in (0.3), (0.4) or in the next assumptions, will identify the class of admissible controls and disturbances. Traditionally $p = 2$, but this choice plays no role in our discussion.

1. Definitions

We now list the basic assumptions we need. In the following we are given the sets A, B, where $A \subset \mathbb{R}^K$ and $0 \in B \subset \mathbb{R}^M$ are closed, the maps $f : \mathbb{R}^N \times A \times B \to \mathbb{R}^N$, $g : \mathbb{R}^N \times A \times B \to \mathbb{R}^P$, $\ell : \mathbb{R}^N \times A \times B \to \mathbb{R}$ are continuous. We assume that the following is satisfied: there are p, q, $0 < q < p$, such that

(1.1) $$\begin{cases} |f(x,a,b) - f(y,a,b)| \leq L(1 + |a|^q + |b|^q)|x - y|, \\ |f(x,a,b)| \leq L(1 + |x| + |a|^q + |b|^q) \quad \text{for all } x,y,a,b, \\ |\ell(x,a,b) - \ell(y,a,b)| \leq L_R(1 + |a|^p + |b|^p)|x - y|, \\ C_R^0 |a|^p - L_R(1 + |a|^q + |b|^q) \leq \ell(x,a,b) \leq L_R(1 + |a|^p + |b|^q), \\ \text{for some } C_R^0 > 0, \text{ for all } |x|, |y| \leq R, \, a \in A, \, b \in B, \, R > 0. \end{cases}$$

Moreover the following sign condition on the cost is assumed

(1.2) $$\ell(x,a,0) \geq 0 \,.$$

The growth conditions we impose on the system lead naturally to consider the following sets of admissible controls and disturbances

$$\begin{cases} \mathcal{A} = L^p_{\text{loc}}(\mathbb{R}_+, A), \\ \mathcal{B} = L^p_{\text{loc}}(\mathbb{R}_+, B) \,. \end{cases}$$

1. DEFINITIONS

In the following by *undisturbed system* we mean the controlled system

(1.3) $$\dot{y} = f(x, a, 0), \qquad y(0) = x,$$

for $a \in \mathcal{A}$. One of the conditions we want on the undisturbed system (1.3) is that it is *open loop Lyapunov stable* at the origin in the following sense: for any neighborhood $\mathcal{U}$ of the origin, there is an open set Ω containing the origin such that for all $x \in \Omega$ there is $a_x \in \mathcal{A}$ satisfying

$$y_x(t; a_x, 0) \in \mathcal{U}, \qquad \text{for all } t \geq 0 \,.$$

As a matter of fact, we will not require the stability of the system in advance but prefer to prove it as a consequence of other assumptions. We will test such stability with the cost function ℓ above, see Definition 1.1 and Remark 1.2. Without other assumptions on the cost ℓ than (1.2), open loop stability is the natural request for the undisturbed system. If instead we want the undisturbed system to be open loop asymptotically stable, then it would be more appropriate to suppose that also the following condition holds true,

(1.4) there are $\varrho, \delta > 0$ such that if $|x| \leq \delta$, $a \in \mathcal{A}$ satisfy
$$|y_x(t, a, 0)| \leq \varrho \text{ for } t \geq 0 \text{ and } \lim_{t \to +\infty} \text{dist}(y_x(t; a, 0), \mathcal{Z}) = 0$$
then $\lim_{t \to +\infty} y_x(t; a, 0) = 0 \,.$

In (1.4) we indicate with $\mathcal{Z} = \{x \in \mathbb{R}^N : \inf_{a \in \mathcal{A}} \ell(x, a, 0) = 0\}$ the degenerate set for the cost. Here and in the following, we implicitly assume that $0 \in \mathcal{Z}$. The condition (1.4) is a modification of the property of controlled systems known as *zero state detectability*. Note that (1.4) holds, in particular, if 0 is isolated in $\mathcal{Z}$.

DEFINITION 1.1. Given $\gamma > 0$, *we solve* the $\mathcal{H}_\infty$ suboptimal control problem with disturbance attenuation level γ if there are a nonnegative function $U : \mathbb{R}^N \to \mathbb{R}$, null at the origin and a family of strategies $\{\alpha_x\}_{x \in \mathbb{R}^N} \subset \Gamma$ (i.e., nonanticipating or causal functionals $\alpha : \mathcal{B} \to \mathcal{A}$ in the sense of Chapter VIII) such that the two following conditions are satisfied:

(i) the controls $\alpha_x[0]$ provide the open loop Lyapunov stability of the undisturbed system,

(ii)

(1.5) $$\int_0^t \ell(y_x, \alpha_x[b], b) \, ds \leq \gamma^p \int_0^t |b|^p \, ds + U(x), \quad \text{for all } x \in \mathbb{R}^N, t \geq 0, b \in \mathcal{B} \,.$$

We say that the problem is *solved strongly* if the following conditions are satisfied. There is a feedback map $a : \mathbb{R}^N \to A$ such that for all $b \in \mathcal{B}$ and $x \in \mathbb{R}^N$ there is a unique absolutely continuous solution of the differential equation

$$\dot{y} = f(y, a(y), b), \quad y(0) = x,$$

this closed loop dynamical system is Lyapunov asymptotically stable at the origin when $b \equiv 0$, and the definition $\alpha_x[b](t) := a(y(t))$, gives the family of strategies solving the problem. ◁

REMARK 1.2. As usual in differential games, in full generality, one has to give some advantage to one of the players in order to have a tractable problem by means of the Dynamic Programming approach. Therefore in the definition above we used strategies for the controller and controls for the disturbance. The problem as defined above, is usually referred to in the literature as the full information problem. When the Isaacs condition on the data holds, then one would prefer to restrict the class of admissible strategies to feedback strategies, or causal functionals of the state. This is referred to as the full state information problem.

When only property (ii) of Definition 1.1 holds, we say that the system has finite L^p *gain*. The terminology is due to the fact that when the cost function is expressed in terms of an output $\bar{z}$, as in the case of linear (0.3) and nonlinear-affine (0.4) systems, then at $x = 0$ the inequality (1.5) can be rewritten as

$$\frac{\|\Gamma_\alpha b\|_{L^p(0,t)}}{\|b\|_{L^p(0,t)}} \leq \gamma, \qquad \text{for all } t \geq 0, \ b \in \mathcal{B},$$

where for the given strategy $\alpha = \alpha_0$ the map $\Gamma_\alpha b = \bar{z}$ is the so called *input-output map*. Observe that if the system is linear and α is given by a linear feedback, then also Γ_α is linear and so γ is an upper estimate for the norm of such functional. Actually in this case the real number $\|\Gamma_\alpha\|$ is the best value of γ one can use in (1.5) for the given strategy α. In this case it is well known that the norm of Γ_α coincides with the $\mathcal{H}_\infty$ norm of the *transfer function* in the Hardy space of analytic functions bounded in the right halfplane.

As stability is a local property, we can define in a suitable way the local $\mathcal{H}_\infty$ suboptimal control problem. We do not address here, however, this part of the theory.

If we solve the $\mathcal{H}_\infty$ problem in the sense of Definition 1.1, then by the stability of the undisturbed system at $x = 0$, it is easy to check that $y_0(\cdot\,;\alpha_0,0) \equiv 0$, therefore $f(0,\alpha_0[0](\cdot),0) = 0$ a.e. Moreover, by (1.5) and (1.2) we also have $\ell(0,\alpha_0[0](\cdot),0) = 0$ a.e. Therefore there is at least one $a \in A$ such that simultaneously $f(0,a,0) = 0$ and $\ell(0,a,0) = 0$. Thus the origin must be an equilibrium point for the controlled system (0.1) when no disturbances are present. This is a necessary condition on our system in order to solve the problem. ◁

We now describe the connection of the problem as formulated above and differential games. It is clear that (1.5) is equivalent to stating that

$$\sup_{b \in \mathcal{B}} \sup_{t \in \mathbb{R}_+} \int_0^t (\ell(y_x, \alpha_x[b], b) - \gamma^p |b|^p)\, ds \leq U(x) \qquad \text{for all } x \in \mathbb{R}^N.$$

We are therefore led to introduce the following (lower) value function for the differential game (0.1)

(1.6) $$V_\gamma(x) = \inf_{\alpha \in \Gamma} \sup_{b \in \mathcal{B}} \sup_{t \in \mathbb{R}_+} \int_0^t (\ell(y_x, \alpha_x, b) - \gamma^p |b|^p)\, ds\,.$$

This value function gives by construction the best value $U(x)$ to be used in (1.5). We skip for the moment the problems related to the existence of the value of the game and limit ourselves to the lower value approach. We will come back to this issue later. Observe that by choosing $0 = t \in \mathbb{R}_+$ in (1.6), we immediately check that $V_\gamma \geq 0$, but this function may attain the value $+\infty$ at some points. The previous discussion leads to the following definition.

DEFINITION 1.3. Given $\gamma > 0$, the $\mathcal{H}_\infty$ suboptimal control problem with level γ is *solvable* if the undisturbed system is open loop Lyapunov stable and the value function V_γ is finite in $\mathbb{R}^N$, and null at the origin. ◁

Clearly Definition 1.1 is more restrictive than Definition 1.3. We will show however in the following that the solution of the problem can be pursued by investigating the properties of the value function V_γ and that the two definitions are in fact equivalent in some cases. As we will see in the examples, the solvability of the problem will depend on the choice of the index γ. As a matter of fact, the main goal of the theory is to compute the real number

$$\gamma^* = \inf\{\gamma > 0 : \text{the problem with index } \gamma \text{ can be solved}\},$$

since it is a measure of the robustness of the system and, on the opposite hand, of the influence that disturbances have on its behavior. Unfortunately, there are no general effective methods to our knowledge, to characterize the value γ^*, and we will only characterize here the solvability of the problem for a fixed index γ.

EXAMPLE 1.4. Consider the following one dimensional system

$$\begin{cases} \dot{y} = a - b, & y(0) = x \in \mathbb{R}, \\ z = y^2 + a^2 \,. \end{cases}$$

In this case $A = B = \mathbb{R}$ and the system is clearly linear according to (0.3), by choosing $D^T = (1\ 0)$, $E^T = (0\ 1)$ (T denotes transposition). We set $p = 2$ and prove that for all $\gamma > 1$ we can solve the $\mathcal{H}_\infty$ suboptimal control problem. Since we have not developed any better method yet, we guess and use the following strategy,

(1.7) $$\alpha[b](t) = -y(t) + b(t)\chi_{\{b:y(t)b\leq 0\}}(b(t)),$$

where $\chi_S(\,\cdot\,)$ denotes the characteristic function of the set S. The second term in the right-hand side of (1.7) cancels bad disturbances that push the system away from the origin, while the first is a stabilizing term. It is easy to check that, using the strategy defined in (1.7), the trajectories of the system satisfy

(1.8) $$|y(t)| \leq |x|\exp(-t),$$

for $t \geq 0$. It is also easy to verify that the function $\phi(b) = (\chi_{\{yb\leq 0\}}(b) - \gamma^2)b^2 - 2by\chi_{\{by\leq 0\}}(b)$ achieves its maximum value at the point $b = y/(1-\gamma^2)$. Therefore we can estimate the value function as follows

$$V_\gamma(x) \leq \sup_{b\in\mathcal{B}} \sup_{t\in\mathbb{R}_+} \int_0^t (2y^2 + (\chi_{\{yb\leq 0\}} - \gamma^2)b^2 - 2by\chi_{\{by\leq 0\}})\,ds$$

$$\le \sup_{t \in \mathbb{R}_+} \int_0^t \left(2y^2 + \frac{y^2}{1-\gamma^2} - \frac{2y^2}{1-\gamma^2}\right) ds$$

$$\le \sup_{t \in \mathbb{R}_+} \int_0^t x^2 \exp(-2t) \frac{2\gamma^2 - 1}{\gamma^2 - 1} ds \le x^2 \frac{2\gamma^2 - 1}{2(\gamma^2 - 1)},$$

from which we get the conclusion. Note that since $V_\gamma \ge 0$, the previous computation verifies that the strategy $\alpha[b] = b$ is optimal at the origin. The same strategy is also optimal at the origin when $\gamma = 1$ by a direct computation, therefore $V_1(0) = 0$. On the other hand the estimate above does not work for $\gamma \le 1$. We will come back later to this example. ◁

2. Linear systems

For linear systems (0.3) the definition we gave in the previous section can be simplified because, as often in the linear case, information at the origin propagates to the whole space. In this section we assume that

(2.1) there is a linear feedback $a(x) = Fx$ such that the undisturbed, closed loop system
$\dot{y} = (C + PF)y$, $y(0) = x$,
is asymptotically stable at the origin

and,

(2.2) for all $b \in \mathcal{B}$, the solution of the differential equation
$\dot{y} = (C + PF)y + Qb$, $y(0) = 0$,
satisfies the L^p gain condition
$$\int_0^\infty |(D + EF)y|^p \, dt \le \gamma^p \int_0^\infty |b|^p \, dt, \text{ for all } b \in L^p(\mathbb{R}_+, B).$$

The two previous conditions will be recalled as (2.2). We will see in the following, more precisely in Remark 6.3, that the $\mathcal{H}_\infty$ suboptimal problem for linear systems can be solved in the strong sense under certain conditions, and feedbacks solving the problem are linear. Therefore the assumption (2.2) is not restrictive in the linear case.

PROPOSITION 2.1. *Assume (2.1) and (2.2). Then the $\mathcal{H}_\infty$ suboptimal control problem is solved strongly by the feedback map $a(x) = Fx$.*

PROOF. We will give the proof only under the following simplifying assumption: the system

(2.3) $$\begin{cases} \dot{y} = Ry + Qb, & y(0) = x, \\ z = |Sy|^p, \end{cases}$$

where we indicated by $R = C + AF$, $S = D + EF$, is controllable to the origin namely for all $x \in \mathbb{R}^N$, there is $b_x \in \mathcal{B}$ such that the solution $y_0(\,\cdot\,; b_x)$ of (2.3) with null initial condition satisfies $y_0(t_x; b_x) = x$ for some $t_x \ge 0$.

Let $x \in \mathbb{R}^N$ be fixed. By the simplifying assumption there are a control $\bar{b} \in \mathcal{B}$ and $T \geq 0$ such that $y_0(T; \bar{b}) = x$. Now let $b \in \mathcal{B}$; we define the admissible control

$$b_1(t) = \begin{cases} \bar{b}(t), & 0 \leq t \leq T, \\ b(t-T), & t > T. \end{cases}$$

Observe that $y_0(t; b_1) = y_x(t-T; b)$ if $t \geq T$. By (2.2) we then get for $t \geq T$

$$0 \geq \int_0^t (|Sy_0(s; b_1)|^p - \gamma^p |b_1(s)|^p) \, ds$$
$$= \int_0^T (|Sy_0(s; \bar{b})|^p - \gamma^p |\bar{b}(s)|^p) \, ds + \int_0^{t-T} (|Sy_x(s; b)|^p - \gamma^p |b(s)|^p) \, ds,$$

therefore the L^p gain condition (1.5) at x is satisfied with the choice of $U(x) = (-\int_0^T (|Sy_0(s; \bar{b})|^p - \gamma^p |\bar{b}(s)|^p) \, ds) \vee 0$. Observe that T and $\bar{b}$ depend only on x. ◂

REMARK 2.2. Unfortunately, at least to our knowledge, a complete proof of the maybe surprising Proposition 2.1 is not trivial. It also requires some results that we will state in the following, namely Theorem 4.5 and the classical statement of linear state space $\mathcal{H}_\infty$ control theory stating that (2.2) is equivalent to the existence of a positive definite solution of the Riccati equation associated with the problem. To give the flavor of the complete result, we can also make some direct computations based on the linearity of (2.3). In particular for all $b \in \mathcal{B}$ the solutions of (2.3) satisfy

$$y_x(t; b) = \exp(Rt)x + y_0(t; b).$$

Therefore

$$|Sy_x(t)|^p = |S\exp(Rt)x + Sy_0(t)|^p \leq 2^{p-1}(|S\exp(Rt)x|^p + |Sy_0(t)|^p).$$

If we now fix any $t \geq 0$, indicate by $U(x) = 2^{p-1}|S|^p \int_0^\infty |\exp(Rt)x|^p \, dt$, for any $b \in \mathcal{B}$ denote $\bar{b} = b\chi_{[0,t]}$, and use (2.2), we get

$$\int_0^t |Sy_x|^p \, ds \leq U(x) + 2^{p-1} \int_0^t |Sy_0|^p \, ds$$
$$\leq U(x) + 2^{p-1}\gamma^p \int_0^\infty |\bar{b}|^p \, ds = (2^{1-1/p}\gamma)^p \int_0^t |b|^p \, ds + U(x),$$

which ensures that the problem is solved according to Definition 1.1, though with the larger attenuation level index $2^{1-1/p}\gamma$. ◁

The simplified version (2.2) of the definition (with $p = 2$) is the usual definition one can find in the literature. We will verify later that the result of Proposition 2.1 is false in general if (2.2) is not guaranteed by a linear, asymptotically stabilizing feedback control.

EXAMPLE 2.3. We now apply Proposition 2.1 to the system of Example 1.4 and solve the corresponding $\mathcal{H}_\infty$ problem strongly. A linear, asymptotically stabilizing feedback control for the undisturbed system has necessarily to be of the form $a(x) = -cx$, with $c > 0$. In this case, the trajectory solution with zero initial state is given by the formula

$$y_0(t; a, b) = -\int_0^t \exp(-c(t-s))b(s)\,ds.$$

Therefore by the estimates on convolution, we have that, for all $b \in L^2(\mathbb{R}_+, B)$,

$$\int_0^\infty (y^2 + a^2)\,dt = (1 + c^2)\int_0^\infty y^2\,dt$$

$$\leq (1 + c^2)\left(\int_0^\infty |b|^2\,dt\right)\left(\int_0^\infty \exp(-ct)\,dt\right)^2 = \frac{1+c^2}{c^2}\int_0^\infty |b|^2\,dt.$$

From Proposition 2.1 it follows that for any $\gamma > 1$, choosing the constant c satisfying $c \geq 1/(\gamma^2 - 1)^{1/2}$, the feedback above will solve the problem. Actually we find infinitely many feedbacks solving the problem, which is due to the fact that we are not requiring their optimality when the system starts at a point different from the origin. We will see in Example 6.4 that, when doing this, the selection will become much more stringent. Note that the strategy at the point $x \in \mathbb{R}^N$ corresponding to a choice of c is given by

$$\alpha_x^c[b](t) = a(y(t)) = c\left(\int_0^t \exp(-c(t-s))b(s)\,ds - \exp(-ct)x\right),$$

which, in particular, is optimal at the origin and we can compare with the guess (1.7).

Observe also that even if $V_1(0) = 0$, the computation above does not guarantee the assumptions of Proposition 2.1 and therefore we cannot deduce in general the stronger estimate (1.5) when $\gamma = 1$. ◁

3. $\mathcal{H}_\infty$ control and differential games

In this section we motivate the introduction of the value function V_γ and show that the existence of a solution of the $\mathcal{H}_\infty$ problem can be established by studying V_γ. The next results also show some simple conditions on V_γ under which Definitions 1.1 and 1.3 are equivalent and how stability of the undisturbed system can be proved by the properties of the value function V_γ. Here and in the following we indicate, as in Chapter V, by $(V_\gamma)_*(x) = \lim_{r \to 0} \inf_{|y-x| \leq r} V_\gamma(y)$, the lower semicontinuous envelope of the value function V_γ.

LEMMA 3.1. *Assume (1.1) and (1.2). Suppose that the system has finite L^p gain and the value function is continuous at the origin. Assume that there is $\sigma > 0$ such that for all $x \in B(0, \sigma) \smallsetminus \{0\}$ we have $V_{\gamma*}(x) > 0$, and V_γ has an optimal strategy at the origin. Then we can find a family of strategies $\{\alpha_x\}_{x \in \mathbb{R}^N}$ and a* function U continuous at the origin, solving the $\mathcal{H}_\infty$ suboptimal control problem. If in addition (1.4) holds, then the controls $\{\alpha_x[0]\}_{x \in \mathbb{R}^N}$ provide open loop local asymptotic stability of the system.

PROOF. 1. In the assumptions of the statement, we consider a basis of open and bounded neighborhoods of the origin $\mathcal{F} = \{\mathcal{U}_n\}_{n \in \mathbb{N}}$ such that $\mathcal{U}_{n+1} \subset\subset \mathcal{U}_n \subset B(0, \sigma)$, for all $n \in \mathbb{N}$. Given $m_0 = 1$, we define $\varepsilon_1 = \inf\{V_\gamma(x) : x \in \partial \mathcal{U}_{m_0}\}$. Note that $\varepsilon_1 > 0$. This follows from the assumption $V_{\gamma*} > 0$ on $B(0, \sigma) \smallsetminus \{0\}$, the fact that $0 \notin \partial \mathcal{U}_{m_0}$ and $V_{\gamma*}$ is lower semicontinuous. Then set $\Omega_{\varepsilon_1} = \{x \in \mathcal{U}_{m_0} : V_\gamma(x) < \varepsilon_1\}$ which is a neighborhood of the origin by the continuity assumption. Next given Ω_{ε_1}, we set $m_1 = \min\{k : \mathcal{U}_k \subset \Omega_{\varepsilon_1}\} \vee (m_0 + 1)$.

In the general case, given $m_n \in \mathbb{N}$, we proceed and define ε_{n+1}, $\Omega_{\varepsilon_{n+1}}$ and m_{n+1} as above with $\mathcal{U}_{m_n}$ instead of $\mathcal{U}_{m_0}$. The family $\mathcal{F} = \{\Omega_{\varepsilon_{n+1}}\}_{n \in \mathbb{N}}$ is a nested basis of neighborhoods of the origin. Now we construct the strategies and the function solving the suboptimal problem. For $x \in \mathbb{R}^N \smallsetminus \Omega_{\varepsilon_1}$, $\varepsilon > 0$, let α_x be such that

$$\sup_{b \in \mathcal{B}} \sup_{t \in \mathbb{R}_+} \int_0^t (\ell(y, \alpha_x, b) - \gamma^2 |b|^2)\, ds \leq V_\gamma(x) + \varepsilon,$$

and define U in $\mathbb{R}^N \smallsetminus \Omega_{\varepsilon_1}$ by setting $U = V_\gamma + \varepsilon$. For $x \in \Omega_{\varepsilon_n} \smallsetminus \Omega_{\varepsilon_{n+1}}$, we choose α_x such that

(3.1) $$\sup_{b \in \mathcal{B}} \sup_{t \in \mathbb{R}_+} \int_0^t (\ell(y, \alpha_x, b) - \gamma^2 |b|^2)\, ds < \varepsilon_n,$$

and set $U(x) = \varepsilon_n$. Finally at $x = 0$ we choose α_0 as an optimal strategy for $V_\gamma(0)$ and set $U(0) = 0$. Clearly by construction $\varepsilon_n \to 0$ as $n \to +\infty$, therefore U is continuous at the origin.

2. We have left to prove that the family of strategies $\{\alpha_x\}_{x \in \mathbb{R}^N}$ provides open loop stability, or open loop asymptotic stability of the undisturbed system if (1.4) holds. We start with the stability of the system. Given a neighborhood $\mathcal{V}$ of the origin, we choose m_n such that $\mathcal{U}_{m_n} \subset \mathcal{V}$, then for any $x \in \Omega_{\varepsilon_{n+1}} \subset \mathcal{U}_{m_n}$, from (3.1) and the definition of $V_\gamma(y_x(t))$, it follows that the chosen α_x also satisfies

$$\int_0^t (\ell(y_x, \alpha_x, b) - \gamma^2 |b|^2)\, ds + V_\gamma(y_x(t)) < \varepsilon_{n+1}, \quad \text{for all } t \geq 0,\, b \in \mathcal{B}.$$

Therefore choosing $b \equiv 0$ and using the assumption $\ell(z, a, 0) \geq 0$, we get

$$V_\gamma(y_x(t; \alpha_x[0], 0)) < \varepsilon_{n+1} \quad \text{for all } t \geq 0.$$

We conclude that there is no $\bar{t}$ such that $y_x(\bar{t}; \alpha_x[0], 0) \in \partial \mathcal{U}_{m_n}$, as required.

3. To prove the local asymptotic stability, we only need to show that for any x in a suitably small neighborhood $\mathcal{V}$ of the origin we have

$$\lim_{t \to +\infty} \operatorname{dist}(y_x(t; \alpha_x, 0), \mathcal{Z}) = 0, \quad \text{and} \quad |y_x(t, \alpha_x, 0)| \leq \varrho,\quad t \geq 0.$$

We then reach the conclusion by (1.4) and the stability.

Let $x \in \Omega_{\varepsilon_1}$. By definition of α_x and (1.2) we have that

$$\int_0^{+\infty} \ell(y_x, \alpha_x, 0)\, ds < +\infty,$$

so we already know that $\liminf_{t\to+\infty} \operatorname{dist}(y_x(t; \alpha_x[0], 0), \mathcal{Z}) = 0$ and by part 2 of the proof $y_x(t) \in \mathcal{U}_0$ for all $t \geq 0$. Let us assume by contradiction that for some $x \in \Omega_{\varepsilon_1}$ we have $\limsup_{t\to+\infty} \operatorname{dist}(y_x(t; \alpha_x, 0), \mathcal{Z}) = 3\varrho > 0$. Since $\mathcal{U}_0$ is bounded, we can find an increasing sequence $t_n \to +\infty$ such that $y_x(t_n) \to \overline{x}$, $\operatorname{dist}(\overline{x}, \mathcal{Z}) = 3\varrho$. We can then construct another increasing sequence $s_n \to +\infty$ such that $|y_x(s_{2n}) - \overline{x}| = \varrho$, $|y_x(s_{2n+1}) - \overline{x}| = 2\varrho$, $|y_x(t) - \overline{x}| \in (\varrho, 2\varrho)$ for $t \in (s_{2n}, s_{2n+1})$. Let $C_\varrho = \inf\{\ell(z, a, 0)/(1 + |a|^2) : |z - \overline{x}| \leq 2\varrho, a \in A\} > 0$ by the coercivity of ℓ in (1.1). Therefore we can estimate

$$\sum_n C_\varrho \int_{s_{2n}}^{s_{2n+1}} (1 + |\alpha_x[0]|^2)\, dt \leq \sum_n \int_{s_{2n}}^{s_{2n+1}} \ell(y_x, \alpha_x[0], 0)\, dt$$
$$\leq \int_0^{+\infty} \ell(y_x, \alpha_x[0], 0)\, dt < +\infty,$$

while on the other hand for each $n \in \mathbb{N}$ we have

$$0 < \varrho \leq |y_x(s_{2n+1}) - y_x(s_{2n})| \leq \int_{s_{2n}}^{s_{2n+1}} |f(y_x, \alpha_x[0], 0)|\, dt$$
$$\leq L \int_{s_{2n}}^{s_{2n+1}} (1 + |\overline{x}| + 2\varrho + |\alpha_x[0]|^2)\, dt \leq L(1 + |\overline{x}| + 2\varrho) \int_{s_{2n}}^{s_{2n+1}} (1 + |\alpha[0]|^2)\, dt.$$

Combining the two estimates, we deduce that

$$+\infty = \sum_n \varrho \leq L(1 + |\overline{x}| + 2\varrho) \sum_n \int_{s_{2n}}^{s_{2n+1}} (1 + |\alpha[0]|^2)\, dt$$
$$\leq L(1 + |\overline{x}| + 2\varrho)/C_\varrho \int_0^{+\infty} \ell(y_x, \alpha_x[0], 0)\, dt < +\infty,$$

hence a contradiction.

To prove the second statement we need, we can take any $x \in \Omega_{\varepsilon_{n+1}}$ where n is sufficiently large so that $\mathcal{U}_{m_n} \subset \mathcal{V} \cap B(0, \delta)$, and use part 2 of this proof. ◂

It is not hard to show that the sign condition on the value function V_γ in the statement of Lemma 3.1 follows from (1.2) and (1.1) if the origin is an isolated point of $\mathcal{Z}$, as it happens in many applications. A precise statement is the following.

LEMMA 3.2. *Assume* (1.1), (1.2). *The lower semicontinuous envelope $V_{\gamma*}$ may vanish only on the degenerate set $\mathcal{Z}$.*

PROOF. Observe that by definition we can compute, using (1.2)

$$V_\gamma(x) \geq \inf_{a \in \mathcal{A}} \int_0^{+\infty} \ell(y_x, a, 0)\, dt.$$

Now note that if $\bar{x} \notin \mathcal{Z}$, then by the assumptions on ℓ (1.1) and (1.2), we have that there is $\sigma \in (0,1)$ sufficiently small such that

$$\inf_{a \in \mathcal{A}} \frac{\ell(x,a,0)}{1+|a|^2} \geq \sigma > 0, \qquad \text{for all } x \in B(\bar{x}, 2\sigma) .$$

For $\varepsilon > 0$ let now $V_{\gamma*}(\bar{x}) < +\infty$, $x_n \to \bar{x}$ be such that $x_n \in B(\bar{x},\sigma)$ and $V_\gamma(x_n) \to V_{\gamma*}(\bar{x})$, and $a_n \in \mathcal{A}$ be a control such that $V_\gamma(x_n) + \varepsilon \geq \int_0^{+\infty} \ell(y_{x_n}, a_n, 0) \, dt$. Let $s_n > 0$ be the first time such that $\sigma = |y_{x_n}(s_n) - x_n|$ (and choose $s_n = +\infty$ if this never happens), then we use (1.1) for f and get, for $s \in [0, s_n[$ and n large,

$$|y_{x_n}(s) - x_n| \leq \int_0^s L(2 + |\bar{x}| + |a_n|^q + |y_{x_n}(r) - x_n|) \, dr$$

$$\leq L(|\bar{x}| + 2)s + L\left(\int_0^s |a_n|^2 \, dr\right)^{q/2} s^{1-q/2} + \int_0^s |y_{x_n}(r) - x_n| \, dr$$

$$\leq L(|\bar{x}| + 2)s + L((V(x_n) + \varepsilon)/\sigma)^{q/2} s^{1-q/2} + \int_0^s |y_{x_n}(r) - x_n| \, dr \, .$$

Hence from Gronwall's Lemma, if $s_n < +\infty$, we obtain that

$$\sigma = |y_{x_n}(s_n) - x_n| \leq L[(2 + |\bar{x}|)s_n + ((V(x_n) + \varepsilon)/\sigma)^{q/2} s_n^{1-q/2}] e^{Ls_n},$$

and therefore we can find $t > 0$ independent of n such that $s_n \geq t$ for all n. We then conclude that $V_\gamma(x_n) + \varepsilon \geq \int_0^t \sigma(1 + |a_n|^2) \, dr \geq \sigma t$. As ε, n are arbitrary it follows that $V_{\gamma*}(\bar{x}) \geq \sigma t > 0$. ◂

The following result is an immediate consequence of the two lemmas above and the definitions.

THEOREM 3.3. *Assume (1.2) and (1.1). The system has finite L^p gain if and only if the value function V_γ is finite, $V_\gamma(0) = 0$ and V_γ has an optimal strategy at the origin. Let moreover the origin be an isolated point of $\mathcal{Z}$. Then the $\mathcal{H}_\infty$ control problem has a solution with a function U continuous at the origin if and only if the system has finite L^p gain and V_γ is continuous at the origin.*

REMARK 3.4. Note that Definition 1.1 does not require the strategies to be optimal at $x \neq 0$. However, as we already mentioned, if we use an optimal strategy, we achieve the best possible estimate in (1.5) with the given disturbance attenuation level γ by definition of V_γ. ◁

4. Dynamic Programming equation

In this section we will study the properties of the value function V_γ, that we need in order to solve the $\mathcal{H}_\infty$ problem, by PDE methods. We start introducing the Hamiltonian related to the $\mathcal{H}_\infty$ control problem and the value function (1.6). Precisely we consider the function

$$\mathcal{H}(x,p) = \inf_{b \in B} \sup_{a \in \mathcal{A}} \{-f(x,a,b) \cdot p - \ell(x,a,b) + \gamma^p |b|^p\} \, .$$

The following result is a variant of similar results in Chapter VIII. The proof is therefore omitted.

PROPOSITION 4.1. *Assume (1.1). Then the Hamiltonian $\mathcal{H}$ is locally Lipschitz continuous.*

There are several technical difficulties with the Hamiltonian $\mathcal{H}$ and the corresponding Hamilton-Jacobi equation

$$(4.1) \qquad \mathcal{H}(x, DU(x)) = 0, \quad \text{in } \mathbb{R}^N .$$

We mention the unboundedness of the sets of controls and disturbances, which we can take care of in part using the coercivity of the running cost $\ell(x, a, b) - \gamma^p |b|^p$, see (1.1), and also the unboundedness of the vector field f with respect to the parameters a and b. These facts do not allow Gronwall estimates of the trajectories starting at a given point which are uniform in controls and disturbances. The latter is one of the main technical points when we study the existence of solutions of (4.1) by game theoretic methods. Moreover it is clear that (4.1) does not have a unique solution, since it is invariant by adding a constant to the function U. Even requiring the solution to be nonnegative and null at the origin there is no guarantee that it is unique (in either classical or viscosity sense), due to the weakness of the sign assumption (1.2). The regularity of the Hamiltonian, as deduced from Proposition 4.1 is also very poor to apply the standard theory of the book.

Despite all the difficulties mentioned above, the following results can be proved about the equation (4.1).

THEOREM 4.2. *Assume (1.1) and (1.2) and let $\Omega \subset \mathbb{R}^N$ be an open set such that V_γ is locally bounded in Ω. Then V_γ is a viscosity solution of*

$$\mathcal{H}(x, DV_\gamma(x)) = 0, \quad \text{in } \Omega .$$

In the assumptions of Theorem 4.2, the value V_γ may not be even lower semicontinuous in general. Therefore solution has to be meant in the sense of noncontinuous viscosity solutions, as in Chapter V. The following more unusual but very general optimality principle for supersolutions holds as well.

THEOREM 4.3. *Assume (1.1) and (1.2) and let $U : \mathbb{R}^N \to \mathbb{R}$ be a continuous viscosity supersolution of (4.1). Then*

$$(4.2) \qquad U(x) = \inf_{\alpha \in \Gamma} \sup_{b \in \mathcal{B}} \sup_{t \in \mathbb{R}_+} \left\{ \int_0^t (\ell(y, \alpha, b) - \gamma^p |b|^p) \, ds + U(y(t)) \right\} .$$

If U is only lower semicontinuous and A is compact, then (4.2) holds if the sets $\{(f(x, a, b), \ell(x, a, b)) : a \in A\}$ are convex for all x, b, or provided we substitute the set of strategies Γ with that of relaxed strategies Γ^r, namely the set of nonanticipating functionals $\alpha : \mathcal{B} \to A^r$, where A^r is the set of relaxed controls.

4. DYNAMIC PROGRAMMING EQUATION

REMARK 4.4. Since the function V_γ is defined as the value function of an optimal stopping time differential game, following Chapter III it could look more natural to consider the following equation,

$$\min\{\mathcal{H}(x, DU), U\} = 0,$$

instead of (4.1). As a matter of fact, due to the sign condition (1.2) they are equivalent for nonnegative solutions, see Exercise 7.3. It could also appear strange that (4.2) is written with the equality sign, while U is only a supersolution. Observe however that one direction ($\geq$) is what one would expect, while the other is obvious by choosing $t = 0$ in the right-hand side. We also note that the corresponding statement holds for subsolutions as well, substituting $\sup_{t \in \mathbb{R}_+}$ with $\inf_{t \in \mathbb{R}_+}$. The result of Theorem 4.3 holds true also if U is a supersolution in an open set Ω by changing $\mathbb{R}_+$ in (4.2) with $[0, t_x(a, b)[$, where $t_x(a, b)$ is the exit-time of the trajectory $y_x(\,\cdot\,; a, b)$ from Ω. ◁

The proofs of Theorems 4.2 and 4.3 are rather technical and will not be given in detail. Let us only sketch the main ideas. First of all, we use a reparametrization procedure. For all $a \in \mathcal{A}$, $b \in \mathcal{B}$, we consider the increasing change of parameter

$$\tau(t) = \tau_{a,b}(t) = \int_0^t (1 + |a|^p + |b|^p)\, ds = t + \|a\|^p_{L^p(0,t)} + \|b\|^p_{L^p(0,t)},$$

and set $t(\,\cdot\,) = t_{a,b}(\,\cdot\,) = \tau^{-1}(\,\cdot\,)$. Observe that the reparametrization $t(\,\cdot\,)$ is a nonanticipating functional of the controls. We will briefly denote in the following the reparametrized measurable functions as $a(\tau) = a(t(\tau))$, $b(\tau) = b(t(\tau))$, and for such controls solve the dynamical system

(4.3) $$y'(\tau) = \bar{f}(y(\tau), a(\tau), b(\tau)), \qquad y(0) \in \mathbb{R}^N,$$

where $\bar{f} : \mathbb{R}^N \times A \times B \to \mathbb{R}^N$, is defined by $\bar{f}(x, a, b) = f(x, a, b)/(1 + |a|^p + |b|^p)$. Observe that trajectories of the systems (0.1) and (4.3) are equivalent by means of the above introduced change of parameter. Indeed if $y(\tau)$ solves (4.3), then

$$y(\tau) = y_x(t(\tau); a, b).$$

By a change of variables in the integral, we can then rewrite the value function as

$$V_\gamma(x) = \inf_{\alpha \in \Gamma} \sup_{b \in \mathcal{B}} \sup_{\tau \in \mathbb{R}_+} \int_0^\tau \frac{\ell(y, \alpha[b], b) - \gamma^p |b|^p}{1 + |\alpha[b]|^p + |b|^p}\, d\tau.$$

It is now easy to check that, by the assumptions (1.1), the trajectories of the system with reparametrized controls satisfy the following condition

$$|y_x(s) - x| \leq \omega(\tau), \qquad \text{for all } |x| \leq R,\ s \in [0, \tau],\ a \in \mathcal{A},\ b \in \mathcal{B},$$

where $\omega : \mathbb{R}_+ \to \mathbb{R}_+$ is continuous, increasing and $\omega(0) = 0$. In particular we have $y_x(\tau; a, b) \to x$ as $\tau \to 0$, uniformly for all $a \in \mathcal{A}$, $b \in \mathcal{B}$ and x bounded. The main advantage of the reparametrization procedure is indeed the fact that our value function corresponds also to another differential game whose trajectories satisfy such a uniform Gronwall type estimate.

The Hamiltonian associated with the reparametrized problem is

$$\mathcal{H}^{\#}(x,p) = \inf_{b \in B} \sup_{a \in A} \left\{ -\bar{f}(x,a,b) \cdot p - \bar{\ell}(x,a,b) + \frac{\gamma^p |b|^p}{1 + |a|^p + |b|^p} \right\},$$

where $\bar{\ell}(x,a,b) = \ell(x,a,b)/(1 + |a|^p + |b|^p)$. It is now almost standard to show that the value function V_γ satisfies the equation $\mathcal{H}^{\#}(x, DU) = 0$, and then conclude the result of Theorem 4.2 with a few simple computations based on the coercivity of the cost ℓ.

To prove the optimality principle of Theorem 4.3, further technical ideas are needed since the reparametrized Hamiltonian is still not regular enough to apply standard comparison theorems. Given a function U as in the statement of Theorem 4.3, we start with a change of variables. Let $\varrho : \mathbb{R} \to \mathbb{R}_+$, be bounded, smooth, such that $M \geq \dot{\varrho} > 0$ and $\varrho(s) \to 0$, as $s \to -\infty$. Define the nonnegative, bounded function $u(x,r) = \varrho(U(x) + r)$. By the standard formulas of change of variables as in Chapter II, u is a viscosity supersolution of

$$\inf_{b \in B} \sup_{a \in A} \{ -F(z,a,b) \cdot D_z u \} \geq 0, \quad \text{in } \mathbb{R}^{N+1},$$

where $z = (x,r)$ and $F(z,a,b) = (\bar{f}(x,a,b), \bar{\ell}(x,a,b) - \gamma^p |b|^p/(1 + |a|^p + |b|^p))$.

We proceed as above and consider the following dynamical system with reparametrized controls

$$z'(\tau) = F(z(\tau), a(\tau), b(\tau)), \quad z(0) = z \in \mathbb{R}^{N+1}.$$

It is clear by the definition of F that this system has a unique solution for any $a \in \mathcal{A}$, $b \in \mathcal{B}$.

For a fixed $\varepsilon > 0$, we now consider a smooth function $\zeta_\varepsilon : \mathbb{R}^{N+1} \to \mathbb{R}$ such that $0 \leq \zeta_\varepsilon(z) = \zeta_\varepsilon(x,0) \leq 1$, $\zeta_\varepsilon = 0$ on $\Sigma^c_{\varepsilon/2}$, $\zeta_\varepsilon = 1$ in Σ_ε, where for $\sigma > 0$ we define $\Sigma_\sigma = \{z = (x,r) \in \mathbb{R}^{N+1} : |x| < 1/\sigma\}$. Therefore u is a supersolution of

$$\mathcal{H}^{\#}_\varepsilon(x, D_z u) = \inf_{b \in B} \sup_{a \in A} \{ -F_\varepsilon(z,a,b) \cdot D_z u(z) \} \geq 0, \quad \text{in } \mathbb{R}^{N+1},$$

where $F = F_\varepsilon$ in $\Sigma_\varepsilon \times A \times B$ and $F_\varepsilon = 0$ in $\Sigma^c_{\varepsilon/2} \times A \times B$, $\Sigma^c = \mathbb{R}^{N+1} \setminus \Sigma$. Finally we conclude that for any $\lambda \in \,]0,1[$, the function $V = u$ is a viscosity solution of

$$\lambda V + \min\{\mathcal{H}^{\#}_\varepsilon(x, D_z V), V - (1+\lambda)u\} = 0, \quad \text{in } \mathbb{R}^{N+1}.$$

To this last equation we can apply a comparison theorem and state that u is the value function of the corresponding differential game. We then need some careful but tedious computations to take the limits as $\lambda \to 0$ first and $\varepsilon \to 0$ next, and complete the proof of the result.

With the above results at hand, it is now easy to derive necessary and sufficient conditions for the solvability of the $\mathcal{H}_\infty$ suboptimal control problem in terms of the equation (4.1). This result completes Theorem 3.3.

4. DYNAMIC PROGRAMMING EQUATION

THEOREM 4.5. *Assume* (1.1) *and* (1.2). *If the $\mathcal{H}_\infty$ suboptimal control problem with level γ is solvable, then there is a nonnegative, lower semicontinuous supersolution, null at the origin of* (4.1). *If, on the other hand, there is a nonnegative, continuous supersolution $U : \mathbb{R}^N \to \mathbb{R}$ of* (4.1), *satisfying in a neighborhood of the origin $U(x) = 0$ only at $x = 0$, then $U \geq V_\gamma$ and the $\mathcal{H}_\infty$ suboptimal control problem is solvable. If V_γ has an optimal strategy at the origin, then there is a solution of the $\mathcal{H}_\infty$ problem.*

PROOF. The necessary part is a direct consequence of Theorem 4.2, using V_{γ_*} as the supersolution. For the sufficiency part observe that from (4.2), the fact that U is nonnegative and the definition of V_γ, we get $U \geq V_\gamma$. We are left to show that the undisturbed system is open loop Lyapunov stable. Let $\{\mathcal{U}_\varepsilon\}_{\varepsilon>0}$ be a nested local base of neighborhoods of the origin made of open bounded sets. We then define the sets $\Omega_\varepsilon = \{x \in \mathcal{U}_\varepsilon : U(x) < \min_{\partial \mathcal{U}} U\}$ and observe that for ε small they are neighborhoods of the origin and, in fact, $\{\Omega_\varepsilon\}_\varepsilon$ is a nested local base of the origin. Fix $\varepsilon > 0$ and let $x \in \Omega_\varepsilon$, by (4.2) we can choose $\delta > 0$ and $\alpha \in \Gamma$ so that

$$\sup_{b \in \mathcal{B}} \sup_{t \in \mathbb{R}_+} \left\{ \int_0^t \left(\ell(y, \alpha[b], b) - \gamma^p |b|^p \right) dt + U(y_x(t)) \right\} \leq U(x) + \delta \leq \min_{\partial \mathcal{U}_\varepsilon} U - \delta .$$

In particular for $b \equiv 0$, and since $\ell(x, a, 0)$ is nonnegative, we get $\sup_{t \in \mathbb{R}_+} U(y_x(t)) \leq \min_{\partial \mathcal{U}_\varepsilon} U - \delta$, and then $y(t) \in \mathcal{U}_\varepsilon$ for all $t \geq 0$, from which the open loop stability immediately follows.

The last statement is a consequence of Lemma 3.1. ◂

REMARK 4.6. Theorem 4.3 shows that, when V_γ is continuous, it is the smallest nonnegative supersolution of (4.1). There is a gap in Theorem 4.5 between the necessary and the sufficient conditions. As a result of Theorem 4.3, if A is compact, in Theorem 4.5 we can relax continuity of U to lower semicontinuity, still requiring U to be null and continuous at the origin and satisfying in a neighborhood of the origin $U(x) = 0$ only at $x = 0$. This can be done by either considering the value function with relaxed strategies or assuming the convexity of the sets $\{(f(x, a, b), \ell(x, a, b)) : a \in A\}$ for all x, b. ◁

EXAMPLE 4.7. We go back to the system of Example 1.4 and compute the corresponding Hamiltonian. It is immediate to check that $\mathcal{H}(x, p) = \frac{1}{4}(1 - 1/\gamma^2)|p|^2 - |x|^2$. Therefore for the choice of $\gamma > 1$, we can check that the function $U_\gamma(x) = \gamma/(\gamma^2 - 1)^{1/2}|x|^2$ is a solution of the equation, while there is clearly no solution if $\gamma \leq 1$. By Theorem 4.5 we can now conclude that $\gamma^* = 1$.

By Proposition 2.1 this implies that, for $\gamma = 1$, there is no linear stabilizing feedback control satisfying (2.2) even if $V_1(0) = 0$. ◁

The following result is concerned with the existence of optimal strategies and therefore with the actual solution of the $\mathcal{H}_\infty$ suboptimal control problem. It completes Theorem 3.3 and Theorem 4.5, and its proof is an adaptation of classical results in differential games theory. The main ideas of this proof can be also found in Propositions 5.5 and 5.7.

PROPOSITION 4.8. *Assume* (1.1) *and let A be compact. If the sets*
$$\{(f(x,a,b), \ell(x,a,b)) : a \in A\}$$
are convex for all x, b, then there is an optimal strategy for $V_\gamma(x)$ at any $x \in \mathbb{R}^N$. Otherwise there is always an optimal relaxed strategy $\alpha \in \Gamma^r$.

The theory of linear systems proves that, when (2.2) holds, there is a classical solution of (4.1) which is nonnegative and null at the origin. The next example motivates the use of viscosity solutions. We want to show that when we depart from linear systems, (super) solutions of (4.1) may be not differentiable, even in a neighborhood of the origin and in special cases often used in the applications as the nonlinear-affine systems (0.4).

EXAMPLE 4.9. Let us consider the following system
$$\begin{cases} \dot{y} = a - b, \quad y(0) = x, \\ z = \ell(x)^2 + |a|^2, \end{cases}$$
where $\ell : \mathbb{R}^N \to \mathbb{R}$ is a nonnegative function null at the origin to be specified. Again it is easy to check that the Hamiltonian is given by
$$\mathcal{H}(x, p) = \frac{1}{4}(1 - 1/\gamma^2)|p|^2 - \ell(x)^2 \ .$$

It is then clear that, when $\gamma > 1$, the Isaacs equation (4.1) is equivalent to

(4.4) $$|DU(x)| = \frac{2\gamma}{(\gamma^2 - 1)^{1/2}} \ell(x),$$

which is the eikonal equation. Our purpose is to show that, with a suitable choice of ℓ, the equation (4.4) does not have a classical nonnegative solution vanishing at the origin. The reader can easily check by control theoretic ideas that if ℓ satisfies (1.4) and we write (4.4) as a Bellman equation, setting $|p| = \max_{|a| \leq 1}\{-a \cdot p\}$, then the value function of the control problem is nonnegative, continuous and vanishes at the origin, see Exercise 7.4. In this case it can also be proven that there is a unique viscosity solution of (4.4) of this sort. We will now show that this solution will not be smooth for a suitable choice of ℓ in any neighborhood of the origin. Consider the function of two variables
$$U(x, y) = (x^2 + y^2)(2 - |y| + y^2) \ .$$
Such a function is differentiable for all (x, y) such that $y \neq 0$ and it is also differentiable at the origin. Note that $U(x, y) > 0$ if $(x, y) \neq (0, 0)$. If $x \neq 0$ we can compute $D^+U(x, 0) = \{(4x, p_2) : p_2 \in [-x^2, x^2]\}$ and $D^-U(x, 0) = \emptyset$. Moreover the function
$$\ell(x, y) = \begin{cases} |DU(x, y)|, & y \neq 0, \\ |x|(16 + x^2)^{1/2}, & y = 0, \end{cases}$$
is immediately seen to be continuous. By the definition of viscosity solution it follows that $|DU| = \ell$ in $\mathbb{R}^N$, in the viscosity sense. ◁

In the case of nonlinear-affine systems, some positive statements concerning the regularity of the value function in a neighborhood of the origin can be found in the literature. We will not give here more details about this matter.

If the value of the differential game we studied so far does not exist, we are possibly not completely satisfied by the solution of the problem in the sense of Definitions 1.1 and 1.3 and we may desire also some estimates for the upper value as well. To this end we note that all the results above can be rephrased for the upper value function

$$\overline{V}_\gamma(x) = \sup_{\beta \in \Delta} \inf_{a \in \mathcal{A}} \sup_{t \in \mathbb{R}_+} \int_0^t (\ell(y, a, \beta[a]) - \gamma^p |\beta[a]|^p) \, ds,$$

where Δ is the set of nonanticipating or causal functionals $\beta : \mathcal{A} \to \mathcal{B}$, and the corresponding Hamiltonian

$$\overline{\mathcal{H}}(x, p) = \sup_{a \in A} \inf_{b \in B} \{-f(x, a, b) \cdot p - \ell(x, a, b) + \gamma^p |b|^p\}.$$

Of course if the Isaacs condition holds (i.e., $\mathcal{H} = \overline{\mathcal{H}}$) the two approaches turn out to be equivalent even if this is unfortunately not enough to prove the existence of value with the results we presented. As a consequence of Theorem 4.3, however, the following result holds.

COROLLARY 4.10. *Assume that* (1.1), (1.2) *and the Isaacs condition hold. If moreover the value functions V_γ and $\overline{V}_\gamma$ are continuous, then the value of the game exists, i.e., $V_\gamma = \overline{V}_\gamma$.*

5. On the partial information problem

In this section we describe some results concerning the partially observed system (0.1)–(0.2). In this new framework the controller is allowed to use a restricted class of strategies based on the available information. We will denote by $\mathcal{W} = \mathcal{M}(\mathbb{R}_+, \mathbb{R}^P)$, the set of measurable *observed outputs*.

DEFINITION 5.1. *Let $x \in \mathbb{R}^N$; an open loop strategy $\alpha \in \Gamma$ is said to be an* output feedback strategy *admissible at x if for all $b_1, b_2 \in \mathcal{B}$, $T \geq 0$ such that*

$$g(y_x(\,\cdot\,; \alpha[b_1], b_1), \alpha[b_1], b_1) = g(y_x(\,\cdot\,; \alpha[b_2], b_2), \alpha[b_2], b_2) \quad \text{a.e. in } [0, T],$$

then $\alpha[b_1] = \alpha[b_2]$ a.e. in $[0, T]$. ◁

We often indicate, with a slight abuse of notation, an output feedback strategy as a functional of the observed output $\alpha = \alpha[w]$ (which is defined in (0.2)), since if we define

$$\mathcal{W}_{x,\alpha} = \{g(y_x(\,\cdot\,; \alpha[b], b), \alpha[b], b) : b \in \mathcal{B}\} \subset \mathcal{W},$$

then α determines a causal functional $\mathcal{W}_{x,\alpha} \to \mathcal{A}$. We will also denote by $\Gamma_x \subset \Gamma$ the set of all output feedback strategies admissible at x.

REMARK 5.2. We give two explicit examples of output feedback strategies. Suppose that $\alpha : \mathcal{W} \to \mathcal{A}$ is a causal functional and that for all $T > 0$ and $b \in \mathcal{B}$ the map $L^p(0,T;A) \to L^p(0,T;A)$ given by

$$a \longmapsto \alpha[g(y_x(\,\cdot\,;a,b),a(\,\cdot\,),b(\,\cdot\,))]$$

has a unique fixed point $\bar{a}$, which is the unique trajectory solution of the problem

(5.1)
$$\begin{cases} \dot{y} = f(y,a,b), \quad y(0) = x, \\ w = g(y,a,b), \\ a = \alpha[w] \,. \end{cases}$$

The functional α can be viewed as an element of Γ_x by the position $\alpha[b] = \bar{a}$.

Another trivial but helpful example is a constant mapping $\alpha[w] \equiv a \in \mathcal{A}$, for all $w \in \mathcal{W}$. In particular, in the case with no available information on the system, i.e., $g(y,a,b) \equiv \text{const.}$, Γ_x reduces exactly to the set of all constant strategies, as easily checked. ◁

REMARK 5.3. Note that the following useful property holds for output feedback strategies. For all $T > 0$, $\alpha \in \Gamma_x$ and $b_1 \in L^2([0,T],B)$ then the position $\overline{\alpha}[b] = \alpha[\bar{b}]$, where

$$\bar{b}(t) = \begin{cases} b_1(t), & [0,T[, \\ b(t-T), &]T,+\infty[, \end{cases}$$

defines an element $\overline{\alpha} \in \Gamma_{y_x(T;\alpha,b)}$. ◁

The previous remark allows us to use the dynamic programming approach for the value function of the partially observed game, namely

(5.2)
$$W_\gamma(x) = \inf_{\alpha \in \Gamma_x} \sup_{b \in \mathcal{B}} \sup_{t \in \mathbb{R}_+} \int_0^t (\ell(y,\alpha_x,b) - \gamma^p |b|^p)\, ds \,.$$

We assume that the initial state of the system is known, while equation (0.2) provides only partial information about the system as the time proceeds. This is the main reason for us to study the value function in (5.2). Observe that if $g(x,a,b) = x$, then the set of admissible strategies Γ_x is the class of feedback strategies, see also Chapter VIII. The notion of solution or of solvability of the $\mathcal{H}_\infty$ suboptimal control problem with partial information can be adapted for this new class of admissible strategies for the controller from Definitions 1.1 and 1.3. If the initial state of the system is also unknown, it seems to be more appropriate to reformulate the problem as a full information problem for an infinite dimensional system. For this approach, we refer the reader to the literature.

To have a feeling of some additional difficulties involved in the partial information problem, observe that since constant output feedback strategies are admissible at any point, the following estimates hold

(5.3)
$$V_\gamma(x) \leq W_\gamma(x) \leq V_\gamma^\#(x) := \inf_{a \in \mathcal{A}} \sup_{b \in \mathcal{B}} \sup_{t \in \mathbb{R}_+} \int_0^t (\ell(y,\alpha_x,b) - \gamma^p |b|^p)\, ds \,.$$

Note that the map $V_\gamma^\#$ actually coincides with the function W_γ when g is constant and clearly $V_\gamma = W_\gamma$, when $g(x,a,b) = b$, therefore the estimate is sharp. On the other hand it is immediately seen that the upper value function satisfies the following inequality

$$\overline{V}_\gamma \leq V_\gamma^\#,$$

where the inequality may be strict in general. The simple argument to prove the previous inequality is as follows. Fix $\varepsilon > 0$, then by definition of upper value there is $\beta \in \Delta$ such that

$$\overline{V}_\gamma(x) - \varepsilon \leq J(x,a,\beta[a]) := \sup_{t \in \mathbb{R}_+} \int_0^t (\ell(y,\alpha_x,b) - \gamma^p |b|^p)\, ds \leq \sup_{b \in \mathcal{B}} J(x,a,b),$$

for all $a \in \mathcal{A}$.

The function $V_\gamma^\#$ is the upper value function of the static game where no information about the dynamics is available to the players as time proceeds. In the very special case that the payoff of the game $J(x,a,b)$ has a saddle point as a functional of control functions $a(\cdot) \in \mathcal{A}$ and disturbances $b(\cdot) \in \mathcal{B}$, then the lower and upper values of the static game coincide and in (5.3) we can substitute inequalities by equalities.

In order to solve the partial information $\mathcal{H}_\infty$ problem, by adapting Theorem 3.3 to this case, we need to show that W_γ is finite, vanishes, is continuous, and has optimal strategies at the origin. The next result follows as Theorem 4.2 by using Remark 5.3, and shows the relationship between W_γ and the Isaacs equation.

THEOREM 5.4. *Assume (1.1) and (1.2). If W_γ is finite, then its lower semicontinuous envelope $(W_\gamma)_*$ is a nonnegative viscosity supersolution of (4.1).*

In general W_γ is not a solution of equation (4.1) nor a subsolution of the upper value equation at the end of Section 4. Therefore the solvability of the problem cannot be proved by the only equation (4.1). Assume we solved the full information problem; a standard but formal way of proceeding would be the approach known in the literature as *worst case certainty equivalence principle*. This means that from an optimal strategy of the full information problem, we hope to construct an optimal output feedback strategy as follows. For any fixed $x \in \mathbb{R}^N$, $T > 0$, control $a(\cdot)$ and output $w(\cdot)$, denote the subset of disturbances

$$\mathcal{B}_{x,T}(a,w) = \{b \in \mathcal{B} : w(\cdot) = g(y_x(\cdot\,;a,b),a,b) \text{ a.e. in } [0,T]\},$$

which are compatible with the given data. Then assume that there is a unique solution $b_{x,T} \in \mathcal{B}$ of the maximization problem

$$M_{x,T}(a,w) = \sup_{b \in \mathcal{B}_{x,T}(a,w)} \left[\int_0^T (h(y,a,b) - \gamma^2 |b|^2)\, ds + V_\gamma(y(T)) \right].$$

The goal, given an optimal strategy $\alpha \in \Gamma$ of the full information value function $V_\gamma(x)$ is to show that the output feedback strategy $\overline{\alpha}$ defined by the implicit relation

$$\overline{\alpha}[w](\cdot) = \alpha[b_{x,T}](\cdot), \quad \text{in } [0,T],$$

is optimal for the value $W_\gamma(x)$. We do not give details about the actual proof of this statement, which can be found in the literature and requires other assumptions that are in practical cases strong and difficult to check. As a matter of fact, if this procedure can be applied, we end up proving that $V_\gamma(x) = W_\gamma(x)$ which is a restrictive property of the system.

In the following we are concerned with the existence of optimal strategies for the value function W_γ, at least if the system satisfies some additional structural conditions.

For the rest of this section, in addition to (1.1) we require that $p = 2$ and for all $R > 0$

(5.4)
$$\begin{cases} |g(x,a,b)| \leq L_R(1+|a|^k+|b|^k), \\ |g(x,a,b) - g(y,a,b)| \leq L_R(1+|a|^k+|b|^k)|x-y|, \\ \ell(x,a,b) \geq C^0|a|^2 - L(1+|a|^q+|b|^2), \\ A \text{ is compact,} \end{cases}$$

for some $k \in [1,2]$, $q \in [1,2[$, $C^0 > 0$, for all $|x|, |y| \leq R$, $a \in A$, $b \in B$, so that if $j = 2/k$ we have $g(y_x(\cdot;a,b),a,b) \in L^j_{\text{loc}}(\mathbb{R}_+, \mathbb{R}^P)$ for all x, a, b. We also indicate the space of measured outputs as $\mathcal{W} = L^j_{\text{loc}}(\mathbb{R}_+, \mathbb{R}^P)$. We start considering the relaxation of the original problem, with class of admissible controls $\mathcal{A}^r = L^\infty(\mathbb{R}_+; A^r)$, where A^r is the set of relaxed controls as in Chapter III, and extend the system defining, for $m \in A^r$,

$$\nu^r(x,m,b) = \int_A \nu(x,a,b)\,dm,$$

where $\nu \in \{f, g, \ell\}$. Observe that (f^r, g^r, ℓ^r) satisfy (1.1) and (5.4). Since A is compact, A^r is metrizable and compact in the weak* topology as a subset of the dual of $\mathcal{C}(A; \mathbb{R})$. For $x \in \mathbb{R}^N$ we consider the set Γ^r of relaxed open loop strategies and the set Γ^r_x of relaxed output feedback strategies admissible at x.

Existence of optimal relaxed feedback strategies can be deduced from the following result.

PROPOSITION 5.5. *Let $x \in \mathbb{R}^N$ and assume (1.1) and (5.4). Suppose that there is a sequence of output feedback strategies $\alpha_n \in \Gamma_x$ which can be extended to an equicontinuous family of causal functionals $\alpha_n : \mathcal{W} \to A$ and such that*

(5.5)
$$\int_0^T (\ell(y,\alpha_n,b) - \gamma^2|b|^2)\,dt \leq K_x + 1/n, \quad \text{for all } b \in \mathcal{B}, T \in \mathbb{R}_+.$$

Then there is $\alpha^* \in \Gamma^r_x$ such that

$$\int_0^T (\ell^r(y^r,\alpha^*,b) - \gamma^2|b|^2)\,dt \leq K_x, \quad \text{for all } b \in \mathcal{B}, T \in \mathbb{R}_+.$$

PROOF. In the following we denote $w_n(\cdot) = g(y_x(\cdot;\alpha_n,b), \alpha_n[b], b)$ for a fixed $b \in \mathcal{B}$. By the weak* compactness of $L^\infty(0,T;A^r)$, the Tychonov theorem applied

5. ON THE PARTIAL INFORMATION PROBLEM

to $(L^\infty(0,T;A^r))^\mathcal{B} \times (L^\infty(0,T;A^r))^\mathcal{W}$ tells us that this space is compact in the product topology. In particular we can find a subnet $\{\alpha_\lambda^1\}$ of $\{\alpha_n\}$ such that $(\alpha_\lambda^1, \alpha_\lambda^1)$ converges in the product topology to an element

$$(\alpha^{*1}, \alpha^{\sharp 1}) \in (L^\infty(0,1;A^r))^\mathcal{B} \times (L^\infty(0,1;A^r))^\mathcal{W}.$$

In general, given a subnet $\{\alpha_\lambda^k\}$ of $\{\alpha_n\}$ such that

$$(\alpha_\lambda^k, \alpha_\lambda^k) \longrightarrow (\alpha^{*k}, \alpha^{\sharp k}) \in (L^\infty(0,k;A^r))^\mathcal{B} \times (L^\infty(0,k;A^r))^\mathcal{W},$$

we can extract a subnet $\{\alpha_\lambda^{k+1}\}$ of $\{\alpha_\lambda^k\}$ such that

$$(\alpha_\lambda^{k+1}, \alpha_\lambda^{k+1}) \longrightarrow (\alpha^{*(k+1)}, \alpha^{\sharp(k+1)}) \in (L^\infty(0,k+1;A^r))^\mathcal{B} \times (L^\infty(0,k+1;A^r))^\mathcal{W}.$$

Define $\alpha^*[b](t) = \alpha^{*(k+1)}[b](t)$ if $t \in [k,k+1[$ and $\alpha^\sharp$ similarly, then, by the construction, for all $T > 0$ we can find some subnet $\{\alpha_\lambda\}$ of $\{\alpha_n\}$ such that for all $b \in \mathcal{B}$ and $w \in \mathcal{W}$ we have $\alpha_\lambda[b] \to \alpha^*[b]$ and $\alpha_\lambda[w] \to \alpha^\sharp[w]$ weak* in $L^\infty(0,T;A^r)$. In particular α^* and $\alpha^\sharp$ are indeed causal functionals of their arguments. By definition of weak* convergence, for all $T > 0$ and $\varphi \in L^1(0,T;\mathcal{C}(A))$ we get

$$\int_0^T \varphi(t, \alpha_\lambda[w_\lambda](t))\, dt = \int_0^T \varphi(t, \alpha_\lambda[b](t))\, dt \to \int_0^T \left(\int_A \varphi(t,a)\, d\alpha^*[b](t)(a)\right) dt,$$

$$\int_0^T \varphi(t, \alpha_\lambda[w](t))\, dt \to \int_0^T \left(\int_A \varphi(t,a)\, d\alpha^\sharp[w](t)(a)\right) dt,$$

for all $b \in \mathcal{B}$ and $w \in \mathcal{W}$. In particular this leads to the following facts for the relaxed trajectory $y_x^r(\,\cdot\,;\alpha^*[b],b)$ for any fixed $b \in \mathcal{B}$ (note that (1.1), (5.4), particularly the coercivity of ℓ, and the assumption (5.5), give, for all fixed $T > 0$, $b \in \mathcal{B}$, a uniform bound for the norm of $\alpha_\lambda[b]$ in $L^2(0,T;A)$):

$$y_x(\,\cdot\,;\alpha_\lambda[w_\lambda],b) \longrightarrow y_x^r(\,\cdot\,;\alpha^*[b],b), \quad \text{locally uniformly,}$$

$$\int_0^T (h^r(y^r, \alpha^*[b], b) - \gamma^2 |b|^2)\, dt \le K_x.$$

Moreover for all $\varphi \in L^{j'}_{\text{loc}}(\mathbb{R}_+, \mathbb{R})$, $j' = j/(j-1)$, we also have

$$\int_0^T \varphi(t) g(y_x, \alpha_\lambda[w_\lambda], b)\, dt \longrightarrow \int_0^T \varphi(t) g^r(y_x^r, \alpha^*[b], b)\, dt,$$

and therefore, if we indicate $w(t) := g^r(y^r(t;\alpha^*[b],b)\alpha^*[b](t), b(t))$, for the particular choice $\varphi = \text{sgn}(w_\lambda - w)|w_\lambda - w|^{j-1}$, we get

$$\int_0^T |w_\lambda - w|^j\, dt \longrightarrow 0.$$

By the equicontinuity of the sequence α_n, we deduce that $\alpha_\lambda[w_\lambda] - \alpha_\lambda[w] \to 0$, so $\alpha^*[b] = \alpha^\sharp[w]$ and the limit strategy α^* depends only on the observed output. Therefore it is a relaxed output feedback strategy admissible at x, i.e., $\alpha^* \in \Gamma_x^r$. ◂

REMARK 5.6. Concerning the technical assumption in Proposition 5.5, regarding the family of strategies, observe that if the sequence α_n is given by constant strategies, then it is obviously extendable to a family of constant (therefore equicontinuous) functionals $\mathcal{W} \to \mathcal{A}$. We need to be careful when passing to limits since it is not true in general that disturbances producing the same output with a given strategy will also enjoy the same property with other strategies. ◁

In certain cases, the optimal output feedback strategy provided in Proposition 5.5 can be represented by means of non-relaxed strategies. This fact can be shown by adapting classical selection arguments. To this end we further assume that

(5.6)
$$\begin{cases} g(x,a,b) = g_1(x,b)g_2(a), \\ \nu(x,a,b) = \nu_1(x,b)\nu_2(g(x,a,b),a) & \text{if } \nu \in \{f,\ell\}, \\ A \text{ is compact}, \mathcal{E}(w) = \operatorname{co}\mathcal{E}(w) & \text{for all } w, \end{cases}$$

where $\mathcal{E}(w) = \{(f_2(w,a), \ell_2(w,a), g_2(a)) : a \in A\}$, g_1, f_1, ℓ_1 are matrix valued, g_2, f_2, ℓ_2 are vector valued and all of them are suitably defined and continuous. With $\operatorname{co} X$, we indicate the convex hull of the set X. It is easy to check that if $f(x,a,b), \ell(x,a,b), g(x,a,b)$ are nonlinear in x and affine in the controls a, b, and A is convex and compact, then (5.6) is satisfied. As a matter of fact, any function of the type $\nu(x,a,b) = \nu^1(x)\nu^2(a) + \nu^3(x,b)$, where ν^1 is matrix valued and ν^2, ν^3 are vector valued and $\nu^2(A)$ is convex, is allowed by (5.6).

The following result holds.

PROPOSITION 5.7. *Let $x \in \mathbb{R}^N$. If the assumptions of Proposition 5.5 and (5.6) hold, then there is an output feedback strategy $\overline{\alpha} \in \Gamma_x$ such that*

$$\int_0^T (\ell(y, \overline{\alpha}, b) - \gamma^2 |b|^2) \, dt \leq K_x, \quad \text{for all } b \in \mathcal{B}, T \in \mathbb{R}_+ .$$

PROOF. The goal here is to define an output feedback strategy with the same behavior of the relaxed strategy α^* constructed in Proposition 5.5. Note that by that proof, α^* is in fact a causal functional $\mathcal{W} \to \mathcal{A}$.

1. Fix $w(\,\cdot\,) \in \mathcal{W}$, $w \in \mathbb{R}^P$ and $t > 0$, and define the vector valued function

$$e(w,t;w(\,\cdot\,)) = \int_A (f_2(w,a), \ell_2(w,a), g_2(a)) \, d\alpha^*[w](t)(a) \in \operatorname{co}\mathcal{E}(w) .$$

Then define the measurable map $\Phi : \mathbb{R}_+ \to \mathcal{P}(A)$ by setting

$$\Phi_{w(\,\cdot\,)}(t) = \{a \in A : (f_2(w(t),a), \ell_2(w(t),a), g_2(a)) = e(w(t),t;w(\,\cdot\,))\} .$$

Of course by the assumptions and the construction, $\Phi_w(\,\cdot\,)$ is compact and nonempty valued. Therefore by Filippov's Selection Theorem, we can find a measurable selection $a(t) \in \Phi(t)$, thus $a \in \mathcal{A}$. We set $\overline{\alpha}[w] = a$.

2. We now proceed with the construction of a causal functional of the output. Assume that a nonanticipating functional has been defined in the set $\mathcal{W}_1 \subset \mathcal{W}$. For $w \in \mathcal{W} \smallsetminus \mathcal{W}_1$ we consider

$$T = \sup\{t : w = w_1 \text{ a.e. in } [0,t], \text{ for some } w_1 \in \mathcal{W}_1\}.$$

If $T < +\infty$, for any $n \in \mathbb{N}$, $n \geq 2$, we determine $w_n \in \mathcal{W}_1$ such that $w = w_n$ a.e. in $[0, (1-1/n)T]$, then set

$$\overline{\alpha}[w] = \overline{\alpha}[w_n], \quad \text{a.e. in } [(1-1/(n-1)T), (1-1/n)T],$$

and define $\overline{\alpha}[w]$ in $[T, +\infty[$ by using the argument of part 1. If $T = +\infty$, the procedure is similar and left to the reader. Zorn's lemma guarantees that indeed $\overline{\alpha}$ can be defined as a nonanticipating functional $\mathcal{W} \to \mathcal{A}$.

3. We can now check that the process leads to what we wanted. Let $b \in \mathcal{B}$, $x \in \mathbb{R}^N$. Denote

$$e(w, t; w(\,\cdot\,)) = (e_f(w, t; w(\,\cdot\,)), e_\ell(w, t; w(\,\cdot\,)), e_g(w, t; w(\,\cdot\,))),$$
$$w(\,\cdot\,) = g(y_x(\,\cdot\,; \alpha^*, b), \alpha^*, b),$$

and define the causal functional by setting $\alpha^\#[b] = \overline{\alpha}[w]$. Any relaxed trajectory of α^* is by construction a trajectory of $\alpha^\#$ and vice versa, since

$$(5.7) \quad \begin{cases} \dot{\hat{y}} = f^r(\hat{y}, \alpha^*[b], b) = f_1(\hat{y}, b)\, e_f(w, t) = f(\hat{y}, \overline{\alpha}[w], b) = f(\hat{y}, \alpha^\#[b], b), \\ h^r(\hat{y}, \alpha^*[b], b) = \ell_1(\hat{y}, b) e_\ell(w, t) = \ell(\hat{y}, \overline{\alpha}[w], b) = \ell(\hat{y}, \alpha^\#[b], b), \\ w = g^r(\hat{y}, \alpha^*[b], b) = g_1(\hat{y}, b) e_g(t) = g(\hat{y}, \overline{\alpha}[w], b) = g(\hat{y}, \alpha^\#[b], b)\,. \end{cases}$$

Moreover by its definition $\alpha^\#$ depends only on the output, so it is admissible at x. The final inequality of Proposition 5.5 gives the result. ◂

6. Solving the problem

In Section 4 we stated that, under suitable assumptions, optimal strategies for the value function of the full information problem exist. However the proof is not constructive and therefore not particularly useful from the practical point of view of the applications. As in the case of standard control theory, one can produce explicit formulas for optimal feedback controls, at least formally. In differential games theory however, this is unfortunately not always likely to succeed. We will consider here only the case of the full information problem. For any $(x, q) \in \mathbb{R}^{2N}$ we denote

$$F_{x,q}(a, b) = -f(x, a, b) \cdot q - \ell(x, a, b) + \gamma^p |b|^p,$$

so that $\mathcal{H}(x, q) = \inf_{b \in B} \sup_{a \in A} F_{x,q}(a, b)$ and $\overline{\mathcal{H}}(x, q) = \sup_{a \in A} \inf_{b \in B} F_{x,q}(a, b)$. We look for necessary conditions, supposing that the value function is of class C^1, which is true for linear systems (0.3) and for the nonlinear-affine systems (0.4) with smooth data, at least in a neighborhood of the origin. Assume that the feedback $a(x)$ solves strongly the $\mathcal{H}_\infty$ suboptimal control problem and provides optimal strategies for V_γ, then by Definition 1.1 we have

$$\int_0^t (\ell(y, a(y), b) - \gamma^p |b|^p)\, ds \leq V_\gamma(x), \quad \text{for all } b \in \mathcal{B},\, t \geq 0.$$

Therefore by the definition of V_γ, for any fixed $t > 0$, we get

$$\int_0^t (\ell(y, a(y), b) - \gamma^p |b|^p)\, ds + V_\gamma(y(t)) \leq V_\gamma(x), \quad \text{for all } b \in \mathcal{B},\, t \geq 0.$$

Dividing by t and letting $t \to 0^+$, we conclude

$$0 \leq F_{x, DV_\gamma(x)}(a(x), b), \quad \text{for all } b \in B,$$

and finally we must have

(6.1) $$0 \leq \inf_{b \in B} F_{x, DV_\gamma(x)}(a(x), b) \leq \overline{\mathcal{H}}(x, DV_\gamma(x)) \leq \mathcal{H}(x, DV_\gamma(x)) = 0.$$

Therefore equalities hold in (6.1) and in particular V_γ has to be also a solution of the upper Isaacs equation. Of course, this is automatically true when the Isaacs condition holds. Moreover, it certainly holds when the value of the game exists.

In view of the necessary condition, we proceed and assume that U is a nonnegative, differentiable supersolution of the upper Isaacs equation. Then we define the possibly multivalued map

(6.2) $$A(x, q) = \arg\max_{a \in A} \{ \inf_{b \in B} F_{x,q}(a, b) \},$$

and assume

(6.3) $$\begin{cases} \text{there is a selection } a(x) \in A(x, DU(x)) \text{ such that the system} \\ \dot{y} = f(y, a(y), b), \quad y(0) = x, \\ \text{has a unique absolutely continuous solution for all } b \in \mathcal{B} \text{ and } x \in \mathbb{R}^N. \end{cases}$$

PROPOSITION 6.1. *Assume that U is a differentiable, positive definite supersolution of the upper Isaacs equation. Then a feedback control $a(x)$ satisfying (6.3) solves the $\mathcal{H}_\infty$ suboptimal control problem. If the assumptions are satisfied by $U = V_\gamma$, then $a(x)$ is optimal.*

PROOF. Let us prove that the position $\alpha[b](t) = a(y_x(t))$ defines a strategy satisfying (1.5). It is clear by the definition of $a(\cdot)$ that

$$\inf_{b \in B} F_{x, DU(x)}(a(x), b) = \overline{\mathcal{H}}(x, DU(x)) \geq 0,$$

(if $U = V_\gamma$, the necessary condition (6.1) then holds by construction) and therefore for all $b \in \mathcal{B}$ the solution guaranteed by (6.3) satisfies

$$0 \leq -f(y, \alpha[b], b) \cdot DU(y) - \ell(y, \alpha[b], b) + \gamma^p |b|^p.$$

6. SOLVING THE PROBLEM

Integrating on $[0, t]$, we obtain

$$U(y(t)) + \int_0^t (\ell(y, a, b) - \gamma^p |b|^p)\, ds \leq U(x), \quad \text{for all } b \in \mathcal{B},\, t \geq 0,$$

and we conclude the proof since U is nonnegative. If $U = V_\gamma$ the same inequality shows the optimality of a for $V_\gamma(x)$. ◀

REMARK 6.2. To our knowledge, there are no general results concerning the existence of optimal feedback controls, i.e., under which conditions (6.3) is satisfied. However, one can call the map $A(x) = A(x, DV_\gamma(x))$ a multivalued feedback synthesis, in the sense that any solution of the differential inclusion

$$\dot{y} \in f(y, A(y), b), \quad y(0) = x,$$

satisfies the gain condition (to prove this, just apply the argument of Proposition 6.1 to a selection $a(t) \in A(y(t))$ such that $\dot{y} = f(y, a, b)$). Moreover, since in general the value function is not differentiable, we should really use generalized gradients in some suitable sense, i.e., define the feedback control as a selection $a(x) \in A(x, D^{\pm} V_\gamma(x))$ for which (6.3) holds, as in §III.2.5. This is also an open problem. Otherwise $A(x) = A(x, D^{\pm} V_\gamma(x))$ will again be a multivalued feedback synthesis. ◁

REMARK 6.3. In the case of nonlinear-affine systems, when the matrix $\ell_2^T \ell_2$ is invertible at each point and $p = 2$, the above map A is luckily single valued and the following explicit formula holds

$$A(x, q) = -(\ell_2^T(x)\ell_2(x))^{-1}(f_2^T(x)q/2 + \ell_2^T(x)\ell_1(x)) \,.$$

If, in particular, the system is linear and the control sets are finite dimensional vector spaces, it is not restrictive to assume that $E^T E = I$ and $E^T D = 0$ (otherwise we change coordinates in the control variable by defining the new control u as $u = (E^T E)^{1/2} a + (E^T E)^{-1/2} E^T D x$) and it is then known that the value function is quadratic, i.e., $V_\gamma(x) = x^T F x$, for some nonnegative definite symmetric matrix F. Therefore the optimal feedback control is given by the formula

$$a(x) = -P^T F x,$$

and so it is linear. ◁

EXAMPLE 6.4. We want to apply the previous Remark 6.3 and compute the optimal linear stabilizing feedback control for the system of Example 1.4 and $\gamma > 1$. In this case $E = (0\ 1)^T$, $D = (1\ 0)^T$, $P = (1)$ and $F = (\gamma/(\gamma^2 - 1)^{1/2})$ as we computed in Example 4.7. Therefore the explicit formula for the feedback control is

$$a(x) = -x \frac{\gamma}{(\gamma^2 - 1)^{1/2}},$$

which is of course one of those allowed by the computations of Example 2.3. ◁

7. Exercises

7.1. Complete the computations of Example 1.4.

7.2. Prove Proposition 4.1.

7.3. Prove that if $U : \mathbb{R}^N \to \mathbb{R}$ is a nonnegative locally bounded function and (1.2) holds, then the following statements are equivalent

(i) U is a viscosity solution of (4.1);

(ii) U is a viscosity solution of $\min\{\mathcal{H}(x, DU), U\} = 0$, in $\mathbb{R}^N$.

7.4. As outlined in Example 4.9, prove that the value function of the control problem related to the eikonal equation (4.4) is continuous if (1.4) holds.

7.5. Compute super- and subdifferentials of the function U in Example 4.9.

7.6. Consider the one dimensional, linear system ($A = B = \mathbb{R}$)

$$\begin{cases} \dot{y} = -y + a - b, \quad y(0) = x, \\ z = y^2 + a^2 \, . \end{cases}$$

Prove that the Hamilton-Jacobi-Isaacs equation of the $\mathcal{H}_\infty$ suboptimal control problem has a solution if and only if $\gamma \geq 2^{-1/2}$. [Hint: look for quadratic solutions and note in particular that there are two such solutions if $\gamma \in (2^{-1/2}, 1)$.] Therefore $\gamma^* = 1/\sqrt{2}$. Construct the optimal linear, stabilizing feedbacks associated with each acceptable disturbance attenuation level.

7.7. Consider the system as in the previous exercise, now with bounded control set $A = [-r, r]$, $r > 0$. Compute the Hamilton-Jacobi-Isaacs equation and show that for all $\gamma \geq 1$ there is a supersolution of the form $u(x) = cx^2$. Prove that for $\gamma < 1$, $V_\gamma \equiv +\infty$, hence $\gamma^* = 1$. [Hint: for $\gamma < 1$ and fixed $x \geq 0$, try with the disturbance $b(t) = -(t + x) - (r + 1)$.]

7.8. Consider the two dimensional linear system (perturbed harmonic oscillator) with $A = B = \mathbb{R}$

$$\begin{cases} \dot{x} = y, \\ \dot{y} = x + a - b, \\ z = x^2 + y^2 + a^2 \, . \end{cases}$$

Compute the Hamiltonian associated with the $\mathcal{H}_\infty$ control problem. Prove that for any $\gamma > 1$ there is a solution of the Isaacs equation of the form $U(x, y) = cx^2 + 2dxy + ey^2$. When $\gamma = 1$, represent the Isaacs equation by a system without controls and then use Theorem 4.3 to prove that there is no nonnegative solution. Therefore $\gamma^* = 1$. Compute the linear stabilizing feedbacks solving the problem for $\gamma > 1$.

7.9. Consider the nonlinear-affine system

$$\begin{cases} \dot{y} = -y + f(y)a - b, \\ z = x^2 + a^2, \end{cases}$$

where $A = B = \mathbb{R}$, $N = 1$, and $f : \mathbb{R} \to \mathbb{R}$ satisfies

$$\inf\{\,(f(x))^2 : x \in \mathbb{R}\,\} = k^2 > 0\,.$$

Compute the Hamilton-Jacobi equation and show that, if $\gamma^2 \geq 1/(1+k^2)$, there is a supersolution of the form $u(x) = cx^2$. Compute the corresponding feedback solving the $\mathcal{H}_\infty$ problem. Show that, if $\gamma^2 < 1/(1+k^2)$, then there is no solution to the Hamilton-Jacobi equation, hence $\gamma^* = 1/(1+k^2)$. [Hint: by contradiction, use Remark 4.4 to show that there is no solution of

$$xu' - \frac{\sigma+1}{4}(u')^2 \geq x^2,$$

in an open set, if $\sigma > 0$.]

8. Bibliographical notes

The $\mathcal{H}_\infty$ control problem was proposed by Zames in [Za81] for linear systems in the frequency domain. For the solution of the problem for linear systems we refer to [GD88, Sch89] and [DGKF89], also in the output feedback case. General surveys on the linear problem can be found in the books [Fra87] and [Sto92]. The problem for nonlinear-affine systems was studied, assuming smoothness of the solution of the Hamilton-Jacobi equation in [vdS92] and [BHW93]. In [vdS92] a regularity result for the solution of the Hamilton-Jacobi equation is also proved. For an introductory and a fairly complete treatment, see also [vdS93]. The partial information problem for nonlinear affine systems was studied in [IA92a] and [IA92b] again assuming the existence of smooth supersolutions of Hamilton-Jacobi equations. A different approach, based on the introduction of an output state infinite dimensional system was introduced and studied by [JBE94] and [JB96]. The differential games approach to the $\mathcal{H}_\infty$ problem was proposed and studied in the book by Başar-Bernhard [BaBe91]. A study of the problem using the theory of viscosity solutions in the case of dissipative systems (no control dependence of the dynamics), can be found in [Jam93]. The proofs of the statements in Sections 3 and 4, and of other results in this appendix appear in [Sor96a] and [Sor97a]. The results in Section 5 are new, see also [Sor94b]. Numerical studies to compute the $\mathcal{H}_\infty$ norm of a nonlinear system appear in [JY95, Cam96]. For the risk sensitive approach, originally introduced by Jacobson [Jac73] for linear systems and formally by Whittle [Wh90] in the nonlinear case, the reader can refer to [FM92a, FM95] and [Jam92]. The $\mathcal{H}_\infty$ problem for infinite dimensional systems is studied in various generalities in [Bar93b, Bar95a, Bar95b, vKPC93, vK93, KSor96].

Bibliography

[AT87] R.A. ADIATULLINA AND A.M. TARAS'YEV. A differential game of unlimited duration. *J. Appl. Math. Mech.*, 51:415–420, 1987.

[Ak90] A.M. AKIAN. Analyse de l'algorithme multigrille FMGH de resolution d'equations de Hamilton-Jacobi-Bellman. In A. Bensoussan and J.L. Lions, editors, *Analysis and Optimization of Systems*, volume 144 of *Lecture Notes in Control and Inform. Sci.*, pages 113–122. Springer, Berlin, 1990.

[AGLM93] L. ALVAREZ, F. GUICHARD, P.L. LIONS, AND J.M. MOREL. Axioms and fundamental equations of image processing. *Arch. Rat. Mech. Anal.*, 123:199–257, 1993.

[Alv97a] O. ALVAREZ. Bounded from below solutions of Hamilton-Jacobi equations. *Differential Integral Equations*, to appear.

[Alv97b] O. ALVAREZ. Homogenization of Hamilton-Jacobi equations in perforated sets. Preprint, 1997.

[ALL97] O. ALVAREZ, J.M. LASRY, AND P.L. LIONS. Convex viscosity solutions and state constraints. *J. Math. Pures Appl.*, 76:265–288, 1997.

[AlT96] O. ALVAREZ AND A. TOURIN. Viscosity solutions of nonlinear integro-differential equations. *Ann. Inst. H. Poincaré Anal. Non Linéaire*, 13:293–317, 1996.

[AL94] B. ALZIARY AND P.L. LIONS. A grid refinement method for deterministic control and differential games. *Math. Models Methods Appl. Sci.*, 4(6):899–910, 1994.

[Al91] B. ALZIARY DE ROQUEFORT. Jeux différentiels et approximation numérique de fonctions valeur. 1re partie: étude théorique, 2e partie: étude numérique. *RAIRO Modél. Math. Anal. Numér.*, 25:517–560, 1991.

[ACS93] L. AMBROSIO, P. CANNARSA, AND H.M. SONER. On the propagation of singularities of semi-convex functions. *Ann. Scuola Norm. Sup. Pisa*, (IV) 20:597–616, 1993.

[An96] F. ANTONIALI. Problemi di controllo ottimo con variabili veloci e vincoli di stato: equazioni di Bellman e perturbazione singolare. Thesis, Università di Padova, 1996.

[Ar95a]   M. ARISAWA. Ergodic problem for the Hamilton-Jacobi-Bellman equation — Existence of the ergodic attractor. Cahiers de mathematiques de la decision, CEREMADE, 1995.

[Ar95b]   M. ARISAWA. Ergodic problem for the Hamilton-Jacobi-Bellman equation II. Cahiers de mathematiques de la decision, CEREMADE, 1995.

[ArL96]   M. ARISAWA AND P.L. LIONS. Continuity of admissible trajectories for state constrained control problems. *Discrete Continuous Dynamical Systems*, 2:297–305, 1996.

[ArT96]   M. ARISAWA AND A. TOURIN. Regularizing effects for a class of first-order Hamilton-Jacobi equations. *Nonlinear Anal.*, 26, 1996.

[Au81]   J.P. AUBIN. Contingent derivatives of set-valued maps and existence of solutions to nonlinear inclusions and differential inclusions. In L. Nachbin, editor, *Mathematical Analysis and Applications, Part A*, volume 7A of *Advances in Math. Supplementary Studies*, pages 160–232. Academic Press, 1981.

[Au91]   J.P. AUBIN. *Viability theory*. Birkhäuser, Boston, 1991.

[AC84]   J.P. AUBIN AND A. CELLINA. *Differential inclusions*. Springer, Berlin, 1984.

[AE84]   J.P. AUBIN AND I. EKELAND. *Applied nonlinear analysis*. Wiley, New York, 1984.

[AF90]   J.P. AUBIN AND H. FRANKOWSKA. *Set valued analysis*. Birkhäuser, Boston, 1990.

[Say91]   SAYAH AWATIF. Équations d'Hamilton-Jacobi du premier ordre avec termes intégro-différentiels: I. Unicité des solutions de viscosité: II. Existence de solutions de viscosité. *Comm. Partial Differential Equations*, 16:1057–1093, 1991.

[Bac86]   A. BACCIOTTI. *Fondamenti geometrici della teoria della controllabilità*. Number 31 in Quaderni U.M.I. Pitagora, Bologna, 1986.

[BBi90]   A. BACCIOTTI AND R.M. BIANCHINI. The linear time optimal problem revisited: regularization and approximation. *Boll. Un. Mat. Ital.*, (7) 4-B:245–253, 1990.

[Bad94]   M. BADIALE. Geometrical properties of solutions of fully nonlinear equations and an application to singularities. *J. Differential Equations*, 112:33–52, 1994.

[BadB90]   M. BADIALE AND M. BARDI. Symmetry properties of solutions of Hamilton-Jacobi equations. *Nonlinear Anal.*, 15:1031–1043, 1990.

[BadB92]   M. BADIALE AND M. BARDI. Asymptotic symmetry of solutions of nonlinear partial differential equations. *Comm. Pure Appl. Math.*, 45:899–921, 1992.

[Bag93]   F. BAGAGIOLO. Soluzioni di viscosità vincolate di equazioni di Bellman e perturbazione singolare in problemi di controllo ottimo. Thesis, Università di Padova, 1993.

[BagB96]   F. BAGAGIOLO AND M. BARDI. Singular perturbation of a finite horizon problem with state-space constraints. Preprint 43, Università di Padova, 1996.

[BBCD94]   F. BAGAGIOLO, M. BARDI, AND I. CAPUZZO DOLCETTA. A viscosity solutions approach to some asymptotic problems in optimal control. In G. Da Prato and J.P. Zolesio, editors, *Partial differential equation methods in control and shape analysis (Pisa 1994)*, pages 29–39, New York, 1997. Marcel Dekker.

[BHW93] J.A. BALL, J.W. HELTON, AND M.L. WALKER. H_∞ control for nonlinear systems with output feedback. *IEEE Trans. Automat. Control*, 38:546–559, 1993.

[Bar93a] V. BARBU. *Analysis and control of nonlinear infinite dimensional systems.* Academic Press, Boston, 1993.

[Bar93b] V. BARBU. H_∞ control for semilinear systems in Hilbert spaces. *Systems Control Lett.*, 21:65–72, 1993.

[Bar95a] V. BARBU. H_∞ boundary control with state feedback: the hyperbolic case. *SIAM J. Control Optim.*, 33:684–701, 1995.

[Bar95b] V. BARBU. The H_∞ problem for infinite dimensional semilinear systems. *SIAM J. Control Optim.*, 33:1017–1027, 1995.

[BDaP83] V. BARBU AND G. DA PRATO. *Hamilton-Jacobi equations in Hilbert spaces*, volume 86 of *Research Notes in Mathematics*. Pitman, 1983.

[B85] M. BARDI. Geometric properties of solutions of Hamilton-Jacobi equations. *J. Differential Equations*, 58:364–380, 1985.

[B87] M. BARDI. An asymptotic formula for the Green's function of an elliptic operator. *Ann. Sc. Norm. Sup. Pisa*, (IV) 14:569–586, 1987.

[B89] M. BARDI. A boundary value problem for the minimum time function. *SIAM J. Control Optim.*, 26:776–785, 1989.

[B92] M. BARDI. Viscosity solutions of Isaacs' equations and existence of a value. In G. Ricci and C. Torricelli, editors, *Proceedings of the Summer School on Dynamic Games*, 1992.

[B90] M. BARDI. Homogenization of quasilinear elliptic equations with possibly superqaudratic growth. In A. Marino and M.K.V. Murthy, editors, *Nonlinear variational problems and partial differential equations (Isola d'Elba, 1990)*, volume 320 of *Pitman Res. Notes Math. Ser.*, pages 44–56. Longman Sci. Tech., Harlow, 1995.

[B95] M. BARDI. Some applications of viscosity solutions to optimal control and differential games. In I. Capuzzo Dolcetta and P.L. Lions, editors, *Viscosity solutions and applications (Montecatini, 1995)*, volume 1660 of *Lecture Notes in Mathematics*, pages 44–97. Springer, Berlin, 1997.

[BB95] M. BARDI AND S. BOTTACIN. Discontinuous solutions of degenerate elliptic boundary value problems. Preprint 22, Università di Padova, 1995.

[BBF95] M. BARDI, S. BOTTACIN, AND M. FALCONE. Convergence of discrete schemes for discontinuous value functions of pursuit-evasion games. In G.J. Olsder, editor, *New Trends in Dynamic Games and Application*, pages 273–304. Birkhäuser, Boston, 1995.

[BDL96] M. BARDI AND F. DA LIO. On the Bellman equation for some unbounded control problems. *NoDEA Nonlinear Differential Equations Appl.*, to appear.

[BE84] M. BARDI AND L.C. EVANS. On Hopf's formulas for solutions of Hamilton-Jacobi equations. *Nonlinear Anal.*, 8:1373–1381, 1984.

[BFag96] M. BARDI AND S. FAGGIAN. Hopf-type estimates and formulas for non-convex non-concave Hamiton-Jacobi equations. *SIAM J. Math. Anal.*, to appear.

[BF90a]   M. BARDI AND M. FALCONE. An approximation scheme for the minimum time function. *SIAM J. Control Optim.*, 28:950–965, 1990.

[BF90b]   M. BARDI AND M. FALCONE. Discrete approximation of the minimal time function for systems with regular optimal trajectories. In A. Bensoussan and J.L. Lions, editors, *Analysis and Optimization of Systems*, volume 144 of *Lecture Notes in Control and Inform. Sci.*, pages 103–112. Springer, Berlin, 1990.

[BFS94]   M. BARDI, M. FALCONE, AND P. SORAVIA. Fully discrete schemes for the value function of pursuit-evasion games. In T. Basar and A. Haurie, editors, *Advances in Dynamic Games and Applications*, pages 89–105. Birkhäuser, Boston, 1994.

[BFS97]   M. BARDI, M. FALCONE, AND P. SORAVIA. Numerical methods for pursuit-evasion games via viscosity solutions. In M. Bardi, T. Parthasarathy, and T.E.S. Raghavan, editors, *Stochastic and differential games: Theory and numerical methods*. Birkhäuser, Boston, to appear.

[BG97]   M. BARDI AND P. GOATIN. A Dirichlet type problem for nonlinear degenerate elliptic equations arising in time-optimal stochastic control. Preprint, SISSA-ISAS, Trieste, 1997.

[BKS96]   M. BARDI, S. KOIKE, AND P. SORAVIA. On the pursuit evasion game with state constraints. Preprint, 1996.

[BOs91]   M. BARDI AND S. OSHER. The nonconvex multi-dimensional Riemann problem for Hamilton-Jacobi equations. *SIAM J. Math. Anal.*, 22:344–351, 1991.

[BP90]   M. BARDI AND B. PERTHAME. Exponential decay to stable states in phase transitions via a double log-transformation. *Comm. Partial Differential Equations*, 15:1649–1669, 1990.

[BP91]   M. BARDI AND B. PERTHAME. Uniform estimates for some degenerating quasilinear elliptic equations and a bound on the Harnack constant for linear equations. *Asymptotic Anal.*, 4:1–16, 1991.

[BSa91a]   M. BARDI AND C. SARTORI. Approximations and regular perturbations of optimal control problems via Hamilton-Jacobi theory. *Appl. Math. Optim.*, 24:113–128, 1991.

[BSa91b]   M. BARDI AND C. SARTORI. Differential games and totally risk-averse optimal control of systems with small disturbances. In R.P. Hamalainen and H.K. Ethamo, editors, *Differential Games — Developments in modeling and computation*, volume 156 of *Lecture Notes in Control and Inform. Sci.*, pages 91–99. Springer, Berlin, 1991.

[BSa92]   M. BARDI AND C. SARTORI. Convergence results for Hamilton-Jacobi-Bellman equations in variable domains. *Differential Integral Equations*, 5:805–816, 1992.

[BS89]   M. BARDI AND P. SORAVIA. A PDE framework for differential games of pursuit-evasion type. In T. Basar and P. Bernhard, editors, *Differential games and applications*, volume 119 of *Lecture Notes in Control and Inform. Sci.*, pages 62–71. Springer, Berlin, 1989.

[BS91a]   M. BARDI AND P. SORAVIA. Hamilton-Jacobi equations with singular boundary conditions on a free boundary and applications to differential games. *Trans. Amer. Math. Soc.*, 325:205–229, 1991.

[BS91b]   M. BARDI AND P. SORAVIA. Approximation of differential games of pursuit-evasion by discrete-time games. In R.P. Hamalainen and H.K. Ethamo, editors, *Differential Games — Developments in modelling and computation*, volume 156 of *Lecture Notes in Control and Inform. Sci.*, pages 131–143. Springer, Berlin, 1991.

[BS91c]   M. BARDI AND P. SORAVIA. Time-optimal control, Lie brackets, and Hamilton-Jacobi equations. Technical report, Università di Padova, 1991.

[BS92]   M. BARDI AND P. SORAVIA. Some remarks on the Bellman equation of the minimum time function with superlinear vector fields. Technical report, Università di Padova, 1992.

[BS94]   M. BARDI AND P. SORAVIA. A comparison result for Hamilton-Jacobi equations and applications to some differential games lacking controllability. *Funkcial. Ekvac.*, 37:19–43, 1994.

[BSt93]   M. BARDI AND V. STAICU. The Bellman equation for time-optimal control of non-controllable nonlinear systems. *Acta Appl. Math.*, 31:201–223, 1993.

[Ba84]   G. BARLES. Existence results for first-order Hamilton-Jacobi equations. *Ann. Inst. H. Poincaré Anal. Non Linéaire*, 1:325–340, 1984.

[Ba85a]   G. BARLES. Deterministic impulse control problems. *SIAM J. Control Optim.*, 23:419–432, 1985.

[Ba85b]   G. BARLES. Quasi-variational inequalities and first-order Hamilton-Jacobi equations. *Nonlinear Anal.*, 9:131–148, 1985.

[Ba85c]   G. BARLES. Asymptotic behavior of viscosity solutions of first order Hamilton-Jacobi equations. *Ricerche Mat.*, 2:227–260, 1985.

[Ba90a]   G. BARLES. An approach of deterministic control problems with unbounded data. *Ann. Inst. H. Poincaré Anal. Non Linéaire*, 7:235–258, 1990.

[Ba90b]   G. BARLES. Uniqueness and regularity results for first-order Hamilton-Jacobi equations. *Indiana Univ. Math. J.*, 39:443–466, 1990.

[Ba90c]   G. BARLES. Regularity results for first-order Hamilton-Jacobi equations. *Differential Integral Equations*, 3:103–125, 1990.

[Ba93]   G. BARLES. Discontinuous viscosity solutions of first order Hamilton-Jacobi equations: a guided visit. *Nonlinear Anal.*, 20:1123–1134, 1993.

[Ba94]   G. BARLES. *Solutions de viscosité des équations de Hamilton-Jacobi*, volume 17 of *Mathematiques et Applications*. Springer, Paris, 1994.

[BES90]   G. BARLES, L.C. EVANS, AND P.E. SOUGANIDIS. Wavefront propagation for reaction-diffusion systems. *Duke Math. J.*, 61:835–858, 1990.

[BaL91]   G. BARLES AND P.L. LIONS. Fully nonlinear Neumann type boundary conditions for first-order Hamilton-Jacobi equations. *Nonlinear Anal.*, 16:143–153, 1991.

[BaP87]   G. BARLES AND B. PERTHAME. Discontinuous solutions of deterministic optimal stopping time problems. *RAIRO Modél. Math. Anal. Numér.*, 21:557–579, 1987.

[BaP88]   G. BARLES AND B. PERTHAME. Exit time problems in optimal control and vanishing viscosity method. *SIAM J. Control Optim.*, 26:1133–1148, 1988.

[BaP90] G. BARLES AND B. PERTHAME. Comparison principle for Dirichlet-type Hamilton-Jacobi equations and singular perturbations of degenerated elliptic equations. *Appl. Math. Optim.*, 21:21–44, 1990.

[BSS93] G. BARLES, H.M. SONER, AND P.E. SOUGANIDIS. Front propagation and phase field theory. *SIAM J. Control Optim.*, 31:439–469, 1993.

[BaSo91] G. BARLES AND P.E. SOUGANIDIS. Convergence of approximation schemes for fully nonlinear second order equations. *Asymptotic Anal.*, 4:271–283, 1991.

[Bn85] E.N. BARRON. Viscosity solutions for the monotone control problem. *SIAM J. Control Optim.*, 23:161–171, 1985.

[Bn90] E.N. BARRON. Differential games with maximum cost. *Nonlinear Anal.*, 14:971–989, 1990.

[Bn93] E.N. BARRON. Averaging in Lagrange and minimax problems of optimal control. *SIAM J. Control Optim.*, 31:1630–1652, 1993.

[BEJ84] E.N. BARRON, L.C. EVANS, AND R. JENSEN. Viscosity solutions of Isaacs' equation and differential games with Lipschitz controls. *J. Differential Equations*, 53:213–233, 1984.

[BI89] E.N. BARRON AND H. ISHII. The Bellman equation for minimizing the maximum cost. *Nonlinear Anal.*, 13:1067–1090, 1989.

[BJ80] E.N. BARRON AND R. JENSEN. Optimal control problems with no turning back. *J. Differential Equations*, 36:223–248, 1980.

[BJ86] E.N. BARRON AND R. JENSEN. The Pontryagin maximum principle from dynamic programming and viscosity solutions to first-order partial differential equations. *Trans. Amer. Math. Soc.*, 298:635–641, 1986.

[BJ87] E.N. BARRON AND R. JENSEN. Generalized viscosity solutions for Hamilton-Jacobi equations with time-measurable Hamiltonians. *J. Differential Equations*, 68:10–21, 1987.

[BJ89] E.N. BARRON AND R. JENSEN. Total risk aversion, stochastic optimal control, and differential games. *Appl. Math. Optim.*, 19:313–327, 1989.

[BJ90] E.N. BARRON AND R. JENSEN. Semicontinuous viscosity solutions of Hamilton-Jacobi equations with convex Hamiltonians. *Comm. Partial Differential Equations*, 15:1713–1742, 1990.

[BJ91] E.N. BARRON AND R. JENSEN. Optimal control and semicontinuous viscosity solutions. *Proc. Amer. Math. Soc.*, 113:397–402, 1991.

[BJ95] E.N. BARRON AND R. JENSEN. Relaxed minimax control. *SIAM J. Control Optim.*, 33:1028–1039, 1995.

[BJL96] E.N. BARRON, R. JENSEN, AND W. LIU. Hopf-Lax-type formulas for $u_t + H(x, Du) = 0$. *J. Differential Equations*, 126:48–61, 1996.

[BJM93] E.N. BARRON, R. JENSEN, AND J.L. MENALDI. Optimal control and differential games with measures. *Nonlinear Anal.*, 21:241–268, 1993.

[BL96a] E.N. BARRON AND W. LIU. Optimal control of the blowup time. *SIAM J. Control Optim.*, 34:102–123, 1996.

[BL96b]   E.N. BARRON AND W. LIU. Semicontinuous solutions for Hamilton-Jacobi equations and the L^∞-control problem. *Appl. Math. Optim.*, 34:325–360, 1996.

[BL97]   E.N. BARRON AND W. LIU. Semicontinuous and continuous blowup and minimal time functions. *J. Differential Equations*, to appear.

[BaBe91] T. BASAR AND P. BERNHARD. H^∞-*optimal control and related minimax design problems*. Birkhäuser, Boston, 2 edition, 1995.

[BaH94]   T. BASAR AND A. HAURIE, editors. *Advances in Dynamic Games and Applications*, volume 1 of *Ann. Internat. Soc. Dynam. Games*, Boston, 1994. Birkhäuser.

[BaO82]   T. BASAR AND G.J. OLSDER. *Dynamic non-cooperative game theory*. Academic Press, New York, 1982.

[Bat78]   G.R. BATES. Lower closure and existence theorems for optimal control problems with infinite horizon. *J. Optim. Theory Appl.*, 24:639–649, 1978.

[Bel57]   R. BELLMAN. *Dynamic Programming*. Princeton University Press, Princeton, NJ, 1957.

[Bel71]   R. BELLMAN. *Introduction to the Mathematical Theory of Control Processes*, volume 2. Academic Press, New York, 1971.

[BelD62] R. E. BELLMAN AND S.E. DREYFUS. *Applied Dynamic Programming*. Princeton University Press, Princeton, NJ, 1962.

[BM96]   PH. BÉNILAN AND M. MALIKI. Approximation de la solution de viscosite d'un probleme d'Hamilton-Jacobi. *Rev. Mat. Univ. Complut. Madrid*, 9:369–383, 1996.

[Ben88]   A. BENSOUSSAN. *Perturbation methods in optimal control*. Gauthier-Villars, Paris, 1988.

[BDDM93] A. BENSOUSSAN, G. DA PRATO, M.C. DELFOUR, AND MITTER S.K. *Representation and control of infinite dimensional systems*, volume 2. Birkhäuser, Boston, 1993.

[BL82]   A. BENSOUSSAN AND J.L. LIONS. *Contrôle impulsionnel et inéquations quasi variationnelles*. Dunod, Paris, 1982.

[Bt77]   S. BENTON. *The Hamilton-Jacobi equation: a global approach*. Academic Press, New York, 1977.

[Be64]   L.D. BERKOVITZ. A differential game with no pure strategy solution. In M. et al. Dresher, editor, *Advances in Game Theory*, volume 52 of *Annals of Mathematical Studies*, pages 175–194. Princeton University Press, Princeton, NJ, 1964.

[Be85]   L.D. BERKOVITZ. The existence of value and saddle point in games of fixed duration. *SIAM J. Control Optim.*, 23:172–196, 1985. Errata and addenda in SIAM J. Control Optim. 26:740–742, 1988.

[Be88]   L.D. BERKOVITZ. Characterizations of the values of differential games. *Appl. Math. Optim.*, 17:177–183, 1988.

[Be89]   L.D. BERKOVITZ. Optimal feedback controls. *SIAM J. Control Optim.*, 27:991–1006, 1989.

[Be94]   L.D. BERKOVITZ. A theory of differential games. In T. Basar and A. Haurie, editors, *Advances in dynamic games and applications*, pages 3–22. Birkhäuser, Boston, 1994.

[Ber76]   P. BERNHARD. *Commande optimale, décentralization et jeux dynamiques*. Dunod, Paris, 1976.

[Ber77]   P. BERNHARD. Singular surfaces in differential games. In P. Hagedorn, Knobloch H.W., and Olsder G.J., editors, *Differential games and applications*, volume 3 of *Lecture Notes in Control and Inform. Sci.*, pages 1–33. Springer, 1977.

[Ber92]   P. BERNHARD. Differential games: lectures notes on the Isaacs-Breakwell theory. In G. Ricci and C. Torricelli, editors, *Proceedings of the Summer School on Dynamic Games*, 1992.

[BiS90]   R.M. BIANCHINI AND G. STEFANI. Time-optimal problem and time-optimal map. *Rend. Sem. Mat. Univ. Pol. Torino*, 48:401–429, 1990.

[BiS93]   R.M. BIANCHINI AND G. STEFANI. Controllability along a trajectory: a variational approach. *SIAM J. Control Optim.*, 31:900–927, 1993.

[Bi83]   R.M. BIANCHINI TIBERIO. Instant controllability of linear autonomous systems. *J. Optim. Theory Appl.*, 39:237–250, 1983.

[Bla97]   A.P. BLANC. Deterministic exit time problems with discontinuous exit cost. *SIAM J. Control Optim.*, 35:399–434, 1997.

[Bol66]   V.G. BOLTYANSKII. Sufficient conditions for optimality and the justification of the Dynamic Programming Principle. *SIAM J. Control*, 4:326–361, 1966.

[BLP97]   B. BONNARD, G. LAUNAY, AND M. PELLETIER. Classification generique de synthese temps minimales avec cible de codimension un et applications. *Ann. Inst. H. Poincaré Anal. Non Linéaire*, 14:55–102, 1997.

[Bon69]   J.-M. BONY. Principe du maximum, inegalite de Harnack et unicite du probleme de Cauchy pour les operateurs elliptiques degeneres. *Ann. Inst. Fourier Grenoble*, 19,1:277–304, 1969.

[Bo93]   S. BORTOLETTO. The Bellman equation for constrained deterministic optimal control problems. *Differential Integral Equations*, 6:905–924, 1993.

[Bot94]   N.D. BOTKIN. Approximation schemes for finding the value functions for differential games with nonterminal payoff functional. *Analysis*, 14:203–220, 1994.

[Br89]   J.V. BREAKWELL. Time-optimal pursuit inside a circle. In T. Basar and P. Bernhard, editors, *Differential games and applications*, volume 119 of *Lecture Notes in Control and Inform. Sci.*, pages 72–85. Springer, Berlin, 1989.

[BrB90]   J.V. BREAKWELL AND P. BERNHARD. A simple game with a singular focal line. *J. Optim. Theory Appl.*, 64:419–428, 1990.

[Bren89]   Y. BRENIER. Un algorithme rapide pour le calcul de transformées de Legendre-Fenchel discretes. *C. R. Acad. Sci. Paris Sér. I Math.*, 308:587–589, 1989.

[Bre80]   A. BRESSAN. On two conjectures by Hàjek. *Funkcial. Ekvac.*, 23:221–227, 1980.

[Bre93]   A. BRESSAN. Lecture notes on the mathematical theory of control. International School for Advanced Studies, Trieste, 1993.

[BPi95]   A. BRESSAN AND B. PICCOLI. A Baire category approach to the bang-bang property. *J. Differential Equations*, 116:318–337, 1995.

[Bru80]   P. BRUNOVSKY. Existence of regular synthesis for general problems. *J. Differential Equations*, 38:317–343, 1980.

[BH87]   A.E. BRYSON AND Y.C. HO. *Applied Optimal Control*. Hemisphere, Washington DC, 1987.

[CafCa95]   L. CAFFARELLI AND X. CABRE. *Fully nonlinear elliptic equations*, volume 43 of *Colloquium Publ.* Amer. Math. Soc., Providence, RI, 1995.

[CCKS96]   L. CAFFARELLI, M.G. CRANDALL, M. KOCAN, AND A. ŚWIECH. On viscosity solutions of fully nonlinear equations with measurable ingredients. *Comm. Pure Appl. Math.*, 49:365–397, 1996.

[Caf89]   L.A. CAFFARELLI. Interior a priori estimates for solutions of fully non-linear equations. *Ann. Math.*, 130:189–213, 1989.

[Cam96]   F. CAMILLI. Computation of the $\mathcal{H}_\infty$ norm for nonlinear systems: a convergence result. *Systems Control Lett.*, 28:139–150, 1996.

[CFa95a]   F. CAMILLI AND M. FALCONE. An approximation scheme for the optimal control of diffusion processes. *RAIRO Modél. Math. Anal. Numér.*, 29:97–122, 1995.

[CFa95b]   F. CAMILLI AND M. FALCONE. Approximation of control problems involving ordinary and impulsive controls I. Preprint, Università di Roma "La Sapienza", 1995.

[CFa96]   F. CAMILLI AND M. FALCONE. Approximation of optimal control problems with state constraints: estimates and applications. In B.S. Mordukhovic and H.J. Sussman, editors, *Nonsmooth analysis and geometric methods in deterministic optimal control*, volume 78 of *I.M.A. Volumes in Applied Mathematics*, pages 23–57. Springer, New York, 1996.

[CFLS94]   F. CAMILLI, M. FALCONE, P. LANUCARA, AND A. SEGHINI. A domain decomposition method for Bellman equations. In D.E. Keyes and Xu J., editors, *Domain Decomposition methods in Scientific and Engineering Computing*, volume 180 of *Contemp. Math.*, pages 477–483. Amer. Math. Soc., 1994.

[CamS95]   F. CAMILLI AND A. SICONOLFI. Discontinuous solutions of a Hamilton-Jacobi equation with infinite speed of propagation. Preprint, Università di Roma "La Sapienza", 1995.

[CDaP89]   P. CANNARSA AND G. DA PRATO. Nonlinear optimal control with infinite horizon for distributed parameter systems and stationary Hamilton-Jacobi equations. *SIAM J. Control Optim.*, 27:861–875, 1989.

[CDaP90]   P. CANNARSA AND G. DA PRATO. Some results on nonlinear optimal control problems and Hamilton-Jacobi equations in infinite dimensions. *J. Funct. Anal.*, 90:27–47, 1990.

[CF91]   P. CANNARSA AND H. FRANKOWSKA. Some characterizations of optimal trajectories in control theory. *SIAM J. Control Optim.*, 29:1322–1347, 1991.

[CGS91]   P. CANNARSA, F. GOZZI, AND H.M. SONER. A boundary value problem for Hamilton-Jacobi equations in Hilbert spaces. *Appl. Math. Optim.*, 24:197–220, 1991.

[CGS93]   P. CANNARSA, F. GOZZI, AND H.M. SONER. A dynamic programming approach to nonlinear boundary control problems of parabolic type. *J. Funct. Anal.*, 117:25–61, 1993.

[CSi95a] P. CANNARSA AND C. SINESTRARI. Convexity properties of the minimum time function. *Calc. Var.*, 3:273–298, 1995.

[CSi95b] P. CANNARSA AND C. SINESTRARI. On a class of minimum time problems. *Discrete Continuous Dynamical Systems*, 1:285–300, 1995.

[CS87]   P. CANNARSA AND H.M. SONER. On the singularities of the viscosity solutions to Hamilton-Jacobi-Bellman equations. *Indiana Univ. Math. J.*, 36:501–524, 1987.

[CS89]   P. CANNARSA AND H.M. SONER. Generalized one-sided estimates for solutions of Hamilton-Jacobi equations and applications. *Nonlinear Anal.*, 13(3):305–323, 1989.

[CT96]   P. CANNARSA AND M.E. TESSITORE. Infinite dimensional Hamilton-Jacobi equations and Dirichlet boundary control problems of parabolic type. *SIAM J. Control Optim.*, 34:1831–1847, 1996.

[CD83]   I. CAPUZZO DOLCETTA. On a discrete approximation of the Hamilton-Jacobi equation of dynamic programming. *Appl. Math. Optim.*, 10:367–377, 1983.

[CDE84]   I. CAPUZZO DOLCETTA AND L.C. EVANS. Optimal switching for ordinary differential equations. *SIAM J. Control Optim.*, 22:1133–1148, 1984.

[CDF89]   I. CAPUZZO DOLCETTA AND M. FALCONE. Viscosity solutions and discrete dynamic programming. *Ann. Inst. H. Poincaré Anal. Non Linéaire*, 6 (Supplement):161–183, 1989.

[CDG86]   I. CAPUZZO DOLCETTA AND M.G. GARRONI. Oblique derivative problems and invariant measures. *Ann. Sc. Norm. Sup. Pisa*, 13:689–720, 1986.

[CDI84]   I. CAPUZZO DOLCETTA AND H. ISHII. Approximate solutions of the Bellman equation of deterministic control theory. *Appl. Math. Optim.*, 11:161–181, 1984.

[CDL90]   I. CAPUZZO DOLCETTA AND P.L. LIONS. Hamilton-Jacobi equations with state constraints. *Trans. Amer. Math. Soc.*, 318:643–683, 1990.

[CDL97]   I. CAPUZZO DOLCETTA AND P.L. LIONS, editors. *Viscosity solutions and applications (Montecatini, 1995)*, volume 1660 of *Lecture Notes in Mathematics*, Berlin, 1997. Springer.

[CDM81]   I. CAPUZZO DOLCETTA AND M. MATZEU. On the dynamic programming inequalities associated with the deterministic optimal stopping problem in discrete and continuous time. *Num. Funct. Anal. Optim.*, 3, 1981.

[CDMe86] I. CAPUZZO DOLCETTA AND J.L. MENALDI. On the deterministic optimal stopping time problem in the ergodic case. In C.I. Byrnes and A. Lindquist, editors, *Theory and Applications of Nonlinear Control Systems*, pages 453–460. North-Holland, 1986.

[CDMe88]   I. CAPUZZO DOLCETTA AND J.L. MENALDI. Asymptotic behavior of the first order obstacle problem. *J. Differential Equations*, 75(2):303–328, 1988.

[CDP96]   I. CAPUZZO DOLCETTA AND B. PERTHAME. On some analogy between different approaches to first order PDE's with non smooth coefficients. *Adv. Math. Sci. Appl.*, 6:689–703, 1996.

[Ca35]   C. CARATHÉODORY. *Calculus of variations and partial differential equations of the first order.* Teubner, Berlin, 1935. (In German). Second (revised) English edition: Chelsea, New York, 1982.

[CQS94a]   P. CARDALIAGUET, M. QUINCAMPOIX, AND P. SAINT-PIERRE. Some algorithms for differential games with two players and one target. *RAIRO Modél. Math. Anal. Numér.*, 28:441–461, 1994.

[CQS94b]   P. CARDALIAGUET, M. QUINCAMPOIX, AND P. SAINT-PIERRE. Temps optimaux pour des problèmes de controle avec contraintes et sans controlabilité locale. *C. R. Acad. Sci. Paris Sér. I Math.*, 318:607–612, 1994.

[CQS97]   P. CARDALIAGUET, M. QUINCAMPOIX, AND P. SAINT-PIERRE. Set-valued numerical analysis for optimal control and differential games. In M. Bardi, T. Parthasarathy, and T.E.S. Raghavan, editors, *Stochastic and differential games: Theory and numerical methods.* Birkhäuser, Boston, to appear.

[Car93]   F. CARDIN. On viscosity and geometrical solutions of Hamilton-Jacobi equations. *Nonlinear Anal.*, 20:713–719, 1993.

[CH87]   A.D. CARLSON AND A. HAURIE. *Infinite horizon optimal control*, volume 290 of *Lecture Notes in Economics and Mathematical Systems.* Springer, 1987.

[Cas92]   V. CASELLES. Scalar conservation laws and Hamilton-Jacobi equations in one-space variable. *Nonlinear Anal.*, 18:461–469, 1992.

[Ce83]   L. CESARI. *Optimization — Theory and Applications.* Springer, New York, 1983.

[CGG91]   Y.G. CHEN, Y. GIGA, AND S. GOTO. Uniqueness and existence of viscosity solutions of generalized mean curvature flow equations. *J. Differential Geom.*, 33:749–786, 1991.

[ChSu95]   A.G. CHENTSOV AND A.I. SUBBOTIN. Iterative procedure for constructing minimax and viscosity solutions of Hamilton-Jacobi equations. Preprint, 1995. Russian version in Dokl. Akad. Nauk 348:736–739, 1996.

[Cla75]   F.H. CLARKE. Generalized gradients and applications. *Trans. Amer. Math. Soc.*, 205:247–262, 1975.

[Cla83]   F.H. CLARKE. *Optimization and nonsmooth analysis.* Wiley, New York, 1983.

[Cla89]   F.H. CLARKE. *Methods of Dynamic and Nonsmooth Optimization*, volume 57 of *CBMS-NSF Regional Conference Series in Applied Math.* SIAM, Philadelfia, 1989.

[ClaL94]   F.H. CLARKE AND YU.S. LEDYAEV. Mean value inequalities in Hilbert space. *Trans. Amer. Math. Soc.*, 344:307–324, 1994.

[CLSW95]   F.H. CLARKE, YU.S. LEDYAEV, R.J. STERN, AND P.R. WOLENSKI. Qualitative properties of trajectories of control systems: a survey. *J. Dynamical Control Systems*, 1:1–48, 1995.

[CLS97]   F.H. CLARKE, YU.S. LEDYAEV, AND A.I. SUBBOTIN. Universal feedback strategies for differential games of pursuit. *SIAM J. Control Optim.*, 35:552–561, 1997.

[ClaV87]   F.H. CLARKE AND R.B. VINTER. The relationship between the maximum principle and dynamic programming. *SIAM J. Control Optim.*, 25:1291–1311, 1987.

[Col89]   F. COLONIUS. Asymptotic behaviour of optimal control systems with low discount rates. *Math. Oper. Res.*, 14:309–316, 1989.

[Co85]   R. CONTI. *Processi di controllo in $\mathbb{R}^n$*, volume 30 of *Quaderni U.M.I.* Pitagora, Bologna, 1985.

[Cor96]   L. CORRIAS. Fast Legendre–Fenchel transform and applications to Hamilton–Jacobi equations and conservation laws. *SIAM J. Numer. Anal.*, 33:1534–1558, 1996.

[C95]   M.G. CRANDALL. Viscosity solutions: a primer. In I. Capuzzo Dolcetta and P.L. Lions, editors, *Viscosity solutions and applications (Montecatini, 1995)*, volume 1660 of Lecture Notes in Mathematics, pages 1–43. Springer, Berlin, 1997.

[CEL84]   M.G. CRANDALL, L.C. EVANS, AND P.L. LIONS. Some properties of viscosity solutions of Hamilton-Jacobi equations. *Trans. Amer. Math. Soc.*, 282:487–502, 1984.

[CIL87]   M.G. CRANDALL, H. ISHII, AND P.L. LIONS. Uniqueness of viscosity solutions revisited. *J. Math. Soc. Japan*, 39:581–596, 1987.

[CIL92]   M.G. CRANDALL, H. ISHII, AND P.L. LIONS. User's guide to viscosity solutions of second order partial differential equations. *Bull. Amer. Math. Soc.*, 27:1–67, 1992.

[CKSS96]   M.G. CRANDALL, M. KOCAN, P. SORAVIA, AND A. ŚWIECH. On the equivalence of various weak notions of solutions of elliptic PDEs with measurable ingredients. In A. et al. Alvino, editor, *Progress in elliptic and parabolic partial differential equations*, volume 350 of *Pitman Research Notes*, pages 136–162. Longman, Harlow, 1996.

[CL81]   M.G. CRANDALL AND P.L. LIONS. Conditions d'unicité pour les solutions generalises des equations d'Hamilton-Jacobi du premier ordre. *C. R. Acad. Sci. Paris Sér. I Math.*, 292:487–502, 1981.

[CL83]   M.G. CRANDALL AND P.L. LIONS. Viscosity solutions of Hamilton-Jacobi equations. *Trans. Amer. Math. Soc.*, 277:1–42, 1983.

[CL84]   M.G. CRANDALL AND P.L. LIONS. Two approximations of solutions of Hamilton-Jacobi equations. *Math. Comp.*, 43:1–19, 1984.

[CL85]   M.G. CRANDALL AND P.L. LIONS. Hamilton-Jacobi equations in infinite dimensions. Part I: Uniqueness of solutions. *J. Funct. Anal.*, 62:379–396, 1985.

[CL86]   M.G. CRANDALL AND P.L. LIONS. On existence and uniqueness of solutions of Hamilton-Jacobi equations. *Nonlinear Anal.*, 10:353–370, 1986.

[CL87]   M.G. CRANDALL AND P.L. LIONS. Remarks on the existence and uniqueness of unbounded viscosity solutions of Hamilton-Jacobi equations. *Illinois J. Math.*, 31:665–688, 1987.

[CL94a]   M.G. CRANDALL AND P.L. LIONS. Viscosity solutions of Hamilton-Jacobi equations in infinite dimensions. Part VI: Nonlinear A and Tataru's method refined. In P. Clément and G. Lumer, editors, *Evolution equations, control theory, and biomathematics*, volume 155 of *Lect. Notes Pure Appl. Math.*, pages 51–89. Marcel Dekker, New York, 1994.

[CL94b]   M.G. CRANDALL AND P.L. LIONS. Viscosity solutions of Hamilton-Jacobi equations in infinite dimensions. part VII: The HJB equation is not always satisfied. *J. Funct. Anal.*, 125:111–148, 1994.

[CL96]   M.G. CRANDALL AND P.L. LIONS. Convergent difference schemes for nonlinear parabolic equations and mean curvature motion. *Numer. Math.*, 75:17–41, 1996.

[CLS89]   M.G. CRANDALL, P.L. LIONS, AND P.E. SOUGANIDIS. Maximal solutions and universal bounds for some partial differential equations of evolution. *Arch. Rat. Mech. Anal.*, 105:163–190, 1989.

[CN85]   M.G. CRANDALL AND R. NEWCOMB. Viscosity solutions of Hamilton-Jacobi equations at the boundary. *Proc. Amer. Math. Soc.*, 94:283–290, 1985.

[Cu69]   J. CULLUM. Discrete approximations to continuous optimal control problems. *SIAM J. Control Optim.*, 7:32–49, 1969.

[Cut93]   A. CUTRÌ. Some remarks on Hamilton-Jacobi equations and non convex minimization problems. *Rend. Mat. Appl.*, 13:733–749, 1993.

[DL96]   F. DA LIO. On the Bellman equation for infinite horizon problems with unbounded cost functional. Preprint, Università di Padova, 1996.

[DaM96]   B. DACOROGNA AND P. MARCELLINI. Theoremes d'existence dans les cas scalaire et vectoriel pour les equations de Hamilton-Jacobi. *C. R. Acad. Sci. Paris Sér. I Math.*, 322:237–240, 1996.

[DM93]   G. DAL MASO. *An Introduction to Γ-convergence*. Birkhäuser, Boston, 1993.

[DeG92]   E. DE GIORGI. New ideas in calculus of variations and geometric measure theory. In G. Buttazzo and A. Visintin, editors, *Motion by mean curvature and related topics (Trento, 1992)*, pages 63–69. de Gruyter, Berlin, 1994.

[DeG90]   E. DE GIORGI. New conjectures on flow by mean curvature. In A. Marino and M.K.V. Murthy, editors, *Nonlinear variational problems and partial differential equations (Isola d'Elba, 1990)*, volume 320 of *Pitman Res. Notes Math. Ser.*, pages 120–128. Longman Sci. Tech., Harlow, 1995.

[DeM71]   V.F. DEM'YANOV AND V.N. MALOZEMOV. On the theory of nonlinear minimax problems. *Russian Math. Surveys*, 26:57–115, 1971.

[DMG95]   S. DI MARCO AND R.L.V. GONZALEZ. Une procedure numerique pour la minimisation du cout maximum. *C. R. Acad. Sci. Paris Sér. I Math.*, 321:869–873, 1995.

[Di92]   G. DÍAZ. Blow-up time involved with perturbed Hamilton-Jacobi equations. *Proc. Roy. Soc. Edinburgh*, 122 A:17–44, 1992.

[DR93]   G. DÍAZ AND J.M. REY. Finite extinction time for some perturbed Hamilton-Jacobi equations. *Appl. Math. Optim.*, 27:1–33, 1993.

[D1877]   U. DINI. Sulle funzioni finite continue di variabili reali che non hanno mai derivata. *Atti R. Acc. Lincei*, (3) 1:130–133, 1877.

[DZ93]   A.L. DONTCHEV AND T. ZOLEZZI. *Well-Posed Optimization Problems*, volume 1543 of *Lecture Notes in Mathematics*. Springer, 1993.

[Dou61]   A. DOUGLIS. The continuous dependence of generalized solutions of non-linear partial differential equations upon initial data. *Comm. Pure Appl. Math.*, 14:267–284, 1961.

[DGKF89] J.C. DOYLE, K. GLOVER, P.P. KHARGONEKAR, AND B.A. FRANCIS. State-space solutions to standard H_2 and H_∞ control problems. *IEEE Trans. Automat. Control*, 34:831–847, 1989.

[DI90]   P. DUPUIS AND H. ISHII. On oblique derivative problems for fully nonlinear second-order elliptic partial differential equations on nonsmooth domains. *Nonlinear Anal.*, 15:1123–1138, 1990.

[DI91]   P. DUPUIS AND H. ISHII. On oblique derivative problems for fully nonlinear second-order elliptic PDEs on domains with corners. *Hokkaido Math. J.*, 20:135–164, 1991.

[DIS90]   P. DUPUIS, H. ISHII, AND H.M. SONER. A viscosity solution approach to the asymptotic analysis of queueing systems. *Ann. Probab.*, 18:226–255, 1990.

[DME97]   P.G. DUPUIS AND W.M. MCENEANEY. Risk-sensitive and robust escape criteria. *SIAM J. Control Optim.*, to appear.

[Ei87]   A. EIZENBERG. The vanishing viscosity method for Hamilton-Jacobi equations with a singular point of attracting type. *Comm. Partial Differential Equations*, 12:1–20, 1987.

[Ei90]   A. EIZENBERG. The exponential leveling in elliptic singular perturbation problems with complicated attractors. *J. Analyse Math.*, 55:229–249, 1990.

[Ei93]   A. EIZENBERG. Elliptic perturbations for a class of Hamilton-Jacobi equations. *Asymptotic Anal.*, 7:251–285, 1993.

[El77]   R.J. ELLIOTT. Feedback strategies in deterministic differential games. In P. Hagedorn, Knobloch H.W., and Olsder G.J., editors, *Differential games and applications*, volume 3 of *Lecture Notes in Control and Inform. Sci.*, pages 136–142. Springer, 1977.

[El87]   R.J. ELLIOTT. *Viscosity solutions and optimal control*, volume 165 of *Pitman Research Notes in Math.* Longman, Harlow, 1987.

[EK72]   R.J. ELLIOTT AND N.J. KALTON. The existence of value in differential games. *Mem. Amer. Math. Soc.*, 126, 1972.

[EK74]   R.J. ELLIOTT AND N.J. KALTON. Cauchy problems for certain Isaacs-Bellman equations and games of survival. *Trans. Amer. Math. Soc.*, 198:45–72, 1974.

[En86]   H. ENGLER. On Hamilton-Jacobi equations in bounded domains. *Proc. Roy. Soc. Edinburgh*, 102 A:221–242, 1986.

[EL91]   H. ENGLER AND S.M. LENHART. Viscosity solutions for weakly coupled systems of Hamilton-Jacobi equations. *Proc. London Math. Soc.*, (3) 63:212–240., 1991.

[EM94]   J.R. ESTEBAN AND P. MARCATI. Approximate solutions to first and second order quasilinear evolution equations via nonlinear viscosity. *Trans. Amer. Math. Soc.*, 342:501–521, 1994.

[E78]   L.C. EVANS. A convergence theorem for solutions of nonlinear 2nd order elliptic equations. *Indiana Univ. Math. J.*, 27:875–887, 1978.

[E80]   L.C. EVANS. On solving certain nonlinear partial differential equations by accretive operator methods. *Israel J. Math.*, 36:225–247, 1980.

[E83]   L.C. EVANS. Nonlinear systems in optimal control theory and related topics. In J.M. Ball, editor, *Systems of nonlinear PDEs*, pages 95–113. D. Reidel, Dordrecht, 1983.

[E89]   L.C. EVANS. The perturbed test function technique for viscosity solutions of partial differential equations. *Proc. Roy. Soc. Edinburgh*, 111 A:359–375, 1989.

[E90]   L.C. EVANS. *Weak convergence methods for nonlinear partial differential equations*, volume 74 of *CBMS Regional Conf. Series in Math.* Amer. Math. Soc., 1990.

[E92]   L.C. EVANS. Periodic homogenization of certain fully nonlinear partial differential equations. *Proc. Roy. Soc. Edinburgh*, 120 A:245–265, 1992.

[EG92]   L.C. EVANS AND R.F. GARIEPY. *Measure theory and fine properties of functions*. CRC Press, Boca Raton, FL, 1992.

[EI84]   L.C. EVANS AND H. ISHII. Differential games and nonlinear first order PDE in bounded domains. *Manuscripta Math.*, 49:109–139, 1984.

[EI85]   L.C. EVANS AND H. ISHII. A PDE approach to some asymptotic problems concerning random differential equations with small noise intensities. *Ann. Inst. H. Poincaré Anal. Non Linéaire*, 2:1–20, 1985.

[EJ89]   L.C. EVANS AND M.R. JAMES. The Hamilton-Jacobi-Bellman equation for time optimal control. *SIAM J. Control Optim.*, 27:1477–1489, 1989.

[ESS92] L.C. EVANS, H.M. SONER, AND P.E. SOUGANIDIS. Phase transitions and generalized motion by mean curvature. *Comm. Pure Appl. Math.*, 45:1097–1123, 1992.

[ES84]   L.C. EVANS AND P.E. SOUGANIDIS. Differential games and representation formulas for solutions of Hamilton-Jacobi equations. *Indiana Univ. Math. J.*, 33:773–797, 1984.

[ES89]   L.C. EVANS AND P.E. SOUGANIDIS. A PDE approach to geometric optics for certain semilinear parabolic equations. *Indiana Univ. Math. J.*, 38:141–172, 1989.

[ESp91] L.C. EVANS AND J. SPRUCK. Motion of level sets by mean curvature. I. *J. Differential Geom.*, 33:635–681, 1991.

[ESp95] L.C. EVANS AND J. SPRUCK. Motion of level sets by mean curvature. IV. *J. Geom. Anal.*, 5:77–114, 1995.

[Fa87]   M. FALCONE. A numerical approach to the infinite horizon problem of deterministic control theory. *Appl. Math. Optim.*, 15:1–13, 1987. Corrigenda in Appl. Math. Optim. 23: 213–214, 1991.

[Fa94] M. FALCONE. The minimum time problem and its applications to front propagation. In A.Visintin and G.Buttazzo, editors, *Motion by mean curvature and related topics*, pages 70–88. De Gruyter verlag, Berlin, 1994.

[FaF94] M. FALCONE AND R. FERRETTI. Discrete time high-order schemes for viscosity solutions of Hamilton-Jacobi-Bellman equations. *Numer. Math.*, 67:315–344, 1994.

[FaF90] M. FALCONE AND R. FERRETTI. High order approximations for viscosity solutions of Hamilton-Jacobi-Bellman equations. In A. Marino and M.K.V. Murthy, editors, *Nonlinear variational problems and partial differential equations (Isola d'Elba, 1990)*, volume 320 of *Pitman Res. Notes Math. Ser.*, pages 197–209. Longman Sci. Tech., Harlow, 1995.

[FaF96] M. FALCONE AND R. FERRETTI. A class of fully discrete high-order schemes for advection equations. *SIAM J. Numerical Analysis*, to appear.

[FG94] M. FALCONE AND T. GIORGI. An approximation scheme for evolutive Hamilton-Jacobi equations. Technical report, Università di Roma "La Sapienza", 1994.

[FGL94] M. FALCONE, T. GIORGI, AND P. LORETI. Level sets of viscosity solutions: some applications to fronts and rendez-vous problems. *SIAM J. Appl. Math.*, 54:1335–1354, 1994.

[FLS94] M. FALCONE, P. LANUCARA, AND A. SEGHINI. A splitting algorithm for Hamilton-Jacobi-Bellman equations. *Appl. Numer. Math.*, 15:207–218, 1994.

[FH94] G. FERREYRA AND O. HIJAB. A simple free boundary problem in $\mathbb{R}^d$. *SIAM J. Control Optim.*, 32:501–515, 1994.

[F61] W.H. FLEMING. The convergence problem for differential games. *J. Math. Anal. Appl.*, 3:102–116, 1961.

[F64] W.H. FLEMING. The convergence problem for differential games. II. In M. et al. Dresher, editor, *Advances in Game Theory*, volume 52 of *Annals of Mathematical Studies*, pages 195–210. Princeton University Press, Princeton, NJ, 1964.

[F69] W.H. FLEMING. The Cauchy problem for a nonlinear first order partial differential equation. *J. Differential Equations*, 5:515–530, 1969.

[FM92a] W.H. FLEMING AND W.M. MCENEANEY. Risk sensitive optimal control and differential games. In *Stochastic theory and adaptive control*, volume 184 of *Lecture Notes on Control and Inform. Sci.*, pages 185–197. Springer, 1992.

[FM92b] W.H. FLEMING AND W.M. MCENEANEY. Risk sensitive control with ergodic cost criteria. In *Proc. 31st IEEE Conference on Decision and Control*, pages 2048–2052. 1992.

[FM95] W.H. FLEMING AND W.M. MCENEANEY. Risk-sensitive control on an infinite time horizon. *SIAM J. Control Optim.*, 33:1881–1915, 1995.

[FR75] W.H. FLEMING AND R.W. RISHEL. *Deterministic and stochastic optimal control*. Springer, New York, 1975.

[FS93] W.H. FLEMING AND H.M. SONER. *Controlled Markov processes and viscosity solutions*. Springer, New York, 1993.

[FSo86]   W.H. FLEMING AND P.E. SOUGANIDIS. Asymptotic series and the method of vanishing viscosity. *Indiana Univ. Math. J.*, 35:425–448 and 925F, 1986.

[FSo89]   W.H. FLEMING AND P.E. SOUGANIDIS. On the existence of value functions of two-players, zero-sum stochastic differential games. *Indiana Univ. Math. J.*, 38:293–314, 1989.

[Fra87]   B.A. FRANCIS. *A course in H_∞ theory*, volume 89 of *Lecture Notes Control Inform. Sci.* Springer, 1987.

[Fr87]   H. FRANKOWSKA. Equations d'Hamilton-Jacobi contingentes. *C. R. Acad. Sci. Paris Sér. I Math.*, 304:295–298, 1987.

[Fr89a]   H. FRANKOWSKA. Hamilton-Jacobi equations: viscosity solutions and generalized gradients. *J. Math. Anal. Appl.*, 141:21–26, 1989.

[Fr89b]   H. FRANKOWSKA. Optimal trajectories associated with a solution of the contingent Hamilton-Jacobi equation. *Appl. Math. Optim.*, 19:291–311, 1989.

[Fr93]   H. FRANKOWSKA. Lower semicontinuous solutions of Hamilton-Jacobi-Bellman equations. *SIAM J. Control Optim.*, 31:257–272, 1993.

[FPR94]   H. FRANKOWSKA, S. PLASKACZ, AND T. RZEZUCHOWSKI. Measurable viability theorems and Hamilton-Jacobi-Bellman equation. *J. Differential Equations*, 116:265–305, 1995.

[Fri71]   A. FRIEDMAN. *Differential games*. Wiley, New York, 1971.

[Fri74]   A. FRIEDMAN. *Differential games*, volume 18 of *CBMS Regional Conf. Series in Math.* American Math. Soc., Providence, R.I., 1974.

[FrSo86]   A. FRIEDMAN AND P.E. SOUGANIDIS. Blow-up of solutions of Hamilton-Jacobi equations. *Comm. Partial Differential Equations*, 11:397–443, 1986.

[Gh91]   K.H. GHASSEMI. Differential games of fixed duration with state constraints. *J. Optim. Theory Appl.*, 68:513–537, 1991.

[GT83]   D. GILBARG AND N.S. TRUDINGER. *Elliptic partial differential equations of second order*. Springer, Berlin, 2nd edition, 1983.

[Gi85]   F. GIMBERT. Problemes de Neumann quasilineaires. *J. Funct. Anal.*, 62:65–72, 1985.

[GD88]   K. GLOVER AND J. DOYLE. State-space formulas for all stabilizing controllers that satisfy an H_∞ norm bound and relations to risk-sensitivity. *Systems Control Lett.*, 11:167–172, 1988.

[Go76]   R. GONZALEZ. Sur l'existence d'une solution maximale de l'équation de Hamilton-Jacobi. *C. R. Acad. Sci. Paris Sér. I Math.*, 282:1287–1290, 1976.

[GR85]   R. GONZALEZ AND E. ROFMAN. On deterministic control problems:an approximation procedure for the optimal cost, I and II. *SIAM J. Control Optim.*, 23:242–285, 1985.

[GS90]   R. GONZALEZ AND C. SAGASTIZABAL. Un algorithme pour la résolution rapide d'équations discretes de Hamilton-Jacobi-Bellman. *C. R. Acad. Sci. Paris Sér. I Math.*, 311:45–50, 1990.

[GTi90]   R. GONZALEZ AND M.M. TIDBALL. Fast solutions of discrete Isaacs' inequalities. Technical Report 1167, INRIA, 1990.

[GTi92] R.L.V. GONZALEZ AND M.M. TIDBALL. Sur l'ordre de convergence des solutions discrétisées en temps et en espace de l'équation de Hamilton-Jacobi. *C. R. Acad. Sci. Paris Sér. I Math.*, 314:479–482, 1992.

[Gru97] L. GRUNE. An adaptive grid scheme for the discrete Hamilton-Jacobi-Bellman equation. *Numer. Math.*, 75:319–337, 1997.

[Gy84] E. GYURKOVICS. Hölder continuity for the minimum time function of linear systems. In Thoft-Chistensen, editor, *System modelling and optimization, Proc. 11th IFIP Conference*, pages 383–392. Springer, 1984.

[Ha75] O. HÀJEK. *Pursuit games*. Academic Press, New York, 1975.

[Ha91] O. HÀJEK. Viscosity solutions of Bellman's equation. In S. Elaydi, editor, *Differential equations*, pages 191–200. Marcel Dekker, New York, 1991.

[HH70] G.W. HAYNES AND H. HERMES. Nonlinear controllability via Lie theory. *SIAM J. Control Optim.*, 8:450–460, 1970.

[He88] H. HERMES. Feedback synthesis and positive, local solutions to Hamilton-Jacobi-Bellman equations. In C.I. Byrnes, C.F. Martin, and R.E. Saeks, editors, *Analysis and Control of Nonlinear Systems*, pages 155–164. North Holland, Amsterdam, 1988.

[He91] H. HERMES. Nilpotent and high-order approximations of vector field systems. *SIAM Review*, 33:238–264, 1991.

[HL69] H. HERMES AND J.P. LASALLE. *Functional Analysis and time optimal control*. Academic Press, New York, 1969.

[Hes66] M.R. HESTENES. *Calculus of Variations and Optimal Control Theory*. Wiley, New York, 1966.

[Hi91] O. HIJAB. Control of degenerate diffusions in $\mathbb{R}^d$. *Trans. Amer. Math. Soc.*, 327:427–448, 1991.

[H65] E. HOPF. Generalized solutions of non-linear equations of first order. *J. Math. Mech.*, 14:951–973, 1965.

[Hop86] R. HOPPE. Multigrid methods for Hamilton-Jacobi-Bellman equations. *Numer. Math.*, 49:235–254, 1986.

[HI97] K. HORIE AND H. ISHII. Homogenization of Hamilton-Jacobi equations on domains with small scale periodic structure. Preprint 1, Tokyo Metropolitan University, 1997.

[Hor68] L. HÖRMANDER. Hypoelliptic second order differential equations. *Acta Math.*, 119:147–171, 1968.

[Ho60] R.A. HOWARD. *Dynamic Programming and Markov processes*. Wiley, New York, 1960.

[Hr78] M.M. HRUSTALEV. Necessary and sufficient conditions in the form of Bellman's equation. *Soviet Math. Dokl.*, 19:1262–1266, 1978.

[Is65] R. ISAACS. *Differential games*. Wiley, New York, 1965.

[I83] H. ISHII. Remarks on existence of viscosity solutions of Hamilton-Jacobi equations. *Bull. Facul. Sci. Eng. Chuo Univ.*, 26:5–24, 1983.

[I84] H. ISHII. Uniqueness of unbounded viscosity solutions of Hamilton-Jacobi equations. *Indiana Univ. Math. J.*, 33:721–748, 1984.

[I85a] H. ISHII. Hamilton-Jacobi equations with discontinuous Hamiltonians on arbitrary open sets. *Bull. Facul. Sci. Eng. Chuo Univ.*, 28:33–77, 1985.

[I85b] H. ISHII. On representation of solutions of Hamilton-Jacobi equations with convex Hamiltonians. In K. Masuda and M. Mimura, editors, *Recent topics in nonlinear PDE II*, volume 8 of *Lecture Notes in Numerical and Applied Analysis*, pages 15–52. Kinokuniya-North Holland, Tokyo, 1985.

[I86] H. ISHII. Existence and uniqueness of solutions of Hamilton-Jacobi equations. *Funkcial Ekvac.*, 29:167–188, 1986.

[I87a] H. ISHII. Perron's method for Hamilton-Jacobi equations. *Duke Math. J.*, 55:369–384, 1987.

[I87b] H. ISHII. A simple, direct proof of uniqueness for solutions of Hamilton-Jacobi equations of eikonal type. *Proc. Amer. Math. Soc.*, 100:247–251, 1987.

[I88a] H. ISHII. Representation of solutions of Hamilton-Jacobi equations. *Nonlinear Anal.*, 12:121–146, 1988.

[I88b] H. ISHII. Lecture notes on viscosity solutions. Brown University, Providence, RI, 1988.

[I89] H. ISHII. A boundary value problem of the Dirichlet type for Hamilton-Jacobi equations. *Ann. Sc. Norm. Sup. Pisa*, (IV) 16:105–135, 1989.

[I91] H. ISHII. Fully nonlinear oblique derivative problems for nonlinear second-order elliptic PDEs. *Duke Math. J.*, 62:633–661, 1991.

[I92a] H. ISHII. Viscosity solutions for a class of Hamilton-Jacobi equations in Hilbert spaces. *J. Funct. Anal.*, 105:301–341, 1992.

[I92b] H. ISHII. Perron's method for monotone systems of second-order elliptic PDEs. *Differential Integral Equations*, 5:1–24, 1992.

[I97] H. ISHII. A comparison result for Hamilton-Jacobi equations without growth condition on solutions from above. Preprint 2, Tokyo Metropolitan University, 1997.

[IK91a] H. ISHII AND S. KOIKE. Remarks on elliptic singular perturbation problems. *Appl. Math. Optim.*, 23:1–15, 1991.

[IK91b] H. ISHII AND S. KOIKE. Viscosity solutions of a system of nonlinear second-order elliptic PDEs arising in switching games. *Funkcial. Ekvac.*, 34:143–155, 1991.

[IK96] H. ISHII AND S. KOIKE. A new formulation of state constraints problems for first order PDE's. *SIAM J. Control Optim.*, 36:554–571, 1996.

[IMZ91] H. ISHII, J.L. MENALDI, AND L. ZAREMBA. Viscosity solutions of the Bellman equation on an attainable set. *Problems Control Inform. Theory*, 20:317–328, 1991.

[IR95] H. ISHII AND M. RAMASWAMY. Uniqueness results for a class of Hamilton-Jacobi equations with singular coefficients. *Comm. Partial Differential Equations*, 20:2187–2213, 1995.

[IA92a]   A. ISIDORI AND A. ASTOLFI. Nonlinear H_∞ control via measurement feedback. *J. Math. Systems Estimation Control*, 2:31–44, 1992.

[IA92b]   A. ISIDORI AND A. ASTOLFI. Disturbance attenuation and H_∞ control via measurement feedback in nonlinear systems. *IEEE Trans. Automat. Control*, 37:1283–1293, 1992.

[IzK95]   S. IZUMIYA AND G. T. KOSSIORIS. Semi-local classification of geometric singularities for Hamilton-Jacobi equations. *J. Differential Equations*, 118:166–193, 1995.

[Jac73]   D.H. JACOBSON. Optimal stochastic linear systems with exponential criteria and their relation to deterministic differential games. *IEEE Trans. Automat. Control*, 18:121–134, 1973.

[Jam92]   M.R. JAMES. Asymptotic analysis of nonlinear stochastic risk-sensitive control and differential games. *Math. Control Signals Systems*, 5:401–417, 1992.

[Jam93]   M.R. JAMES. A partial differential inequality for dissipative nonlinear systems. *Systems Control Lett.*, 21:315–320, 1993.

[JB96]   M.R. JAMES AND J.S. BARAS. Partially observed differential games, infinite-dimensional Hamilton-Jacobi-Isaacs equations, and nonlinear H_∞ control. *SIAM J. Control Optim.*, 34:1342–1364, 1996.

[JBE94]   M.R. JAMES, J.S. BARAS, AND R.J. ELLIOTT. Risk-sensitive control and dynamic games for partially observed discrete-time nonlinear systems. *IEEE Trans. Automat. Control*, 39:780–792, 1994.

[JY95]   M.R. JAMES AND S. YULIAR. Numerical approximation of the H_∞ norm for nonlinear systems. *Automatica J. IFAC*, 31:1075–1086, 1995.

[Ja77]   R. JANIN. On sensitivity in an optimal control problem. *J. Math. Anal. Appl.*, 60:631–657, 1977.

[J88]   R. JENSEN. The maximum principle for viscosity solutions of fully nonlinear 2nd order partial differential equations. *Arch. Rat. Mech. Anal.*, 101:1–27, 1988.

[J89]   R. JENSEN. Uniqueness criteria for viscosity solutions of fully nonlinear elliptic partial differential equations. *Indiana Univ. Math. J.*, 38:629–667, 1989.

[Kal63]   R.E. KALMAN. The theory of optimal control and the calculus of variations. In R. Bellman, editor, *Mathematical optimization techniques*, pages 309–331. University of California Press, Berkeley, CA, 1963.

[Kam84]   S. KAMIN. Exponential descent of solutions of elliptic singular perturbation problems. *Comm. Partial Differential Equations*, 9:197–213, 1984.

[Kam86]   S. KAMIN. Singular perturbation problems and the Hamilton-Jacobi equation. *Integral Equations Operator Theory*, 9:95–105, 1986.

[Kam88]   S. KAMIN. Elliptic perturbation for linear and nonlinear equations with a singular point. *J. Analyse Math.*, 50:241–257, 1988.

[Ka95]   M.A. KATSOULAKIS. A representation formula and regularizing properties for viscosity solutions of 2nd order fully nonlinear degenerate parabolic equations. *Nonlinear Anal.*, 24:147–158, 1995.

[Kaw90] M. KAWSKI. High-order small-time controllability. In H.J. Sussmann, editor, *Nonlinear controllability and optimal control*, chapter 14, pages 431–467. Marcel Dekker, New York, 1990.

[Kle93] A.F. KLEIMENOV. *Nonantagonistic positional differential games*. Nauka, Ekaterinburg, 1993. (in Russian).

[KSor96] M. KOCAN AND P. SORAVIA. Differential games and nonlinear H_∞ control in infinite dimensions. *SIAM J. Control Optim.*, to appear.

[KSS97] M. KOCAN, P. SORAVIA, AND A. ŚWIECH. On differential games for infinite dimensional systems with nonlinear, unbounded operators. *J. Math. Anal. Appl.*, 211:395–423, 1997.

[KSw95] M. KOCAN AND A. ŚWIECH. Second order unbounded parabolic equations in separated form. *Studia Math.*, 115:291–310, 1995.

[K91] S. KOIKE. On the rate of convergence of solutions in singular perturbation problems. *J. Math. Anal. Appl.*, 157:243–253, 1991.

[K95] S. KOIKE. On the state constraint problem for differential games. *Indiana Univ. Math. J.*, 44:467–487, 1995.

[K96a] S. KOIKE. On the Bellman equations with varying controls. *Bull. Australian Math. Soc.*, 53:51–62, 1996.

[K96b] S. KOIKE. Semicontinuous viscosity solutions for Hamilton-Jacobi equations with a degenerate coefficient. *Differential Integral Equations*, to appear.

[KN97] S. KOIKE AND I. NAKAYAMA. Uniqueness of lower semicontinuous viscosity solutions for the minimum time problem. Preprint, Saitama University, 1997.

[Kok84] P.V. KOKOTOVIC. Applications of singular perturbation techniques to control problems. *SIAM Review*, 277:1–42, 1984.

[Ko93a] G.T. KOSSIORIS. Propagation of singularities for viscosity solutions of Hamilton-Jacobi equations in one space variable. *Comm. Partial Differential Equations*, 18:747–770, 1993.

[Ko93b] G.T. KOSSIORIS. Propagation of singularities for viscosity solutions of Hamilton-Jacobi equations in higher dimensions. *Comm. Partial Differential Equations*, 18:1085–1108, 1993.

[KK95] A.N. KRASOVSKII AND N.N. KRASOVSKII. *Control under lack of information*. Birkhäuser, Boston, 1995.

[KS74] N.N. KRASOVSKII AND A.I. SUBBOTIN. *Positional differential games*. Nauka, Moscow, 1974. (In Russian.) French translation: Jeux differentiels, Mir, Moscou, 1979. Revised English edition: Game-theoretical control problems, Springer, New York, 1988.

[Kr60] S.N. KRUŽKOV. The Cauchy problem in the large for certain nonlinear first order differential equations. *Sov. Math. Dokl.*, 1:474–477, 1960.

[Kr64] S.N. KRUŽKOV. The Cauchy problem in the large for nonlinear equations and for certain quasilinear systems of the first order with several variables. *Soviet Math. Dokl.*, 5:493–496, 1964.

[Kr66a] S.N. KRUŽKOV. On solutions of first-order nonlinear equations. *Soviet Math. Dokl.*, 7:376–379, 1966.

[Kr66b]   S.N. KRUŽKOV. The method of finite differences for a first-order non-linear equation with many independent variables. *USSR Comput. Math. Math. Phys.*, 6(5):136–151, 1966.

[Kr67]   S.N. KRUŽKOV. Generalized solutions of nonlinear first order equations with several independent variables II. *Math. USSR Sbornik*, 1:93–116, 1967.

[Kr70]   S.N. KRUŽKOV. First order quasilinear equations in several independent variables. *Math. USSR Sbornik*, 10:217–243, 1970.

[Kr75]   S.N. KRUŽKOV. Generalized solutions of the Hamilton-Jacobi equations of eikonal type I. *Math. USSR Sbornik*, 27:406–445, 1975.

[Kry80]   N.V. KRYLOV. *Controlled diffusion processes*. Springer, New York, 1980.

[Kry87]   N.V. KRYLOV. *Nonlinear elliptic and parabolic equations of the second order*. Reidel, Dordrecht, 1987.

[KP72]   L.A. KUN AND YU.F. PRONOZIN. Sufficient conditions for smoothness of the Bellman function in linear time-optimal problems. *Engrg. Cybernetics*, 10:6–11, 1972.

[Ku86]   J. KURZWEIL. *Ordinary Differential Equations*. Elsevier, Amsterdam, 1986.

[Kus77]   H.J. KUSHNER. *Probability methods for approximations in stochastic control and for elliptic equations*. Academic Press, New York, 1977.

[KD92]   H.J. KUSHNER AND P.G. DUPUIS. *Numerical methods for stochastic control problems in continuous time*. Springer, New York, 1992.

[La67]   R.E. LARSON. *State increment Dynamic Programming*. American Elsevier, New York, 1967.

[LL86]   J.M. LASRY AND P.L. LIONS. A remark on regularization in Hilbert spaces. *Israel J. Math.*, 55:257–266, 1986.

[LM67]   E.B. LEE AND L. MARKUS. *Foundations of optimal control theory*. Wiley, New York, 1967.

[Lei66]   G. LEITMANN. *An introduction to optimal control*. McGraw-Hill, New York, 1966.

[Le87]   S.M. LENHART. Viscosity solutions associated with switching control problems for piecewise-deterministic processes. *Houston J. Math.*, 13:405–426, 1987.

[Le89]   S.M. LENHART. Viscosity solutions associated with impulse control problems for piecewise-deterministic processes. *Internat. J. Math. Math. Sci.*, 12:145–157, 1989.

[LeY91]   S.M. LENHART AND N. YAMADA. Perron's method for viscosity solutions associated with piecewise-deterministic processes. *Funkcial. Ekvac.*, 34:173–186, 1991.

[LeY92]   S.M. LENHART AND N. YAMADA. Viscosity solutions associated with switching game for piecewise-deterministic processes. *Stochastics Stochastics Rep.*, 38:27–47, 1992.

[Lew94]   J. LEWIN. *Differential games*. Springer, London, 1994.

[LiY95]   X. LI AND J. YONG. *Optimal control theory for infinite dimensional systems*. Birkhäuser, Boston, 1995.

[L82]   P.L. LIONS. *Generalized solutions of Hamilton-Jacobi equations*. Pitman, Boston, 1982.

[L83a]   P.L. LIONS. Optimal control of diffusion processes and Hamilton-Jacobi-Bellman equations. Part 1: The dynamic programming principle and applications; Part 2: Viscosity solutions and uniqueness. *Comm. Partial Differential Equations*, 8:1101–1174 and 1229–1276, 1983.

[L83b]   P.L. LIONS. On the Hamilton-Jacobi-Bellman equations. *Acta Appl. Math.*, 1:17–41, 1983.

[L83c]   P.L. LIONS. Optimal control of diffusion processes and Hamilton-Jacobi-Bellman equations, Part III: Regularity of the optimal cost function. In *Nonlinear PDE's and Applications*, volume V of *College de France Seminar*, pages 95–205. Pitman, Boston, 1983.

[L83d]   P.L. LIONS. Existence results for first-order Hamilton-Jacobi equations. *Ricerche Mat.*, 32:1–23, 1983.

[L85a]   P.L. LIONS. Optimal control and viscosity solutions. In I. Capuzzo Dolcetta, W.H. Fleming, and T. Zolezzi, editors, *Recent Mathematical Methods in Dynamic Programming*, volume 1119 of *Lecture Notes in Mathematics*, pages 94–112. Springer, 1985.

[L85b]   P.L. LIONS. Neumann type boundary condition for Hamilton-Jacobi equations. *Duke Math. J.*, 52:793–820, 1985.

[L85c]   P.L. LIONS. Equation de Hamilton-Jacobi et solutions de viscosité. In *Ennio De Giorgi colloquium (Paris 1983)*, volume 125 of *Res. Notes in Math.*, pages 83–97. Pitman, Boston, 1985.

[L85d]   P.L. LIONS. Regularizing effects for first-order Hamilton-Jacobi equations. *Applicable Anal.*, 20:283–307, 1985.

[L86]   P.L. LIONS. Optimal control of reflected diffusion processes: an example of state constraints. In *Stochastic differential systems (Bad Honnef, 1985)*, volume 78 of *Lecture Notes in Control and Inform. Sci.*, pages 269–276. Springer, Berlin, 1986.

[L88]   P.L. LIONS. Viscosity solutions of fully nonlinear second-order equations and optimal stochastic control in infinite dimensions. Part I: the case of bounded stochastic evolutions. *Acta Math.*, 161:243–278, 1988.

[L89]   P.L. LIONS. Viscosity solutions of fully nonlinear second-order equations and optimal stochastic control in infinite dimensions. Part III: Uniqueness of viscosity solutions for general second-order equations. *J. Funct. Anal.*, 86:1–18, 1989.

[L92]   P.L. LIONS. Viscosity solutions and optimal control. In *Proceedings of ICIAM 91*, pages 182–195. SIAM, Philadelphia, 1992.

[LMr80] P.L. LIONS AND B. MERCIER. Approximation numerique des equations de Hamilton-Jacobi-Bellman. *RAIRO Anal. Numer.*, 14:369–393, 1980.

[LPV86] P.L. LIONS, G. PAPANICOLAOU, AND S. VARADHAN. Homogenization of Hamilton-Jacobi equations. unpublished paper.

[LP86]   P.L. LIONS AND B. PERTHAME. Quasi-variational inequalities and ergodic impulse control. *SIAM J. Control Optim.*, 24:604–615, 1986.

[LP87] P.L. LIONS AND B. PERTHAME. Remarks on Hamilton-Jacobi equations with measurable time-dependent Hamiltonians. *Nonlinear Anal.*, 11:613–621, 1987.

[LR86] P.L. LIONS AND J.-C. ROCHET. Hopf formula and multi-time Hamilton-Jacobi equations. *Proc. Amer. Math. Soc.*, 96:79–84, 1986.

[LRT93] P.L. LIONS, E. ROUY, AND A. TOURIN. Shape-from-shading, viscosity solutions and edges. *Numer. Math.*, 64:323–353, 1993.

[LS85] P.L. LIONS AND P.E. SOUGANIDIS. Differential games and directional derivatives of viscosity solutions of Bellman's and Isaacs' equations. *SIAM J. Control Optim.*, 23:566–583, 1985.

[LS86] P.L. LIONS AND P.E. SOUGANIDIS. Differential games and directional derivatives of viscosity solutions of and Isaacs' equations II. *SIAM J. Control Optim.*, 24:1086–1089, 1986.

[LS93] P.L. LIONS AND P.E. SOUGANIDIS. Fully nonlinear 2nd order degenerate elliptic equations with large zeroth-order coefficients. *Indiana Univ. Math. J.*, 42:1525–1543, 1993.

[LS95] P.L. LIONS AND P.E. SOUGANIDIS. Convergence of MUSCL and filtered schemes for scalar conservation laws and Hamilton-Jacobi equations. *Numer. Math.*, 69:441–470, 1995.

[LSV87] P.L. LIONS, P.E. SOUGANIDIS, AND J.L. VAZQUEZ. The relation between the porous medium and the eikonal equation in several space dimensions. *Rev. Mat. Iberoamericana*, 3:275–310, 1987.

[Liv80] A.A. LIVEROVSKII. Some properties of Bellman's functions for linear and symmetric polysystems. *Differential Equations*, 16:255–261, 1980.

[Lo86] P. LORETI. Approssimazione di soluzioni viscosità dell'equazione di Bellman. *Boll. Un. Mat. Ital.*, (6) 5-B:141–163, 1986.

[Lo87] P. LORETI. Some properties of viscosity solutions of Hamilton-Jacobi-Bellman equations. *SIAM J. Control Optim.*, 25(5):1244–1252, 1987.

[LoS93] P. LORETI AND A. SICONOLFI. A semigroup approach for the approximation of a control problem with unbounded dynamics. *J. Optim. Theory Appl.*, 79:599–610, 1993.

[LoT94] P. LORETI AND M.E. TESSITORE. Approximation and regularity results on constrained viscosity solution of Hamilton-Jacobi-Bellman equations. *J. Math. Systems Estimation Control*, 4:467–483, 1994.

[MS82] J. MACKI AND A. STRAUSS. *Introduction to optimal control theory*. Springer, New York, 1982.

[MSou94] A.J. MAJDA AND P.E. SOUGANIDIS. Large-scale front dynamics for turbulent reaction-diffusion equations with separated velocity scales. *Nonlinearity*, 7:1–30, 1994.

[M79] K. MALANOWSKI. On convergence of finite difference approximation to optimal control problems for systems with control appearing linearly. *Archivum Automatyki Telemechaniki*, pages 155–170, 1979.

[ME95] W.M. MCENEANEY. Uniqueness for viscosity solutions of nonstationary HJB equations under some a priori conditions (with applications). *SIAM J. Control Optim.*, 33:1560–1576, 1995.

[Mel94]   A.A. MELIKYAN. Singular paths in differential games with simple motion. In T. Basar and A. Haurie, editors, *Advances in Dynamic Games and Applications*, pages 125–135. Birkhäuser, Boston, 1994.

[Me80a]   J.L. MENALDI. On the optimal stopping time problem for degenerate diffusions. *SIAM J. Control Optim.*, 18:697–721, 1980.

[Me80b]   J.L. MENALDI. On the optimal impulse control problem for degenerate diffusions. *SIAM J. Control Optim.*, 18:722–739, 1980.

[Me82]   J.L. MENALDI. Le probleme de temps d'arret optimal deterministe et l'inequation variationelle du premier ordre associee. *Appl. Math. Optim.*, 8:131–158, 1982.

[Me89]   J.L. MENALDI. Some estimates for finite difference approximations. *SIAM J. Control Optim.*, 27:579–607, 1989.

[MeR86]   J.L. MENALDI AND E. ROFMAN. An algorithm to compute the viscosity solution of the Hamilton-Jacobi-Bellman equation. In C.I. Byrnes and A. Lindquist, editors, *Theory and Applications of Nonlinear Control Systems*, pages 461–468. North-Holland, 1986.

[MP86]   F. MIGNANEGO AND G. PIERI. On the sufficiency of the Hamilton-Jacobi-Bellman equation for optimality of the controls in a linear optimal-time problem. *Systems Control Lett.*, 6:357–363, 1986.

[Mi88]   S. MIRICA. Inégalités différentielles impliquant l'équation de Bellman-Isaacs et ses généralisations dans la théorie du controle optimal. *Anal. Univ. Bucuresti Mat.*, 37:25–35, 1988.

[Mi90a]   S. MIRICA. *Control optimal*. Editura Stiintifica, Bucharest, 1990. In rumanian, english translation to appear.

[Mi90b]   S. MIRICA. Some generalizations of the Bellman-Isaacs equation in deterministic optimal control. *Studii Cercetari Mat.*, 42:437–447, 1990.

[Mi92a]   S. MIRICA. A proof of Pontryagin's minimum principle using dynamic programming. *J. Math. Anal. Appl.*, 170:501–512, 1992.

[Mi92b]   S. MIRICA. Differential inequalities and viscosity solutions in optimal control. *Anal. Univ. Iaşi*, 38:89–102, 1992.

[Mi93a]   S. MIRICA. Nonsmooth fields of extremals and constructive Dynamic Programming in optimal control. Preprint 18, Dipartimento di Matematica, Università di Padova, 1993.

[Mi93b]   S. MIRICA. Optimal feedback control in closed form via Dynamic Programming. Preprint, 1993.

[Mo66]   J.J. MOREAU. Fonctionelles convexes. Technical report, Seminaire Leray-College de France, 1966.

[Mos76]   U. MOSCO. Implicit variational problems and quasi variational inequalities. In *Nonlinear operators and the calculus of variations*, volume 543 of *Lect. Notes Math.*, pages 83–156. Springer, 1976.

[Mo95]   M. MOTTA. On nonlinear optimal control problems with state constraints. *SIAM J. Control Optim.*, 33:1411–1424, 1995.

[MR96a] M. MOTTA AND F. RAMPAZZO. Dynamic Programming for nonlinear systems driven by ordinary and impulsive controls. *SIAM J. Control Optim.*, 34:199–225, 1996.

[MR96b] M. MOTTA AND F. RAMPAZZO. The value function of a slow growth control problem with state constraints. *J. Math. Systems Estim. Control*, to appear.

[MR96c] M. MOTTA AND F. RAMPAZZO. State-constrained control problems with neither coercivity nor L^1 bounds on the controls. *Appl. Math. Optim.*, to appear.

[Ol95] G.J. OLSDER, editor. *New Trends in Dynamic games and Applications*, volume 3 of *Ann. Internat. Soc. Dynam. Games*, Boston, 1995. Birkhäuser.

[OSe88] S. OSHER AND J.A. SETHIAN. Fronts propagating with curvature dependent speed: Algorithms on Hamilton-Jacobi formulation. *J. Comp. Phys.*, 79:12–49, 1988.

[OS91] S. OSHER AND C.W. SHU. High-order essentially nonoscillatory schemes for Hamilton-Jacobi equations. *SIAM J. Numer. Anal.*, 28:907–922, 1991.

[PBKTZ94] V.S. PATSKO, N.D. BOTKIN, V.M. KEIN, V.L. TUROVA, AND M.A. ZARKH. Control of an aircraft landing in windshear. *J. Optim. Theory Appl.*, 83:237–267, 1994.

[PS88] B. PERTHAME AND R. SANDERS. The Neumann problem for nonlinear second order singular perturbation problems. *SIAM J. Math. Anal.*, 19:295–311, 1988.

[Pes94a] H.J. PESCH. A practical guide to the solution of real-life optimal control problems. *Control Cybernetics*, 23:7–60, 1994.

[Pes94b] H.J. PESCH. Solving optimal control and pursuit-evasion game problems of high complexity. In R. Bulirsch and D. Kraft, editors, *Computational Optimal Control*, pages 43–64. Birkhäuser, Basel, 1994.

[PB94] H.J. PESCH AND R. BULIRSCH. The Maximum Principle, Bellman's equation, and Carathéodory's work. *J. Optim. Theory Appl.*, 80:199–225, 1994.

[Pe68] N.N. PETROV. Controllability of autonomous systems. *Differential Equations*, 4:311–317, 1968.

[Pe70a] N.N. PETROV. On the Bellman function for the time-optimal process problem. *J. Appl. Math. Mech.*, 34:785–791, 1970.

[Pe70b] N.N. PETROV. The continuity of Bellman's generalized function. *Differential Equations*, 6:290–292, 1970.

[Pe79] N.N. PETROV. On the continuity of the Bellman function with respect to a parameter. *Vestnik Leningrad Univ. Math.*, 7:169–176, 1979.

[PSus96] B. PICCOLI AND H.J. SUSSMANN. Regular synthesis and sufficiency conditions for optimality. Preprint 68, SISSA-ISAS, Trieste, 1996.

[P66] L.S. PONTRYAGIN. On the theory of differential games. *Russian Math. Surveys*, 21:193–246, 1966.

[P90] L.S. PONTRYAGIN, editor. *Optimal control and differential games*, volume 185 of *Proc. Steklov Inst. Math.* Amer. Math. Soc., 1990.

[PBGM62] L.S. PONTRYAGIN, V.G. BOLTYANSKII, R.V. GAMKRELIDZE, AND E.F. MISHCHENKO. *The mathematical theory of optimal processes.* Interscience, New York, 1962.

[Q80] J.P. QUADRAT. Existence de solutions et algorithme de resolution numerique des problemes stochastiques degenerees ou non. *SIAM J. Control Optim.*, 18:199–226, 1980.

[QSP95] M. QUINCAMPOIX AND P. SAINT-PIERRE. An algorithm for viability kernels in Hölderian case: approximation by discrete viability kernels. *J. Math. Systems Estimation Control*, 5:115–118, 1995. (Summary).

[Ra95] F. RAMPAZZO. Differential games with unbounded versus bounded controls. *SIAM J. Control Optim.*, to appear.

[RS96] F. RAMPAZZO AND C. SARTORI. The minimum time function with unbounded controls. *J. Math. Systems Estim. Control*, to appear.

[Ra82] M. RANGUIN. Propriete Hölderienne de la fonction temps minimal d'un systeme lineaire autonome. *RAIRO Automatique Syst. Anal. Control*, 16:329–340, 1982.

[Ro83] M. ROBIN. Long run average control of continuous time Markov processes: a survey. *Acta Appl. Math.*, 1:281–299, 1983.

[R70] R.T. ROCKAFELLAR. *Convex analysis.* Princeton University Press, Princeton, NJ, 1970.

[R81] R.T. ROCKAFELLAR. *The theory of subgradients and its applications to problems of optimization.* Heldermann, Berlin, 1981.

[Rou92] E. ROUY. Numerical approximation of viscosity solutions of first order Hamilton-Jacobi equations with Neumann type boundary conditions. *Math. Models Methods Appl. Sci.*, 2:357–374, 1992.

[RT92] E. ROUY AND A. TOURIN. A viscosity solutions approach to shape-from-shading. *SIAM J. Numer. Anal.*, 29:867–884, 1992.

[RV91] J.D.L. ROWLAND AND R.B. VINTER. Construction of optimal feedback controls. *Systems Control Lett.*, 16:357–367, 1991.

[Ro69] E. ROXIN. Axiomatic approach in differential games. *J. Optim. Theory Appl.*, 3:153–163, 1969.

[RS88] I. ROZYEV AND A.I. SUBBOTIN. Semicontinuous solutions of Hamilton-Jacobi equations. *J. Appl. Math. Mech.*, 52:141–146, 1988.

[Ru86] W. RUDIN. *Real and complex analysis.* McGraw-Hill, New York, 3 edition, 1986.

[SP94] P. SAINT-PIERRE. Approximation of the viability kernel. *Appl. Math. Optim.*, 29:187–209, 1994.

[Sa90] C. SARTORI. On the continuity with respect to a parameter of the value function in some optimal control problems. *Rend. Sem. Mat. Univ. Pol. Torino*, 48:383–399, 1990.

[Sch89] C. SCHERER. H_∞ control by state feedback: an alternative algorithm and characterization of high-gain occurrence. *Systems Control Lett.*, 12:383–391, 1989.

[Si95] A. SICONOLFI. A first order Hamilton-Jacobi equation with singularity and the evolution of level sets. *Comm. Partial Differential Equations*, 20:277–308, 1995.

[Sin95] C. SINESTRARI. Semiconcavity of solutions of stationary Hamilton-Jacobi equations. *Nonlinear Anal.*, 24:1321–1326, 1995.

[S86] H.M. SONER. Optimal control problems with state-space constraints I and II. *SIAM J. Control Optim.*, 24:551–561 and 1110–1122, 1986.

[S88] H.M. SONER. Optimal control of jump-Markov processes and viscosity solutions. In W. Fleming and P.L. Lions, editors, *Stochastic differential systems, stochastic control theory and applications*, volume 10 of *IMA*, pages 501–511. Springer, New York, 1988.

[S93] H.M. SONER. Singular perturbations in manufacturing. *SIAM J. Control Optim.*, 31:132–146, 1993.

[S95] H.M. SONER. Controlled Markov processes, viscosity solutions and applications to mathematical finance. In I. Capuzzo Dolcetta and Lions P.L., editors, *Viscosity solutions and applications (Montecatini 1995)*, volume 1660 of *Lecture Notes in Mathematics*, pages 134–185. Springer, Berlin, 1997.

[Son90] E.D. SONTAG. *Mathematical control theory: Deterministic finite dimensional systems*. Springer, New York, 1990.

[Sor92a] P. SORAVIA. Hölder continuity of the minimum time function for C^1 manifold targets. *J. Optim. Theory Appl.*, 75:401–421, 1992.

[Sor92b] P. SORAVIA. The concept of value in differential games of survival and viscosity solutions of Hamilton-Jacobi equations. *Differential Integral Equations*, 5:1049–1068, 1992.

[Sor93a] P. SORAVIA. Pursuit-evasion problems and viscosity solutions of Isaacs equations. *SIAM J. Control Optim.*, 31:604–623, 1993.

[Sor93b] P. SORAVIA. Discontinuous viscosity solutions to Dirichlet problems for Hamilton-Jacobi equations with convex Hamiltonians. *Comm. Partial Differential Equations*, 18:1493–1514, 1993.

[Sor94a] P. SORAVIA. Generalized motion of a front along its normal direction: a differential games approach. *Nonlinear Anal.*, 22:1247–1262, 1994.

[Sor94b] P. SORAVIA. Differential games and viscosity solutions to study the H_∞ control of nonlinear, partially observed systems. In Breton M. and G. Zaccour, editors, *Preprint volume of the 6th International Symposium on Dynamic Games and Applications*, Montreal, 1994.

[Sor95] P. SORAVIA. Stability of dynamical systems with competitive controls: the degenerate case. *J. Math. Anal. Appl.*, 191:428–449, 1995.

[Sor96a] P. SORAVIA. H_∞ control of nonlinear systems: Differential games and viscosity solutions. *SIAM J. Control Optim.*, 34:1071–1097, 1996.

[Sor96b] P. SORAVIA. Optimality principles and representation formulas for viscosity solutions of Hamilton-Jacobi equations: I. Equations of unbounded and degenerate control problems without uniqueness; II. Equations of control problems with state constraints. *Advances Differential Equations*, to appear.

[Sor97a]   P. SORAVIA. Equivalence between nonlinear H_∞ control problems and existence of viscosity solutions of Hamilton-Jacobi-Isaacs equations. *Appl. Math. Optim.*, to appear.

[Sor97b]   P. SORAVIA. Estimates of convergence of fully discrete schemes for the Isaacs equation of pursuit-evasion differential games via maximum principle. *SIAM J. Control Optim.*, to appear.

[SSou96]   P. SORAVIA AND P.E. SOUGANIDIS. Phase-field theory for FitzHugh-Nagumo-type systems. *SIAM J. Math. Anal.*, 27:1341–1359, 1996.

[Sou85a]   P.E. SOUGANIDIS. Approximation schemes for viscosity solutions of Hamilton-Jacobi equations. *J. Differential Equations*, 57:1–43, 1985.

[Sou85b]   P.E. SOUGANIDIS. Max-min representations and product formulas for the viscosity solutions of Hamilton-Jacobi equations with applications to differential games. *Nonlinear Anal.*, 9:217–257, 1985.

[Sou86]   P.E. SOUGANIDIS. A remark about viscosity solutions of Hamilton-Jacobi equations at the boundary. *Proc. Amer. Math. Soc.*, 96:323–329, 1986.

[Sou95]   P.E. SOUGANIDIS. Front propagation: theory and applications. In I. Capuzzo Dolcetta and P.L. Lions, editors, *Viscosity solutions and applications (Montecatini, 1995)*, Lecture Notes in Mathematics, pages 186–242. Springer, 1997.

[Sta89]   V. STAICU. Minimal time function and viscosity solutions. *J. Optim. Theory Appl.*, 60:81–91, 1989.

[St91]   G. STEFANI. Regularity properties of the minimum-time map. In C.I. Byrnes and A. Kurzhansky, editors, *Nonlinear systems*, pages 270–282. Birkhäuser, Boston, 1991.

[Sto92]   A.A. STOORVOGEL. *The H_∞ control problem*. Prentice Hall, New York, 1992.

[Su80]   A.I. SUBBOTIN. A generalization of the basic equation of the theory of differential games. *Soviet Math. Dokl.*, 22:358–362, 1980.

[Su84]   A.I. SUBBOTIN. Generalization of the main equation of differential game theory. *J. Optim. Theory Appl.*, 43:103–133, 1984.

[Su91a]   A.I. SUBBOTIN. *Minimax inequalities and Hamilton-Jacobi equations*. Nauka, Moscow, 1991. In Russian.

[Su91b]   A.I. SUBBOTIN. Existence and uniqueness results for Hamilton-Jacobi equations. *Nonlinear Anal.*, 16:683–699, 1991.

[Su93a]   A.I. SUBBOTIN. Discontinuous solutions of a Dirichlet type boundary value problem for first order partial differential equations. *Russian J. Numer. Anal. Math. Modelling*, 8:145–164, 1993.

[Su93b]   A.I. SUBBOTIN. On a property of the subdifferential. *Math. USSR Sbornik*, 74:63–78, 1993.

[Su95]   A.I. SUBBOTIN. *Generalized solutions of first order PDEs: The Dynamic Optimization Perspective*. Birkhäuser, Boston, 1995.

[Su97]   A.I. SUBBOTIN. Constructive theory of positional differential games and generalized solutions to Hamilton-Jacobi equations. In M. Bardi, T. Parthasarathy, and T.E.S. Raghavan, editors, *Stochastic and differential games: Theory and numerical methods*. Birkhäuser, Boston, to appear.

[SS78]   A.I. SUBBOTIN AND N.N. SUBBOTINA. Necessary and sufficient conditions for a piecewise smooth value of a differential game. *Soviet Math. Dokl.*, 19:1447–1451, 1978.

[SS83]   A.I. SUBBOTIN AND N.N. SUBBOTINA. On justification of dynamic programming method in an optimal control problem. *Izv. Acad. Nauk SSSR Tehn. Kibernet.*, 2:24–32, 1983. (In Russian).

[ST86]   A.I. SUBBOTIN AND A.M. TARASYEV. Stability properties of the value function of a differential game and viscosity solutions of Hamilton-Jacobi equations. *Problems Control Inform. Theory*, 15:451–463, 1986.

[Sua89]   N.N. SUBBOTINA. The maximum principle and the superdifferential of the value function. *Problems Control Inform. Theory*, 18:151–160, 1989.

[Sua91]   N.N. SUBBOTINA. The method of Cauchy characteristics and generalized solutions of the Hamilton-Jacobi-Bellman equation. *Soviet Math. Dokl.*, 44:501–506, 1992.

[Sun93]   M. SUN. Domain decomposition algorithms for solving Hamilton-Jacobi-Bellman equations. *Numer. Funct. Anal. Appl.*, 14:143–166, 1993.

[Sus83]   H.J. SUSSMANN. Lie brackets, real analiticity and geometric control. In R.W. Brockett, R.S. Millman, and H.J. Sussmann, editors, *Differential geometric control theory*, pages 1–116. Birkhäuser, Boston, 1983.

[Sus87]   H.J. SUSSMANN. A general theorem on local controllability. *SIAM J. Control Optim.*, 25:158–194, 1987.

[Sus89a]   H.J. SUSSMANN. Optimal control. In H. Nijmeijer and J.M. Schumacher, editors, *Three decades of mathematical system theory*, volume 135 of *Lect. Notes Control Inform. Sci.*, pages 409–425. Springer, 1989.

[Sus89b]   H.J. SUSSMANN. Lecture notes on topics in system theory. Rutgers University, 1989.

[Sus90]   H.J. SUSSMANN. Synthesis, presynthesis, sufficient conditions for optimality and subanalytic sets. In H.J. Sussmann, editor, *Nonlinear controllability and optimal control*, chapter 1, pages 1–19. Marcel Dekker, New York, 1990.

[Sw96a]   A. ŚWIECH. Sub- and superoptimality principles of Dynamic Programming revisited. *Nonlinear Anal.*, 26:1429–1436, 1996.

[Sw96b]   A. ŚWIECH. Another approach to the existence of value functions of stochastic differential games. *J. Math. Anal. Appl.*, 204:884–897, 1996.

[Ta94]   A.M. TARAS'YEV. Approximation schemes for constructing minimax solutions of Hamilton-Jacobi equations. *J. Appl. Math. Mech.*, 58:207–221, 1994.

[Ta97]   A.M. TARASYEV. Optimal control synthesis in grid approximation schemes. Technical Report 12, I.I.A.S.A., Laxenburg, Austria, 1997.

[T92]   D. TATARU. Boundary value problems for first order Hamilton-Jacobi equations. *Nonlinear Anal.*, 19:1091–1110, 1992.

[T94] D. TATARU. Viscosity solutions for Hamilton-Jacobi equations with unbounded nonlinear term: a simplified approach. *J. Differential Equations*, 111:123–146, 1994.

[Te95] M.E. TESSITORE. Optimality conditions for infinite horizon control problems. *Boll. Un. Mat. Ital.*, (7) 9-B:785–814, 1995.

[Ti91] M.M. TIDBALL. *Sobre la resolucion numerica de la ecuacion de Hamilton-Jacobi-Bellman*. PhD thesis, Universidad Nacional de Rosario, Argentina, 1991.

[Ti95] M.M. TIDBALL. Undiscounted zero sum differential games with stopping times. In G.J. Olsder, editor, *New Trends in Dynamic games and Applications*, pages 305–322. Birkhäuser, Boston, 1995.

[TiG94] M.M. TIDBALL AND R.L.V. GONZALEZ. Zero sum differential games with stopping times: some results about their numerical resolution. In T. Basar and A. Haurie, editors, *Advances in dynamic games and applications*, pages 106–124. Birkhäuser, Boston, 1994.

[Tou92] A. TOURIN. A comparison theorem for a piecewise Lipschitz continuous Hamiltonian and application to Shape-from-Shading problems. *Numer. Math.*, 62:75–85, 1992.

[Val64] F.A. VALENTINE. *Convex sets*. McGraw-Hill, New York, 1964.

[vdS92] A.J. VAN DER SHAFT. L_2 gain analysis for nonlinear systems and nonlinear H_∞ control. *IEEE Trans. Automat. Control*, 37:770–784, 1982.

[vdS93] A.J. VAN DER SHAFT. Nonlinear state space H_∞ control theory. In H.L. Trentelman and J.C. Willems, editors, *Perspectives in Control*. Birkhäuser, Boston, 1993.

[vK93] B. VAN KEULEN. H_∞-control with measurement-feedback for linear infinite-dimensional systems. *J. Math. Systems Estim. Control*, 3:373–411, 1993.

[vKPC93] B. VAN KEULEN, M. PETERS, AND R. CURTAIN. H_∞-control with state-feedback: the infinite-dimensional case. *J. Math. Systems Estim. Control*, 3:1–39, 1993.

[Va67] P.P. VARAIYA. On the existence of solutions to a differential game. *SIAM J. Control*, 5:153–162, 1967.

[Ve96] V.M. VELIOV. On the Lipschitz continuity of the value function in optimal control. *J. Optim. Theory Appl.*, to appear.

[VW90a] R.B. VINTER AND P. WOLENSKI. Hamilton-Jacobi theory for optimal control problems with data measurable in time. *SIAM J. Control Optim.*, 90:1404–1419, 1990.

[VW90b] R.B. VINTER AND P.R. WOLENSKI. Coextremals and the value function for control problems with data measurable in time. *J. Math. Anal. Appl.*, 153:37–51, 1990.

[Wa72] J. WARGA. *Optimal control of differential and functional equations*. Academic Press, New York, 1972.

[Wh90] P. WHITTLE. *Risk-sensitive optimal control*. Wiley, New York, 1990.

[W25] N. WIENER. Note on a paper of O. Perron. *J. Math. Physics (M.I.T.)*, 4:21–32, 1925.

[Wi96] S.A. WILLIAMS. Local behavior of solutions of Hamilton-Jacobi equations. *Nonlinear Anal.*, 26:323–328, 1996.

[WZ97] P.R. WOLENSKI AND YU ZHUANG. Proximal analysis and the minimal time function. *SIAM J. Control Optim.*, to appear.

[Y89] J. YONG. Systems governed by ordinary differential equations with continuous, switching and impulse controls. *Appl. Math. Optim.*, 20:223–235, 1989.

[Y90] J. YONG. A zero-sum differential game in a finite duration with switching strategies. *SIAM J. Control Optim.*, 28:1234–1250, 1990.

[Y94] J. YONG. Zero-sum differential games involving impulse controls. *Appl. Math. Optim.*, 29:243–261, 1994.

[Yo68] J.A. YORKE. Extending Liapunov's second method to non-Lipschitz Liapunov functions. *Bull. Amer. Math. Soc.*, 74:322–325, 1968.

[You69] L.C. YOUNG. *Lectures on the Calculus of Variations*. Chelsea, New York, 2 edition, 1980.

[Za81] G. ZAMES. Feedback optimal sensitivity: model reference transformations, multiplicative seminorms, and approximate inverses. *IEEE Trans. Automat. Control*, 26:301–320, 1981.

[Z90] X.Y. ZHOU. Maximum principle, dynamic programming, and their connection in deterministic control. *J. Optim. Th. Appl.*, 65:363–373, 1990.

[Z93] X.Y. ZHOU. Verification theorems within the framework of viscosity solutions. *J. Math. Anal. Appl.*, 177:208–225, 1993.

[ZYL97] X.Y. ZHOU, J. YONG, AND X. LI. Stochastic verification theorems within the framework of viscosity solutions. *SIAM J. Control Optim.*, 35:243–253, 1997.

[Zo92] T. ZOLEZZI. Wellposedness and the Lavrentiev phenomenon. *SIAM J. Control Optim.*, 30:787–789, 1992.

Index

A
acceleration algorithm, 500
adjoint system, 14, 138
adjoint vector, 13, 168, 174
admissible controls, 272
approximation, 17, 20, 118, 154
    by discrete time games, 457
    by vanishing viscosity, 20, 383
    scheme, 472
    semidiscrete, 359
Ascoli-Arzelà theorem, 219
attainable set, 282
averaging, 430

B
bang-bang principle, 113
Barles-Perthame procedure, 357
barrier, 302
Barron-Jensen discontinuous solution, 285
Bellman functional equation, 389
Bellman's function, 99
bilateral solution, 120, 382
    subsolution, 120
    supersolution, 79, 120, 349
        non-continuous, 342
Bolza problem, 147
boundary condition, 227, 347
    at infinity, 245
    Dirichlet, 256
        in the viscosity sense, 317
    for state constrained problems, 277
    for the minimal time function, 241
    Neumann, 281
    nonlinear oblique derivative, 281

boundary layer, 323
Bump Lemma, 303

C
certainty equivalence principle, 467, 523
chain rule, 135
characteristic system, 86, 181
characteristics, 77, 96, 181
chattering controls, 113
Chow theorem, 233
Clarke's gradient, 63, 85
comparison principle, 7, 9, 27, 50, 107, 109, 191, 215, 256, 262, 278, 281, 286, 326, 328, 329, 389
    for Cauchy problems, 56, 152, 156
    for Cauchy-Dirichlet problems, 295
    for convex Hamiltonian, 82
    for time-dependent HJB equations, 182
    local comparison
        for Dirichlet problems, 266
        for initial value problems, 156, 295
compatibility of the boundary data, 284
compatible sets of strategies, 454
compatible terminal cost, 251
complementarity system, 196
complete solution, 107, 151, 244, 256, 287, 334
computational methods, 17, 471
conditions of optimality, 133
    necessary, 134, 171
    necessary and sufficient, 138, 172
    sufficient, 136, 172
cone of dependence, 156
constrained viscosity solution, 277, 399

constraints on the endpoint of the trajectories, 348
contingent derivatives, 125
continuity of the value function, 99, 146, 185, 194, 200, 211, 248, 251, 260, 268, 274
control, 1
    closed loop, 111
    law, 13, 110
        multivalued, 143
    open loop, 98
    optimal, 103, 368
    piecewise constant, 99
    space, 1
    suboptimal, 17
    system, 1, 97
controllability, 227, 398
    matrix, 231
    small-time, 228, 268
        local, 228, 331
convexified problem, 112
cost functional, 1, 99
costate, 13, 168, 174

D

Danskin theorem, 95
detectability, 507
differentiability
    of solutions of HJ equations, 80
    of solutions of ODEs with respect to the initial data, 222
    of the value function, 167
differential games, 431, 512
    N-person, 470
    positional, 456
    stochastic, 470
    two-person zero-sum, 431
differential inequalities, 86
directional derivatives
    Dini, 64, 162
        lower, 125
        upper, 125
    generalized, 63
    regularized, 63
Dini solution, 125, 163
    generalized, 127
    lower, 131
    subsolution, 162
        generalized, 127
    supersolution, 127, 162

Dirichlet problem, 51, 84, 254, 260, 277, 305, 308, 316, 342
    subsolution, 287
    supersolution, 287
    v-subsolution, 318, 347
    v-supersolution, 318
discontinuous equations, 319
discount factor, 2
discrete time
    Dynamic Programming, 17, 388
    game, 457
        fair, 458
    infinite horizon problem, 388
    system, 388
dissipativity, 397, 403
distance function, 34, 42, 47
disturbances, 506
domain decomposition, 502
Dynamic Programming, 1
    equation, 389, 458, 473, 515
    Principle, 2, 102, 149, 155, 182, 187, 194, 201, 211, 239, 254, 276, 388, 437, 451, 465
        Backward, 119, 160, 344
        discrete time, 18, 389
dynamics, 1

E

e-solution, see envelope solution
eikonal equation, 47, 82, 84, 241
entry time, 228
envelope
    solution, 309, 347
    subsolution, 309, 319
    supersolution, 309, 319
ergodic problems, 397
escape controllability domain, 237, 245, 265
escape time, 237
estimate of the error, 464, 475
existence
    of an optimal control, 368
    of an optimal strategy, 520
    of a value, 442
    of solutions of HJ equations, 302
    of trajectories
        global, 220
        local, 219
exit time, 123, 228
exterior normal, 48
    generalized, 48, 269

F
falsification function, 112, 151
fast components, 420
feedback map, 111, 387
  admissible, 111, 144
  memoryless, 111
  multivalued
    admissible, 141
    fully optimal, 142
    weakly optimal, 142
feedback strategies, 431, 435
finite horizon problem, 147, 348
free boundary, 199, 241, 246, 263, 265
fundamental matrix, 222

G
game
  differential, 431, 512
  discrete time, 457
  majorant, 458
  minorant, 458
  pursuit-evasion, 469, 494
generalized gradient, 63
generalized solution, 10, 32, 77
geometric optics, 430
geometric properties of solutions, 96
geometric theory of controllability, 283
Gronwall inequality, 218
guaranteed outcome, 99

H
$\mathcal{H}_\infty$ control, 505
$\mathcal{H}_\infty$ norm, 508
Hamilton-Jacobi equation, 25, 516
Hamilton-Jacobi-Bellman equation, 2, 3, 104, 150, 188, 196, 203, 212, 240, 255, 414
  backward, 154
  evolutive, 156, 307
Hamilton-Jacobi-Isaacs equation, 438
  lower, 439, 453
  upper, 439, 453
Hamiltonian system, 181
Hausdorff distance, 116, 376
high-order methods, 501
homogenization, 430
Hopf's formulas, 96

I
imperfect state measurement, 467
implicit obstacle, 212
impulse, 183, 200
  control, 200
  cost, 200
inf-convolution, 43, 72, 345
infinite horizon discounted regulator, 1
infinite horizon problem, 99, 271, 376, 397, 435, 472
  with constraints, 227
information pattern, 431, 448, 458
input, 506
integrodifferential equations, 226
interest rate, 99
intrinsically classical solution, 128, 132, 163, 170
Isaacs' condition, 444
Ishii discontinuous solution, 285
Ito stochastic differential equation, 384

J
Jacobi bracket, 233

K
K-strategy, 449
Kružkov transform, 243, 263

L
L^p gain, 508
Lagrange problem, 147
large deviations, 396
Lavrentiev phenomenon, 223
Lewy-Stampacchia type inequality, 216
Lie bracket, 233, 276
linear system, 224, 231, 506, 510
linearized system, 231
Lipschitz continuity, 62
  of subsolutions, 294
Lipschitz control, 425, 427
local grid refinements, 501
long run averaged cost, 397, 400
lower semicontinuous envelope, 296
LQ problem, 224
Lyapunov stable, 507
  asymptotically, 507

M
marginal function, 34, 42, 65, 163
maximum principle, 27
Mayer problem, 147, 160
minimal interior time function, 228, 240
minimal time function, 228, 313
  continuity, 230
  Hölder continuity, 230, 233, 235

minimal time function (*continued*)
    Lipschitz continuity, 230, 234, 269
    regularization, 379
minimax solution, 125, 127, 358
minimizing sequence, 363
Minimum Principle, 135, 170, 173
modulus, 98
monotone acceleration procedure, 477
monotone controls, 99
    problem, 183, 426
    value function, 184
motions of the game, 456

N

Nash equilibrium, 447
nonanticipating strategy, 431
nonautonomous problems, 182
nonautonomous systems, 147, 154, 159
nonconvex problems in the Calculus of Variations, 96
nonlinear $\mathcal{H}_\infty$ control, 505
nonlinear systems, 98, 431
    affine, 112, 444, 506
numerical tests, 481, 496

O

observed output, 506, 521
observed state, 467
optimal control, 2, 103
    approximate, 362
    existence, 368
    in a differential game, 446
    law, 110
optimal feedback
    control, 393
    map, 2, 11, 199
    multivalued, 142
optimal point, 119, 160
optimal relaxed
    control, 366
    feedback strategies, 524
    strategy, 446, 520
        existence, 520
optimal result function, 99
optimal stopping, 193, 402
optimal strategy, 520
optimal switching, 210, 411
optimal trajectory, 169
optimizing sequence of strategies, 466
output, 467, 506
    state feedback strategy, 466

P

parallel algorithms, 502
partial information, 521
partial observation, 467
penalization, 414
    of state constraints, 417
    of stopping costs, 414
    of unilateral constraints, 414
perfect memory, 449
perfect state measurement, 449
Perron's method, 96, 304, 311, 357
perturbation, 118, 154, 244, 279, 291, 320, 333, 336
    of the domain, 323, 333
    regular, 333, 376
    singular, 21, 322, 333, 384, 420, 421
Petrov theorem, 231, 269
pointwise monotone limits, 306
Pontryagin Maximum Principle, 13, 135, 138, 168, 173, 355
    extended, 179
    for relaxed controls, 175
positive basis, 238
    condition, 232
presynthesis, 110
    multivalued, 143
        maximal optimal, 143
    of strategies, 441
problems with exit times, 247, 342
problems with state constraints, 271, 484
projections, 47
propagation of fronts, 96, 430
proximal normal, 48, 50
pursuit-evasion games, 469, 494

Q

quasivariational inequality, 183, 203
queuing systems, 430

R

Rademacher theorem, 63
Radon measure, 114
rate of convergence, 369, 372, 385, 408, 464, 476
reachable set, 228, 263
reaction-diffusion equations, 430
reduced order modeling, 421
regularity, 80, 163, 224
regularizing effect, 96
relaxation, 110, 154, 244, 279

relaxed controls, 113, 175, 354, 444
representation formula, 96, 260
Riccati equation, 394
risk-sensitive control, 505
robust attenuation of disturbances, 505
rocket railroad car, 242, 490
running cost, 1, 99

S

saddle, 444
    point, 454
Schauder fixed point theorem, 219
semiconcave, 11, 65, 77, 355
    -convex, 164
    locally, 163
semiconcavity, 372, 387, 391
semidifferentials, 29
semigroup property, 155
separation property, 444
shape-from-shading, 96
signed distance, 234, 272
simplex method, 477
singular perturbation, 21, 322, 333, 384, 420, 421
singular surfaces, 468
singularities of solutions, 96
solution in the extended sense, 85
solution operator, 175
solvability of the small game, 444
stability, 9, 35, 116, 289, 291, 306, 333, 336, 354
    of v-solutions, 320
    with respect to the first player, 456
    with respect to the second player, 456
stable bridge, 457
state constraints, 271, 397, 417, 484
state equation, 1
stochastic control, 20, 383
stochastic differential equation, 20
stopping cost, 193
stopping time, 183, 193, 200
strategy
    causal, 433
    closed-loop, 449
    feedback, 448
    memoryless, 449
    nonanticipating, 433
        discrete, 458
    optimal, 446, 454, 520
        approximate, 462

output feedback, 521
playable, 449
sampled-data, 449
slightly delayed, 442
with finite memory, 452
subdifferential, 5, 29, 95
suboptimality principle
    global, 124, 161
    local, 122, 161
sup-convolution, 72
superdifferential, 5, 29
superoptimality principle
    global, 124, 257
    local, 122, 161
switching, 183, 425, 426
    cost, 211
symmetric system, 233, 276
symplectic mechanics, 96
synthesis, 2, 10, 19, 111, 140, 209, 215, 382, 390
    multivalued, 143
        maximal optimal, 143
    of feedback controls, 478, 502
    of optimal strategies, 462
system of quasivariational inequalities, 212, 411
systems of HJB equations, 226

T

target set, 227
    smooth, 233
    with piecewise smooth boundary, 236
tenet of transition, 102
terminal cost, 147, 247
    compatible, 251
    lower semicontinuous, 349
test functions, 5
time optimal control, 227, 313, 333, 345, 488
trajectory, 98
    backward, 119

U

unbounded costs, 156
undiscounted value, 406
unique evolution property, 155
uniqueness result, 7, 50, 107, 152, 159, 197, 206, 213, 241, 244, 246, 256, 260, 278, 330
    in the sense of the graph, 336

upper semicontinuous envelope, 296
usable part of the boundary, 331

V

value function, 2, 99
    approximate, 17
    continuity, 99, 146, 185, 194, 200, 211, 248, 251, 260, 268, 274
    differentiability, 167
    discontinuous, 298
    vector valued, 212
value (of a differential game), 434
    Berkovitz, 456
    existence, 442
    feedback, 450
    Fleming, 461
    Friedman, 465
    Krasovskiĭ and Subbotin, 456
    lower, 434
    relaxed, 446
    upper, 434
    Varaiya, Roxin, Elliott and Kalton, 434, 468
vanishing discount, 397, 402
vanishing switching costs, 411
vanishing viscosity, 20, 322, 337, 383

variational inequality, 183, 185, 188
    of obstacle type, 193, 196
verification function, 12, 151, 171, 316, 447
    generalized, 112
verification theorems, 110, 154, 244, 279, 287, 316, 446
viability, 275, 358, 470
viscosity solution, 1, 5, 25
    non-continuous, 296, 318
    subsolution, 5, 25, 31
        lower semicontinuous, 286
        of Dirichlet boundary condition, 318
    supersolution, 5, 25, 31, 286, 318
        upper semicontinuous, 286
        of Dirichlet boundary condition, 318

W

weak limits, 10, 321, 359, 360
    lower, 288
    upper, 288

Z

Zermelo Navigation Problem, 300, 307, 357